The Large Hadron Collider

Thomas Schörner-Sadenius
Editor

The Large Hadron Collider

Harvest of Run 1

Springer

Editor
Thomas Schörner-Sadenius
Deutsches Elektronen-Synchrotron (DESY)
Hamburg
Germany

ISBN 978-3-319-15000-0 ISBN 978-3-319-15001-7 (eBook)
DOI 10.1007/978-3-319-15001-7

Library of Congress Control Number: 2015933362

Springer Cham Heidelberg New York Dordrecht London

Cover art: Jorge Cham—www.phdcomics.com

Printed on acid-free paper

Springer International Publishing AG Switzerland is part of Springer Science+Business Media
(www.springer.com)

Foreword

The Large Hadron Collider is the largest scientific experiment mankind ever devised, and already the first period of data-taking was a tremendous success. The accelerator, detector and computing Grid performance surpassed all expectations. Only 4 years after the start-up the first major milestone was reached: Everybody who witnessed the discovery of a Higgs boson, the messenger of the Brout–Englert–Higgs field, in the year 2012 and appreciates the importance of its existence will certainly agree. The 2013 Nobel Prize in physics to Francois Englert and Peter Higgs is a testimony to this breakthrough result. This discovery and many other outstanding achievements of the LHC raise great hopes: The imminent restart of the physics programme in 2015 will increase the sensitivity by an order of magnitude while pushing the energy frontier to unprecedented values. We are looking forward to unravelling new mysteries that the universe may have in store for us.

Technologically, the LHC was—and continues to be—a significant challenge, demanding numerous innovations and breakthroughs in areas so diverse as magnet development, detector sensitivity and robustness, and large-scale computing—to mention only a few. So far, these challenges have been well met, thanks to the ingenuity and perseverance of the staff at CERN and at the many other institutions around the world involved in the realisation of the LHC and its detectors.

The LHC is not only a scientific and technological success: Rarely before has a scientific endeavour raised so much public interest and has received so much attention in the media. The LHC fascinates the young and the old alike, and it increases the awareness for scientific and technical questions. The public at large is convinced of the importance—and the rewards—of fundamental research for today's societies. It is my particular pleasure to see that it attracts pupils and students to get involved with the natural sciences, a prerequisite to providing a next generation of well-educated and responsible experts that can tackle the demanding problems the twenty-first century will bring along.

Finally, and perhaps most importantly, the LHC is a political triumph. It is a truly global endeavour, supported by thousands of scientists, engineers and technicians from all over the world who work on a common project based, to a large extent, on a shared vision and driven by curiosity and enthusiasm. The LHC bridges cultural and political boundaries and economic disparities, and it shows what can be achieved when human minds are driven by a common goal. The collaborative spirit of the LHC is unparalleled.

Last year, 2014, CERN turned 60. Sixty years, during which CERN has always managed to enthral scientists and non-scientists alike. Today, more than ever, CERN is a truly global laboratory, with now 21 member states reaching beyond Europe and a significant number of associate and observer states or applicants (Brazil, Pakistan, Russia, Turkey, Ukraine and others). These countries form an impressive global network of scientific spirit, which is an ideal basis for scientific success and progress, and which indirectly fosters understanding and peace among all nationalities involved. Science can and should be a prime example for worldwide coexistence and cooperation and may serve as a stronghold in international development.

The LHC may still be in its infancy—but the long shutdown that will soon come to an end constituted a significant milestone. This book tries to compile the essence of our knowledge gained or corroborated at the LHC at this specific point in time, and to present it in a way that keeps its value independent of newer and still more exiting results rolling in. I am confident that young and also more senior physicists will find it an educating and fascinating reading.

Geneva Rolf-Dieter Heuer

Preface

Early in 2013, shortly after the beginning of the first long LHC shutdown and after the discovery of a Higgs-like particle, there seemed to be a common desire to take stock of the scientific harvest of Run 1 and to aim for a modest extrapolation into the future. It was clear from the start that such a summary could only be a snapshot—in spite of its size, inertia is small at CERN and discussions are moving fast at the LHC—and that most results presented would very quickly be "outdated" after their publication—at the latest once first results from Run 2 would start to appear.

We—the authors and the editor—therefore decided to take a slightly relaxed view, concentrating on the impact of the LHC for its main fields of investigation, on the most striking results of Run 1, and on the basic methods and techniques that were used to achieve the latter—methods and techniques that were independent of time and would still be in use for later data-taking periods. All this should be presented in a modestly pedagogical way, taking a slightly historical perspective (e.g. by comparing LHC achievements with results achieved previously at the Tevatron or elsewhere), and garnishing everything with the relevant references. We hoped that a few "text book" results could also be included, and time will show.

In short, we wanted to write a book that was comprehensive, easy and fun to read, and useful both for younger scientists with a wish to familiarise them with certain aspects of LHC physics and for more senior physicists who were looking for an overview on specific topics or for a rather complete set of references. It is now up to the reader to decide whether or not we succeeded with this ambitious goal.

The structure of the book is as follows: A first part, consisting of Chaps. 1–3, describes the basics of the LHC: The first chapter—"The Large Hadron Collider—Background and History"—discusses the motivation for and the genesis of the LHC project. It also sketches the history of proton–proton (or proton–antiproton) collider physics and the relevant predecessor machines and projects, and it gives an overview of the LHC financing and of the history of the experimental LHC collaborations. The second chapter—"A Journey to the Heart of the LHC"—presents the technical side of the LHC construction, commissioning and operation. The third chapter—"The LHC Detectors"—discusses the involved technologies and the performances of the main LHC experiments.

The following Chaps. 4–11 cover the main physics topics at the LHC, from the well-established (the electroweak Standard Model and QCD, Higgs physics, top, flavour and heavy-ion physics) to the more speculative (searches for supersymmetry and searches for other, more exotic physics beyond the Standard Model). The book concludes with an outlook chapter—"Perspectives on the Energy Frontier"—that tries to bundle the conclusions of all other chapters and to translate them into a look into the future of our field of high energy physics.

A few technical remarks: We tried to be as consistent in the notation throughout the whole book as possible—failure to achieve this is entirely due to the editor, as are all other shortcomings and mistakes that might have escaped the editing and proofreading process (please send any errors you find to thomas.schoerner@desy.de). Sticking to the guidelines set by the LHC experimental collaborations, and in order to prepare a book which presents final results that will not be obsolete tomorrow, we decided to only use published and publicly available results as references. This rule has only been violated in very few, well-motived places, e.g. in cases where journal publications do not exist and are also not foreseen. Throughout the book, the convention $\hbar = c = 1$ is used.

Writing a book like the present one is a major challenge, and it involves the engagement and goodwill of many people who deserve deepest appreciation and gratefulness. First and foremost, I would like to thank all authors who—despite their numerous other demanding commitments and responsibilities—have shown great enthusiasm and a strong will to endure the inconveniences imposed upon them by their editor until the end of the project. The editor is in particular indebted to the following persons for their support, information and critical comments (in alphabetical order): Eckhard Elsen, Lyn Evans, Peter Jenni, Burton Richter, Herwig Schopper, Volker Sörgel, Florian Sonnemann. Then, of course, a lot of technical and organisational support is required. I would specifically like to thank Claus Ascheron from Springer Publishing for his readiness to constantly answer questions. Ian Brock (Bonn University) provided his LaTeX framework, which has served me extremely well for the third book project in a row. Thanks Ian! I am very grateful for Kati Brock's superb work on the layout and style of the figures—Kati, thanks a lot for our by now well-established cooperation! A very warm "thank you" goes to Jorge Cham for his intriguing title picture. I finally would like to thank my family for their endurance during the final editing of the book, which did not always perfectly match our family agenda.

Hamburg Thomas Schörner-Sadenius

Contents

About the Authors

Reyes Alemany-Fernandez graduated in physics at the University of Valencia where she received her Ph.D. in experimental physics in 1999 for her work on the search for supersymmetric particles with the DELPHI experiment. Afterwards she worked as a research physicist at CERN and at the Experimental Physics Institute (LIP) in Lisbon, where she contributed to the design of the CMS tracker alignment system and of the electromagnetic calorimeter data acquisition and trigger system. Since 2006 she is CERN staff member in the Accelerator Department where she works as "Engineer In Charge of LHC". She is at the heart of the commissioning and operation of the accelerator and the responsible person for setting-up the LHC as a proton-nucleus collider.

Ralf Averbeck studied physics in Münster and Giessen where he received his Ph. D. in 1996 on neutral-meson measurements with the TAPS experiment. After working as a post-doctoral researcher on the FOPI experiment at GSI, in 1999 he was awarded a Feodor Lynen fellowship of the Alexander von Humboldt Foundation at Stony Brook University and Brookhaven National Laboratory where he joined the PHENIX experiment at RHIC. In 2002 he became a member of the faculty of Stony Brook University as a research assistant professor. In 2008 he accepted a position as senior staff scientist at GSI where he currently works on the ALICE experiment at the CERN LHC. Being interested in many aspects of heavy-ion collisions at high energies, his particular expertise is on heavy-flavour production in such collisions.

Philip Bechtle studied at the Universities of Dortmund and Hamburg, where he worked on the HERA-B and OPAL experiments, on studies for the ILC, and on interpretations of supersymmetry. As a post-doctoral researcher, he worked for the BABAR experiment at SLAC before joining DESY as a Young Investigator Group Leader to work on SUSY searches at ATLAS and on further phenomenological interpretations and ILC studies. He habilitated at the University of Bonn and is still occupied there with the above-mentioned topics.

Maarten Boonekamp studied physics at the University of Orsay and joined DAPNIA (now IRFU) for his Ph.D., which involved Higgs-boson searches with the DELPHI experiment at LEP. He authored several prospective studies on the observation of the Higgs boson in diffraction at the LHC, and then specialised on electroweak physics. He is a member of the ATLAS experiment, where he participated in the construction and calibration of the electromagnetic calorimeter and performed measurements of *W*-boson and *Z*-boson production. His current main project is a precise measurement of the *W*-boson mass.

Tancredi Carli studied physics and philosophy at the Universities of Göttingen, Munich and Paris where he also obtained his Ph.D. working on the H1 experiment in 1992. He worked on calorimeters, QCD measurements and searches for new particles at the H1 and ZEUS experiments at the electron–proton collider HERA. Since 2002 he is research staff scientist at CERN where he works for the ATLAS experiment at the LHC. He worked on the ATLAS electromagnetic and hadronic barrel calorimeters, on QCD measurements, and on top physics.

Stefan Dittmaier studied physics at the University of Würzburg where he received his Ph.D. in theoretical physics in 1993. He worked as a scientist at the Universities of Bielefeld and Vienna, at CERN and at DESY before becoming a senior scientist at the MPI for Physics in Munich in 2002. Since 2009 he holds a professorship for theoretical physics at the University of Freiburg. His field of research comprises precision calculations for electroweak and QCD processes at high-energy colliders as well as techniques and concepts in perturbative quantum field theory.

Frank Ellinghaus studied physics at the University of Münster, working in nuclear physics at the KVI Groningen for his diploma. For his Ph.D. he moved to DESY in order to work on the HERMES experiment at HERA, where he received his Ph.D. from Humboldt University Berlin in 2004. As a post-doctoral researcher at the University of Colorado he moved to Brookhaven National Laboratory to work on the PHENIX experiment at RHIC. He joined the University of Mainz and the ATLAS experiment in 2008 with a fellowship from the German Helmholtz Alliance "Physics at the Terascale". His research focuses on Standard Model precision measurements and searches for physics beyond the Standard Model.

Eckhard Elsen was awarded a Feodor Lynen fellowship of the Alexander von Humboldt Foundation at SLAC (Stanford) after earning his Ph.D. on the JADE experiment at PETRA in Hamburg in 1981 with a study on 3-jet production. Initial studies on the production of heavy quarks in e^+e^- collisions were continued in 1984 at Heidelberg with a measurement of the electroweak asymmetry of the b quark. Joining the H1 experiment in 1990, he concentrated on advancing the hardware trigger of the experiment and continued with the study of electroweak physics at DESY. As the spokesperson, he led the H1 experiment through its luminosity upgrade phase before spending a 1-year sabbatical on the BABAR experiment at SLAC. Since 2004 he is promoting the case for the International Linear Collider (ILC), which includes the advancement of the superconducting technology for the collider. He is a professor at the University of Hamburg.

Ingrid-Maria Gregor is an expert in detector development for high-energy physics. She completed two diplomas, one in physics engineering (1994) and one in physics (1998), both at Wuppertal University, where she also received her Ph.D. in 2001. She worked on detector projects for the particle physics experiments DELPHI, HERMES, ZEUS and ATLAS, mostly in the field of silicon-tracking detectors. As post-doctoral researcher she developed and constructed the silicon-strip recoil detector for the HERMES target region (2002–2005). Since 2005 she is a senior staff scientist at DESY, where she first coordinated the ZEUS uranium calorimeter group. During the EUDET project, she led the development of the high-resolution pixel telescope that is now being used for many R&D studies for pixel and strip detectors around the world. Currently, Gregor is the project leader of the ATLAS strip detector for the HL-LHC, organising the efforts towards the replacement of the ATLAS inner detector.

Christophe Grojean is, since 2012, a research professor at the Institució Catalana de Recerca i Estudis Avançats (ICREA) working at the Institut de Fisica d'Altes Energies (IFAE) in Barcelona. He got his Ph.D. at the University Paris XI, Orsay, and has worked at CEA-Saclay as a permanent research staff member. He worked for 2 years at the University of California at Berkeley as a post-doctoral researcher and spent 1 year at the University of Michigan at Ann Arbor as a visiting professor. He spent 7 years as a junior staff in the theory unit of the physics department at CERN from 2006 to 2012. His topic of research includes Higgs physics, various aspects of physics beyond the Standard Model, and astroparticles.

Stephanie Hansmann-Menzemer studied at the Universities of Karlsruhe and Grenoble and received her Ph.D. in 2003 at Karlsruhe University on track reconstruction for the CDF vertex detector. She then got a DFG post-doctoral fellowship and joined the MIT group at Fermilab as a guest scientist. There she was involved in the discovery of the $B_s^0 - \overline{B}_s^0$ oscillation frequency. In 2006, Stephanie moved to the Physikalische Institut of Heidelberg University as Emmy Noether junior research group leader. There she joined the LHCb Experiment. In 2009, she became professor in Heidelberg. Her research focuses on precision measurements in the flavour sector.

John W. Harris is professor of physics at Yale University and fellow of the American Physical Society. He received his BS in physics from the University of Washington (Seattle) and Ph.D. from Stony Brook University. His primary field of research is relativistic heavy-ion physics. He was a senior scientist at Lawrence Berkeley Laboratory before moving to Yale and has worked on various experiments at Brookhaven Laboratory and CERN. He was awarded a Humboldt fellowship and a Humboldt senior award with stays at GSI, Frankfurt University, and CERN.

Bernhard Holzer studied physics at the Universities of Heidelberg and Grenoble and specialised soon on the design of accelerators and storage rings. He took part in the construction and commissioning of the heavy ion storage ring TSR at the Max Planck Institute in Heidelberg where he also got his diploma. For his Ph.D. he moved to DESY to design a new high-luminosity e^+e^- collider, a so-called B factory. Being just in time for the final design and optimisation of the new HERA collider, he took over the responsibility for the HERA lattice and beam optics, and became soon coordinator and responsible optics physicist of the superconducting ring. After the shutdown of HERA he moved to CERN and took part in the LHC start-up and Run 1 as coordinator of the collider. In parallel he is involved in several future accelerator projects at CERN and teaching in a large number of accelerator schools.

Kerstin Hoepfner studied physics and crystallography in Berlin and obtained her Ph.D. in particle physics while at CERN working on the CHORUS neutrino oscillation experiment. After a Leopoldina fellowship at the Technion Haifa, Israel, she accepted a post-doctoral position at DESY Hamburg to coordinate the vertex detector upgrade for the HERA-B experiment. In 2001 she joined the CMS experiment and moved to RWTH Aachen, where she now holds the position of a senior researcher. In between, from 2005 to 2007, a research stay brought her back to CERN and to the University of Göttingen as a visiting professor. After 6 years of leading the Aachen muon detector construction and commissioning effort, she transitioned to searches for new physics and LHC phase-II upgrade studies.

Karl Jakobs is professor of experimental particle physics at the University of Freiburg in Germany. He studied physics at the University of Bonn and obtained his Ph.D. at the University of Heidelberg. After a research fellowship and staff positions at CERN and at the Max Planck Institute for Physics in Munich he was appointed professor at the University of Mainz (1996–2003) and at the University of Freiburg (since 2003). He has been engaged in experiments at CERN (UA2, ALEPH, ATLAS) and Fermilab (DØ). His main research activities are the study of the properties of the Higgs boson, the search for supersymmetric particles, and research and development activities on semiconductor detectors.

Kevin Kröninger studied at the University of Bonn, where he worked on top-quark physics with the DØ experiment. His Ph.D. work, conducted at the Max Planck Institute for Physics in Munich, was on novel experimental techniques involving germanium detectors used in searches for neutrinoless double-beta decay with the GERDA experiment. He is involved in the ATLAS experiment and habilitated with his research on top-quark properties at the University of Göttingen in 2012. He was appointed professor of experimental particle physics at the TU Dortmund in 2014.

Andreas B. Meyer studied physics at Hamburg University where he received his Ph.D. in 1997. After post-doctoral positions in Berkeley on the BABAR experiment and Hamburg University on the H1 experiment, he became senior scientist at DESY and "Privatdozent" at Hamburg University in 2005. Since 2006 he has been working on the CMS experiment, where he has held responsibilities for the CMS data quality monitoring systems (2007–2010) and in the data analysis of top-quark physics (since 2012).

Matthias Mozer studied physics at Ohio University and the University of Heidelberg, where he obtained his Ph.D. on diffractive physics with the H1 experiment. He moved to the Free University of Brussels to start working on the

CMS experiment. In 2010 he moved to CERN as a fellow and began studies involving electroweak bosons. Since then he has taken up a position at KIT, where his work in electroweak physics continues.

Ulrich Nierste studied physics at the University of Würzburg and SUNY at Stony Brook. In 1995 he received his Ph.D. from TU Munich and later was post-doctoral researcher at DESY, Fermilab, and CERN. In 2002 he moved to a tenure-track position at Fermilab and became a tenured scientist in 2005. In the same year he was appointed to a professorship at the University of Karlsruhe, which in 2009 merged with another institution to the Karlsruhe Institute of Technology. His main research interest is flavour physics, with emphasis on the search for new physics in flavour-changing processes of B, D and K mesons.

Thorsten Ohl studied physics at TH Darmstadt and received his Ph.D. in theoretical physics there in 1990. He worked as a post-doctoral researcher at Darmstadt, DESY, Harvard and Würzburg. Since 2007 he is a senior scientist at Würzburg University and was appointed supernumerary professor of theoretical physics there the following year. His research activities cover the range from Monte Carlo event-generator construction and the collider phenomenology of physics beyond the Standard Model to mathematical physics.

Tilman Plehn is professor for theoretical physics at Heidelberg University. He studied physics in Heidelberg, Zürich, and Hamburg/DESY. After post-doctoral positions at Madison/Wisconsin, CERN, and the Max Planck Institute for Physics, he became a lecturer/reader in Edinburgh. In 2008 he moved to Heidelberg. He has been working on many aspects of LHC physics, including supersymmetry, Higgs signals and property measurements, top-quark identification, and QCD.

Günter Quast received his Ph.D. at University of Siegen on the "Measurement of Direct CP Violation in Neutral Kaon Decays" with the NA31 experiment at CERN. He then worked as a CERN fellow and DESY research associate on the OPAL experiment. He wrote his habilitation on "Measurement of Z Boson Parameters at LEP" at Mainz University, where he worked on the experiments ALEPH and ATLAS. In 2001, he became a professor at the University of Karlsruhe, now Karlsruhe Institute of Technology (KIT), where he now works on the CMS experiment. Among his responsibilities is the Grid computing for the LHC; his research interests are electroweak physics, QCD and Higgs physics.

Klaus Rabbertz obtained his Ph.D. in 1998 at the RWTH Aachen for research performed within the H1 experiment at the electron-proton collider HERA. As a CERN research fellow he worked within the OPAL experiment at the e^+e^- collider LEP. Since 2002, he is a member of the CMS collaboration at the LHC as senior scientist for the University of Karlsruhe, now Karlsruhe Institute of Technology (KIT). Following his convenership of the CMS working group on QCD from 2007–2008, he concentrated on precision measurements with the LHC jet data and the determination of the strong coupling constant.

Christian Sander studied physics at the University of Karlsruhe where he also received his Ph.D. in astro-particle physics. After a post-doctoral position at Hamburg University, he became a junior professor for experimental particle physics in 2011. His main research interest is physics beyond the Standard Model, in particular searches for supersymmetry at the LHC.

Björn Schenke studied physics at Justus Liebig University in Giessen and Goethe University in Frankfurt, where he received his Ph.D. in 2008. He worked as a Tomlinson fellow at McGill University in Montreal, and Goldhaber fellow in the Nuclear Theory Group at Brookhaven National Laboratory, Upton. In 2014 he became associate scientist at Brookhaven National Laboratory. His main interests are the theory and phenomenology of heavy-ion physics, ranging from colour-glass condensate effective theory to relativistic fluid dynamics to thermal field theory and Monte Carlo simulations of jet quenching.

Thomas Schörner-Sadenius studied physics at the Universities of Hamburg and Munich. He held post-doctoral positions in Munich, at CERN and in Hamburg, working on a number of different experiments (OPAL, H1, ATLAS, ZEUS, CMS). In 2008 he joined DESY where he currently is acting as Scientific Manager of the German Helmholtz Alliance "Physics at the Terascale".

Steffen Schumann studied physics at the TU Dresden where he also received his Ph.D. in theoretical physics in 2008. He worked as a post-doctoral researcher in Edinburgh and Heidelberg before moving to Göttingen where, in 2011, he became a junior professor for theoretical particle physics and phenomenology. His main field of interest is the development of Monte Carlo event generators for high-energy collider experiments, with a particular focus on parton-shower simulations and multi-particle final states.

Arno Straessner studied physics at the RWTH Aachen where he received his Ph. D. in the year 2000. After a CERN research fellowship he joined the University of Geneva as maître assistant in 2003. In 2008 he became junior professor at the TU Dresden and in 2014 professor for experimental particle physics, also in Dresden. He worked on the L3 and ATLAS experiments at CERN, mainly on electroweak and Higgs physics, as well as on detector development.

Peter Uwer studied physics at the RWTH Aachen where he also received his Ph.D. in theoretical physics in 1998. He worked as a scientist in Saclay, Karlsruhe and at CERN before he became professor for theoretical particle physics at the Humboldt University, Berlin, in 2008. His main research interests are QCD and top-quark physics.

Georg Weiglein is leading scientist in the theory group at DESY, Hamburg. He studied physics at the University of Würzburg and SUNY at Stony Brook. After post-doctoral positions at Bielefeld, Karlsruhe and CERN, he was appointed as a lecturer at the IPPP, University of Durham, in 2001. At Durham he was promoted to reader and full professor before moving to DESY in 2009. His field of research is the phenomenology of the electroweak and strong interactions, in particular Higgs physics, physics beyond the Standard Model and electroweak precision physics.

Contributors

Reyes Alemany-Fernandez CERN, Geneva 23, Switzerland

Ralf Averbeck EMMI, GSI Helmholtzzentrum für Schwerionenforschung, Darmstadt, Germany

Philip Bechtle Physikalisches Institut, Universität Bonn, Bonn, Germany

Maarten Boonekamp CEA, IRFU, Gif-sur-Yvette Cedex, France

Tancredi Carli CERN, Geneva 23, Switzerland

Stefan Dittmaier Albert-Ludwigs-Universität Freiburg, Freiburg, Germany

Frank Ellinghaus Institut für Physik, Johannes Gutenberg-Universität Mainz, Mainz, Germany

Eckhard Elsen DESY, Hamburg, Germany

Ingrid-Maria Gregor DESY, Hamburg, Germany

Christophe Grojean ICREA and IFAE, Universitat Autònoma de Barcelona, Bellaterra, Barcelona, Spain

Stephanie Hansmann-Menzemer Physikalisches Institut, Universität Heidelberg, Heidelberg, Germany

John W. Harris Department of Physics, Yale University, New Haven, CT, USA

Kerstin Hoepfner Rheinisch-Westfälische Technische Hochschule Aachen, Aachen, Germany

Bernhard Holzer CERN, Geneva 23, Switzerland

Karl Jakobs Albert-Ludwigs-Universität Freiburg, Freiburg, Germany

Kevin Kröninger Technische Universität Dortmund, Dortmund, Germany

Andreas B. Meyer DESY, Hamburg, Germany

Matthias Mozer Institut für Experimentelle Kernphysik, KIT, Karlsruhe, Germany

Ulrich Nierste Institut für Theoretische Teilchenphysik, KIT, Karlsruhe, Germany

Thorsten Ohl Universität Würzburg, Würzburg, Germany

Tilman Plehn Institut für Theoretische Physik, Universität Heidelberg, Heidelberg, Germany

Günter Quast Institut für Experimentelle Kernphysik, KIT, Karlsruhe, Germany

Klaus Rabbertz Institut für Experimentelle Kernphysik, KIT, Karlsruhe, Germany

Christian Sander Institut für Experimentalphysik, Universität Hamburg, Hamburg, Germany

Björn Schenke Physics Department, Brookhaven National Laboratory, Upton, NY, USA

Steffen Schumann II. Physikalisches Institut, Georg-August-Universität Göttingen, Göttingen, Germany

Thomas Schörner-Sadenius DESY, Hamburg, Germany

Arno Straessner Institut für Kern- und Teilchenphysik, Technische Universität Dresden, Dresden, Germany

Peter Uwer Humboldt-Universität zu Berlin, Berlin, Germany

Georg Weiglein DESY, Hamburg, Germany

Chapter 1
The Large Hadron Collider—Background and History

Thomas Schörner-Sadenius

Abstract This introductory chapter gives a short account of the history of the Large Hadron Collider (LHC) project, i.e. it describes the rationale for the LHC, the situation of high energy physics in the period in which the LHC was initially conceived, and the development of the project from first ideas to first beams in the machine. In doing so, some emphasis is put on the comparison of the LHC with other pp or $p\bar{p}$ collider projects, which are also discussed from a historical point of view. Finally, the development of the LHC experimental collaborations is sketched.

1.1 The LHC—A Marvel in Every Respect

The Large Hadron Collider (LHC) is clearly a "marvel of technology".[1] The collider itself, the cryogenics installations and the experiments were all ground-breaking endeavours at the technological frontier, and the sheer size and complexity of the machines and their intrinsic beauty fascinate scientists and laypersons alike. Also the necessary civil-engineering work posed numerous awe-inspiring challenges. But the LHC is remarkable also in many other, less technical, respects:

- The LHC lifespan from first ideas to the last publication will, according to plan, amount to about six decades—enough to touch the careers of four generations of scientists.
- The number of people involved in the creation or the exploitation of the project easily reaches 10,000, and the fact that the management of the project relies chiefly on common sense and commitment without strong hierarchy and only lean formalised

[1] The excellent book of the same name edited by Evans [1] provides an abundance of useful information about the subject. A short discussion especially of the transition from LEP to LHC is provided by Schopper [2].

T. Schörner-Sadenius (✉)
DESY, Notkestr. 85, 22607 Hamburg, Germany
e-mail: thomas.schoerner@desy.de

T. Schörner-Sadenius (ed.), *The Large Hadron Collider*,
DOI 10.1007/978-3-319-15001-7_1

responsibility sets a guiding example for other projects involving people from different backgrounds, cultures and nationalities. "The common cause seems to be a very strong motivator in keeping individual institutes on track [1]".

In order to understand and fully appreciate the immense effort that made the LHC a reality, a little digression into history is indicated.

1.1.1 The Origins of the LHC

The LHC—or the option of a hadron collider in the tunnel of the Large Electron-Positron Collider (LEP) at CERN—was reportedly first mentioned [3] by former CERN director general Sir John Adams who, in 1977, suggested that a potential LEP tunnel be made wide enough to accommodate a superconducting proton collider of above 3 TeV beam energy [4]. The late 1970 s were a period busy with exciting physics results and, at CERN, with LEP preparations. In fact, one of the arguments for a relatively large circumference for the LEP machine—which was conceived 2 years earlier, in 1975, and approved by CERN Council in 1981—was to avoid compromises to the energy of a potential hadron-collider successor of the electron-positron machine[2] [5].

By 1977, electron-positron colliders were well established, but hadron accelerators had so far exclusively been working in fixed-target mode—except for the Intersecting Storage Rings (ISR) at CERN, which, since 1971, were colliding protons with beam energies of up to 31.4 GeV (see Sect. 1.2.2).

Lepton colliders had already provided interesting results, like the co-discovery (together with a fixed-target hadron machine) of the J/ψ and thus of charm in 1974. However, the wish of particle physicists to go to ever higher centre-of-mass energies posed a severe problem to lepton colliders: The energy loss due to synchrotron radiation increases with the fourth power of a particle's energy. The fact that the energy loss per turn also scales with the inverse of the bending radius favours large accelerators—like LEP. Since there are, naturally, restrictions to the possible size of accelerators, also the achievable energy is limited for circular lepton colliders. However, the energy loss also goes with the inverse of the particle mass to the fourth power. So one way to realise collisions at higher energies is to choose protons as beam particles. Hadron colliders—so it seemed—are the only way towards the discovery of new physics phenomena at highest energies. And ideas for such new phenomena abounded already in the late 1970s!

---

[2] When, in 1981, the decision about location and circumference of the LEP tunnel had to be taken in the light of geologically dangerous ground beneath the Jura mountains, then CERN director general Herwig Schopper argued that the suggested smaller circumference of 22 km would make a successful pp collider in the LEP tunnel impossible [2].

1.1.2 The Picture of the Microcosm Around 1977

By 1977, a large fraction of what came to be known as the Standard Model (SM) of particle physics was well established.[3] At the same time, numerous questions remained unanswered, which were at the top of the research agenda of high energy physics.

- Already in the 1960s, the "zoo" of strongly interacting particles had been organised with the invention of the "eightfold way" and of quarks by Gell-Mann, Zweig, Ne'eman and others. Gell-Mann was awarded the 1969 Nobel Prize in Physics for his contributions.
- There was a model for the generation of mass for gauge bosons—the BEH mechanism invented around 1964 by Brout, Englert, Guralnik, Hagen, Higgs and Kibble, which led to the 2013 Nobel Prize in Physics for Englert and Higgs. There was, however, no direct experimental evidence of the existence of a Higgs particle that was a necessary ingredient of the theory.
- Around 1967, electroweak interactions and the BEH mechanism had been merged by Glashow, Salam and Weinberg (GSW)[4] into a renormalisable gauge theory (rewarded by the 1979 Nobel Prize in Physics). The gauge bosons of this theory ($W^{\pm}$, Z^0) were only discovered in 1983 at the Sp$\bar{\text{p}}$S—although charged-current interactions were already well established.
- The GIM mechanism (Glashow–Iliopoulos–Maiani, 1970) had postulated the existence of a fourth quark beyond the well established u, d and s quarks; the discovery of the fourth—the c or "charm"—quark through the measurement of J/ψ mesons in 1974 in both e^+e^- collisions and fixed-target experiments beautifully confirmed this hypothesis (1976 Nobel Prize for Richter and Ting).
- In 1973, the GARGAMELLE experiment at CERN had discovered neutral-current interactions in neutrino experiments, thus indirectly confirming the existence of heavy neutral gauge bosons (Z^0 bosons) as predicted by the GSW theory. GARGAMELLE also discovered that only about 50 % of the proton's momentum is carried by its charged constituents, the quarks.
- Another important discovery of the year 1973 was that of asymptotic freedom—a key ingredient of QCD—by Gross, Politzer and Wilczek (2004 Nobel Prize).
- Also in 1973, at the CERN ISR collider, high-p_{T} particles had been observed. This and other breakthroughs in strong-interaction physics made quantum chromodynamics (QCD), as formulated in 1973, a serious contender for a gauge theory of strong interactions. Gluons, the postulated gauge bosons of QCD, were only discovered at the PETRA e^+e^- collider at DESY in Hamburg in 1978.
- A third charged lepton, the τ lepton, was discovered in 1975 by Perl and collaborators at SPEAR (1995 Nobel Prize for Perl).

[3] See the excellent book by Cahn and Goldhaber [6] for a historical account of particle physics.

[4] See the two articles [7, 8] for a historical perspective.

- At Fermilab in 1977, L. Lederman and collaborators obtained evidence for a fifth (the *b* or "beauty") quark. The 1979 proceedings of the LEP Summer Study 1978 consequently stated that "with a little theoretical help, we can already take for granted" [9] the existence of the remaining particles of this third family (the *t* or "top" quark and the τ neutrino, which finally were discovered in 1995 and 2000 by the Tevatron experiments and the DONUT collaboration at Fermilab, respectively).

1.1.3 Arguments for the LHC and First Design Parameters

So there remained a lot to do before even the Standard Model would be fully established—not to talk of the many ideas about alternatives to or extensions of the Standard Model that were already around in the 1970s. At the 1984 ECFA-CERN workshop on a "Large Hadron Collider in the LEP Tunnel" [10], therefore, the main arguments for a multi-TeV hadron collider were the need to investigate the origin of mass (i.e. the role of the BEH mechanism) and to search for signs of unification beyond the Standard Model (i.e. to understand the true nature of the recently observed W and Z bosons).

Consequently, on the agenda of the LHC would be the search for the Higgs boson,[5] the understanding of the mechanism of electroweak symmetry breaking [12], the search for supersymmetry[6] "at a scale of 1 TeV, or below", as a "necessary and sufficient condition for [...] cancellations to occur" [11], the investigation of the phenomenology of *b* and *t* quarks, and the investigation of new forms of matter and, potentially, the quark-gluon plasma [12], among others.

The tool to achieve these goals was to be a proton-proton collider[7] of centre-of-mass energy between 10 and 20 TeV (1 TeV at constituent level) and with a luminosity of up to $10^{33}\,\mathrm{cm}^{-2}\,\mathrm{s}^{-1}$. It was understood that such an ambitious machine, which was to be housed in the LEP tunnel, required an extensive R&D programme especially on the magnets, for which a maximum field strength of 10 T was assumed.

After having established the physics agenda of the LHC and its basic properties (see also Chap. 2), as perceived in the late 1970s and early 1980s, we will now turn to a discussion of pp and $p\bar{p}$ colliders that preceded the LHC or that were planned or conceived as competitors—see Fig. 1.1.

[5] "The Higgs mechanism works, but it can hardly represent the whole truth: it's implementation [...] is far too ugly and arbitrary" [11].

[6] It is interesting to note that in the literature of that time, there is no connection drawn between supersymmetry and the phenomenon of dark matter, the existence of which had been postulated since the 1930s.

[7] The proton-antiproton option was also studied, but it was quickly understood that the necessary luminosity would be difficult to achieve with antiproton beams.

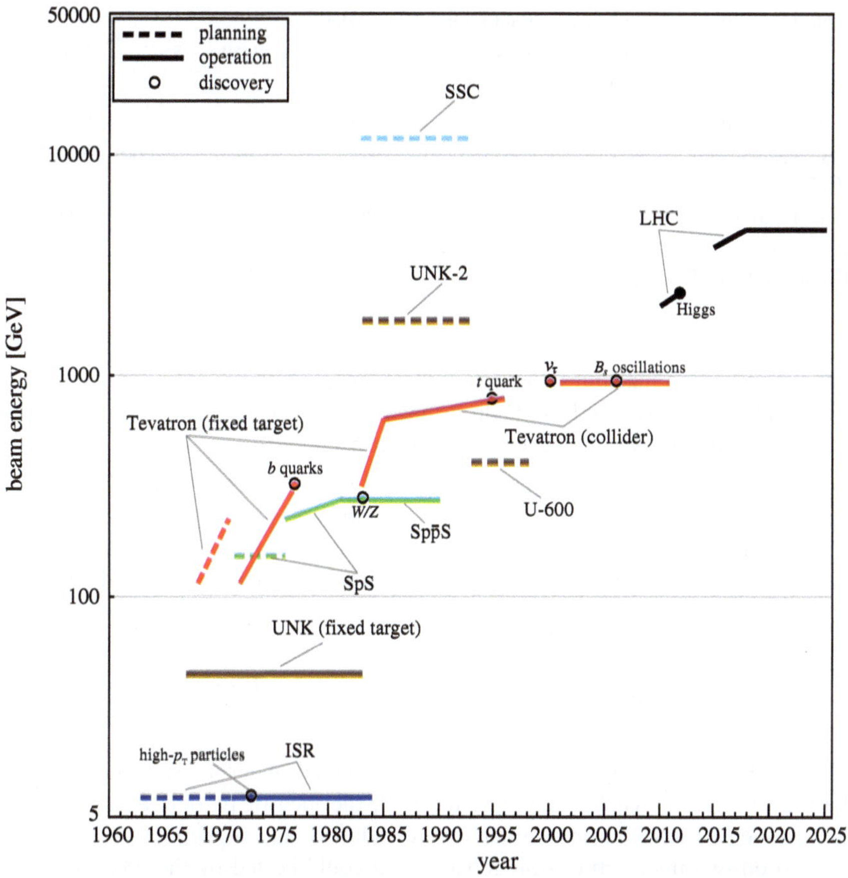

Fig. 1.1 An overview of pp and $p\bar{p}$ colliders, their beam energies and major achievements

1.2 $p\bar{p}$ and pp Colliders Before the LHC

1.2.1 Fixed-Target Experiments Versus Colliders

Fixed-target experiments had been experimenters' choice for many decades. The first mention of colliding-beam experiments is reportedly due to Wideröe who—not even working in the field of particle physics at that time—put forward the idea in 1943 and even registered a patent, which he finally received in 1953 [13, 14]. However, although people had of course realised the advantage of colliding beams with respect to fixed-target collisions in terms of usable energy, the particle densities obtained in accelerators in these day made colliders seem a very unrealistic option. This changed in 1957, when the idea of stacking particles into circular accelerators was first put forward by Kerst and collaborators [15]. Although the first thoughts about colliders

Table 1.1 Main parameters of important pp and $p\bar{p}$ colliders

	ISR	$Sp\bar{p}S$	Tevatron	UNK	SSC	LHC
Operation	1971–83	1981–89	1985–2011	–	–	Since 2008
Diameter [km]	0.3	2.2	2.18	6.61	27.7	8.49
Max. beam energy [GeV]	28/31.4	273/315 (450)	980	3000	20,000	7000
Max. luminosity [$10^{30}\,\text{cm}^{-2}\,\text{s}^{-1}$]	140	6	430	–	–	$\approx$8000
Dipole strength [T]	1.2	1.3	4.2	5	6.6	8.3

focused on hadron machines, the first realisations used electrons or, later, electrons and positrons. The research on hadron machines, which posed much more severe technical problems, was, however, always continued.[8]

In the following, the relevant precursor pp and $p\bar{p}$ machines to the LHC will briefly be discussed. A comparison of a few technical numbers is given in Table 1.1; Fig. 1.2 sheds light on the physics reach (in terms of parton luminosity versus reachable energy scale) of hadron colliders operating at different centre-of-mass energies.

1.2.2 Intersecting Storage Rings (ISR, 1971–1984)

Also at CERN, colliding hadron beams were pursued. After the successful start of the CERN Proton Synchrotron (PS) in 1959, the idea of particle acceleration in the foreseen CERN hadron collider was abandoned, and conceptual work instead focused on two intersecting storage rings that could be fed by the 28 GeV proton beams of the PS.[9]

In parallel, design work for the "CERN Electron Storage and Accumulation Ring" CESAR had started in 1960, and first beams were captured in December 1963. With CESAR, many important insights into the storage and stacking of particles could be gained, so that in 1964 a proposal for the proton-proton "Intersecting Storage Rings" (ISR) was finally put forward to CERN Council and approved in December 1965; construction began in early 1966 on a site in France, just across the border from the the CERN Swiss site. In January 1969, the ISR Committee was set up; it's task was to review and select proposals for experiments to be conducted at the ISR.

The two ISR rings were ready in October 1970 and January 1971, respectively; the official inauguration of the machine (and the begin of regular operations) took place on 16 October 1971, after first collisions had already been produced on 27

[8] See the article [16] about a discussion, from 1973, of fixed target versus colliding beams.

[9] See the review article on CERN synchrotrons by Brianti [3] for more details on the CERN machines since the 1950s.

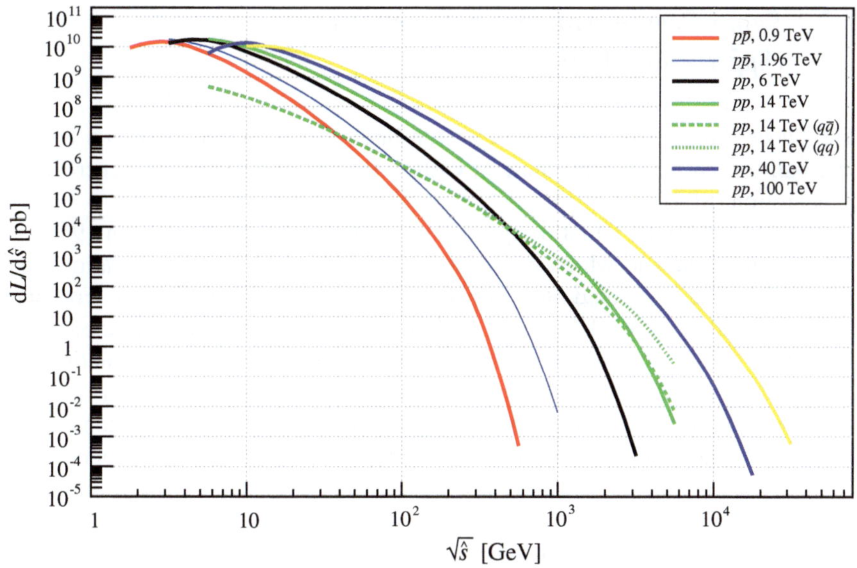

Fig. 1.2 Parton-luminosity comparisons for various pp and $p\bar{p}$ scenarios. Material from A. Cooper-Sakar

January 1971. ISR operation stopped with the dump of the last proton beams in the morning of 23 December 1983. See [17] for a thorough overview of the ISR machine and physics.

During the ISR's lifespan, important steps were taken in understanding pp colliders—among them the first tests of stochastic cooling in the first half of the 1970s (the idea of stochastic cooling had been brought forward by S. van de Meer as early as 1968). In many respects, the ISR served as a test plant for numerous concepts that were useful for all subsequent hadron colliders. "The accelerator physicists learned how to build proton storage rings that overcame the lack of synchrotron radiation damping [18]". Table 1.1 presents some of the important parameters of the ISR.

On the physics side, the experiments at the ISR—which could be set up at eight points where the counter-rotating beams crossed each other—brought many exciting insights into the physics of strong interactions [17], the most spectacular probably being the discovery of single high-p_T particles in the momentum range 2–9 GeV in 1972 [19–21] with ensuing discussions about the existence of hadronic "jets". At the time, the high-p_T phenomenon caused considerable interest because it could readily be explained in the parton model that had received support from deep-inelastic scattering experiments [22]; it was also in line with findings in e^+e^- collisions at SPEAR. However, in hadron collisions, there was, for a considerable time, no unambiguous and universally supported evidence. See [23] for a thorough discussion of the subject. A further important finding of the ISR and its experiments is the continued logarithmic rise of the pp cross section, when it was expected to flatten out at ISR energies.

1.2.3 Super Proton-Antiproton Synchrotron (Sp$\bar{p}$S, 1981–1989)

The Super Proton-Antiproton Synchrotron Sp$\bar{p}$S was a proton-antiproton collider that built on an existing machine at CERN, the proton accelerator Super Proton Synchrotron (SPS) [24].

The SPS, of which a first design was suggested to CERN Council in 1964 as a 300 GeV machine and which was approved in 1971, had been in operation since 1976.[10] The machine could ultimately deliver proton beams of 450 GeV. For many years it served as the work horse for CERN's fixed-target physics programme.

In June 1976, three quite different proposals for proton-proton or antiproton-proton colliders had been handed in to the Fermilab Program Committee [14]. One of these proposals [25], was based on colliding beams of antiprotons and protons each at about 1 TeV. At Fermilab, none of the three proposals was immediately supported—although finally, the one by Cline and collaborators evolved into the Tevatron and the CERN Sp$\bar{p}$S. The Sp$\bar{p}$S, as detailed for the first time in [26], was designed as an extension of the existing SPS; it was intended as a tool for the search for the massive weak vector bosons that were predicted by the Glashow–Salam-Weinberg theory (and for other high-p_T phenomena, since the ISR had raised interest in high-p_T physics). The plan, however, was also criticised: In 1977, when the Sp$\bar{p}$S was first discussed as a CERN project, the SPS—which had already been quite an expensive project—had only been in operation for a few years; at the same time the planning for a large e^+e^- collider had just started in earnest, driven by an ECFA recommendation in May 1977 to the CERN SPC that "an electron-positron storage ring of about 200 GeV c.m. energy, possibly with an initial phase of 140 GeV, be considered by the high-energy physics community as the prime candidate for a major European project in the 1980s".[11] To some, for example John Adams, trying to squeeze in the Sp$\bar{p}$S as a slightly less expensive project seemed like asking too much from the funding agencies.

The decision to convert the SPS machine—a process requiring the first large-scale underground excavations for experiments at CERN—was nevertheless taken by CERN Council in 1979, and first collisions were observed in the experiments on 10 July 1981.

The Sp$\bar{p}$S provided $p\bar{p}$ collisions at centre-of-mass energies of 546 and 630 GeV, mainly to the UA1 and UA2 experiments. For some exceptional runs, when the Sp$\bar{p}$S rings were pushed to 450 GeV each, collisions could also be seen at 900 GeV. The main user of these was the UA5 experiment that, among other things, studied hadron production in high-energy collisions. UA1 recorded a few minimum-bias data during the 900 GeV runs.

---

[10] The SPS first managed to deliver beams at 400 GeV on 17 June 1976, which would have been a world record had not the Tevatron achieved 500 GeV just 4 weeks before, on 14 May 1976, see Sect. 1.2.4.

[11] The same report still talks about a "hypothetical site" for LEP construction.

The Sp$\bar{\text{p}}$S had been designed and built for the discovery of the weak gauge bosons, and it fulfilled its purpose within the remarkably short time of less than 3 years: In a seminar on 20 January 1983, Rubbia (for the UA1 collaboration) and Luigi Di Lella (for UA2) presented 6 and 4 W boson candidates, respectively, and at a press conference on 25 January, the discovery of charged weak gauge bosons was announced. The discovery of the neutral boson, the Z^0, followed soon after [27–30], as did the 1984 Nobel Prize that was awarded to Rubbia and Simon van der Meer, the inventor of stochastic cooling.

Operation of the Sp$\bar{\text{p}}$S continued after the discoveries and the Nobel Prize. Among other things, the experiments were searching for top quarks and for supersymmetry. All in all, the machine was a great success, and—as Richter [14] put it—it culminated "with [...] an essential confirmation of the Standard Model. van der Meer's invention made it possible, and Rubbia's drive and determination brought it about." For a short time, in the years before the Tevatron at Fermilab resumed operations as a $p\bar{p}$ collider in 1985 (Sect. 1.2.4), the leadership in hadron collisions had been in Europe.

1.2.4 Tevatron (1983/1985–2011)

Like the Sp$\bar{\text{p}}$S, which was based on the fixed-target proton accelerator SPS, the Tevatron $p\bar{p}$ collider has its roots in a proton accelerator, the "Main Ring" of the US National Accelerator Laboratory (NAL), which today is known as Fermilab or FNAL.

The NAL Main Ring [31], designed as a proton accelerator with a maximum energy of 200 GeV, had been in construction since 3 October 1969 (informal ground breaking by NAL director R. Wilson). The final of its 1014 dipole magnets was put in place on 16 April 1971, just two and a half years later. The first proton beam was steered through the ring on 30 June 1971 and accelerated to 7 GeV, and on 1 March 1972 the design energy of 200 GeV was reached. The maximum energy of 0.5 TeV was reached on 14 May 1976. The Main Ring went out of fixed-target physics operation in mid-1982 [32]. It continued its life as injector of 150 GeV protons for the later "Energy Doubler" (or Tevatron, as especially the colliding-beam machine was called) until September 1997 [33]. A historical account of the history of the Main Ring can be found in [31]. See also [34] for a history of FNAL accelerators.

The upgrade of the Main Ring to a (in fact the world's first) superconducting machine [35, 36] with about twice the beam energy (hence the name "Energy Doubler") and subsequently to a $p\bar{p}$ colliding-beam machine [37] (the Tevatron) was approved at Fermilab in 1978, with the design goals of a centre-of-mass energy of 1.8 TeV and a luminosity of $10^{30}\,\text{cm}^{-2}\,\text{s}^{-1}$. The Department of Energy (DOE) authorised Fermilab to build the machine on 5 July 1979. A major milestone was reached when, on 3 July 1983, protons were accelerated to 512 GeV, which at that time constituted a new world record. In parallel to the construction of the Energy Doubler, the antiproton source was developed. Routine fixed-target operation at 400 GeV started again on 1 October 1983, and on 16 February 1984 the new record beam energy of 800 GeV was achieved. The antiproton source was commissioned in 1985, and first $p\bar{p}$ collisions were observed in the then operational parts of the CDF detector on 13 October of that year, with a centre-of-mass energy of 1600 GeV. First 900 GeV beams were produced on 21 October, and on 30 November first collisions at 1800 GeV took place. The first substantial physics run between June 1988 and June 1989 allowed CDF to collect about $5\,\text{pb}^{-1}$ of data at 1800 TeV. Luminosity at that time was limited to about $1.6 \times 10^{30}\,\text{cm}^{-2}\,\text{s}^{-1}$.

The following years brought many improvements to core parts of the Tevatron machine, together leading to drastic increases in luminosity. During Run 1, from August 1992 to the beginning of a long shutdown in February 1996, about $180\,\text{pb}^{-1}$ where delivered to both DØ[12] and CDF (typical peak luminosity: $1.6 \times 10^{31}\,\text{cm}^{-2}\,\text{s}^{-1}$). By then, work on the new main injector (approved in October 1991) that would replace the old Main Ring had started. The machine, which was complemented by the "Recycler" for the storage of larger stashes of antiprotons than the old antiproton accumulator could hold, was ready for physics in 1999. In 2000, the Tevatron fixed-target programme came to an end [38, 39].

Tevatron Run 2 started in 2001, then with a centre-of-mass energy of 1960 GeV, and lasted until 2011, showing peak luminosities of more than $4 \times 10^{32}\,\text{cm}^{-2}\,\text{s}^{-1}$. Interestingly, the luminosity record for pp or $p\bar{p}$ colliders had been held by the ISR for around 20 years; the Tevatron took over on 16 July 2004, with more than $1 \times 10^{32}\,\text{cm}^{-2}\,\text{s}^{-1}$. By the end of its life, the Tevatron had accumulated of the order of $11\,\text{fb}^{-1}$ for both DØ and CDF. The early view on the Tevatron and a summary of its life from a machine perspective can be found in [32, 40], respectively.

From a physics point of view, the Tevatron and its predecessors at FNAL were extremely successful machines. One highlight of the fixed-target days was the discovery of b quarks by Lederman and collaborators in 1977. Others were the observation of direct CP violation by KTeV (1999) and the discovery of the last missing fermion of the Standard Model, the τ neutrino, by the DONUT experiment (2000).

The greatest success of the Tevatron, however, was the discovery of the top quark by the CDF and DØ experiments in 1995 [41, 42] and the precise determination of the mass of this last fermion of the Standard Model. After the discovery of already five quarks (until 1977) and five leptons, and after the confirmation of the Standard Model through the results of the Sp$\bar{\text{p}}$S and LEP, nobody had really doubted the existence of this particle. Indirect experimental evidence from electroweak fits, however, had hinted to a rather large mass, which meant small cross sections and difficult and time-consuming analysis.

The story of discoveries at the Tevatron did not end with the t quark: Both CDF and DØ continued to make important contributions: In 1998 the B_c meson was discovered, followed by the Σ_b baryon (with quark content uub and ddb) in 2006, the ssb baryon Ω_b (2008) and the dsb and usb Ξ_b baryons (2007, 2011). In addition, B_s oscillations were observed at the Tevatron for the first time in 2006, and in 2009, the first measurements of single top quark production were published.

[12] The DØ experiment had been operational since 1992.

The Tevatron, however, did no only succeed in the quark and hadron sector of the Standard Model—it also made important contributions to electroweak physics and beyond, e.g. by providing a very precise value for the mass of the W boson, by providing stringent limits to the mass of the Higgs boson or by narrowing down the parameter space for models of new physics.

1.2.5 UNK

The UNK was conceived in 1983 as a 21 km proton-proton storage ring at the Institute for High Energy Physics in Protvino near Serpukhov and Moscow, on a site on which since 1960 the U-70 synchrotron (in its days for some time the strongest proton synchrotron in the world) was constructed. The U-70 went into operation on 29 August 1967 and delivered beams of up to 76 GeV to numerous important experiments. Among the discoveries of the time are the phenomenon of scale invariance in hadronic interactions and the rising cross sections for collisions between pions and protons and other hadrons.

UNK aimed for two circulating beams—one of 400 GeV in a normal-conducting machine (UNK-1), and one of 3 TeV using superconducting magnets (UNK-2). Protons were to be delivered by U-70 via a 2.7 km transfer tunnel; the transfer beam line was actually commissioned in March 1994 with protons of 65 GeV. There were even ideas of porting the UA1 detector to UNK after it had fulfilled its mission by discovering the weak bosons. However, due to financial problems, already in January 1993 the focus had been shifted, by the Scientific Programme Committee of the Russian National Scientific Programme "High Energy Physics", from the initially prioritised 3 TeV machine UNK-2 to the "earliest possible commissioning" [43] of UNK-1 with beams of up to 600 GeV ("U-600"). The work on the superconducting machine was to be kept at R&D level.

In subsequent years, civil engineering on the UNK tunnel was finalised; about a quarter of the ring is, in principal, ready for machine installation, and another half ready to be equipped with services [44]. As of 1998, about 75 % of all dipoles for U-600 were available, with many of them tested. 17 km of vacuum chamber are in store, as are many other parts necessary for building the accelerator (power supplies, RF generators, corrector magnets, etc.). In 1998, manufacturing of UNK components was stopped, and since then the tunnel and the already procured equipment are kept under more or less safe conditions.[13]

[13] The websites [45, 46] show a few impressions from the abandoned tunnel.

1.2.6 Super-Conducting Super Collider (SSC, 1983–1993)

The SSC was certainly set to "regain leadership" in high energy physics—as the US President's science advisor demanded after the W/Z boson shock [2, 5] ("Europe 3, US not even Z-Zero" [47]). The SSC was first formally discussed in the US National Reference Designs Study in 1983—the same year that also the first design ideas for the LHC were brought up [48]. Not surprisingly, the early days of the LHC were dominated by "sometimes acrimonious competition and comparison" between the two projects [5].

However, in the US, the early 1980s were a time full of discussion about future directions, and numerous proposals for e^+e^- or $pp/p\bar{p}$ colliders were on the table [49]:

- At Brookhaven National Laboratory (BNL), there were mature plans for ISABELLE—a 800 GeV centre-of-mass energy pp collider that had been proposed in 1976, endorsed by HEPAP (the US High Energy Physics Advisory Panel) in 1977 and approved by Congress in 1978. By 1981, the tunnel for ISABELLE (later called the Colliding Beam Accelerator CBA) was essentially finished, and first beams were supposed to be possible in October 1987.
- In early 1983—just before the restart of operations of the now superconducting Tevatron —Fermilab put forward a proposal for a "Dedicated Collider" (DC) with a centre-of-mass energy of 4 TeV for $p\bar{p}$ physics, based on Tevatron technology.
- Brookhaven also considered a "Sandatron" of 10–30 TeV, with the CBA as injector.
- SLAC proposed a e^+e^- linear collider in the 1 TeV range.
- And there were first workshops for a 20 TeV proton-proton collider—later to become the SSC. At the first of these workshops, held from 28 March to 2 April 1983, also a preliminary cost estimate (without R&D, contingency, escalation etc.) of about 1.720 billion US dollar was made.

Then, later in 1983, under the impression of the discoveries at the Sp$\bar{p}$S at CERN and despite the enormous costs (which triggered many heated discussions in the US), the responsible HEPAP Subpanel on New Facilities for the US High Energy Physics Program recommended "the immediate initiation of a multi-TeV high luminosity proton-proton collider". It also stated that it recommends "Fermilab not proceed at this time with the Dedicated Collider" and that "the [CBA] project at Brookhaven not be approved[14]" [50]. The recommendations were endorsed by HEPAP, and first funds for preliminary SSC studies were transferred from CBA. A study, completed in 1984, estimated the total costs of the collider complex to be of the order of 3–4 billion dollar.

The SSC plan was quite daring: The parameters foreseen in 1984 by the Central Design Group suggested building a pp collider of 20 TeV beam energy with a luminosity of $10^{33}\ \mathrm{cm}^{-2}\ \mathrm{s}^{-1}$ and six experimental areas. A conceptual design report with these benchmarks was delivered to the DOE on 1 April 1986. On 30 January 1987,

[14] Today, the tunnel constructed for ISABELLE houses the heavy-ion collider RHIC.

President Reagan approved the SSC at a total cost estimate of 4.4 billion dollar, and first funds (still dedicated to R&D, since the funding situation in the US was not easy) were provided by Congress in 1988.

The SSC site selection was a delicate process during which important political supporters of the SSC were turned into opponents of the project—a fact with "long-range repercussions" [49]. Initially 45 site proposals were submitted, out of which seven sites were considered to meet all selection criteria. These were ranked, and on 18 January 1989, it was finally decided to site the SSC in Texas.

The building-up of the SSC Laboratory (SSCL) and the construction of the machine could start after first construction funds had been approved by Congress in autumn 1989. In parallel to these activities, a site-specific design report was prepared, which (involving external expertise) in July 1990 led to a new cost estimate of 8.6 billion dollar. Ideas were circulated to decrease the centre-of-mass energy as the only means to significantly reduce costs, but they were refuted with the argument that the physics potential would be seriously compromised [51]. A report by a DOE committee delivered in August 1993 led to the "claim that the SSCL costs went up to over" 11 billion dollar [49]; in addition it was reported that most technical systems were late by up to 1 year and that the magnet construction posed a serious cost risk. These developments finally, after significant political "roller-coaster" [49], led to the termination of the SSC project by the US Houses in October 1993. By then, about 2 billion dollar had been spent, and 23.5 km of SSC tunnel had been constructed. 640 million dollar were approved for termination costs of the project. In the end, the close-out of the SSCL took 2 years and cost 736 million dollar. In 1994, the responsible HEPAP subpanel recommended US participation in and contribution to the LHC [52].

The reasons for the failure of the SSC project have been discussed in many places. The sheer size and cost of the project were problematic—the LHC, relying on existing infrastructure like the LEP tunnel and a system of pre-accelerators, had a clear advantage in this respect; and the fact that the LHC allegedly could be realised at significantly lower costs certainly was a serious (psychological) burden for the SSC.[15] Also the strong increase in the project costs from start (1,72 billion US dollar for the accelerator complex) to end (11 billion dollar total project costs) was a problem, not least because it was taken as further evidence for the bad management of the project.

Also the physics potential of the SSC—in comparison to the LHC—was an issue: The design centre-of-mass energies differed by a factor of almost 3—but the design luminosity of the LHC was a factor of 10 higher than that of the SSC, making the LHC competitive for almost all relevant processes (see the discussion in the next section) and thus weakening the SSC physics case.

Finally, international collaboration was heavily debated during SSC planning: The SSC was designed to "restore US pre-eminence" in high energy physics [49],

[15] Note, however, that the cost estimates for building the SSC close to Fermilab, and thus using the Tevatron and its facilities as injector, were, according to most sources, not significantly lower than for the Texas site.

a fact which at first sight precluded international contributions to the project.[16] Nevertheless, various models for international collaboration were discussed, ranging from the usual model of commonly operated experiments over what later became to be known as the "HERA model" (in which roughly 20 % by cost of the accelerator's components had been produced by countries other than Germany) up to a purely financial involvement in SSC construction. The focus later was on international involvement in all R&D aspects, with purely national SSC construction, also because there were strong opinions against (in-kind) contributions that ostensibly would transfer US jobs to abroad. Still, many connections on laboratory and also government level were established between the USA and other countries; in the end they led to nothing, also because the new Clinton administration that stepped into office in January 1993 ordered a "go slow" policy and planned for a 3 year delay in the completion of the machine [51]. This, in turn, discouraged foreign partners: "Major foreign participation has remained elusive because of uncertainty about the US commitment to the project, yet our own commitment has wavered in large part because of the absence of foreign funding" [51].

All in all, the SSC probably did not seem very inviting to other nations ("The President has decided to build such a machine and you have the option to join the project or leave it." [2]), and when later the SSC went into serious funding problems and other nations were asked to contribute at the billion-dollar level, this attitude was well remembered [5].

1.3 LHC Development and Timelines

1.3.1 From First Ideas to First Approval

As described above, the possibility of a hadron collider in the LEP tunnel was first mentioned in the notebook of John Adams. Serious discussions of a large proton collider started with the first internal notes in 1983 [48] and the first CERN–ECFA workshop in March 1984 [53], under CERN director general Herwig Schopper. The 1984 workshop clearly was a reaction to the developments in the US, where the SSC project was maturing quickly after the $W^{\pm}/Z^0$ shock. And already in May 1984, Schopper presented plans to the meeting of the AAAS, the American Association for the Advancement of Science, to build a superconducting proton collider with 5 TeV beam energy on top of the LEP magnets. A price of only 500 million US dollar was mentioned [49], and the existing LEP tunnel was introduced as a decisive advantage in terms of cost and time for the LHC over the SSC.

[16] As an example, in the site-selection process, a proposal locating the SSC across the US–Canadian border was rejected partly because of its not being fully national.

Wrong as the absolute numbers may have been (see Sect. 1.3.4 for a discussion of LHC costs), the tone for the ensuing competition between SSC and LHC was set: The LHC might be realised faster and cheaper than the SSC (there were initial estimates of the cost relation between SSC and LHC as high as 3:1 or higher); the LHC might even be regarded as an intermediate project before the realisation of the SSC. The SSC, on the other hand, claimed a clear physics advantage on the grounds of the significantly higher centre-of-mass energy. To this, CERN again reacted by pointing out the higher LHC luminosity (after careful studies of the magnet potential, the design value had been changed to $10^{34}\,\mathrm{cm}^{-2}\,\mathrm{s}^{-1}$).[17]

For physical reasons, any new proton-proton collider was required to achieve collisions with centre-of-mass energies, at the constituent level, of at least 1 TeV. Taking this energy as a reference, Fig. 1.2 shows that indeed the LHC, with a ten times higher luminosity compared to the SSC, has similar parton luminosities both for gluon-gluon induced processes and for those induced by quark-antiquark pairs. The plot shows, as a function of $\sqrt{\hat{s}} = \sqrt{s \cdot x_1 \cdot x_2}$, the gluon-gluon parton luminosity $\frac{\mathrm{d}L}{\mathrm{d}\sqrt{\hat{s}}}$ for proton-proton collisions at different centre-of-mass energies. For the LHC design energy of 14 TeV also the quark-antiquark and quark-quark parton luminosities are shown.[18] The LHC can even compete with the SSC for slightly higher energy scales, up to maybe 5 TeV. The higher centre-of-mass energy of the SSC clearly becomes relevant for the highest scales of 10 TeV where, however, the overall statistics for any rare process will be tiny even at the SSC.[19]

The next years were very eventful: In 1985, a "Long-Range Planning Committee" (LRPC) was installed at CERN, chaired by Carlo Rubbia. 1987 was a particularly eventful year: In January, US President Reagan approved the SSC—a step that very much put the LHC project in doubt.[20] A second general workshop on the "Physics at Future Accelerators" was held in La Thuile, at which the three subpanels of the LRPC discussed their findings [54]; they supposed the LHC to have realisable luminosities between 10^{33} and $10^{34}\,\mathrm{cm}^{-2}\,\mathrm{s}^{-1}$ and a maximum beam energy of 8 TeV. This idea was put to CERN Council in 1987, together with an R&D programme for the 10 T dipole magnets, which was started soon after. In 1987, finally, also the Texas site was selected for the SSC.

In 1989, Rubbia became CERN director general. Since the beginning of his mandate, his line of argument for the LHC was fixed (see for example his presentation

---

[17] It is interesting to read, in parallel, accounts of the SSC–LHC competition written by SSC advocates [49] and by LHC supporters [2]. Even when talking about the same events, e.g. a US Congress hearing in April 1987, they seem to be telling rather different stories. Especially the evaluation of efforts for international collaboration on the SSC is very different—even in texts written 15 years after SSC cancellation.

[18] The quark-quark parton luminosity here is obtained by using the quark-antiquark parton luminosity for $p\bar{p}$ collisions.

[19] Note that, due to the large total proton-proton cross section, luminosities beyond values of around $10^{36}\,\mathrm{cm}^{-2}\,\mathrm{s}^{-1}$ do not result in an increased physics potential, but only in massively reduced beam lifetimes.

[20] "It was only the resilience and conviction of Carlo Rubbia [...] that kept the project alive" [18].

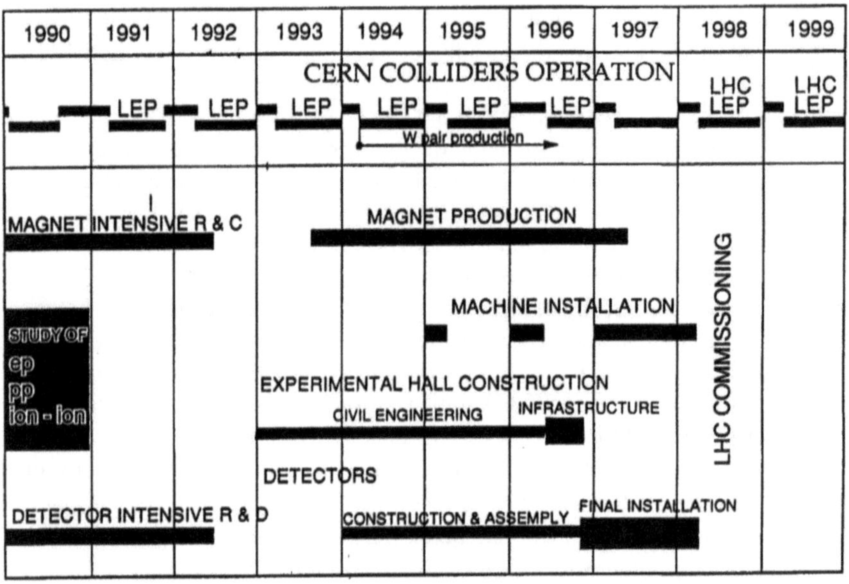

Fig. 1.3 LHC construction schedule, as proposed by Rubbia in 1990. Adapted from [12]

to the fourth general "Large Hadron Collider Workshop" held in Aachen, Germany, in 1990 [12]):

- Hadron collisions at the "energy frontier" and with a constituent-level centre-of-mass energy of 1 TeV are the correct approach to finding new, interesting physics.
- The LHC is healthy competition for the SSC, at moderate cost and an advantageous timescale.
- Because of the possibility of a heavy-ion programme and the potential to have concurrent running with LEP and thus also the *ep* option, the LHC is much more versatile than SSC.[21]
- The LHC disadvantage in centre-of-mass energy was, by Rubbia, seen as compensated by its higher luminosity.

Rubbia, at the 1990 workshop, announced the goal of having the first full LHC magnet string ready by 1992. The machine parameters he presented then suggested a beam energy of 7.7 TeV, a maximum luminosity of up to $5 \times 10^{34}\,\mathrm{cm}^{-2}\,\mathrm{s}^{-1}$ and—very optimistically—first operations in 1998 (see the timeline plot in Fig. 1.3)!

[21] Note that in the La Thuile workshop [54] even the possibility of also having $p\bar{p}$ collisions—at much reduced luminosity—in the LHC is mentioned. Also, from very early days of LHC discussions on, the complementarity of "high-precision" e^+e^- instruments like LEP and of "high-energy" machines like the LHC has been stressed [12]. In Aachen, Rubbia also discussed the necessity of linear e^+e^- colliders "such as CLIC": "In particular the LHC has to be conceived as a machine precursory to CLIC [...] As often in the past a first 'exploratory' phase with hadron collisions is necessary precursory to the second 'consolidating' phase using electrons and positrons".

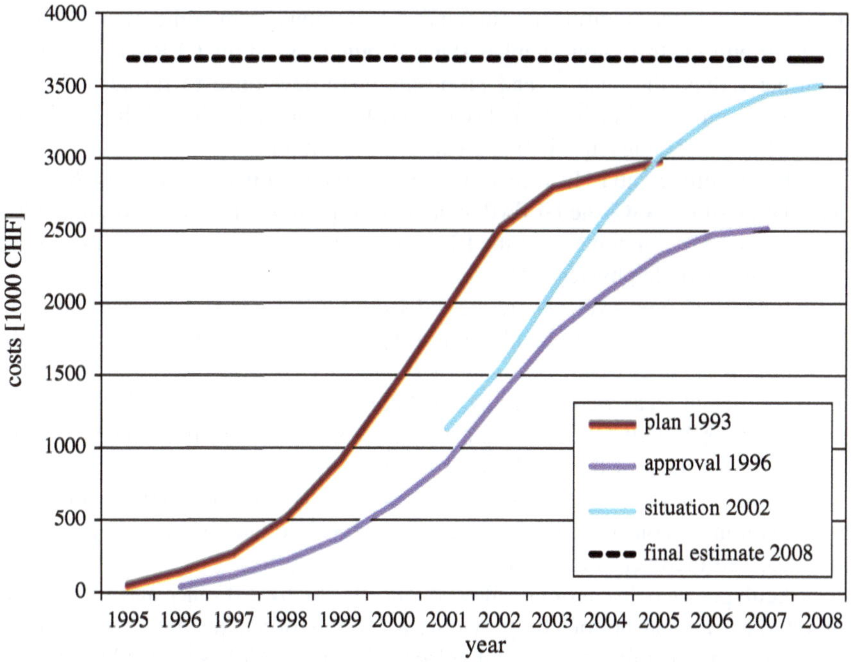

Fig. 1.4 LHC funding scenarios at different stages of the project. *Source* CERN

Then, in 1991, the LHC started its way through the CERN approval procedure: In December, the LHC was first presented in detail to CERN Council, and Council adopted the project, recognising the LHC as "the right machine for the advance of the subject and the future of CERN". Rubbia was asked to prepare a full proposal, including a cost estimate, by the end of 1993 [5]. In early 1993—when, in the light not only of the SSC developments, the LHC did not seem "inevitable" [4]—Rubbia handed over the responsibility for producing a complete proposal to the designated next director general, Chris Llewellyn-Smith. Prospects then were not encouraging: "[...] costing was significantly bigger than previous estimates [...]; attitudes towards high-energy physics were hardening in several CERN member states; and the CERN Council had just agreed to a temporary reduction in Germany's contribution on the grounds that reunification was proving very costly" [5].

In October 1993, the SSC was cancelled. Llewellyn-Smith later wrote that "I do not think that the LHC would have been approved if the SSC had not been cancelled [...]" [4]. In December 1993, Llewellyn-Smith presented his proposal for the LHC to CERN Council, based on a two-in-one machine (see Chap. 2 for technical details) installed above the LEP accelerator[22] and to be commissioned in 2002. The overall proposed budget (see Fig. 1.4) relied on a mixture of a general budget

[22] This idea was only given up in 1995, when it was decided to dismantle LEP in order to gain space for the LHC. However, nobody had really pursued the idea of *ep* physics in the LEP/LHC tunnel seriously.

increase for CERN, additional voluntary contributions from some member states and contributions from non-member states. Council asked the CERN management to further reduce the costs,[23] and proposals were developed to further delay the machine commissioning by 1 or 2 years and to reduce all other CERN activities (except the CERN flagship LEP) to an absolute minimum.

In the meantime, the LHC received (moral) support from other sources: After the cancellation of the SSC, the HEPAP in the US suggested to join the LHC, and ICFA issued a supportive statement for the LHC, which they considered to be "the correct next step for particle physics" [4].

In June 1994, approval was requested from CERN Council for a machine of 14 TeV centre-of-mass energy with luminosity up to $10^{34}\ \mathrm{cm}^{-2}\ \mathrm{s}^{-1}$. The cost for the machine was given as 2230 million swiss francs (MCHF) [55]. Additional costs for the experimental areas were stated as 210 MCHF, and CERN's contributions to the experiments as 220 MCHF. The vote on the proposal and in fact the Council meeting itself were, for the first time in Council history, left open because it was clear that the UK and Germany would not have accepted the plan; the two countries demanded further voluntary contributions from France and Switzerland who, in their view, as host states profited very much from the LHC.

Difficult political discussions followed, during which e.g. France and Switzerland agreed to extra contributions by providing a 2 % inflation adjustment (compared to 1 % for all other member states), and in December 1994, CERN Council approved a two-stage procedure for LHC construction [56]: A "missing-magnet" machine that left out one third of all magnets was to be constructed for commissioning in 2002 and operations at 9–10 TeV centre-of-mass energy in 2004, with an upgrade to 14 TeV envisaged for 2008. The plan saved about 300 MCHF in direct construction costs for the first stage; the staged approach, however, came with additional managerial and other costs, so that the net effect was unclear. The plan was to be reviewed latest in 1997 on the grounds of additional non-member state funds (which were to be used not to reduce member state contributions, but to speed up the project), and depending on the budget conditions required by the UK and Germany, including the contribution rebate that Germany had received on the grounds of the expensive reunification.

1.3.2 From First Approval to First Beams

In the following 2 years, several non-member states became observers to the LHC project: Japan, India and Russia; in December 1997, also an agreement about LHC contributions could be signed with the USA. Also Canada promised substantial contributions to the LHC. These developments made the CERN management confident that a single-stage machine was still possible. Then, quite unexpectedly, Germany announced their intention to reduce CERN contributions (and in fact all international

[23] In fact, mainly the UK and Germany opposed the plan and even managed to get the voting procedures in Council changed to a scheme that favoured the larger contributors.

scientific contributions) by around 9 %. The UK chimed in. The US agreement was not yet signed at this point, and CERN felt quite anxious that the German and UK plans might scare away the non-member states, which had been assured that their contributions would not be used to reduce the contributions of the member states.

In the following discussions (see Llewellyn-Smith's vivid account of the time [4]), there were rumours of Germany and UK even wanting to leave CERN; on the other hand CERN made it clear that the suggested budget cuts would kill the LHC project. The solution that was found was unique for scientific projects: It was decided that CERN was to be allowed to take out loans (the German statement was that "a greater degree of risk would inevitably have to accompany the LHC"). With this risky solution, LHC construction in a single stage (a necessary condition to keep the non-member states on board) was approved by CERN Council in December 1997 [57], with commissioning at full energy foreseen for 2005, the start of operations in 2006 and design performance in 2007. The corresponding budget cuts were marginally smaller than demanded by Germany, and in addition a 1 year "crisis-levy" on all CERN salaries was imposed. The plan was to be reviewed half-way.

So in late 1997, the LHC was approved—but, as Lyn Evans, then LHC Project Leader, says [18]: "With all contingency removed, it was inevitable that a financial crisis would occur at some time, and this was indeed the case when the cost estimate was revised upwards by 18 % in 2001." "[...] the budgetary position was extremely fragile. The deficit financing of the LHC was hyper-sensitive to small changes in the timing of the contracts" [5].

In 2001, it was communicated by CERN management that the LHC would become significantly more expensive than expected (see above), a statement that came as a shock to the public and was not well prepared. The main drivers of this increase were the dipole magnets and the necessary erection of a second cryogenics line. The solution to this problem—which "should [...] not have been a surprise" [5]—was to increase the level of borrowing and to further extend the construction period "which was necessary anyway on technical grounds for both the machine and the detectors" [18]. The rest of the story is quickly told (and told in more detail in Chap. 2 and in [1]): LHC was commissioned in 2008, with first beams travelling around the machine on 10 September, producing beam-splash and beam-gas events.

1.3.3 Evolution of High Energy Physics Since 1977

When first physics collisions were recorded in the LHC experiments, the world of high energy physics had changed (if only slightly) with respect to 1977, when the LHC was first mentioned (see Sect. 1.1.2):

- LEP, Tevatron, HERA, the *B* factories and many other facilities and experiments had convinced us that the Standard Model works extremely well at the one-loop level. Furthermore, it had become clear that, if nature were supersymmetric, the gauge couplings of the three gauge groups forming the Standard Model would unify at very high energy scales, thus sharpening the case for searches for supersymmetric extensions of and new physics beyond the Standard Model.

- With the top quark and the τ neutrino, the last two fermions of the Standard Model had been discovered. The very large top-quark mass opened room for speculations about new physics phenomena in a regime that LHC, with its high energy and high luminosity, could cover exceedingly well.
- LEP had demonstrated that there are only three light neutrinos; furthermore it had been learned that neutrinos had small, but non-zero masses.
- The scientific case for SUSY had further sharpened through its natural connection with dark matter, which, initially, had not been part of the case for the LHC.
- However, supersymmetry or other extensions of the Standard Model, and also the Higgs boson of the Standard Model had not yet been discovered—so that even after more than 20 years after the first outline of the LHC physics menue this menue was still valid!
- Observations of small deviations between Standard Model predictions and measurements—i.e. of B decays, of $(g-2)_{mu}$ or of the effective electroweak mixing angle—are not necessarily significant; but they do raise considerable interest and are, partly, open for scrutiny at the LHC.
- On the experimental side, the largest steps forward had been in tracking detectors. First medium-scale semiconductor detectors had already been in use in the LEP experiments and at the Tevatron (for the top-quark discovery!) for secondary vertexing. However, at the time of John Adams nobody would have dreamt of building a radiation-hard all-silicon tracker with an area the size of a tennis court and some 76 million channels!

So, all in all, during the more than 30 years between the first ideas about LHC in 1977 and the start of operation, only small changes to the objectives of the LHC had become necessary.

1.3.4 LHC Funding and Construction Timelines

As has become clear, the genesis of the LHC was a complex story—also politically and financially. It is not easy to summarise the costs of the LHC and to follow the funding proposals over the years between first plan (December 1993) and commissioning (2008). This difficulty is due to several effects:

- Firstly, the numbers from different years often refer to price levels of different years, i.e. they are not always corrected for inflation effects.
- Secondly, the definition of the LHC project costs varied. A few examples: inclusion of personnel or not; inclusion of generic R&D costs; inclusion of test and pre-operation costs; inclusion of in-kind contributions; inclusion of escalation etc. In the following, only material costs to completion will be discussed—which will lead to differences to the numbers specified in the text in Sects. 1.3.1 and 1.3.2. The numbers quoted below include the machine and tunnel construction,

the construction of the experimental areas and, at least for the later years, also in-kind contributions, machine R&D, tests and pre-operations.
- Thirdly, the accounting schemes at CERN underwent significant changes over the years.

For all these reasons, the numbers given below (and also above in the text) and also in Fig. 1.4 have to be taken with great care. Figure 1.4 shows the following funding profiles for the LHC:

- the December 1993 plan with commissioning foreseen for 2002. The total material costs were given as 2988 MCHF (in 1993 prices);
- the budget plan approved in 1996 which—after having collected sufficiently many non-memberstate contributions and after having economised on the CERN programme—foresaw commissioning at full energy in 2005. This plan included reductions due to more economies on the CERN side and a stretched schedule;
- the budget estimate from 2002 that was developed after the increase in cost had come to light in 2001. With respect to the 1996 plan (which in 2001 had already been significantly violated), a materials cost increase of around 1150 MCHF can be inferred (see the comments below);
- the final 2008 materials cost sum. Not included in the discussion here are costs for repair after the 2009 incidence (see Sect. 2.3.4) and for increased quality control and monitoring etc.

Difficult as a direct comparison of the various scenarios may be, it is worth looking a bit closer at the approved number from 1996 (2530 MCHF) and the 2008 cost of the project (3685 MCHF). The first of these numbers is given in 1996 prices, i.e. assuming (completely unrealistically) no inflation. The 2008 number, in contrast, is given in actual costs. Assuming an average annual indexation on materials costs of 2 %, the 1996 number would already become over 3200 MCHF, leaving a gap of under half a billion MCHF and thus representing a cost increase of roughly 13 %. The author is tempted to say "only" 13 % since—as a brief look at current large-scale public projects easily shows—budget overruns of even several 100 % seem not to be the exception but rather the silently accepted rule. As Llewellyn-Smith [5] cites from a 2003 "The Times" supplement: "If those involved didn't lie about the cost, they [the large construction projects (ed.)] would never be built."

So, all in all, it seems fair to say that—for a project of this size and complexity, and for a project that basically was approved before the necessary R&D was completed–the LHC stuck remarkably well to schedule and budget. This is particularly notable since at the time of approval, many future developments were not yet foreseeable (e.g. exchange rates, inflation, etc.). Figure 1.3 shows Rubbia's 1990 expectation that—with civil engineering starting in 1993—first beams should be delivered after 5 years, in 1998. In fact, civil engineering started in 2000, and first beams entered the machine after 8 years, in September 2008 (see Sect. 2.3.3).

1.4 Superconducting Magnets for Particle Physics

The history of superconducting (SC) magnets for particle accelerators [58, 59] begins in the 1960s, when first R&D programmes were started. A first collaborative effort aimed at providing SC magnets for the SPS (Sect. 1.2.3) was, however, abandoned because of technical problems, and the SPS became a machine with resistive magnets (commissioned in 1976). At CERN, the SC idea was further pursued for the ISR which received, in 1980, powerful SC quadrupoles as part of a luminosity increase.

The SC magnet story was continued by efforts towards increasing the energy of the NAL Main Ring (Sect. 1.2.4), and towards the design of the ISABELLE or CBA 4–5 T dipole magnets (Sect. 1.2.6). The development of the latter, after a promising start, proved more complicated and time-consuming than initially expected, and after the discovery of the weak bosons at the Sp$\bar{p}$S and the progress in the Tevatron programme, CBA was considered superfluous and finally abandoned.

At the Tevatron, thanks to the new superconducting magnets of 4.2 T, in 1983 a new energy world record was reached, with a beam energy of 512 GeV. The Tevatron magnet design still had a warm return yoke—in contrast to ISABELLE, where the yoke would have been part of the cold mass kept at 4.2 K. A big advantage of the Tevatron magnets—and a concept that from then one became integral part of all SC accelerator dipole magnets—was the "collar system" that helped to contain the radial forces exerted by the electromagnet fields on the coils. A further ingenious achievement was the acceleration of both protons and antiprotons in one common beampipe, which saved a second magnet ring. All in all, the Tevatron used more than 700 SC dipole magnets.

The proton ring of the HERA electron-proton collider was developed in the 1980s. Similar in size to the Tevatron ring (see Table 1.1), it was based on a few conceptual improvements: The two most important ones were probably that, first, the iron of the return yoke was integrated into the cold mass—a solution that avoids problems concerning coil centering and alignment. Second, the approximately 500 5.5 T SC magnets for the HERA ring were produced by industry and not manufactured at the laboratory. This led to a significant cost reduction.

Also the SC magnets for the heavy-ion collider RHIC—which was under constant financial pressure—were to a large extent produced by industry. RHIC profited from experience gained earlier at other machines—also in the development of the magnets of the SSC, which was to be built in parallel. SSC was a "leading project for SCM development, backed by a joint effort of all major US laboratories" [58]. Numerous studies had led to the final design of the SSC magnets for 6.6 T at 4.4 K, necessary to meet the design goal of 20 TeV beams. Also SSC foresaw steel collars and cold iron. Among the many key developments for the SSC was that of new SC cables that showed a very high critical current. However, some technology choices of the SSC showed to not have been optimal. One drawback—also in comparison to the LHC design—was the use of single-bore magnets that necessitated the construction of two separate rings. Others were the ambitious choice of the working point and a later increase of the bore.

The history and design of the LHC magnets is documented in many places (see e.g. [1]). The LHC, relying substantially on previous R&D, took a few innovative design choices (two-in-one magnets, cooling to 1.9K and thus work with superfluid helium) and "pushed the $NbTi$ technology to its extreme" [1], finally resulting in the known dipole fields of 8.3T.

1.5 Forming the Collaborations

One of the first large-scale dedicated meetings for LHC detectors[24] was an ECFA "Study Week on Instrumentation Technology for High-Luminosity Hadron Colliders" in Barcelona in 1989 [62, 63]. Here, also the first precursor of experimental collaborations—EAGLE ("Experiment for Accurate Gamma, Lepton and Energy measurements")—started forming.

The next important step was the 1992 CERN–ECFA workshop "Towards the LHC Experimental Programme" in Evian [64]. It took place soon after CERN Council's unanimous December 1991 vote that the LHC is "the right machine for the advance of the subject and for the future of CERN" [5]. During this workshop, proto-collaborations presented "expressions of interest" describing the respective detector plans. The interest in contributing to the LHC experimental programme was large: all in all, 12 proposals were made in Evian:

- There were four proposals for general-purpose experiments: EAGLE, ASCOT ("Apparatus with SuperCOnducting Toroids"), CMS ("Compact Muon Solenoid") and L3+1 (or L3P, an upgrade of the L3 LEP experiment for the LHC).
- Three *b* physics experiments entered into the competition for approval. One concept was based on *pp* collisions in the LHC ("COBEX"); a group which called themselves the "LHB collaboration" would use a beam extracted from the LHC by crystal channeling for fixed-target operation; the third wanted to use a gas-jet target for the proton beam in one of the straight LHC sections ("GAJET").
- Similarly, three proposals were made for heavy-ion experiments; the one later called ALICE; one that wanted to use the DELPHI detector from LEP, and one that suggested a heavy-ion programme for the CMS detector.
- In addition there were two proposals for neutrino experiments, one brought in by the NOMAD collaboration.

It was probably clear from the beginning that only two general-purpose experiments would be accepted at the LHC, one of them potentially being a toroidal apparatus like ASCOT and EAGLE. Therefore, these two proto-collaborations, in a voluntary move, merged to form the new ATLAS ("A Toroidal LHC ApparatuS") collaboration in spring 1992. Over the summer of 1992, the ATLAS, CMS and L3+1

[24] Detailed technical discussions of the existing large LHC experiments can be found in Chap. 3. An overview of the organisation, funding and management of large collaborations can be found in [60, 61].

collaborations wrote "Letters of Intent" (LoIs) that, on 1 October 1992, were handed in to the newly formed CERN LHC Experiments Committee (LHCC) [1, 65, 66].

Among the remaining three general-purpose experiments, ATLAS and CMS were invited to provide [67] detailed technical proposals, which they did in 1994. CMS and ATLAS were finally approved in January 1996, and "green light" for construction was given on 31 January 1997, with an expenditure ceiling of 475 MCHF (1995 currency rate). In fact, up to the LHC start of operation, the requested funding of ATLAS and CMS amounted to 540.9 and 566.3 MCHF, respectively [60], meaning rather moderate cost overruns.[25] Construction started in 1997, the ATLAS cavern was inaugurated on 4 June 2003, and the last large pieces of ATLAS and CMS were lowered into the experimental caverns on 29 February and 23 July 2008, respectively.

The situation for the *b* physics experiments was slightly more involved: COBEX, GAJET and LHB submitted LoIs in October 1993. At that time, the new *B* factories at SLAC and KEK were already under construction. However, the LHCC came to the conclusion that the large data statistics available at the LHC would allow a dedicated experiment to gain physics results beyond what was possible at the factories. Therefore, it was recommended in January 1994 to foresee a dedicated *b* physics experiment at the LHC, and the LHCC asked the three submitters to form a single collaboration, based on the collider-mode suggestion. In February 1998, the newly formed LHCb collaboration submitted a technical proposal, and in September 1998 it was accepted.

The ALICE LoI was submitted in March 1993, the technical proposal followed in 1996, and the experiment was approved in February 1997. The DELPHI proposal was turned down.

The four big experiments ALICE, ATLAS, CMS and LHCb were followed by three smaller, more focused proposals for experiments: The TOTEM experiment (LoI 1997) is investigating the total *pp* cross section, elastic *pp* scattering and diffraction dissociation; MoEDAL (LoI 1998) is searching for magnetic monopoles and other exotic phenomena; LHCf (LoI 2003), finally, uses very forward particles created in LHC's *pp* collisions to simulate cosmic rays.

1.6 Conclusion

The history of the LHC cannot be told without telling the history of other collider projects and—in fact—of the greatest part of particle physics in the last five or six decades. Beyond the obvious topics in the field of collider physics, many aspects would have to be discussed: politics, finance, sociology, management. The full history of our field, seen from these points of view, still needs to be written.

[25] These numbers do not contain the extraordinary efforts from laboratories and universities around that world that contributed to the experiments using national funding sources.

References

1. L. Evans (ed.), *The Large Hadron Collider: A Marvel of Technology* (EPFL Press, Lausanne, 2009)
2. H. Schopper, *LEP: The Lord of the Collider Rings at CERN 1980–2000: The Making, Operation and Legacy of the World's Largest Scientific Instrument* (Springer, Heidelberg, 2009)
3. G. Brianti, in *CAS-CERN Accelerator School: 50 years of Synchrotrons*, CERN, Switzerland, 1996, ed. by J.D. Lawson, G. Brianti, p. 29
4. C. Llewellyn-Smith, Phil. Trans. Roy. Soc. Lond. A **370**, 995–1004 (2012)
5. C. Llewellyn-Smith, Nature **448**, 281 (2007)
6. R. Cahn, G. Goldhaber, *The Experimental Foundations of Particle Physics*, 2nd edn. (Cambridge University Press, Cambridge, 2009)
7. S. Weinberg, Eur. Phys. J. C **34**, 5–13 (2004)
8. G. 't Hooft, Nature **448**, 271–273 (2007)
9. M. Jacob, in *Proceedings of the LEP Summer Study 1978*, Les Houches, France, and CERN, Switzerland, 1978, ed. by M. Jacob et al., p. 29
10. G. Brianti et al. in *Proceedings of the ECFA-CERN Workshop "Large Hadron Collider in the LEP Tunnel"*, Lausanne/Geneva, Switzerland, 1984, ed. by M. Jacob, pp. 1–16
11. C. Llewellyn-Smith, in *Proceedings of the ECFA-CERN Workshop "Large Hadron Collider in the LEP Tunnel*, Lausanne/Geneva, Switzerland, 1984, ed. by M. Jacob, pp. 27–48
12. C. Rubbia, in *Proceedings of the ECFA Large Hadron Collider Workshop*, Aachen, Germany, 1990, ed. by G. Jarlskog, D. Rein, pp. 9–39
13. V. Shiltsev, arXiv:1307.3116 (2013)
14. B. Richter, in *Proceedings of the Conference The Rise of the Standard Model: Particle Physics in the 1960s and 1970s*, Stanford, USA, 1992, ed. by L.H. Hoddeson et al., p. 261
15. D. Kerst et al., Phys. Rev. **102**, 590–591 (1956)
16. W. Panofsky, Proc. Natl. Acad. Sci. **70**, 611 (1973)
17. M. Jacob, K. Johnsen, *A Review of Accelerator and Particle Physics at the CERN Intersecting Storage Rings*. CERN-84-13 (1984)
18. L. Evans, Phil. Trans. Roy. Soc. Lond. A **370**, 831–858 (2012)
19. British-Scandinavian ISR Collaboration, Phys. Lett. B **44** 521–526 (1973)
20. M. Banner et al., Phys. Lett. B **44**, 537–540 (1973)
21. F. Büsser et al., Phys. Lett. B **46**, 471–476 (1973)
22. S. Berman, J. Bjorken, J.B. Kogut, Phys. Rev. D **4**, 3388 (1971)
23. L. Di Lella, Phys. Rept. **403–404**, 147–164 (2004)
24. CERN Timelines—CERN Accelerators. http://timeline.web.cern.ch/timelines/CERN-accelerators. Accessed 26 June 2014
25. D. Cline, et al. *Proposal 493. Search for New Phenomena Using Very High-Energy p p and anti-p p Colliding Beams Devices at Fermilab*. FERMILAB-PROPOSAL-0493, NAL-1976-030 (1976)
26. C. Rubbia, P. McIntyre, D. Cline, in *Proceedings of the International Neutrino Conference*, Aachen, Germany, 1976, ed. by H. Faissner et al., p. 175
27. UA1 Collaboration, Phys. Lett. B, **122**, 103–116 (1983)
28. UA1 Collaboration, Phys. Lett. B **126** 398–410 (1983)
29. UA2 Collaboration, Phys. Lett. B **122** 476–485 (1983)
30. UA2 Collaboration, Phys. Lett. B **129** 130–140 (1983)
31. Fermilab History and Archives Project; Accelerator History—Main Ring. http://history.fnal.gov/main_ring.html. Accessed 17 June 2014
32. H. Edwards, Ann. Rev. Nucl. Part. Sci. **35**, 605–660 (1985)
33. Historic Main Ring Bids Farewell Forever. FermiNews 19/1997
34. Tevatron Interactive Time-Line. http://www.fnal.gov/pub/tevatron/milestones/interactive-timeline.html. Accessed 18 June 2014
35. L. Hoddeson, *The First Large—Scale Application of Superconductivity: The Fermilab Energy Doubler* 1972–1983, FPRINT-87-01 (1987)

36. Fermilab History and Archives Project; Accelerator History—Main Ring Transition to Energy Doubler/Saver. http://history.fnal.gov/transition.html. Accessed 18 June 2014
37. Fermilab History and Archives Project; The Lederman Years, 1979–1989—The Tevatron. http://history.fnal.gov/tevatron.html. Accessed 18 June 2014
38. In Celebration of the Fixed Target Program with the Tevatron. http://conferences.fnal.gov/tevft/book/. Accessed 18 June 2014
39. J.A. Appel, C.N. Brown, P.S. Cooper, H.B. White, arXiv:hep-ex/0008076
40. S. Holmes, R.S. Moore, V. Shiltsev, JINST **6**(T08), 001 (2011)
41. DØ Collaboration, Phys. Rev. Lett. **74**, 2632–2637 (1995)
42. CDF Collaboration, Phys. Rev. Lett. **74** 2626–2631 (1995)
43. G. Gurov, in *Proceedings of the Conference Particle Accelerator (PAC95)*, Dallas, USA, 1995 pp. 416–419
44. K. Myznikov, E. Troyanov, in *Proceedings of the 17th Conference on "High-Energy Accelerators" (HEACC98)*, Dubna, Russia, 1998 pp. 35–39
45. Protvino—Hadron Collider: Accelerator. http://a-a-ah.com/accelerator. Accessed 18 June 2014
46. Large Twin of Hadron Collider. http://englishrussia.com/2011/02/14/large-twin-of-hadroncollider/. Accessed 18 June 2014
47. The New York Times, *Europe 3, U.S. Not Even Z-Zero*. 6 June 1983
48. S. Myers, W. Schnell, *Preliminary Performance Estimates for a LEP Proton Collider*. CERN-LEP-NOTE-440 (1983)
49. S. Wojcicki, Rev. Accel. Sci. Tech. **1**, 259–302 (2008)
50. Report of the 1983 HEPAP Subpanel on New Facilities for the U.S. High Energy Physics Program. DOE/ER-0169, 1983
51. S. Wojcicki, Rev. Accel. Sci. Tech. **2**, 265–301 (2009)
52. Report of the HEPAP Subpanel on Vision for the Future of High Energy Physics. DOE/ER-0614P, 1994
53. M. Jacob (ed.), *Proceedings of the ECFA-CERN Workshop "Large Hadron Collider in the LEP Tunnel"*, Lausanne, Switzerland, 1984
54. J. Mulvey (ed.), *Proceedings of the Workshop on "Physics at Future Accelerators"*, La Thuile, France, 1987
55. CERN Council, *The Large Hadron Collider Project*. CERN/2039 (1994)
56. CERN Council, *Funding and Approval of the LHC*. CERN/CC/2071 (1994)
57. CERN Council, *Resolution Concerning the Construction of the LHC and the Funding of the Organization*. CERN/2179 (1996)
58. L. Rossi, L. Bottura, Rev. Accel. Sci. Tech. **05**(30), 003 (2012)
59. A. Tollestrup, E. Todesco, Rev. Accel. Sci. Tech. **1**, 185–210 (2008)
60. K. Ehret, Funding of High Energy Physics, in *Physics at the Terascale*, ed. by I. Brock, T. Schörner-Sadenius (Wiley-VCH, Weinheim, 2011) pp. 415–430
61. R. Barnett, M. Nordberg, The Sociology and Management of Terascale Experiments, in *Physics at the Terascale*, ed. by I. Brock, T. Schörner-Sadenius, (Wiley-VCH, Weinheim, 2011) pp. 401–414
62. E. Fernandez, G. Jarlskog (eds.), *Proceedings of the ECFA Study Week on "Instrumentation Technology for High-Luminosity Hadron Colliders"* (Vol. 1), Barcelona, Spain, 1989
63. E. Fernandez, G. Jarlskog (eds.), *Proceedings of the ECFA Study Week on "Instrumentation Technology for High-Luminosity Hadron Colliders"* (Vol. 2), Barcelona, Spain, 1989
64. CERN-ECFA workshop "Towards the LHC Experimental Programme (LHC92)", Evian, France, 1992
65. CERN Timelines—Building the LHC. http://timeline.web.cern.ch/timelines/The-Large-Hadron-Collider. Accessed 3 March 2015
66. M. Della Negra, P. Jenni, T. Virdee, Science **338**, 1560–1568 (2012)
67. D. Froidevaux, P. Sphicas, Ann. Rev. Nucl. Part. Sci. **56**, 375–440 (2006)

Chapter 2
A Journey to the Heart of the LHC

Bernhard Holzer and Reyes Alemany-Fernandez

Abstract This chapter provides a summary of the most important design parameters of the LHC, and it gives an introduction to the layout and the optics for the different beam energies. Based on this, the luminosity reach of LHC is discussed, together with the most important performance limits encountered during Run 1, like space charge in the injectors, beam-beam interaction effects and the electron-cloud effect. Aspects of the LHC operation and the careful staged approach to final luminosity production are described. Proton-proton and ion-ion commissioning and operation are documented, as are other, more exotic ways of running the LHC, e.g. as an ion-proton collider. The chapter finishes by looking into the future of high-energy hadron colliders, covering the high-luminosity LHC upgrade and the "Future Circular Colliders" of 100 km circumference.

2.1 Introduction and Basics of the LHC Machine

The LHC as a proton-proton storage ring is standing in a tradition of hadron colliders that, since the startup of the first hadron collider—the CERN ISR—in the beginning of 1971, have been built as discovery machines (see Sect. 1.2). Building on the experience gained in machines like the Tevatron [1], HERA [2] and Sp$\bar{p}$S [3], the LHC has been designed to operate at the highest reachable beam energies, limited only by the size of the machine, which was dictated by the existing LEP tunnel, and by the technically realisable magnetic fields.

B. Holzer (✉) · R. Alemany-Fernandez
CERN, 1211 Geneva 23, Switzerland
e-mail: bernhard.holzer@cern.ch

R. Alemany-Fernandez
e-mail: reyes.alemany.fernandez@cern.ch

T. Schörner-Sadenius (ed.), *The Large Hadron Collider*,
DOI 10.1007/978-3-319-15001-7_2

Fig. 2.1 The LHC storage ring in the former LEP tunnel. *Source* CERN

2.1.1 Design Parameters

The LHC is a two-ring, superconducting accelerator and collider with a centre-of-mass collision energy of up to 14 TeV installed in the 27 km long tunnel formerly used by LEP (Fig. 2.1).[1] It has been designed and constructed with the aim of discovering the Higgs particle and for the investigation of rare high energy physics events.

The high beam intensities required for a luminosity of $10^{34}\,\mathrm{cm}^{-2}\,\mathrm{s}^{-1}$ exclude the use of antiprotons and, consequently, a design as single-ring $p\bar{p}$ collider where both beams are stored in a common magnet structure and vacuum chamber (as it had been the case e.g. at the Tevatron). Due to the limited geometrical space in the tunnel, however, the two LHC rings had to be designed as "twin-aperture" magnets; with only 3.76 m tunnel diameter the installation of two separate storage rings was technically impossible. As a consequence, the superconducting magnets of the two rings are housed in a common return yoke and cryostat, thus achieving a compact design. Correspondingly, both rings have to be operated in a synchronised way during particle injection, acceleration and luminosity operation.

While traditional single-ring colliders are limited in the number of bunches, i.e. the number of groups of particles captured within one radio-frequency bucket, that can be carried in the machine, the two-ring design of the LHC allows the operation with multiple bunch trains, which are brought into collision in four interaction points (IPs). Unprecedented beam intensities can thus be stored in the machine, grouped and injected in trains of up to 288 proton bunches. Given the design bunch distance

[1] The tunnel is precisely 26,658 m in circumference. For simplicity, we shall call it 27 km in the rest of the book.

Table 2.1 Nominal LHC parameters for the injection and collision energies

	Injection	Collision
Proton energy (GeV)	450	7000
Circumference (m)	26658.883	
Particles/bunch (10^{11})	1.15	
Number of bunches	2808	
Bunch distance (ns)	25	
Beam current (mA)	584	
Norm. emittance (x and y) (μmrad)	3.5	3.75
Stored energy per beam (MJ)	23.3	362
Rms beam size at IP1 and IP5 (μm)	375	17
Rms beam size at IP2 and IP8 (μm)	280	71
Peak luminosity ($cm^{-2}\,s^{-1}$)		1.0×10^{34}

of 25 ns or 7.5 m within a train, up to 2808 bunches with a single-bunch intensity in the order of 1.2×10^{11} protons can be stored in each LHC beam and brought into collision at the IPs.

The main parameters of the machine are listed separately in Table 2.1 for the injection and collision energies [4].

To collide two beams of equally charged particles requires opposite magnetic dipole fields, as indicated in Fig. 2.2. The same requirement of an alternating field configuration between the two storage ring magnets has to be fulfilled for the complete focusing structure of the arc lattice. This requirement sets strict boundary conditions

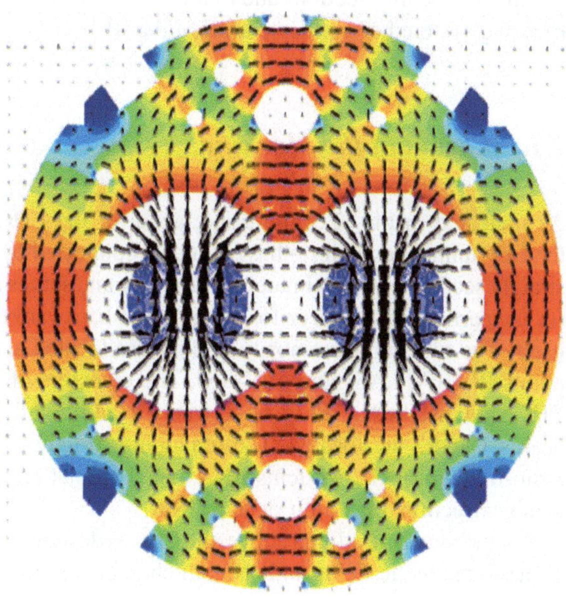

Fig. 2.2 Field configuration of the LHC twin-aperture dipole magnets. *Source* CERN

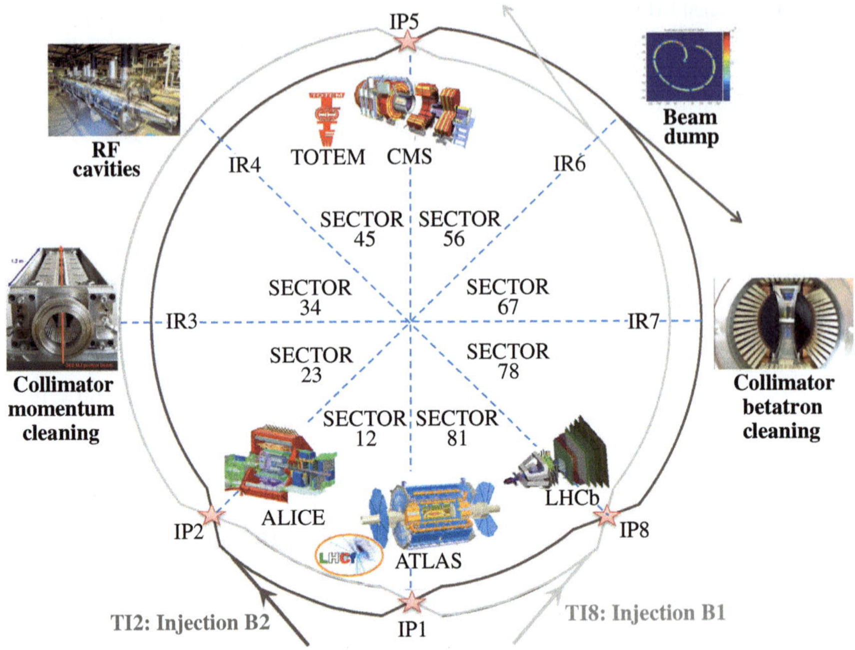

Fig. 2.3 Geometry of the LHC with eight arcs and eight straight sections. The two beams cross at the interaction points 1, 2, 5 and 8. *Source* CERN

for the layout of the interaction region where the two beams pass through special so-called matching sections and finally share a common vacuum chamber and triplet focusing system before they collide at the IP.

2.1.2 Layout of the Machine

As indicated in Fig. 2.3, the basic layout of the LHC machine—while following the LEP tunnel geometry—has eight arcs connected by eight straight sections. Each straight section is approximately 530 m long and serves as insertion region (IR) either for the particle detectors or for the machine hardware equipment of the storage ring itself. The two high-luminosity regions are located at the two opposite interaction points IP 1 (ATLAS) and 5 (CMS). The detectors of the experiments ALICE and LHCb are installed at IPs 2 and 8, where in addition the beam-injection schemes for beams 1 and 2 are located. The remaining straight sections are equipped with the radio-frequency (RF) systems (IR 4), the collimation system (IRs 3 and 7), and the beam extraction and dump system (IR 6).

For the reasons outlined above, the LHC is designed as proton-proton collider with separate magnetic fields and vacuum chambers in the main arcs and with common

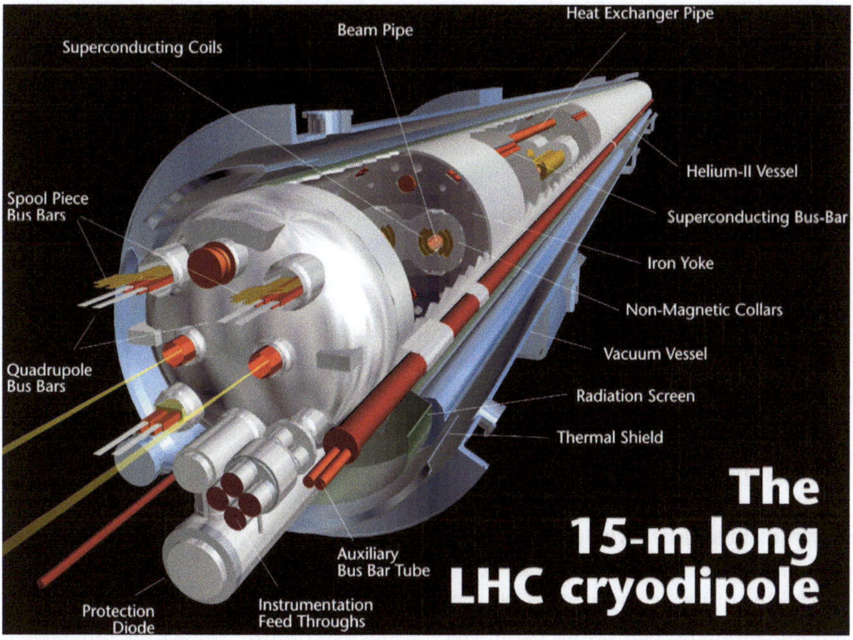

Fig. 2.4 Schematical view of the 15 m long twin-aperture dipoles, housed in a common cryostat. *Source* CERN

lattice structures only at those insertion regions where both beams are brought into collision. In these long straight sections the two beams share the same beampipe of approximately 130 m length. In order to maintain an equal circumference of the two beams, they cross over at IPs 1, 2, 5 and 8, as indicated schematically in Fig. 2.3.

The maximum beam momentum (or energy) that can be carried in the storage ring is determined by the field of the dipoles and their effective magnetic length in the LHC arcs. In a large design and optimisation campaign, the super conducting cable, coil geometry and cryostat design had been optimised to obtain dipole fields considerably higher than achieved in any superconducting storage ring before. The nominal magnetic field strength of 8.3 T corresponds to a maximum beam energy of 7 TeV. In Run 1, technical boundary conditions limited the achievable field, and the highest possible proton energy was 4 TeV per beam. The schematic layout of the LHC magnets is shown in Fig. 2.4.

2.1.3 Beam Optics and Magnet Lattice

The beam optics of the LHC is based on a standard "FODO lattice", i.e. a magnet structure that contains, in a periodic manner, quadrupole lenses that alternately focus the particles in the horizontal and vertical plane. The basic cell of the LHC consists

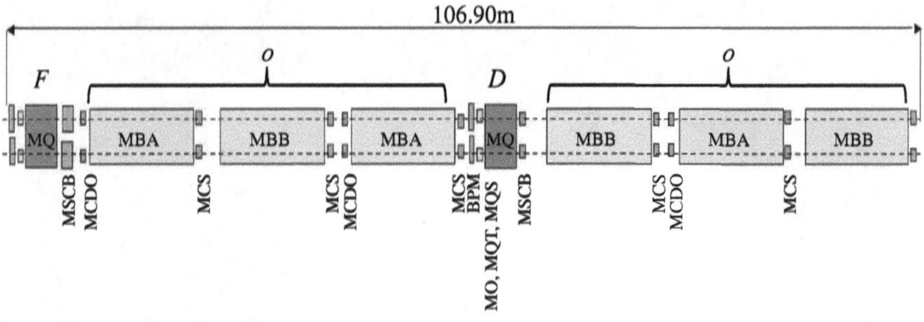

Fig. 2.5 Fundamental magnet structure of the LHC storage ring: Two quadrupoles MQ and six dipoles MBA, MBB are grouped to obtain the FODO structure that represents the basic cell in the LHC arc. *Source* CERN

of a focusing and a defocusing quadrupole and six dipole magnets (Fig. 2.5), all of which are superconducting. Given the magnetic length of 14.3 m for the dipoles, the length of this basic structure adds up to 107 m. As in most proton storage rings a phase advance of 90° per cell has been chosen.

The resulting beam optics is presented in Fig. 2.6a. The figure shows the beam amplitude function or "Twiss parameter" β that determines—under the influence of the focusing fields—the beam envelope. The plot refers to the injection case where, due to the larger beam emittance and the aperture limitations that go along with it, the β function in the straight sections has to be limited to values below 500 m. The periodic optics in the arc of LHC is clearly visible. Typical values of $\beta = 180$ m are obtained. Given the normalised design beam emittance of $\varepsilon_{\text{norm}} = 3.75\,\mu$mrad, we obtain an absolute value at injection energy of $\varepsilon_{450\text{GeV}} = 7.3 \times 10^{-9}\,\mu$mrad and a beam size of

$$\sigma = \sqrt{\varepsilon * \beta} \approx 1.1\text{mm}.$$

It should be emphasised that this value refers to the injection energy and will—due to the adiabatic shrinking of the emittance—be reduced during beam acceleration.

In addition to this basic layout of two quadrupole lenses and six dipole magnets per cell, a number of correction coils is included in the arc lattice, also designed in superconducting technology. Specific dipole correctors keep the beam orbit within the tight tolerances required by machine performance and safety; sextupole magnets are used for chromatic corrections; octupoles provide Landau damping.

These main building blocks of the LHC lattice are complemented by an ensemble of multipole compensation coils, designed as so-called "spool-piece correctors", which are necessary because—like superconducting magnets in general—the LHC dipole and quadrupole magnets are far from being perfect. Especially at low field, i.e. at beam injection and during the initial stage of the ramp, unwanted multipole errors of the LHC main magnets have a detrimental influence on the beam stability. The compensation magnets are designed to keep the lattice functions within tight

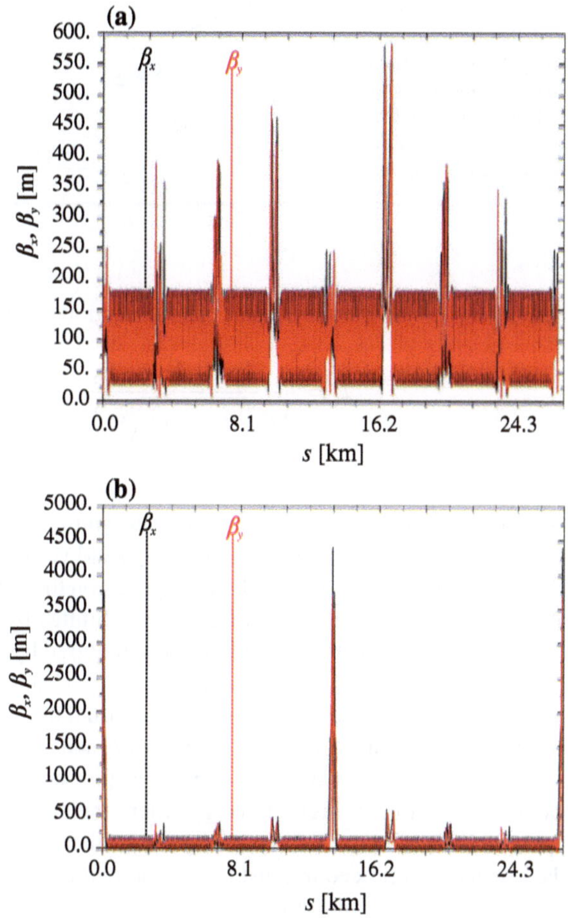

Fig. 2.6 **a** Beam optics of the LHC at 450 GeV injection energy. Due to the large emittance at this energy, the maximum β function has to follow strict limits in the arcs and in the straight sections to guarantee sufficient aperture. **b** Luminosity optics of the LHC. *Source* CERN

tolerance limits by counter-acting the most prominent multipoles errors like skew quadrupoles normal and skew sextupoles, normal and skew octupoles, and 10- and 12-pole coefficients. The spool-piece correctors are powered octant-wise.

2.1.4 Luminosity

The number of events generated in particle collisions, N_{event}, depends on the cross section of the event category considered, σ_{event}, and the luminosity, $\mathcal{L}$, that is a function of the storage ring parameters:

$$N_{\text{event}} = \mathcal{L} \times \sigma_{\text{event}}.$$

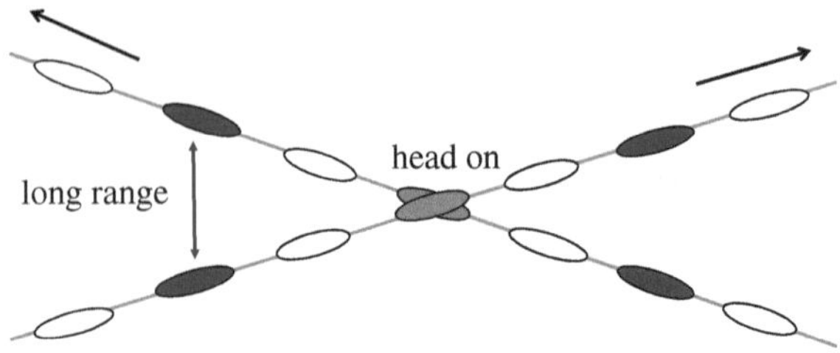

Fig. 2.7 A finite crossing angle has to be applied at each interaction point to avoid detrimental secondary collisions between the bunches spaced at 25 ns. *Source* CERN

For a Gaussian transverse particle distribution, $\mathcal{L}$ can be written as

$$\mathcal{L} = \frac{N_1 N_2 f_{\text{rev}} n_{\text{b}}}{2\pi\sqrt{\sigma_{1x}^2 + \sigma_{2x}^2}\sqrt{\sigma_{1y}^2 + \sigma_{2y}^2}} \cdot F \cdot W \tag{2.1}$$

Here, N_1 and N_2 refer to the number of particles in the n_{b} colliding bunches, f_{rev} describes the revolution frequency in the ring and σ stands for the transverse beam size at the interaction point in the horizontal (x) and vertical (y) plane for beam 1 and 2. As the minimum beam size at the IP depends on the β function that can be obtained at that position, the maximum luminosity is limited by the aperture requirements of the final triplet magnets and the free space required by the installation of the particle detectors. The two luminosity reduction factors F and W in (2.1) describe the effect of a final crossing angle of the two beams (F) and the transverse offset at the collision point (W). For ideal conditions the factors F and W are equal to one.

By design, a value of $\beta^* = 55$ cm is foreseen in the LHC, which, however, can only be set up at highest energy where the emittance is small enough to avoid aperture limitations. The luminosity optics therefore is applied at "flat top", after the acceleration procedure, just before the two beams are brought into collision. In Fig. 2.6b, the beam amplitude function at luminosity optics is shown in comparison to the more relaxed optics of Fig. 2.6a at injection energy.

As shown schematically in Fig. 2.7, the large number of stored bunches and the resulting small bunch distance of 25 ns create unwanted collisions every 3.75 m. If ignored, up to 34 of these collison points are obtained upstream and downstream of the interaction points. This would have a detrimental effect on the beam stability and on the quality of the detector data analysis. A crossing angle of $2 \times 142\,\mu$rad in IPs 1 and 5 is needed to avoid these unwanted collisions. A consequence of this choice of crossing angle is, however, that the bunches do not collide anymore head on at the IPs and that the geometrical reduction factor F becomes smaller than one. F depends on the bunch length σ_{s}, on the transverse beam size in the plane of crossing, and on the size of the crossing angle ϕ:

$$F = \frac{1}{\sqrt{1 + 2\frac{\sigma_{\text{s}}^2}{\sigma_{1x}^2 + \sigma_{2x}^2}\tan^2\frac{\phi}{2}}}. \tag{2.2}$$

For the values listed in Table 2.1, we obtain the unavoidable luminosity reduction of $F = 0.836$, which, in a sense, is the price to pay for the high luminosity gained by the large bunch number that can be stored in the machine. Using the LHC specifications of Table 2.1, a peak luminosity of the machine of $\mathcal{L}_{\text{peak}} = 1.0 \times 10^{34}\,\text{cm}^{-2}\,\text{s}^{-1}$ is obtained.

In LHC Run 1, the beam energy was limited to 3.5–4 TeV [5, 6], and as a consequence the maximum achievable value for β^* was (due to the higher emittance at that energy) 60 cm. Still, a peak luminosity $\mathcal{L}_{\text{peak}}$ of about $7 \times 10^{33}\,\text{cm}^{-2}\,\text{s}^{-1}$ has been obtained, thanks to the high-quality beams that had been produced in the pre-accelerator chain.

A special challenge in this context arises from the quite different running conditions of the four LHC experiments. In IPs 1 and 5, the high-luminosity experiments ATLAS and CMS have been installed that are capable of analysing the large number of events, created in LHC collisions. During Run 1, this number included up to 35 pile-up events per bunch crossing. In contrast to ATLAS and CMS, the ALICE and LHCb detectors are optimised for more specific purposes and cannot be operated at highest luminosities. During LHC operation, therefore, the luminosity delivered in their IPs had to be reduced to a tolerable level. Typical values of $\mathcal{L} = 10^{32}\,\text{cm}^{-2}\,\text{s}^{-1}$ in IP 8 and $\mathcal{L} = 10^{27}\,\text{cm}^{-2}\,\text{s}^{-1}$ in IP 2 have been established. To achieve this, the beam collisions were set up with a different beam optics at these interaction points and, in addition, with a transverse offset between the beams that corresponds to about 2–3 σ of the transverse beam size. The reduction factor

$$W = e^{\frac{-(d_1-d_2)^2}{2(\sigma_1^2+\sigma_2^2)}} \tag{2.3}$$

leads to the necessary event rate reduction as indicated in (2.1) ($d_{1,2}$ and $\sigma_{1,2}$ are the distances and the beam sizes in the direction of the beam offset, respectively). To optimise these special operation conditions, a "luminosity levelling" system was commissioned in 2010 and used in routine operation since 2011 [7]. Luminosity levelling is achieved by an automatic accelerator-experiment real-time feedback system. The experiment online system computes and monitors the instantaneous luminosity and the online detector performance. Based on that it establishes a maximum target luminosity, which is communicated to an LHC application that in turn adjusts the beam overlap at the interaction point to reach the required luminosity. This luminosity control has been a direct tool to maximise the LHCb physics yield. It allowed the operation of the experiment at extremely stable data-taking conditions with more than 95 % of the luminosity collected within 3 % of the target, and a total integrated luminosity that is three times higher than foreseen. In Fig. 2.8, the luminosity evolution for ATLAS, CMS and LHCb is shown for an arbitrarily chosen fill. While in ATLAS and CMS the luminosity decays exponentially once the beams are brought into collision, in LHCb the luminosity is continuously levelled to the target luminosity proposed by the experiment, until the levelling potential is exhausted. At that moment both beams collide head on, without any transverse separation, and the luminosity begins to decay exponentially.

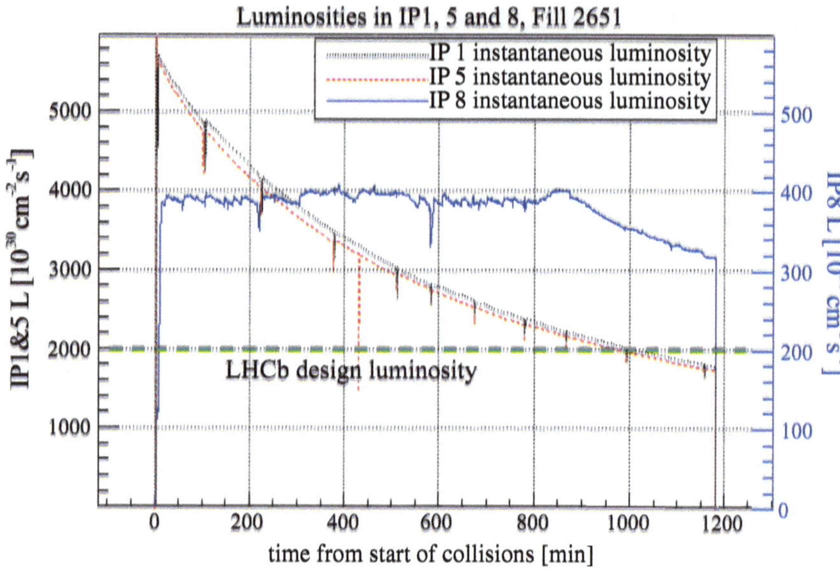

Fig. 2.8 Luminosity evolution in ATLAS, CMS and LHCb during an LHC fill. See the text for more details

2.2 LHC Performance Limits in Run 1

During beam operation in LHC Run 1, the performance of the storage ring and its pre-accelerators has been improved and, step by step, pushed to the maximum achievable values set by fundamental limits in energy, beam current and luminosity. It is worth discussing at least the most important of these limits and their consequences.

2.2.1 Space-Charge Effect

The single bunch intensity obtained in the LHC is limited by several effects. The most obvious one is the space-charge effect that plays a major role at low energies, i.e. in the pre-accelerator chain.

In a highly charged proton bunch, as they are injected into the PS Booster or the PS Synchrotron, every single proton will feel the space charge of the surrounding particles. This space-charge effect has a strong defocusing effect on the bunches. Like in the case of a defocusing quadrupole lens, the transverse oscillation frequency (tune) of the particles is affected; it can even be influenced strongly enough to push the protons towards destructive resonances. In linear approximation, the resulting tune shift ΔQ depends on the bunch population and the energy of the particles:

$$\Delta Q = -2\pi \frac{N_b}{\varepsilon_0} * \frac{r_0}{\beta^2 \gamma^3}, \qquad (2.4)$$

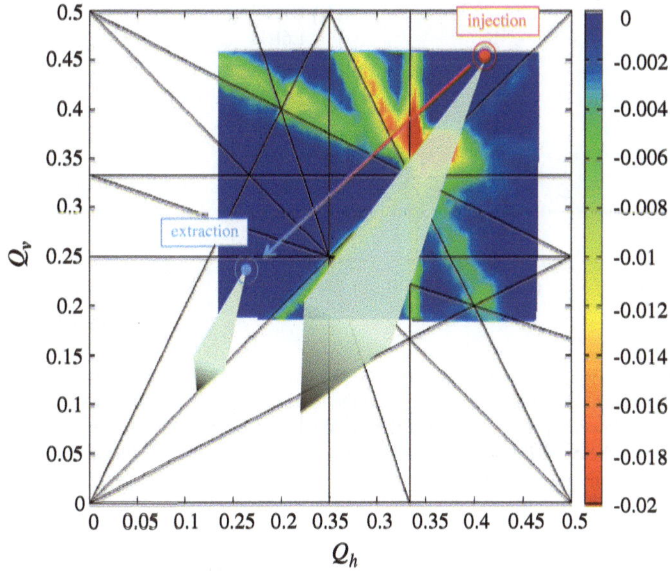

Fig. 2.9 Space charge effect in the PS Booster during injection and acceleration. Due to the energy-dependent space-charge effect the working point of the beam covers a large area in the working diagram during beam acceleration. *Source* CERN

where N_b and ε_0 are the number of particles in the bunch and the absolute emittance of the opposing bunch, r_0 stands for the classical particle radius, and β and γ are the usual relativistic parameters. Figure 2.9 shows the simulated result of the space-charge effect during injection and acceleration in the CERN PS Booster [8].

The tune shift depends strongly on the relativistic parameters β and γ and therefore is most severe at low energies. At present, a new linear accelerator, Linac 4 [9], is under construction that will replace the old Linac 2. The top energy of 160 MeV (compared to 50 MeV in Linac 2) will allow the space-charge limit to be pushed up by a factor of two, thus paving the way to future higher beam intensities.

2.2.2 Beam-Beam Effect

While the space-charge effect is most critical at low energies and determines the beam quality in the pre-accelerator chain, a similar limitation is observed even at highest energies, as soon as the two beams are brought into collision. During collision, the space charge of the bunch in beam 1 has a strong defocusing effect on the protons of beam 2, and vice versa. As a consequence, the particles have effectively different tunes, $Q = Q_{\text{ideal}} + \Delta Q_{\text{beam-beam}}$, which depend on the number of particles per bunch, N_b, the optical function β, the energy of the particle beam in units of rest mass, γ, the beam size at the interaction point, $\sigma_{x(y)}$ and the classical particle radius, r_0:

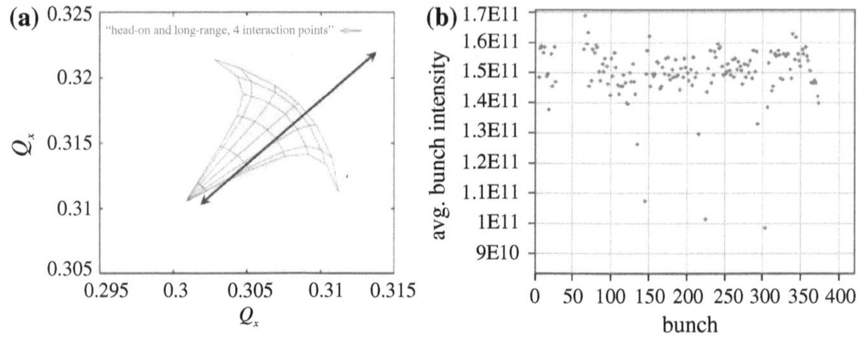

Fig. 2.10 **a** Effect of the beam-beam force on the working point of the protons in the LHC. The *straight line* points to the unperturbed beam on the *upper right side* and the maximum obtained tune distortion on the *lower left* of the plot. **b** Effect of the beam-beam force observed during luminosity operation. While the overall tune of the machine is optimised, a certain number of bunches within the train suffers from reduced lifetime and rapidly loses intensity. *Source* CERN

$$\Delta Q_{x beam-beam} = \frac{\beta_x^* r_0 N_b}{2\pi\gamma(\sigma_x + \sigma_y)\sigma_x} \tag{2.5}$$

The expression refers to the beam–beam tune shift in the horizontal plane; the corresponding equation for the vertical one is obtained via obvious change of the indices. Figure 2.10a shows the result of a simulation for the LHC: under the influence of the beam-beam force, the tune of the particle ensemble is no longer a single value but covers a wide area in the working diagram. The oscillation of the different particles is affected in quite a non-linear manner due to the non-linearity of the beam-beam force. The arrow on the plot shows the overall detuning, compared to the ideal working point of a particle without any beam-beam effect [10, 11].

During Run 1, the machine had been pushed to higher and higher single-bunch intensities until first signature of beam-beam limitations had been observed. Figure 2.10b shows, as an example, the single-bunch intensity for about 350 bunches during standard luminosity operation. Depending on the number of interactions per turn, a certain fraction of the bunches is pushed to the limit, suffering more from the beam-beam force and thus losing particles rapidly, while other bunches remain basically unaffected.

2.2.3 *Electron-Cloud Effect*

The last limiting effect mentioned here is the so-called "electron-cloud problem". Triggered by primary protons that are hitting the vacuum chamber wall, or even more by photons emitted by the circulating proton beam, electrons are desorbed from the vacuum chamber wall and get accelerated by the positive beam potential

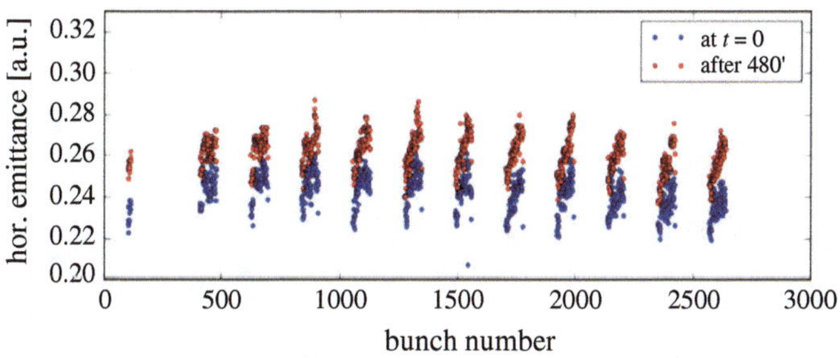

Fig. 2.11 Electron-cloud effect observed during LHC commissioning. The electron cloud builds up as the bunches pass by and can lead to considerable emittance blow-up towards the end of the bunch train. *Source* CERN

of the protons. Some of these electrons will hit the opposite side of the vacuum chamber and create secondary electrons. In case the "secondary-emission yield" is larger than one (in the LHC a value of 1.3 has been determined), an avalanche-like effect can occur, and more and more electrons are kept within the positive potential of the stored proton beam, accumulating around the closed orbit of the LHC beam. As a result instabilities and a blow-up of the beam emittance are observed. The size of the effect depends on the bunch intensity and the energy of the stored beam, on the bunch pattern in a bunch train, and on the material properties and geometry of the vacuum chamber wall.

LHC operation has been optimised to limit this effect. For the 50 ns bunch distance that had been used for luminosity production in Run 1, the observed emittance blow-up during a standard fill could be kept on a tolerable level (see Fig. 2.11).

2.3 LHC Operation

The LHC is an unprecedented machine in many aspects; it hosts proton beams with the highest stored energy ever achieved; it is a 27 km circumference cryogenic installation with record energy stored in the magnets when powered at 7 TeV; sophisticated systems like the beam injection, beam extraction, collimators, superconducting radio-frequency systems, beam instrumentation, etc., work all together in an orchestrated way to assure a reliable and secured machine operation. But the accelerator operation of such a machine had to be learnt on the way as well. This section highlights some of the most important aspects of the LHC operation like the machine protection, the cryogenics operation, the staged approach to the beam commissioning before physics production starts, and last but not least, it summarises the ion operation and the proton-ion run.

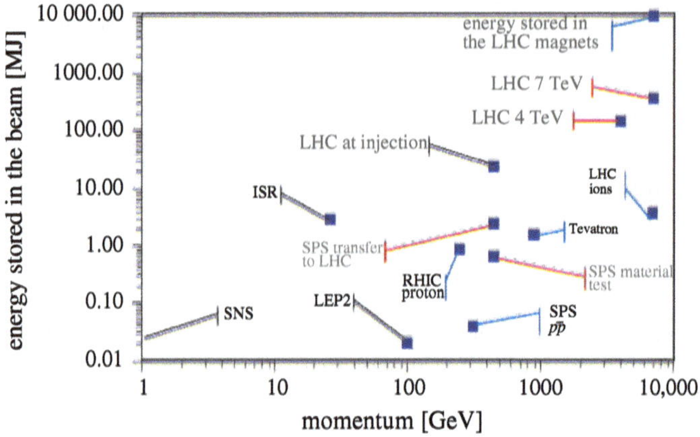

Fig. 2.12 Energy stored in the LHC proton beams, compared to previous accelerators. *Source* CERN

2.3.1 Protecting the LHC from Itself

The high-intensity beams circulating in the LHC and their unprecedented energy content—362 MJ as indicated in Fig. 2.12 and thus enough to melt 500 kg of copper—bring demanding challenges. Most importantly it has to be guaranteed that, in case of problems, the stored energy can be released in a safe and fast manner. The complexity of the accelerator requires a machine protection system that relies on several interdependent modules with twofold priority: to protect the accelerator equipment from beam damage and to prevent the superconducting magnets from quenching while still keeping LHC availability uncompromised.

The pivotal systems engaged in the LHC protection are the collimators [12], the beam-loss monitors (BLMs) [13, 14], the beam controls, the beam injection system and the LHC beam dump system (LBDS) [15]. The master plan for machine protection, which involves all of these systems, is based on several fundamental principles [16, 17].

In a system as complex and heterogeneous as the LHC, many types of malfunctions can occur. Dealing with high-energy particle beams, one of the most crucial parameters of a failure scenario is the time scale since it determines the response time required by the protection system. The fastest and most serious failures occur on the timescale of a single turn or less, for example during beam injection or beam extraction. High-reliability systems are designed to minimise the probability for such events; they are complemented by other safety measures like interlocking the injection elements whenever they are not needed. Despite these design precautions, malfunctions—such as incorrect firing of the kicker magnets—cannot completely be excluded. Active protection based on the detection of a fault and an appropriate reaction is not possible on these short timescales. Therefore, machine safety in this case has to rely on passive protection using beam absorbers and collimators that are positioned with high accuracy close to the beam and that capture accidentally deflected particles. Moreover, a high-intensity beam cannot be injected into an empty machine. Only a so-called single "safe beam" bunch or "pilot bunch" with strongly reduced intensity (which is highly unlikely to damage equipment) is allowed to be injected into an empty ring. This single bunch of $5-10 \times 10^9$ protons is used to measure and adjust the most critical beam parameters like tune, chromaticity, coupling, trajectory, injection oscillations and energy matching between the SPS and the LHC. Once this is accomplished, bunch trains of high intensity can be injected into LHC.

In contrast to problems on the shortest timescales, the majority of possible equipment breakdowns lead to beam instabilities generating fast movements of the orbit or growth in beam size that must be detected on a timescale of 1 ms. Protection against such events relies on fast monitoring of the beam position and of beam losses. The LHC is equipped with of the order of 4000 BLMs distributed along its circumference that are used to protect all elements against excessive beam loss. Equipment monitoring, like quench detectors and monitors for failures of magnet powering, provides redundancy for the most critical failure scenarios.

On the longest timescales, there will be unavoidable beam losses around the LHC machine during all phases of normal operation. Most of these losses will be captured in the collimation sections IRs 3 and 7 (see Fig. 2.3), where the beam losses and the heat load at collimators are monitored; when these become unacceptably high, the beam is dumped.

2.3.2 *The LHC Cryogenic System: One of the Coldest Places on Earth*

The LHC is the largest cryogenic system in the world and one of the coldest places on Earth. The LHC main magnets operate at a temperature of 1.9 K (–271.3 °C). In order to produce the magnetic field of 8.3 T that is needed to keep particle beams on course around the 27 km ring, the coils of the LHC main magnets are supplied with a current of 11,850 A. The use of superconducting material (niobium-titanium, NbTi) is the only way to achieve these values. 120 t of liquid helium are needed to keep the NbTi wires at their operational temperature. The layout of the LHC magnet cooling system is based on five cryogenic islands with eight cryogenic plants in total, which distribute the cooling fluid and transport kilowatts of cooling power over several kilometers. Each cryogenic plant combines an 18 kW refrigerator at 4.5 K and a 2.4 kW one at 1.8 K (Fig. 2.13).

Helium was a natural choice of coolant since it allows components of large-scale installations to be kept cool over long distances. At atmospheric pressure, gaseous helium becomes liquid at around 4.2 K (–269.0 °C). However, if cooled below 2.17 K (–271.0 °C), it passes from the ordinary fluid to the superfluid state. Superfluid helium has remarkable properties: very high thermal conductivity, high heat capacity and low viscosity. Thanks to the low viscosity, superfluid helium permeates the NbTi windings where it acts as a thermal barrier against temperature transients due to its high specific heat. The excellent thermal conductivity enables it to conduct heat without mass

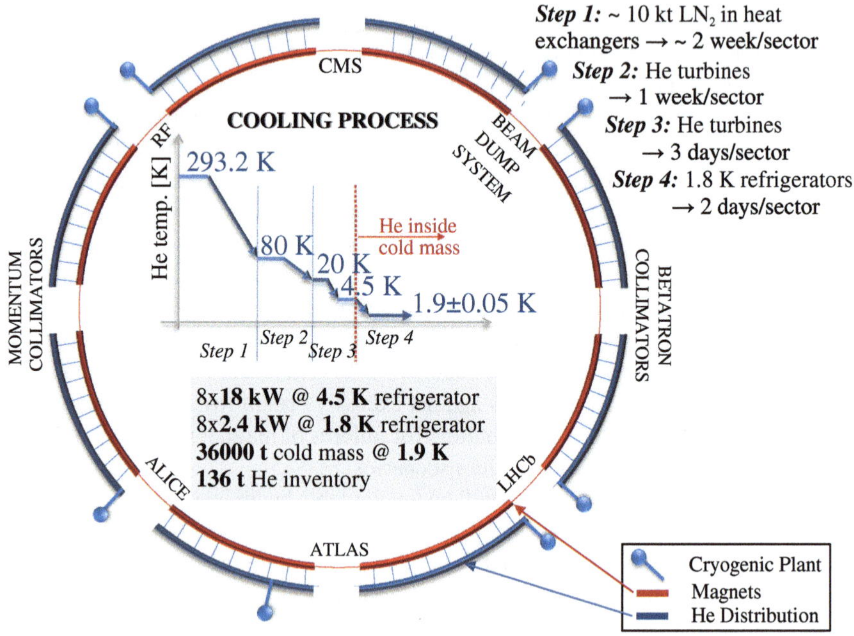

Fig. 2.13 The cryogenic distribution system of the LHC. *Source* CERN

transport, i.e. without the need for fluid circulation or pumps. Consequently, LHC magnets operate in static baths of pressurised superfluid helium (note in Fig. 2.14 that the magnet is embedded in a static bath of pressurised He II) close to atmospheric pressure continuously cooled by heat exchange with saturated helium flowing in a heat exchanger tube as sketched in Fig. 2.14. Two consecutive magnets are connected by a superconducting interconnection busbar, indicated in Fig. 2.14 as "SC BUS

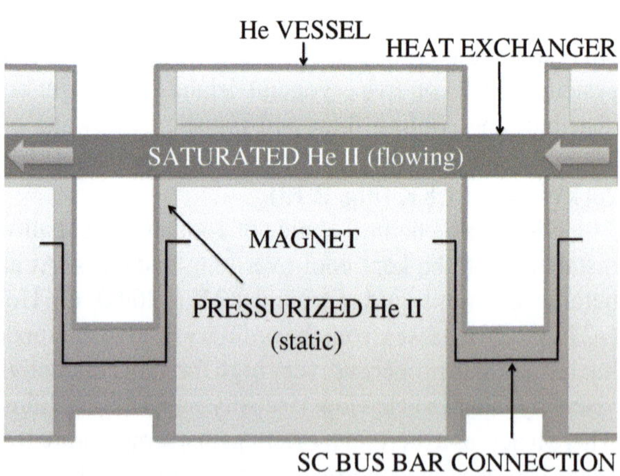

Fig. 2.14 LHC magnet cooling scheme. *Source* CERN

BAR CONNECTION", made of NbTi cables surrounded by copper. The bus bar interconnection is also embedded in the static bath of superfluid helium.

The cooling process of one sector takes around four weeks and it is divided in four steps as indicated in Fig. 2.14 and explained in [18].

2.3.3 *Start-Up of the LHC in 2008*

Almost 25 years of design and construction preceded the first injection of a single proton bunch into the LHC on 10 September 2008 (see Chap. 1 for a short historical account). Important technological choices had been taken over this quarter century; prototypes were built and tested to asses the feasibility of the technology; thousands of magnets were produced; and finally the underground installation of all accelerator equipment was accomplished. The last dipole was installed in the tunnel in April 2008.

The complexity of the LHC required a staged approach to the beam commissioning. Before aiming for a full commissioning of the machine at injection energy, a series of injection tests were carried out. The first LHC sector that saw beam was sector 23 (Fig. 2.3). Between 8 and 11 August 2008, beam 1 was successfully injected and stopped with the momentum collimators at IR 3. The next injection test took place from 22–25 August; this time beam 2 was injected for the first time in LHC and stopped with the collimators at IR 7 [19, 20].

Beam commissioning started on 10 September 2008. On the first day the LHC saw, for the first time ever, both beams traveling around the 27 km machine exceeding all expectations. Some hours later, the anti-clock wise beam was successfully bunched (or captured) by the radio-frequency system, and circulated for over half an hour before being safely extracted from the LHC. The next step consisted to repeat the process for beam 1, but a failure of a power transformer on 12 September in point 8, switched off the main compressors of the cryogenics for sectors 78 and 81. In this conditions, injection into LHC is not possible and the beam commissioning was put on hold for several days waiting for the recovery of the beam conditions. On the unaffected sectors, powering test to validate the machine for 7 TeV continued.

During all these tests single bunches of about 5×10^9 protons were injected to guarantee safe beam conditions.

2.3.4 *A Forced Break*

The series of successful commissioning steps was interrupted by a severe incident on 19 September 2008. During the powering test of sector 34 (the sector between the momentum collimators and the radio-frequency system, see Fig. 2.3) without beam in the machine, a faulty superconducting interconnection busbar between dipole C24 and quadrupole Q24 developed a resistance of around $\sim$200 nΩ (the nominal value is of the order of 0.3 nΩ). This resulted in a voltage difference of around 2 mV while a current of 9000 A was circulating. The resulting heat load of about 16 W could not be cooled down by the helium circuit, thus provoking a thermal runaway. The transition from superconducting to normalconducting state (quench) of the interconnection busbar, which was not in the quench protection system (QPS) [21, 22] of the magnets, melted and boiled away the conductor. As a result, an electrical arc developed between the two exposed cable ends. This arc punctuated the helium vessel, releasing approximately two tons of helium into the cryostat. The existing overpressure valves could not cope with the released mass flow of 20 kg/s. As a consequence, a pressure of seven bars developed on the vacuum barriers (which are designed for 1.5 *bar*), causing extensive mechanical damage, i.e. magnets were physically displaced, many interconnections were destroyed, the beam vacuum was contaminated, support jacks were ripped out of their concrete sockets, etc. In total 53 magnets (39 dipole and 14 quadrupoles) had to be brought to surface. Ten of them had to be replaced, the others could be repaired and/or cleaned.

Following the incident, numerous consolidation and compensatory measures were devised and implemented:

- The QPS extended its functionality to detect and localise potentially faulty splices in the main dipoles and quadrupoles. If the system measures a voltage of more than 300 μV in a splice for more than 10 s, the system will interlock the concerned circuit, for example all the dipoles of one sector, by initiating a fast discharge through the activation of the energy extraction system. After this, a complete and controlled energy ramp-down of the rest of the machine takes place to bring the magnets back to injection current.
- When a magnet quenches, heat transfer between the magnets of both beam apertures can sometimes trigger a quench. This was difficult to detect with the old QPS, therefore, an additional layer of protection electronics was added to the QPS to ensure the safe detection of such aperture-symmetric quenches.
- New overpressure relief valves have been installed on all dipoles, on the standalone magnets in the long straight sections and on various current feed boxes.
- New support jacks, able to withstand the forces generated in a similar incident, have been installed on the main quadrupoles, with vacuum barriers preventing lateral movement and knock-on damage. These vacuum barriers lead to the sectorisation of the insulation vacuum of the LHC magnet cryostats, housing the superconducting magnets. Each vacuum barrier is a leak-tight austenitic stainless-steel thin-wall structure mounted on the main quadrupoles between the magnet helium enclosure and the vacuum vessel.
- A campaign for the detection of resistive splices was initiated in order to get a comprehensive map of the quality of the splices in the whole machine. During the long shutdown 2013–14, all splices have been measured and, if necessary, repaired in order to prepare the machine for safe operation at 6.5 TeV in Run 2.

2.3.5 LHC Proton Run

After fourteen months of intense repair work, LHC Run 1 re-started on 20 November 2009. Three main periods can be identified: beam commissioning (2009–2010), exploring the machine limits (2011) and beam operation (2012). The quality of the superconducting interconnection splices limited the maximum achievable energy to 3.5 TeV until 2011 and to 4 TeV during 2012.

The beam commissioning period in 2009 saw several milestones listed below in chronological order [23]:

- injection and capture of both beams;
- establishment of circulating beams at 450 GeV;
- first beam collisions (no stable beams declared yet) at injection energy with pilot bunches;
- first successful energy ramp up to 1.18 TeV with one pilot bunch per beam;
- first stable beams declared at 450 GeV with four pilot bunches per beam;
- first successful energy ramp to 1.18 TeV with two pilot bunches per beam and first collisions in ATLAS. A week after a second trial brought collisions in all four experiments;
- stable beams at injection energy with 16 pilot bunches per beam;
- first squeeze of the β function at the interaction point up to 7 m with four bunches per beam at 1.18 TeV. Collisions in all four experiments were established afterwards.

2010 was the first full year of LHC commissioning, with several operational breakthroughs realised [24]. First collisions with unsqueezed beams were delivered on 30 March 2010 at 3.5 TeV with a β function at the IPs down to 2 m. Summer 2010 was devoted to increasing the single-bunch intensity, and from September onwards, bunch train commissioning started. Already then the single-bunch intensity was high enough to cause instabilities so that the beams had to be stabilised using octupole magnets when brought into collisions. A staged increase in total intensity was carried out throughout autumn as can be seen in Fig. 2.15a, with operational and machine protection validation at each increase of the number of bunches. The crowning event of the 2010 proton run was the achievement of 368 bunches per beam of around 1.2×10^{11} protons per bunch colliding at 3.5 TeV per beam at 3.5 m β^* at all IPs. A peak luminosity of $2 \times 10^{32}\mathrm{cm}^{-2}\,\mathrm{s}^{-1}$ was attained with an integrated luminosity of 6 pb^{-1} per day delivered in the final week of proton operations. Figure 2.15a summarises the luminosity evolution, taking as example the peak luminosity measured in ATLAS. The figure also shows how the number of bunches per beam was increased over the year. During 2010, the stored beam energy evolved from 2 MJ in August to more than 20 MJ per beam, as indicated in Fig. 2.16.

All in all, in 2010 the LHC achieved an availability of 65 % of the scheduled beam time—an impressive value for the first full year of machine exploitation.

At the end of October a single fill with 50 ns was tried. The electron cloud produced induced significant heat load in the cold regions and enough vacuum activity in

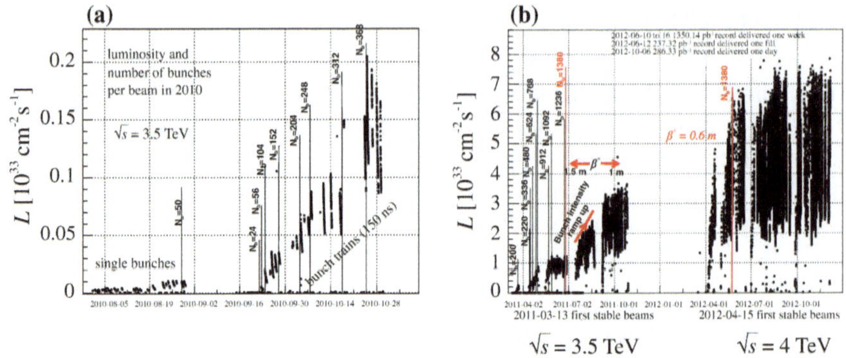

Fig. 2.15 **a** Evolution of luminosity and number of bunches per beam during 2010. The peak luminosity as measured by ATLAS is plotted; CMS luminosities are very similar. The plot also indicates the staged bunch train commissioning at 150 ns performed during the last quarter of the year. **b** Evolution of the ATLAS peak luminosity during 2011 and 2012. The main commissioning steps are indicated in the plot. *Source* CERN

Fig. 2.16 Evolution of the stored beam energy during Run 1. *Source* CERN

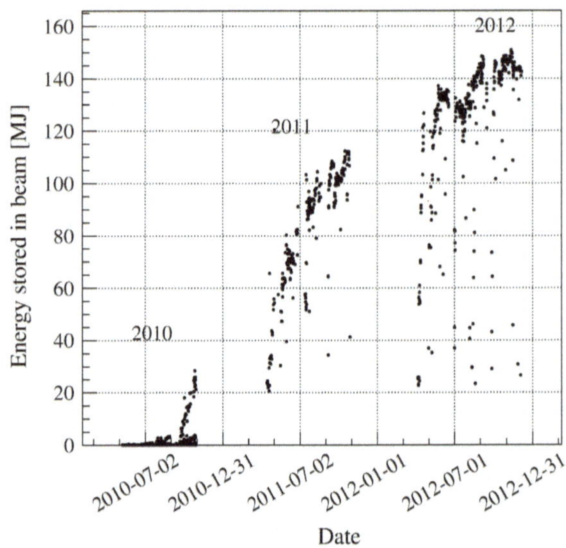

the warm regions to bring the vacuum valves in and dump the beams. Instabilities and beam size growth were observed, as was surface conditioning (scrubbing) with associated gas-desorption effects and a drop in secondary electron emission yield.

In 2011 the LHC moved from commissioning into an "exploring the limits" phase, with the objective of delivering 1 fb^{-1} per high-luminosity experiment [25]. The re-commissioning of the machine after the Christmas shutdown was very fast, and circulating beams were re-established on 19 February 2011. On 21 March 2011 the peak luminosity was back to $1 \times 10^{32}\ cm^{-2}\ s^{-1}$. By the end of June, peak luminosities of $1.26 \times 10^{33}\,\mathrm{cm^{-2}\,s^{-1}}$ had been achieved, with already $1\,\mathrm{fb^{-1}}$ of integrated luminosity available for the summer conferences.

The operational energy remained at 3.5 TeV, and the first stable-beam period of the year took place at 75 ns bunch spacing, with a maximum of 200 bunches per beam. At the end of March 2011, five days were dedicated to running at 1.38 TeV per beam. ATLAS and CMS accumulated around $300\,\mathrm{nb^{-1}}$ of data. This energy per beam corresponds to the energy per nucleon in a lead-lead collision at 3.5 TeV per lead beam, i.e. $(Z/A) \cdot 3.5\,\mathrm{TeV} = (82/208) \cdot 3.5\,\mathrm{TeV} = 1.38\,\mathrm{TeV}$.

The encouraging results of the electron cloud (concerning the secondary emission yield) obtained during the ten days of scrubbing run and the 50 ns injection commissioning in March and April supported the decision of operating at 50 ns bunch spacing with up to 1380 bunches per beam. The staged bunch train commissioning at 50 ns performed at the beginning of 2011 is indicated in Fig. 2.15b. Once the maximum number of bunches (1380) was reached at the end of June 2011, the rest of the summer was devoted to the bunch intensity ramp-up with the corresponding steady increase of luminosity. On 21 April 2011 the LHC broke the luminosity record for hadron colliders so far held by the Tevatron with a peak luminosity of $4.6 \times 10^{33}\mathrm{cm^{-2}\,s^{-1}}$. The drastic luminosity increase brought up issues of machine availability due to single-event errors, and the beam intensity ramp-up had to proceed slowly due to beam instabilities and beam losses. Nevertheless, the performance of the machine was pushed up further thanks to the injector's performance, and the LHC was able to achieve beam intensities of up to 1.35×10^{11} protons per bunch and transverse emittances of $2.2\text{–}2.4\,\mu\mathrm{m}$, around 40 % smaller than the design emittance quoted in (2.2). A dedicated aperture study showed that sufficient aperture was available to operate the machine at even lower β function in ATLAS and CMS, and therefore, since September the squeeze was carried out until 1 m, as indicated in Fig. 2.15b. The figure also shows the corresponding luminosity increase by a factor 1.5.

The start of operations in 2012 was characterised by an increase of beam energy to 4 TeV, with the immediate aims of gaining luminosity and increasing the event yields of a potential Higgs boson. The decision was driven by the positive experience in the previous run years and the absence of beam-induced quenches at 3.5 TeV. The measured aperture at the interaction regions showed to be bigger than expected and, together with the use of tight collimator settings, allowed a β^* of 60 cm in ATLAS and CMS to be reached—which is close to the design value of 55 cm. The bunch spacing remained at 50 ns, with approximately 1380 bunches stored in each beam.

First beam injections took place on 14 March 2012, and both beams were circulating and captured in less than two hours. Two days later the pilot bunches were ramped to 4 TeV for the first time. On 18 March 2012 both beams were squeezed down to 60 cm in ATLAS and CMS. Luminosity production began on 4 May 2012 and lasted until 17 December 2012. The peak luminosity evolution during 2012 can be seen in Fig. 2.15b.

The bunch intensity at injection peaked at 1.7×10^{11} protons per bunch, $\approx 50\,\%$ more than the design bunch intensity quoted in (2.2). While running far above the design bunch intensities, and thus taking machine performance to its extreme, beam

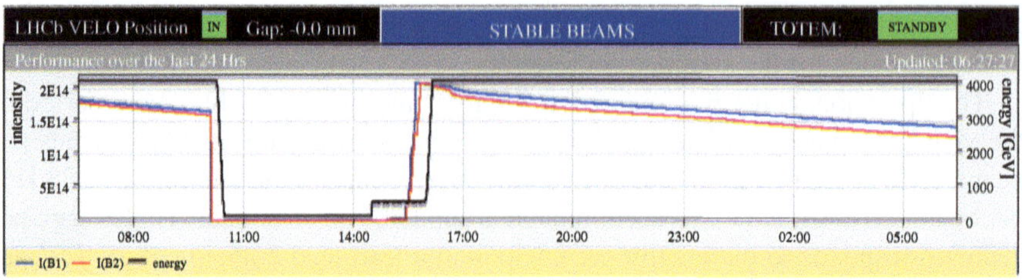

Fig. 2.17 A typical LHC fill including cycle time for the machine preparation, injection plateau, ramp and stable beams. *Source* CERN

instabilities during the squeeze and before collisions were a serious concern and sometimes compromised the operational efficiency. The most effective cure was the head on beam-beam tune spread and the corresponding Landau damping effect [26].

Once in collision, however, the LHC running conditions turned out to be very stable and reproducible, including the luminosity levelling system in the experiments LHCb and ALICE. In Fig. 2.17 a typical LHC fill is shown, starting from beam injection and going over acceleration to stable data-taking conditions.

An increase in beam intensity is expected to enhance the sources for beam-induced heating. In fact, several temperature and pressure diagnostics installed in near-beam LHC equipment revealed significant temperature increase when proton intensity was accumulated and/or when the beams were ramped. As a consequence, some equipment provoked un-programmed beam dumps, and other components were even damaged. There are several reasons for this phenomenon: exposure to synchrotron radiation generated by the beam, impact of protons lost from the beam, impact of photo-electrons generated by the beam, and interaction of the electromagnetic field generated by the beam ("wake field") with surrounding equipment [27]. Other external sources of cooling and heating are also present and set limits to the overall intensity that can be stored in the machine.

All in all, in the year 2012, 201 days were dedicated to proton physics operation. Stable beams lasted an average of 6.1 h (as in 2011), and out of the 201 days of operation 36.5 % were spent in stable beams, delivering an integrated luminosity of 23.3 fb^{-1} to both ATLAS and CMS.

In Fig. 2.18, the integrated luminosity delivered to CMS is presented for the years 2010–2012. After increases in bunch intensity and in the number of stored bunches in the machine, and after the reduction of the beam size at the interaction point when decreasing the β^* the luminosity could be optimised to finally reach a maximum value of $\mathcal{L} = 7.7 \times 10^{33} \text{cm}^{-2}\,\text{s}^{-1}$, close to design.

The LHC parameters reached in the three years of Run 1 are summarised in Table 2.2.

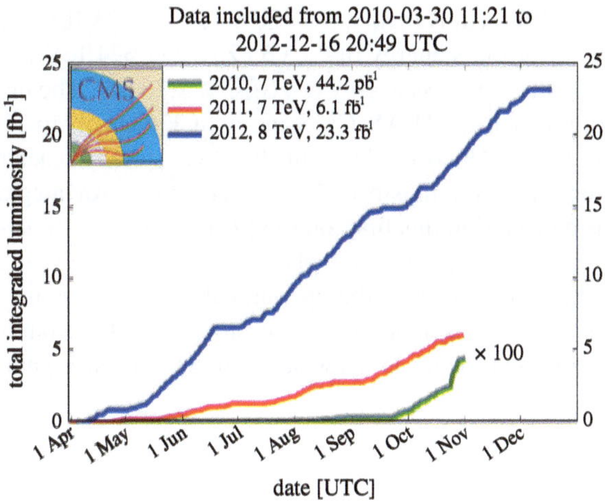

Fig. 2.18 Integrated luminosity delivered to CMS in the years 2010–2012. *Source* CERN

Table 2.2 The LHC main parameters obtained in the three run years 2010–2012

Parameter	2010	2011	2012	Design value
Beam energy (TeV)	3.5	3.5	4	7
β^* IPs 1 and 5 (m)	2.0/3.5	1.5/1.0	0.6	0.55
Bunch spacing (ns)	150	75/50	50	25
Max. num. of bunches per beam	368	1380	1380	2808
Max. bunch intensity (protons per bunch) (10^{11})	1.2	1.45	1.7	1.15
Normalized emittance at start of injection (μmrad)	2	2.4	2.5	3.75
Peak luminosity (10^{33} cm^{-2} s^{-1})	0.21	3.7	7.7	10
Max. mean number of inelastic collisions per bunch crossing	4	17	37	19
Stored beam energy (MJ)	28	110	140	362

2.4 Special LHC Runs

2.4.1 Heavy-Ion Runs

Besides the standard operational mode in proton-proton collisions, heavy-ion collisions are the second major building block of the LHC physics programme and were included in the conceptual design of the accelerator from an early stage. In the evening of 4 November 2010, the first fully stripped lead ions (Pb^{82}_{208}) circulated in the LHC, using the same magnetic configuration as the protons, i.e. with a beam energy of $3.5 \cdot Z$ TeV or, equivalently, $3.5 \cdot Z/A$ TeV $= 1.38$ TeV per nucleon. This yields a total centre-of-mass energy of $2 \cdot 3.5 \cdot Z$TeV $= 574$ TeV or, equivalently, a nucleon-nucleon centre-of-mass energy of $\sqrt{s_{\rm NN}} = 2.76$TeV. The first stable heavy-ion beams with collisions in ATLAS, CMS and ALICE were obtained after four days of beam commissioning. Up to 137 lead bunches were stored in each beam with a β^* of 3.5 m, providing a peak luminosity of $3 \times 10^{25}\,\mathrm{cm}^{-2}\,\mathrm{s}^{-1}$. An integrated luminosity of $9.7\,\mu\mathrm{b}^{-1}$ was delivered in this first ion run [28].

Another heavy-ion (Pb^{82}_{208}) run took place at the end of the year 2011, the commissioning of which profited from the operational experience gained with protons. This time β^* in ALICE was squeezed further to 1 m, and the number of bunches per beam was increased up to 358, resulting in a luminosity gain of a factor of roughly 15 [29].

2.4.2 Proton-Lead Run

After the successful heavy-ion runs in the previous years, a previously unexplored mode of collider operation that was also not foreseen in the baseline programme of the LHC was set up in 2013: collisions of protons with lead ions. After a first feasibility test on 31 October 2011 [30], stable beams were provided for these exotic running conditions [31].

At injection energy, due to their different mass-to-charge ratios, the radio frequencies for protons and lead ions differ by about 5 kHz for the same magnetic settings for both beams, which were the same as in the proton run. Since—due to their higher mass-to-charge ratio—lead ions are slower than protons, the leading bunches of both beams cross each other at a different position in the accelerator. For example, at injection energy when both leading bunches meet each other in IP 1, they meet again there only every 7.7 s. However, the revolution frequencies of protons and lead ions converge as the energy increases during the ramp. At flat top, the frequency difference between the different particle beams is of the order of only 70 Hz. With the ions being a bit faster now, the bunches cross each other in ATLAS every 10 min. The moment when they are very close together around IP 1 is used to lock the frequencies together. The locking of the frequencies is performed in several steps, and the whole process can take few seconds or up to 10 min, depending how close the leading bunches are to IP 1. When the radio frequencies of both beams are locked together, the particles are forced to run at a different frequency than the natural value; the ions, being always slower than the protons, will compensate the running at higher frequency by moving the orbit to the inner side of the accelerator. The protons, on the other hand, will move to the outer side of the circumference. The closed orbits for the beams can be seen in Fig. 2.19 for both protons and ions.

With this machine configuration the first stable beams, with 13 individually injected bunches per beam, were declared on 20 January 2013. At the following fill the first bunch trains were injected and stable beams were declared with 96 lead bunches on 120 proton bunches. The full filling scheme with 338 on 338 bunches

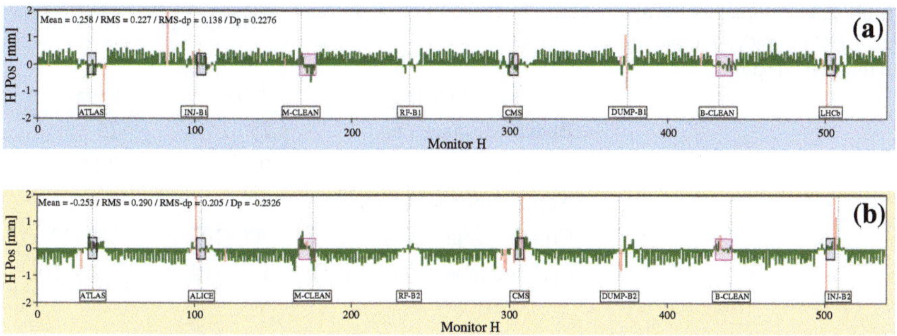

Fig. 2.19 Horizontal beam positions for closed orbits for protons (*top*) and ions (*bottom*) at top energy when the radio frequencies are locked to the same value. The horizontal axis represents the position in the 27 km circumference where the orbit is measured. Each bar represents a beam position monitor reading. *Source* CERN

was injected and successfully ramped on 21 January with record lead intensity in LHC—a great success of the injectors that provide the high-quality bunches.

Since 24 January 2013, stable beams were routinely declared twice per day providing a peak luminosity at the beginning of the collisions of the order of $10^{29}\text{cm}^{-2}\,\text{s}^{-1}$. The LHC produced an integrated luminosity of about 2 nb^{-1} per experiment and per day. The proton and ion bunches had an average intensity of 1.5×10^{10} charges. The total integrated luminosity was above the request from the experiments of 30 nb^{-1}. During this mode of LHC operation, all seven LHC experiments took data together: ATLAS, CMS, ALICE, LHCb, TOTEM, ALFA and LHCf.

2.5 LHC Upgrade Plans and the High-Luminosity LHC

The operation of the LHC in the near and mid-term future foresees, as first step, a jump in energy to values close to the design energy of 14 TeV. After the consolidation work on the superconducting connections in the main magnet chain during the long shutdown 2013/2014, the storage ring will first operate at an energy of 6.5 TeV per beam. Since the higher energy will lead to a reduction of emittance with respect to Run 1, a smaller β^* can be achieved, leading to a first increasing step in luminosity.

A major luminosity enhancement is planned after the long shutdown LS 3, currently scheduled to end in 2024. New triplet magnets will be installed to deliver at the same time the necessary aperture and the considerable step in focusing strength and luminosity. Based on Nb_3Sn technology, higher critical fields—and thus stronger gradients—can be achieved. The parameters of the improved machine, the "high-luminosity LHC" (HL-LHC) [32], are listed in Table 2.3.

The luminosity upgrade project of the LHC is based on a strong focusing scheme that is designed to reach lowest values of the β function at the collision points. It aims at producing a total integrated luminosity of approximately 3000 fb^{-1}. To achieve

Table 2.3 Parameters for the HL-LHC upgrade, compared to the LHC design

	LHC design	HL-LHC
Particles/bunch (10^{11})	1.15	2.2
Number of bunches	2808	2808
Bunche distance (ns)	25	25
β^* (m)	0.55	0.15
Norm. emittance (μmrad)	3.5	2.5
Crossing angle (μrad)	300 (10 σ)	590 (12 σ)
Peak luminosity ($cm^{-2}\,s^{-1}$)	1.2×10^{34}	2.2×10^{35}

this ambitious goal, a considerable reduction of the β function at the high-luminosity IPs is needed, and values as low as $\beta^* = 15$ cm are aimed for. The fundamental concept is based on an "achromatic telescopic squeezing" scheme [33], which has been proposed to enable both the matching of very low β^* optics and the correction of the induced chromatic aberrations.

The key issue for the optics design and the lattice modifications that are required is to establish a smooth matching between the small Twiss parameters at the IPs and the periodic arc structure. New large-aperture magnets will be needed in the inner triplet of the straight section due to the strong beam divergence that is created by the extremely low β^* values. The optics for the foreseen value of β^* of 15 cm is shown in Fig. 2.20.

One of the most crucial features of the HL-LHC upgrade project is the use of crab cavities that compensate for the geometric luminosity loss related to the crossing angle of the two beams. In the luminosity formula (2.1), for a limited beam intensity the emittances of the beams and the transverse beam sizes at the IP are the only ingredients available for optimisation.

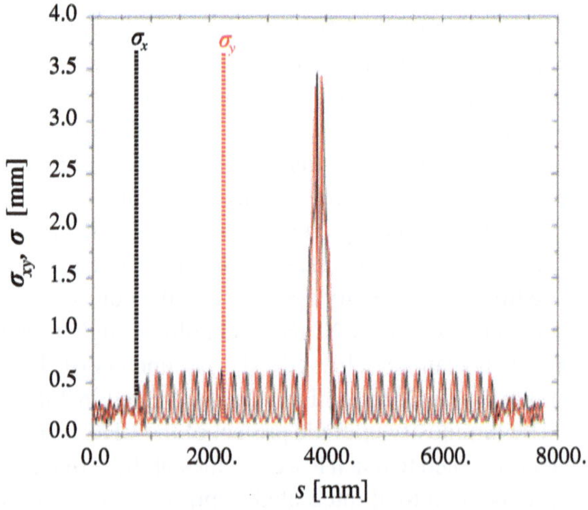

Fig. 2.20 Beam optics for the HL-LHC upgrade. *Source* CERN

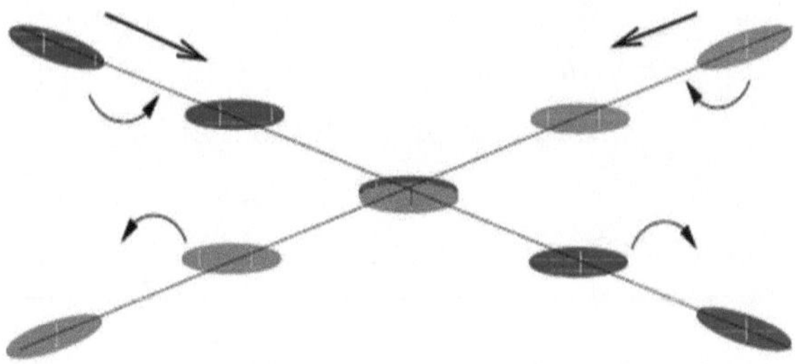

Fig. 2.21 Crab effect to compensate the geometric loss factor resulting from interactions under a large crossing angle. *Source* CERN

However, given the design parameters of the HL-LHC optics, a considerable crossing angle will be needed to avoid parasitic encounters of the bunches spaced at 25 ns in the machine. Unlike the present LHC design, where a crossing angle of 285 μrad is sufficient to establish a 10 σ separation at each parasitic encounter, in the upgrade design a much larger angle of 590 μrad has to be established at each high-luminosity IP, and the loss factor F (which is mainly related to the geometric effect of the crossing angle—see (2.2)) amounts to 0.31.

The installation of crab cavities that compensate this reduction, therefore, is indispensable, and the beam optics in the new interaction regions has to be optimised accordingly. Schematically, the problem is shown in Fig. 2.21. Before and after each IP, transversely deflecting crab cavities [34] will be used to create a local shearing of the bunches.

2.6 Plans for Future Colliders at CERN

Looking even farther into the future, several new projects are under study in a world-wide collaboration: Precise measurements of the parameters of the new Higgs boson as well as the search for new physics beyond the Standard Model define the baseline for the next generation colliders. ILC [35], CLIC [36] and a new 100 km long circular electron-positron collider [37] on one side and a new future circular hadron collider (FCC-hh) are the alternative projects for the long-term future. Especially the future hadron machine, as successor for the LHC collider, is being studied in detail at CERN (see Fig. 2.22).

The current design of this machine, FCC-hh, is based on a 90° FODO structure with 12 arcs and a racetrack-like geometry, where the usual long straight sections needed for beam injection, extraction, collimation, radio frequency etc. are combined into two very long straights that even house the foreseen four particle detectors. A first layout of the beam optics considers using a 200 m long basic cell housing twelve 14 m long dipole and 2 quadrupole magnets. The storage ring design (see Table 2.4) is based on Nb_3Sn as new superconducting high-field magnet material: A magnetic induction of 16 T seems to be in reach. It would allow beam energies of 50 TeV to

Fig. 2.22 Schematic view of a potential future 100 km collider in the Geneva region. *Source* CERN

Table 2.4 Parameters for the FCC-hh collider

Particle	LHC	FCC-hh
Circumference (km)	27	100
Centre-of-mass energy (TeV)	14	100
Dipole field (T)	8.33	16
Number of straight sections	8	12
Normalised emittance (25 ns) (μm)	3.75	2.2
Normalised emittance (5 ns) (μm)		0.44
Amplitude function at IP β^* (m)	0.55	1.1
Number of bunches (25 ns)	2808	10,600
Number of bunches (5 ns)		53,000
Particles/bunch (25 ns) (10^{11})	1.15	1.0
Particles/bunch (5 ns) (10^{11})		0.2
Peak luminosity ($10^{34}\,\mathrm{cm}^{-2}\,\mathrm{s}^{-1}$)	1.0	5.0

be reached in a 100 km tunnel. As for the dipoles, the quadrupole magnets will need a considerable step in gradient and thus will be designed in Nb_3Sn technology [38].

A first version of the luminosity optics, matched for an amplitude function of $\beta^* = 1.1$ m is presented in Fig. 2.23.

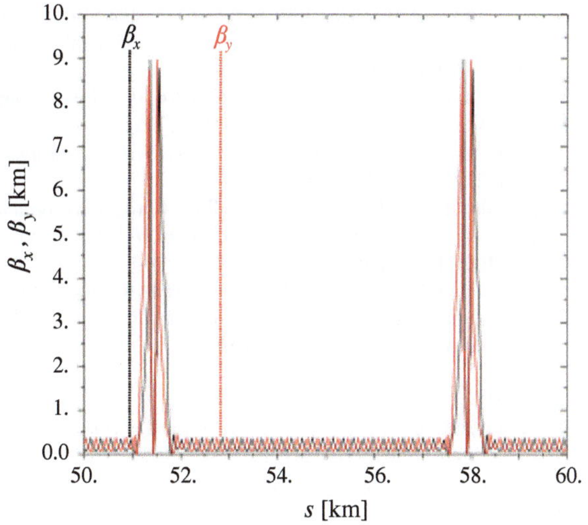

Fig. 2.23 Luminosity optics in the two adjacent interaction regions of the FCC-hh. *Source* CERN

At present it is foreseen to deliver a technical design report in about 4 years from now (autumn 2014). Still, however, the restart of the LHC after the present shutdown is eagerly awaited, and the existing machine will be optimised for operation at higher energy. The results of the foreseen operation at a centre-of-mass energy of 14 TeV will show the way to go.

References

1. TeVI Group, Design Report Tevatron 1 Project. FERMILAB-DESIGN-1984-01 (1984)
2. B.H. Wiik (ed.), HERA—a proposal for a large electron proton colliding beam facility at DESY. DESY-HERA-81-10 (1981)
3. J. Adams, The 400GeV proton synchrotron. CERN annual report 1976 (1976)
4. O. Brüning et al., LHC design report. CERN-2004-003 (2004)
5. A. Siemko et al., in *Proceedings of the Chamonix 2012 Workshop on LHC Performance*, ed. by C. Carli, Chamonix, France (2012), p. 170
6. M. Koratzinos et al., in *Proceedings of the 1st International Particle Accelerator Conference (IPAC 2010)*, Kyoto, Japan (2010), p. 373
7. R. Alemany et al., in *Proceedings of the 4th International Particle Accelerator Conference (IPAC 2013)*, Shanghai, China (2013), p. 1346
8. K. Schindl, Space charge. CERN-2006-002 (2006)
9. L. Arnaudon et al., in *Proceedings of the 2nd International Particle Accelerator Conference (IPAC 2011)*, San Sebastian, Spain (2011), p. 900
10. W. Herr et al., in *Proceedings of the 2nd International Particle Accelerator Conference (IPAC 2011)*, San Sebastian, Spain (2011), p. 1936
11. W. Herr, Part. Accel. **50**(1–3), 69 (1995)
12. B. Salvachua et al., in *Proceedings of the 4th International Particle Accelerator Conference (IPAC 2013)*, Shanghai, China (2013), p. 1002

13. E.B. Holzer et al., Beam loss monitoring system for the LHC. CERN-AB-2006-009 (2006)
14. E.B. Holzer et al., Commissioning and optimization of the LHC BLM system. CERN-BE-2010-031 (2010)
15. B. Goddard, in *Proceedings of LHC Days 2003*, Les Diablerets, Switzerland (2003), p. 99
16. R. Schmidt et al., New J. Phys. **8**, 290 (2006)
17. R. Assmann et al., in *Proceedings of the 3rd International Particle Accelerator Conference (IPAC 2012)* (2012), p. 4062
18. F. Millet et al., in *Proceedings of the 2007 Cryogenic Engineering Conference and International Cryogenic Materials Conference (CEC-ICMC)*, Chattanooga, USA (2007), p. 1403
19. J. Uythoven et al., in *Proceedings of the 11th European Particle Accelerator Conference* (EPAC 08), eds. by I. Andrian, C. Petit-Jean-Genaz (Genoa, Italy, 2008), p. 2668
20. M. Lamont et al., in *Proceedings of the 23rd Particle Accelerator Conference (PAC2009)*, Vancouver, Canada (2008), p. 4254
21. L. Coull et al., in *Proceedings of 13th International Conference on Magnet Technology (MT-13)* (1994), pp. 1742–1745
22. F. Formenti et al., in *Proceedings of the 1st International Particle Accelerator Conference (IPAC 2010)*, Kyoto, Japan (2010), p. 696
23. K. Fuchsberger et al., in *Proceedings of the 1st International Particle Accelerator Conference (IPAC 2010)*, Kyoto, Japan (2010), p. 456
24. M. Lamont, in *Proceedings of the 2nd International Particle Accelerator Conference (IPAC 2011)*, San Sebastian, Spain (2011), p. 11
25. J. Wenninger et al., in *Proceedings of the 3rd International Particle Accelerator Conference (IPAC 2012)*, New Orleans, Louisiana (2012), p. 3767
26. T. Pieloni, in *Proceedings of the 4th International Particle Accelerator Conference (IPAC 2013)*, Shanghai, China (2013), p. 1418
27. B. Salvant et al., in *Proceedings of the 4th International Particle Accelerator Conference (IPAC 2013)*, Shanghai, China (2013), p. 1646
28. J. Jowett et al., in *Proceedings of the 2nd International Particle Accelerator Conference (IPAC 2011)*, San Sebastian, Spain (2011), p. 1837
29. J.M. Jowett et al., in *Proceedings of the Chamonix 2011 Workshop on LHC Performance*, ed. by C. Carli (Chamonix, France, 2011), p. 276
30. R. Alemany, others, First proton-nucleus collisions in the LHC: the p_-Pb pilot physics run. CERN-ATS-Note-2012-094 MD (2012)
31. J.M. Jowett et al., in *Proceedings of the 4th International Particle Accelerator Conference (IPAC 2013)*, Shanghai, China (2013), p. 49
32. B. Holzer et al., in *Proceedings of the 4th International Particle Accelerator Conference (IPAC 2013)*, Shanghai, China (2013), p. 1385
33. S. Fartoukh, in *Proceedings of the 2nd International Particle Accelerator Conference (IPAC 2011)*, San Sebastian, Spain (2011), p. 2088
34. B. Xiao et al., in *Proceedings of the 4th International Particle Accelerator Conference (IPAC 2013)*, Shanghai, China (2013), p. 2417
35. C. Adolphsen et al., The international linear collider technical design report. ILC-Report-2013-040
36. M. Aicheler et al., Multi-TeV linear collider based on CLIC technology: CLIC conceptual design report. CERN-2012-007 (2012)
37. F. Zimmermann, in *Proceedings of the 5th International Particle Accelerator Conference (IPAC 2014)*, Dresden, Germany (2014), p. 1
38. A. Ball et al., Future circular collider study: hadron collider parameters. FCC-ACC-SPC-0001 (2014)

Chapter 3
The LHC Detectors

Ingrid-Maria Gregor and Arno Straessner

Abstract The physics results described in this book are made possible by the excellent performances of the LHC machine (see Chap. 2) and the four main detectors ALICE, ATLAS, CMS, and LHCb. These experiments are the most complex and advanced detectors for particle physics ever built. The planning and detector R&D started in the late 1980s. The actual construction and installation took place between 2000 and 2008 and involved thousands of scientists, engineers and technicians. This chapter will give an overview of the main design features of the experiments as well as a summary of key performance indicators.

3.1 The LHC Detectors—The Big Picture

3.1.1 General Requirements

The four large detectors at the LHC—ALICE, ATLAS, CMS, LHCb—were optimised for their respective specific physics requirements and experimental conditions, which differ from experiment to experiment [1–4]. ATLAS and CMS are designed to cover the full solid angle—a goal that is achieved using a cylindrical configuration with a central barrel and end-caps on both sides that also detect particles traversing the detector under a shallow angle with respect to the beamline (in the so-called "forward" direction). LHCb and ALICE are conceptualised for studies at well-defined solid-angle ranges: Both consist of a main detector section and an extended forward

I.-M. Gregor (✉)
DESY, Notkestr. 85, 22607 Hamburg, Germany
e-mail: ingrid.gregor@desy.de

A. Straessner
Institut für Kern- und Teilchenphysik, Technische Universität Dresden,
01062 Dresden, Germany
e-mail: arno.straessner@tu-dresden.de

T. Schörner-Sadenius (ed.), *The Large Hadron Collider*,
DOI 10.1007/978-3-319-15001-7_3

arm in one direction. Within the chosen solid angle of an experiment—be it the full solid angle or only a fraction of it—no particle should escape detection, except for weakly-interacting particles (such as e.g. neutrinos) that do not interact on the length scale of a typical detector.

The following detector descriptions are based on the conventions which are common for all four experiments: The interaction point is defined as the origin of the coordinate system (Fig. 3.1). The z axis runs along the beamline. The x–y plane is perpendicular to the beamline and is referred to as the "transverse plane". The positive x axis points from the interaction point to the centre of the LHC ring; the positive y axis points upward to the surface of the Earth. In case of the cylindrical detectors ATLAS, CMS and ALICE, also cylindrical coordinates are used. The transverse plane is often described in terms of r–ϕ coordinates. The azimuthal angle ϕ is measured from the x axis, around the beam. The radial dimension—the radius r—measures the distance from the beamline. The polar angle θ is defined as the angle from the positive z axis. The polar angle is often also given in terms of the pseudo-rapidity, defined as $\eta = -\ln\tan(\theta/2)$ and illustrated in Fig. 3.1.

The pseudo-rapidity is the massless approximation of the rapidity variable y defined for a particle with energy E, momentum $\boldsymbol{p}$ and longitudinal momentum $p_\mathrm{L} = |\boldsymbol{p}|\sin\theta$ along the beam as

$$y = \frac{1}{2}\ln\frac{E + p_\mathrm{L}}{E - p_\mathrm{L}} \approx \eta = \frac{1}{2}\ln\frac{|\boldsymbol{p}| + p_\mathrm{L}}{|\boldsymbol{p}| - p_\mathrm{L}} = -\ln\tan\left(\frac{\theta}{2}\right).$$

The usage of rapidity, respectively pseudo-rapidity, at hadron colliders is motivated by the fact that the longitudinal momenta of the colliding partons inside the protons and nuclei are a priori unknown in each individual collision. Rapidity intervals Δy are Lorentz invariant under boosts along the beam direction, so that Δy and differential cross sections $\mathrm{d}\sigma/\mathrm{d}y$ remain independent of the unknown longitudinal momentum sum of the partons. Furthermore, the particle density in inelastic proton-proton collisions is approximately constant in the central region of the detectors when measured in equal rapidity intervals. Therefore, the LHC detector components are typically segmented in such intervals to achieve a constant particle occupancy for the dominant fraction of collision events.

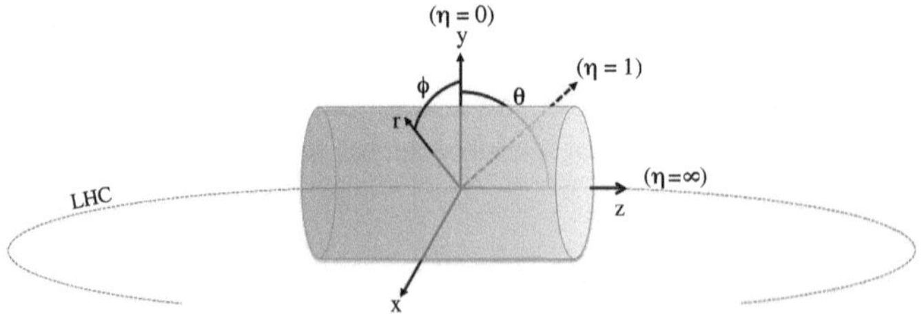

Fig. 3.1 Right-handed coordinate system used by all LHC experiments. Cartesian and cylindrical coordinates are used. The relation of the pseudo-rapidity variable η and the polar angle θ in the r–z plane is also illustrated

All LHC detectors pursue the “onion” concept, according to which a sequence of different sub-detector components provide detailed measurements of different properties of the particles emerging from the interaction point. For the cylindrical experiments, the detectors are arranged concentrically in layers of increasing radius around the beamline—thus the comparison to an onion. The combination of various detectors such as tracking detectors and calorimeters provides the maximum information on the traversing particles and hence allows their full reconstruction to be accomplished.

Independent of cylindrical or forward configuration, the main components are typically as follows: At smallest radii, tracking detectors are placed within a strong magnetic field to measure the directions and momenta of charged particles originating from the interaction point (Sect. 3.2). At increasing radii, electromagnetic and hadronic calorimeters measure the energies of electrons, photons and hadrons (Sect. 3.3). The calorimeters are followed by chambers providing measurements of high-energy muons escaping the calorimeters (muon chambers, Sect. 3.4). A trigger system that selects the most relevant events in real-time is interleaved throughout all detector subsystems (Sect. 3.6).

3.1.2 Identifying Physics Objects

The main physics objects that shall be reconstructed using the detector information are photons, electrons, muons, τ leptons, individual hadrons, jets, total energy and missing transverse energy. For unstable particles like τ leptons or short-lived hadrons, the specific decay patterns are exploited. The topology of interesting physics events at the LHC can be illustrated using Fig. 3.2, which shows the decay of a Z boson into a pair of muons, $Z \rightarrow \mu^+\mu^-$, recorded by the ATLAS detector. The muons have large transverse momentum, p_{T}, with respect to the beam axis and are shown in the upper hemisphere of the detector. They traverse all detector layers and are mainly measured in the outer muon spectrometer chambers and in the innermost tracking detectors. If the muon track is not fully reconstructed in the muon chambers, e.g. due to limited detector acceptance or efficiency, the calorimeter information is used to complete the muon identification.

In the inner tracking detectors a large number of tracks are measured, as can be seen in Fig. 3.2. These are from low-momentum hadrons that are produced both from the remains of the proton-proton collision that produced the Z boson, but also from additional pp interactions in the same collision of the proton bunches, the so-called pile-up. The vertices of the colliding pp pairs can be reconstructed by interpolating the tracks of charged particles to a common intersection point, as shown at the bottom of Fig. 3.2. Also the muon tracks from the Z-boson decay can be traced back to such a primary vertex, which can be used to separate signal muon tracks from possible background.

One challenge for the LHC detectors is the large number of particle tracks and high occupancy and energy density in each detector segment. In particular for heavy-ion

Fig. 3.2 Event display of a Z-boson decay, $Z \rightarrow \mu^+\mu^-$, with two muons detected in the muon spectrometer. At the *bottom* the reconstructed tracks of charged particles are displayed which clearly show that there is more than one interaction vertex in each proton-proton collision. *Source* CERN

collisions the number of reconstructed particle tracks per event reaches values of >1500 per unit interval in pseudo-rapidity [5]. The ALICE detector is designed in particular for such conditions, and Fig. 3.3 shows an example event from the first lead-ion collisions in 2011.

In crossings of proton-proton bunches, the pile-up also leads to a high particle density. The mean number of interactions per bunch crossing, μ, is determined by the inelastic proton-proton cross section, σ_{inel}, the instantaneous luminosity, $\mathcal{L}$, and the mean time interval between two bunch crossings, $\langle t \rangle$:

$$\mu = \sigma_{\text{inel}} \times \mathcal{L} \times \langle t \rangle .$$

The distribution of μ during the 2012 data-taking as determined by CMS is shown in Fig. 3.4. The time interval between two filled bunches had been 50 ns during the first LHC run, which is two times the LHC design value of 25 ns. Thus, to reach the same instantaneous luminosity twice the number of protons need to be filled in each LHC bunch. This leads to an increase in pile-up with mean values of $\mu \approx 40$. The multi-purpose detectors ATLAS and CMS were thus already operated in conditions beyond their design, which was optimised for a mean number of 25 pile-up events.

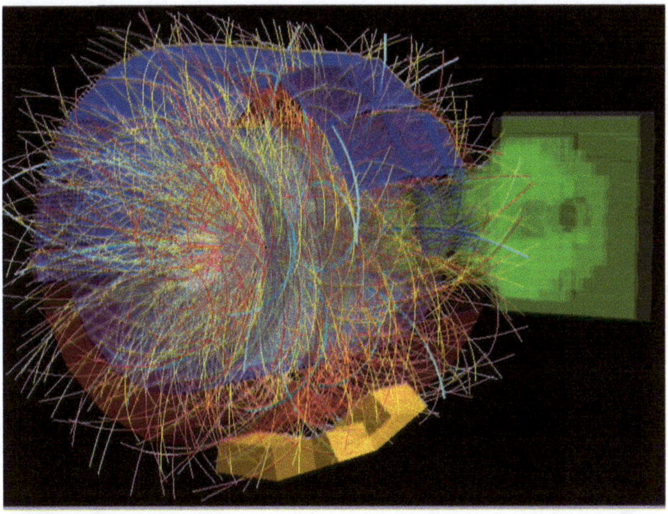

Fig. 3.3 Event recorded by the ALICE experiment during the first lead-ion run in 2011 at a centre-of-mass energy of 2.76 TeV per nucleon pair. *Source* CERN

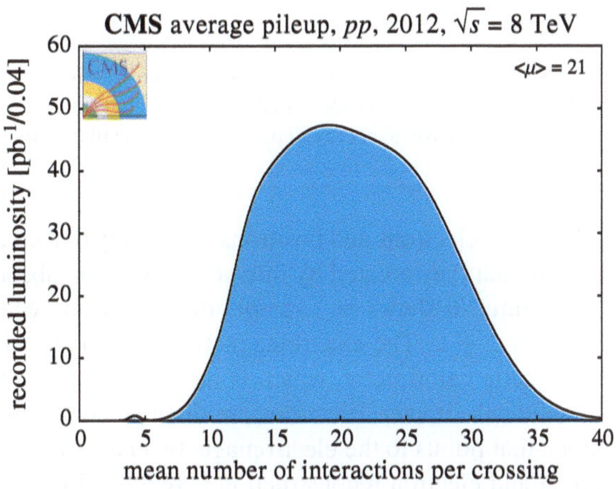

Fig. 3.4 Mean number of pile-up events, as determined during the 2012 data taking at 8 TeV centre-of-mass energy by CMS. *Source* CERN

In a large fraction of proton-proton collision events, the hadrons form collimated particle jets, which are produced from the fragmentation of primary quarks or gluons. Figure 3.5 shows a dijet event measured in the CMS detector. Such jets are reconstructed both from the energy depositions in the calorimeters, which are shown as rectangular columns with sizes proportional to the deposited energy, and from the momenta of charged particles tracked in the inner detectors. Iterative procedures, like the anti-k_t algorithm [6], allow the well-defined reconstruction of hadronic jets with given resolution parameters (see Sect. 5.3).

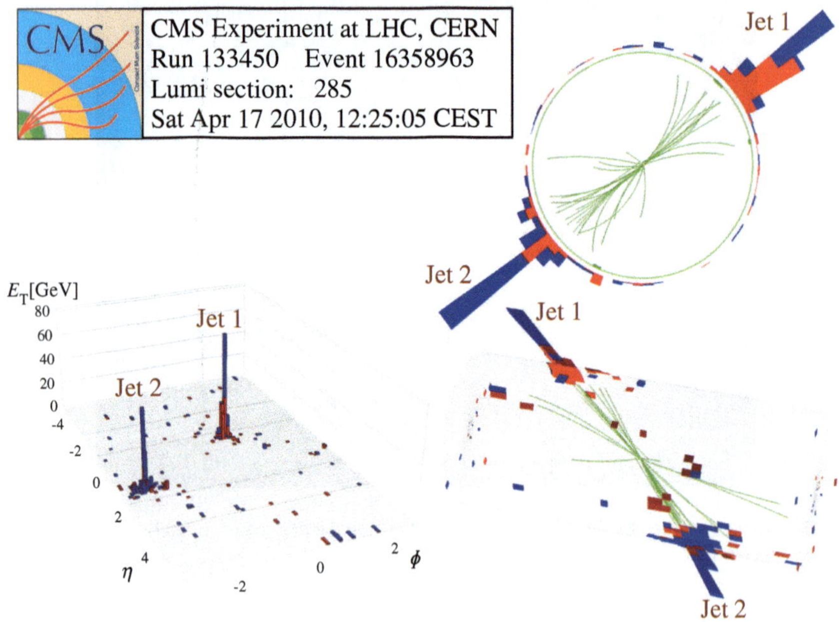

Fig. 3.5 Hadrons produced from quark or gluon fragmentation usually form collimated particle jets like in this dijet event measured by CMS, shown in the $r\phi$ and rz views together with the reconstructed transverse energy deposits in the η−ϕ plane. *Source* CERN

Photons, electrons and positrons are mainly detected through the electromagnetic showers that they created by interacting with the absorber material of the calorimeters. Figure 3.6 shows an example of a Z-boson decay accompanied by a photon, $Z \rightarrow e^+e^- + \gamma$. The electromagnetic energy is nearly fully deposited in the electromagnetic calorimeter, which is installed closer to the interaction point than the hadronic calorimeter. Electrons are distinguished from photons by the presence of a track that points to the electromagnetic energy deposit. Hadron background in the photon and electron reconstruction is reduced by a veto against energy deposits in the hadronic calorimeter.

While electrons and positrons are stable and muons have a lifetime much longer than their flight time across the detector, τ leptons all decay inside the detector volume. The leptonic decays $\tau^\pm \rightarrow e^\pm + 2\nu$ and $\tau^\pm \rightarrow \mu^\pm + 2\nu$, which amount to 35 % of all τ decays, can therefore not be easily distinguished from primary electrons, positrons and muons. A dedicated τ identification is only possible for hadronic τ decays, for which an example is shown in Fig. 3.7, which is a display of a W-boson decay, $W \rightarrow \tau\nu$. The τ decay products visible in the detector are one or three charged hadrons (mostly $\pi^\pm$ and $K^\pm$) accompanied by a given number of neutral hadrons (mostly π^0). The reconstruction algorithms try to identify the individual hadrons as hadronic energy deposits, distinguishing charged from neutral particles by the

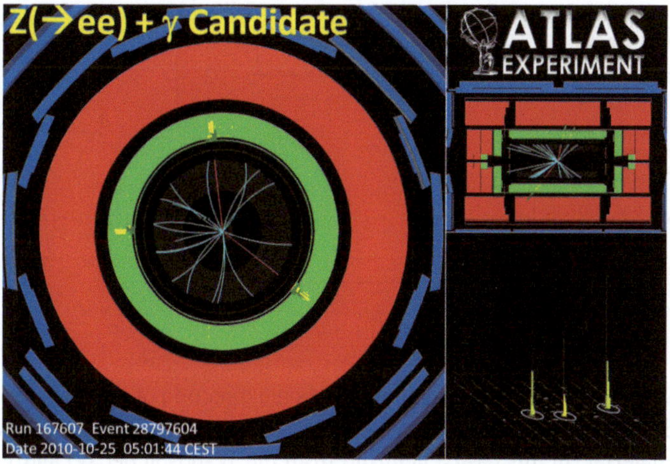

Fig. 3.6 A Z-boson decay, $Z \rightarrow e^+e^-$, accompanied by a photon, as measured in the ATLAS detector. *Source* CERN

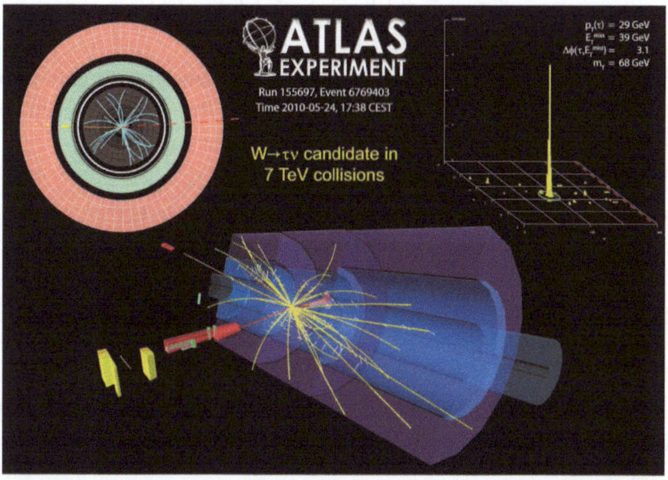

Fig. 3.7 Decay of a W boson, $W \rightarrow \tau\nu$, recorded with the ATLAS detector. Hadronic τ decays are identified by one or more track in the central detector pointing to collimated energy deposits in the calorimeters. The characteristic shape and distribution of the energy depositions are used to distinguish τ decays from electrons or from quark and gluon jets. *Source* CERN

presence of a track pointing to them. In this way, hadronic τ decays are identified and energy and direction are assigned to the τ candidates.

Since charged leptons also appear in hadronic jets, e.g. in the decay chain of heavy-flavour hadrons, primary leptons are typically required to have a certain angular separation from jets, a property which is usually measured in terms of the so-called "isolation".

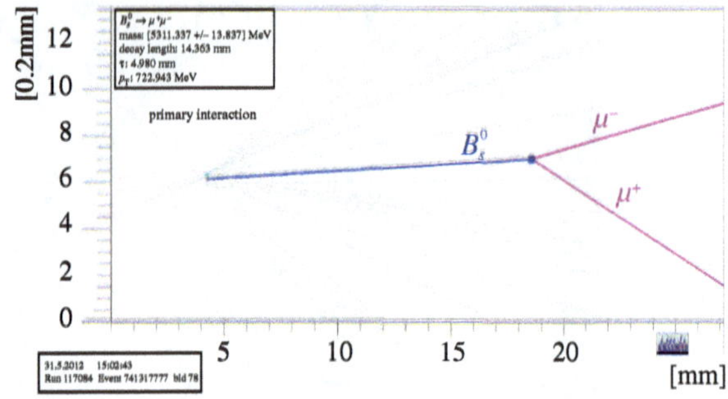

Fig. 3.8 Candidate event of a $B_s \to \mu^+\mu^-$ decay measured in the LHCb detector. The event is identified by the dimuon vertex that is clearly separated from the primary vertex due to the non-zero lifetime of the B_s meson. *Source* CERN

The event display of Fig. 3.7 also indicates the direction of the transverse missing energy, $E_{\mathrm{T}}^{\mathrm{miss}}$, which is a signature for particles that only weakly interact with the detector material (in this case a neutrino from the W-boson decay) and which thus remain invisible otherwise. The quantity $E_{\mathrm{T}}^{\mathrm{miss}}$ is reconstructed as

$$E_{\mathrm{T}}^{\mathrm{miss}} = \sqrt{\left(E_{\mathrm{T},x}^{\mathrm{miss}}\right)^2 + \left(E_{\mathrm{T},y}^{\mathrm{miss}}\right)^2},$$

where the energy sums $E_{\mathrm{T},x}^{\mathrm{miss}}$ and $E_{\mathrm{T},y}^{\mathrm{miss}}$ in the two transverse directions x and y is given by the negative sum of the energy flow in these directions determined from all detected objects i in the calorimeters and muon spectrometers:

$$\begin{pmatrix} E_{\mathrm{T},x}^{\mathrm{miss}} \\ E_{\mathrm{T},y}^{\mathrm{miss}} \end{pmatrix} = -\sum_i \begin{pmatrix} E_i \sin\theta_i \cos\phi_i \\ E_i \sin\theta_i \sin\phi_i \end{pmatrix}.$$

The details to which the detectors need to measure the LHC collision events are in general very complex, as indicated by Fig. 3.8 of the reconstructed, very rare decay of a B_s meson into a pair of muons measured by LHCb. A very good experimental precision optimized for the LHC operating conditions is thus required which had driven the design of the LHC detectors.

3.1.3 The Four Main Detectors

In Fig. 3.9, the drawings of the four LHC detectors show the large difference in size and layout. These differences mainly arise from the different physics goals. The layouts are, however, also based on preferences of the experimental collaborations that designed and built the machines, and on cost considerations. The effect of these considerations is most strikingly visible in the layout of ATLAS and CMS—two detectors with exactly the same physics goals, nevertheless completely different in size, weight and magnetic field configuration. The main parameters of the four experiments are summarised in Table 3.1.

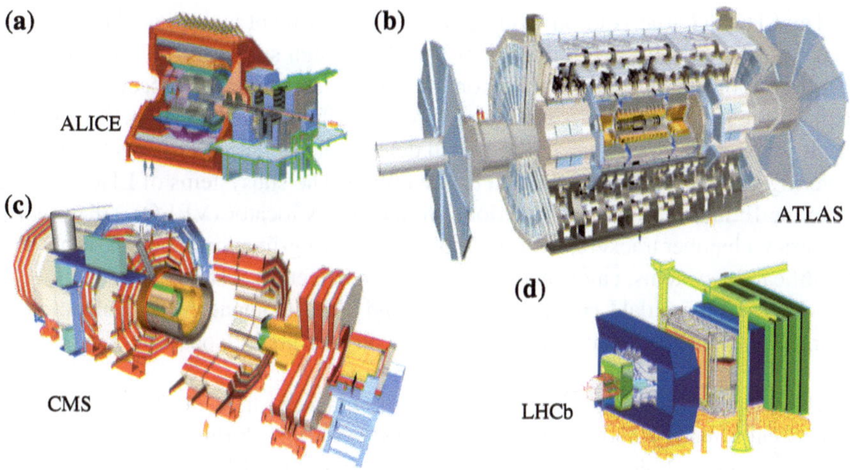

Fig. 3.9 The four LHC experiments ALICE, ATLAS, CMS and LHCb. The sizes of the drawings roughly reflect the relative proportions of the detectors. *Source* CERN

Table 3.1 The main design parameters for ATLAS, CMS, ALICE and LHCb [1–4]

Parameter	ATLAS	CMS	ALICE	LHCb
Total weight (tons)	7000	12,500	10,000	5600
Overall length (m)	46	20	26	21
Overall diameter (m)	22	15	16	–
Width × height (m × m)	–	–	–	13 × 10
Magnet	Toroid/solenoid	Solenoid	Solenoid/dipole	Dipole
Magnetic field for tracking (T)	1/2	3.8	0.5/0.67	1.1

ALICE (**A** **L**arge **I**on **C**ollider **E**xperiment) is a general-purpose, heavy-ion detector optimised for studies of the physics of strongly interacting matter and of the quark-gluon plasma at extreme values of energy density and temperature in nucleus-nucleus collision (see Chap. 9 for a discussion of heavy-ions physics at the LHC) [3]. Most challenging in this environment is the extreme particle multiplicity in central lead-lead (Pb-Pb) collisions that is up to three orders of magnitude larger than in pp interactions at the same energy. ALICE concentrates on physics at or close to mid-rapidity in the plane orthogonal to the beamline (i.e. at $\eta \approx 0$). It is therefore designed as a central-barrel detector with an additional single-arm muon spectrometer covering the small-angle area between $-4.0 < \eta < -2.5$. This forward spectrometer serves for studies of the complete spectrum of heavy quarkonia via their decay in the $\mu^+\mu^-$ channel in the forward direction.

From inside (i.e. close to the beamline) to outside, ALICE features the inner tracking system, a time projection chamber (TPC), a transition radiation detector (TRD), a time-of-flight detector (TOF), electromagnetic calorimeters, a high momentum particle identification system (HMPID) and a photon detector. The forward region features a muon arm and zero-degree calorimeters.

The LHCb (**L**arge **H**adron **C**ollider **b**eauty) experiment is dedicated to measure heavy-flavour physics [4]. It does so primarily through studies of *CP* violation and rare decays of beauty and charm hadrons. It is a single-arm forward spectrometer that exploits the fact that $b\bar{b}$ pairs are mainly produced in forward direction. About 35 % of all $b\bar{b}$ pairs produced are in the LHCb detector acceptance. The spectrometer is covering production angles close to the beamline. The subsystems of LHCb (going from the IP into the forward direction) are the vertex locator (VELO), silicon-strip and straw-chamber trackers, a warm dipole magnet, ring-imaging Cherenkov particle identification systems, calorimeters, and a muon detection system.

ATLAS (**A T**oroidal **LHC A**pparatu**S**) [1] and CMS (**C**ompact **M**uon **S**olenoid) [2] are multi-purpose experiments with extensive physics goals. Consequently, both are configured as cylindrical detectors with a barrel and two end-cap sections. Both detectors have an inner tracking system immersed in a magnetic field, electromagnetic and hadron calorimeters, and muon chambers outside of the calorimeter system.

The essential choice affecting the overall design of the LHC experiments is that of the magnetic field configuration used to bend charged particles and thus to enable measurements of their momenta. An additional driving parameter for ATLAS and CMS is the momentum resolution for muons. The muon momentum resolution, $\Delta p_{\mathrm{T}}/p_{\mathrm{T}}$, of the LHC detectors should be better than 10 % for muons of up to 1 TeV. It is proportional to $B^{-1}L^{-2}$, where B is the strength of the magnetic field and L is the distance from the interaction point to the muon momentum measurement [7]. In order to achieve the optimal momentum resolution, one can choose a strong magnetic field B or a long lever arm L or both. ATLAS and CMS adopted different approaches: While ATLAS chose a large lever arm L, CMS pursued the concept of a very compact detector (small L) with a strong magnetic field B.

CMS decided to use a superconducting solenoid with a 3.8 T field that encloses the tracking system and the calorimeters. This configuration leads to a large homogeneous field inside the coil and a weak opposite direction field in the return yoke. Disadvantages are the limited size (due to cost reasons) and the relatively high material budget introduced by the solenoid insulation. ATLAS, on the other hand, chose a combination of an extended toroidal magnetic field of about 1 T for the muon system and an additional 2 T solenoid surrounding the tracking detectors in the inner region of the detector. The advantage of the toroid is that the magnetic field is always perpendicular to the particle trajectories and thus the calculation of momentum is simplified. However, the field of a toroid is rather non-uniform. In the case of ATLAS, it has peak fields of up to 4 T—a fact that complicates the track reconstruction. Furthermore, the complex skeleton structure of the toroidal magnet is limiting the accessibility to the ATLAS detector for repairs and upgrades. In the case of LHCb, the forward orientation of the experiment led to the choice of a warm-coil dipole magnet for the tracking system. ALICE features a large solenoid magnet (the one that was formerly used in the L3 experiment at LEP) in the central region and a dipole magnet on the single-arm forward muon detector.

Besides the design considerations concerning coverage, magnetic field and lever arm, a lot of thought went into the technology choices for the different subsystems. All detector systems need to be highly efficient, even after 10 years of operation in the harsh LHC environment. A main concern for all four collaborations were the extreme radiation levels, even for the muon systems that are located on the far ends of the experiments. Only ALICE has somewhat more relaxed requirements in terms of radiation tolerance due to the reduced luminosity. All chosen technologies had to be qualified for the harsh environment, a requirement that explains the long R&D time needed for the LHC experiments. Not only the active sensor material had to be qualified, but every single item in the cavity, including screws, bolts, cables, glue and other non-sensing materials, had to be certified for the operation in the LHC environment.

In the following, an overview of the various detector subsystems is given, from the innermost layers in the tracking section to the outermost muon systems. Furthermore, the trigger concepts and the different luminosity measurements are described.

3.2 Tracking Detectors

The first detectors encountered by the particles emerging from the interaction point are the tracking detectors. They need to provide good pattern recognition and the exact measurement of particle trajectories to provide precise track reconstruction at very small radii. For the efficient identification of heavy flavours, a precise measurement of the secondary vertex and the impact parameter (see below) is required. In addition, tracking systems should enable the identification of electrons by matching tracks to clusters in the electromagnetic calorimeters. During LHC Run 1, with its peak luminosities of $10^{34}\,\mathrm{cm}^{-2}\,\mathrm{s}^{-1}$, events with a few thousand tracks had to be disentangled by the tracking system. Therefore all systems had to be designed with high granularity and fast response to identify the trajectories with high precision.

The tracking detectors are systems in many layers in which different technologies can be employed, from high-precision pixel detectors in the innermost region to large-area outer tracker systems build of silicon or gas detectors at larger radii. The first layers are often referred to as "vertex detector". (In principle, most muon spectrometers are also tracking systems, but will be discussed separately in Sect. 3.4.)

The number of layers in a tracking system is a trade-off between performance, the amount of material, and the cost. While a higher number of hits, i.e. a larger number of layers, improves the overall tracking performance, the amount of material within the tracking system should be kept as small as possible in order to minimise the effects of multiple scattering. In Table 3.2 the tracking subsystems of each LHC experiment are listed, including the detector type, the maximum number of hits, the number of channels, the coverage and the cell size (either the pixel dimension, the strip pitch or the diameter of straw tubes).

All pixel detectors at the LHC are built of hybrid pixels where the sensor and the front-end readout chip are sandwiched to one system via the use of small solder ball connections, the so-called "bump bonds". This high-integration process enables the development of the sensor and the front-end electronics in different technologies and thus allows each to be optimised for their respective task. The disadvantage of

Table 3.2 Main parameters for the tracking systems of ALICE, ATLAS, CMS and LHCb [1–4]. For each detector the basic detector type, the maximum number of hits in the central tracker, the number of channels, the cell size (pixel size or strip pitch) and the area of silicon is given

	Tracker	Type	# of hits	# of channels	Angular coverage η, ϕ	Cell size (μm)	Size Si: m^2 gas: m^3
ATLAS	Pixel	Silicon	3	80 M	±2.5, full	50 × 400	1.7
	Strips	Silicon	8	6.2 M	±2.5, full	80	60
	TRT	Gaseous	35	0.3 M	±2.0, full	4000	12
CMS	Pixel	Silicon	3	66 M	±2.5, full	100 × 150	1
	Strips	Silicon	13	9.6 M	±2.5, full	80–205	200
LHCb	VELO	Silicon	3–5	180 k	1.6–4.9	40–100	0.3
	Trigger tracker	Silicon	4	145 k	1.6–4.9	200	8.4
	Inner tracker	Silicon	4	129 k	4.0–4.9	200	4.0
	Outer tracker	Gaseous	12	54 k	1.6–4.9	4900	29.4
ALICE	Pixel	Silicon	2	9.8 M	±2, ±1.4	50 × 425	0.21
	Inner strip	Silicon	2	133 k	±0.9, ±0.9	350	1.31
	Outer strip	Silicon	2	2.6 M	±0.97, ±0.97	95	5.0
	TPC	Gaseous	159	558 k	<0.9, full	>4000 × 7500	90

this technology is the relatively high material budget—both sensor and readout chip come with a certain thickness, and also the solder bump (which is made of higher Z material) is adding to the material budget.

The ATLAS inner detector (ID) is formed of three subsystems that use different tracking technologies. The silicon-pixel detector is located closest to the beampipe, with $50 \times 400\,\mu\text{m}^2$ pixels. The pitch of $50\,\mu$m was chosen for the r–ϕ plane in order to have the best spatial resolution in this plane. The $400\,\mu$m for the second direction is then a consequence of the area required for the front-end electronics to read out one pixel cell. The ATLAS pixel detector provides a spatial resolution in r–ϕ of $10\,\mu$m and in z a resolution of $115\,\mu$m. The smallest building blocks of the pixel detector are hybrid pixel modules which are arranged into three concentric layers of silicon hybrid pixel modules in the barrel and disks in the end-caps. The pixel detector is surrounded by the semi-conductor tracker (SCT), a silicon-strip detector with $80\,\mu$m pitch. The strip sensors are arranged back-to-back with a 40 mrad stereo angle for better 2D measurements. The two sensors constitute one module. The SCT consists of four double-sided layers in the barrel and nine layers in the end-caps. The outermost tracking detector in the ATLAS ID is the transition radiation tracker (TRT). It is a gaseous tracking detector which is also used for particle identification (Sect. 3.5). It consists of about 300,000 gas-filled straw tubes of 4 mm diameter with a central sense wire of $30\,\mu$m diameter. The maximum straw length is 144 cm. In the barrel each straw is divided in two halves at the center and read out at both ends to reduce the occupancy. Each channel provides a drift-time measurement, giving an intrinsic spatial resolution of $<120\,\mu$m for the straw tubes. The straws are surrounded by radiators in which relativistic charged particles produce transition radiation (TR) when they cross the interface of two media of different refraction indices. The TR photons will then be absorbed and produce large signals in the straw tubes. The TRT readout electronics provides two thresholds, a lower threshold for detecting signals from ionisation in the gas and a higher threshold for detecting the transition radiation. The electronics will also give the time-over-threshold (ToT) for the lower threshold. This additional information has been shown to help in particle identification.

CMS has the currently largest all-silicon tracking detector in the world, with 13 layers in the barrel, 14 layers in the end-caps and an overall area of about $205\,\text{m}^2$ of silicon. The innermost three layers are pixel detectors with $100 \times 150\,\mu\text{m}^2$ cell size to provide good vertex reconstruction. The next four layers consist of $180\,\mu$m wide strips of 10 cm length, followed by six layers of $180\,\mu$m wide strips of 25 cm length. The third coordinate is given by the sensor plane position. In total, 1440 hybrid pixel modules cover an area of about $1\,\text{m}^2$ with 66 million pixel cells. The strip tracker has about 15,000 silicon modules, which in total cover an active area of about $198\,\text{m}^2$ with 9.3 million strips. Also the strip detector of CMS is equipped with double sided modules incorporating a stereo angle of 100 mrad. This layout results in a spatial resolution of $15\,\mu$m in the r–ϕ plane and $45\,\mu$m in the r–θ plane.

The LHCb tracking system includes a high precision silicon-strip vertex detector (VELO) surrounding the interaction region, a large-area silicon-strip detector upstream of the dipole magnet (Trigger Tracker, TT), and three tracking stations downstream (T_1–T_3). The latter are a combination of a large area with straw-tube drift chambers at more central rapidity (outer tracker, OT) and silicon-strip detectors close to the beamline (inner tracker, IT). The LHCb VELO is actually located inside the beam vacuum. The silicon-strip sensors with a pitch between 40 and 100 μm allow a measurement of the r and ϕ coordinates of the tracks to be performed. The detector consists of two halves which can be retracted from the beam during injection to reduce possible radiation damage. The TT is located between the VELO and the spectrometer magnet and is used for triggering and momentum measurement. It is a silicon-strip detector with four detection planes of silicon sensors with a pitch of 183 μm. The orientation of the strips for the first plane is vertical, the following plane has its strips rotated by $-5°$ around the beam axis; the third plane is rotated by $+5°$. The orientation of the final plane is again vertical. The IT close to the beampipe covers a 120 cm wide and 40 cm high cross shaped area built of strip detectors with a strip pitch of 200 μm. The OT is designed as an array of individual, gas-tight straw-tube modules of 5 mm diameter. As a counting gas a mixture of Argon (70 %) and CO_2 (30 %) was chosen for its fast drift time.

The innermost tracking detector in the ALICE barrel section is the inner tracking system (ITS), which is equipped with six layers of silicon: two innermost layers of pixel detectors with a pixel size of 50 μm $\times$ 425 μm, two layers of silicon drift detectors with a pitch of 350 μm, and, at the largest radii, two silicon-strip detector layers with a pitch of 95 μm. The ITS is surrounded by the largest time projection chamber (TPC) ever built functioning as main tracking detector of the central barrel. The TPC is composed of a cylindrical gas volume of 88 m^3 divided into two halves, separated by a 30 μm thick high-voltage central cathode that generates a uniform drift field along the z axis. The inner (outer) radius of the cylinder is about 80 cm (280 cm); the overall length is 500 cm. Multi-wire proportional chambers with pad readout are attached at both end-plates. The pads are $4 \times 7.5\ mm^2$ ($r \times \phi$) for the inner chambers and 6×10 and $6 \times 15\ mm^2$ for the outer chambers. The readout provides 1000 samples per trigger resulting in $\sim 5.6 \times 10^8$ three-dimensional "pixels".

3.2.1 Silicon Detectors and the Harsh LHC Environment

The main challenge for silicon tracking detectors is the harsh environment of the LHC which requires a certain radiation tolerance of the sensor material. The unit used to give the expected radiation level is typically chosen to be 1 MeV equivalent neutrons (n_{eq}) per cm^2 based on the non-ionising energy loss (NIEL) hypothesis which states that the radiation damage is proportional to the NIEL damage cross section [8, 9]. The silicon sensors in the strip and pixel detectors are junctions of differently doped silicon materials (n-type or p-type silicon) that form a diode. The n-type silicon is doped with elements from the group V in the periodic table (such as arsenic or phosphorus) that add an additional valence electron that becomes unbonded from individual atoms and allow the compound to be an electrically conductive n-type semiconductor. In contrast, p-type material is doped with elements from group III (e.g. boron) that add an additional hole. The doping concentration is typically 10^{12}–$10^{16} cm^{-3}$. At the time

Table 3.3 Silicon sensor choices and expected fluences for the various pixel and strip detectors

	System	Sensor type	Fluence (1 MeV n_{eq}/cm^2)
ATLAS	Pixel	n-in-n	1×10^{15}
	Strips	p-in-n	2×10^{14}
CMS	Pixel	n-in-n	3×10^{15}
	Strips	p-in-n	1.6×10^{14}
LHCb	VELO	n-in-n, n-in-p	1.3×10^{14}
ALICE	Pixel	p-in-n	1×10^{13}
	Drift	p-in-n	1.5×10^{12}
	Strips	p-in-n	1.5×10^{12}

of the design of the main silicon detectors (end of the 1990s), the technologies proven to be radiation tolerant for the LHC were p-in-n, n-in-p as single-sided processes and n-in-n as double-sided process. Table 3.3 summarises the sensor choices and the expected fluences for the various silicon systems. The highest fluences and thus highest radiation-damage levels are expected for the innermost layers of ATLAS and CMS because of the envisaged high integrated luminosities for both detectors and the small distance from the collision region at which these layers are placed.

All silicon sensors at the LHC suffer from the harsh radiation environment [10, 11]. The effects were studied in detail during the R&D phase [8, 9], and mitigation techniques were developed in order to keep the effects of the radiation damage in the sensors small during the expected lifetime of the tracking detectors. Radiation-induced effects are cumulative in time and essentially irreversible. Bulk damage due to non-ionising energy loss of the particles passing through the sensor material results in increased leakage current. Thus it is possible to monitor the radiation damage in the systems by closely monitoring the leakage current over time. The current increase, ΔI, can also be calculated using the predicted fluence and the measured temperature:

$$\Delta I = \alpha \Phi V_{\mathrm{Si}},$$

where α is the annealing parameter in A/cm, which depends on the temperature history, V_{Si} is the volume in cm^3, and Φ is the fluence in cm^{-2} [9]. This so-called "Hamburg model" allows the current increase to be predicted very reliably. Indeed, the increases in the leakage currents observed in most LHC silicon detectors during Run 1 are in agreement with the predicted behaviour for the the accumulated luminosities. Figure 3.10 shows the measured current increase in the silicon-strip sensors of the LHCb VELO detector as a function of time, matched with the integrated luminosity delivered to LHCb. The prediction from simulation is in agreement with the mean measured current and shows that the damages caused by non-ionising energy loss is well understood. Similar studies were performed for the other LHC silicon detectors, and they typically show equally good agreement between predictions and measurements [11].

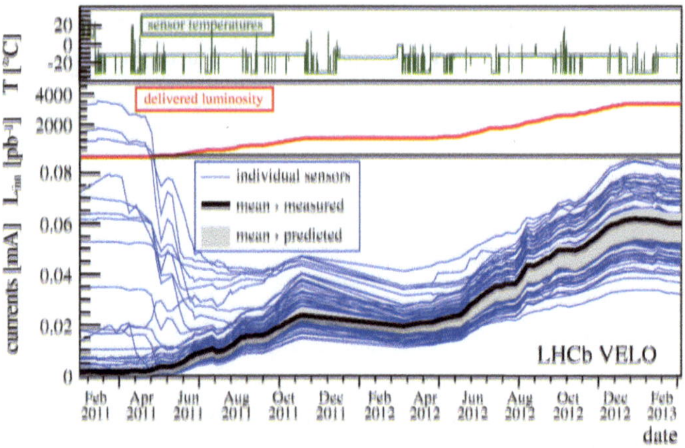

Fig. 3.10 Currents measured for each sensor in the LHCb VELO detector as a function of time. The evolution of the mean current increase is in agreement with the prediction from simulation. Adapted from [11]

3.2.2 Performance of the Tracking Systems

The intrinsic hit efficiency, ϵ, of individual layers is typically measured by extrapolating well-reconstructed tracks through the entire tracking detector and then comparing the number of hits counted in a specific layer, N_{measured}, with the overall number of tracks, N_{expected}:

$$\epsilon = \frac{N_{\text{measured}}}{N_{\text{expected}}}.$$

The term "well-reconstructed tracks" refers to tracks that are based on a certain (minimum) number of hits in the respective subsystem. As an example, the efficiency for the layers in the ATLAS SCT is measured based on tracks with a minimum of six hits in the SCT. In Fig. 3.11 the intrinsic hit efficiency for each layer of the ATLAS SCT measured with proton-proton data at $\sqrt{s} = 8$ TeV in 2012 is shown. In addition, the fraction of bad strips in the system is given. It is important for tracking detectors to have a high intrinsic hit efficiency as well as a low number of non-operating channels to ensure an overall high tracking efficiency. In ATLAS, for the pixel detector, an intrinsic efficiency of >97 % was measured during Run 1, while the SCT shows an efficiency of better than 99.5 % in all layers. In CMS the efficiencies were even slightly larger and better than 99.5 % (pixels) and 99.8 % (strips).

The overall track reconstruction efficiency (also referred to as tracking efficiency) is an important parameter of the tracking system as it is an indication for the physics performance of the experiment. All experiments invest a lot of effort in maintaining the tracking efficiency at above 95 %. In Table 3.4 the average track reconstruction efficiencies as measured by all four experiments are given [12–15]. A method to

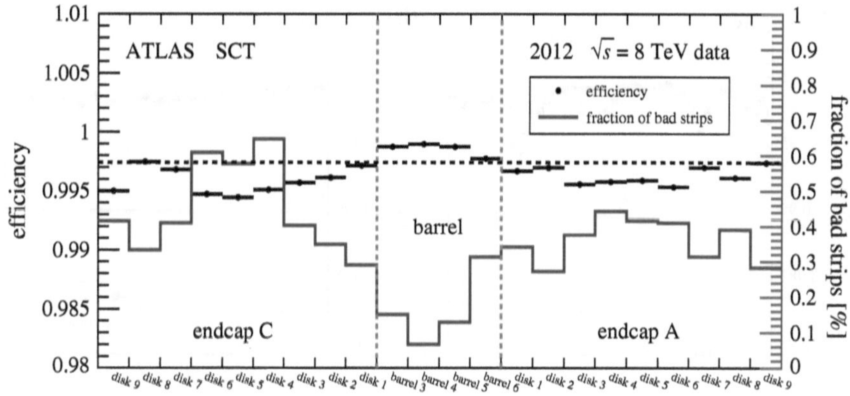

Fig. 3.11 Mean intrinsic hit efficiency for each layer of the ATLAS SCT measured with proton-proton data at $\sqrt{s} = 8$ TeV in 2012. In addition the fraction of bad strips is given (*line*) using the right-hand y axis. Adapted from [10]

Table 3.4 Overview of tracking parameters describing the performance of the tracking systems. For the transverse impact parameter resolution the optimal values are given as it is also depending on the detector region and the track momentum

	Track reconstruction efficiency in central region (%)	Primary-vertex resolution	Transverse impact parameter resolution at p_T = 10 GeV (μm)	Momentum resolution at p_T = 10 GeV (%)
ATLAS	>98	x: 20 μm y: 30 μm	>10	<1
CMS	>99.5	x: 10 μm z: 12 μm	>10	0.8
LHCb	>98	x: 13 μm z: 71 μm	<35	0.4
ALICE	>95	x: 10 μm y: 10 μm	20	<2

measure the tracking efficiency of the tracking system is the "tag&probe" method [14, 16, 17]. Here, the fraction of muon tracks from decays of known resonances are used. An example are $Z \to \mu^+\mu^-$ candidates that are selected by an invariant-mass cut around the Z-boson mass. The fraction of tracks measured in the inner tracker with a matching track in the muon tracker over the number of tracks measured in the muon system only is defining the tracking efficiency. Typically this parameter is shown as a function of the pseudo-rapidity η and of the number of reconstructed primary vertices. In Fig. 3.12 the tracking efficiency of the CMS tracker is shown as an example.

The primary vertex is the spatial location where two particles in the colliding beams interacted. The location, and the associated uncertainty, of all proton-proton interaction vertices in each event using the available reconstructed tracks is measured precisely using the track information from the vertex detector. Typically a split

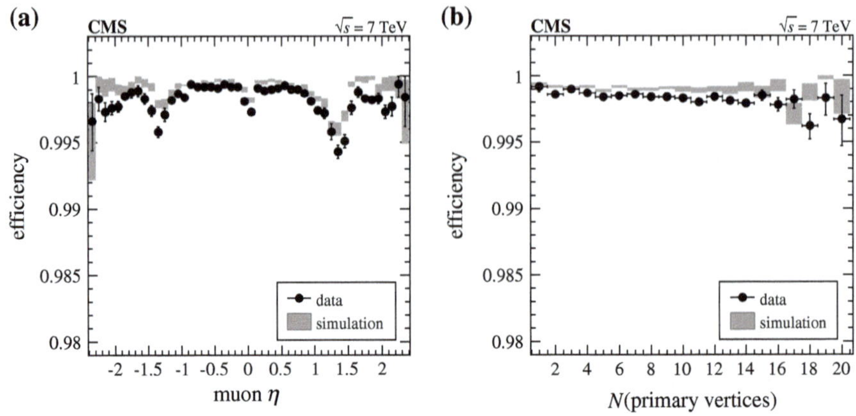

Fig. 3.12 Tracking efficiency of the CMS tracker measured with a tag&probe method, for muons from Z-boson decays, as a function of **a** the muon η and **b** the number of reconstructed primary vertices in the event for data (*dark dots*) and simulation (*shaded bands*). Adapted from [13]

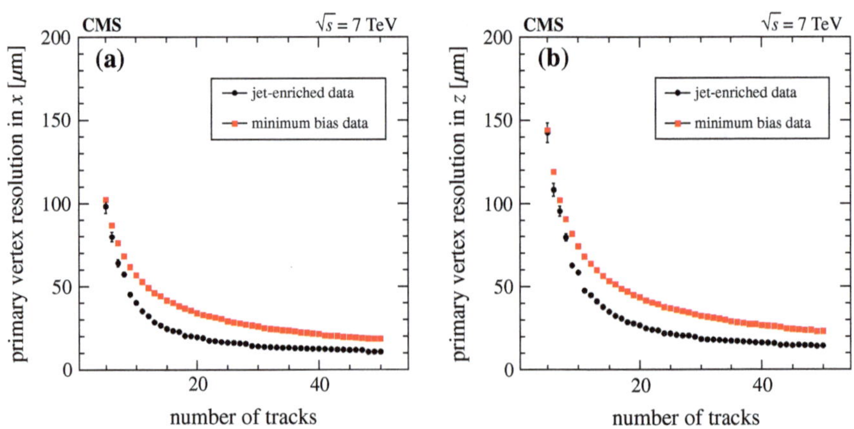

Fig. 3.13 Primary-vertex resolution for x (*left*) and z (*right*) as a function of the number of tracks at the fitted vertex. Adapted from [13]

method is being used to estimate the primary-vertex resolution. Here the group of tracks associated to a given vertex is divided into two sets. The distance between the vertex positions determined from both sets is proportional to the resolution. The vertex resolutions in the transverse plane and along the z direction are shown as a function of the number of fitted tracks for the CMS pixel detector in Fig. 3.13. The jet-enriched samples indicated in the figure are selected by requiring a reconstructed jet with transverse energy $E_T > 20$ GeV in the event. In these samples, the mean p_T of the tracks is higher, resulting in a better vertex resolution [13].

The impact parameter (IP) resolution of a tracking system describes the ability to reconstruct secondary vertices. The IP is defined as the shortest distance between

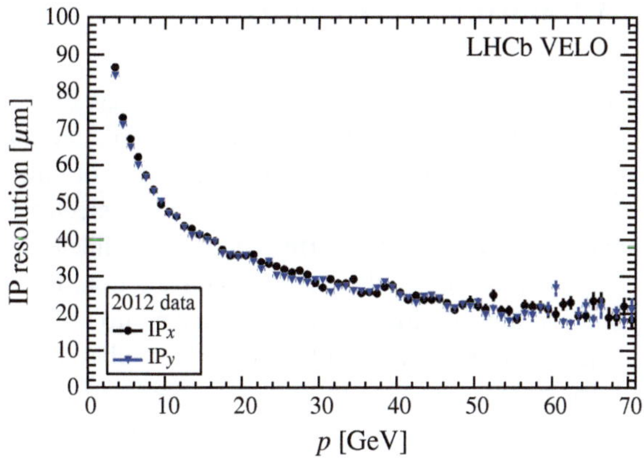

Fig. 3.14 Impact-parameter resolutions IP_x and IP_y are shown as functions of the momentum for the LHCb VELO. Adapted from [11]

the reconstructed track and the primary vertex. The impact parameter resolution is affected by the position resolution of the detector planes, by the multiple scattering of the particles in the detector material, and by the distance of the first hit in the system with respect to the primary vertex. The resolution in the central region of the ATLAS and CMS detector ranges from 10 to 100 μm and is similar for both. In Fig. 3.14 the impact-parameter resolutions IP_x and IP_y are shown as functions of the momentum for the LHCb VELO determined in 2012 data.

The tracking systems of the LHC detectors were also designed to provide excellent momentum resolution for the measured tracks. It depends on the precise knowledge of the material distribution within the tracking detector and the magnetic field, the alignment of the system, and the algorithm used to reconstruct the tracks. The fractional momentum resolution $\sigma(p_\mathrm{T})$, for $p_\mathrm{T} > 20\,\mathrm{GeV}$, can be parametrised to good approximation by the quadratic sum of two terms [18]:

$$\frac{\sigma(p_T)}{p_T} = a \oplus b \cdot p_T,$$

where a is constant describing the multiple scattering contribution, while b describes the intrinsic resolution of the detector components as well as any residual misalignment. In Table 3.4 the momentum resolution for all four experiments is shown in the rightmost column. Here, LHCb has the best performance with $\sigma(p_T)/p_T$ of 0.4 % for the energy range from 2 to 20 GeV. This extremely good performance is mainly owed to the fact that the material and thus the multiple scattering was minimised in the central tracking system of LHCb, for example by placing the VELO within the beam vacuum.

3.3 Calorimetry Detectors

The electromagnetic and hadronic calorimeters of the LHC experiments are used to determine the energy and also the momentum direction of particles that are produced at the interaction points or as decay products of other unstable particles. Typically, both types of calorimeters are built such that electromagnetic and hadronic showers are fully contained in the corresponding calorimeter material in order to optimise the energy resolution. In the energy reconstruction, the material between the collision point and the calorimeter devices is taken into account because the shower development may already have started in detector subsystems upstream of the calorimeters.

The relative energy resolution, $\Delta E/E$, of the calorimeters can typically be described by stochastic, noise and constant terms with coefficients c_s, c_n, and c_c, according to

$$\left(\frac{\Delta E}{E}\right)^2 = \left(\frac{c_s}{\sqrt{E}}\right)^2 + \left(\frac{c_n}{E}\right)^2 + \left(c_c\right)^2. \tag{3.1}$$

The energy dependence of the stochastic term derives from the number of shower particles, which increases with energy, such that the relative uncertainty is proportional to $1/\sqrt{E}$. The detector and electronics noise is independent of the deposited energy, and after dividing by energy a term proportional to $1/E$ appears. At high energies, the constant term thus limits the energy resolution. This term is mainly due to shower tails that are not captured by the calorimeter, inhomogeneities of the detector, limited inter-calibration of detector segments, and non-linearities in the signal response. The resolution parameters are all optimised in the calorimeter designs.

Apart from precise energy and direction measurement of particles, an important functionality of the calorimeters is the generation of trigger signals. All calorimeters provide a fast signal response, which is used for deciding if a collision event is interesting enough to be recorded for further data analysis.

3.3.1 Electromagnetic Calorimeters

For the measurement of electrons and photons, homogeneous and sampling detector technologies are employed. Table 3.5 presents an overview of the different active and passive materials that are used to measure electromagnetic showers produced by the absorbed particles [1–4]. The absorption strength of all electromagnetic calorimeters is more than 20 radiation lengths (X_0) and thus guarantees shower containment. The detector signals are generated by different physics processes in the active material layers, like ionisation of the material or Cherenkov light generation. These signals are amplified and read-out electronically.

In ATLAS [1], sampling calorimeters with liquid argon (LAr) as active material and lead (Pb) or copper (Cu) as absorber are used. The energy deposit is proportional to the ionisation signal in the LAr volume, which is read out in up to three layers

Table 3.5 Overview of the main properties of the electromagnetic calorimeters installed in the main LHC detectors. The relative energy resolution for electrons and photons is given in terms of the constants for the stochastic and constant term, according to (3.1)

	Calorimeter	Material	Number of channels	Angular coverage	Energy resolution	
					c_s (%)	c_c (%)
ATLAS	EM barrel	$LAr + Pb$	109,568	$\|\eta\| < 1.475$	10	0.7
	EM end-cap	$LAr + Pb$	63,744	$1.375 < \|\eta\| < 3.2$	10	0.7
	FCal	$LAr + Cu$	2016	$3.1 < \|\eta\| < 4.9$	28.5	3.5
CMS	ECAL barrel	$PbWO_4$	61,200	$\|\eta\| < 1.479$	2.8	0.3
	ECAL end-cap	$PbWO_4$	14,648	$1.479 < \|\eta\| < 3.0$	2.8	0.3
LHCb	ECAL	Scint. + Pb	6016	$0.756 < \eta_x < 2.19$ $1.037 < \eta_y < 2.19$	9	0.8
ALICE	PHOS	$PbWO_4$	17,920	$\|\eta\| < 0.12$, $220° < \phi < 320°$	3.3	1.1
	EMCal	Scint. + Pb	12,672	$\|\eta\| < 0.7$, $80° < \phi < 187°$	10	2

via finely segmented electrodes. The CMS electromagnetic calorimetry [2] relies on homogeneous $PbWO_4$ crystals in which the charged shower particles produce Cherenkov light. The light is converted to electrical signals by avalanche photo diodes (APD) and vacuum phototriodes (VPT) in the barrel and end-cap sections, respectively.

The LHCb electromagnetic calorimeter [4] is covering a more limited acceptance range than those of the general-purpose detectors ATLAS and CMS. The calorimeter modules are installed in planes perpendicular to the beam axis in such a way that they cover the same angular range as the LHCb tracking system. The LHCb electromagnetic calorimeter is built in lead-scintillator shashlik technology. The scintillator tiles are made of polystyrene with 2.5 % pTP and 0.01 % POPOP admixture. Phototubes convert the scintillator light, which is collected by wavelength-shifting fibers, into amplified electrical signals which are proportional to the energy deposit.

Also ALICE [3] is using electromagnetic energy measurements only in a reduced angular range. The high-resolution photon spectrometer (PHOS) aims at direct photon and $\pi^0 \rightarrow \gamma\gamma$ detection. It is made of $PbWO_4$ crystals equipped with an APD read-out. The electromagnetic calorimeter (EMCal) of ALICE is installed opposite in azimuth to the PHOS detector. It is constructed as a Pb-scintillator sampling device with shashlik geometry. The scintillator tiles are made of polystyrene with 1.5 % pTP and 0.4 % POPOP. The light produced is guided by wavelength-shifting fibers to APD sensors.

A comparison of the detector resolutions is made in Table 3.5. As expected, the sampling devices based on lead as absorber have a slightly worse resolution than the homogeneous crystal calorimeters.

3.3.2 Hadronic Calorimeters

The energy measurements of the electromagnetic calorimeters are complemented by measurements of hadronic calorimeters, which are all designed as sampling devices. The measured hadrons may be produced in quark-initiated or gluon-initiated particle jets, decays of τ leptons, or single hadron decays.

In the central region of ATLAS, a lead-scintillator tile calorimeter is installed, which is complemented by a hadronic end-cap calorimeter (HEC) with a LAr-copper structure. Closer to the beampipe, the forward calorimeter (FCal) also has a hadronic section using a combination of tungsten (W) and LAr, in order to reduce the lateral shower width and improve the resolution. The acceptance ranges and resolution parameters of the LHC calorimeters are summarised in Table 3.6.

The steel/brass-scintillator hadronic calorimeter (HCAL) of CMS is fully contained within the solenoidal magnetic field. It is divided into barrel and end-cap sections (HB and HE) and complemented by an outer hadronic calorimeter (HO), which is integrated in the steel structure of the muon spectrometer and serves as "tail catcher" for the hadronic showers that penetrate the superconducting solenoid. The forward region is equipped with a steel-quartz-fiber detector. The choice of this technology is motivated by the radiation tolerance of the active material.

Table 3.6 Detector technology, acceptance ranges, and energy-resolution parameters for the main hadronic calorimeters of the LHC detectors (3.1)

	Calorimeter	Material	Number of channels	Angular coverage	Energy resolution	
					c_s (%)	c_c (%)
ATLAS	Tile	Scint. + Pb	9852	$\lvert\eta\rvert < 1.7$	52	3
	HEC	$LAr + Cu$	5632	$1.5 < \lvert\eta\rvert < 3.2$	84	–
	FCal	$LAr + W$	1508	$3.1 < \lvert\eta\rvert < 4.9$	94	7.5
CMS	HB	Scint. + steel/brass	2592	$\lvert\eta\rvert < 1.3$	90	9
	HE	Scint. + steel/brass	2592	$1.3 < \lvert\eta\rvert < 3$	90	9
	HO	Scint. + steel	2160	$\lvert\eta\rvert < 1.4$	–	–
	HF	Quartz fibre + steel	1728	$3 < \lvert\eta\rvert < 5.2$	120	–
LHCb	HCAL	Scint. + steel	1488	$\lvert\eta_x\rvert < 1.87$	69	9
				$\lvert\eta_y\rvert < 2.07$		

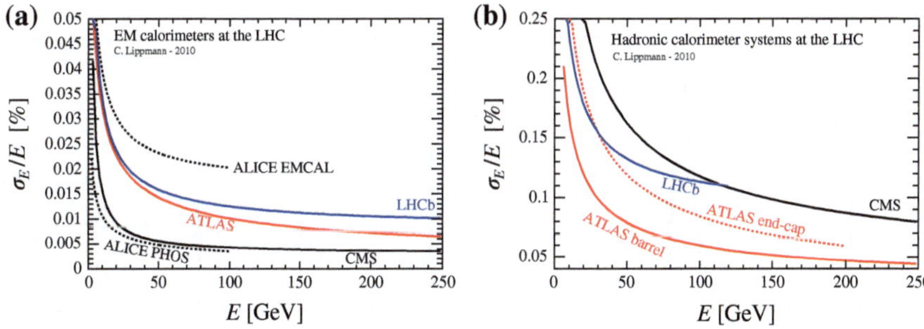

Fig. 3.15 Relative energy resolution of the electromagnetic and hadronic calorimeters installed in the LHC detectors. Adapted from [19]

A comparison of the relative energy resolution of the electromagnetic and hadronic calorimeters of the LHC detectors is compiled in Fig. 3.15 [1–4].

3.3.3 Calibration and Performance

The calibration of the calorimeters is performed using dedicated calibration systems, e.g. precision electronic or light pulsers, or radioactive sources with well-known emission spectra. This initial calibration is further refined exploiting high-statistics physics processes with known kinematics. Figure 3.16 shows an example of the calibration steps of the CMS ECAL using the light-monitoring system and the precisely known Z-boson mass in order to obtain a calibration of the energy of electrons. The final energy scale is calibrated such that the invariant mass spectrum of e^+e^- pairs in $Z \rightarrow e^+e^-$ events matches the line-shape of the Z resonance [20].

Reference signals like Z decays allow the assignment of uncertainties on the energy scales of physics objects, like electrons, hadronically decaying τ leptons and hadronic jets. The transverse momentum, p_{T}, or transverse energy, E_{T}, of these objects, as well as their momentum direction, is, however, not only based on measurements in the calorimeters—tracking information is obviously a crucial ingredient in the object reconstruction. For electrons the available information on track momentum, detection of bremsstrahlung along the particle track, and calorimetric energy measurement is combined in the final electron energy determination. For τ leptons and hadronic jets, the decay and hadronisation products, which can be both charged and neutral, the calorimetric energy measurements are typically weighted with scale factors depending on the assignment of a track, since neutral and charged particles have a different response in the calorimeters.

Figure 3.17a shows an example of the energy-scale factor α for electrons as a function of energy, where the calibrated energy, E_{cali}, is rescaled by a factor $(1+\alpha)$ with respect to reconstructed energy E_{reco}, such that $E_{\mathrm{cali}} = (1+\alpha)E_{\mathrm{reco}}$. A value of α close to zero indicates that the final calibration corrections are not large.

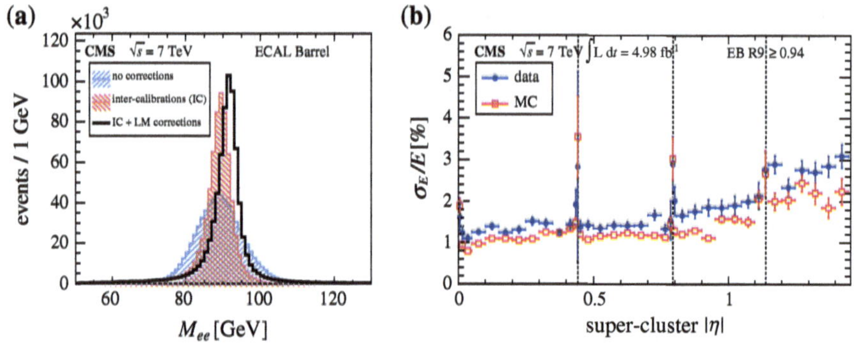

Fig. 3.16 **a** Reconstructed invariant mass, $M_{e^+e^-}$, of electron-positron pairs in $Z \rightarrow e^+e^-$ events measured by CMS at $\sqrt{s} = 7$ TeV after application of channel inter-calibration and light-monitoring corrections. **b** Relative energy resolution determined in $Z \rightarrow e^+e^-$ events after the final calibration step in which the invariant-mass peak is adjusted to the nominal Z-boson mass. The resolution is given for groups of crystals of the CMS calorimeter, so-called super-clusters, which are used in the electron reconstruction. Adapted from [20]

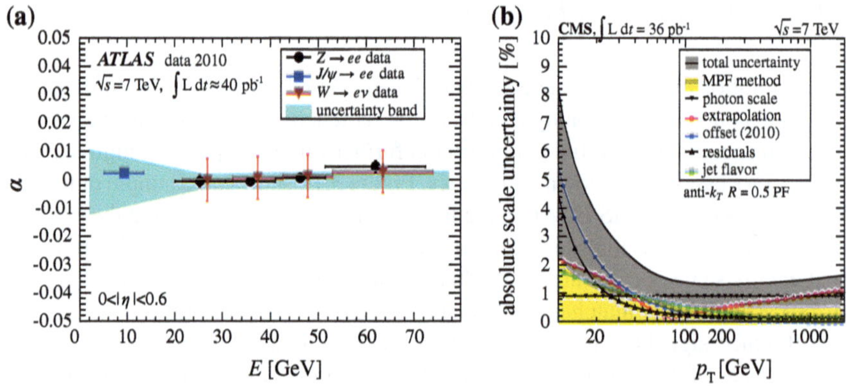

Fig. 3.17 **a** Energy-scale factor α for electrons in the central ATLAS detector region, derived in collision data with Z, J/ψ and W decays, together with systematic uncertainties that are shown as a horizontal band. **b** Uncertainty on the p_{T} measurement of jets in the CMS detector. Adapted from [20, 22]

The systematic uncertainty on α determined from detailed detector studies is also displayed. In ATLAS, the final systematic uncertainty is typically below 10 % and dominated by the knowledge of the exact detector material composition in front of the calorimeters.

The jet energy scale is derived exploiting transverse energy balance in events which contain hadronic jets, like γ+jet and $Z \rightarrow e^+e^-/\mu^+\mu^-$+jet events. Jets in a p_{T} range between 20 and 300 GeV can typically be calibrated and lower or higher p_{T} jet energy scales are extrapolated. Figure 3.17b shows a summary of systematic uncertainties on the jet energy scale for anti-k_t jets [6] with resolution parameter

$R = 0.5$ measured by the CMS experiment in the central detector region [21]. Apart from the uncertainties on the absolute p_{T} scale and the extrapolation, the pile-up modelling, the knowledge of the jet-flavour composition and the time stability of the calorimeters are taken into account. The total uncertainties are about 2 % and increase to 8 % at low values of p_{T}.

Studies of hadronic τ-lepton decays reveal similar sources of uncertainties on the determination of the transverse energy scale as hadronic jets, like pile-up, modelling of detector material and time-dependent detector effects. Also, the single-particle response and the hadronic activity in pp collisions that is not related to the hard scattering (the so-called "underlying event"), play an important role. The τ lepton energy uncertainties are estimated to be about 2–3 % [23, 24]. The possible effect of a variation of the energy scale on the reconstructed mass of the visible τ lepton decays is displayed in Fig. 3.18a for τ decays into three charged pions.

The decay products of τ leptons, but also the final state of many other physics processes, contain neutrinos or, more generally, weakly interacting particles. Because of their small cross section with the detector material, these particles are measured as missing energy. Only the transverse component, $E_{\mathrm{T}}^{\mathrm{miss}}$, is experimentally accessible because the longitudinal momentum component of the hard scattering process of the pp collision is a priori unknown, and because the energy flow can not be fully measured in the forward region close to the beam axis. The $E_{\mathrm{T}}^{\mathrm{miss}}$ resolution along the two transverse directions in the x–y plane, $\sigma(E_{\mathrm{T},x}^{\mathrm{miss}}, E_{\mathrm{T},y}^{\mathrm{miss}})$, obviously scale with $\sqrt{\sum E_{\mathrm{T}}}$ since $E_{\mathrm{T}}^{\mathrm{miss}}$ is calculated as the recoil to the total transverse energy flow in the detector:

$$\sigma(E_{\mathrm{T},x}^{\mathrm{miss}}, E_{\mathrm{T},y}^{\mathrm{miss}}) = \sigma_s \sqrt{\sum E_{\mathrm{T}}},$$

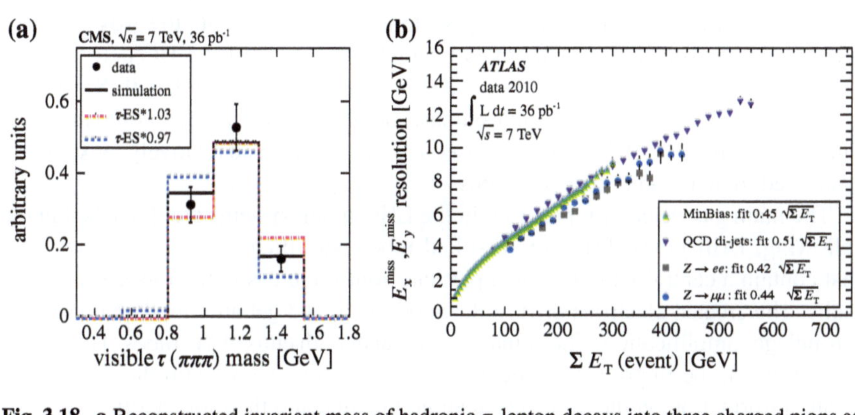

Fig. 3.18 **a** Reconstructed invariant mass of hadronic τ-lepton decays into three charged pions as measured by the CMS experiment, compared to predictions of the simulation with the τ-energy scale varied up and down by 3 % with respect to the best-fit value. **b** $E_{\mathrm{T}}^{\mathrm{miss}}$ resolution in the x and y directions derived by ATLAS from events with leptonic Z decays, in QCD jet events, and in so-called minimum-bias events. Adapted from [23, 25]

with $\sigma_s \approx 0.4 - 0.5$. An example of the dependence of the resolution on $\sum E_{\mathrm{T}}$ is shown in Fig. 3.18. Refined techniques allow a suppression of the pile-up influence on the $E_{\mathrm{T}}^{\mathrm{miss}}$ reconstruction, which is particularly important for measurements of rare processes with $E_{\mathrm{T}}^{\mathrm{miss}}$ signatures.

3.4 Muon Spectrometers

The physics reach of the LHC experiments depends strongly on the trigger efficiency and momentum resolution provided by the muon system in a p_{T} range extending up to the TeV scale. A trigger with muons in a hadron environment is a clean signature for basically all discovery physics. Therefore, highly efficient muon detection systems needed to be developed for the LHC detectors. Each layer in the muon system requires a high detection efficiency (~99 %) to obtain an overall trigger efficiency of >95 %. Additionally a fast response is required to comply with the 25 ns bunch crossing intervals at LHC. In most LHC experiments, the muon spectrometer additionally provides muon tracking with a spatial resolution in the 100 μm range in the r–ϕ plane.

As muon systems are placed behind the calorimeters at large radii or in the forward direction (ALICE and LHCb), they cover extensive areas of the order of a few 100 m^2. Typically an absorber is placed in front of the muon detector to reduce the number of hadrons entering the muon system which might lead to false identifications (hadron punch-through). The choice of absorber in a muon system is critical for the reduction of multiple scattering. The most commonly used absorber is iron/steel which features low cost, high density (and hence compactness), and easy magnetisation. The low occupancy in the muon-detector region allows cell sizes in the mm range. Compared to the inner tracking systems the requirements towards radiation tolerance are relatively low. The main technologies used at the LHC detectors are gas-filled detectors, which have the advantage of a large radiation length X_0 and the possibility to construct large volumes/areas, and thus to optimise the momentum resolution ($\Delta p/p \sim L^{-2}$). In addition, gaseous detectors are also relatively inexpensive compared to many other detector types.

The main gas-detector types used in the LHC muon systems are described in the following. Drift tubes (DT) are gas-filled tubes with diameters of a few ten mm that contain a central wire. Charged particles ionise the gas in the tubes along their paths. The created electrons drift to the anode wire and get amplified near the wire through gas amplification. The signal is read out at both ends of the tubes. The spatial resolution can be improved by recording the drift time along with the drift signal giving information on the hit position along the tube and therefore adding a second coordinate. Drift tubes are implemented in the ATLAS and CMS muon systems.

Multi-wire proportional chambers (MWPC) are more complex systems in which several wires are positioned between planar cathodes operated in proportional mode. The cross-wired configuration yields in the measurement of two coordinates for each passing charged particle. George Charpak showed in 1968 that an array of many closely spaced anode wires in the same chamber could each act as independent proportional counters [26]. Each wire is instrumented with an amplifier to allow a separate readout and thus a position sensing. MWPCs are employed in the LHCb forward muon spectrometer. A further development of MWPCs are cathode-strip chambers (CSC). They consist of arrays of anode wires crossed with copper cathode strips within a gas volume allowing 3D coordinate measurements. These detectors with short drift distances are useful for environments with high rates or inhomogeneous B field as in the CMS forward region or in ATLAS at very high η.

Another gas detector used by all LHC detectors are resistive-plate chambers (RPC). RPCs are gaseous parallel-plate avalanche detectors that implement electrodes made from a material with a high volume resistivity between 10^7 and10^{12} Ωcm. RPCs combine a good spatial and time resolution. Large-area RPCs with 2 mm single gaps operated in avalanche mode provide better than 98 % efficiency and a time resolution of around 1 ns up to a particle flux of several thousand particles per second and cm^2. With slightly varied configurations they are used as trigger or tracking planes in the muon systems.

Micro-pattern gas detectors such as gas electron multipliers (GEM) are the next-generation muon chambers exploiting gas amplification. They are used in regions where better spatial resolutions is required—for example in areas where higher occupancies are expected. GEMs are manufactured using thin perforated 50 μm kapton foils clad on both sides with copper. The two surfaces are maintained at a potential gradient, thus providing the necessary field for electron amplification. Coupled with a drift electrode above and a read-out electrode below, the foil acts as a high-performance electron amplifier. The major advantage of such detectors is that amplification and detection are decoupled and that the read-out is at zero potential. GEMs are used in the LHCb muon system.

In Table 3.7, an overview of the technologies chosen for all fours LHC experiments is given.

Table 3.7 Overview of technology choices for the muon systems giving the purpose of each plane (trigger or muon tracking), the measured efficiencies and the selected gas mixtures

	Type	Coverage	Aim	Efficiency (%)	Gas mixture
ATLAS	RPC	$\|\eta\| < 1.05$	Trigger	>98	$C_2H_2/F_4/$ C_4H_{10}/SF_6
	TGC	$\|\eta\| > 1.05$	Trigger	>96	CO_2/n-pentane
	MDT	$\|\eta\| < 2$	Tracking	>96	Ar/CO_2
	CSC	$2.0 < \|\eta\| < 2.7$	Tracking	>96	Ar/CO_2
CMS	DT	$\|\eta\| < 1.2$	Track/trigger	>96	Ar/CO_2
	CSC	$0.9 < \|\eta\| < 2.4$	Track/trigger	>96	$CF_4/Ar/CO_2$
	RPC	$\|\eta\| < 2.1$	Trigger	>95	$C_2H_2F_4$, C_4H_{10}
LHCb	GEM	$1.9 < \eta < 4.9$	Track/trigger	>96	$Ar/CO_22/CF_4$
	MWPC	–	Track/trigger	>99.5	$CO_2/Ar/CF_4$
ALICE	RPC	$-4.0 < \eta < -2.5$	Trigger	>98	$C_2HF_4/C_4H_{10}/SF_6$
	CPC	$-4.0 < \eta < -2.5$	Tracking	>98	Ar/CO_2

The ATLAS muon spectrometer is designed with two separate sets of detectors: trigger chambers with fast response time and precision chambers for good momentum resolution without loss of trigger efficiency. The precision tracking muon system is composed of two sub-detectors: monitored drift tubes (MDT) and cathode-strip chambers (CSC). The MDTs cover most of the solid angle, with three concentric layers in the barrel and up to $|\eta| = 2.7$ in the end-cap. One MDT chamber consists of three to eight layers of gas-filled drift tubes. In total about 355,000 tubes cover an area of 3650 m^2. Due to the small channel size, cathode-strip chambers (CSC) can function up to a counting rate of 1000 Hz/cm^2 and therefore are better suited for higher particle fluxes at higher pseudo-rapidity. Both detectors use a mixture of Ar and CO_2 gas inside the chambers for particle detection. The system with 70,000 channels provides a resolution of 60 μm. The trigger chambers used for triggering are also divided into two sub-detectors: resistive-plate chambers (RPC) in the barrel and thin-gap chambers (TGC) in the end-cap. The RPC extends to $|\eta| = 1.05$ with 380,000 channels. The TGC covers the range $1.05 < |\eta| < 2.4$ and is a multi-wire proportional chamber with a smaller wire-cathode gap than the wire-wire gap.

The CMS muon spectrometer uses drift tubes (DT) for the momentum measurement (tracking) in the barrel wheels; in the end-cap disks cathode-strip chambers (CSC) are employed to cope with the higher background rates. Additionally, both detector types feature self-triggering capabilities. The main purpose of the resistive-plate chambers (RPC) in both barrel and end-cap regions is to provide a fast and independent trigger. The CMS barrel section of the muon system ($|\eta| < 1.2$) consists of five wheels, each with four different types of DT chambers, called stations, forming concentric cylinders around the beam line. Three iron layers are embedded within the yoke. The first muon station, MB1, is placed in front of the yoke; the next two, MB2 and MB3, are between iron layers. The last station, MB4, is placed outside of the yoke. The combination of the four stations forms a sector in ϕ, and in total 12 sectors make up the concentric cylinder.

The LHCb muon system located in the forward arm is composed of five stations placed along the beam axis (M1–M5) with a total area of 435 m^2 and 122,000 channels. M1 is located in front of the calorimeters and improves the p_T resolution from ~35 to ~25 % with respect to what could be obtained using only the four stations behind the calorimeter [27]. Stations M2 to M5 are placed outside of the calorimeters and are interleaved with 80 cm thick iron absorbers. Each station is equipped with 276 multi-wire proportional chambers, except for the inner region of the first station M1. This region is exposed to the highest radiation level, and therefore 12 GEM detectors were chosen to cope with the harsher environment. All stations consist of two mechanically independent halves that can be moved horizontally to access the beampipe and the detector chambers for installation and maintenance.

The ALICE muon spectrometer in the single forward arm is also designed to trigger and track the muons with good momentum resolution. The first layer is a front absorber made of carbon and concrete to suppress hadrons and photons arriving from the central detector. Next are five muon tracking stations with the third being located in a warm dipole magnet about 7 m from the interaction point with a field of 0.7 T in the horizontal plane in the direction perpendicular to the beam. The tracking system with cathode-pad chambers (CPC) covers a total area of about $100\,m^2$. This muon tracking system fulfils the requirements of a spatial resolution of about $100\,\mu m$ and has the capability to operate in a high particle multiplicity environment. (CPCs are a modification of cathodes-strip chambers). Each tracking station consists of two CPCs with pads installed on both sides of the gas gap in order to measure correlated x–y coordinates in each chamber. The muon trigger system in the ALICE muon spectrometer consists of two stations with two RPC planes each, with a total of 72 chambers. The trigger system exposes a chamber efficiency and a trigger efficiency both close to 100 % with a very good stability.

3.4.1 Performance

Muon objects are used in most of the analyses at the LHC experiments, and several algorithms were developed within the collaborations to ensure robust and high-efficiency muon reconstruction and identification. For this purpose different subsystems of the experiments have to be exploited, including the inner tracking systems, the calorimeters and the muon spectrometers. Typically, muons are reconstructed independently in the tracking system and as standalone muon tracks. Standard candles such as J/ψ, Υ, $Z \rightarrow \mu^+\mu^-$ are used to achieve an unbiased measurement of the trigger and muon reconstruction efficiencies, the muon-momentum scale and the muon-momentum resolution.

The intrinsic efficiency of a muon system can be evaluated by comparing the number of hits in a single plane with the number of tracks identified in the remaining muon planes as an external trigger and tracking system. Inefficiencies are introduced by dead channels or small geometrical losses due to the chamber structures of the muon detector. As an example, Fig. 3.19 shows the average efficiency of one ALICE RPC detection plane as a function of time. The observed efficiencies are always >95 % and well within requirements for the systems.

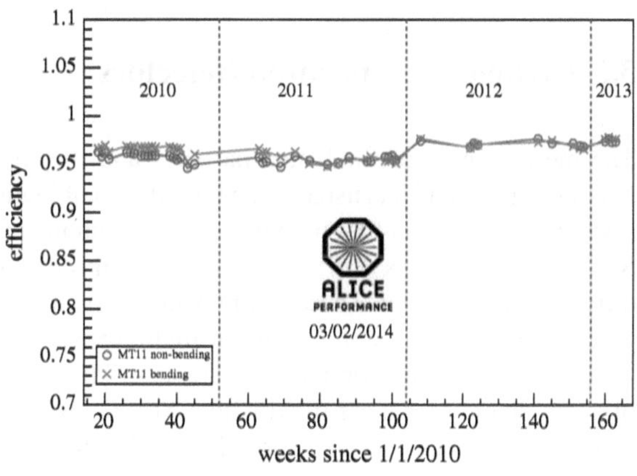

Fig. 3.19 Average efficiency of one ALICE RPC detection plane for both the bending and non-bending planes, as a function of time expressed in weeks since the beginning of 2010. Adapted from [28]

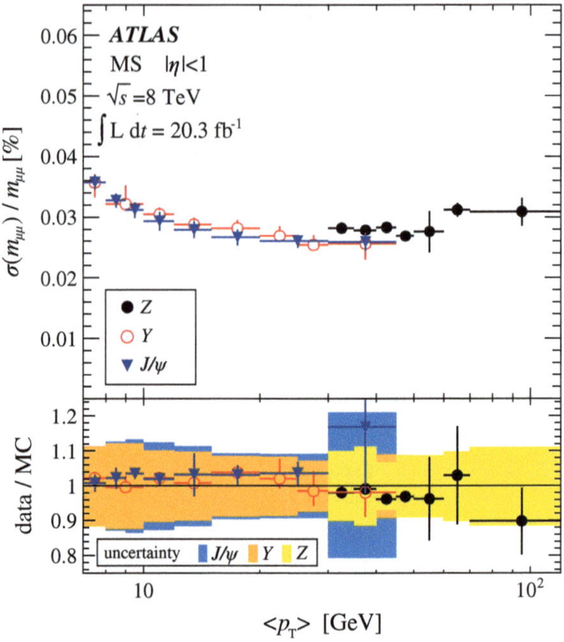

Fig. 3.20 Dimuon invariant-mass resolution for muons reconstructed with the ATLAS muon system only, measured from J/ψ, Υ and Z decays as a function of the average transverse momentum in the barrel section ($|\eta| < 1.0$). Adapted from [18]

The measurement of the transverse momentum p_T for muons is highly sensitive to the alignment of the inner tracking detector and of the muon chambers. Also the composition of material and its distribution inside the tracking volume and the knowledge of the magnetic field inside and outside of the solenoid volume are playing an important role.

In Fig. 3.20 the dimuon invariant-mass resolution for muons reconstructed with the ATLAS muon spectrometer only is shown. The tracks detected in the muon spectrometer are matched to muon tracks detected in the inner tracking system also using the "tag&probe" method (see Sect. 3.2.2).

3.5 Particle-Identification Detectors

The measurement of final-state particles and of physics objects like electrons, τ leptons, muons and jets usually relies on the combination of specific signatures from several sub-detectors like tracking systems, muon spectrometers and calorimeters. Some LHC detectors are complemented with dedicated systems that distinguish different particle types. These components are typically based on transition radiation, Cherenkov-light emission or time-of-flight measurements.

The ATLAS transition radiation tracker (TRT) is built of polyimide straw tubes of 4 mm diameter that are filled with a Xe/CO_2-based gas. A high-voltage potential is applied between the tube cathode and the thin tungsten anode wire in the tube center, at which the signals are read out. The emission of radiation by charged particle appears at the transition region between two materials of different dielectric constants. The transition radiation has a threshold behaviour and sets on at values of the relativistic γ above approximately 1000. This typical photon-emission signature is used to distinguish charged-pion background from electron signals, because the lighter electrons have a greater γ value at the same momentum. In addition, transition radiation photons produce signals with larger amplitudes than those stemming from an ionisation of the TRT gas. Figure 3.21a shows the probability that a particle produces a pre-defined high threshold of the measured TRT signal amplitude for a sample of pions and electrons at different γ values measured during the 0.9 TeV run in 2009 [29]. The high-threshold probability clearly increases with increasing γ and thus yields separation power between electron signal and pion background.

In LHCb, the ring-imaging Cherenkov (RICH) detectors [4] play a key role in the identification of charged hadrons—mainly in the separation of pions, kaons and protons. The RICH detector has two segments: one at 25–300 mrad with respect to the beam-axis optimised for particles of lower momentum between 2 and 40 GeV, and one covering the 15–120 mrad region for particle momenta from 15–100 GeV. Cherenkov light is produced when the velocity β of a charged particle exceeds the velocity of light in a medium with refractive index $n > 1$. Both LHCb detectors are filled with fluorocarbon gases, C_4F_{10} in RICH1 and CF_4 in RICH2, with refractive indices n of 1.0014 and 1.0005, respectively. The refractive index also defines the opening angle θ_C of the light cone into which the Cherenkov light is emitted, fulfilling $\cos\theta_C = 1/(\beta n)$ for charged particles with velocity $\beta > 1/n$. In particular for kaons it is useful to extend the range in which Cherenkov radiation is created to lower momenta. Therefore, the RICH1 section is equipped with silica aerogel with a larger $n = 1.03$ as radiator. The Cherenkov light is reflected by a mirror system to, in total, 484 hybrid photon detector tubes (HPD), each of which has a pixelised silicon photoelectron detector containing 1024 pixels of $2.5 \times 2.5\,\text{mm}^2$ area.

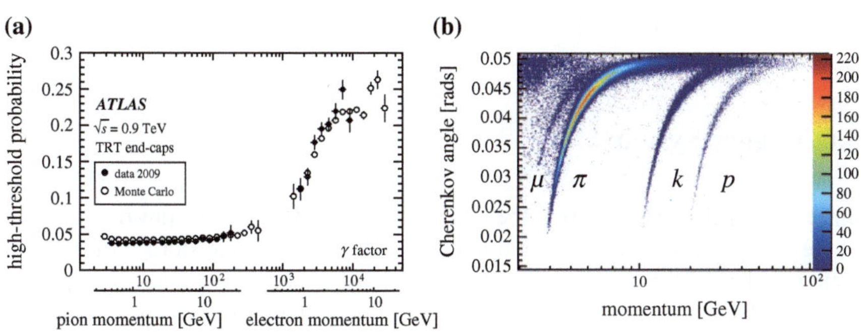

Fig. 3.21 **a** The fraction of high-threshold transition radiation hits on tracks as a function of the relativistic γ factor, facilitating an electron-pion separation over a momentum range from 1–150 GeV. **b** Reconstructed Cherenkov angle as a function of track momentum in the C_4F_{10} radiator for isolated particle tracks measured by LHCb. Adapted from [29, 30]

Particle identification with the RICH system requires a combined measurement of the radius of the Cherenkov ring, and thus θ_C, and of the particle momentum, using the LHCb VELO tracking detector. The RICH1 is therefore installed in direct contact to the VELO detector in order to minimise the material traversed by the charged particles. After calibration, the reconstructed Cherenkov angle is separating pions, kaons and protons very well, as illustrated in Fig. 3.21b. Particle identification in data is performed using a likelihood method in which the data signatures are compared to template distributions for pions, kaons and protons. These are obtained from dedicated event samples of $K_S^0 \to \pi^+\pi^-$, $\Lambda \to p\pi^-$ and $D^{*+} \to D^0(K^-\pi^+)\pi^+$ decays. As an example, at low event pile-up, a kaon identification efficiency of about 80 % is achieved at a pion misidenitification efficiency of 2×10^{-2} [30].

The ALICE detector has implemented three devices for particle identification [3]: a time-of-flight (TOF) detector, a high-momentum particle-identification detector (HMPID), and a transition radiation detector (TRD). These detectors mainly cover the central detector area at $|\eta| < 0.9$ and $|\eta| < 0.6$, respectively. The TOF system is made of gas-filled multi-gap resistive-plate chambers with 160,000 channels and an intrinsic time resolution around 50 ps. It is optimised for π/K and p/K separation in the momentum range up to 2–4 GeV. The HMPID is a ring-imaging Cherenkov detector. Like the TOF system, it aims at a separation of charged hadrons, but at slightly higher momenta of up to 3–5 GeV. The radiator is perfluorohexane (C_6F_{14}) with a refractive index of $n = 1.30$, so that the minimal Cherenkov velocity is $\beta = 0.77$. The Cherenkov photons are traversing a quartz window and are converted into electrical signals using a 300 nm thin-film CsI photocathode, which at the same time represents the cathode pad of a multi-wire pad chamber. The separation of electrons from pions is achieved by measuring transition radiation photons in the TRD. These multi-keV photons create an additional pulse in case a high-velocity electron ($\gamma > 1000$) is crossing the radiator layer constructed as a polypropylene fibre/foam sandwich. The detection of the transition radiation photons and the measurement of the tracks of charged particle is performed in a multi-wire proportional chamber filled with Xe/CO_2. As an example, the pion background efficiency is about 2 % at an electron signal efficiency of 90 % for tracks of 5 GeV.

3.6 Trigger Systems

Since the bunch-crossing rate at the LHC is about 40 MHz and the raw data volumes of the LHC detectors reach several hundred MB per event, the storage of the complete raw data of the detectors is impossible. Furthermore, most interesting physics processes have cross sections several orders of magnitude smaller than the inelastic proton-proton cross section, so that a filtering of collision events is applied before the data are written to permanent storage. Figure 3.22 illustrates the large variation of cross sections at LHC energies, and also at the proton-antiproton collider Tevatron, for comparison.

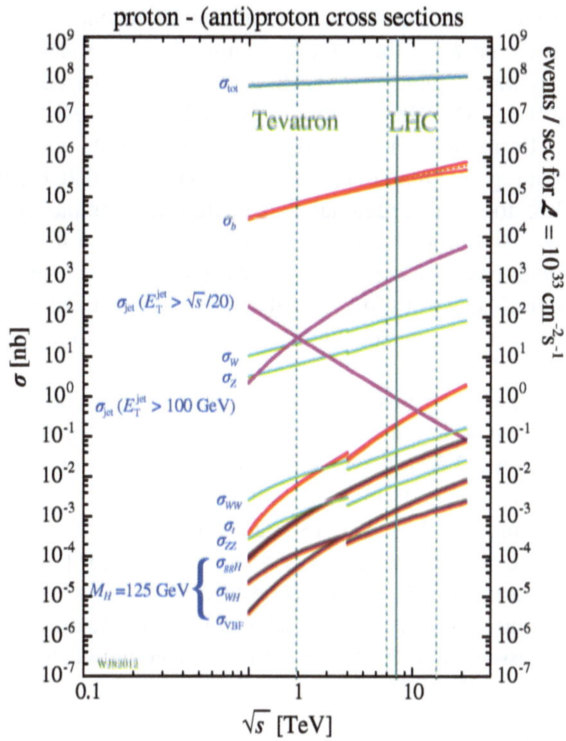

Fig. 3.22 Proton-proton and proton-antiproton cross-sections at LHC and Tevatron collision energies for different processes. Also indicated are the expected rates of collision events at an instantaneous luminosity of 10^{33} cm^2 s^2. *Source* W.J. Stirling

The trigger signatures of the multi-purpose detectors ATLAS and CMS are mainly based on reconstructed physics objects, like electrons, photons, muons, hadronically decaying τ leptons and jets, which pass given thresholds in transverse momentum, p_T. Such signatures indicate that a hard scattering of proton constituents has taken place. Events potentially containing weakly interacting particles, like neutrinos, are triggered by significant missing transverse momentum above a certain threshold. Large total transverse energy flow typically points to the production of heavy particles.

The trigger system of the LHCb experiment is focusing on decays of mesons and baryons with charm-quark or bottom-quark content—signatures that can also be selected by the trigger systems of the multi-purpose detectors. In LHCb, these decays are identified by their decay topology, e.g. by a given number of particle tracks matched to a decay vertex which fall into an appropriately chosen window in invariant mass.

Complicated trigger signatures require information from several detector systems and can typically not be processed at the full LHC collision rate of 40 MHz. Therefore, all the LHC detectors apply a trigger strategy involving at least two levels. The first level is hardware-based, and only the detector components with fast signal response are included, like calorimeters and muon spectrometers. The first trigger level has a response time, or trigger latency, in the order of 2 μs, during which the detector data remain stored in on-detector buffers. The latency time is mainly driven by the cable length between the detector front-end and the off-detector trigger electronics. The hardware triggers of ATLAS, CMS and LHCb accept events at typical rates between 100 and 500 kHz, and about a factor 10 less for ALICE in heavy-ion collision mode.

In the next trigger stages, either the whole detector is read out (LHCb, CMS) or a regional read-out is launched (ATLAS, ALICE). With the information from all detectors in one regional slice or from the complete detector, respectively, particle-reconstruction algorithms can be applied. For example, calorimeter energy measurements can be matched to particle tracks from the central tracking detectors in order to identify electrons. In the final trigger stage of all LHC detectors, reconstruction algorithms nearly identical to those used in the off-line event reconstruction are applied. This allows an event selection triggered by signatures of generally interesting physics processes like those containing high-p_T objects, but also initiated by dedicated decay topologies, e.g. those potentially stemming from B meson decays.

Figure 3.23a shows examples of trigger efficiencies realised at the first (hardware) trigger level and subsequent software trigger levels. Displayed are jet-trigger efficiencies measured by ATLAS at a centre-of-mass energy of 7 TeV, which are detected in the central detector region at $|\eta| < 0.8$ [31]. The efficiency is calculated with respect to anti-k_t jets [6] reconstructed offline with a resolution parameter $R = 0.4$. All curves reach a plateau value close to 100 % efficiency when the trigger threshold is passed. An example for a trigger selecting dedicated particle-decay signatures is given in Fig. 3.23b for the LHCb detector. The first-level hardware trigger is tuned to identify $B^+ \to J/\psi(\mu^+\mu^-)K^+$ decays by single-muon or dimuon signatures.

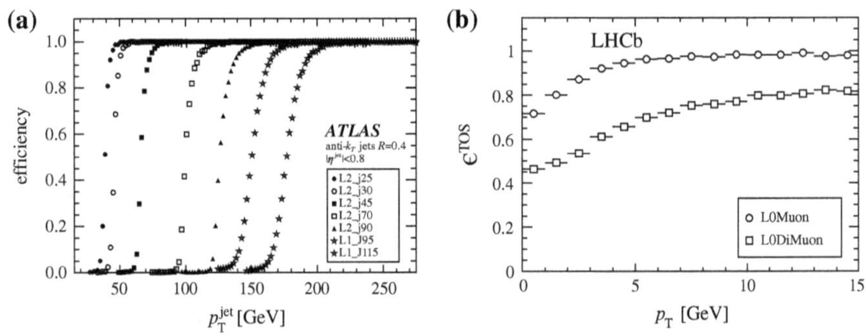

Fig. 3.23 **a** Trigger efficiency for hadronic jets passing different thresholds in transverse momentum, p_T, as a function of p_T reconstructed offline, as measured in the ATLAS detector at a centre-of-mass energy of 7 TeV. The trigger signatures marked with "L1" indicate the first-level hardware trigger, and those marked with "L2" are selected by the second-level software trigger, which uses regional detector information. **b** Efficiency of the LHCb hardware triggers for $B^+ \to J/\psi(\mu^+\mu^-)K^+$ decays as a function of the J/ψ p_T for single-muon and dimuon trigger signatures. Adapted from [31, 32]

3.7 Luminosity Measurement

The measurement of cross sections of physics processes requires the knowledge of the luminosity integrated over the data taking period (2.1.4). The instantaneous luminosity, $\mathcal{L}$, varies with time and needs to be measured in real-time during data-taking. At the LHC, two methods are applied to determine $\mathcal{L}$: either directly from the beam parameters and the beam profile, or by using a reference physics process with a theoretically well-known cross section calculable to high precision.

According to 2.1.4, the average instantaneous luminosity of each pair of colliding particles in the LHC is proportional to

$$\mathcal{L} \propto \frac{W}{2\pi \cdot \sqrt{\sigma_{x1}^2 + \sigma_{x2}^2} \cdot \sqrt{\sigma_{y1}^2 + \sigma_{y2}^2}},$$

where W depends on the Gaussian widths, $\sigma_{x1}, \sigma_{x2}, \sigma_{y1}, \sigma_{y2}$, of the transverse profiles of the two beams. This expression can be easily derived from the overlap integral Ω of the transverse particle densities, $\rho_1(x, y)$ and $\rho_2(x, y)$:

$$\mathcal{L} \propto \Omega(\rho_1, \rho_2) = \int \rho_1(x, y)\rho_2(x, y)\, \mathrm{d}x\, \mathrm{d}y\,.$$

All other parameters, like the beam currents and the number of filled bunches, are determined using beam-monitoring devices [33–36].

The overlap integral $\Omega(\rho_1, \rho_2)$ is measured at the interaction point of the LHC experiments using the van der Meer method [37]. This method exploits the fact that the luminosity varies when one of the beams is displaced, e.g. in x and y direction by amounts b_x and b_y, respectively. The normalisation of the beam particle densities ensures that

$$\int \rho_1(x, y)\rho_2(x + b_x, y + b_y)\, \mathrm{d}x\, \mathrm{d}y\, \mathrm{d}b_x\, \mathrm{d}b_y = 1$$

when neglecting variations in z and in time t. A direct consequence is that only a particle rate $R(b_x, b_y) \propto \mathcal{L}(b_x, b_y)$ needs to be measured in a given detector to determine the overlap integral which can be expressed as

$$\Omega(\rho_1, \rho_2) = \frac{R(b_x = 0, b_y = 0)}{\int R(b_x, b_y)\, \mathrm{d}b_x\, \mathrm{d}b_y}.$$

For uncorrelated Gaussian beam profiles in x and y direction, one obtains 2.1.4. However, the overlap integral is defined in a more general way and also applies to non-Gaussian beam profiles, for which the parameters $\sigma_{x1}, \sigma_{x2}, \sigma_{y1}, \sigma_{y2}$ can be understood as effective beam widths. Once the beam profile is known, the remaining beam parameters measured in beam-monitoring devices are used to calculate the luminosity according to 2.1.4.

The sub-detectors used for the particle-rate measurement primarily are the LUCID detector [1] in ATLAS [33], the hadronic forward (HF) calorimeter and the pixel-tracking detector in CMS [34, 38], the VELO tracking detector in LHCb [35], and two scintillator and Cherenkov detector arrays, V0 and T0 [3], in ALICE [36]. Figure 3.24a shows an example of the particle-rate profiles determined during a van der Meer scan by the LHCb VELO detector [35].

Detailed studies of systematic uncertainties on the luminosity measurements are carried out by all collaborations. These studies investigate in great detail e.g. the length-scale calibration of the beam-displacement measurement, as shown in Fig. 3.24b, the fit procedure used to extract the profile peak and integral, bunch-to-bunch variations, and beam-position jitter. For the determination of the currents of proton and ion beams, the charge distribution within the bunches is determined, taking into account so-called ghost and satellite charges in bunches which are nominally not filled (Chap. 2). Statistical uncertainties in the rate measurements are negligible. The luminosity precision reached by the LHC experiments varies for the different data-taking periods and depends on the types of the colliding particles. When applying the van der Meer method for luminosity determination, the LHCb detector reaches the best precision of all LHC experiments because of the acceptance range of the VELO detector, which is positioned close to the beam directions. A 1.4 % relative precision is obtained by LHCb with the van der Meer method [35], in agreement with an alternative beam-gas interaction procedure to determine the beam profile. ATLAS, CMS and ALICE obtain slightly less precise relative luminosity uncertainties between 1.8 and 3.5 % [33, 34, 36, 38, 39], for colliding proton-proton, proton-ion, and ion-ion beams.

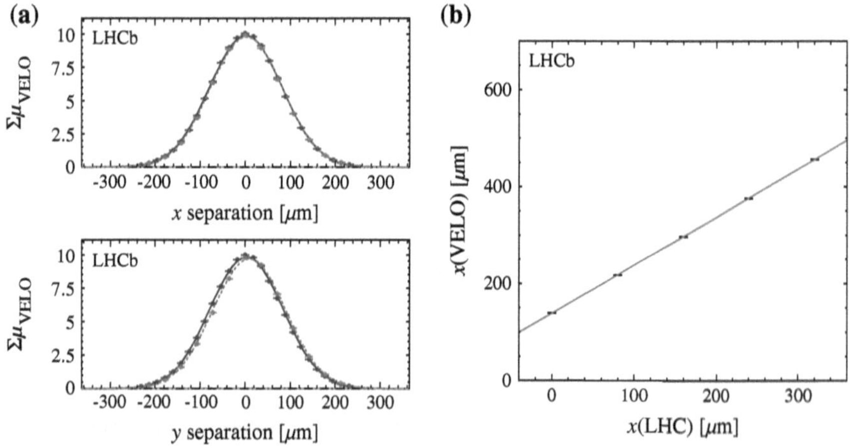

Fig. 3.24 **a** Mean number of interactions per bunch-crossing, μ_{VELO}, summed over all colliding bunches, measured by the LHCb VELO detector during the van der Meer scan of the beam position in x and y directions. The profiles are fitted with a Gaussian to determine peak position and integral. **b** Linear calibration curve of the length scale measured by the LHCb VELO detector with respect to the LHC beam displacement. Adapted from [35]

3.8 Conclusions and Outlook

The hard work of thousands of physicists and engineers to design, construct and commission the main LHC detectors ATLAS, CMS, LHCb and ALICE is already paying off, even though we are only at the beginning of the adventure to run such a complex accelerator and its experiments. The wealth of physics results measured with these tools will be discussed in the following chapters. But this does not mean that the detector designers are out of job: detector upgrades will be needed in the next decades. The LHC luminosity is planned to be increased in steps up to $7 \times 10^{34}\,\mathrm{cm}^{-2}\,\mathrm{s}^{-1}$ and the total luminosity accumulated is foreseen to reach $3000\,\mathrm{fb}^{-1}$ by 2035. Increased radiation levels and doses will therefore exceed the specifications of the current LHC detectors. The future LHC programme requires replacements and complements of the central tracking detectors, of parts of the calorimetry, of the muon spectrometers and of the read-out and trigger electronics. A full R&D programme is already in place in order to keep and improve the performance of all LHC detectors for the upcoming LHC runs at highest luminosities.

References

1. ATLAS Collaboration, JINST **3**, S08003 (2008)
2. CMS Collaboration, JINST **3**, S08004 (2008)
3. ALICE Collaboration, JINST **3**, S08002 (2008)
4. LHCb Collaboration, JINST **3**, S08005 (2008)
5. ALICE Collaboration, Phys. Rev. Lett. **106**, 032301 (2011)
6. M. Cacciari, G.P. Salam, G. Soyez, JHEP **04**, 063 (2008)
7. W.R. Leo, *Techniques for Nuclear and Particle Physics Experiments*, 2nd revised edn. (Springer, Berlin, 1994)
8. R. Wunstorf, A Systematic investigation of the radiation hardness of silicon detectors for high-energy physics experiments, Ph.D. thesis, Hamburg University, DESY-FH1K-92-01 (1992)
9. M. Moll, Radiation damage in silicon particle detectors: Microscopic defects and macroscopic properties, Ph.D. thesis, Hamburg University, DESY-THESIS-1999-040 (1999)
10. ATLAS Collaboration, JINST **9**(10), P08009 (2014)
11. LHCb VELO Group, JINST **9**, 09007 (2014)
12. ATLAS Collaboration, New J. Phys. **13**, 053033 (2011)
13. CMS Collaboration, JINST **9**(10), P10009 (2014)
14. LHCb Collaboration, JINST **10**, P02007 (2015)
15. ALICE Collaboration, PoS Vertex 2012, 048 (2013)
16. ATLAS Collaboration, J. Phys. Conf. Ser. **219**, 032023 (2010)
17. CMS Collaboration, JHEP **10**, 132 (2011)
18. ATLAS Collaboration, Eur. Phys. J. C **74**, 3130 (2014)
19. C. Lippmann, Nucl. Instrum. Meth. A **666**, 148–172 (2012)
20. CMS Collaboration, JINST **8**, P09009 (2013)
21. CMS Collaboration, JINST **6**, P11002 (2011)
22. ATLAS Collaboration, Eur. Phys. J. C **72**, 1909 (2012)
23. CMS Collaboration, JINST **7**, P01001 (2012)
24. ATLAS Collaboration, Phys. Lett. B **719**, 242–260 (2013)
25. ATLAS Collaboration, Eur. Phys. J. C **72**, 1844 (2012)

26. G. Charpak et al., Nucl. Instrum. Meth. **62**, 262–268 (1968)
27. A. Alves Jr et al., JINST **8**, P02022 (2013)
28. M. Fontana (2014), arXiv:1406.3437
29. ATLAS Collaboration, JHEP **09**, 056 (2010)
30. LHCb RICH Group, Eur. Phys. J. C **73**, 2431 (2013)
31. ATLAS Collaboration, Eur. Phys. J. C **72**, 1849 (2012)
32. R. Aaij et al., JINST **8**, P04022 (2013)
33. ATLAS Collaboration, Eur. Phys. J. C **73**, 2518 (2013)
34. CMS Collaboration, Luminosity Calibration for the 2013 Proton-Lead and Proton-Proton Data Taking. CMS-PAS-LUM-13-002, 2013
35. LHCb Collaboration. JINST **7**, P01010 (2012)
36. ALICE Collaboration, Eur. Phys. J. C **73**, 2456 (2013)
37. S. van der Meer, Calibration of the Effective Beam Height in the ISR. CERN-ISR-PO-68-31, 1968
38. CMS Collaboration, CMS Luminosity Based on Pixel Cluster Counting—Summer 2013 Update. CMS-PAS-LUM-13-001, 2013
39. ALICE Collaboration (2013), arXiv:1305.7044

Chapter 4
Electroweak Standard Model Physics

Maarten Boonekamp, Stefan Dittmaier and Matthias Mozer

Abstract In its first running period, the LHC has produced more W bosons than any accelerator before. This makes the LHC experiments hotbeds of electroweak physics. Although the ultimate precision in the measurement of electroweak parameters in many cases is still held by LEP, SLC or the Tevatron, the LHC has opened new views on measurements at high scales, especially in the field of multi-boson couplings. Furthermore, at the LHC the electroweak gauge bosons are used in a wide range of measurements that constrain the parton densities of the proton in ways that are complementary to previous results. The future promises further exciting results, among them W-boson mass measurements with unprecedented precision and the detailed study of vector-boson scattering.

4.1 Introduction

Precision measurements of electroweak (EW) parameters at the LHC are commonly limited in precision by the high pile-up and by the more complicated initial state compared to lepton colliders. Nevertheless, measurements of EW parameters are still of interest, especially for parameters that have a scale dependence because the high centre-of-mass energy of the LHC allows for studying previously inaccessible values of the scale. The measurements will typically have considerable uncertainties due to the underlying proton parton distribution functions (PDFs) and to other aspects of

M. Boonekamp (✉)
CEA, IRFU, 91191 Gif-sur-Yvette Cedex, France
e-mail: maarten.boonekamp@cea.fr

S. Dittmaier
Albert-Ludwigs-Universität Freiburg,
Hermann-Herder-Str. 3, 79104 Freiburg, Germany
e-mail: stefan.dittmaier@physik.uni-freiburg.de

M. Mozer
Institut für Experimentelle Kernphysik, KIT,
Wolfgang-Gaede-Str. 1, 76131 Karlsruhe, Germany
e-mail: Matthias.Mozer@kit.edu

T. Schörner-Sadenius (ed.), *The Large Hadron Collider*,
DOI 10.1007/978-3-319-15001-7_4

QCD. While QCD effects may therefore limit the access to the EW parameters, it is possible to invert this relation and measure the underlying QCD dynamics under the assumption of the SM EW parameters. In this case the EW bosons serve as a probe to study the proton structure and related QCD effects. Measurements that have been interpreted in this approach are studies of the total and differential cross sections of W and Z bosons as well as the production of W/Z bosons in association with jets (see Sect. 4.8). In addition, the large centre-of-mass energy of the LHC gives access to multi-boson production, which had limited cross sections at previous colliders. Measurements of multi-boson production provide direct access to the triple and quartic vertices of the Standard Model (SM). Physics beyond the SM may lead to modifications in these couplings, which are searched for in terms of anomalous couplings. No hints for anomalous couplings have so far been found, but thanks to the large centre-of-mass energy of the LHC compared to previous colliders, current LHC limits on anomalous couplings surpass previous limits, and further improvements are expected from the increase of the LHC's beam energy.

4.2 The Standard Model Lagrangian

Before we start a discussion of the phenomenology of electroweak processes at the LHC, we briefly review the general structure of the SM. In the following we define the full SM Lagrangian in a uniform way, including both electroweak and strong interactions, in order to also set the stage for the subsequent sections of this book. The description closely follows the review [1]; more details can be found in standard textbooks [2–6].

4.2.1 The Gauge Structure of the Standard Model

The Lagrangian $\mathcal{L}_{\mathrm{SM}}$ of the SM can be divided into four different parts,

$$\mathcal{L}_{\mathrm{SM}} = \mathcal{L}_{\mathrm{YM}} + \mathcal{L}_{\mathrm{ferm}} + \mathcal{L}_{\mathrm{H}} + \mathcal{L}_{\mathrm{Yuk}}\,,$$

which will be discussed one by one in the following. We start with the first two, which contain the dynamics of the gauge bosons and matter fermions as predicted by the unbroken gauge symmetry, before we turn to the latter two which involve the Higgs field.

The Yang–Mills part $\mathcal{L}_{\mathrm{YM}}$ describes the genuine dynamics of the gauge fields, i.e. their free propagation as well as their self-interactions,

$$\mathcal{L}_{\mathrm{YM}} = -\frac{1}{4} W^i_{\mu\nu} W^{i,\mu\nu} - \frac{1}{4} B_{\mu\nu} B^{\mu\nu} - \frac{1}{4} G^a_{\mu\nu} G^{a,\mu\nu}\,, \tag{4.1}$$

where

$$W^i_{\mu\nu} = \partial_\mu W^i_\nu - \partial_\nu W^i_\mu - g\epsilon^{ijk} W^j_\mu W^k_\nu, \qquad i, j, k = 1, 2, 3,$$

$$B_{\mu\nu} = \partial_\mu B_\nu - \partial_\nu B_\mu,$$

$$G^a_{\mu\nu} = \partial_\mu G^a_\nu - \partial_\nu G^a_\mu - g_s f^{abc} G^b_\mu G^c_\nu, \qquad a, b, c = 1, \dots, 8,$$

are the field-strength tensors of the gauge fields W^i_μ for the $SU(2)_I$ group of the weak isospin I^i_w, B_μ for the $U(1)_Y$ of weak hypercharge Y_w, and G^a_μ for the $SU(3)_c$ of colour. The respective gauge couplings of these groups are denoted g, g', and g_s, and the structure constants of the non-Abelian groups SU(2) and SU(3) are ϵ^{ijk} and f^{abc}, following the usual notation. The $\mathcal{L}_{YM}$ of (4.1) contains gauge-boson self-interactions between three and four gauge bosons, which carry the information about the underlying gauge group. The experimental studies of these triple and quartic interactions among the EW gauge bosons γ, Z, and $W^\pm$ will be discussed in some detail in later sections.

The interaction of the gauge fields with the fermions is encoded in

$$\mathcal{L}_{\text{ferm}} = \mathrm{i}\overline{\Psi}_L \not{D}\Psi_L + \mathrm{i}\overline{\psi}_{\ell_R} \not{D}\psi_{\ell_R} + \mathrm{i}\overline{\Psi}_Q \not{D}\Psi_Q + \mathrm{i}\overline{\psi}_{u_R} \not{D}\psi_{u_R} + \mathrm{i}\overline{\psi}_{d_R} \not{D}\psi_{d_R},$$

where $L = (\nu_{\ell_L}, \ell_L)^T$ are the left-handed $SU(2)_I$ doublets of charged leptons $\ell = e, \mu, \tau$ and neutrinos $\nu_\ell = \nu_e, \nu_\mu, \nu_\tau$, $Q = (u_L, d_L)^T$ the left-handed $SU(2)_I$ doublets of up-type quarks $u = u, c, t$ and down-type quarks $d = d, s, b$, and ℓ_R, u_R, d_R are the respective right-handed $SU(2)_I$ singlets.[1] The actual interaction is contained in the covariant derivative

$$D_\mu = \partial_\mu + \mathrm{i}g I^i_w W^i_\mu + \mathrm{i}g' Y_w B_\mu + \mathrm{i}g_s T^a_c G^a_\mu, \qquad (4.2)$$

where I^i_w, Y_w, and T^a_c are the generators of the respective gauge groups in the representation of the fermions they act on. In detail, $I^i_w = \sigma^i/2$ (with the Pauli matrices σ^i) for the left-handed $SU(2)_I$ doublets and $I^i_w = 0$ for the right-handed singlets, the weak hypercharge Y_w (= numbers) is related to the relative electric charge Q by the Gell-Mann–Nishijima relation $Q = I^3_w + Y_w/2$, and $T^a_c = \lambda^a/2$ (λ^a = Gell-Mann matrices) for $SU(3)_c$ quark triplets and $T^a_c = 0$ for the leptons. The requirement that the coupling structure of the photon be parity-blind and proportional to $Q\overline{\psi} \not{A}\psi$, as in quantum electrodynamics, identifies the photon field A_μ and the Z-boson field Z_μ as linear combinations of W^3_μ and B_μ obtained by the rotation

$$\begin{pmatrix} Z_\mu \\ A_\mu \end{pmatrix} = \begin{pmatrix} c_w & -s_w \\ s_w & c_w \end{pmatrix} \begin{pmatrix} W^3_\mu \\ B_\mu \end{pmatrix}$$

[1] Right-handed neutrinos are omitted, since they do not play any role at high-energy colliders.

with the weak mixing angle θ_w and electric unit charge e fixed by

$$\cos\theta_\mathrm{w} = c_\mathrm{w} = \sqrt{1-s_\mathrm{w}^2} = \frac{g}{\sqrt{g^2+(g')^2}}, \qquad e = \frac{gg'}{\sqrt{g^2+(g')^2}}.$$

Once the photon is identified among the gauge fields, it is easy to see that the fields

$$W_\mu^\pm = (W_\mu^1 \mp \mathrm{i} W_\mu^2)/\sqrt{2}$$

correspond to the charged weak gauge bosons $W^\pm$ of charge $\pm e$.

Note that the part of the SM described so far—encoded in $\mathcal{L}_\mathrm{YM} + \mathcal{L}_\mathrm{ferm}$—does not involve any mass terms. Naive gauge-boson mass terms, such as $W_\mu^i W^{i,\mu}$ for the W bosons, obviously violate gauge invariance. Since left-handed and right-handed fermions transform differently under $\mathrm{SU}(2)_\mathrm{I} \times \mathrm{U}(1)_\mathrm{Y}$ gauge transformations, naive fermion-mass terms $\propto (\overline{\psi}_{f\mathrm{L}}\psi_{f\mathrm{R}} + \overline{\psi}_{f\mathrm{R}}\psi_{f\mathrm{L}})$ for a fermion f are also ruled out by gauge invariance. Introducing particle masses without destroying the gauge invariance of the dynamics of the model requires extensions of the theory.

4.2.2 Electroweak Symmetry Breaking and the Higgs Mechanism

The Higgs part

$$\mathcal{L}_\mathrm{H} = (D_\mu \Phi)^\dagger (D^\mu \Phi) - V(\Phi) \tag{4.3}$$

of the Lagrangian extends the previously introduced particle content by the complex scalar $\mathrm{SU}(2)_\mathrm{I}$ doublet $\Phi = (\phi^+, \phi^0)^\mathrm{T}$ of weak hypercharge $Y_{\mathrm{w},\Phi} = 1$, so that ϕ^+ carries charge $+e$ and ϕ^0 is neutral. In total, Φ involves four real degrees of freedom. The self-interaction of Φ is described by the potential

$$V(\Phi) = -\mu^2(\Phi^\dagger\Phi) + \frac{\lambda}{4}(\Phi^\dagger\Phi)^2,$$

the form of which is constrained by the requirements of gauge invariance and renormalisability of the model. The latter is guaranteed by the polynomial structure of degree four in the field components, the former by the Φ dependence via the combination $\Phi^\dagger\Phi$, leaving only the two real free parameters μ^2 and λ in V. While $\lambda > 0$ is required by vacuum stability, the sign of μ^2 is deliberately taken positive in order to enforce a non-vanishing vacuum expectation value (vev) Φ_0 of Φ. Minimising V yields the condition

$$\Phi_0^\dagger\Phi_0 = \frac{v^2}{2}, \qquad v = 2\sqrt{\frac{\mu^2}{\lambda}}.$$

The requirement that the vev be electrically neutral forces the upper component of Φ_0 to vanish, i.e. Φ_0 is fixed up to a phase, with the usual choice $\Phi_0 = (0, v/\sqrt{2})^{\mathrm{T}}$. This freedom in choosing the vev Φ_0 of Φ reflects the spontaneous breakdown of the $\mathrm{SU(2)_I} \times \mathrm{U(1)_Y}$ symmetry down to the remaining electromagnetic $\mathrm{U(1)_{em}}$ invariance. Splitting off the vev from Φ,

$$\Phi = \begin{pmatrix} \phi^+ \\ \phi^0 = (v + H + \mathrm{i}\chi)/\sqrt{2} \end{pmatrix}, \tag{4.4}$$

we reparametrise Φ in terms of the real physical Higgs field H and the unphysical would-be Goldstone boson fields ϕ^+ and χ, which are complex and real, respectively. That ϕ^+ and χ do not correspond to physical states can already be seen by the fact that they are connected to the vev by gauge transformations; in fact one can always find a gauge, known as the "unitary gauge", in which ϕ^+ and χ vanish. Making use of this gauge and inserting the parametrisation (4.4) of Φ and the covariant derivative (4.2) with $I^i_{\mathrm{w},\Phi} = \sigma^i/2$, $Y_{\mathrm{w},\Phi} = 1$, $T^a_{\mathrm{c},\Phi} = 0$ into the Higgs Lagrangian (4.3), we find

$$\begin{aligned} \mathcal{L}_{\mathrm{H,U\text{-}gauge}} &= \frac{1}{2}(\partial H)^2 + \frac{g^2}{4}(v+H)^2 W^+_\mu W^{-,\mu} + \frac{g^2}{8c_{\mathrm{w}}^2}(v+H)^2 Z_\mu Z^\mu \\ &\quad + \frac{\mu^2}{2}(v+H)^2 - \frac{\lambda}{16}(v+H)^4, \end{aligned}$$

which in particular contains bilinear terms in the gauge fields $W^\pm$, Z and in the Higgs field H, i.e. mass terms for the corresponding weak gauge bosons $W^\pm$ and Z as well as for the Higgs boson H. Identifying these masses according to

$$M_W = \frac{gv}{2}, \qquad M_Z = \frac{M_W}{c_{\mathrm{w}}}, \qquad M_H = \sqrt{2\mu^2},$$

we can eliminate the parameters μ^2, λ, v completely and get

$$\begin{aligned} \mathcal{L}_{\mathrm{H,U\text{-}gauge}} &= \frac{1}{2}(\partial H)^2 - \frac{1}{2}M_H^2 H^2 + M_W^2 W^+_\mu W^{-,\mu} + \frac{1}{2}M_Z^2 Z_\mu Z^\mu \\ &\quad + gM_W H W^+_\mu W^{-,\mu} + \frac{g^2}{4} H^2 W^+_\mu W^{-,\mu} \\ &\quad + \frac{gM_Z}{2c_{\mathrm{w}}} H Z_\mu Z^\mu + \frac{g^2}{4c_{\mathrm{w}}^2} H^2 Z_\mu Z^\mu \\ &\quad - \frac{gM_H^2}{4M_W} H^3 - \frac{g^2 M_H^2}{32 M_W^2} H^4 + \text{const.}, \end{aligned}$$

where we have not spelled out an irrelevant constant. In summary, the SM makes the following important phenomenological predictions in the Higgs sector which can be tested experimentally:

- Associated to the Higgs field H, a physical neutral, spinless particle of mass M_H is postulated—the Higgs boson. Since it can be viewed as some kind of vacuum excitation, it carries the quantum numbers of the vacuum and is thus even with respect to *CP* symmetry. Note that M_H is the only free SM parameter that is tied to a property of the Higgs boson, while the other parameters are fixed by the masses of the weak gauge boson masses and the gauge couplings.
- The ratio $\rho = M_W^2/(c_w^2 M_Z^2)$ [7][2] is equal to one, which is a non-trivial relation among the masses of the weak bosons and the gauge couplings $g = e/s_w$ and $g' = e/c_w$. While $\rho = 1$ in this form is used to define $c_w = M_W/M_Z$ in the process of renormalisation to all orders, it nevertheless has important phenomenological consequences. For instance, on top of the Z-boson resonance, as measured at LEP and SLD via $e^+e^- \to Z \to f\bar{f}$, an effective weak mixing angle θ_{eff}^f can be defined for each fermion species f, usually quantified via $\sin^2\theta_{\mathrm{eff}}^f$, which can be measured from various asymmetries. The ratios $\rho_f = M_W^2/(\cos^2\theta_{\mathrm{eff}}^f M_Z^2)$ thus are predicted to be equal to one up to radiative corrections, a fact that is experimentally confirmed at a level of better than 10^{-3} [8].
  The lowest-order property $\rho = 1$ is not shared by all possible scalar sectors that can be employed to lend masses to the $W^\pm$ and Z bosons. If not accidental or forced by fine-tuned parameters, it is rather a consequence of a "custodial symmetry" [9] of the scalar sector. In the SM this is an SO(4) symmetry of the special form of $\mathcal{L}_{\mathrm{H}}$ with respect to the exchange of the four real components of Φ, which holds up to $\mathrm{U(1)_Y}$ gauge interactions and differences in the fermion masses within $\mathrm{SU(2)_I}$ doublets, both inducing custodial-symmetry-breaking effects, however, only in higher orders.
- The model predicts couplings of the Higgs boson to a massive weak gauge boson $V = W, Z$ proportional to M_V^2/v, because the $HV^\dagger V$ couplings originate from the factor $(1 + H/v)^2$ multiplying the gauge-boson mass term in $\mathcal{L}_{\mathrm{H}}$. Owing to the square in this factor also quartic couplings of two Higgs bosons and two gauge bosons are predicted, which are proportional to M_V^2/v^2.
- Finally, triple and quartic Higgs-boson self-interactions are predicted, both scaling with M_H^2. Since these couplings are in one-to-one correspondence with the shape of the Higgs potential that drives EW symmetry breaking, an experimental reconstruction of these couplings from an analysis of scattering processes would be part of an ultimate phenomenological confirmation of the Higgs mechanism. However, since the relevant processes involve multi-Higgs-boson final states with very low cross sections, the LHC will at best be able to give qualitative results here. A future high-energy e^+e^- collider should make more precise studies possible.

[2] The ratio ρ is called β in [7].

4.2.3 Yukawa Couplings and Fermion Masses

Renormalisability and gauge invariance of the SM Lagrangian allow for so-called Yukawa couplings of the Higgs doublet Φ to all fermions. The most general form of these interactions is

$$\mathcal{L}_{\mathrm{Yuk}} = -\overline{\Psi}_L G_\ell \psi_{\ell_{\mathrm{R}}} \Phi - \overline{\Psi}_Q G_u \psi_{u_{\mathrm{R}}} \tilde{\Phi} - \overline{\Psi}_Q G_d \psi_{d_{\mathrm{R}}} \Phi + \mathrm{h.c.},$$

where "h.c." means hermitian conjugate and $\tilde{\Phi} = \mathrm{i}\sigma^2 \Phi^* = ((\phi^0)^*, -\phi^-)^{\mathrm{T}}$ denotes the charge-conjugate Higgs doublet with quantum numbers opposite to Φ. The matrices G_f $(f = \ell, u, d)$ represent arbitrary complex 3×3 matrices, i.e. at first sight $\mathcal{L}_{\mathrm{Yuk}}$ involves a large number of free parameters. However, most of them turn out to be not physically relevant and can be transformed to canonical values or eliminated by appropriate field redefinitions. We first note that each term in $\mathcal{L}_{\mathrm{Yuk}}$ involves terms that are bilinear in the fermion fields, because Φ and $\tilde{\Phi}$ contain a constant piece in the form of the vev v. More precisely, the non-diagonal elements of G_f mix the left- and right-handed parts of the different generations of fermion type $f = \ell, u, d$, where ℓ generically stands for charged leptons, u for up-type quarks, and d for down-type quarks. Owing to these mixing terms, a fermion of flavour f_i $(i = 1, 2, 3)$ of the ith generation oscillates into f_j of the other generations $(j \neq i)$ even during a free propagation in space and time. This oscillation can be removed upon transforming the existing "flavour basis" $(\psi_{f_{\tau,1}}, \psi_{f_{\tau,2}}, \psi_{f_{\tau,3}})$ of left-handed and right-handed fields $(\tau = \mathrm{L}, \mathrm{R})$ into a "mass basis" $(\hat{\psi}_{f_{\tau,1}}, \hat{\psi}_{f_{\tau,2}}, \hat{\psi}_{f_{\tau,3}})$ with a unitary matrix U,

$$\hat{\psi}_{f_{\tau,i}} = U_{ij}^{f_\tau} \psi_{f_{\tau,j}}, \qquad f = \ell, u, d, \qquad \tau = \mathrm{L}, \mathrm{R},$$

where in this process the matrices G_f receive a diagonal form,

$$U^{f_{\mathrm{L}}} G_f (U^{f_{\mathrm{R}}})^\dagger = \frac{\sqrt{2}}{v} \operatorname{diag}\{m_{f_1}, m_{f_2}, m_{f_3}\}. \tag{4.5}$$

The diagonal value m_{f_i}, which can be chosen non-negative by convention, is the mass of the fermion f_i. For completeness we mention that the left-handed neutrino fields are transformed with the same unitary matrix as their charged counterparts; this is possible as long as we work with mass-degenerate (such as massless) neutrinos. The effect of this field redefinition on the whole SM Lagrangian can easily be summarised: (i) The coupling matrices G_f are replaced by their diagonal form (4.5); (ii) all fermion fields $\psi_{f_{\tau,i}}$ are replaced by their counterparts $\hat{\psi}_{f_{\tau,i}}$ of the mass basis; (iii) the only remnant of the U matrices is the appearance of the Cabibbo–Kobayashi–Maskawa matrix $V = U^{u_{\mathrm{L}}} (U^{d_{\mathrm{L}}})^\dagger$ in fermion chains of type $\overline{\hat{\psi}}_{u_{\mathrm{L}}} \ldots V \hat{\psi}_{d_{\mathrm{L}}}$ and of $V^\dagger$ in $\overline{\hat{\psi}}_{d_{\mathrm{L}}} \ldots V^\dagger \hat{\psi}_{u_{\mathrm{L}}}$. In other words, only charged-current interactions receive modifications by V, while neutral currents remain unchanged. In the following we adopt the common convention to omit the clumsy hats on fermionic fields and assume the use of the mass basis.

In the unitary gauge, where the would-be Goldstone fields are absent, the Yukawa Lagrangian takes the simple final form

$$\mathcal{L}_{\text{Yuk,U-gauge}} = -\sum_f m_f\,(\overline{\psi}_{f\text{L}}\psi_{f\text{R}} + \overline{\psi}_{f\text{R}}\psi_{f\text{L}})\left(1 + \frac{H}{v}\right),$$

where the sum over f runs over all fermion flavours of all generations. This form shows a distinctive footprint of the Higgs mechanism in the fermionic sector: The Higgs boson couples to each fermion f of mass m_f with the strength $y_f = m_f/v$. Moreover, the coupling is the one of a pure scalar, i.e. the coupling to fermions does not have any pseudo-scalar admixture proportional to γ_5. Testing these features offers a possibility to empirically tell the observed Higgs state from scalar particles predicted by other models. Alternative models with non-minimal Higgs sectors often predict new pseudoscalars as well, or even scalar particles without definite *CP* quantum numbers. Moreover, the strict proportionality of the Yukawa coupling strength to the fermion masses might be broken, as it is for instance the case in (type-II) Higgs-doublet models, where the proportionality factor between y_f and m_f is different for up- and down-type fermions.

4.2.4 The Input Parameters of the Standard Model

As it is obvious from the construction of the model, the free input parameters of the SM are the gauge couplings g, g', g_s, the parameters μ^2 and λ of the Higgs sector, the fermion masses m_f, and the CKM matrix V. For phenomenology it is much more convenient to take instead the following parameters as input: the electromagnetic coupling $\alpha = e^2/(4\pi)$, the strong couplings constant $\alpha_s = g_s^2/(4\pi)$, the weak gauge-boson masses M_W and M_Z, the Higgs-boson mass M_H, and finally m_f and V. The masses can all be defined as "pole masses", defined from the locations of the particle poles in the respective propagators. For the heavy quarks, however, it is often useful to switch to a running mass at some appropriate scale. The couplings α and α_s have to be chosen thoughtfully. The strong coupling is usually defined as running coupling $\alpha_s(\mu_R)$ in the so-called $\overline{\text{MS}}$ scheme. Very often the value of α_s at the Z pole, $\alpha_s(M_Z)$, is used as numerical input and transferred to some other value μ_R using the renormalisation group equation.

For the electromagnetic coupling α, the choice is basically between three different values: the fine-structure constant $\alpha(0) \approx 1/137$, the effective value $\alpha(M_Z) \approx 1/129$, where $\alpha(0)$ is evolved via renormalisation group equations from zero-momentum transfer to the Z pole, and an effective value derived from the Fermi constant G_F leading to $\alpha_{G_\text{F}} = \sqrt{2}G_\text{F}M_W^2(1 - M_W^2/M_Z^2)/\pi \approx 1/132$, defining the so-called "G_F scheme". The various values of α differ by 2–6 %. Very often the actual value of α can be adjusted in such a way that large universal EW corrections are already absorbed into the lowest-order prediction. Even different values of α can be appropriate in one calculation, but care has to be taken that the same choice is taken within gauge-invariant subsets of diagrams. An external photon (virtuality $Q^2 = 0$) always effectively couples with $\alpha(0)$, while internal photons with virtuality Q^2 at a high energy scale effectively couple with $\alpha(Q)$, which is much closer to $\alpha(M_Z)$. On the other hand, the couplings of W and Z bosons should be parametrised with α_{G_F}, which does not only take into account the running from $Q = 0$ to the EW scale, but also universal effects from the ρ parameter. Following these rules, in particular, avoids perturbative instabilities due to the appearance of light-quark masses, as e.g. discussed in [10, 11].

4.3 Higher-Order Electroweak Effects

In the coming years, the LHC will explore phenomena at the TeV scale using its design centre-of-mass energy of $\sqrt{s} = 13{-}14\,\mathrm{TeV}$ with continuously increasingly luminosity, so that numerous processes can be investigated at the accuracy level of several percent. This perspective, and already several measurements at 7 or 8 TeV in Run 1, are calling for theoretical predictions at the level of few percent, i.e. both QCD and EW radiative corrections have to be taken into account in predictions beyond the level of crude approximations. In general, QCD corrections are substantial in predictions for hadronic collisions; their structure and general features are discussed in Sect. 4.8 in more detail. In the following, we provide a brief survey of salient features of EW higher-order corrections, which are typically less pronounced than QCD corrections. Particular attention is paid to those EW effects that can enhance next-to-leading order (NLO) EW corrections (expected to be of the small size α/π) to a size that can compete with or even exceed the size of QCD corrections in specific situations (e.g. final-state radiation, EW corrections at high energies). For more details and a discussion of some subtle aspects in the calculation of EW corrections, such as photon-jet separation or the treatment of unstable particles, we refer to the "EW dictionary" of [11] and the original literature.

4.3.1 Electroweak Corrections at High Energies

At high energies, where scattering processes involve large scales $Q^2 \gg M_W^2$, EW corrections develop large logarithmic contributions such as $(\alpha/s_\mathrm{w}^2)\ln^2(Q^2/M_W^2)$ and $(\alpha/s_\mathrm{w}^2)\ln(Q^2/M_W^2)$ at NLO, and powers of these beyond NLO. These mass-singular corrections originate from soft and/or collinear exchange of EW gauge bosons in loop diagrams, as illustrated in Fig. 4.1.

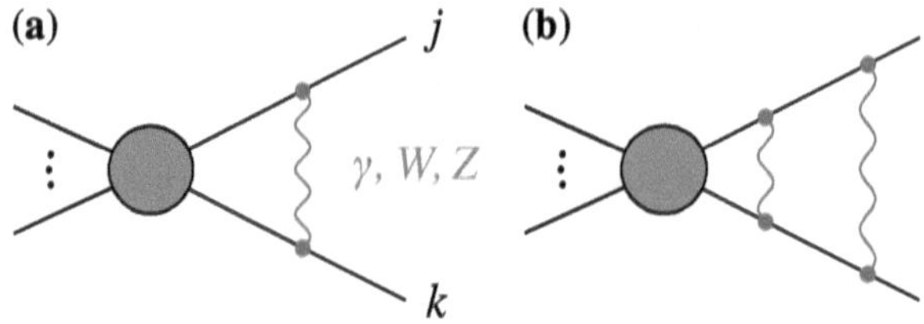

Fig. 4.1 Diagrammatic illustration of soft/collinear EW gauge-boson exchange at high scattering energies

In the TeV energy range, the corresponding logarithms grow to ten percent or more, i.e. EW corrections become very significant at high energies.

The kinematic regime in which such EW corrections are most pronounced is characterised by the situation that all invariants $s_{ij} = 2k_i \cdot k_j$ for pairs of particles' four-momenta $k_{i,j}$ become large ($s_{ij} \gg M_W^2$); it is known as "Sudakov regime". The structure of EW corrections in this domain has been investigated in detail at $\mathcal{O}(\alpha)$ and beyond by several groups (see [12–22] and references therein). As described for example in [16, 19, 22], the leading EW logarithmic corrections, which are enhanced by large factors $L = \ln(s_{ij}/M_W^2)$, can be derived in an effectively SU(2) × U(1)-symmetric theory in which W bosons, Z bosons and photons have a common mass M_W. The leading EW Sudakov corrections $\propto (\alpha L^2)^n$ are obtained to all orders from the respective NLO result via exponentiation.

While the high-energy structure of EW corrections was studied in the literature in great detail for the Sudakov regime, there is only little knowledge on EW corrections beyond NLO in more general kinematical situations where not all invariants s_{ij} are large. Note that there are many cross sections that are in fact not dominated by the Sudakov regime in the high-energy limit, including all processes that are dominated by t-channel diagrams. For example, unless specifically designed cuts are applied, reactions like W-boson pair production via e^+e^-, pp or $\gamma\gamma$ collisions are dominated by the "Regge limit", where the Mandelstam variable t remains small while s gets large. Moreover, it often depends on the specific observable which regime is probed in high-energy tails of kinematical distributions. Taking Drell–Yan processes (see e.g. [10, 23, 24]) and dijet production [25] at the LHC as examples, differential distributions in the transverse momenta of the produced leptons or jets probe the Sudakov regime in the high-momentum tails. On the other hand, the distributions in the invariant masses of leptons or jets of these processes are not dominated by this regime at high scales, so that the EW high-energy logarithms derived in the Sudakov regime do not approximate the EW corrections well in these observables. It is, thus, advisable to make use of full NLO EW corrections, i.e. without applying expansions for high energies, whenever possible for a given process.

Since the EW high-energy logarithmic corrections are associated with virtual soft and/or collinear weak-boson or photon exchange, they all have counterparts in real weak-boson or photon-emission processes which can partially cancel the large negative virtual corrections (but not completely, see [14]). This cancellation

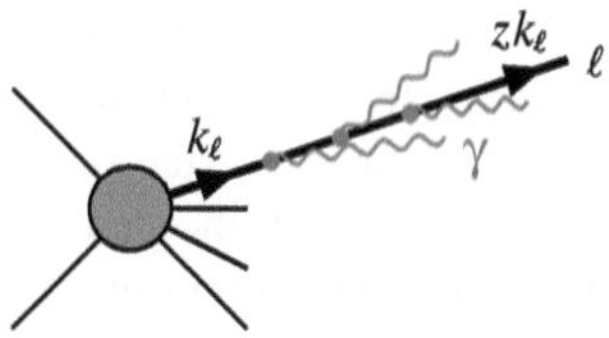

Fig. 4.2 Diagrammatic illustration of collinear FSR off a lepton ℓ

will not be complete, since SU(2) doublets are in general not treated inclusively in EW corrections—a fact that is by some abuse of language called "Bloch–Nordsieck violation".[3] To which extent the cancellation occurs depends on the experimental capabilities to separate final states with or without weak bosons or photons. The general issue and specific examples have been discussed for example in [26–29].

4.3.2 Photonic Final-State Radiation Off Leptons

The emission of photons collinear to incoming or outgoing charged leptons ℓ leads to corrections that are enhanced by large logarithms of the form $\alpha^n \ln^n(m_\ell^2/Q^2)$ with Q denoting a characteristic scale of the process. For lepton colliders, such as a high-energy e^+e^- collider, this collinear initial-state radiation (ISR) leads to pronounced photonic corrections, which are particularly large whenever the underlying total or differential cross section shows strong variations (e.g. near resonances or thresholds). In the following we focus on final-state radiation (FSR) off leptons, a situation that is also relevant at hadron colliders. Pictorially this effect is illustrated in Fig. 4.2.

For both ISR and FSR, the logarithmically enhanced corrections are universal in the sense that they do not change the nature of the cross section of the underlying hard scattering process, but depend only on the type and kinematics of the incoming or outgoing charged particles. In contrast to ISR, FSR only influences the kinematics and acceptance of the outgoing particles. The first-order logarithm $\alpha \ln(m_\ell^2/Q^2)$ is, of course, contained in a full (process-dependent) NLO EW $\mathcal{O}(\alpha)$ correction, and likewise for higher orders, so that Q is unambiguously fixed in any completely calculated order. In a fixed perturbative order that is not completely taken into account, but where ISR or FSR is included in logarithmic accuracy, the ambiguity in the scale Q is part of the remaining theoretical uncertainty.

The universal logarithmic corrections can be evaluated in the so-called "structure-function approach" (see e.g. [30, 31]), or by photonic parton showers (see e.g. [32–34]). For FSR in a process in which a lepton ℓ with momentum k_ℓ is produced, the incorporation of the mass-singular logarithms takes the form of a convolution integral over the leading-order (LO) cross section σ^{LO},

$$\sigma_{\text{LLFSR}} = \int \mathrm{d}\sigma^{\text{LO}}(k_\ell) \int_0^1 \mathrm{d}z\, \Gamma_{\ell\ell}^{\text{LL}}(z, Q^2)\, \Theta_{\text{cut}}(zk_\ell)\,,$$

[3] The Bloch–Nordsieck theorem simply does not apply to non-Abelian gauge theories.

where $\Gamma_{\ell\ell}^{\mathrm{LL}}$ is the structure function describing the radiation in logarithmic accuracy. The variable $(1-z)$ is the momentum fraction of the respective lepton lost by collinear (single or multiple) photon emission, and the step function Θ_{cut} is equal to 1 if the event passes the cuts on the rescaled lepton momentum zk_ℓ, and 0 otherwise. The structure function is known to $\mathcal{O}(\alpha^5)$ in the literature, including the resummation of soft-photon effects. At NLO, i.e. in $\mathcal{O}(\alpha)$, the structure function has the well-known form

$$\Gamma_{\ell\ell}^{\mathrm{LL},1}(z, Q^2) = \frac{\beta_\ell}{4}\left(\frac{1+z^2}{1-z}\right)_+ \tag{4.6}$$

with the variable

$$\beta_\ell = \frac{2\alpha(0)Q_\ell^2}{\pi}\left[\ln\left(\frac{Q^2}{m_\ell^2}\right) - 1\right]$$

quantifying the large logarithm (Q_ℓ denotes the electric charge of the lepton ℓ).

For FSR the Kinoshita–Lee–Nauenberg (KLN) theorem [35, 36] guarantees that these logarithms cancel if collinear lepton-photon systems are treated fully inclusively, like in a total cross section, defined without any phase-space cuts. Such observables are called "collinear safe". In the presence of phase-space cuts and in differential cross sections, in general, mass-singular contributions survive, leading to enhanced radiation effects, since the necessary inclusiveness for their compensation is disturbed. At NLO, these features can be easily understood from the explicit analytic form shown in (4.6), where the integral over the plus distribution vanishes when taken over the full z range (i.e. $\Theta_{\mathrm{cut}}(zk_\ell) = 1$ for all z). For differential observables the level of inclusiveness necessary for collinear safety can be restored by a procedure known as "photon recombination", which treats collinear lepton-photon systems as one quasi-particle. This procedure is similar to the application of a jet algorithm in QCD. For final-state electrons, photon recombination automatically is involved in their reconstruction from electromagnetic showers detected in calorimeters. Muons, on the other hand, can be observed as "bare" leptons from their tracks in the muon chambers, but in order to reduce large FSR corrections, observed muons are sometimes also reconstructed as "dressed" leptons via photon recombination, as e.g. described in [37] for an ATLAS analysis. Working with dressed leptons, where mass-singular FSR effects cancel, has the advantage that the resulting cross section does not depend on the mass (and thus on the flavour) of the charged lepton, i.e. the reconstructed lepton looks universal (at least electrons and muons).

4.3.3 Photonic Corrections to the Initial State

The inclusion of NLO EW $\mathcal{O}(\alpha)$ corrections to hadronic cross sections conceptually proceeds along the same lines as the incorporation of NLO QCD corrections, with the slight generalisation that also the photon appears as a parton inside the proton. The $\mathcal{O}(\alpha)$-corrected parton cross sections contain mass singularities which are due to collinear photon radiation off the initial-state quarks or due to a collinear splitting $\gamma \to q\bar{q}$ for initial-state photons. In complete analogy to factorisation in NLO QCD calculations, these collinear singularities are absorbed into the quark and photon distributions. The explicit form of the redefinition of PDFs can be found e.g. in [24].

In complete analogy to the pure QCD case, a factorisation scale $\mu_{\text{fact,QED}}$ determines up to which scale particles resulting from collinear QED splitting processes in the initial state are considered to be part of the proton remnant. The absorption of the collinear singularities of $\mathcal{O}(\alpha)$ into PDFs requires the inclusion of the corresponding $\mathcal{O}(\alpha)$ corrections into the Dokshitzer–Gribov–Lipatov–Altarelli–Parisi (DGLAP) evolution of these distributions, which describe the dependence of the PDFs on the factorisation scales.

The inclusion of the photon in the set of partons inside hadrons leads to so-called "photon-induced processes", i.e. partonic channels with photons in the initial state, in addition to the partonic channels of QCD. At NLO EW level, contributions from photon-induced processes always appear as as crossed counterparts of photonic bremsstrahlung corrections. For instance, quark-initiated $qq, q\bar{q}, \bar{q}\bar{q}$ channels always receive (real) $\mathcal{O}(\alpha)$ corrections from $q\gamma$ and/or $q\bar{q}\gamma$ scattering, where the additional q or $q\bar{q}$ in the final state leads to an additional jet with respect to the LO signature, similar to real NLO QCD corrections. For specific final states with charged particles, but without net electric charge, there is also a contribution from $\gamma\gamma$ scattering with LO kinematics and without additional partons in the final state. This is, for instance, the case for $\mu^+\mu^-$ or W^+W^- production.

Since the photon PDF is smaller than the gluon PDF by about two orders of magnitude in the typical kinematical range of LHC processes, but similar in shape, γ-induced processes typically contribute only at the percent level to hadronic cross sections if the γ-induced channels have gluon-induced counterparts. More significant contributions from photon-induced processes can only arise if the photonic channels involve diagrams without QCD counterparts, i.e. diagrams where the photon couples to colour-neutral charged particles like muons or W bosons. Recently discussed examples for enhanced photon-induced contributions are the channels $\gamma\gamma \to \ell^+\ell^-$ [24, 38, 39] and $\gamma\gamma \to W^+W^-$ [40–43], where the $\gamma\gamma$ channels comprise more than 10 % in certain regions of phase space.

4.3.4 Combining QCD and Electroweak Corrections

A naive comparison of coupling strengths suggests that—when considering NLO $\mathcal{O}(\alpha_s)$ and NNLO $\mathcal{O}(\alpha_s^2)$ QCD corrections—it should be sufficient to take NLO EW corrections of $\mathcal{O}(\alpha) \sim \mathcal{O}(\alpha_s^2)$ into account. However, as explained in the previous sections, EW $\mathcal{O}(\alpha)$ corrections can be significantly enhanced compared to the small value of α/π by large logarithms and/or kinematical effects, raising the question about mixed QCD-EW corrections of $\mathcal{O}(\alpha_s\alpha)$. In spite of the great progress of recent years in NNLO QCD calculations, the required multi-scale NNLO calculations needed at $\mathcal{O}(\alpha_s\alpha)$ are still out of reach at present. As long as this is the case, we have to rely on approximations—an issue that already comes up in the combination of NLO QCD and EW corrections ($\delta_{\mathrm{QCD}}^{\mathrm{NLO}}$ and $\delta_{\mathrm{EW}}^{\mathrm{NLO}}$) in cross-section predictions. Schematically, the choice is between the two extreme variants of adding, $1 + \delta^{\mathrm{NLO}} = 1 + \delta_{\mathrm{QCD}}^{\mathrm{NLO}} + \delta_{\mathrm{EW}}^{\mathrm{NLO}}$, or multiplying relative corrections, $1 + \delta^{\mathrm{NLO}} = (1 + \delta_{\mathrm{QCD}}^{\mathrm{NLO}}) \times (1 + \delta_{\mathrm{EW}}^{\mathrm{NLO}})$, and variants in between these two extreme cases. The differences between different variants are of $\mathcal{O}(\alpha_s\alpha)$, and the optimal choice should minimise the remaining corrections of this order for the most important observables.

Since long-distance effects such as soft or collinear parton emission off quarks or gluons and collinear final-state photon radiation off leptons are known to factorise from the actual hard scattering, the variant of factorising QCD and EW corrections seems to be preferable in many cases. For the prediction of a cross section which is differential in some observable x at the parton level, this idea of factorisation can be translated into the recipe

$$\frac{\mathrm{d}\sigma}{\mathrm{d}x} = \frac{\mathrm{d}\sigma_{\mathrm{QCD}}}{\mathrm{d}x} \times \left[1 + \delta_{\mathrm{EW}}^{\mathrm{NLO}}(x)\right] + \frac{\mathrm{d}\sigma_\gamma}{\mathrm{d}x},$$

where σ_{QCD} stands for the best available QCD prediction for the cross section, $\delta_{\mathrm{EW}}^{\mathrm{NLO}}(x)$ is the relative EW correction differential in x, and σ_γ denotes the contribution from photon-induced processes. This approach is e.g. useful if the various contributions are calculated independently, possibly even by independent programs. For instance, the state-of-the-art predictions for Higgs production via vector-boson fusion and Higgs-strahlung provided by the LHC Higgs Cross Section Working Group [44–46] are calculated in this way.

Similar factorisation approaches, which apply differential reweighting factors to include EW corrections, have been suggested for event generation within QCD-based event generators. Examples presented in the literature (see section 19 in [47, 48]) comprise, for instance, the inclusion of EW corrections to Higgs production via vector-boson fusion [49, 50] and to diboson production [51, 52] upon reweighting events generated by the HERWIG Monte Carlo generator [51–53]. Before applied in practice, however, the applicability and reliability of any reweighting approach should be checked by a comparison to fully differential results.

One of the basic problems in any approach based on the factorisation idea lies in the fact that QCD and EW corrections of the two different factors are in general defined on different phase spaces. Only a complete calculation of $\mathcal{O}(\alpha_s\alpha)$ corrections can fully solve this problem. Since matrix elements and phase spaces for jet and/or photon emission factorise in the soft and/or collinear limits of the radiated particle, there is at least the possibility to correctly describe the leading soft and/or collinear $\mathcal{O}(\alpha_s\alpha)$ effects at the fully differential level. Based on this idea, Monte Carlo generators such as HERWIG++ [51, 52] and SHERPA [54, 55] already dress QCD-based predictions with soft and/or collinear photon radiation (at least off final-state leptons), in order to catch some leading $\mathcal{O}(\alpha_s\alpha)$ effects and even higher-order effects in α. Likewise, dedicated Monte Carlo programs for Drell–Yan processes such as ResBos-A [56], Horace [57], and WinHac [58] proceed similarly.

Finally, we comment on the issue of "matching" fixed-order NLO calculations of $\mathcal{O}(\alpha_s)$ and $\mathcal{O}(\alpha)$ with parton showers that take into account multiple jet and/or photon emission in some leading logarithmic approximation. Here the central issue is to avoid double-counting effects at the level of one-parton and one-photon emission and to keep full NLO accuracy after the matching. In pure QCD, two matching prescriptions are widely in use: MC@NLO [59] and Powheg [60, 61]. When matching NLO QCD+EW calculations to a pure QCD parton shower (as e.g. done in [62] for Drell–Yan processes in the Powheg framework), thus, improves the NLO prediction not only in the pure QCD sector, but also in $\mathcal{O}(\alpha_s\alpha)$ by dressing $\mathcal{O}(\alpha)$ effects with soft/collinear QCD radiation. If the matched shower even includes photon radiation, where the Powheg matching has to be generalised as described in [63, 64] for Drell–Yan processes, then the effectively included $\mathcal{O}(\alpha_s\alpha)$ effects account for $\mathcal{O}(\alpha_s)$ effects dressed with soft/collinear photon radiation.

4.3.5 Treatment of W/Z Resonances

For an inclusion of the W/Z decays in predictions, the simplest (but somewhat crude) way is to employ the "narrow-width approximation" (NWA), which treats the W/Z bosons as "stable intermediate states", i.e. the full process is decomposed into on-shell W/Z production and on-shell W/Z decays. This decomposition results from the limit $\Gamma_V \to 0$ in the squared matrix element of the full resonance process, where the squared propagator factor (momentum transfer k) behaves like

$$\frac{1}{|k^2 - M_V^2 + \mathrm{i}M_V\Gamma_V|^2} \;\widetilde{\Gamma_V \to 0}\; \frac{\pi}{M_V\Gamma_V}\,\delta(k^2 - M_V^2).$$

The $1/\Gamma_V$ factor on the right-hand side is part of the well-known branching ratio which emerges after the inclusive integration over the V decay phase space. If cuts are imposed on the decay products, or if distributions in kinematical variables of those are considered, in general effects of spin correlations between W/Z production and decay appear. The naive NWA can be easily improved to include these correlations by properly combining production and decay parts for definite polarisation states. Note that $\mathcal{O}(\alpha)$ corrections to a cross section in NWA do not only consist of corrections to the V-production cross section and to the relevant branching ratio, but also comprise the off-shell effects of $\mathcal{O}(\Gamma_V/M_V)$ mentioned above.

A detailed description of a resonance process, keeping the full differential information of the kinematics of the decay products, has to be based on complete matrix elements for the full process, including both resonant and non-resonant diagrams. Note that in standard perturbation theory, particle propagators do not include decay widths. The decay widths rather appear in the propagator denominators only after a Dyson summation of self-energy diagrams (or at least the imaginary parts thereof near resonances). Depending on the details of the field-theoretical definition of the unstable particle's mass, different results are obtained for the resonant propagator. Writing generically

$$P_V(k) = \frac{1}{k^2 - M_V^2 + \mathrm{i}M_V\Gamma_V(k^2)} \tag{4.7}$$

for the V-propagator factor, two frequently used versions are:

- Fixed width (FW): $\Gamma_V(k^2) = \Gamma_V = \text{const.}$
  In this parametrisation, the complex squared mass $M_V^2 - \mathrm{i}M_V\Gamma_V$ plays the role of the complex location of the pole in the propagator, i.e. it is a field-theoretically sound, gauge-invariant quantity resulting from an all-order definition (see [65, 66] and older references therein). The real and imaginary parts of this complex quantity define the "pole mass" and its associated "pole width".
- Running width (RW): $\Gamma_V(k^2) = \Gamma_V \times k^2/M_V^2 \times \theta(k^2)$.
  This behaviour results from the so-called "on-shell definition" (OS) of the gauge-boson masses (see e.g. [67]), M_V^{OS}, which are tied to the zeroes of the real parts of the respective self-energies. This mass definition has the drawback that it is gauge dependent at the two-loop level and beyond. The values for the masses of the Z and W bosons have, however, been determined within this scheme at LEP and the Tevatron.

For reasons of theoretical consistency, the pole definition of mass and width are clearly preferable. But as long as they are only used in the resonance region, both the FW and the RW prescriptions represent different, legal parametrisations of resonances which can be used to fit mass and width, although the resulting resonance parameters correspond to field-theoretically different definitions. Fortunately, there is a simple translation from one scheme to the other [68, 69]:

$$M_V = \left.\frac{M_V}{\sqrt{1 + \Gamma_V^2/M_V^2}}\right|_{\mathrm{OS}}, \qquad \Gamma_V = \left.\frac{\Gamma_V}{\sqrt{1 + \Gamma_V^2/M_V^2}}\right|_{\mathrm{OS}},$$

so that $M_{Z,\mathrm{OS}} - M_Z \approx 34\,\mathrm{MeV}$ and $M_{W,\mathrm{OS}} - M_W \approx 27\,\mathrm{MeV}$. For precision EW physics, in particular for a precision M_W measurement, it is important to be consistent in the use of a scheme and the respective input.

Care has to be taken in predictions that cover both resonant and non-resonant regions, because neither the FW, nor the RW prescription represents a consistent scheme to calculate cross sections based on full matrix elements with resonant and non-resonant contributions. Both in general lead to gauge-dependent results. This is due to the fact that using (4.7) for propagator factors necessarily mixes different orders in perturbation theory in practice, since the width term in the propagator results from partial all-order resummations, but the cross-section calculation stops at some finite loop level. Proper cross-section predictions should be based on consistent, gauge-invariant schemes, the description of which goes far beyond the scope of this book. We, therefore, just list the most prominent methods and refer to the original literature for details:

- The "pole scheme" exploits the fact that both the location of the V propagator pole and its residue in amplitudes are gauge-independent quantities. The idea [70, 71] is, thus, to first isolate the residue for the considered resonance and subsequently to introduce a finite decay width only in the gauge-independent resonant part. Often only a "pole approximation" is kept upon taking only the leading contribution into account, which, however, restricts the validity of the result to the resonance region.
- "effective field theories" [72–74] deliver a field-theoretically elegant way to carry out pole expansions owing to their formulation via Lagrangians and effective actions. Like pole approximations, their validity is restricted to the resonance region, but they offer the combination with further expansions, e.g. around thresholds, and suggest better possibilities to carry out dedicated resummations.
- In the "complex-mass scheme" [75, 76] the squared W-boson and Z-boson masses are consistently identified with the complex values $M_V^2 - \mathrm{i} M_V \Gamma_V$ $(V = W, Z)$, not only in the V propagators, but also in the couplings. NLO calculations deliver uniform predictions with NLO accuracy everywhere in phase space, i.e. both in resonant and non-resonant regions.

More details and references can again be found e.g. in the "EW dictionary" of [11].

4.4 Experimental Techniques

Decays of W and Z bosons to electrons and muons are studied most commonly, as these channels provide the best resolution in the reconstruction of the final-state kinematics and the largest background rejection. In order to identify a Z boson, two oppositely charged leptons are required in the event. The invariant mass of the lepton pair is required to be compatible with the nominal Z-boson mass of 91 GeV. In the case of the W boson, the decay lepton is accompanied by a neutrino, which leaves the detector undetected, leaving only an imbalance in the transverse momenta of the detected particles. The lepton four-momentum and missing transverse energy (MET, assumed to represent the transverse momentum components of the neutrino) are combined to form the so-called transverse mass $M_\mathrm{T} = \sqrt{2 \cdot p_{\mathrm{T,lepton}} \cdot MET \cdot (1 - \cos\phi)}$, where ϕ is the angle between the lepton and MET in the transverse plane. In case that only one neutrino is present in the final state, the neutrino momentum may be completely reconstructed from the MET and lepton momentum by imposing the condition that the invariant mass of the neutrino-lepton pair is compatible with the known W mass.

Leptons may also be produced in QCD processes, e.g. in decays of heavy quarks, or imitated by misinterpreted detector signals (so-called "fake leptons"). Due to their association with QCD processes, these background leptons are typically accompanied by a number of hadrons, while the leptons from Z or W decays are most often geometrically separated from other particles. This is exploited by requiring the leptons to be "isolated", i.e. by imposing a requirement on the maximum energy of particles reconstructed around the leptons. With the increasing luminosity of the LHC, pile-up creates randomly distributed tracks and calorimeter deposits, which reduce the efficiency of the isolation criteria. To ameliorate this problem, the isolation quantities are corrected for pile-up effects before selection cuts are applied. To avoid constructing a Z boson from two leptons created in to separate proton-proton interactions, the particles are required to originate from the same reconstructed collision vertex.

Decays with τ leptons are more complicated compared to the electron and muon channels, as the different decay channels of the τ multiply the number of possible final states. The presence of additional neutrinos in the τ decay further complicates the reconstruction, so that measurements of EW physics using τ leptons are generally less precise than competing measurements with electrons or muons. Nevertheless, the LHC experiments have gone to great length in observing W and Z bosons also in their τ decay channels, mostly as a testbed for the validation of τ-lepton reconstruction which is of importance in the search for the Higgs boson (see Chap. 6).

Hadronic decays of W and Z bosons are not suitable for use in precision measurements, due to the very high backgrounds from QCD multi-jet processes and the better kinematic resolution of leptons compared to jets. However, the hadronic decays have a niche in studies with multi-boson final states at very high momenta, as they are used in searches for anomalous couplings. At these high scales, the QCD background is kinematically suppressed and the small signal cross sections make it impossible to rely solely on the few leptonic final states, which have small branching fractions.

In order to exploit these final states for measurements, the response of the LHC detectors to leptons and MET needs to be understood to high precision. The accuracy of the detector modelling is evaluated comparing real data and simulated events for well known resonances, in particular the J/ψ, Υ, and the Z boson itself. These narrow states constitute excellent probes of the detector resolution and the uniformity of their response. Detector calibration methods and results can be found in Chap. 3 and in [77–81].

4.5 Drell–Yan Processes and EW Precision Observables

4.5.1 Theoretical Preliminaries

In lowest perturbative order (LO), hadronic collisions produce single W or Z bosons via quark-antiquark annihilation, as diagrammatically shown in Fig. 4.3, where the experimentally observable leptonic gauge-boson decays are also included.

We did not indicate partonic channels with incoming photons here, which actually contribute to dilepton pairs via $\gamma\gamma \to \ell\ell$ in LO (see discussion in Sect. 4.3.3). The $\gamma\gamma$ channel was not yet taken into account in experimental analyses because of its small contribution to the production rates near the gauge-boson resonances. In future

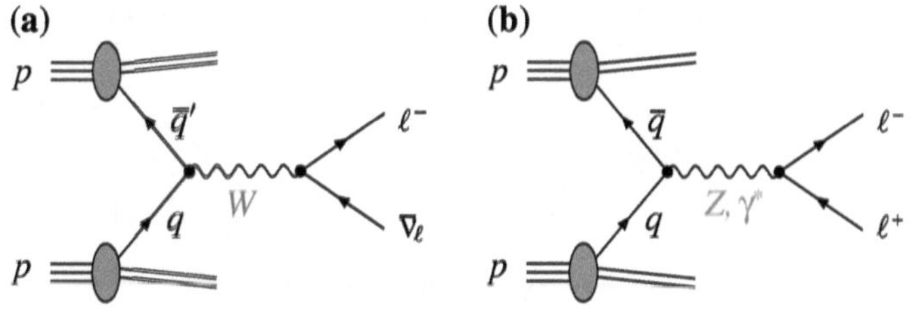

Fig. 4.3 LO diagrams for single hadronic W-boson or Z-boson production

analyses, which reach deep into the TeV range, such photon-induced channels should, however, be taken into account carefully.

Since the clean leptonic signatures can be experimentally reconstructed with high precision and since the production rates are high, theoretical predictions have to include higher-order corrections, both from strong and EW interactions. Owing to the absence of recoiling jets or photons, the produced EW gauge bosons do not receive a transverse momentum p_T in LO predictions. The theoretical description of the p_T spectra of the weak bosons, thus, starts at NLO. In fact, the region of small boson p_T cannot be described in any finite order of perturbation theory, but requires a resummation of soft-gluon radiation to all orders.

In the past, great efforts were made in the theory community to deliver precise predictions matching the required accuracy. QCD corrections are known up to NNLO [82–87], and EW corrections up to NLO [10, 24, 38, 88–93]. Both on the QCD and on the EW sides, there are further refinements by including leading higher-order effects, γ-induced channels, etc. The impact of QCD and EW corrections is illustrated in two examples in Figs. 4.4 and 4.5.

The former shows the sequence of perturbative predictions at LO, NLO QCD, and NNLO QCD for the rapidity distribution of the Z and $W^\pm$ bosons, nicely revealing the convergence of the prediction and the reduction of theoretical scale uncertainties in the first few orders. EW corrections to this observable are at the level of few percent only. EW corrections, for example, are more pronounced in differential cross sections that exhibit the resonance structure, such as the transverse-mass distribution of the W boson, as shown in Fig. 4.5. Here, their dominant contribution results from final-state radiation off the charged lepton, as discussed in Sect. 4.3.2, which explains the differences between EW corrections for bare electrons ($e^+\nu_e$), bare muons ($\mu^+\nu_\mu$), and dressed leptons (γ recomb.). The curves labelled "PA" represent "pole approximations" to the corrections, as briefly discussed in Sect. 4.3.5.

The region of small transverse gauge-boson momenta is described by a resummation of large logarithms through matched parton showers [94–96] or dedicated calculations supplemented with fits to non-perturbative functions [97–101]. First approaches to the combination of QCD and EW corrections can be found in [56, 62–64, 102, 103]. In view of fixed-order calculations, the largest missing piece seems to be the mixed QCD-EW corrections of $\mathcal{O}(\alpha_s\alpha)$. These are a prerequisite to answer the question how to properly combine QCD and EW corrections in predic-

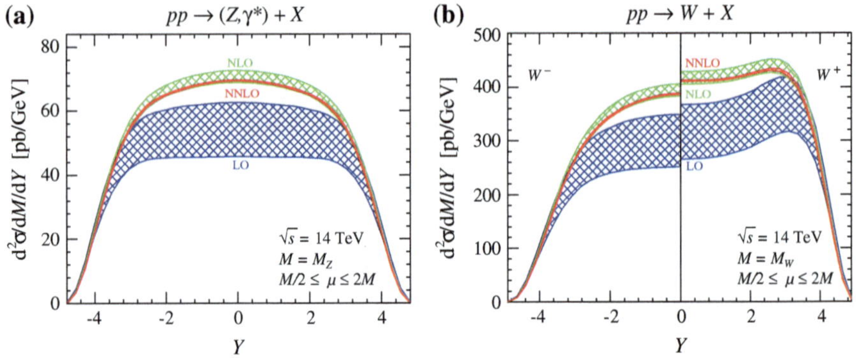

Fig. 4.4 Z-boson and $W^{\pm}$-boson rapidity distributions in Drell–Yan production at the LHC, evaluated at LO, NLO QCD, and NNLO QCD, including corresponding QCD scale-uncertainty bands. Adapted from [84]

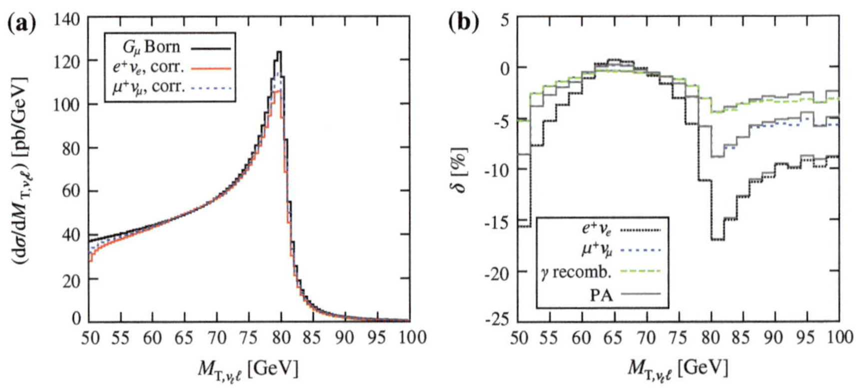

Fig. 4.5 Transverse-mass distribution of the $\ell^+\nu_\ell$ pair in W^+ production at the LHC and corresponding relative EW corrections δ for different leptons. Adapted from [10]

tions (see discussion in Sect. 4.3.4). First steps towards this direction have already been taken in the literature. In [104], a strategy for calculating these corrections in the resonance region has been set up, and first results of this pole approximation can be found in [105]. In view of the full calculation with off-shell W/Z bosons, the two-loop matrix elements [106–108] are available as well as the one-loop corrections to the subcontributions with either hard-jet or hard-photon emission, which comprise the $\mathcal{O}(\alpha)$ EW corrections to W/Z+jet production [109–114], and the $\mathcal{O}(\alpha_s)$ QCD corrections to $W/Z + \gamma$ production [111, 115–118]. More references pointing also to earlier literature can e.g. be found in [104].

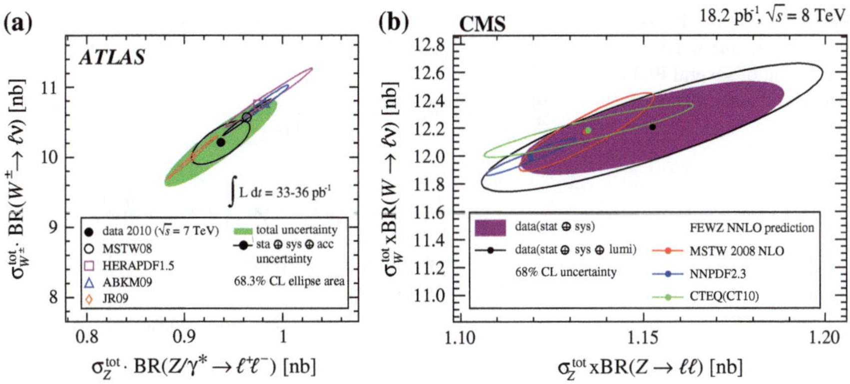

Fig. 4.6 Measured inclusive W and Z production cross sections at the LHC at **a** 7 TeV and **b** 8 TeV centre-of-mass energy. Adapted from [119, 121]

4.5.2 *Total and Differential Cross Sections*

Using the methods discussed in Sect. 4.4, the experiments at the LHC have measured the total and differential cross sections of EW boson production. Results for the inclusive cross sections from the ATLAS [119] and CMS [120, 121] collaborations are shown in Fig. 4.6.

The results are in good agreement with each other and with the predicted cross sections, which depend on the underlying parton densities. The accuracy of the measurement of the total cross section is severely limited by systematic uncertainties. Experimentally, uncertainties on the lepton reconstruction and on the MET reconstruction are particularly important. The uncertainty on the integrated luminosity of the measurements cannot be neglected in the absolute cross-section measurements, but cancels when ratios of W and Z production are measured. Additional theoretical uncertainties come into play when extrapolating the number of bosons visible in the fiducial volumes of the detector to the full phase space.

While the inclusive cross sections provide a handle to constrain PDFs in future fits, much more information can be gained from differential cross sections. Of particular interest is the rapidity of the vector boson y, which is directly related to the momentum fractions of the partons in the initial state via the relation $x_{1,2} = M_{\ell\ell}/s \cdot \exp(\pm y)$, where $x_{1,2}$ are the momentum fractions of the interacting partons, $M_{\ell\ell}$ is the mass of the dilepton system for Z-boson production and the mass of the lepton-neutrino system for W-boson production, and $\sqrt{s}$ is the centre-of-mass energy of the collider. Measurements of this quantity can be found in [119, 122, 123]. Figure 4.7 shows the measured distribution compared to predictions based on several different sets of parton densities, demonstrating the value of such measurements for the future improvement of PDF fits.

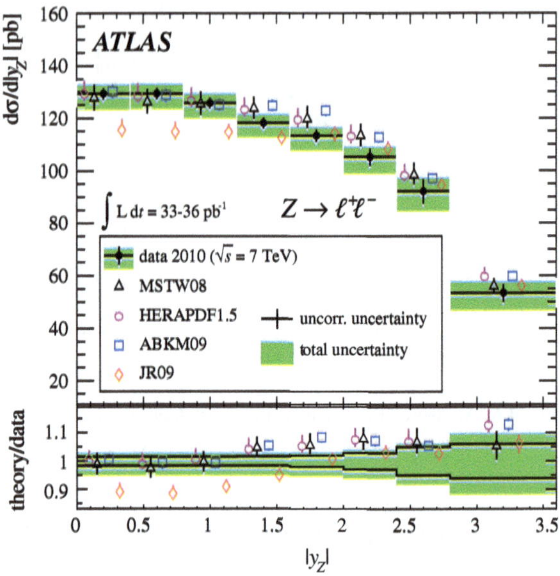

Fig. 4.7 Z-boson rapidity distribution compared to the prediction of several PDFs. Adapted from [119]

Of special interest in this regard are differences in the differential cross sections in y for W^+ and W^- production, which directly access the difference between quark and antiquark densities and the production of W bosons in association with b quarks, which gives access to the otherwise difficult to observe charm-quark density. These measurements are discussed in more detail in Sect. 4.8.

The transverse-momentum distribution of vector bosons is driven by higher-order diagrams in perturbation theory, in particular by jet emission, and is an important ingredient in the measurement of the W mass as discussed below. Due to the uncertainties on predictions of this quantity it is important to measure this distribution to validate the corresponding computational tools. The escaping neutrino of leptonic W decays prevents a direct measurement of the W transverse-momentum distribution with the desired accuracy. Instead, the more precise leptonic Z decays are used to verify theory predictions, which are in turn used to predict the W transverse-momentum distribution. Both ATLAS and CMS have measured this quantity [124–126], where the ATLAS experiment does not only directly measure the transverse momentum, but also studies a related quantity which can be determined from the lepton angles only, avoiding uncertainties from calorimetric measurements. While reasonable agreement between experiment and predictions is observed, there is still a need to improve our knowledge of the vector-boson production dynamics before a competitive measurement of the W mass at the LHC is feasible.

4.5.3 Effective Weak Mixing Angle

At first sight, the symmetric initial state at the LHC seems to preclude a measurement of the forward-backward asymmetry A_{FB} in the Drell–Yan (DY) process. However, on average the momentum carried by quarks in the proton is larger than that of antiquarks, due to the contributions of the proton's valence quarks. Thus the rapidity of the outgoing Z boson is significantly correlated to the direction of the initial-state quark. This correlation can be used to define an average forward-backward asymmetry for the extraction of the effective weak mixing angle, $\sin^2\theta^{\ell}_{\mathrm{eff}}$, though the resulting asymmetry is diluted compared to experiments where the initial state is known exactly (i.e. LEP, SLAC). The LHC experiments exploit this correlation between the initial-state quark and Z boson direction by observing the Drell–Yan decay angle as function of the Z-boson rapidity and invariant mass in order to extract the forward-backward asymmetry A_{FB}, which can be translated into a measurement of the weak mixing angle $\sin^2\theta^{\ell}_{\mathrm{eff}}$ on the Z-boson resonance.

CMS has performed two complementary measurements: a direct measurement of $\sin^2\theta^{\ell}_{\mathrm{eff}}$ at the Z mass [127] and a differential measurement of A_{FB} as function of the dilepton invariant mass [128]. To extract $\sin^2\theta^{\ell}_{\mathrm{eff}}$, a three-dimensional fit to the distributions of the dilepton invariant mass, rapidity, and decay angle are performed, taking into account the detector efficiency and acceptance as well as the dilution in asymmetry due to the unknown initial state. The result of $\sin^2\theta^{\ell}_{\mathrm{eff}} = 0.2287 \pm 0.0020(\mathrm{stat}) \pm 0.0025(\mathrm{syst})$ is compatible with previous results [129], but does not reach the precision achieved at lepton colliders. Already with the data sample of 1 fb^{-1} used in this analysis, the statistical uncertainty is reduced to the size of the systematic uncertainties. Of particular concern are the uncertainties connected to the proton PDFs (see Sect. 4.8), which govern the correlation between the initial quark direction and the Z-boson rapidity. The precision of future measurements of $\sin^2\theta^{\ell}_{\mathrm{eff}}$ thus critically depends on improvements in our understanding of the proton structure, a feature shared with the measurement of W-boson mass discussed in detail below. LHC data themselves may help to drive substantial progress towards the reduction of these uncertainties, as is described in more detail in Sect. 4.8.

Off the Z resonance, the direct correspondence between A_{FB} and $\sin^2\theta^{\ell}_{\mathrm{eff}}$ is lost, so that measurements that probe the scale dependence of the EW couplings are more suitable performed on A_{FB} directly. The measurement of A_{FB} as function of dilepton invariant mass by the CMS collaboration proceeds more simply than the extraction of $\sin^2\theta^{\ell}_{\mathrm{eff}}$ at the Z peak: The lepton asymmetry is measured in the Collins–Soper frame [130] under the assumption that the initial-state particle with the smaller momentum fraction is the antiquark. The results are corrected for detector effects as well as for the effects of final-state radiation and agree well with SM predictions (see Fig. 4.8). Although the measurements of A_{FB} at the LHC cannot compete with previous LEP results in precision, they achieve the direct observation of the scale dependence of the EW interaction, up to a scale of 400 GeV. Future analyses with larger data sets and higher centre-of-mass energy will allow the extension of these studies to even higher scales.

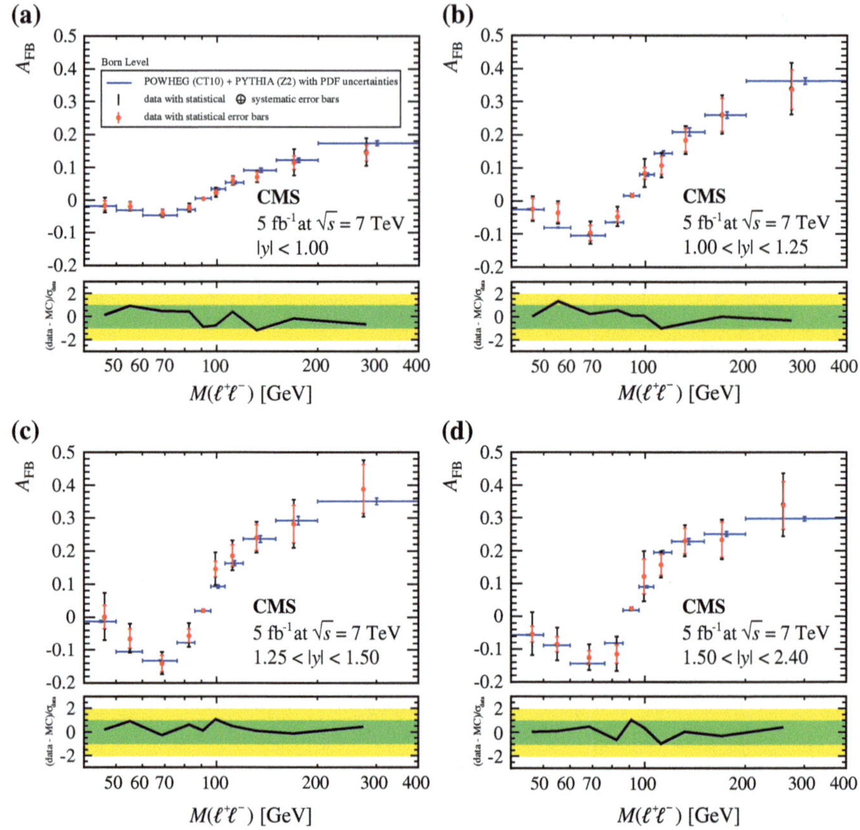

Fig. 4.8 Measurement of A_{FB} in four regions of dilepton rapidity as function of the dilepton invariant mass, clearly showing the evolution of A_{FB} with the mass. Adapted from [128]

4.5.4 *The W-Boson Mass*

The mass of the W boson (M_W) is intricately linked to the masses of other SM particles, which makes the measurement of this quantity of great interest in combination with other mass measurements as discussed in the next section. Great efforts have been spent at previous accelerators in order to measure M_W as precisely as possible, using a variety of techniques. At LEP, M_W has been determined with a scan of the beam energy over the W-pair production threshold and by means of full kinematic reconstruction [131], while the Tevatron and LHC experiments [132] use techniques described in detail below.

At the time of writing, LHC measurements of M_W are still missing from the harvest. While these measurements are actively pursued, the challenges are significant; before competitive measurements can be performed, the ATLAS and CMS collaborations need to achieve an exquisite understanding of the performance of their detectors,

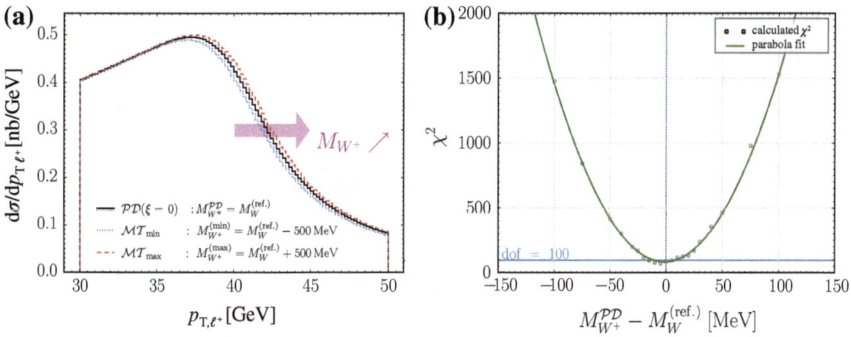

Fig. 4.9 **a** Lepton transverse momentum distribution in $W \to \ell + \nu$ events, for a reference mass (*full line*), and mass shifts of ±500 MeV (*dashed lines*). **b** χ^2 between the shifted and reference distribution, as a function of the mass shift. Adapted from [133]

and the uncertainties in the physics of W-boson production and decay need to be strongly constrained. The main issues are described below.

Since the W final state comprises (at least) a charged lepton and a neutrino, the invariant mass of the boson cannot be reconstructed for each event. Nevertheless, the distributions of the lepton transverse momentum p_T and the transverse mass M_T display peaks which carry information on M_W. The principle of the measurement is simple: Using Monte Carlo simulation, models of the p_T and M_T distributions are constructed for different hypotheses on the W mass, and compared to the data; the measured value is given by the model (or template) closest to the data. The agreement between each template distribution and the data can be quantified using a simple χ^2. This procedure is illustrated in Fig. 4.9 [133].

The p_T and M_T distributions, however, reflect many other effects beyond the W-boson mass itself. Firstly, the observed position of the peak is subject to detector miscalibration. An accurate calibration can be achieved using known resonances, as mentioned in Sect. 4.4; these issues will not be discussed here in detail. The distributions also reflect the production and decay mechanisms: The rapidity (y) distribution of the W boson is given by the proton PDFs; the interactions of the partons initiating the W-production process induce the emission of multiple initial-state gluons and give rise to the transverse-momentum distribution. Both effects are in part non-perturbative and cannot be predicted from first principles only; complementary measurements are needed to constrain the associated uncertainties. The convolution of the decay-lepton distributions with the W-boson y and p_T distributions impacts the shape of the Jacobian peaks; any uncertainty in these effects induces a potential bias on the M_W measurement. This is illustrated in Fig. 4.10, where the impact of a 30 MeV variation in the W-boson mass on the lepton p_T spectrum is compared with that of a change in the theoretical treatment of the W p_T spectrum; the effects are similar; if no attention is paid, a mismodelling of the W p_T distribution can easily occasion a biased measurement of M_W.

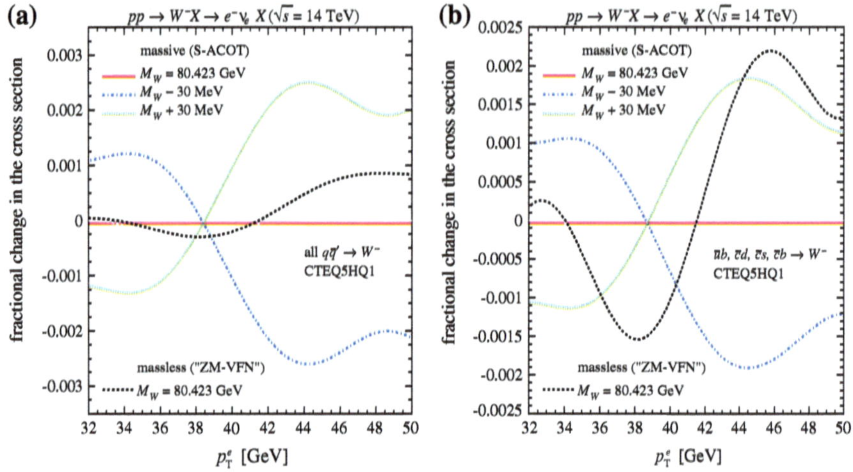

Fig. 4.10 Comparison of the effects of a 30 MeV variation in M_W with that of a change in the theoretical treatment of the mass of the initial quarks, for **a** inclusive W^- production and **b** the subprocesses involving heavy quarks. Adapted from [134]

Measurements of the W-boson and Z-boson cross sections and rapidity distributions are confronted with the current PDF models and help to improve their accuracy. As the W boson p_T distribution cannot be accurately measured due the incomplete kinematic reconstruction, this aspect is most efficiently probed using Z-boson events. As the W boson p_T distribution constitutes one of the principal issues, and the interplay between PDF and p_T modelling uncertainties are quite complex, we will describe it in some detail.

The inclusive Z boson p_T distribution results from the sum of the various partonic subprocesses $q\bar{q} \to Z, q = d, u, s, c, b$. The individual contributions have different distributions, due in part to quark-mass effects. As illustrated in Fig. 4.11 [134], the $b\bar{b} \to Z$ subprocess, which contributes about 4 % of the overall production rate, shows a peak in the p_T distribution at around 15 GeV; in the case of $c\bar{c} \to Z$, representing about 15 %, the peak is around 7 GeV. The light flavours typically have $p_T(uu, dd, ss \to Z) \sim 5$ GeV. The observed inclusive distribution is used to adjust the parameters governing the multiple gluon emissions mentioned above, assuming definite values for the relative contributions of the different quark flavours.

W production, on the other hand, is mostly composed of $d\bar{u} \to W^-$ and $s\bar{c} \to W^-$ (plus the CKM-suppressed channels and the charge-conjugate processes). In particular, the latter involves only one charm quark; there is no contribution from initial b quarks to this process. If our assumption of the c and b quark fractions contribution to Zproduction are wrong, the interpretation of the Z p_T spectrum in terms of multiple gluon radiation effects is biased, and this bias will be transported to the model for $p_T(W)$.

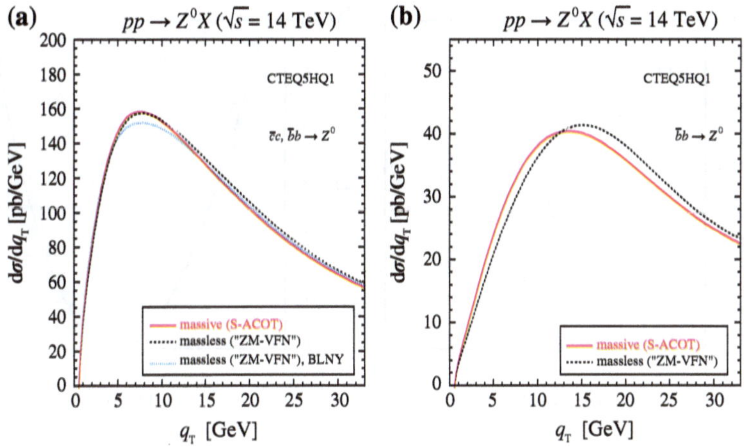

Fig. 4.11 **a** Z boson p_T distribution, for the $c\bar{c} \to Z$ and $b\bar{b} \to Z$ contributions. **b** The same quantity for $b\bar{b} \to Z$ only. Adapted from [134]

Converting the Z-boson p_T distribution measurement into a prediction for the W p_T distribution thus carries an uncertainty that is a function of the second- and third-generation quark PDF uncertainties. The set of measurements required to fully constrain the W-production distributions thus includes:

- W and Z cross sections and rapidity distributions, mainly for PDF constraints;
- cross sections for the associated production of W and Z boson with heavy flavours (c and b quarks), to constrain the s, c, and b quark densities in the proton;
- the Z-boson transverse momentum distribution which, in concurrence with the flavour decomposition mentioned above, allows the W p_T spectrum to be fully specified.

Measurements for all these observables are available (see Sect. 4.8). Their interpretation is ongoing and, once completed, will provide a complete prediction of the W distributions and a sound basis for the ensuing M_W measurement. Results from ATLAS and CMS are expected before the start of the LHC Run 2.

4.5.5 Global Fits

The rigid structure of the SM means that many measurable quantities are derived from only a small number of free parameters, so that the consistency of the SM can be tested by comparing an overconstrained set of independent measurements. The core set of observations used in these compatibility studies are precision measurements of Z-boson production and decay from the LEP and SLC e^+e^- colliders as well as LEP

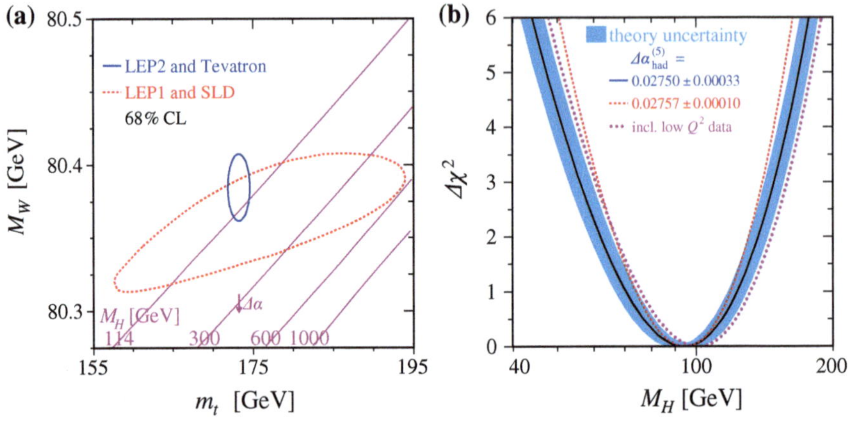

Fig. 4.12 Results derived from electroweak precision measurements [131]. **a** 68 % CL contours of the indirect constraints on M_W and m_t from precision measurements at LEP I, SLD and low-energy experiments in comparison with the direct measurements at LEP II and Tevatron and with the SM prediction for different values of M_H. **b** Constraints on the mass of the Higgs boson within the SM from precision measurements. The line shows the increase in the χ^2 value with respect to its minimum value obtained from a global fit. The band represents the estimated theoretical uncertainties from unknown higher-order corrections. Adapted from [131]

W-pair data [129, 131]. The experiments at the Tevatron contribute measurements of the W-boson and top-quark masses, although the LHC experiments have recently started to compete with the Tevatron in measurements of the latter (see Sect. 7.3).

Historically, the combination of these results has been used to determine the mass of the (then undiscovered) Higgs boson [131] indirectly from the overconstrained system of measurements. The main constraint in this case arises from the fact that the masses of the W boson, top quark and Higgs boson depend on each other through loops in their propagators. Figure 4.12a shows the prediction for M_W within the SM as function of the top-quark mass, m_t, for different values of M_H. This prediction is compared with the experimental measurements of M_W and m_t, indicated by a 68 % confidence-level ellipse, and with indirect predictions for M_W and m_t, which are obtained from a fit where the experimental input on those two observables has been removed. The indirect and direct determinations are in good agreement, but even with the inclusion of the direct measurements for M_W and m_t a large range of possible Higgs masses remains consistent with the data. This is demonstrated in Fig. 4.12b, which shows the χ^2 distribution of a global fit to electroweak precision data shortly before the Higgs boson was discovered (see Sect. 6.8.1). Neglecting limits from direct searches, the fit constrains the Higgs mass to be $M_H < 152$ GeV at 95 % confidence level [131].

With the discovery of a Higgs boson at the LHC, it is now possible to use the observed Higgs-boson mass to test the SM more strictly for consistency, as is done in [135]. The results show that the measurements are consistent with each other under the assumption of the SM with a χ^2 probability of 21 %, where the largest discrepancies originate with a b-quark related measurements from LEP. The addition of the Higgs-mass measurement to the fit inputs increases both the number of degrees of freedom of the fit and χ^2 by about one, so that the overall compatibility remains the same.

As the measurements of the EW sector do not show large deviations from the SM expectation, they may also be used to constrain possible extensions of the SM. Depending on the particular model under investigation, it may be useful to include additional data sets, as is done for example in [136], which adds results from flavour physics, dark-matter searches, and direct SUSY searches at the LHC to constrain free parameters of possible SUSY scenarios (see Sect. 10.6.1 for a more detailed discussion).

In the search for new physics, the measurements of the masses of the Higgs boson, the W boson and the top quark are of particular importance, as these quantities are related to each other through loop effects in the particle propagators. Therefore, deviations from the SM expectations in the relations between these three masses can be interpreted in terms of additional exotic particles that also contribute to the particles masses via loop effects. With the improving detector understanding and increasing integrated luminosity, prospects are good for improved measurements from the LHC in this sector.

4.6 Diboson Production and Anomalous Triple Gauge Couplings

Apart from directly confronting data with SM predictions and looking for possible differences, an alternative approach consists in adding non-standard ("anomalous") couplings to the SM Lagrangian and considering their effect on predictions as corrections induced by new physics. If the SM agrees with data, the experimental results can, thus, be translated into bounds on the size of the anomalous couplings. If significant deviations from SM predictions emerge, data could favour or disfavour specific non-standard coupling structures, pointing to specific classes of SM extensions. A systematic construction of the effective operators representing the non-standard couplings in the Lagrangian can proceed via counting the mass dimensions of these operators. Any dimension beyond 4 leads to an inverse mass factor $1/\Lambda$ in the respective coupling, where Λ typically is of the order of the mass of an undiscovered heavy particle causing the coupling modification. The higher the dimension of a new operator, the higher is the expected suppression of the corresponding coupling.

Since the SM—as a renormalisable field theory—exhausts all dimension-4 operators compatible with its symmetry, the dominant effective operators are of dimension 5 or 6. However, the only possible dimension-5 operator, the so-called "Weinberg operator", is merely relevant for neutrino physics. We are, thus, left with dimension-6 operators, which were completely classified in [137, 138], revealing 59 independent coupling structures compatible with the SM gauge group. Since this number is too high for a comprehensive, global analysis, in the past the number of effective operators was reduced using either additional symmetry assumptions or restrictions to specific classes of coupling modifications, motivated by the accessible processes.

The most prominent example of such a restriction is provided by the investigation of anomalous triple couplings among weak gauge bosons, based on analyses of diboson (W^+W^-, ZZ, $W\gamma$, WZ) production at LEP, the Tevatron, and the LHC. In these analyses, anomalous triple gauge-boson couplings (aTGCs) were based on general phenomenological parametrisations of the AWW, ZWW, AAA, AAZ, AZZ and ZZZ couplings. In the SM, only the $W^+W^-\gamma$ and ZWW vertices occur as triple couplings, while couplings of three neutral bosons are forbidden.

4.6.1 The Effective Lagrangian

The most general phenomenological parametrisation of the VWW vertex was introduced in [139] (see also [140]). The corresponding effective Lagrangian reads (note our sign convention $\epsilon^{0123} = +1$):

$$\begin{aligned}\mathcal{L}_{VWW}/(-eg_{VWW}) = {} & \mathrm{i}g_1^V(W^\dagger_{\mu\nu}W^\mu V^\nu - W^{\mu\nu}W^\dagger_\mu V_\nu) + \mathrm{i}\kappa^V W^\dagger_\mu W_\nu V^{\mu\nu} \\ & + \mathrm{i}\frac{\lambda^V}{M_W^2}W^\dagger_{\rho\mu}W^\mu{}_\nu V^{\nu\rho} - g_4^V W^\dagger_\mu W_\nu(\partial^\mu V^\nu + \partial^\nu V^\mu) \\ & + g_5^V \epsilon^{\mu\nu\rho\sigma}(W^\dagger_\mu \partial_\rho W_\nu - W_\nu \partial_\rho W^\dagger_\mu)V_\sigma \\ & + \frac{\mathrm{i}}{2}\tilde{\kappa}^V \epsilon^{\mu\nu\rho\sigma} W^\dagger_\mu W_\nu V_{\rho\sigma} \\ & + \frac{\mathrm{i}}{2}\frac{\tilde{\lambda}^V}{M_W^2}\epsilon^{\mu\nu\rho\sigma} W^\dagger_{\nu\lambda} W^\lambda{}_\mu V_{\rho\sigma} \, . \end{aligned} \tag{4.8}$$

Here V^μ stands for either the photon or the Z-boson field, W^μ is the W^--boson field, $W_{\mu\nu} = \partial_\mu W_\nu - \partial_\nu W_\mu$, and $V_{\mu\nu} = \partial_\mu V_\nu - \partial_\nu V_\mu$. The operators in (4.8) exhaust all possible Lorentz structures which do not involve the scalar components of any of the three vector bosons ($\partial_\mu V^\mu = \partial_\mu W^\mu = 0$). The Lagrangian (4.8) comprises the most general Lorentz-invariant VWW vertex observable in processes where the vector bosons couple to massless fermions. The general parametrisation (4.8) violates electromagnetic gauge invariance, a concept that is too well established to tamper with. This can be cured by considering appropriate higher-dimensional operators. The overall coupling constants g_{VWW} are fixed to their SM values $g_{AWW} = 1$ and $g_{ZWW} = \frac{c_W}{s_W}$.

Assuming Lorentz and electromagnetic gauge invariance, the most general effective dimension-6 Lagrangian for AAA, AAZ, AZZ, and ZZZ couplings can be written as [141, 142]:

$$\mathcal{L}_{VVV} = \frac{e}{M_Z^2}\Big[[f_4^\gamma(\partial_\mu F^{\mu\nu}) + f_4^Z(\partial_\mu Z^{\mu\nu})] Z_{\nu\rho} Z^\rho \\ + [f_5^\gamma(\partial_\mu F^{\mu\nu}) + f_5^Z(\partial_\mu Z^{\mu\nu})] \tilde{Z}_{\nu\rho} Z^\rho \\ - [h_1^\gamma(\partial_\mu F^{\mu\nu}) + h_1^Z(\partial_\mu Z^{\mu\nu})] F_{\nu\rho} Z^\rho \\ - [h_3^\gamma(\partial_\mu F^{\mu\nu}) + h_3^Z(\partial_\mu Z^{\mu\nu})] \tilde{F}_{\nu\rho} Z^\rho \Big], \tag{4.9}$$

with

$$\tilde{F}_{\mu\nu} = \frac{1}{2}\epsilon^{\mu\nu\rho\sigma} F_{\rho\sigma}\,. \tag{4.10}$$

Again, the operators in (4.9) exhaust all possible Lorentz structures (respecting $\partial_\mu Z^\mu = \partial_\mu A^\mu = 0$).

The scattering cross sections induced by these additional couplings rise with the interaction energy, leading to violations of unitarity at sufficiently high energies even for arbitrarily small anomalous couplings. This effect is not physical—the additional operators in the Lagrangian only represent an effective low-energy approximation of unknown high-scale new physics, which is presumably unitary. Conceptually, the mechanism is similar to the Fermi theory of the weak interaction which describes low-energy β decays well, but is not suitable at EW energy scales, where the full SM is needed.

There is no widely agreed method to regularise these anomalous cross sections. Common approaches in the past have been to introduce ad-hoc form factors [143] or the so-called k-matrix scheme [144]. Alternatively, results have been shown using no regularisation at all, comparing experimental limits to bounds derived from unitarity conditions. While all these approaches allow a model-independent interpretation of the data, they hinder the comparison obtained in different schemes. The regularisation can be regarded as the effect of further undiscovered new physics at high scales. However, at the energies probed at the LHC, the scale of the unitarisation scheme needs to be chosen so low that the mechanism of unitarisation itself should be in experimental reach and the question arises why it has not yet been discovered.

4.6.2 *Predictions for Diboson Production*

In leading order, EW diboson production at hadron colliders receives contributions from the Feynman diagrams shown in Fig. 4.13, where a potential aTGC is highlighted by a blob.

As before, we again do not consider γ-induced partonic channels here, which actually contribute to W-boson pairs via $\gamma\gamma \to WW$. Owing to its small contribution to total production rates, this contribution was not taken into account in experimental analyses yet, but future analyses of differential cross section reaching deeper into the TeV range should included it carefully.

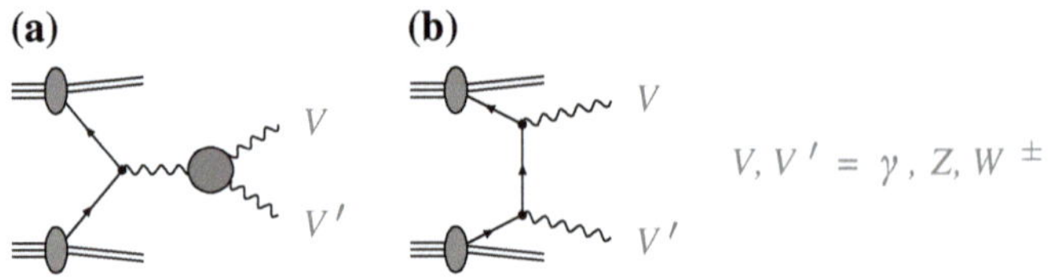

Fig. 4.13 LO diagrams for EW diboson production in hadronic collisions

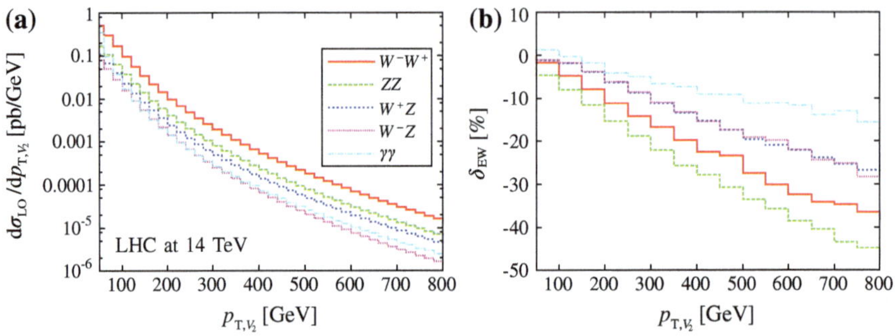

Fig. 4.14 Transverse-momentum distributions for the various diboson production channels at the LHC and corresponding relative EW corrections $\delta_{\rm EW}$. Adapted from [41]

In order to match a generic accuracy of some percent, theoretical predictions for hadronic diboson production are based on higher-order calculations comprising both QCD and EW corrections. Very recently, QCD predictions were pushed to the NNLO level for the $Z\gamma$, ZZ, and W^+W^- production channels [145–147], while the other channels are known to NLO, as e.g. provided by MCFM [116]. EW corrections for all diboson signatures are available at the NLO level [40–43, 148, 149]. Figure 4.14 illustrates the impact of EW corrections in the transverse-momentum distributions, where these EW effects become very large owing to the so-called EW Sudakov logarithms, as discussed in Sect. 4.3.1.

Beyond fixed-order predictions, the NLO-based predictions were also matched to QCD parton showers [59, 150], and dedicated QCD resummations [151, 152] were worked out. Apart from $q\bar{q}$-initiated production, electrically neutral gauge-boson pairs can also be produced in loop-mediated gluon-scattering processes, $gg \to VV'$, as discussed at the one-loop level in [153]. Although formally of higher order in the strong coupling, these reactions can contribute a non-negligible number of events because of the large gluon luminosity at the LHC, in particular if specific search cuts are applied. The most comprehensive QCD-based prediction for (off-shell) weak-diboson production at the LHC has been presented in [154] in a study of these reactions as background to Higgs-boson production.

4.6.3 Diboson Measurements and Limits on Triple Couplings

The anomalous triple gauge couplings are tested by measuring the diboson production cross sections, i.e. W^+W^-, ZZ, WZ, $W\gamma$, and $Z\gamma$. The number of possible diboson pairs combined with the number of different decay modes for the W and Z bosons leads to a large number of different experimental signatures. So far, not all of these have been analysed by the LHC experiments, and preference has been given to channels that are backgrounds in searches for the Higgs boson. The CMS and ATLAS experiments have investigated the ZZ channel in leptonic decays [155, 156] as well as the lepton-neutrino final state [156].

At the LHC, QCD processes can result in the production of two weak bosons without the involvement of the triple gauge coupling and such processes dominate the diboson production cross sections. Nevertheless, contributions from the triple gauge couplings can be isolated where the outgoing bosons have very high transverse momenta or where the pair has a very high invariant mass.

As an example we will discuss here the CMS analysis of ZZ production using the decay to four leptons [155]. The reconstruction of the Z bosons follows the outline given above. Notably, this analysis contains results from final-state τ leptons, even though these turn out to have little contribution to the final results. While the all-leptonic channel has a very small branching fraction ($\sim$0.5 %), it retains good sensitivity, because it is practically background-free. As the ZZ final state probes the ZZZ and $ZZ\gamma$ vertices, which are forbidden in the SM, there is also no contribution from the SM triple vertex. Only the independent production of two Z bosons contributes as background, further enhancing the sensitivity. To constrain anomalous couplings, the invariant mass of the ZZ system is checked for deviations from the SM prediction at high mass (see Fig. 4.15).

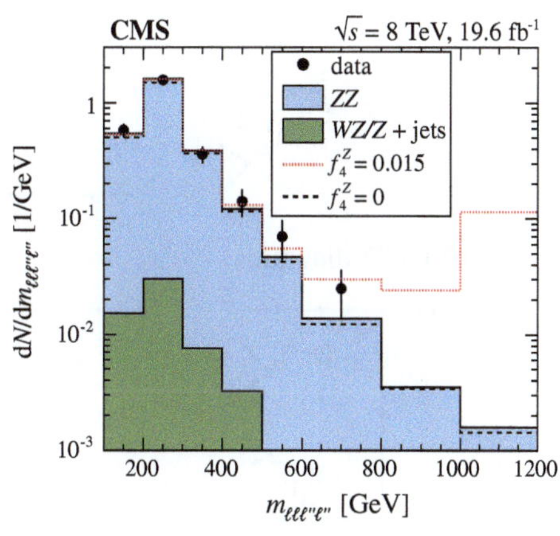

Fig. 4.15 Measurement of the ZZ invariant mass for events with two leptonically decaying Z bosons with the CMS experiment. The expectations for aTGCs are shown in addition to the SM expectation. Adapted from [155]

4.7 Triple Gauge-Boson Production and Vector-Boson Scattering

Anomalous quartic gauge-boson couplings (aQGCs) can be experimentally probed by investigating triboson production and vector-boson scattering (VBS). Similar to the triple gauge-boson couplings, vertices involving four neutral bosons are forbidden in the SM, so that only the $W^+W^-\gamma\gamma$, $W^+W^-Z\gamma$, $WWZZ$, and $WWWW$ vertices are allowed.

Some LO diagrams contributing to the signatures of triple-gauge-boson production and VBS are shown in Fig. 4.16.

4.7.1 Anomalous Quartic Gauge-Boson Couplings

Unlike most of the studies of aTGCs, previous analyses of aQGCs were not based on a general parametrisation of the quartic gauge-boson vertex. On the one hand, there this is a proliferation of free coupling parameters in the general vertex parametrisation; on the other hand, there are only less precise measurements constraining those couplings. In the past, $\gamma\gamma WW$ and $\gamma\gamma ZZ$ aQGCs were experimentally accessible via $WW\gamma$, $Z\gamma\gamma$, and $\nu\bar{\nu}\gamma\gamma$ production at LEP (see e.g. [157] and references therein) and via elastic W^+W^- production at the Tevatron [158] and the LHC [159, 160]. Data from these analyses were used to set bounds on the two coupling parameters a_0 and a_c of the operator basis [161–163]

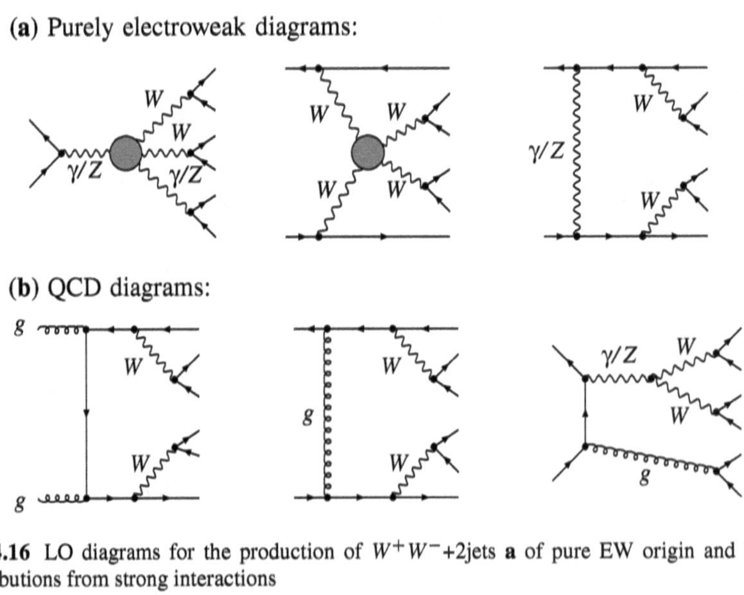

Fig. 4.16 LO diagrams for the production of W^+W^-+2jets **a** of pure EW origin and **b** with contributions from strong interactions

$$
\begin{aligned}
\mathcal{L}_0 &= -\frac{e^2}{16\Lambda^2}\, a_0\, F^{\mu\nu} F_{\mu\nu} \overline{\mathbf{W}}_\alpha \overline{\mathbf{W}}^\alpha, \\
\mathcal{L}_c &= -\frac{e^2}{16\Lambda^2}\, a_c\, F^{\mu\alpha} F_{\mu\beta} \overline{\mathbf{W}}^\beta \overline{\mathbf{W}}_\alpha, \\
\mathcal{L}_n &= -\frac{e^2}{16\Lambda^2}\, a_n\, \epsilon_{ijk} F^{\mu\nu} \overline{W}^i_{\mu\alpha} \overline{W}^j_\nu \overline{W}^{k,\alpha}, \\
\tilde{\mathcal{L}}_0 &= -\frac{e^2}{16\Lambda^2}\, \tilde{a}_0\, F^{\mu\nu} \tilde{F}_{\mu\nu} \overline{\mathbf{W}}_\alpha \overline{\mathbf{W}}^\alpha, \\
\tilde{\mathcal{L}}_n &= -\frac{e^2}{16\Lambda^2}\, \tilde{a}_n\, \epsilon_{ijk}\, \tilde{F}^{\mu\nu} \overline{W}^i_{\mu\alpha} \overline{W}^j_\nu \overline{W}^{k,\alpha},
\end{aligned}
\tag{4.11}
$$

where $F^{\mu\nu}$ is the usual photon field-strength tensor, $\tilde{F}^{\mu\nu}$ its dual, and the combination $\overline{\mathbf{W}}_\mu = (W^1_\mu, W^2_\mu, Z_\mu/c_w)$ with $\overline{W}^{i,\mu\nu} = \partial^\mu \overline{W}^{i,\nu} - \partial^\nu \overline{W}^{i,\mu}$ is inspired by custodial $SU(2)_c$ symmetry. The operator set (4.11) includes only genuine aQGCs of dimension 6 that respect electromagnetic gauge invariance and $SU(2)_c$, i.e. it omits all operators that contribute also to aTGCs, such as the quadrilinear part of the well-known operator $F^{\mu\nu} W^{+,\rho}_\nu W^-_{\rho\mu}$. The operators $\mathcal{L}_n$, $\tilde{\mathcal{L}}_0$, $\tilde{\mathcal{L}}_n$ either break C or P symmetry, or both, and were not yet confronted with data.

The LHC will be the first collider to explore aQGCs involving only massive weak bosons, so that the set of aQGC operators has to be extended. A first (though not fully systematic) extension was adopted in [160] where anomalous γZWW operators were introduced by replacing one photonic field-strength tensor $F_{\rho\sigma}$ in $\mathcal{L}_0$ and $\mathcal{L}_c$ by $Z_{\rho\sigma}$ from the Z-boson field (thereby giving up custodial symmetry). Moreover, one specific dimension-8 operator with coupling $f_{T,0}$ was introduced.

However, in accordance with the model-independent search for new-physics effects in other process classes, such as in analyses of the Higgs boson, a preferred operator basis is provided by the aQGCs included in the 59 operators of dimension 6 [137, 138] that respect the SM $SU(3) \times SU(2) \times U(1)$ gauge symmetry.

4.7.2 Triple Gauge-Boson Production

The aQGCs can be accessed in triple gauge-boson production similar to the way that the aTGCs are probed in diboson production. However, triple-boson production is plagued by even lower production cross sections than diboson production, in particular if all three final state bosons are massive (W or Z), which significantly reduces the available phase space.

At the energies and luminosities reached up to now, the experimental study of three massive weak gauge bosons, $pp \to WWW$, WWZ, etc., was not yet feasible. At present, predictions are globally based on NLO QCD [164–166]; only for WWZ production NLO EW corrections [167] are known as well. For first measurements of total cross sections, the accuracy of $\sim$10 % achieved in this way should be sufficient, but for analyses at the LHC design luminosity the predictions should be further improved. This, in particular, concerns the inclusion of the decays of the W/Z bosons and the evaluation of NLO EW corrections.

The CMS experiment has studied the production of W^+W^- and WZ pairs in association with a photon to put bounds on anomalous $W^+W^-\gamma\gamma$ and $W^+W^-Z\gamma$ couplings [160]. The analysis searches for a semileptonic signature with a leptonically decaying W boson and a dijet pair compatible with the W or Z mass in addition to the photon, as the high hadronic branching fractions of the W and Z guarantee a good overall signal efficiency. The presence of one leptonic decay provides an improved background rejection compared to an all-hadronic analysis. The dijet mass resolution of the detector is not sufficient to distinguish between hadronically decaying W and Z bosons, so that possible deviations from the SM cannot be uniquely attributed to a specific anomalous coupling. Due to large backgrounds from the production of $W\gamma$ final states in association with QCD-induced jets, it is not possible to measure the strength of the SM quartic couplings (see Fig. 4.17). Nevertheless, anomalous couplings could lead to an excess of the observed cross section at high photon transverse momentum. No such excess is observed, and limits on the anomalous couplings are extracted.

Additionally, the quartic vertices may be probed in processes where two bosons are produced from an initial state that indirectly involves vector bosons as well. The quartic vertex with four W bosons is of special interest in the SM and is discussed in the next section. Complementary to the analysis described above, which studies the $W^+W^-\gamma\gamma$ vertex in events with two W bosons and one photon, it is possible to probe this vertex in the photoproduction of two W bosons. Such an analysis has been performed by CMS using events with two leptonically decaying W bosons [159]. The photoproduction process is identified by the absence of fragmentation products

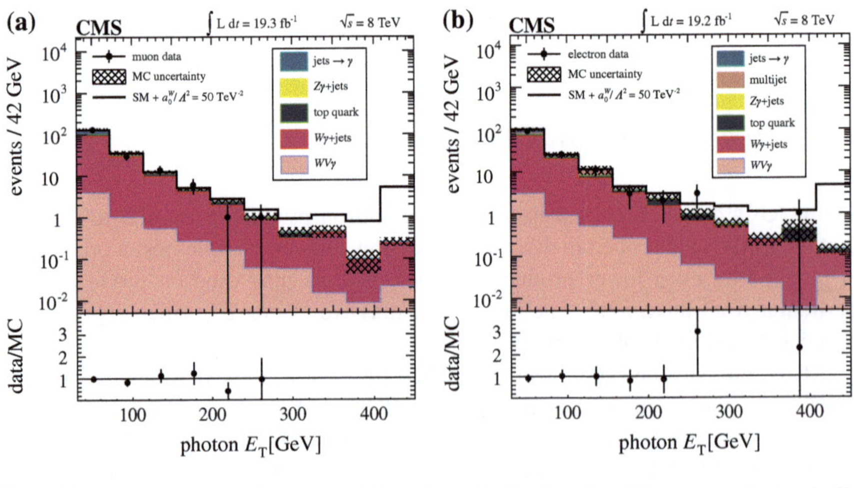

Fig. 4.17 Photon E_T spectra in events with one leptonically decaying W boson, one hadronically decaying W or Z boson and a photon; **a** muon channel; **b** electron channel. Adapted from [160]

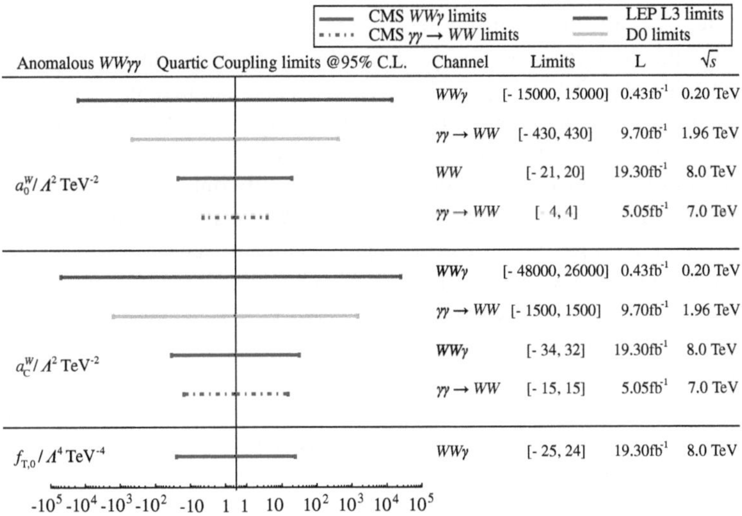

Fig. 4.18 Comparison of limits on aQGCs from different experiments and final states. Adapted from [158–160, 168]

of the initial-state protons. This signature, together with the clean leptonic signature of the W decays, leads to a very pure set of events at the end of the analysis chain, producing more stringent limits on the quartic couplings than obtained in [160], even though the results are derived from a data set of smaller integrated luminosity (see Fig. 4.18 for a comparison of the limits also with previous results from LEP [168] and the Tevatron [158]). This type of analysis is quite sensitive to the presence of pile-up, as the efficiency of the requirement on the absence of proton-fragmentation products drops quickly as function of the number of simultaneous interactions.

4.7.3 Vector-Boson Scattering

The interactions of massive vector-bosons are intricately linked to the mechanism of EW symmetry breaking. In the SM, the amplitude for the scattering of longitudinally polarised W bosons increases with the centre-of-mass energy $\sqrt{s}$ and violates the unitarity bound at around 1 TeV in the absence of the Higgs boson, which regularises the cross section through negative interference [169–171]. While a suitable Higgs-boson candidate has been found at the LHC (see Sect. 6.5), there is a number of scenarios beyond the SM, such as a composite Higgs boson or new high-mass resonances, which would lead to deviations from the SM prediction in VBS processes [172, 173].

Since the experimental signature for massive weak VBS is $WW/WZ/ZZ$ $+2$jets $\to$ 4leptons+2jets, precise theoretical predictions are very complicated. Already at leading order, various production mechanisms, not all related to VBS, contribute, as is illustrated in Fig. 4.16. Some of the diagram types even involve gluons, so that LO cross-section predictions receive contributions from different orders in EW or QCD couplings ($\alpha^4\alpha_s^2$, $\alpha^5\alpha_s$, α^6), but predictions based on full sets of LO diagrams are available. At next-to-leading order, the number of perturbative orders contributing to the cross section further proliferates ($\alpha^4\alpha_s^3$, $\alpha^5\alpha_s^2$, $\alpha^6\alpha_s$, α^7), and not all of them have been evaluated yet. Basically only the NLO QCD corrections are known to the (from the EW point of view) most interesting diagram types shown as category (a) in Fig. 4.16 in the approximation of neglecting colour exchange between different quark lines [174–178]. Also known are QCD corrections to category (b) [179–183], which carry the highest power in α_s. These QCD corrections are consistently matched to a QCD parton shower at NLO [184–186]. For quantum corrections that mix QCD and EW contributions as well as for genuine EW corrections, no results exist yet, although effects at the level of some 10 % can be expected for momentum transfers in the TeV range, owing to the general high-energy pattern of higher-order EW gauge-boson exchange or emission (see Sect. 4.3.1).

At the LHC, VBS is only indirectly accessible in event topologies similar to the vector-boson fusion (VBF) process studied with the Higgs boson (see Sect. 6.3.1). In the LHC Run 1, VBS cross sections are quite small, because the effective boson-boson centre-of-mass energy in the VBF process is reduced compared to the pp centre-of-mass energy of 7 or 8 TeV. Thus, the LHC experiments have so far focused on preparatory measurements that explore analysis techniques planned to be used in the Run 2, where the increased centre-of-mass energy and luminosity promise access to the kinematic region of interest.

A good example for these preparatory studies are measurements of Z bosons in association with forward jets, as this final state has a contribution from VBF diagrams similar to VBS, but replacing the quartic with a triple-gauge coupling (see Fig. 4.16a, middle diagram). Measurements by the CMS and ATLAS collaborations [187, 188] proceed along similar strategies: The leptonically decaying Z is reconstructed as described above, and two additional jets are required to be present in the event. To isolate the signal contribution from the large number of background events, both experiments use the unique kinematics of the dijet system in VBF processes: The jets are expected to be widely separated in rapidity and to have large invariant mass (see Fig. 4.19).

A first glimpse of VBS has been obtained in direct observation [189, 190]. The analysis searches for the signature of two leptonic W-boson decays in association with to jets with large separation in rapidity, similar to the above preparatory studies. The background of QCD-induced diboson pair production is reduced by only studying events in which the two W bosons carry the same electric charge, a configuration which is not permitted in processes with gluons in the initial state. Figure 4.20 shows the invariant-mass distribution of the dijet system and the clear excess of observed events that would arise without accounting for VBS. Evidence of the presence of this process is observed with more than $3\,\sigma$ significance, and the corresponding cross section is compatible with the SM. The results are additionally interpreted in terms of aQGCs, using the k-matrix scheme discussed above. At 8 TeV collision energy

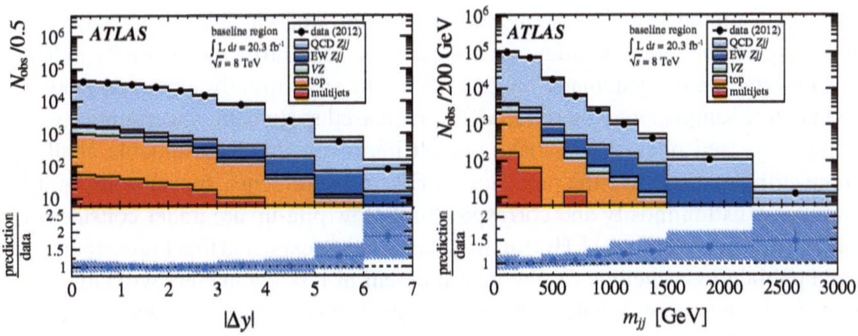

Fig. 4.19 Dijet rapidity difference $|\Delta y|$ and dijet invariant mass for Z+2jet events from different production mechanisms. Adapted from [188]

Fig. 4.20 Dijet invariant-mass (M_{jj}) distribution of W^+W^-+2jets events. Adapted from [189]

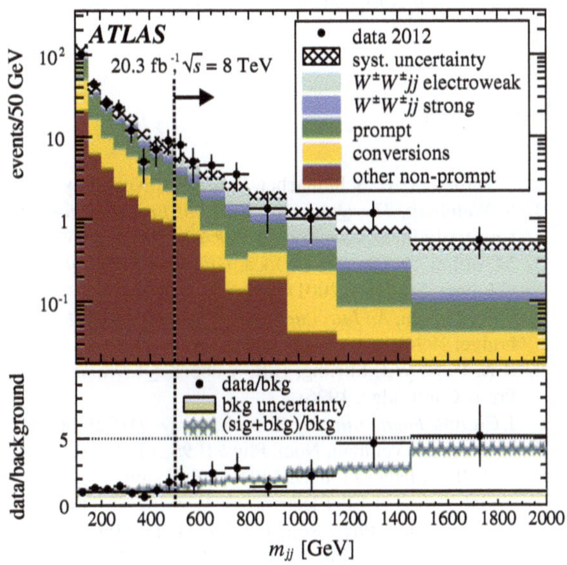

and with the limited Run 1 data sample, constraints on the quartic vertex are still rather weak, and the larger data sets and higher beam energy of Run 2 are likely to improve these results substantially.

4.8 Outlook to Run 2

The upcoming Run 2 at the LHC promises an increase of the beam energy to 13 TeV as well as a substantial increase in instantaneous and integrated luminosity. This improved performance comes at the price of a larger amount of pile-up events. Precision measurements of EW parameters, the accuracy of which is already largely limited by systematic uncertainties, are not expected to improve substantially in Run 2. Even though some systematic uncertainties that are currently extracted from limited control samples may be reduced, the increased systematic uncertainties related to the increased pile-up will make overall improvements a formidable challenge. To nevertheless improve the precision of EW measurements, special runs with low instantaneous luminosity and correspondingly low pile-up are under consideration. Independently of the exact LHC running scenario it is expected that the currently very rapid publication schedule for the measurement of EW parameters will slow down, as a more and more detailed understanding of systematic uncertainties becomes necessary. Measurements with multi-boson final states such as triple weak-boson production or VBS, on the other hand, are likely to profit substantially from the conditions in Run 2, as the higher beams energy disproportionally increases the parton luminosity at high invariant masses.

References

1. S. Dittmaier, M. Schumacher, Prog. Part. Nucl. Phys. **70**, 1–54 (2013)
2. S. Weinberg, *The Quantum Theory of Fields. Volume 2: Modern Applications* (Cambridge University Press, Cambridge, 1996)
3. M. Bohm, A. Denner, H. Joos, *Gauge Theories of the Strong and Electroweak Interaction* (Teubner, Stuttgart, 2001)
4. G.F. Sterman, *An Introduction to Quantum Field Theory* (Cambridge University Press, Cambridge, 1994)
5. R.K. Ellis, W.J. Stirling, B.R. Webber, *QCD and Collider Physics* (Cambridge University Press, Cambridge, 1996)
6. J. Collins, *Foundations of Perturbative QCD* (Cambridge University Press, Cambridge, 2013)
7. D. Ross, M. Veltman, Nucl. Phys. B **95**, 135 (1975)
8. ALEPH Collaboration, DELPHI Collaboration, L3 Collaboration, OPAL Collaboration, SLD Collaboration, LEP Electroweak Working Group, SLD Electroweak Group, SLD Heavy Flavour Group, Phys. Rep. **427**, 257–454 (2006)
9. P. Sikivie, L. Susskind, M.B. Voloshin, V.I. Zakharov, Nucl. Phys. B **173**, 189 (1980)
10. S. Dittmaier, M. Krämer, Phys. Rev. D **65**, 073007 (2002)
11. J. Butterworth et al. (2014), arXiv:1405.1067
12. V.S. Fadin, L. Lipatov, A.D. Martin, M. Melles, Phys. Rev. D **61**, 094002 (2000)
13. J.H. Kühn, A. Penin, V.A. Smirnov, Eur. Phys. J. C **17**, 97–105 (2000)
14. M. Ciafaloni, P. Ciafaloni, D. Comelli, Phys. Rev. Lett. **84**, 4810–4813 (2000)
15. M. Hori, H. Kawamura, J. Kodaira, Phys. Lett. B **491**, 275–279 (2000)
16. A. Denner, S. Pozzorini, Eur. Phys. J. C **21**, 63–79 (2001)
17. M. Melles, Eur. Phys. J. C **24**, 193–204 (2002)
18. W. Beenakker, A. Werthenbach, Nucl. Phys. B **630**, 3–54 (2002)
19. A. Denner, M. Melles, S. Pozzorini, Nucl. Phys. B **662**, 299–333 (2003)
20. B. Jantzen, J.H. Kühn, A.A. Penin, V.A. Smirnov, Phys. Rev. D **72**, 051301 (2005)
21. B. Jantzen, J.H. Kühn, A.A. Penin, V.A. Smirnov, Nucl. Phys. B **731**, 188–212 (2005)
22. A. Denner, B. Jantzen, S. Pozzorini, Nucl. Phys. B **761**, 1–62 (2007)
23. S. Brensing, S. Dittmaier, M. Krämer, A. Mück, Phys. Rev. D **77**, 073006 (2008)
24. S. Dittmaier, M. Huber, JHEP **01**, 060 (2010)
25. S. Dittmaier, A. Huss, C. Speckner, JHEP **11**, 095 (2012)

26. P. Ciafaloni, D. Comelli, JHEP **09**, 055 (2006)
27. U. Baur, Phys. Rev. D **75**, 013005 (2007)
28. G. Bell, J. Kühn, J. Rittinger, Eur. Phys. J. C **70**, 659–671 (2010)
29. M. Chiesa, G. Montagna, L. Barz, M. Moretti, O. Nicrosini et al., Phys. Rev. Lett. **111**, 121801 (2013)
30. W. Beenakker et al. (1996), arXiv:hep-ph/9602351
31. A. Arbuzov, Phys. Lett. B **470**, 252–258 (1999)
32. W. Placzek, S. Jadach, Eur. Phys. J. C **29**, 325–339 (2003)
33. C. Carloni Calame, G. Montagna, O. Nicrosini, M. Treccani, Phys. Rev. D **69**, 037301 (2004)
34. P. Golonka, Z. Was, Eur. Phys. J. C **50**, 53–62 (2007)
35. T. Kinoshita, J. Math. Phys. **3**, 650–677 (1962)
36. T. Lee, M. Nauenberg, Phys. Rev. **133**, B1549–B1562 (1964)
37. ATLAS Collaboration, Phys. Lett. B **705**, 415–434 (2011)
38. C. Carloni Calame, G. Montagna, O. Nicrosini, A. Vicini, JHEP **10**, 109 (2007)
39. R. Boughezal, Y. Li, F. Petriello, Phys. Rev. D **89**, 034030 (2014)
40. A. Bierweiler, T. Kasprzik, J.H. Kühn, S. Uccirati, JHEP **11**, 093 (2012)
41. A. Bierweiler, T. Kasprzik, J.H. Kühn, JHEP **12**, 071 (2013)
42. J. Baglio, L.D. Ninh, M.M. Weber, Phys. Rev. D **88**, 113005 (2013)
43. M. Billoni, S. Dittmaier, B. Jäger, C. Speckner, JHEP **12**, 043 (2013)
44. LHC Higgs Cross Section Working Group (2011), arXiv:1101.0593
45. LHC Higgs Cross Section Working Group (2012), arXiv:1201.3084
46. LHC Higgs Cross Section Working Group (2013), arXiv:1307.1347
47. SM and NLO Multileg Working Group (2010), arXiv:1003.1241
48. S. Gieseke, T. Kasprzik, J.H. Kühn, Eur. Phys. J. C **74**, 8, 2988 (2014)
49. M. Ciccolini, A. Denner, S. Dittmaier, Phys. Rev. Lett. **99**, 161803 (2007)
50. M. Ciccolini, A. Denner, S. Dittmaier, Phys. Rev. D **77**, 013002 (2008)
51. M. Bahr, S. Gieseke, M. Gigg, D. Grellscheid, K. Hamilton et al., Eur. Phys. J. C **58**, 639–707 (2008)
52. K. Arnold et al. (2012), arXiv:1205.4902
53. G. Corcella et al. (2002), arXiv:hep-ph/0210213
54. T. Gleisberg et al., JHEP **02**, 007 (2009)
55. S. Höche, S. Schumann, F. Siegert, Phys. Rev. D **81**, 034026 (2010)
56. Q.H. Cao, C. Yuan, Phys. Rev. Lett. **93**, 042001 (2004)
57. G. Balossini et al., JHEP **01**, 013 (2010)
58. W.P. Baczek, S. Jadach, M. Krasny, Acta Phys. Polon. B *44*(11), 2171–2178 (2013)
59. S. Frixione, B.R. Webber, JHEP **06**, 029 (2002)
60. P. Nason, JHEP **11**, 040 (2004)
61. S. Frixione, P. Nason, C. Oleari, JHEP **11**, 070 (2007)
62. C. Bernaciak, D. Wackeroth, Phys. Rev. D **85**, 093003 (2012)
63. L. Barze et al., JHEP **04**, 037 (2012)
64. L. Barze et al., Eur. Phys. J. C **73**, 2474 (2013)
65. P. Gambino, P.A. Grassi, Phys. Rev. D **62**, 076002 (2000)
66. P.A. Grassi, B.A. Kniehl, A. Sirlin, Phys. Rev. D **65**, 085001 (2002)
67. A. Denner, Fortsch. Phys. **41**, 307–420 (1993)
68. D.Y. Bardin, A. Leike, T. Riemann, M. Sachwitz, Phys. Lett. B **206**, 539–542 (1988)
69. W. Beenakker et al., Nucl. Phys. B **500**, 255–298 (1997)
70. R.G. Stuart, Phys. Lett. B **262**, 113–119 (1991)
71. A. Aeppli, G.J. van Oldenborgh, D. Wyler, Nucl. Phys. B **428**, 126–146 (1994)
72. M. Beneke, A. Chapovsky, A. Signer, G. Zanderighi, Phys. Rev. Lett. **93**, 011602 (2004)
73. M. Beneke, A. Chapovsky, A. Signer, G. Zanderighi, Nucl. Phys. B **686**, 205–247 (2004)
74. A.H. Hoang, C.J. Reisser, Phys. Rev. D **71**, 074022 (2005)
75. A. Denner, S. Dittmaier, M. Roth, D. Wackeroth, Nucl. Phys. B **560**, 33–65 (1999)
76. A. Denner, S. Dittmaier, M. Roth, L. Wieders, Nucl. Phys. B **724**, 247–294 (2005)
77. CMS Collaboration, JINST **8**, P09009 (2013)

78. CMS Collaboration, JINST **8**, P11002 (2013)
79. ATLAS Collaboration, Eur. Phys. J. C **74**, 11, 3130 (2014)
80. ATLAS Collaboration, Eur. Phys. J. C **74**(10), 3071 (2014)
81. ATLAS Collaboration, Eur. Phys. J. C **72**, 1844 (2012)
82. R. Hamberg, W. van Neerven, T. Matsuura, Nucl. Phys. B **359**, 343–405 (1991)
83. R.V. Harlander, W.B. Kilgore, Phys. Rev. Lett. **88**, 201801 (2002)
84. C. Anastasiou, L.J. Dixon, K. Melnikov, F. Petriello, Phys. Rev. D **69**, 094008 (2004)
85. K. Melnikov, F. Petriello, Phys. Rev. D **74**, 114017 (2006)
86. S. Catani et al., Phys. Rev. Lett. **103**, 082001 (2009)
87. R. Gavin, Y. Li, F. Petriello, S. Quackenbush, Comput. Phys. Commun. **182**, 2388–2403 (2011)
88. U. Baur et al., Phys. Rev. D **65**, 033007 (2002)
89. U. Baur, D. Wackeroth, Phys. Rev. D **70**, 073015 (2004)
90. A. Arbuzov et al., Eur. Phys. J. C **46**, 407–412 (2006)
91. C. Carloni Calame, G. Montagna, O. Nicrosini, A. Vicini, JHEP **12**, 016 (2006)
92. V. Zykunov, Phys. Rev. D **75**, 073019 (2007)
93. A. Arbuzov et al., Eur. Phys. J. C **54**, 451–460 (2008)
94. S. Frixione, B.R. Webber (2006), arXiv:hep-ph/0612272
95. S. Alioli, P. Nason, C. Oleari, E. Re, JHEP **07**, 060 (2008)
96. K. Hamilton, P. Richardson, J. Tully, JHEP **10**, 015 (2008)
97. C. Balazs, C. Yuan, Phys. Rev. D **56**, 5558–5583 (1997)
98. F. Landry, R. Brock, P.M. Nadolsky, C. Yuan, Phys. Rev. D **67**, 073016 (2003)
99. G. Bozzi et al., Phys. Lett. B **696**, 207–213 (2011)
100. S. Mantry, F. Petriello, Phys. Rev. D **83**, 053007 (2011)
101. T. Becher, M. Neubert, D. Wilhelm, JHEP **02**, 124 (2012)
102. P. Richardson et al., JHEP **06**, 090 (2012)
103. Y. Li, F. Petriello, Phys. Rev. D **86**, 094034 (2012)
104. S. Dittmaier, A. Huss, C. Schwinn, Nucl. Phys. B **885**, 318–372 (2014)
105. S. Dittmaier, A. Huss, C. Schwinn, PoS LL **2014**, 045 (2014)
106. A. Kotikov, J.H. Kuhn, O. Veretin, Nucl. Phys. B **788**, 47–62 (2008)
107. W.B. Kilgore, C. Sturm, Phys. Rev. D **85**, 033005 (2012)
108. R. Bonciani, PoS **EPS-HEP2011**, 365 (2011)
109. J.H. Kühn, A. Kulesza, S. Pozzorini, M. Schulze, Nucl. Phys. B **727**, 368–394 (2005)
110. J.H. Kühn, A. Kulesza, S. Pozzorini, M. Schulze, Nucl. Phys. B **797**, 27–77 (2008)
111. W. Hollik, T. Kasprzik, B. Kniehl, Nucl. Phys. B **790**, 138–159 (2008)
112. A. Denner, S. Dittmaier, T. Kasprzik, A. Muck, JHEP **08**, 075 (2009)
113. A. Denner, S. Dittmaier, T. Kasprzik, A. Muck, JHEP **06**, 069 (2011)
114. A. Denner, S. Dittmaier, T. Kasprzik, A. Muck, Eur. Phys. J. C **73**, 2297 (2013)
115. L.J. Dixon, Z. Kunszt, A. Signer, Nucl. Phys. B **531**, 3–23 (1998)
116. J.M. Campbell, R.K. Ellis, Phys. Rev. D **60**, 113006 (1999)
117. D. De Florian, A. Signer, Eur. Phys. J. C **16**, 105–114 (2000)
118. J.M. Campbell, R.K. Ellis, C. Williams, JHEP **07**, 018 (2011)
119. ATLAS Collaboration, Phys. Rev. D **85**, 072004 (2012)
120. CMS Collaboration, JHEP **10**, 132 (2011)
121. CMS Collaboration, Phys. Rev. Lett. **112**, 191802 (2014)
122. CMS Collaboration, JHEP **12**, 030 (2013)
123. ATLAS Collaboration, Phys. Lett. B **725**, 223–242 (2013)
124. CMS Collaboration, Phys. Rev. D **85**, 032002 (2012)
125. ATLAS Collaboration, Phys. Lett. B **720**, 32–51 (2013)
126. ATLAS Collaboration, JHEP **1409**, 145 (2014)
127. CMS Collaboration, Phys. Rev. D **84**, 112002 (2011)
128. CMS Collaboration, Phys. Lett. B **718**, 752–772 (2013)
129. The ALEPH, DELPHI, L3, OPAL, SLD Collaborations, the LEP Electroweak Working Group, the SLD Electroweak and Heavy Flavour Groups, Phys. Rep. **427**, 257 (2006)

130. J.C. Collins, D.E. Soper, Phys. Rev. D **16**, 2219 (1977)
131. The ALEPH, DELPHI, L3, OPAL Collaborations, the LEP Electroweak Working Group, Phys. Rep. **532**, 119 (2013)
132. CDF and DØ Collaborations, Tevatron Electroweak Working Group (2012), arXiv:1204.0042
133. F. Fayette (2009), arXiv:0906.4260
134. S. Berge, P.M. Nadolsky, F.I. Olness, Phys. Rev. D **73**, 013002 (2006)
135. Gfitter Group, Eur. Phys. J. C **74**, 3046 (2014)
136. O. Buchmüller et al., Eur. Phys. J. C **74**(6), 2922 (2014)
137. W. Buchmüller, D. Wyler, Nucl. Phys. B **268**, 621–653 (1986)
138. B. Grzadkowski, M. Iskrzynski, M. Misiak, J. Rosiek, JHEP **10**, 085 (2010)
139. K. Hagiwara, R. Peccei, D. Zeppenfeld, K. Hikasa, Nucl. Phys. B **282**, 253 (1987)
140. G. Gounaris et al. (1996), arXiv:hep-ph/9601233
141. G. Gounaris, J. Layssac, F. Renard, Phys. Rev. D **61**, 073013 (2000)
142. G. Gounaris, J. Layssac, F. Renard, Phys. Rev. D **62**, 073012 (2000)
143. O.J. Eboli, M. Gonzalez-Garcia, S. Lietti, S. Novaes, Phys. Rev. D **63**, 075008 (2001)
144. A. Alboteanu, W. Kilian, J. Reuter, JHEP **11**, 010 (2008)
145. M. Grazzini, S. Kallweit, D. Rathlev, A. Torre, Phys. Lett. B **731**, 204–207 (2014)
146. F. Cascioli et al., Phys. Lett. B **735**, 311–313 (2014)
147. T. Gehrmann et al., Phys. Rev. Lett. **113**(21), 212001 (2014)
148. W. Hollik, C. Meier, Phys. Lett. B **590**, 69–75 (2004)
149. E. Accomando, A. Denner, C. Meier, Eur. Phys. J. C **47**, 125–146 (2006)
150. K. Hamilton, JHEP **01**, 009 (2011)
151. M. Grazzini, JHEP **01**, 095 (2006)
152. S. Dawson, I.M. Lewis, M. Zeng, Phys. Rev. D **88**(5), 054 028 (2013)
153. T. Binoth, M. Ciccolini, N. Kauer, M. Kramer, JHEP **03**, 065 (2005)
154. F. Cascioli et al., JHEP **01**, 046 (2014)
155. CMS Collaboration, Phys. Lett. B **740**, 250–272 (2015)
156. ATLAS Collaboration, JHEP **03**, 128 (2013)
157. ALEPH, DELPHI, L3 and OPAL Collaborations, LEP Electroweak Working Group (2006), arXiv:hep-ex/0612034
158. DØ Collaboration, Phys. Rev. D **88**, 012005 (2013)
159. CMS Collaboration, JHEP **07**, 116 (2013)
160. CMS Collaboration, Phys. Rev. D **90**, 032008 (2014)
161. G. Belanger, F. Boudjema, Phys. Lett. B **288**, 201–209 (1992)
162. O.J. Eboli, M. Gonzalez-Garcia, S. Novaes, Nucl. Phys. B **411**, 381–396 (1994)
163. A. Denner, S. Dittmaier, M. Roth, D. Wackeroth, Eur. Phys. J. C **20**, 201–215 (2001)
164. A. Lazopoulos, K. Melnikov, F. Petriello, Phys. Rev. D **76**, 014001 (2007)
165. V. Hankele, D. Zeppenfeld, Phys. Lett. B **661**, 103–108 (2008)
166. T. Binoth, G. Ossola, C. Papadopoulos, R. Pittau, JHEP **06**, 082 (2008)
167. D.T. Nhung, L.D. Ninh, M.M. Weber, JHEP **12**, 096 (2013)
168. L3 Collaboration, Phys. Lett. B **527**, 29–38 (2002)
169. M. Veltman, Acta Phys. Polon. B **8**, 475 (1977)
170. B.W. Lee, C. Quigg, H. Thacker, Phys. Rev. Lett. **38**, 883–885 (1977)
171. B.W. Lee, C. Quigg, H. Thacker, Phys. Rev. D **16**, 1519 (1977)
172. D. Espriu, B. Yencho, Phys. Rev. D **87**(5), 055017 (2013)
173. J. Chang, K. Cheung, C.T. Lu, T.C. Yuan, Phys. Rev. D **87**(9), 093005 (2013)
174. B. Jäger, C. Oleari, D. Zeppenfeld, JHEP **07**, 015 (2006)
175. B. Jäger, C. Oleari, D. Zeppenfeld, Phys. Rev. D **73**, 113006 (2006)
176. B. Jäger, C. Oleari, D. Zeppenfeld, Phys. Rev. D **80**, 034022 (2009)
177. G. Bozzi, B. Jäger, C. Oleari, D. Zeppenfeld, Phys. Rev. D **75**, 073004 (2007)
178. A. Denner, L. Hosekova, S. Kallweit, Phys. Rev. D **86**, 114014 (2012)
179. T. Melia, K. Melnikov, R. Rontsch, G. Zanderighi, JHEP **12**, 053 (2010)
180. T. Melia, K. Melnikov, R. Rontsch, G. Zanderighi, Phys. Rev. D **83**, 114043 (2011)
181. N. Greiner et al., Phys. Lett. B **713**, 277–283 (2012)

182. F. Campanario, M. Kerner, L.D. Ninh, D. Zeppenfeld, Phys. Rev. Lett. **111**, 052003 (2013)
183. F. Campanario, M. Kerner, L.D. Ninh, D. Zeppenfeld, Phys. Rev. D **89**, 054009 (2014)
184. T. Melia, P. Nason, R. Rontsch, G. Zanderighi, Eur. Phys. J. C **71**, 1670 (2011)
185. T. Melia, P. Nason, R. Rontsch, G. Zanderighi, JHEP **11**, 078 (2011)
186. B. Jäger, G. Zanderighi, JHEP **11**, 055 (2011)
187. CMS Collaboration, Eur. Phys. J. C **75**(2), 66 (2015)
188. ATLAS Collaboration, JHEP **04**, 031 (2014)
189. ATLAS Collaboration, Phys. Rev. Lett. **113**, 141803 (2014)
190. CMS Collaboration, Phys. Rev. Lett. **114**, 051801 (2015)

Chapter 5
Studies of Quantum Chromodynamics at the LHC

Tancredi Carli, Klaus Rabbertz and Steffen Schumann

Abstract A successful description of hadron-hadron collision data demands a profound understanding of quantum chromodynamics. Inevitably, the complexity of strong-interaction phenomena requires the use of a large variety of theoretical techniques—from perturbative cross-section calculations up to the modelling of exclusive hadronic final states. Together with the unprecedented precision of the data provided by the experiments in the first running period of the LHC, a solid foundation of hadron-hadron collision physics at the TeV scale could be established that allowed the discovery of the Higgs boson and that is vital for estimating the background in searches for new phenomena. This chapter presents the advances in theoretical methods side-by-side with related key measurements in an integrated approach.

5.1 Introduction

Quantum chromodynamics (QCD) is the well-established quantum field theory of the strong interaction [1, 2] and one cornerstone of the Standard Model (SM) of particle physics. Like the electromagnetic and the weak force, QCD belongs to the

T. Carli (✉)
CERN, 1211 Geneva 23, Switzerland
e-mail: tancredi.carli@cern.ch

K. Rabbertz
Institut für Experimentelle Kernphysik, KIT, Wolfgang-Gaede-Str. 1,
76131 Karlsruhe, Germany
e-mail: klaus.rabbertz@kit.edu

S. Schumann
II. Physikalisches Institut, Georg-August-Universität Göttingen,
Friedrich-Hund-Platz 1, 37077 Göttingen, Germany
e-mail: steffen.schumann@phys.uni-goettingen.de

T. Schörner-Sadenius (ed.), *The Large Hadron Collider*,
DOI 10.1007/978-3-319-15001-7_5

class of gauge field theories. The assumption that the corresponding gauge symmetry is an exact symmetry of nature results in truly massless force carriers of the strong force, the gluons. However, as a consequence of the non-Abelian character of the SU(3) QCD gauge group, the gluons carry a strong or so-called "colour" charge and interact amongst themselves. This is a striking difference to the electromagnetic force, which is mediated by electrically uncharged photons, and it induces—amongst other effects—the confinement of strongly interacting particles at low energies. In the following, the colour-charged constituents of hadrons, i.e. quarks, antiquarks, and gluons, will generically be denoted as "partons".

In hadronic collisions like at the LHC, QCD effects are omnipresent, and their detailed understanding is indispensable for the interpretation of collider data, whether to search for new phenomena or to perform precision studies of model parameters. Despite complications through the gluon self-interactions, the perturbative expansion of QCD (pQCD) that describes interactions with large momentum exchanges ("hard interactions") in terms of parton-parton scatterings remains the most powerful theoretical technique. This technique allows quantitative predictions at parton level for observables ranging from inclusive production rates to shape observables, which are sensitive to details of the QCD radiation pattern. Nevertheless the theory's confining nature obliges us to address non-perturbative aspects. This includes a reliable understanding of the short-distance parton structure of the initial-state protons, the fragmentation of final-state partons into hadrons, or the modelling of soft proton interactions.

The unprecedented experimental precision achieved with the new detectors at the LHC requires equally accurate theoretical predictions and has sparked rapid progress in the field of perturbative calculations using both analytical methods and modern Monte Carlo (MC) event generators. The interplay between experiment and theory enforced the development of new observables and novel techniques to match the challenges arising on both sides. In summary, the improved theoretical understanding of the QCD dynamics including the ability to precisely predict even complicated high-multiplicity final states and the excellent performance and understanding of the LHC machine and detectors together with sophisticated analysis techniques reveal a more refined and detailed picture of QCD than ever before.

For the presentation of QCD-related experimental results obtained during LHC Run 1 and the underlying theoretical developments, an integrative approach is chosen—reflecting the productive and fruitful interplay of the two communities. As a consequence, compromises on the content had to be taken, and only a selection of the most important measurements is discussed—omissions in the presentation of theoretical methods were unavoidable.

The chapter at hands is structured as follows: After a brief reminder of the basics of the QCD theory and the central aspects of perturbative QCD, the discussion focuses on various approximations in the modelling of scattering processes, including parton-shower simulations and parton-level predictions at next-to-leading and next-to-next-to-leading order in the strong coupling. Then the discussion turns to the observation of multi-jet final states, the successful description of which requires the combination of both fixed-order calculations and all-order (i.e. parton-shower) techniques. This is followed by a section on analytical methods for the resummation of large logarithms; these are exemplified using gap-fraction and jet-substructure observables. The chapter closes with a presentation of various phenomena and measurements sensitive to non-perturbative aspects of QCD.

5.2 Basic Elements of QCD

The discussion shall begin with a brief reminder of the ingredients of the QCD Lagrangian that defines the Feynman rules required for a perturbative analysis of QCD. The classical QCD Lagrangian is composed out of the free Dirac Lagrangians for the six quark fields and the kinetic and self-interaction terms for the gluon fields A_μ^a, labelled by a colour index $a = 1, \ldots, 8$. These two parts get minimally coupled through a gauge covariant derivative D_μ:

$$\mathcal{L}_{\text{QCD}} = \mathcal{L}_{\text{gauge}} + \mathcal{L}_{\text{quarks}} ,$$

where

$$\mathcal{L}_{\text{gauge}} = -\frac{1}{4} F_{\mu\nu}^a F_a^{\mu\nu} , \quad \text{with} \quad F_{\mu\nu}^a = \partial_\mu A_\nu^a - \partial_\nu A_\mu^a - g_s f_{abc} A_\mu^b A_\nu^c$$

the gluon field-strength tensor, and

$$\mathcal{L}_{\text{quarks}} = \sum_{q \in \{u,d,s,c,b,t\}} \bar{q} \left(i\gamma^\mu D_\mu - m_q \right) q , \quad \text{with} \quad D_\mu = \partial_\mu + i g_s t^a A_\mu^a$$

the QCD covariant derivative (see also the introduction to the SM Lagrangian in Sect. 4.2). Quark masses are denoted by m_q. The SU(3) generator matrices introduced here obey the algebra

$$[t^a, t^b] = i f_{abc} t^c ,$$

defining the QCD structure constants f_{abc}. The classical QCD Lagrangian exhibits the property of local gauge invariance, i.e. invariance under a simultaneous redefinition of the quark and gluon fields. As a consequence of this internal symmetry, it is impossible to define the gluon field propagator without explicitly specifying a choice of gauge. A Lorentz-covariant way to fix the gauge is given by the class of R_ξ gauges, imposed by adding a term

$$\mathcal{L}_{\text{gauge-fixing}} = -\frac{1}{2\xi} (\partial^\mu A_\mu^a)^2$$

to the classical Lagrangian. Because of the non-Abelian character of the QCD gauge group, the full Lagrangian of the quantum field theory features a further contribution, the ghost Lagrangian

$$\mathcal{L}_{\text{ghost}} = \partial_\mu \eta^{a\dagger} \left(D^\mu_{ab} \eta^b \right) ,$$

that represents the field-dependent Faddeev–Popov determinant. The ghost fields η^a are represented by anti-commuting scalar fields. This completes the Lagrangian for a consistent version of a quantum field theory of the strong interaction. Accordingly one can read off the QCD Feynman rules, featuring three-point quark-quark-gluon and ghost-ghost-gluon interactions as well as triple and quartic gluon self-interactions. All of these interaction vertices are proportional to the strong charge g_s. This is also the relevant parameter when applying the method of perturbation theory to QCD. Defining the QCD counterpart of the QED fine-structure constant $\alpha_s = g_s^2/4\pi$, one can expect a truncation of the power-series expansion for a given observable $\mathcal{O}$, i.e.

$$\mathcal{O} = \mathcal{O}_0 + \mathcal{O}_1 \alpha_s + \mathcal{O}_2 \alpha_s^2 + \cdots ,$$

to yield meaningful estimates as long as $\alpha_s \ll 1$.

A prime example of a quantity evaluated in perturbation theory is the QCD β function. It determines the running of the coupling constant α_s through the renormalisation group equation

$$Q^2 \frac{\partial \alpha_s}{\partial Q^2} = \beta(\alpha_s) , \quad \text{with} \quad \beta(\alpha_s) = -\alpha_s^2 (b_0 + b_1 \alpha_s + b_2 \alpha_s^2 + \mathcal{O}(\alpha_s^3)) , \qquad (5.1)$$

and

$$b_0 = \frac{33 - 2n_F}{12\pi} , \quad b_1 = \frac{153 - 19n_F}{24\pi^2} , \quad b_2 = \frac{77139 - 15099 n_F + 325 n_F^2}{3456\pi^3} . \qquad (5.2)$$

n_F denotes the number of quark flavours with masses m_q smaller than the scale Q. Note that the higher coefficients b_2 and b_3 (see [3]) are renormalisation-scheme dependent. Here b_2 is quoted in the $\overline{\text{MS}}$ scheme. Retaining only the leading term b_0, (5.1) is solved by

$$\alpha_s(Q^2) = \frac{\alpha_s(\mu^2)}{1 + b_0 \ln\left(Q^2/\mu^2\right) \alpha_s(\mu^2)} , \qquad (5.3)$$

which relates the strength of the coupling at a scale Q to the one at scale μ, assuming both scales to be in the perturbative regime. The non-Abelian nature of QCD manifests itself in the negative sign of the β function. Thus, as long as $n_F < 17$, the coupling becomes weaker at higher scales Q, or, in other words, the QCD colour charge decreases when the distance decreases. For high scales Q, QCD becomes almost a free theory—a property known as "asymptotic freedom". It is this weakly coupled regime where perturbative methods can successfully be applied and quantitative predictions for hard scattering processes can be made. The world average

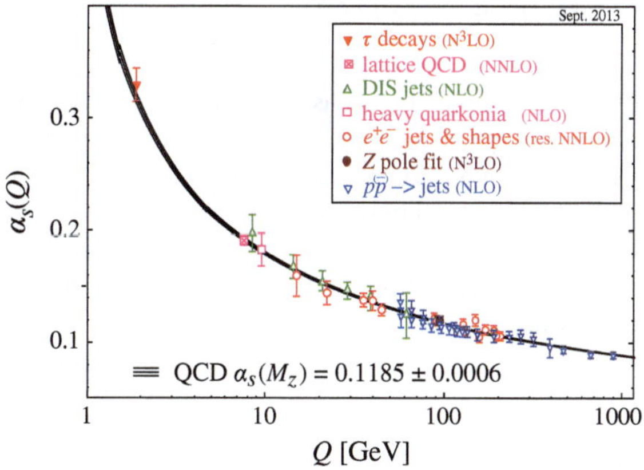

Fig. 5.1 Summary of measurements of the strong coupling α_s as a function of the respective energy scale Q. By including new LHC data, the range in Q could be extended to the TeV scale. Adapted from [4]

value of the strong coupling as of 2014, quoted at the scale of the Z-boson mass M_Z, is given by

$$\alpha_s(M_Z) = 0.1185 \pm 0.0006\,,$$

derived from hadronic τ-lepton decays, lattice QCD calculations, deep-inelastic scattering data, electron-positron annihilation processes, and electroweak precision fits [4]. Figure 5.1 shows a summary of measurements of the strong coupling at energy scales ranging from the mass of the τ-lepton of $M_\tau \approx 1.8\,\text{GeV}$ up to the TeV scale thanks to newly included LHC data. The historical development of α_s determinations is discussed in Fig. 12.2 in Chap. 12.

The dynamical behaviour of $\alpha_s(Q^2)$ implies an increase of the QCD coupling at small momentum transfer, i.e. large distances. When the coupling approaches unity, perturbation theory is not valid anymore. The parameter Λ_{QCD} is defined as the scale, where $\alpha_s(Q^2)$ formally diverges. With this definition, (5.3) can be rewritten as $\alpha_s(Q^2) = \left(b_0 \ln\left(Q^2/\Lambda^2_{\text{QCD}}\right)\right)^{-1}$. For $n_F = 5$ flavours in the $\overline{\text{MS}}$ scheme, Λ_{QCD} roughly amounts to 214 MeV and represents the dividing line below which one is in the manifestly non-perturbative regime of QCD. It is the growth of the coupling at small scales that makes QCD a theory of the strong interaction—the fundamental force that confines the quarks and gluons into ordinary hadronic matter, e.g. the protons and neutrons. For the purpose of LHC physics, one has to account for this phenomenon of "confinement" when modelling the transition from free quarks and gluons to the bound-state hadrons observed in the detectors. Lacking a first-principles understanding of this process, one mostly has to rely on MC models for this aspect. Even more fundamentally, the partonic content of the colliding protons needs to be parametrised in order to allow for a description of LHC collision events through partonic scattering processes.

5.3 Perturbative QCD

Quantitative predictions based on the non-Abelian QCD Lagrangian can be obtained either with lattice methods employing a discretised space-time or using perturbation theory. Given the complexity of the final states produced in proton-proton collisions with high momentum transfer, lattice techniques are of no practical importance for the prediction of LHC events at present. Instead one has to rely on perturbative methods, which are possibly supplemented by models for the transition of partons to hadrons.

5.3.1 *Cross-Section Predictions*

In perturbative QCD, the cross section for a hard scattering process at a hadron-hadron collider can be written in the following factorised form

$$\sigma_{H_1H_2\to X} = \sum_{i,j} \int \mathrm{d}x_1 \mathrm{d}x_2 \; f_{i/H_1}(x_1, \mu_F) f_{j/H_2}(x_2, \mu_F) \\ \times \hat{\sigma}_{ij\to X}\left(x_1 P_1, x_2 P_2, \alpha_s(\mu_R), \frac{Q}{\mu_F}\right), \tag{5.4}$$

where the sum extends over all contributing initial-state partons $i, j \in \{q, \bar{q}, g\}$. It is assumed here that the scale associated with the hard process, Q, is much larger than Λ_{QCD}, the delimiting scale for the applicability of perturbative methods to QCD. In this high-energy limit, effects related to the binding of the partons in the initial-state protons can be neglected. As a consequence, the cross section simply factorises into a product of parton distribution functions (PDFs) $f_{i/H}(x, \mu_F)$ of non-perturbative origin and the perturbatively calculable partonic cross section. The parton distribution functions model the probability to find a parton of flavour i in the incoming hadron H (protons at the LHC) with a fraction x of the hadron's momentum P. The resulting squared partonic centre-of-mass energy is given by $\hat{s} = x_1 x_2 s$, with $s = (P_1 + P_2)^2$ the squared hadronic centre-of-mass energy.

The collinear factorisation ansatz underlying (5.4) is the key to quantitative predictions in the framework of QCD that can be compared to actual LHC collision data. Based on the property of asymptotic freedom of QCD, the desired cross section can be expanded as a power series of the coupling constant α_s. The lowest-order coefficient is denoted as leading order (LO), the subsequent ones as next-to-leading order (NLO) and next-to-next-to-leading order (NNLO), respectively.

When calculating the LO, NLO or NNLO estimate for a partonic cross section, the appropriate QCD evolution of the parton distribution functions has to be used.

Leading-order cross-section calculations are fully automated by means of tree-level matrix-element generator programs such as ALPGEN [5], AMEGIC [6], COMIX [7], HELAC/ PHEGAS [8], MADGRAPH [9] or WHIZARD [10]. These codes are capable of providing integrated cross sections and parton-level events for almost arbitrary Standard Model final states, with multiplicities ranging up to ten particles. In particular for high-multiplicity final states, implementations relying on recursive algorithms for the generation of the expressions for the amplitudes, e.g. Berends and Giele recursion [11], prove most efficient [12, 13].

Over the past years there has been enormous progress in the evaluation of processes at NLO and NNLO in the strong coupling. These developments and related precision measurements will be addressed in detail in later sections of this chapter.

5.3.2 Fragmentation and Hadronic Jets

One entity from (5.4) that has not yet been discussed is the final state X of a collision. The simplest reaction that can be considered is the Drell–Yan process [14], where a quark and an antiquark annihilate to produce a lepton pair: $\hat{\sigma}\left(q\bar{q} \to \ell^+\ell^-\right)$. In this case there are no strongly interacting particles in the final state, and the theory prediction can directly be compared to the measured leptons. Merely the proton remnants, which fragment into hadrons along the beam lines, have to be described by non-perturbative models in MC event generators. At high transverse momenta, the two leptons are well separated from any such proton debris and high-precision comparisons with theory even at NNLO become possible. This is discussed in more detail later in this chapter.

However, in the vast majority of reactions at least some colour-charged partons are produced so that a further step covering the transition from the partonic final state to measurable particles, the so-called "particle level", is needed. Here, "measurable" refers to colour-neutral particles with mean decay lengths such that $c\tau > 10\,\text{mm}$, where c is the speed of light and τ the lifetime of a particle. One possibility to account for this transition is to reuse the concept underlying the PDFs that describe the partonic content of a hadron, only in an inverted sense. The necessary functions $D_{k\to h}(z, \mu_F)$ are called fragmentation functions (FFs) and are the final-state analogues of the PDFs. They parametrise the probability of finding a hadron h within the fragmentation products of parton k, carrying the fraction z of the parton momentum. Like the PDFs, fragmentation functions depend on a non-physical resolution or fragmentation scale μ_F. Again, these functions can currently not be determined by first principles in QCD, but once they have been measured (for example under the experimentally more favourable conditions of e^+e^- collisions at the LEP collider), they are universally valid.

A second possibility to account for the transition to measurable particles makes use of the concept of hadronic jets. Instead of looking into the detailed production of identified particles—an experimentally very challenging endeavour—for the majority of processes it is sufficient to know how much energy and momentum is carried away by hadrons. QCD predicts that large-distance non-perturbative (NP) effects are mostly decoupled from the hard reaction so that highly energetic partons fragment into a collimated stream or "jet" of hadrons, which inherits energy and momentum

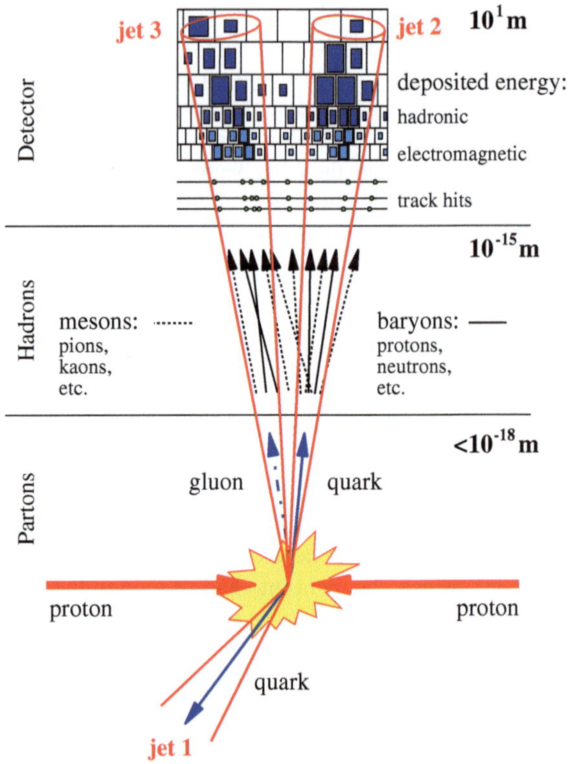

Fig. 5.2 Illustration of a jet to which bundles of partons, hadrons, or detector measurements are grouped together

from its parent parton. To define what "collimated" means, a prescription is required that, given some distance measure, unambiguously decides which objects belong to a jet. As one wants to compare predictions by pQCD with measured particles, tracks, or energy depositions as illustrated in Fig. 5.2, a jet algorithm is needed that is applicable to theoretical calculations as well as to measurements from different experiments.

Most importantly, to deal with the cancellation of collinear and soft singularities appearing in pQCD, a jet algorithm must be collinear- and infrared-safe. This means that the outcome of a jet-clustering procedure depends neither on the splitting or merging of collinear parton four-vectors nor on the addition of arbitrarily soft partons to the list of objects to be clustered. The first description of a collinear- and infrared-safe jet algorithm, which grouped partons or particles together that are inside an angular cone around a specific direction, was given by Sterman and Weinberg in 1977 [15].

The most important requirements for jet algorithms from the experimental side are (i) independence of detector details, (ii) maximal reconstruction efficiency, (iii) minimal resolution smearing, (iv) computational efficiency, and (v) ease of calibration. The extension of the original jet definition, which was specialised to 2-jet events in

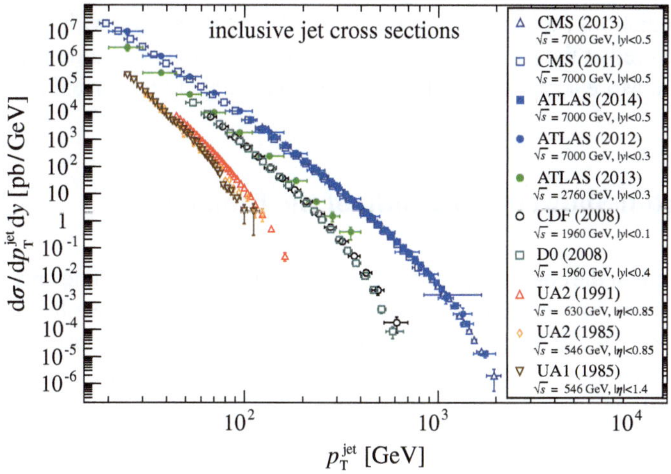

Fig. 5.3 Summary of inclusive jet cross-section measurements at pp or $p\bar{p}$ colliders. Data taken from [24–34]

e^+e^- collisions, to all kind of reactions and the partially conflicting requirements lead to various new propositions that were tried and discussed in several workshops [16–18]. Two classes of jet algorithms emerged:

1. cone algorithms that geometrically assign objects to the leading energy-flow objects in an event;
2. sequential-recombination algorithms that iteratively combine the closest pairs of objects [19].

Many cone algorithms need starting points with a minimum energy or momentum for the cone directions—so-called "seeds". These spoil the condition of collinear safety. A seedless infrared-safe cone algorithm exists in the form of the SISCONE algorithm [20], which avoids this problem. For reasons of computational efficiency, though, the method of choice employed at the LHC is the anti-k_t clustering algorithm [21] as implemented in the FASTJET package [22]. For the jet size parameter R (the equivalent to the cone size in cone algorithms) the LHC collaborations chose the values 0.4 and 0.6 for ATLAS respectively 0.5 and 0.7 for CMS. An extensive overview of jet definitions in QCD and their history is presented in [23].

The most fundamental quantity of jet production that can be investigated is the inclusive single-jet cross section as a function of the jet p_T and rapidity y, $d^2\sigma/dp_T^{jet}dy$, where every jet of an event contributes. Figure 5.3 presents a summary of inclusive jet cross section measurements, performed with various jet algorithms, from pp or $p\bar{p}$ collisions at centre-of-mass energies from 546 GeV up to 7 TeV at central rapidity. The measured cross sections stretch over eleven orders of magnitude and range up to 200 GeV at SPS, 600 GeV at the Tevatron, and 2 TeV at the LHC.

LO predictions (not shown) correctly describe the steep drop with increasing $p_{\mathrm{T}}^{\mathrm{jet}}$, but fail to predict the absolute values more precisely than in an order-of-magnitude estimation. More accurate calculations at NLO are able to improve this situation.

5.4 Parton Showers: The Bulk of the Emissions

The strength of the collinear factorisation ansatz allowing us to define universal PDFs and FFs lies in its repeated applicability. It enables us to define evolution equations such as the DGLAP equations [35–38], which determine the variation of the PDFs and FFs when the factorisation scale μ_F is changed. For the case of the initial state PDFs one derives

$$\mu_F^2 \frac{\partial f_i(x,\mu_F)}{\partial \mu_F^2} = \sum_{j=\{q,\bar{q},g\}} \int_x^1 \frac{dz}{z} \frac{\alpha_s}{2\pi} P_{ij}(z) f_{j/H}(x/z,\mu_F)\,, \tag{5.5}$$

with P_{ij} the (plus-prescription) regularised Altarelli–Parisi splitting functions

$$\begin{aligned} P_{gq}(z) &= C_F\left(\frac{1+(1-z)^2}{z}\right), \quad P_{qq}(z) = C_F\left(\frac{1+z^2}{(1-z)_+} + \frac{3}{2}\delta(1-z)\right), \\ P_{qg}(z) &= T_R\left(z^2 + (1-z^2)\right), \\ P_{gg}(z) &= 2C_A\left(\frac{z}{(1-z)_+} + \frac{1-z}{z} + z(1-z)\right) + \delta(1-z)\frac{11C_A - 4n_F T_R}{6}. \end{aligned} \tag{5.6}$$

Furthermore, it should be noted that $P_{g\bar{q}} = P_{gq}$ and $P_{\bar{q}\bar{q}} = P_{qq}$. While the LO, i.e. one-loop, approximation is quoted here, the QCD splitting functions are known up to NNLO, i.e. three-loop accuracy [39, 40]. The LO DGLAP evolution allows for a simple interpretation by means of branching processes. A parton i resolved at scale μ_F may have originated from a branching of parton j resolved at some higher scale. This transition of parton j to i is accompanied by the emission of an additional QCD parton. When applying the DGLAP equations to solve for the scale evolution of PDFs or FFs, these emitted particles get ignored, by considering inclusive processes only.

However, in parton-shower MC programs these emissions are made explicit, and the subsequent branching of initial-state and final-state partons results in a cascade-like picture, modelling the initial- and final-state evolution of the hard process' parton configuration into an ensemble of many QCD partons. In essence these parton-shower emissions approximate higher-order real-emission corrections to the hard scattering process. Shower simulations form an integral part of MC event generators—simulation tools to describe the fully exclusive hadronic final states of individual scattering events [41]. They link the hard scattering at some high scale Q^2 to an exclusive set of partons with typical separation scales of order $Q_0^2 \approx 1\,\text{GeV}^2$, the cut-off scale of the shower evolution. Through this parton-shower link it is possible to invoke universal models for the hadronisation of partons into hadrons, independent of the hard process scale Q^2. Physically, parton showers account for the intra-jet evolution of jets by accounting for the emission of almost collinear partons, and they model the inter-jet QCD activity through the emission of soft, wide-angle partons.

5.4.1 Colour Coherence

In particular the correct incorporation of soft-gluon emissions needs special consideration. In contrast to collinear emissions, which factorise at the cross-section level, soft-gluon emissions factorise at the level of individual QCD amplitudes. One should therefore consider soft gluons to be emitted by the scattering process as a whole, given by the squared sum of all contributing amplitudes. At first glance this approach seems to spoil the parton-shower picture of independent subsequent emissions. However, soft colour-coherence effects originating from the interference of individual amplitudes can be incorporated into parton-shower simulations by suitable choices for the shower-evolution variable. Most prominently this can be achieved by using the emission's opening angle as ordering parameter, as employed in the HERWIG and HERWIG++ generators [42, 43].

However, angular ordering is not the only option to include colour-coherence effects. A lot of effort went into the construction of new parton-shower algorithms based on subtraction formalisms as they are used in NLO QCD calculations. These new shower schemes, either based on Catani–Seymour dipole factorisation [44–46] or antenna subtraction [47, 48], implement soft-gluon coherence through a partial fractioning of the QCD radiator antennas, thereby introducing the notion of an emitter, i.e. splitter, and an associated spectator parton that accompanies the splitting process.

In order to study QCD coherence effects one needs to devise observables sensitive to rather soft emissions. This can be achieved by selecting final states with at least three jets exhibiting a sizeable spread in transverse momentum between the hardest and the softest jet. One such analysis was presented by the CMS collaboration in [49], based on a data set with an integrated luminosity of 36 pb^{-1} collected in 2010. The analysis inspects events with at least three anti-k_t jets using a distance parameter of $R = 0.5$, ordered in transverse momentum such that $p_{\text{T},1} > p_{\text{T},2} > p_{\text{T},3}$. The event selection criteria are given by

$$
\begin{aligned}
&p_{\text{T},1} > 100\,\text{GeV}\,,\quad p_{\text{T},3} > 30\,\text{GeV}\,,\quad |\eta_{1,2}| < 2.5 \\
&M_{12} > 220\,\text{GeV}\,,\quad 0.5 < \Delta R_{23} < 1.5\,.
\end{aligned}
$$

The observable considered to probe colour-coherence effects, called β, is defined as the azimuthal angle of the third jet with respect to the second jet, i.e.

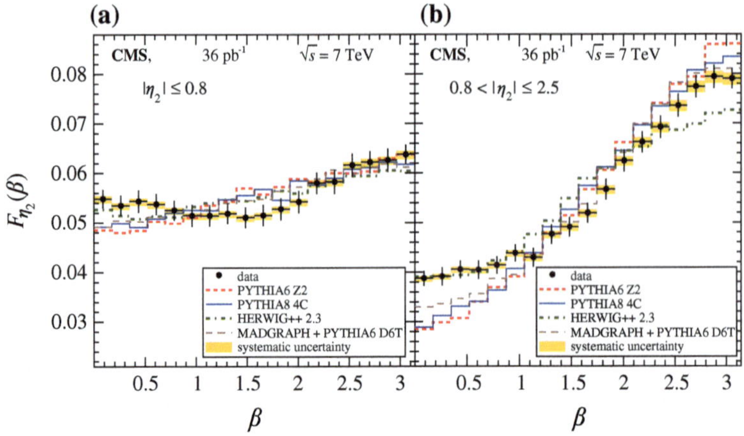

Fig. 5.4 Measured normalised β distribution, corrected for detector effects, in comparison to various MC predictions. The *error bars* indicate the statistical uncertainties, while the *shaded bands* correspond to the total systematic uncertainty. Adapted from [49]

$$\tan\beta = \frac{|\Delta\phi_{23}|}{\Delta\eta_{23}} .$$

In the presence of QCD coherence, the emission of the parton initiating the third jet is expected to preferentially lie in the event plane defined by the emitting parton and the beam axis. Figure 5.4 shows the normalised β distributions measured for two regions of the second-jet pseudo-rapidity, i.e. central $|\eta_2| \leq 0.8$ and forward $0.8 < |\eta_2| \leq 2.5$. The observable is thereby defined as $F_{\eta_2,i}(\beta) = N_{\eta,i}/N_\eta$, with N_η the total number of events in the respective η_2 region and $N_{\eta,i}$ the number of events in the ith β bin. The data are compared to various particle-level MC predictions, i.e. specific tunes of PYTHIA6, PYTHIA8 [50] and HERWIG++, all based on LO $2 \to 2$ matrix elements plus parton showers, and a combination of MADGRAPH using exact $2 \to 2$ and $2 \to 3$ tree-level matrix elements supplemented by PYTHIA6 parton showers. It has been shown in [49] that the data clearly support the inclusion of coherence effects in the simulation; the regions of small β and $\beta \approx \pi/2$ are particularly sensitive. However, none of the generators used in the analysis describes the data satisfactorily over the entire phase space. HERWIG++ models the data best but the agreement is rather poor in the forward region around $\beta \approx \pi$. The inclusion of the exact $2 \to 3$ matrix element from MADGRAPH slightly improves the pure PYTHIA6 parton-shower prediction.

5.4.2 Azimuthal Decorrelation

Clearly, the inclusion of parton showers is indispensable when attempting to simulate event configurations sensitive to the emission of multiple partons. Exemplary quantities are event shapes [51–53], or sub- or intra-jet-related observables [54–57]. Here, the measurement of dijet azimuthal decorrelations as performed by ATLAS and CMS [58, 59] shall briefly be discussed. The analyses evaluate 36 and $2.9\,\text{pb}^{-1}$ of pp collision data at $\sqrt{s} = 7\,\text{TeV}$ by ATLAS and CMS, respectively. The observable considered is the azimuthal angle $\Delta\phi$ between the two jets leading in jet p_T. For events with exactly two high-p_T jets and nothing else, the azimuthal angle between the two jets is fully correlated through momentum conservation in the transverse plane and $\Delta\phi \approx \pi$. Multi-jet production disturbs this balance so that the two leading jets become decorrelated in azimuthal angle and $\Delta\phi < \pi$. Hence, the observable $\Delta\phi$ is indicative of multi-jet production, while at the same time only the azimuthal angles of the two leading jets are measured, avoiding the large uncertainties associated to the jet energy calibration.

With $\Delta\phi$ it is possible to probe complementary aspects of perturbative QCD. The region $\Delta\phi \approx \pi$ is sensitive to multiple soft and collinear emissions and thus requires a proper resummation of soft and collinear logarithms. The region of large azimuthal decorrelations, i.e. $\Delta\phi \ll \pi$, indirectly probes the production mechanism for additional hard jets and requires the inclusion of higher-order matrix elements. Figure 5.5a shows a comparison of the normalised distribution of $\Delta\phi$ for five slices in leading-jet p_T as measured by CMS [59] to perturbative predictions for three-parton production at LO and NLO, respectively, in the region $2\pi/3 < \Delta\phi < \pi$. The leading order has predictive power only in a very narrow range of $\Delta\phi$. The NLO calculation improves this substantially, but fails when approaching $\Delta\phi \approx \pi$, and below $2\pi/3$ matrix elements for 4-jet production are required.

In Fig. 5.5b the normalised $\Delta\phi$ distributions as measured by ATLAS [58] are compared for nine slices in leading-jet p_T to predictions from the MC event generators PYTHIA6, HERWIG, and SHERPA. PYTHIA6 and HERWIG are based on LO $2 \to 2$ QCD matrix elements matched with parton showers. SHERPA additionally includes tree-level matrix elements for $2 \to 3$–6 jet production properly matched to its parton-shower algorithm [44, 60]. All three generators describe the data well over the measured range of $\pi/2 < \Delta\phi < \pi$ and all $p_T^{\max}$ slices. This includes both the region $\pi/2 < \Delta\phi < 2\pi/3$ where multi-jet contributions are significant and $\Delta\phi \approx \pi$ that cannot be described by a fixed-order calculation.

5.5 NLO: The New Standard

For most observables, NLO predictions in the strong coupling represent the first accurate theoretical estimate that allows an assessment of associated theoretical uncertainties. The evaluation of NLO cross sections, however, is more involved and requires to

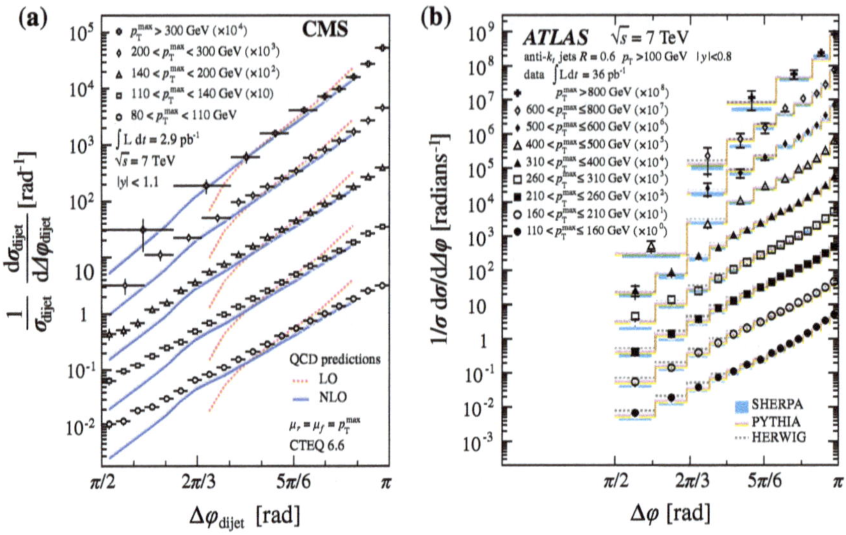

Fig. 5.5 Normalised differential cross section $(1/\sigma)(d\sigma/d\Delta\phi)$ for multiple slices in leading-jet p_T: **a** CMS data compared to fixed-order predictions in pQCD; **b** ATLAS data compared to predictions by the MC event generators PYTHIA6, HERWIG, and SHERPA. Adapted from auxiliary material provided with [58, 59]

consider real-emission and virtual one-loop corrections that are individually singular in the infrared region. While these divergences cancel in the sum of both contributions for sufficiently inclusive observables, they render the numerical evaluation of such cross sections difficult. Several state-of-the-art techniques exist that allow the exact cancellation of the divergences separately for the real and virtual contributions through the introduction of suitable subtraction terms. Most widely used is the dipole-factorisation method by Catani and Seymour [61]; alternatives are provided by the Frixione–Kunszt–Signer (FKS) method [62], antenna subtraction [63], or the recently developed Nagy–Soper formalism [64].

The enormous progress recently experienced in the field of NLO QCD calculations was sparked by two important developments: (i) the introduction of fast and efficient methods for the calculation of virtual amplitudes, see for instance [65–70]; (ii) the organisation and implementation of complete NLO calculations in the framework of parton-level Monte Carlo event generators such as HELAC/PHEGAS [71], MADGRAPH [9] or SHERPA [72, 73]. All these approaches employ automated subtraction-term generators, see [74–78], that implement the Catani–Seymour or FKS subtraction formalism. The real-emission corrections as well as the phase-space integration are handled by tree-level matrix-element generators such as AMEGIC [6], COMIX [7], MADGRAPH [9] or HELAC [8]. Virtual amplitudes, typically provided by specialised one-loop generators such as BLACKHAT [70], GOSAM [79], HELAC- 1LOOP [80], MADLOOP [81], NJET [82], OPENLOOPS+COLLIER [83] or RECOLA [84] can be incorporated via the universal BLHA interface [85, 86] or dedicated solutions. Examples of recent NLO calculations that have been performed using a combination of the tools listed above include: W+4, 5 jets [87, 88], $Z + 4$ jets [89], 4-jet and 5-jet production [90–92], $t\bar{t} + 2$ jets [93] and $\gamma\gamma + 3$ jets [94]. Most of these new tools are readily available to perform NLO QCD event generation for use in LHC data analyses.

5.5.1 Jet Counting

With NLO predictions at hand, more precise comparisons of data and theory become possible. The production of single inclusive jets and of dijets constitute the most basic QCD processes at hadron colliders. Figure 5.6 shows a comparison between data and theory at NLO, i.e. $\mathcal{O}(c_1\alpha_s^2 + c_2\alpha_s^3)$, for (a) inclusive jets from CMS [31] and (b) dijets from ATLAS [95], both at 7 TeV centre-of-mass energy. Recently, also trijet cross sections were measured [96, 116]. The inclusive jet cross section has been measured as a function of $p_{\mathrm{T}}^{\mathrm{jet}}$ and y in ranges of $0.1 \leq p_{\mathrm{T}} < 2.0$ TeV and $|y| < 2.5$ (see [31, 32]). The dijet cross section is presented as a function of the dijet mass, m_{12}, and rapidity separation, $y^* = |y_1 - y_2|/2$, of the two highest-p_{T} jets in regions of $0.3 \leq m_{12} < 5.0$ TeV and $y^* < 3.0$, respectively. In both cases the NLO theory has been corrected for non-perturbative effects, and for the more recent ATLAS publication also electroweak (EW) corrections have been considered. While EW effects start being sizeable at high $p_{\mathrm{T}}^{\mathrm{jet}}$ and $|y|$ or large m_{12} [97], non-perturbative effects, estimated via their modelling in MC event generators, become negligible. The data are well described by theory over many orders of magnitude

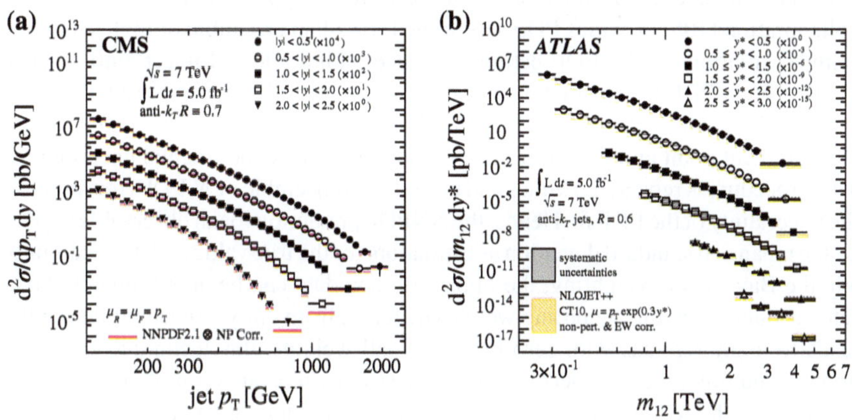

Fig. 5.6 **a** Inclusive jet cross section from CMS and **b** dijet cross section from ATLAS, both at 7 TeV centre-of-mass energy, in comparison to theory at NLO including non-perturbative (CMS) or non-perturbative and electroweak corrections (ATLAS). The anti-k_t algorithm has been applied with the jet-size parameter R being 0.7 for CMS and 0.6 for ATLAS. Adapted from [31, 95]

in cross section and in wide kinematic ranges so that fits of parameters entering the pQCD calculations become feasible.

5.5.2 Jets and the Gluon PDF

In the following, PDF fits are considered as a first example of determining parameters entering pQCD calculations. Many choices are possible in terms of data set selection, theoretical ingredients and approximations, parameterisation, or fitting techniques and criteria. There is no unique approach to PDF fits. The PDF sets in use at the LHC are produced by six PDF fitting groups: ABM, CTEQ, (G)JR, HERAPDF, MSTW, and NNPDF [98–103], where the quoted publications refer to pre-LHC determinations. The fundamental experimental input central to all PDF determinations are deep-inelastic scattering (DIS) data collected by the H1 and ZEUS collaborations at the *ep* collider HERA. The HERA collaborations' own PDF fits based solely on their data are presented as HERAPDF sets. The ABM group extends the choice of data to other DIS measurements and fixed-target Drell–Yan data, while the (G)JR group also considers high-p_T jet production, which cannot yet be described at NNLO, to better fix the gluon PDF. The trio of CTEQ, MSTW, and NNPDF finally try to incorporate as many data sets as possible and to provide truly global fits. New PDF sets including LHC data became available recently or are in preparation for the next start-up of the LHC with 13 TeV center-of-mass energy.

It is possible to judge the potential impact of new data, e.g. from the LHC, on a given PDF set without a complete refit. The technique of "Bayesian reweighting" was first implemented in the context of the NNPDF approach [104]. The NNPDF collaboration [105] provides PDF sets in the form of an ensemble of replicas, which sample variations in the PDF parameter space within allowed uncertainties. Their method fits a PDF to smeared data points, where each data point is sampled from a multi-Gaussian distribution. The central value of the distribution is equal to the measurement point and the covariance is taken from the experimental uncertainties. This procedure is repeated N times and provides an ensemble of N PDFs representing the uncertainty of the PDF fit. Hence, the NNPDF prediction for an observable is given by the mean and standard deviation as estimators for the true value and its uncertainty. Via the Bayesian reweighting, the impact of new data can be approximately taken into account by giving according weights different from unity to each replica. If this leads to many replicas with zero or very small weights, this technique cannot be applied and refits must be performed. Otherwise the re-evaluation of an observable while using the re-weighted set of replicas approximately accounts for changes in the PDFs due to the additional data.

In a similar way to the mean and standard deviation as estimators for the main prediction and uncertainty of an observable, the NNPDF ensemble of replicas can be used to investigate the correlation between an observable and the PDFs as a function of energy scale Q and momentum fraction x. As an example, Fig. 5.7a shows the correlation coefficient ρ between the inclusive jet cross section as measured by

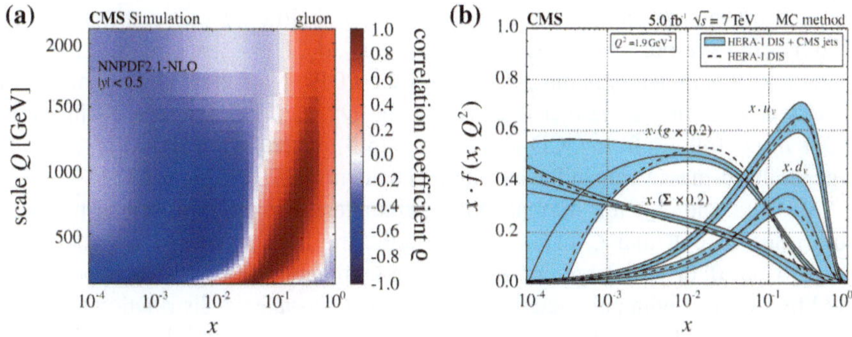

Fig. 5.7 **a** Correlation coefficient ρ between the inclusive jet cross section and the gluon PDF as a function of energy scale Q and momentum fraction x. **b** PDFs as determined from fits to HERA DIS data alone or in combination with CMS jet data. The uncertainties on the gluon PDF at high x are reduced compared to a fit with DIS data alone (uncertainties not shown). Adapted from [106] and auxiliary material

CMS [106] and the gluon PDF. For $x > 0.01$ a large positive correlation is exhibited between the two.

In case of the other five PDF sets, the method of diagonalisation of the Hessian matrix [107] (also called "eigenvector method") is employed to express uncertainty estimates for their respective choices of PDF parameterisation. Their prediction for an observable in pQCD can thus be evaluated from one central PDF set with an uncertainty given by quadratic addition of all deviations obtained while using additional PDFs that correspond to variations along the directions of the eigenvectors. A Bayesian reweighting was shown to be feasible here as well [108], and an alternative approach is studied in [109].

Employing an open-source tool like the HERAFITTER project [101, 110, 111], which provides fitting code, data sets, and theory predictions in a common framework, it is even possible to perform complete PDF fits. The CMS collaboration used this tool to study the impact of their inclusive jet measurements on the gluon PDF [106] following the fitting setup and using the same DIS data as described in [101]. The shape of the PDFs at a starting scale Q_0 is assumed to be:

$$
\begin{aligned}
xg(x) &= A_g x^{B_g}(1-x)^{C_g} - A'_g x^{B'_g}(1-x)^{C'_g}\,,\\
xu_v(x) &= A_{u_v} x^{B_{u_v}}(1-x)^{C_{u_v}}(1+E_{u_v}x^2)\,,\\
xd_v(x) &= A_{d_v} x^{B_{d_v}}(1-x)^{C_{d_v}}\,,\\
x\overline{U}(x) &= A_{\overline{U}} x^{B_{\overline{U}}}(1-x)^{C_{\overline{U}}}\,,\\
x\overline{D}(x) &= A_{\overline{D}} x^{B_{\overline{D}}}(1-x)^{C_{\overline{D}}}\,.
\end{aligned}
$$

Here, $xg(x)$ is the gluon distribution, $xu_v(x)$ and $xd_v(x)$ represent the valence quarks with $u_v(x) = u(x) - \overline{u}(x)$, $d_v(x) = d(x) - \overline{d}(x)$, and $\overline{U}(x)$, $\overline{D}(x)$ are the up- and down-type antiquark distributions $\overline{U}(x) = \overline{u}(x)$, $\overline{D}(x) = \overline{d}(x) + \overline{s}(x)$. The indexed symbols A_i, B_i, C_i with $i \in \{g, u_v, d_v, \overline{U}, \overline{D}\}$, A'_g, B'_g, C'_g, and E_{u_v} stand for 19 parameters in the definition of the PDFs. Note that there are no heavy quarks at a starting scale that is chosen to be $Q_0^2 = 1.9\,\text{GeV}^2$. The evolution of PDFs in Q^2 follows from QCD according to (5.5), where in the HERAFITTER setup a generalised-mass variable flavour number scheme [112, 113] is used.

A_g, A_{u_v}, and A_{d_v} are normalisation parameters that are constrained by QCD sum rules. The B and C parameters describe the limiting behaviour when very small, $x \to 0$, or very large, $x \to 1$, proton momentum fractions are approached. Additional terms with parameters for the gluon and u-valence distribution allow for some more flexibility in shape, while constraints of $B_{\overline{U}} = B_{\overline{D}}$ and $A_{\overline{U}} = A_{\overline{D}}(1-f_s)$ are applied to ensure the same normalisation for the $\bar{u}$ and $\bar{d}$ densities at high x. The strangeness fraction is set to $f_s = 0.31$, as obtained from neutrino-induced dimuon production [114]. The parameter C'_g is fixed to 25 [102, 113] and the strong coupling constant to $\alpha_s(M_Z) = 0.1176$ so that in total 13 free parameters remain to be determined.

The PDF fit result using HERA DIS data alone or in combination with CMS jet data is presented in Fig. 5.7b. As expected, a clear impact on the gluon PDF for momentum fractions $x > 0.01$ and a reduction of uncertainties (not shown) is exhibited.

5.5.3 Jets, Cross-Section Ratios, and the Strong Coupling

Absolute jet cross sections as used in the above CMS example are subject to many experimental sources of uncertainty, and in particular to details of the jet energy calibration. Since the jet-p_T spectrum is steeply falling (see Fig. 5.6), an uncertainty in the jet energy scale (JES) of 5 % would translate into an uncertainty on the measured cross section of about 25–30 %. A means to at least partially cancel such experimental uncertainties was investigated by ATLAS in the form of the ratio of inclusive jet cross sections measured for two different center-of-mass energies (2.76 and 7 TeV) [34]. Figure 5.8 shows the considerably reduced experimental uncertainties of this ratio compared to the cross sections themselves, thus demonstrating the possibility to provide stronger constraints on PDFs. For comparison, the PDF uncertainties for various PDF sets are indicated as well.

Finally, the possibility to extract the strong coupling constant $\alpha_s(M_Z)$ from inclusive jet production data is studied. As can be seen from Fig. 5.7 and the DGLAP equations (5.5), the appearance of gluons and the strength of $\alpha_s(Q)$ are coupled. Therefore, a simultaneous fit of PDFs and $\alpha_s(M_Z)$ with DIS data alone is not possible, which is also the reason why the value of $\alpha_s(M_Z)$ is fixed in the above-mentioned PDF fit of [101]. The inclusion of jet data facilitates such a simultaneous fit of PDFs and the strong coupling as demonstrated by CMS [106]. In the following, the focus shall be, however, on the running behaviour of $\alpha_s(Q)$ according

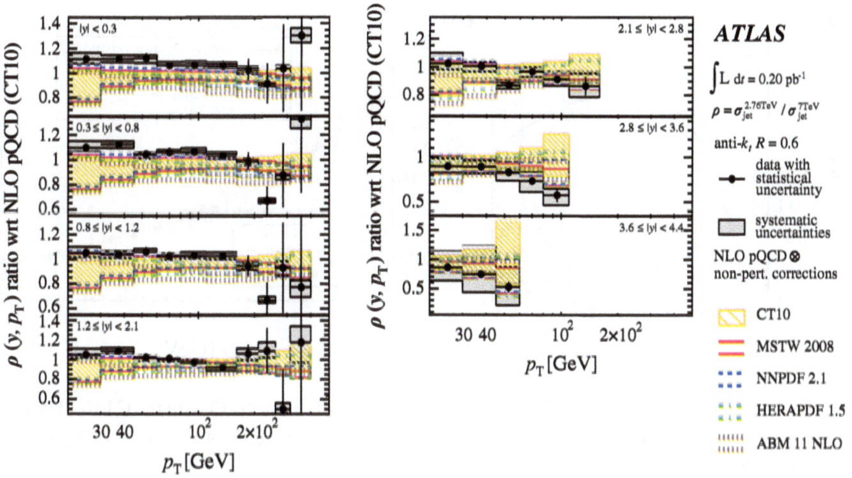

Fig. 5.8 The ratio of the inclusive jet cross sections at $\sqrt{s} = 2.76$ and 7 TeV as measured by ATLAS divided by a pQCD prediction at NLO (with CT10 PDFs) of the same ratio. Experimental uncertainties are indicated by *error bars* (statistical) and *gray boxes* (systematic). The uncertainties inherent to the PDFs are represented by bands. Adapted from [34]

to (5.3). To eliminate or reduce numerous experimental sources of uncertainty—such as those related to the luminosity measurement or the jet energy scale—again a ratio is used: that of the inclusive 3-jet over the inclusive 2-jet production cross section, R_{32}. In this quantity, also theoretical uncertainties like those stemming from the PDFs are reduced. Since 3-jet production requires one real emission more than the usual dijets, the numerator is approximately proportional to α_s^3, while the denominator is $\propto \alpha_s^2$. R_{32} is therefore a direct measure for the strength of the strong coupling. The CMS collaboration has investigated this ratio with the result $\alpha_s(M_Z) = 0.1148 \pm 0.0014\,(\text{exp}) \pm 0.0018(\text{PDF}) \pm 0.0050(\text{theo})$ [115]. By performing the analysis separately for different regions of the average p_T of the two leading jets, the running could be tested up to scales of 1.39 TeV. Figure 5.9a shows the ratio R_{32} measured by CMS in comparison to predictions at NLO obtained with the CT10 PDF set for a series of assumptions on $\alpha_s(M_Z)$. Figure 5.9b presents a summary of extractions of $\alpha_s(Q)$ including the latest results achieved at the LHC with scales Q reaching beyond 1 TeV.

5.6 NNLO: The Quest for Precision

For a few benchmark processes and observables, the experimental accuracy of QCD measurements at the LHC is aiming for the few-percent level or even better. Examples are inclusive single-jet or dijet production, vector boson (pair) production, and top-quark or Higgs-boson production. Clearly, for these channels theoretical predictions

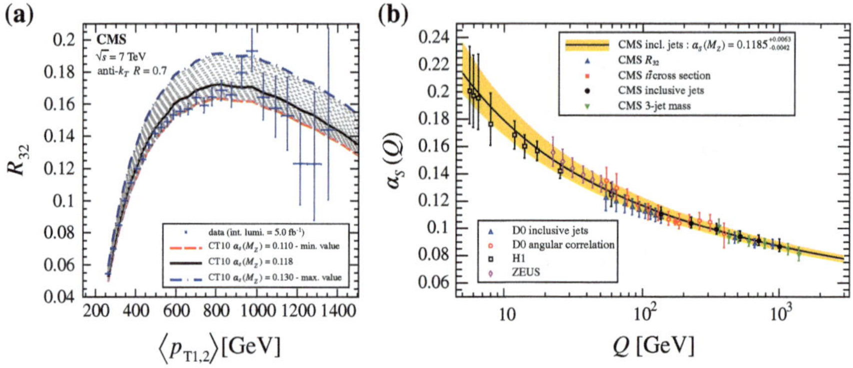

Fig. 5.9 **a** R_{32} as measured by CMS in comparison to predictions at NLO with the CT10 PDF for a series of assumptions on $\alpha_s(M_Z)$. **b** Summary of extractions of $\alpha_s(Q)$ from low to very high scales Q from various experiments including the latest results achieved at the LHC up to $Q \approx 1$ TeV. Adapted from [115, 116] and auxiliary material

at NNLO in the strong coupling are needed, as otherwise the extraction of Standard Model parameters will be limited by theoretical uncertainties.

The evaluation of NNLO corrections for a given process requires the calculation of three contributions: (i) generic two-loop graphs; (ii) one-loop graphs with one extra final-state parton; (iii) real-emission tree diagrams with two additional partons. All of these contributions exhibit infrared singularities, and only their sum yields finite results. In explicit calculations, the divergences are typically treated in dimensional regularisation. Similar to what was discussed for the treatment of infrared singularities appearing in NLO corrections before, there exist now several subtraction schemes to handle these divergences consistently [117–120].

Concerning the above-mentioned processes in need of precision calculations, full NNLO results exist for inclusive W/Z-boson production [121–123], for diboson [124–126] and diphoton production [127], for top-quark pair production [128] and for various Higgs-boson production channels [129–135]. For the NNLO evaluation of dijet production, results have been achieved for two-gluon final states, initiated by gluon annihilation [136], and for quark-antiquark initial states [137]. For the inclusive jet cross section at $\sqrt{s} = 8$ TeV, considering anti-k_t jets with $R = 0.7$, transverse jet momenta of $p_{\mathrm{T}} > 80$ GeV and jet rapidities $|y| < 4.4$, the authors find an increase of the NNLO prediction with respect to the NLO estimate of about 27–16 %. The corrections are largest for small jet transverse momenta and decrease for larger values of jet p_{T}. The given calculation relies on the MSTW2008-NNLO PDF set [102] and uses a dynamical scale choice of $\mu = \mu_R = \mu_F$ equal to the transverse momentum of the leading jet. The size of these corrections highlights the importance of higher-order calculations for LHC precision observables. In what follows, two standard "candles" (i.e. well-measured processes or processes with theoretically well-understood cross sections) at the LHC—the Drell–Yan processes and diphoton production—shall be discussed in some more detail.

5.6.1 Inclusive Vector-Boson Production

Due to the relatively simple colour-singlet production mechanism and an experimental signature based on leptons that can be measured very accurately, the production of W and Z bosons is well suited for precision tests of QCD and electroweak interactions.

At hadron colliders, massive electroweak bosons are dominantly produced via quark-antiquark annihilation. The production of Z bosons is commonly referred to as Drell–Yan process, since in 1970 Drell and Yan applied the developing parton model to predict the scaling of the $pp \to \gamma/Z \to \mu^+\mu^-$ cross sections $\mathrm{d}\sigma/\mathrm{d}M^2 \propto K\ L(M^2/s)$, with M the invariant boson mass, L the parton luminosity and K a constant that depends on the parton spin and coupling [138]. In modern QCD, the cross section can be expressed as $\mathrm{d}\sigma/\mathrm{d}M^2 = (4\pi\alpha_{\mathrm{EM}}^2)/(3n_c M^2 s) L_{q\bar{q}}(M^2/s)$, where $L_{q\bar{q}}$ is the quark-antiquark luminosity, α_{EM} the electromagnetic coupling, and n_c the number of colours.

Vector-boson processes are relatively simple to calculate, since there is no QCD interaction involving the final-state particles. In fact, the Drell–Yan process was the first hadron-collider channel known to NNLO accuracy [121–123]. The NNLO Drell-Yan production cross sections can be calculated with a precision of 1 % for the renormalisation-scale and factorisation-scale uncertainty. The calculations also include the decay of W and Z bosons to leptons. For inclusive Z-boson production at the LHC, the NLO-to-LO correction is sizeable (about 20 %), while the NNLO-to-NLO correction amounts to 2 % only . Two public codes implement the NNLO QCD calculations for inclusive Z-boson and W-boson production: DYNNLO [123] and FEWZ [139, 140]. Furthermore, the dominant electroweak corrections have been determined [141–146].

The precise measurement of the inclusive W-boson and Z-boson production cross sections at LHC and their successful comparison to precise QCD calculations was one of the first and very important confirmations of pQCD at very high energies and momentum transfers. Measurements at $\sqrt{s} = 7\,\mathrm{TeV}$ have been performed with the 2010 data set corresponding to an integrated luminosity of $36\,\mathrm{pb}^{-1}$ by ATLAS [147] and CMS [148]. CMS also measured the cross sections at $\sqrt{s} = 8\,\mathrm{TeV}$ using a data set of $18.2\,\mathrm{fb}^{-1}$[149].

The cross sections are measured by counting the number of events in the detector acceptance and subtracting the estimated background contributions. The efficiency ϵ is estimated from Monte Carlo simulations and corrected for differences between these simulations and the data. To better separate experimental and modelling uncertainties (in particular from the extrapolation of the detector acceptance to the total phase space), the efficiency is decomposed in a fiducial acceptance, A, and a correction for detector effects, C, i.e. $\epsilon = A \cdot C$.

The fiducial acceptance A is given by the ratio of the number of events passing the kinematic cuts applied on particle-level (i.e. without detector simulation) over the total number of generated events. The kinematic selection cuts are set close to the requirements on the reconstructed objects typically for leptons: $p_{\mathrm{T}} > 20\,\mathrm{GeV}$ and

Table 5.1 Summary of the cross-section results of the inclusive W-boson and Z-boson analyses of ATLAS and CMS. The total uncertainty (statistical, experimental and acceptance modelling added in quadrature) and the luminosity uncertainties are also given. The theory cross sections are calculated with FEWZ [139, 140] with the MSTW2008-NNLO PDF set [102] and contain renormalisation-scale and factorisation-scale uncertainties and PDF uncertainties

	$W^+ \to l^+\nu$ (nb)	$W^- \to l^-\nu$ (nb)	$Z \to \ell^+\ell^-$ (nb)
7 TeV			
ATLAS	$6.05 \pm 0.12 \pm 0.21$	$4.16 \pm 0.10 \pm 0.14$	$0.937 \pm 0.02 \pm 0.03$
CMS	$6.04 \pm 0.10 \pm 0.24$	$4.26 \pm 0.08 \pm 0.17$	$0.974 \pm 0.02 \pm 0.04$
NNLO QCD	5.98 ± 0.3	4.2 ± 0.2	0.991 ± 0.05
8 TeV			
CMS	$7.11 \pm 0.14 \pm 0.18$	$5.09 \pm 0.12 \pm 0.13$	$1.15 \pm 0.02 \pm 0.03$
NNLO QCD	7.12 ± 0.2	5.06 ± 0.13	1.13 ± 0.04

$|\eta| < 2.5$. The detector correction factor C is defined as the number of selected events in the sample including the detector simulation over the number of events passing the fiducial acceptance A. Typically A is 0.45–0.50 and C is 0.7–0.8, depending on the analysis.

A summary of the inclusive Drell–Yan cross-section measurements is shown in Table 5.1. The statistical uncertainty amounts to about 0.3 % for W-boson production and 0.7 % for Z-boson production. The systematic uncertainty is dominated by the knowledge of the integrated luminosity, which at $\sqrt{s} = 7$ TeV results in an uncertainty of 3.4 % (ATLAS) and 4.0 % (CMS) and in 2.5 % at $\sqrt{s} = 8$ TeV. The measurements of ATLAS and CMS agree within their uncertainties, and both agree with the NNLO prediction that has an uncertainty of about 5 % (dominated by the PDF uncertainty).

5.6.2 Differential Vector-Boson Cross Sections

Apart from the total W and Z production cross sections, also some differential cross sections are known to order $\mathcal{O}(\alpha_s^2)$; this includes the dilepton invariant-mass spectrum, the vector-boson's transverse momentum and its rapidity distribution.

As an example, Fig. 5.10 presents a CMS measurement of the Z-boson production cross section as a function of the invariant mass M of the two leptons from the Z decay at $\sqrt{s} = 7$ TeV [150]. The cross section falls over eight orders of magnitudes in the mass range of $15 < M < 1500$ GeV and clearly shows the Z-boson resonance at 90 GeV. The NNLO calculation using the CT10 PDF set describes the data well. Similar measurements were performed by ATLAS [151].

The boson's transverse momentum is an interesting observable to test various regimes in the strong-interaction dynamics. The high-p_T part of the spectrum can be modelled by perturbative QCD calculations considering processes with additional jets ($W/Z + n$-jet processes). Presently calculations can be performed with up to

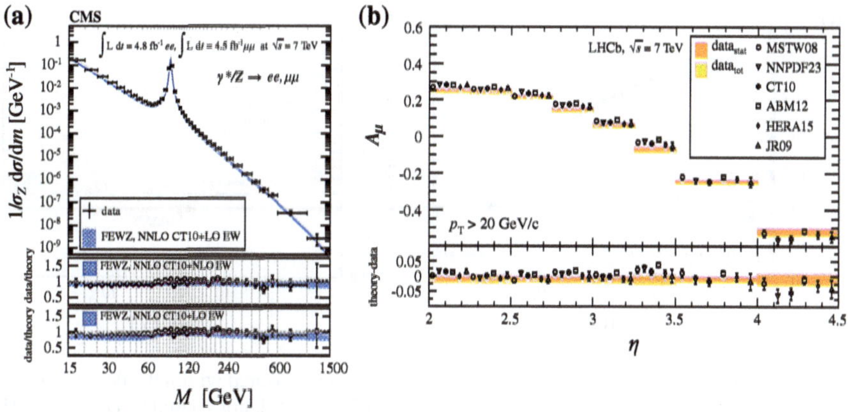

Fig. 5.10 **a** Drell–Yan cross sections at $\sqrt{s} = 7\,\text{TeV}$ as a function of the dilepton invariant mass M. **b** W-boson charge asymmetry as a function of the lepton pseudo-rapidity. Data are compared to NNLO QCD predictions using various PDF sets. Adapted from [150, 152]

$n = 5$ additional partons. When the transverse momentum of the vector boson is much smaller than its invariant mass reconstructed from the two leptons, soft QCD radiation is enhanced and fixed-order perturbation theory has to be supplemented by the resummation of large logarithmic corrections of the form $\alpha_s^n \ln^m(p_T/M)$ to all orders of α_s. The resummed cross sections also include a non-perturbative component at momentum scales below 1 GeV. This region can also be modelled using Monte Carlo generators implementing parton showers and models for hadronisation.

To lowest order, W and Z bosons are produced via quark-antiquark annihilation, i.e. $q\bar{q} \to Z$. Indeed, this process dominates for inclusive Z production within a rapidity range of $|y| < 2.1$. However, already for Z transverse momenta around 25 GeV quark-gluon scattering $qg \to Zq$ is of similar size, and around $p_T = 100\,\text{GeV}$ the latter process constitutes 80 % of the total cross section [153]. For Z-boson transverse momenta of 180 GeV, parton-momentum fractions $x_{1/2} = (M/\sqrt{s})\,\exp(\pm y)$ of about $x = 0.05$ are probed.

The program ResBos [154] implements soft-gluon resummation at low p_T at next-to-next-to-leading-logarithm (NNLL) accuracy and matches to a matrix-element calculation of order $\mathcal{O}(\alpha_s^2)$ at high p_T. It provides fully differential cross sections as a function of the rapidity, the invariant mass and the transverse momentum of the vector boson. It also attempts to estimate non-perturbative corrections [155]. Recently, first calculations for Drell–Yan lepton-pair production at NNLO accuracy including parton-shower effects have been presented [156, 157].

Early measurements of differential cross sections at $\sqrt{s} = 7\,\text{TeV}$ for Z [159] and W [160, 161] bosons were based on an integrated luminosity of 35–40 pb^{-1}. Recently, the ATLAS experiment updated this measurement to a larger data set corresponding to an integrated luminosity of 4.7 fb^{-1}. This measurement reaches Z-boson transverse momenta of 800 GeV for rapidities up to 2.4 [158].

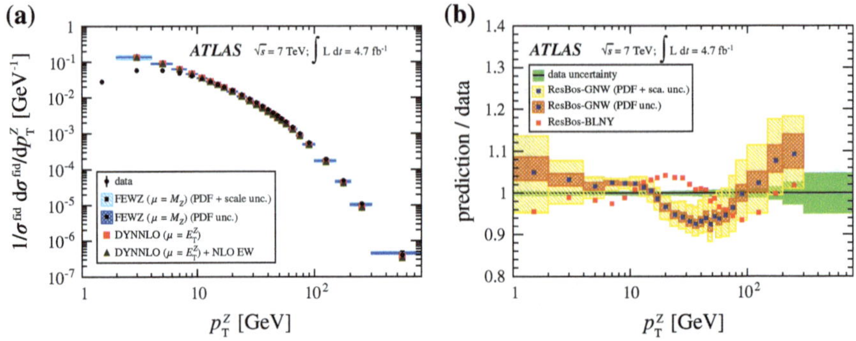

Fig. 5.11 Normalised transverse momentum of the Z boson measured within the detector acceptance at $\sqrt{s} = 7\,\text{TeV}$. **a** ATLAS data compared to NNLO QCD calculations based on DYNNLO [123] and FEWZ [139, 140]) with different renormalisation-scale and factorisation-scale settings. **b** Ratio of an NNLO QCD prediction including NNLL resummation to the data. Shown are calculations that differ in the parametrisation of non-perturbative effects. The bands indicate the respective theory uncertainty estimate. The band around 1 denotes the data measurement uncertainty. Adapted from [158]

The measurement uncertainty amounts to 1 % for $p_T < 150\,\text{GeV}$ and rises to about 5 % for the highest p_T. The leptonic decays of the Z boson to electrons or muons are used in the analysis, and electrons or muons with $p_T > 20\,\text{GeV}$ and rapidity $|y| < 2.4$ are required. The invariant mass of the dilepton pair must lie between $66 < M < 116\,\text{GeV}$. Figure 5.11 shows the comparison of the data to the NNLO QCD predictions based on DYNNLO [123] and FEWZ [139, 140] using the CT10 PDF set. Shown as a band are theory uncertainties that are due to the renormalisation-scale and factorisation-scale variations and associated with the choice of PDFs for both calculations. The measurement uncertainties are below a percent in most bins and are smaller than the theory uncertainties that reach 10 % around 50 GeV. The two calculations are in excellent agreement with each other for $p_T < 100\,\text{GeV}$, but differ at large p_T by about 10 %. In the region $10 < p_T < 100\,\text{GeV}$ the predictions are 10 % lower than the data. For $p_T < 10\,\text{GeV}$ the fixed-order calculations diverge and clearly disagree with the data.

The resummed calculation based on RESBOS using the CTEQ6M PDF [162] describes the data within 10 %. The observed deviations are covered by the theory uncertainties of the standard non-perturbative parametrisation (denoted "BLNY"). A recent new parameterisation denoted "GNW" [155] predicts a somewhat different shape.

An NLO QCD calculation based on POWHEG interfaced with PYTHIA for parton showers and hadronisation describes the data within 5 % for $p_T < 60\,\text{GeV}$, but deviates by up to 20 % over the full p_T range. An NLO QCD calculation based on MC@NLO interfaced with HERWIG for parton shower and hadronisation deviates by up to 40 % at high p_T. Leading order Monte Carlo generators based on multi-leg matrix elements, like SHERPA or ALPGEN, agree with data within 5 % except for the highest p_{T} bin. A dedicated tuning of parton-shower parameters in PYTHIA8 interfaced to POWHEG [158] achieves agreement within 2 % for $p_{\mathrm{T}} < 50\,\mathrm{GeV}$.

Since the dominant $W^{\pm}$ production mechanisms at the LHC are $d\bar{u} \rightarrow W^{-}$ and $u\bar{d} \rightarrow W^{+}$, measurements of $W^{\pm}$-boson production cross sections provide a powerful tool to study the parton density functions of d and u quarks in the proton. In particular, the measurement of the lepton charge asymmetry $A_{\ell} = (\sigma_{W^{+}\rightarrow\ell^{+}\nu} - \sigma_{W^{-}\rightarrow\ell^{-}\bar{\nu}})/(\sigma_{W^{+}\rightarrow\ell^{+}\nu} + \sigma_{W^{-}\rightarrow\ell^{-}\bar{\nu}})$ gives valuable information on the d-to-u ratio and also on the sea-quark and sea-antiquark distributions, in particular of the strange quark. By measuring differential cross sections, for instance as a function of the lepton rapidity, different parton-momentum fraction values x can be probed, since the parton-momentum fraction depends on the vector-boson mass and on the rapidity y: $x_{1/2} = (M_{W,Z}/\sqrt{s})\exp(\pm y)$.

ATLAS published such differential cross-section measurements $\mathrm{d}\sigma_{W^{\pm}/Z}/\mathrm{d}y_{\ell}$ for electrons and muons up to lepton rapidities of $y_{\ell} = 4.9$ based on the 2010 data set at $\sqrt{s} = 7\,\mathrm{TeV}$ [147]. Leptons with a $p_{\mathrm{T}} > 20$–$25\,\mathrm{GeV}$ are selected. For Z bosons, the accuracy reaches about 2 % in the central region of the Z-boson rapidity and 10 % at $y_{Z} = 3.2$. A precision of about 2 % is obtained for W-boson cross sections measured within $|y_{\ell}| < 2.5$. For the A_{ℓ} measurement the accuracy ranges between 4 and 8 %. The data can be described by an NNLO calculation using the HERAPDF1.5 and the MSTW2008-NNLO PDF sets. Further measurements by ATLAS and CMS can be found in [148, 161, 163–166].

The LHCb experiment measured W-boson production in the forward region covering lepton rapidities in the range $2 < |y_{\ell}| < 4.5$. A measurement was performed with the 2011 data set at $\sqrt{s} = 7\,\mathrm{TeV}$ using an integrated luminosity of $1\,\mathrm{fb}^{-1}$. Figure 5.10b shows the measured lepton charge asymmetry for muons with $p_{\mathrm{T}} > 20\,\mathrm{GeV}$ as a function of the muon rapidity compared to NNLO calculations using various PDF sets. The band shows the statistical (inner, dark) and total uncertainty (light) on the data. The experimental accuracy is similar to the theoretical uncertainty estimate, and the predictions are in good agreement with the measurement.

5.6.3 Production of Photon Pairs with Large Invariant Mass

At leading order in pQCD, the production of photon pairs can rather easily be described by quark-antiquark annihilation, ($q\bar{q} \rightarrow \gamma\gamma$), as for the Drell–Yan process before. However, in contrast to the massive vector bosons W and Z, which are predominantly produced in hard scattering processes and can easily be identified via their leptonic decay modes, massless photons can also be radiated off initial-state or final-state partons or may be created in decays. In particular the π^{0} and η mesons, and to some extent also the ω, all of which are copiously produced within jets, have neutral decay modes. Since these particles are boosted within the jets, their decay photons are collimated and difficult to distinguish from a single highly energetic photon. Without applying further selection criteria, the mode of "photon" pair production at large invariant mass via meson decays within two (different) jets is even dominating by several orders of magnitude.

To distinguish the "non-prompt" decay photons within jets from "prompt" photons stemming from the hard scattering, isolation criteria are applied. Typically, isolation demands that the transverse energy not associated to a photon candidate but deposited within a cone of $\Delta R = \sqrt{(\Delta\eta)^2 + (\Delta\phi)^2} = 0.4$ around it may not exceed a few GeV. In addition, a minimal separation of the two photons is required to avoid overlap: $\Delta R_{\gamma\gamma} > 0.40,\ 0.45$ for ATLAS and CMS, respectively. In practice, the experimental selections are more complicated (see e.g. [167–170]), and great care has to be taken to ensure that they can be matched properly to selection criteria applicable in theory (see [171] for a recent discussion on photon isolation).

Finely segmented electromagnetic calorimeters are of great advantage in such photon studies. Investigating the shower shape of energy deposits in these calorimeters helps to differentiate between electromagnetic showers that are caused by a single high-energetic signal photon and those from multiple photons like in $\pi^0 \to \gamma\gamma$ decays. Photon candidates are separated from electron (or positron) candidates by using reconstructed tracks. Depending on the amount of material to be traversed (e.g. the beampipe or the silicon-pixel and silicon-strip detectors), however, highly energetic photons may convert into e^+e^- pairs before reaching the electromagnetic calorimeter. Dedicated reconstruction methods are applied to avoid losing these converted photons, which can make up to 50 % [170] of the total photon yield.

For background subtraction, techniques relying on data are preferred. Detailed studies, necessary to further quantify detector effects on the photon reconstruction and isolation and to unfold the signal yields, are performed with simulated events provided by the LO event generator PYTHIA [50, 172] or by multi-jet-improved event generators that include additional real emissions (extra jets) like SHERPA [73, 173] or MADGRAPH [9]. The latter is combined with PYTHIA for parton showering, hadronisation, and for the modelling of the underlying event, see Sect. 5.9.

At next-to-leading order, the gluon-initiated processes ($gq(\bar{q}) \to \gamma\gamma q(\bar{q})$) join the annihilation process ($q\bar{q} \to \gamma\gamma g$) together with corresponding virtual corrections. Here, a further complication arises from the collinear fragmentation of a hard outgoing parton into a photon. Theoretically, this process is described using fragmentation functions $D_{q\to\gamma}(z, \mu_F)$ and $D_{g\to\gamma}(z, \mu_F)$. In the limit of $\mu_F \approx Q$, where Q is the scale of the hard scattering, this fragmentation process results in a contribution to the cross section that is similar in size to the one of the LO annihilation process [174].

Prompt photons not emerging via fragmentation from the hard process are also called "direct". The effect of the above-mentioned isolation conditions, which aim to suppress fragmentation photons in favour of direct ones, have to be properly estimated not only experimentally for mesons decaying within jets, but also theoretically, e.g. via FFs in the perturbative calculations. The NLO parton-level program RESBOS [175, 176] effectively includes the fragmentation of one quark/gluon to a single photon at leading order and additionally features soft and collinear gluon resummation (see Sect. 5.8). DIPHOX [174] provides parton-level results at NLO taking fully into account up to two fragmentation photons. Although formally an NNLO box graph, the process ($gg \rightarrow \gamma\gamma$) is drastically enhanced at the LHC through the large gluon luminosity. It is comparable in size to the LO terms, and therefore DIPHOX includes this contribution at NLO precision, i.e. up to N^{3}LO corrections in the strong coupling α_s, via GAMMA2MC [177]. Finally, a full NNLO calculation is available in the form of the 2γNNLO [127] program, however without consideration of fragmentation photons.

Four kinematic variables are usually chosen to investigate the differential photon-pair production cross section: the invariant mass $m_{\gamma\gamma}$, the transverse momentum of the photon pair $p_{\mathrm{T},\gamma\gamma}$, the azimuthal angular separation $\Delta\phi_{\gamma\gamma}$, and the cosine of the polar angle in the Collins–Soper reference frame [178] of the diphoton system, $\cos\theta^*_{\gamma\gamma}$. The invariant mass is of obvious interest for resonance searches, where $\cos\theta^*_{\gamma\gamma}$ is useful to examine the spin of diphoton resonances. The transverse momentum $p_{\mathrm{T},\gamma\gamma}$ and the quantity $\Delta\phi_{\gamma\gamma}$ are well-suited for the comparison of specific aspects of the theoretically very challenging description of photon-pair production. In particular, at low $p_{\mathrm{T},\gamma\gamma}$ or for well-balanced photons at $\Delta\phi_{\gamma\gamma} \approx \pi$, where multiple soft-gluon emission becomes important, fixed-order calculations are not expected to work. As an example, Fig. 5.12 shows a comparison between data and parton-shower improved LO predictions by PYTHIA6 and SHERPA for $p_{\mathrm{T},\gamma\gamma}$ and $\Delta\phi_{\gamma\gamma}$ from ATLAS [168].

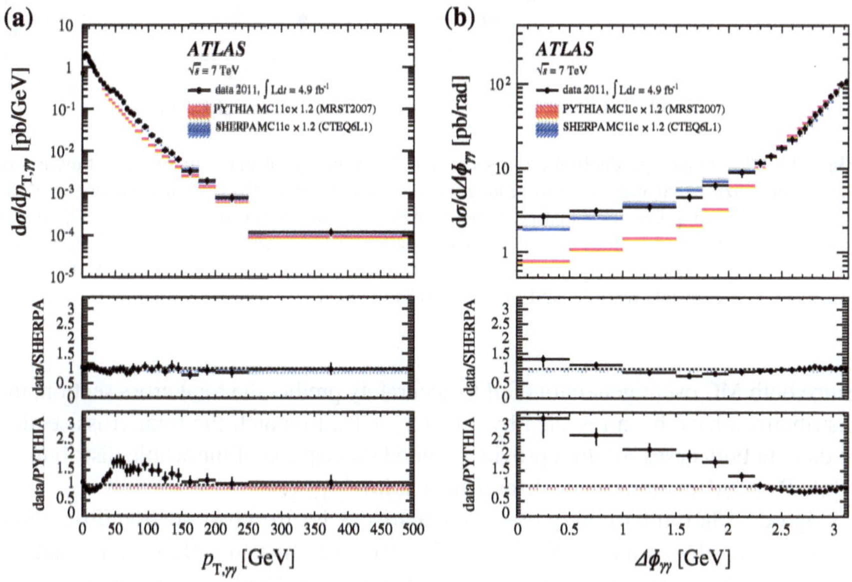

Fig. 5.12 Photon-pair production cross section at 7 TeV centre-of-mass energy as a function of **a** the diphoton transverse momentum $p_{\mathrm{T},\gamma\gamma}$ and **b** the azimuthal angular separation $\Delta\phi_{\gamma\gamma}$ as measured by ATLAS [168]. The data are compared to parton-shower-improved LO predictions by PYTHIA6 and SHERPA. The MC predictions are rescaled by a factor of 1.2 to match the total cross section observed in data. Adapted from [168]

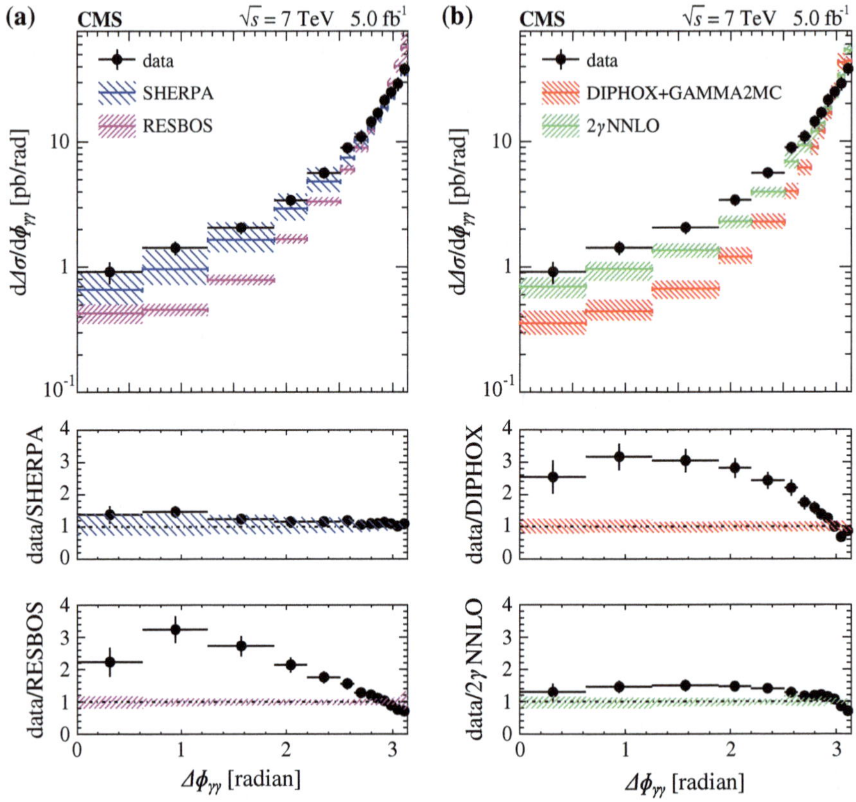

Fig. 5.13 Photon-pair production cross section at 7 TeV centre-of-mass energy as a function of the diphoton azimuthal angular separation $\Delta\phi_{\gamma\gamma}$ as measured by CMS [170]. **a** Comparison of the data to SHERPA using tree-level matrix elements matched to the parton shower and to NLO results from RESBOS that also include soft-gluon resummation. **b** Comparison to full NLO calculations including contributions of photons produced in jet fragmentation from DIPHOX + GAMMA2MC as well as to NNLO results from 2γNNLO. Adapted from [170]

Since both MC event generators fail to accurately predict the total cross section, the distributions have been rescaled by a factor of 1.2 to match the total cross section in data. In both cases SHERPA provides a good description of the whole distribution, while PYTHIA6 exhibits some discrepancy at low $p_{\mathrm{T},\gamma\gamma}$.

Figure 5.13a furthermore presents comparisons for the cross section differential in $\Delta\phi_{\gamma\gamma}$ for SHERPA and to NLO from RESBOS including soft-gluon resummation. Again SHERPA gives a good overall description, while RESBOS predictions are too low by factors of 2–3 at small separation angles. A similar behaviour can be seen in the comparison to NLO from DIPHOX + GAMMA2MC in Fig. 5.13b, although due to resummation RESBOS performs somewhat better around $\Delta\phi_{\gamma\gamma} \approx \pi$. The best description of the data is given by the NNLO result from 2γNNLO (Fig. 5.13b), even though the fragmentation contribution is not included. ATLAS observes similar results when comparing their data to the predictions from DIPHOX + GAMMA2MC and 2γNNLO [168].

5.7 Multi-jets: Precision Meets Multiplicity

Understanding the QCD dynamics governing the emergence of multi-jet final states is of utmost importance at the LHC. Given that typically jet objects with minimal values of their transverse momentum of order 30 GeV are considered, the phase space for producing multiple jets is huge. Accordingly, these jet-production processes need to be described as well as possible, as they constitute severe backgrounds to almost every search for new physics and furthermore have an impact on the appearance of the signals themselves. The modelling of inclusive processes that receive contributions from final states with potentially many jets is the realm of so-called "matching and merging" techniques that combine (multi-parton) fixed-order calculations with parton-shower simulations.

On the one hand, fixed-order calculations provide a well-defined estimate for inclusive production cross sections and reliably account for hard, well-separated parton configurations. On the other hand, parton-shower simulations capture the dominant terms driving the emission of additional soft and/or collinear partons and thus provide the means to perturbatively account for the internal structure of jets and inter-jet energy flows. When combining fixed-order matrix elements with parton-shower resummation beyond leading-order 2-to-2 scattering processes, the obstacle of double-counting configurations that appear in both approaches needs to be resolved. Furthermore, any consistent scheme should neither spoil the logarithmic accuracy of the inherent parton-shower resummation nor destroy the formal precision of the fixed-order part.

There are two basic strategies to distinguish. Tree-level merging techniques correct the hardest emissions of the parton shower off a given core process through exact leading-order QCD matrix elements [179–181]. This is achieved through a slicing of the real-emission phase space in terms of a hardness-measure that regulates any infrared singularities and allows to consistently combine matrix elements of varying parton multiplicity dressed with parton showers into an inclusive sample. Several variants of such leading-order merging techniques exist and are widely used in LHC analyses, see [60, 182–184].

The second ansatz relies on the exact matching of an NLO QCD calculation with a parton-shower cascade off the underlying Born process. Accordingly, the real-emission correction as part of the fixed-order calculation has to be properly synchronised with the first, i.e. hardest, shower splitting. Furthermore, the NLO accuracy with respect to the inclusive production process considered needs to be preserved. Two basic solutions exist to this problem, which are known as MC@NLO [185] and POWHEG [186]. Over the last years there has been tremendous development in implementing these techniques for a wide range of processes and ultimately their automation, see [187–190].

Most recently hybrid solutions emerged that combine NLO plus parton-shower calculations with higher-order tree-level QCD corrections [191, 192] or that even combine parton-shower-matched NLO calculations of varying jet multiplicity [193–196].

In the following section, the main focus shall be on a class of processes that constitutes a prototypical Standard Model background and serves as a test bed for the various types of advanced QCD calculations outlined above: W/Z+jets production.

5.7.1 Weak Bosons and Jets

In the production of W and Z bosons in association with jets (W+jets), very high jet multiplicities can be reached at the LHC. Cross sections for the associated production with up to seven jets with transverse momentum p_{T} larger than 30 GeV have been measured by ATLAS and CMS. Already the leading-order prediction for this process is of order α_s^7. However, this class of processes is extremely important as it constitutes a major background to many other processes with complex final states such as top-quark pair or diboson production or rare signatures from the Standard Model or beyond.

The W+jets processes allow an assessment of the validity of Monte Carlo generators like ALPGEN [5] SHERPA [72, 73] or MADGRAPH [9] that merge tree-level matrix-element calculations with up to five additional partons interfaced with parton showers and hadronisation.

Figure 5.14 shows the W+jets cross section as a function of the leading-jet p_{T} as measured by CMS [197] and ATLAS [198] using proton-proton collision data recorded at $\sqrt{s} = 7$ TeV. In the ATLAS measurement, jets are defined by the anti-k_t jet algorithm using a radius parameter of $R = 0.4$ and measured within $30 \leq p_{\mathrm{T}} < 300$ GeV and $|y| < 4.4$. The results for up to four additional jets are superimposed and are compared to the predictions of SHERPA and ALPGEN. The measured shape of the jet p_{T} distribution is described by the Monte Carlo simulation within the uncertainties.

In the CMS measurement, jets defined by the anti-k_t jet algorithm with $R = 0.5$ are selected with $p_{\mathrm{T}} > 30$ GeV and $|y| < 2.4$. The jet spectrum reaches up to p_{T} values of 800 GeV. The data fall below the leading-order QCD predictions towards high p_{T}. The MADGRAPH (SHERPA) estimate using up to four additional partons is almost 50 % (20 %) higher than the data for $p_{\mathrm{T}} > 200$ GeV.

At the time of the LHC start-up in 2008, NLO QCD calculations with maximally two associated jets were available [199]. Since then, in particular, with the BLACKHAT- SHERPA [87, 200] program significant progress was made in NLO QCD calculations for W or Z production in association with up to five additional partons [88]. Fixed-order calculations, by nature, are not matched to parton showers and hadronisation, so that corrections to account for these effects are derived

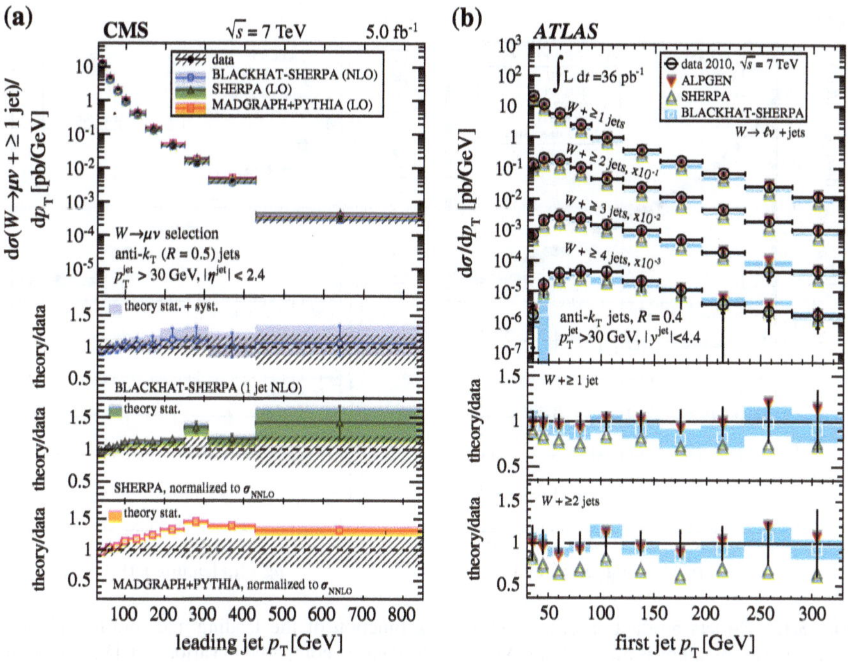

Fig. 5.14 The W+jet cross section as a function of the transverse momentum of the leading jet as measured by CMS (**a**) and ATLAS (**b**). The data are compared to various QCD calculations. For the ATLAS measurement, jet multiplicities up to four jets are shown. The results for higher jet multiplicities are scaled by the factors given in the figure. The leading-order multi-leg calculations are normalised to the inclusive NNLO cross section. Adapted from [197, 198]

from generators like PYTHIA or HERWIG are needed. The resulting predictions from BLACKHAT- SHERPA are in good agreement with the measured distributions within the systematic uncertainties as shown in Fig. 5.14.

Recently, automated computations of NLO QCD cross sections matched to parton-shower simulations for a large variety of processes (with up to four partons) have been developed. The programs SHERPA (in the MEPS@NLO option [7, 193]) and MADGRAPH5_AMC@NLO [190] presently provide the most accurate predictions for multi-jet final states at particle level.

Similar measurements [201] for the Z+jets processes using the full data set at $\sqrt{s}$ = 7 TeV are presented in Fig. 5.15 as a function of the leading-jet p_T and the fourth-leading jet p_T. Jets are defined using the anti-k_t jet algorithm with $R = 0.4$ and are selected if $p_T > 30$ GeV and $|y| < 4.4$. The cross sections are normalised to the inclusive Z cross section. Again, the BLACKHAT- SHERPA predictions are in good agreement with both measured distributions.

An NLO QCD calculation for the inclusive Z cross section merged with parton showers and hadronisation as incorporated in MC@NLO is capable of describing the data for events with one or two low-p_T jets, but falls far below the data towards

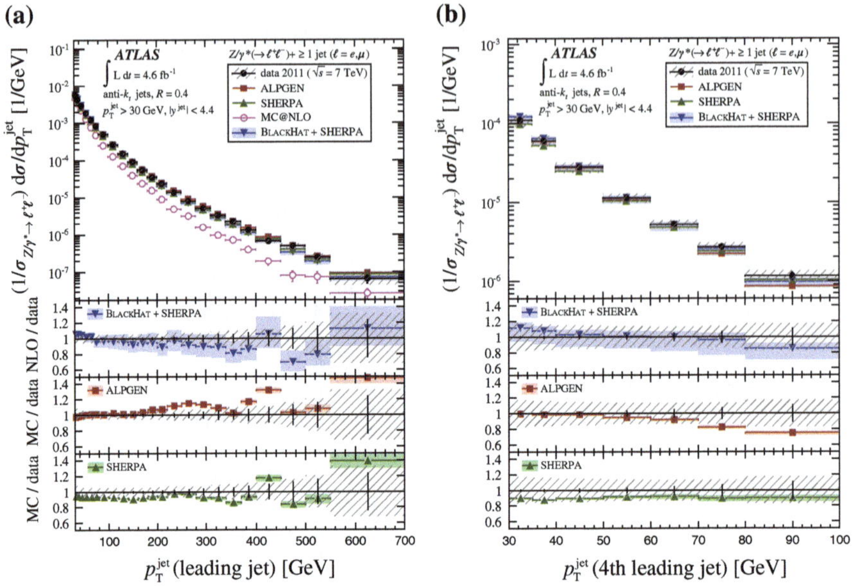

Fig. 5.15 Z+jets production cross section as a function of the leading and fourth-leading jet transverse momentum as measured by ATLAS. The data are compared to various QCD calculations. Adapted from [201]

high values of the leading-jet p_T. This indicates that the fraction of events exhibiting a second jet increases considerably with the p_T of the leading jet and that the parton shower approximation, which is used by MC@NLO for the second jet, fails to describe the data.

The results are also compared to two multi-leg Monte Carlo simulations produced with up to five additional partons. The ALPGEN simulation overestimates the data at high p_T for the leading jets and underestimates the data for the fourth-leading jets. SHERPA is in good agreement with the data considering the experimental uncertainties. The measurements of the Z+jets cross section are limited by the experimental uncertainty on the jet energy measurements. By considering jet-multiplicity ratios, many experimental and theoretical uncertainties cancel, allowing for an even more precise data-to-theory comparison.

The ratio of the Z+jet cross sections for $N_{\text{jet}} + 1$ over N_{jet} ($R_{(N+1)/N} = \sigma(Z+N_{\text{jet}}+1)/\sigma(Z+N_{\text{jet}})$) was measured by ATLAS [201]. An earlier measurement was performed by the CMS experiment on a smaller data set [202]. Figure 5.16 shows the ratio $R_{(N+1)/N}$ for successive exclusive jet multiplicities for all events ("inclusive") and for events with at least one jet with $p_T > 150$ GeV. For the inclusive case, the ratio shows a rather flat, approximately linear dependence on the jet multiplicity with a small slope only, while for events with $p_T > 150$ GeV the measured distribution steeply rises towards low jet multiplicities. These measurements illustrate two limiting cases of scaling patterns, the "staircase scaling" [203] with $R_{(N+1)/N}$ constant and the "Poisson scaling" [204] with $R_{(N+1)/N} = \langle N \rangle / N$, where $\langle N \rangle$ is

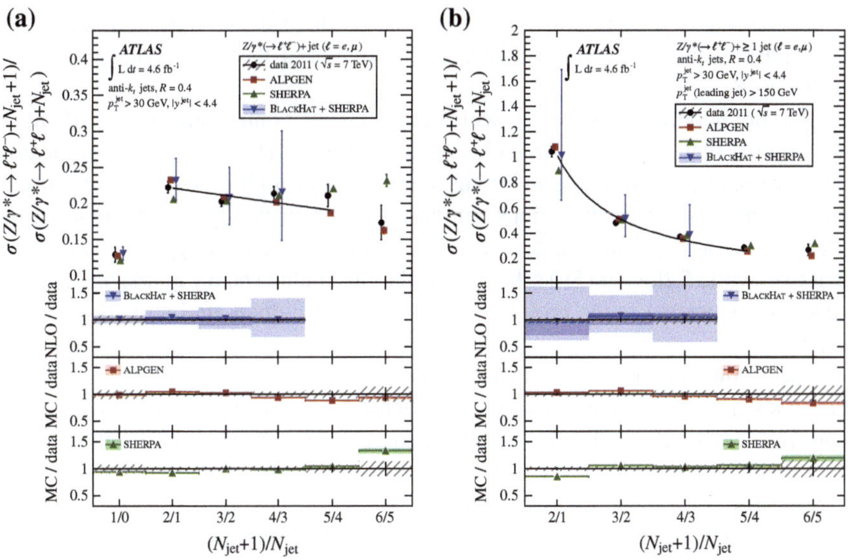

Fig. 5.16 Ratios of Z+jets cross sections for successive exclusive jet multiplicities for **a** inclusive events and **b** for events with at least one jet with $p_T > 150\,\text{GeV}$ and $|y_{\text{jet}}| < 4.4$. The data are compared to various QCD calculations. Adapted from [201]

the average number of jets. These scaling patterns can be qualitatively understood from the expected Poisson-distributed jet multiplicity. For large jet-p_T differences, the Poisson-scaling is directly observed (see Fig. 5.16b). However, for low p_T and small N_{jet}, the emission of additional partons is suppressed by the PDFs, and for high N_{jet} the emission of additional partons no longer follows a Poisson distribution due to the non-Abelian nature of QCD, leading to a proliferation of the number of jets originating from gluon splitting (see Fig. 5.16a).

The scaling patterns observed in data are described by the Monte Carlo simulations BLACKHAT- SHERPA, ALPGEN and SHERPA. The scale uncertainty is only shown for the NLO QCD calculation based on BLACKHAT- SHERPA. The different grey shades indicate the scale uncertainty when it is considered uncorrelated (as proposed in [205]) or correlated across the various jet multiplicities. Assuming the predicted and now observed scaling patterns to be valid, it becomes possible to give estimates for even higher jet multiplicities where there exists no complete NLO prediction yet [206].

5.7.2 Weak Bosons and Jets with Flavour

Several key processes in the Standard Model and beyond—such as the production of single top quarks, top-quark pairs or Higgs bosons—involve heavy quarks. The understanding of their production mechanisms is therefore an important part of the LHC physics programme. For a detailed discussion of flavour physics at the LHC see Chap. 8.

The predictions of processes with heavy-flavour content are more difficult to accomplish than those for the inclusive case. Bottom (b) quarks are significantly heavier than the proton, and in strong interactions at high scales they can only be created in pairs. The calculations can be classified in two types: In the 4-flavour (4F) scheme heavy quarks appear only in the final state and are typically considered massive, while 5-flavour (5F) scheme calculations include heavy quarks in the initial state, i.e. as present in the proton. For this purpose, c-quark and b-quark PDFs are used.

The production of W and Z bosons in association with heavy flavour is an important background for many processes. Several NLO QCD calculations are available [208–218]. Recently, the $W+b\bar{b}$ and $Z+b\bar{b}$ processes (with and without additional jets) have been included in automated NLO calculations matched with parton showers and hadronisation [190].

Z bosons provide a particularly clean experimental signature that allows for precise measurements. Figure 5.17 presents a recent ATLAS measurement of the cross section of Z bosons produced in association with b jets as a function of the p_{T} of the leading jet containing a B hadron. The measurements uses the full 2011 data set. Transverse jet momenta up to 500 GeV are reached. The data are compared to various QCD calculations.

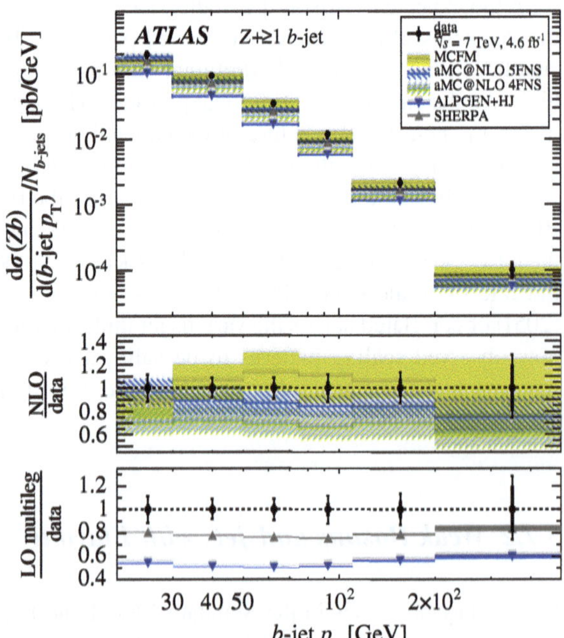

Fig. 5.17 Production cross sections for Z bosons produced in association with jets with heavy-flavour content as a function of the jet p_{T}. Superimposed are various QCD calculations. Adapted from [207]

The MCFM prediction is based on a 5F NLO QCD calculation of $Z + 1b$ jet [211, 213], corrected for hadronisation and effects of multi-parton interactions (see Sect. 5.9.4). Full particle-level predictions with NLO QCD matrix element calculations are also obtained using MADGRAPH5_AMC@NLO [190] in both the 4 F and 5 F schemes. For these calculations, the renormalisation and factorisation scales are set to the transverse Z-boson mass and varied up and down independently by a factor of two to assess the residual dependence on this scale choice. The data are described within the experimental and theoretical uncertainties.

Also shown are the predictions for multi-leg LO QCD calculations based on ALPGEN using the 4 F scheme with up to five partons in the matrix element and SHERPA using the 5 F scheme with up to four partons. The shape of these predictions is in good agreement with the data. However, the LO normalisation of these predictions undershoots the measurement, SHERPA reproducing the normalisation better than ALPGEN.

Further measurements by ATLAS and CMS of vector-boson production in association with heavy-flavour jets can be found in [219–221].

5.8 Resummation: The Realm of Large Logarithms

When discussing the process of vector-boson production before, the necessity to incorporate effects of multiple emissions of initial-state and final-state partons to appropriately model certain observables has been touched upon already. In fact, phase-space regions sensitive to the emission of soft and/or collinear QCD quanta often resist a satisfactory description in terms of fixed-order perturbation theory. Rather, one is forced to identify the relevant enhanced contributions, namely large logarithmic terms, and to reorganise the perturbative expansion for the observable at hand. The aim is to account for the enhanced terms to all orders in perturbation theory. This procedure is referred to as "resummation" and allows an appropriate treatment in a wide range of perturbative QCD phenomena.

Examples for hadron-hadron collider observables sensitive to multiple-emission effects include the transverse-momentum distribution of lepton pairs or gauge bosons at low p_{T} [154, 222, 223], event-shape variables [224–226], or certain jet-substructure observables [227, 228]. Furthermore, resummation techniques become important when the real-emission phase space is constrained, leaving behind finite but possibly large uncancelled QCD corrections. Examples thereof are observables that exhibit thresholds [229–232] or that rely on an explicit veto on QCD activity [233–237]. Resummed expressions for this type of observables might enable us to extract QCD parameters such as the strong coupling, quark masses and the parton distribution functions much more accurately than when having to rely on fixed-order perturbation theory alone. In the following, two concrete examples shall be considered: the application of a veto on additional jet activity and the jet-mass distribution in Z+jet production.

5.8.1 Jet Vetos and Gap Fractions

At first, so called "gap fractions" will be discussed. For this class of observables, used also to select the weak-boson-fusion production channel of Higgs bosons (see Chap. 6), one considers a specific kinematic structure of two hard and widely separated jets. Analysing the QCD activity in-between these hard jets, events are vetoed in case there are additional jets with transverse momentum above a certain veto scale Q_0, filtering out those events that exhibit a gap between the two boundary jets. On the one hand, the presence of the potentially soft jet-veto scale Q_0 induces large logarithms with argument Q/Q_0, where Q denotes the typical hard process scale, e.g. the boundary jets' transverse momenta. This demands for the all-orders treatment of wide-angle soft-gluon radiation [233, 234]. On the other hand, when considering large rapidity separations Δy between the boundary jets, manifestations of BFKL-like dynamics [238, 239] are expected, requiring to sum terms proportional to $\alpha_s^n(\Delta y)^n$ to all orders n [240, 241]. In [242, 243] ATLAS presented measurements of dijet production with a veto on additional jets. In order to gain further insight into the QCD dynamics determining the gap-fraction and gap-jet measurements, [243] considers additional azimuthal-decorrelation observables. In this latter analysis, the data are compared to predictions from POWHEG [188, 244], interfaced to the PYTHIA8 and HERWIG event generators, as well as to the program HEJ [245] with and without invoking the ARIADNE shower model [246]. The two POWHEG simulations provide NLO accuracy for the inclusive dijet production process invoking leading-logarithmic DGLAP resummation through the parton showers attached. The HEJ approach provides a resummation of small-x, BFKL-type logarithmic terms that can be supplemented with DGLAP resummation through, in this case, the ARIADNE parton cascade.

In Fig. 5.18 the gap fraction as a function of the veto scale Q_0 is presented for various slices of the boundary jets' rapidity separation Δy. The fraction of events exhibiting a rapidity gap decreases as the veto scale is lowered or when increasing Δy. This corresponds to the expectation of an increase of the jet activity when considering lower jet transverse-momentum thresholds together with the rise in jet multiplicity when considering larger rapidity intervals. The POWHEG +PYTHIA8 prediction yields a reasonable description of the data. Employing the HERWIG parton shower the agreement worsens, an effect that originates from a prediction of too many jets above the veto scale. For the two predictions from HEJ the agreement with data improves when considering large rapidity separations. Given the high precision of the measured data, no single theoretical prediction is capable of describing the data in all phase-space regions. Considering more observables, including angular variables, [243] establishes the DGLAP-based POWHEG +PYTHIA8 and BFKL-like HEJ +ARIADNE simulations to yield the best overall description of the data. So far no clear-cut statements about the evidence for BFKL effects can be established. However, the small experimental uncertainties allow theoretical models for QCD radiation between widely separated and high-transverse momentum jets to be further constrained and improved [204, 247–250].

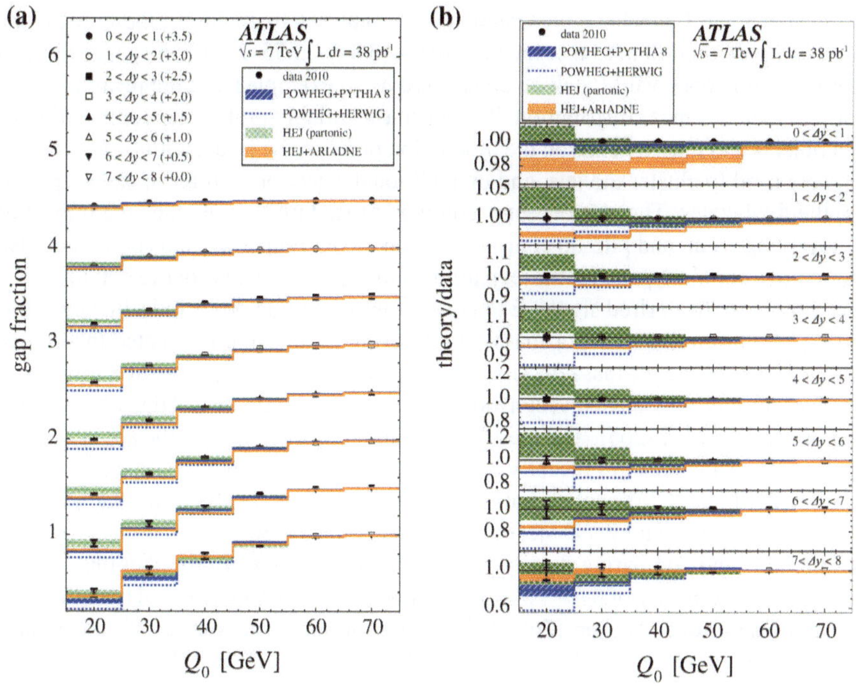

Fig. 5.18 Gap fraction for dijet events as a function of the jet-veto scale Q_0 for various Δy ranges. The data are compared to predictions from POWHEG +PYTHIA8, POWHEG +HERWIG and from the generator HEJ with and without inclusion of the ARIADNE shower algorithm. **a** Direct comparison with the unfolded data (for illustrative purposes the various results are shifted by arbitrary constants); **b** ratio of theory over data. Adapted from auxiliary material provided with [243]

5.8.2 The Jet-Mass Distribution

Many phenomenological studies over the past years have investigated and highlighted the potential of jet-substructure techniques to identify the hadronic decays of boosted heavy objects or to discriminate scenarios of new physics from Standard Model (i.e. QCD) backgrounds [251–253]. The successful application of jet-substructure methods requires a more detailed and even finer understanding and modelling of the inner structure of QCD jets. The new experimental challenges to address include the consideration of various jet algorithms and radii that need to be calibrated, the quantification and possible removal of non-perturbative contributions from underlying event or pile-up activity, or the application of tracking and flavour-tagging methods in busy environments. On the theoretical side, precise and reliable predictions for the new observables at hand are required. This, in particular, asks for resummed calculations, in the form of Monte Carlo parton-shower simulations or as dedicated analytical results.

ATLAS and CMS have presented several analyses that address these questions and compare their data to theoretical models [54–57, 255–258]. One particular and rather simple observable shall be considered as an example here: the invariant mass of jets—the jet-mass distribution. The jet-mass variable probes the composition of jet objects and thus depends on the production process, the jet algorithm and its parameters used to cluster the jets, and on additional selection or filtering steps applied to the final states. The aim of "jet grooming" procedures is to reduce the impact of underlying-event and pile-up contributions on the jet properties and sometimes also to remove certain types of intra-jet parton splittings, all that in order to maximise the sensitivity to the desired signal. Examples of jet-grooming procedures include trimming [259], pruning [254] or mass-drop filtering [260, 261]. Dedicated theoretical predictions for the jet-mass distribution exist for a variety of production processes, including dijet final states, vector-boson plus jet production and Higgs-boson plus jet production [262–265]. The mass distribution of QCD jets after the application of grooming techniques has recently been discussed in [266, 267].

In [255] the CMS collaboration has presented an analysis of the jet-mass observable in dijet and W/Z+jet events, focusing on the effects of various grooming techniques and comparing data to parton-shower Monte Carlo predictions. For an integrated luminosity of 5 fb^{-1} of 7 TeV collision data they considered both the anti-k_t and Cambridge–Aachen [268–270] jet algorithms using various distance parameters, namely $R = 0.7$ for anti-k_t and $R = 0.8$ and $R = 1.2$ for Cambridge–Aachen. The groomed and ungroomed data are compared to Monte Carlo predictions from PYTHIA6 and HERWIG++. As an example, the unfolded jet-mass distribution of $Z(\ell\ell)$+jet events is shown in Fig. 5.19. Depicted there is the jet-mass variable

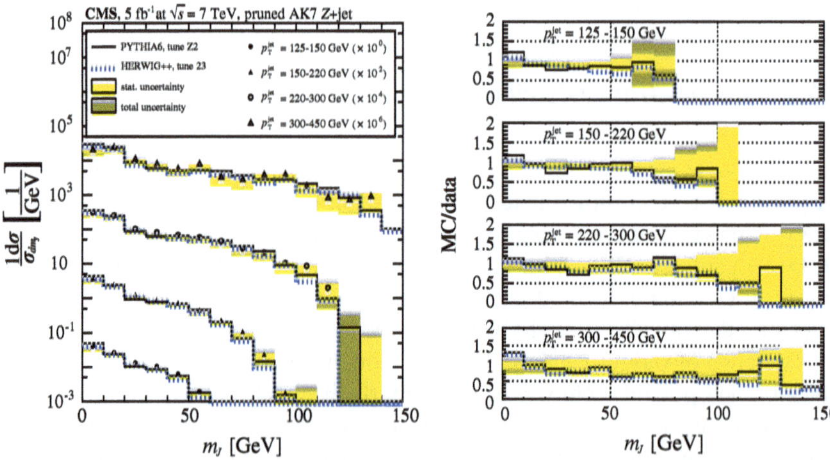

Fig. 5.19 Unfolded jet-mass distribution in $Z(\ell\ell)$+jet events. Jets are reconstructed using the anti-k_t jet algorithm with distance parameter $R = 0.7$ and are groomed using the pruning technique [254]. The data from CMS are compared to Monte Carlo predictions from MADGRAPH +PYTHIA6 (*solid*) and HERWIG++ (*dotted*). The *right panel* shows the ratio of Monte Carlo over data. Adapted from [255]

for anti-k_t jets after a pruning procedure, analysed for four slices in jet transverse momentum (note the arbitrary scaling factors). For all the jet-p_T slices both generators describe the data well. A similar level of congruence is observed for the alternative grooming techniques and combinations of jet algorithm and distance parameter. Somewhat larger deviations between data and Monte Carlo estimates are observed for regions of small jet mass. However, it is in particular this region of phase space that is most affected by non-perturbative corrections from pile-up, underlying event and hadronisation. Attempts to analytically estimate the effects of the underlying event and parton-to-hadron fragmentation on the jet-mass distribution have been presented in [271].

From the analysis presented by CMS and a similar one presented by ATLAS [55], the agreement of jet-mass and substructure observables with modern parton-shower Monte Carlo tools is established. This paves the way to successfully use these techniques in future analyses and particularly in searches for new phenomena.

5.9 Beyond Perturbative QCD

So far the focus has been mainly on measurements that can reliably be addressed using QCD perturbative techniques. Experimentally, however, hadrons are observed in the detectors, and accordingly the transition of partons into hadrons needs to be accounted for. Furthermore, there is a wealth of effects that require the use of non-perturbative methods and models—most prominently the so-called "underlying event" that accompanies a hard scattering. As an example, the underlying event includes contributions from multiple interactions of the colliding protons' constituents ("multi-parton interactions") that give rise to the production of many additional final-state particles.

In this section these aspects shall be exemplified by presenting LHC measurements of jet-shape variables, of jet-radius ratios and of observables sensitive to soft-particle production and double and multi-parton scattering.

5.9.1 Jet Shapes

Focusing on the internal structure of jets, the profile of the distribution of transverse momentum within a jet can be examined. For this type of observable the term "jet shape" has been coined [272, 273]. The differential jet shape $\rho(r)$ as a function of the distance $r_i = \sqrt{(\Delta_{i,\mathrm{jet}}y)^2 + (\Delta_{i,\mathrm{jet}}\phi)^2}$ to the jet axis is defined as the average fraction of jet p_T contained inside an annulus of inner radius $r_a = r - \Delta r/2$ and outer radius $r_b = r + \Delta r/2$ for an ensemble of N jets:

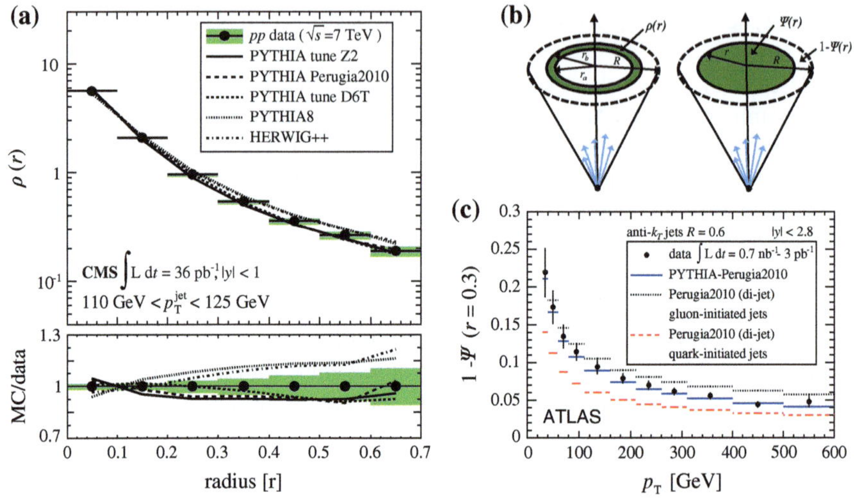

Fig. 5.20 **a** Differential jet shape $\rho(r)$ as a function of the distance r from the jet axis for jets with $110 < p_T < 125\,\text{GeV}$ from CMS. The data are compared to various tunes of the PYTHIA6 and to the PYTHIA8 and HERWIG++ event generators. **b** Illustration of the jet shape observables. **c** Integrated jet shape $1 - \Psi(r = 0.3)$ as a function of the jet p_T from ATLAS [276]. The data are compared to the prediction for gluon-initiated and quark-initiated jets using the PYTHIA6 MC event generator with tune Perugia2010. Adapted from [276, 277]

$$\rho(r) = \frac{1}{N}\sum_{\text{jets}} \frac{1}{\Delta r} \frac{\sum_{r_a \le r_i < r_b} p_{T,i}}{\sum_{r_i \le R} p_{T,i}},$$

where the second sum runs over all jet constituents i. The integrated jet shape $\Psi(r)$ is then given by the integral of the differential jet shape up to a radius r (see the illustrations in Fig. 5.20b). Conventionally, measurements are presented in terms of $1 - \Psi(r = r_{\text{core}})$, where r_{core} is usually taken to be 0.3 [274].

Perturbatively, the internal structure of jets is determined by multiple emissions of gluons and depends on the type of initiating parton, i.e. a quark or gluon. Perturbative QCD predicts that gluon-initiated jets are broader in shape with a higher particle multiplicity on average [275]. To some extent it is therefore possible to differentiate statistically between gluon and quark jets by looking into the jet substructure. Jet shapes, however, are also sensitive to non-perturbative effects of the parton fragmentation and the underlying event. Corresponding measurements are valuable inputs to improve the modelling of such soft contributions in MC event generators like PYTHIA, HERWIG++, or SHERPA.

Figure 5.20a shows a CMS measurement [277] of the differential jet shape, for the example of the jet p_T interval $110 < p_T < 125\,\text{GeV}$, in comparison to several LO MC event generators at particle level. While HERWIG++ and PYTHIA8 predict somewhat broader jet shapes than measured, the PYTHIA6 tunes D6T and Z2 deviate in the opposite direction. The best description of the data is given by PYTHIA6 with the tune Perugia2010, something, which is also observed in a similar investigation by the ATLAS collaboration [276]. By employing this tune for the integrated jet shape, shown in the form of $1 - \Psi(r = 0.3)$ as a function of the jet p_T in Fig. 5.20c, ATLAS demonstrates the sensitivity of this observable with respect to the jet-initiating parton. At small jet p_T the examined inclusive jet sample is predominantly composed of gluon jets, while with increasing jet p_T the quark-initiated component grows. In the figure, the change in the fraction of quark-initiated to gluon-initiated jets is averaged over the jet rapidity up to $|y| < 2.8$. If studied double-differentially, a mild dependence on $|y|$ of this fraction is observed as expected. The sensitivity of jet shapes to non-perturbative effects and modelling uncertainties, however, prevents an extraction of the quark-gluon jet fraction. Instead, they provide valuable input to the tuning of these effects in MC event generators.

In the context of jets initiated by heavy boosted objects, the investigation of jet substructure has attracted much attention lately. Potential new particles like Z' bosons or heavy t' quarks that decay hadronically may be at the origin of jets with two or three, respectively, high-energetic partons fragmenting into only one jet. The required boosts for these hypothetical particles are attainable at the LHC, in particular at 13 TeV centre-of-mass energy as foreseen for the restart in 2015. Searches for these particles profit from a better understanding of the QCD background in the form of quark-initiated or gluon-initiated jets. For some of the investigated observables, e.g. jet mass, resummed predictions are available (see Sect. 5.8).

5.9.2 Jet-Radius Ratio

As described in Sect. 5.5, the inclusive jet cross section can precisely be compared to NLO predictions, and the strong coupling constant can be extracted. A small correction for non-perturbative effects, which is p_T-dependent and increases at small p_T values, has to be taken into account. Looking more closely into the results presented by ATLAS for anti-k_t jet sizes of $R = 0.4$ and 0.6 [29] or by CMS for anti-k_t jet sizes of $R = 0.5$ and 0.7 [30, 31], it is observed that the relative normalisation of the measured cross sections to the theoretical predictions exhibits a dependence on R. Theoretically, this R dependence has been examined in [271, 278], where in a collinear approximation it was found that the impact of perturbative radiation and of the non-perturbative effects of hadronisation and the underlying event on jet transverse momenta scales roughly with $\ln R$, $-1/R$, and R^2 for small R, respectively. By choosing the jet size parameter R, one can therefore steer which aspects of jet formation are emphasised in a jet analysis. In [271] it is suggested, in order to gain insight into the interplay of these effects, to study the relative difference between inclusive jet cross sections that emerge from two different jet definitions:

$$\left(\frac{d\sigma^{\text{alt}}}{dp_T} - \frac{d\sigma^{\text{ref}}}{dp_T}\right) \Big/ \left(\frac{d\sigma^{\text{ref}}}{dp_T}\right) = \mathcal{R}(\text{alt}, \text{ref}) - 1\,.$$

Provided that partons in opposite hemispheres are not clustered together (a condition that is usually fulfilled), the LO two-parton cross sections are identical for arbitrary jet algorithms. Therefore, only partonic final states with three or more partons lead to a numerator different from zero. Hence, Sect. 5.9.2 defines a three-jet observable, $\mathcal{R}$, for which it was shown in [279] that it is calculable to NLO with terms up to α_s^4 with NLOJET++ [280, 281].

Studies in that direction have been performed at the HERA collider by the ZEUS collaboration for two different jet algorithms [282] and by the ALICE experiment for the two different anti-k_t jet sizes of $R = 0.2$ and 0.4 [283]. The results of a CMS analysis for their two jet sizes of 0.5 and 0.7 is presented in [33] as a function of jet p_T and rapidity y. It is expected that QCD radiation reduces this jet-radius ratio $\mathcal{R}$ below unity and that this effect disappears with increasing collimation of jets at high p_T. Figure 5.21 confirms this assumption with a comparison of the measured ratio $\mathcal{R}$ as a function of the jet p_T up to the TeV scale to theoretical predictions at central rapidity $|y| < 0.5$. Figure 5.21a clearly demonstrates that fixed-order calculations up to NLO, even when combined with non-perturbative corrections, are systematically above the data. The LO event generators PYTHIA6 and HERWIG++ lie systematically on either side of the data, as can be seen in Fig. 5.21b. Presumably, this can be improved by including this observable into the tuning of these generators. The best description is given by POWHEG that matches the dijet production process evaluated at NLO with the parton showers and non-perturbative models of PYTHIA6, emphasising the importance of parton showers for the given choice of jet sizes.

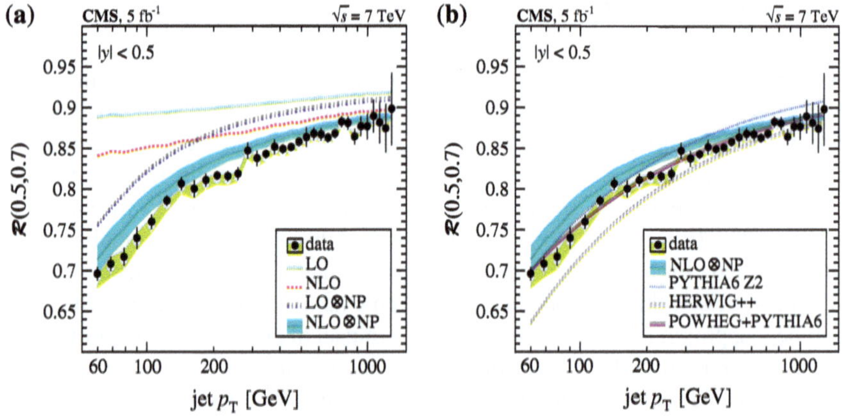

Fig. 5.21 Jet-radius ratio $\mathcal{R}$ as measured by CMS in comparison to **a** fixed-order predictions with and without non-perturbative corrections and to **b** particle-level predictions of LO and NLO event generators with matched parton showers and modelling of hadronisation and the underlying event at central rapidity $|y| < 0.5$. Adapted from auxiliary material provided with [33]

5.9.3 Soft Hadron-Hadron Collisions

When two hadrons collide at high energy, their entire volumes or at least a significant part of them participates in the collision. Most of these collisions are caused by so-called "soft" hadronic interactions.

In a simple geometrical model, the total hadron-hadron collision cross section is related to the transverse extension of the scattering system. Therefore, it is expected to be of the order of $\sigma_{tot} \approx \mathcal{O}(2\,\text{fm}^2) \approx \mathcal{O}(40\,\text{mb}^{-1})$. Since the effective range of soft hadron interactions does not change much with energy, only a weak energy dependence is expected. Moreover, if the collision is soft, the scattering cross section at high energy should not depend on the quark structure of the colliding hadrons.

In 1952 Heisenberg used a geometrical model where hadrons are modelled as Lorentz-contracted disks with a given energy density. The interaction takes place in the disk-overlap region, and only if there is sufficient energy to create two mesons. In this model the total cross section increases like $\ln^2 s$. A consequence of this model is that the interaction cross sections of all hadrons are similar at high energy. For a recent review on the model giving a $\ln^2 s$ high-energy behaviour see [284].

General arguments based on unitarity, analyticity and factorisation imply a bound on the high-energy behaviour of total hadronic cross sections. This bound is independent of the details of the strong-interaction dynamics and states that the total cross section can not rise faster than $\ln^2 s$ [285].

Soft hadron collisions can not be calculated in perturbative QCD. Instead, they are usually described by phenomenological models where the force between hadrons is assumed to be mediated by mesons, for instance by pions [287]. The basic idea is that the exchanged meson interacts with the colour charges confined in the scattered hadrons. Since many mesons exists, the strong force between hadrons can be quite complicated in such models. The typical range of the strong force is anti-proportional to the mass of the exchanged meson. Consequently, the largest range is provided by the lightest meson which has the appropriate quantum numbers. Such models are usually called "Regge theory". For reviews see [288, 289]. Regge theory aimed at explaining the spectrum of the various hadrons, the forces between them, and the high-energy behaviour of the hadron-hadron scattering cross section. It is based on simple assumptions like that the scattering matrix is analytic and unitary. Mesons are understood as composite quark-antiquark states bound by an effective strong potential that depends on the distance between the quark-antiquark pair r and the orbital angular momentum J. Elastic hadron-hadron scattering is viewed as the exchange of a meson in the s-channel or t-channel. In 1961, Chew et al. [290] made the observation that for $\pi^- p \to \pi^0 n$ scattering there is a linear relation between the mass of the exchanged hadron (m) and its angular momentum J. Each integer value of the slope α—called "Regge trajectory"—corresponds to one particle that can be experimentally identified ("Regge-pole"). In the time-like region where $t > 0$ (s-channel meson exchange) the momentum transfer t corresponds to the hadron mass m. If one continues the linear relation between m and J using the slope α for $t < 0$, i.e. the space-like regime, a good description of the experimental measurements

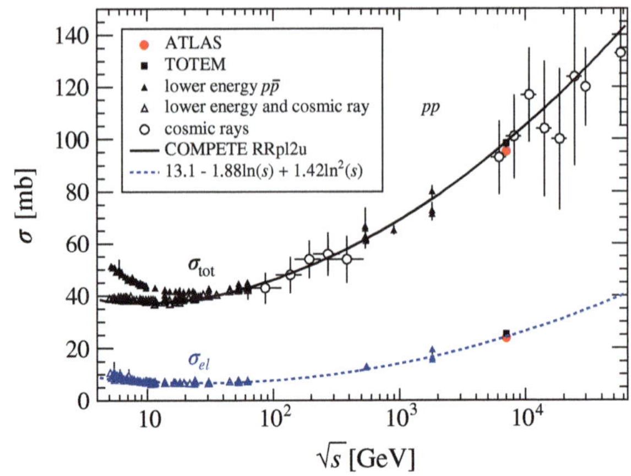

Fig. 5.22 Comparison of total and elastic cross-section measurements in proton-proton and antiproton-proton collisions and in cosmic-ray interactions as a function of the centre-of-mass energy. Adapted from [286]

for α is obtained. In this case, however, t can take any value. This trajectory is called "Pomeron trajectory". The exchanged object carrying the quantum numbers of the vacuum ($C = P = 1$) is called "Pomeron".

The meson exchange leads to a simple power behaviour $s^{\alpha-1}$ of total hadron-hadron scattering cross sections. In 1992 Donnachie and Landshoff [291] made the observation that hadron-hadron total cross sections can be described by two powers, one $\alpha \approx 1.1$ corresponding to Pomeron exchange and one $\alpha = 0.5$ corresponding to the exchange of mesons like ρ, ω, f_2 and a_2. Therefore at low-energies, where the Reggeon exchange dominates, a behaviour $\sigma \sim s^{-0.5}$ is expected, while at high energies $\sigma \sim s^{0.1}$ should prevail.

Such a behaviour was indeed experimentally observed. Figure 5.22 shows the total inclusive proton-proton cross section as a function of the centre-of-mass energy $\sqrt{s}$ for various experiments. In the range 2–20 GeV the proton-proton cross section decreases as measured by fixed-target experiments, while at higher energies it increases towards larger $\sqrt{s}$. The difference between proton-proton and proton-antiproton cross section decreases towards high energy; at about $\sqrt{s} \approx 30$ GeV, the two cross sections are very similar. The rise of the proton-antiproton cross section for $\sqrt{s} \gtrsim 30$ GeV was first observed at the SPS, confirmed by Tevatron and LHC, where the ATLAS [292], CMS [293] and TOTEM [294] experiments measured at $\sqrt{s} = 7$ TeV (TOTEM measured also at $\sqrt{s} = 8$ TeV [295]). Data from cosmic-ray experiments [296–298] are able to provide useful measurements at energies up to 60 TeV that are not covered by colliders. Overlayed in Fig. 5.22 as curve is a parameterisation [299] based on the $\ln^2 s$ behaviour that describes the data well at high energy. The power-law behaviour can also describe the present data [300]. Also shown in Fig. 5.22 are recent measurements of the elastic proton-proton cross section [286, 294].

5.9.4 The Underlying Event and Multi-Parton Interactions

An important aspect of hadron-hadron collisions, neglected in our discussion so far, is the observation that additional mostly low-energetic (soft) particles are produced. The simple picture of a hard scattering between one parton from each hadron with subsequent showering and hadronisation at least needs to be complemented by a treatment of the beam-beam remnants, which is beyond the capabilities of pQCD. In addition, the primary interaction may be accompanied by the production of further particles, a circumstance which is usually modelled in the form of multi-parton interactions (MPI). The concepts employed for MPI by PYTHIA and HERWIG++, for example, are described in [301–303] and [304], respectively. A very useful overview of general-purpose MC event generators including all aspects of event generation for the LHC is given in [41].

An unambiguous association of a specific particle to the reaction from which it originates is impossible. Therefore, the extra activity in a hadron-hadron collision cannot be uniquely separated from effects like initial-state or final-state radiation. One thus conventionally defines the "underlying event" (UE) as everything except for the hard process of interest. Two strategies are applied to measure the UE: The first evaluates events that are collected requiring only minimalistic trigger conditions such that even very soft collisions are registered. These so-called "minimum bias" (MB) measurements are supposed to be dominated by underlying-event physics.

In the second approach, reactions are considered that have a high rate for the production of a leading particle or jet with reasonably high p_{T} (for triggering purposes). The event is then separated into hard and soft contributions through a physically motivated definition of phase-space regions that are dominated by either the hard or soft component of a collision. Traditionally, this is achieved by geometrically subdividing the event into different regions ("towards", "away", and "transverse") with respect to the direction of the triggering particle or jet (Fig. 5.23a). Reactions that are typically chosen for this second approach are dijet or Drell–Yan lepton-pair production. Implicitly, it is assumed that the soft particle production is uncorrelated to details of the hard process—an assumption that can be tested experimentally.

As an example for such a strategy Fig. 5.23b shows a measurement by ATLAS [305] of the charged-particle density in the transverse region as a function of the transverse momentum that is chosen as scale of the event hardness. Three event types are compared: minimum bias, dijet production, and inclusive Z production, with event scales chosen to be the p_{T} of the leading track, the leading jet, and the Z boson, respectively. An approximately continuous transition from MB to jet events is visible in the inset plot. In the comparison between jet and Z events of the soft-particle density in the transverse region, which is attributed to the UE, differences are exhibited, which are even more pronounced in the average summed p_{T} density of all tracks (not shown). These can be explained in terms of a bias between the definitions of the event scale. In the case of the jet events further jets with transverse momenta larger than the p_{T} of the leading jet are forbidden by definition, while for the inclusive Z events additional jet activity with $p_{\mathrm{T}}^{\mathrm{jet}} > p_{\mathrm{T}}^{Z}$ is possible. Taking

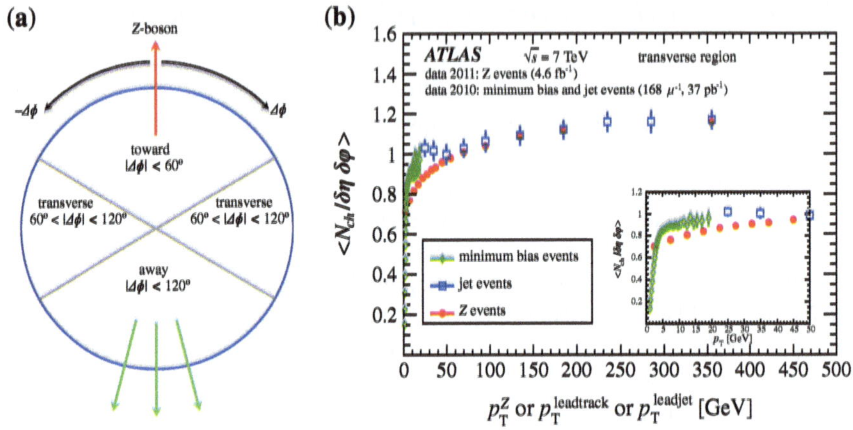

Fig. 5.23 **a** Sketch of the geometrical subdivision of events in azimuthal angle for UE measurements. **b** Average density of charged particles in the transverse region as a function of the leading-jet or leading-track p_T, or the p_T of the reconstructed Z boson. Adapted from [305]

this bias into account, the conjecture that the UE activity is independent of the hard process is supported by the ATLAS measurement and by similar measurements from CMS [306].

Another quantity that is not predictable by perturbative methods is the average density of charged particles produced in MB events for example at central rapidity. Corresponding measurements usually must be performed early on at the beginning of a data-taking period, when the instantaneous luminosity is small and the probability of multiple hadron-hadron collisions in the same or adjacent bunch crossings, the so-called pile-up, is negligible. It should be emphasised that such measurements are of utmost importance for a precise understanding of LHC data as they are used to model individual pile-up collisions. At later stages of the data-taking and with high instantaneous luminosity, the occurrence of up to 10–100 of such pile-up collisions simultaneously to the hard reaction of interest demands that these MB-like collisions are very well understood and modelled in simulations. Otherwise it becomes very difficult to precisely calibrate jet energies (see Chap. 3) or to estimate the background to searches for new physics. The charged-particle density is also of much interest for heavy-ion physics, and its dependence on the centre-of-mass energy is shown in Fig. 9.1 of the corresponding Chap. 9 for various combinations of colliding hadron-hadron beams.

5.9.5 Double-Parton Scattering

The basis for all perturbative results in the previous sections (see in particular Sect. 5.3) is the assumption that collinear factorisation holds and that the cross section of interest can be calculated following (5.4). To some extent it is a great fortune that such a simple picture with only one interacting parton from each colliding hadron and only longitudinal degrees of freedom describes such a huge amount of measurements. Partially this is due to the fact that one is mostly interested in inclusive processes at large p_T. Looking into the regime of small transverse momenta, however, it becomes apparent that this picture must fail, since otherwise the relevant cross sections would grow beyond all limits for $p_T \to 0$. They would thus violate unitarity, i.e. become even larger than the measured total cross sections. So, somehow, the cross sections have to be regularised. Realising that at p_T values of the order of a few GeV, protons are probed at fractional momenta of 10^{-3} or below, one can interpret the consequently large parton densities in such a way that more than one parton per hadron participates in a scattering process. In terms of the MPI, as explained above, this concept has already been successfully applied to model the part of the soft-particle production that manifests itself in the form of the underlying event. The possibility of two hard or semi-hard reactions with two partons from each of the colliding hadrons participating in their own parton-parton scatter has not yet been considered, though.

This type of process is called "double-parton scattering" (DPS), in contrast to the single parton scattering (SPS) discussed previously. Double-parton scattering is one possibility to dampen the rise of the low-p_T SPS cross sections. In a simplified form, the impact of double-parton scattering on the production rate for a final state $A + B$ can be parametrised by one effective cross section that can be written as

$$\sigma_{\mathrm{eff}} = \frac{m}{2} \frac{\sigma_{\mathrm{A}} \cdot \sigma_{\mathrm{B}}}{\sigma_{\mathrm{A+B}}^{\mathrm{DPS}}}, \tag{5.7}$$

where σ_{A} and σ_{B} are the cross sections for the independent processes A and B, $\sigma_{\mathrm{A+B}}^{\mathrm{DPS}}$ is the cross section for processes A and B to occur simultaneously, and m is a symmetry factor that is equal to unity for indistinguishable final states and two otherwise. In this simple picture, σ_{eff} is a measure of the overlap in transverse size of the parton distributions in the colliding hadrons. A naive geometrical interpretation leads to an estimate of $\sigma_{\mathrm{eff}} \approx \pi \mathrm{R}_{\mathrm{p}}^2 \approx 50\,\mathrm{mb}$, where R_{p} is the proton radius.

Measurements have been performed for a number of processes A and B: 4-jet [307, 308], γ+3-jet [309–311], and W+2-jet [312, 313] production. In each case, the second independent scattering reaction is chosen to have a dijet final state, because dijet production occurs at by far the highest rates. To differentiate double-parton scattering from normal high-multiplicity single parton scattering events, the particular properties of DPS events are exploited: the p_T of the jet pair (both jet pairs in 4-jet production) should be balanced and the azimuthal angle between the vectors of the final-state objects of the leading scattering and the lower-p_T jet pair should be uncorrelated, i.e. randomly oriented. Specialising to W+2-jet events, observables sensitive to DPS can then be defined as

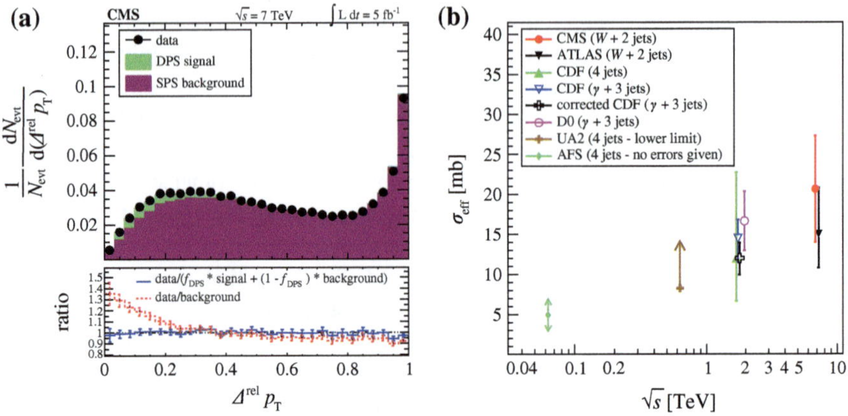

Fig. 5.24 **a** Template fit of SPS background and DPS signal to data for the $\Delta^{\mathrm{rel}} p_{\mathrm{T}}$ distribution in W+2-jet events as measured by CMS. **b** Extracted effective cross section σ_{eff} as a function of the centre-of-mass energy as measured by various experiments. Adapted from [313]

$$\Delta^{\mathrm{rel}} p_{\mathrm{T}} = \frac{|\boldsymbol{p}_{\mathrm{T}}(j_1) + \boldsymbol{p}_{\mathrm{T}}(j_2)|}{|\boldsymbol{p}_{\mathrm{T}}(j_1)| + |\boldsymbol{p}_{\mathrm{T}}(j_2)|} ,$$

$$\Delta S = \arccos\left(\frac{\boldsymbol{p}_{\mathrm{T}}(\mu, E_{\mathrm{T}}^{\mathrm{miss}}) + \boldsymbol{p}_{\mathrm{T}}(j_1, j_2)}{|\boldsymbol{p}_{\mathrm{T}}(\mu, E_{\mathrm{T}}^{\mathrm{miss}})| + |\boldsymbol{p}_{\mathrm{T}}(j_1, j_2)|}\right) ,$$

where $\boldsymbol{p}_{\mathrm{T}}$ are the transverse momentum vectors of the jets j_1, j_2, or of the jet pair (j_1, j_2) and the decay products of the W boson, one muon and missing transverse energy $E_{\mathrm{T}}^{\mathrm{miss}}$ representing the undetected neutrino. Figure 5.24a shows, as an example, the measurement of $\Delta^{\mathrm{rel}} p_{\mathrm{T}}$ by CMS [313] together with the composition of single parton scattering background and double-parton scattering signal, which peaks at small p_{T} differences, as determined in a template fit. Figure 5.24b summarises determinations of σ_{eff} from various experiments as a function of the centre-of-mass energy. The "Corrected CDF" point includes an improved estimate [314, 315] for the fact that (5.7) implies processes A and B to be inclusive such that also higher terms than double scatterings are parametrised by σ_{eff}. The bias of exclusive event selections, when applying (5.8) in the W+2-jet case,

$$\sigma_{\mathrm{eff}} = \frac{\sigma_{W+0\text{-jet}}}{\sigma^{\mathrm{DPS}}_{W+2\text{-jet}}} \cdot \sigma_{2\text{-jet}} , \tag{5.8}$$

has been considered in the presented analysis from CMS [313] and a similar measurement by ATLAS [312]. Otherwise, σ_{eff} would not be independent any more of the choice of scattering processes A and B.

The validity of the presented simplified theoretical approach anyway relies on a couple of additional assumptions. For a recent critical overview see [316].

5.10 Summary and Outlook

The excellent performance of the LHC machine and of the four major detectors around the ring made it possible to carry out precise measurements at very high particle, i.e. jet transverse momenta, high invariant masses, and for very complex event topologies with large particle multiplicities. The high luminosities and the centre-of-mass energies of 7 and 8 TeV reached for proton-proton collisions are unprecedented. These conditions allowed to assess the validity of the QCD theory in the TeV regime and to stress-test our current understanding of particle-production processes at such high energies. Many measurements in particular of vector-boson and jet-production processes reached an experimental accuracy of a few percent uncertainty only—true QCD precision tests.

To meet the experimental challenges, theoretical predictions are needed that match the precision and complexity requirements. Accordingly, enormous efforts have been undertaken to provide better and better theoretical predictions accompanied and driven by exciting new theoretical developments and insights. This includes largely automated NLO and many new NNLO calculations, improved analytical and parton-shower resummation techniques, the merging (i.e. matching) of parton showers and exact matrix elements, and attempts to gain a better understanding of the low-energy (i.e. non-perturbative) regime of the strong interaction. A major role is played by the Monte Carlo event generators that have been developed and optimised for the analysis and interpretation of LHC measurements.

Calculations at NNLO in the strong coupling provide predictions for inclusive production cross sections with an estimated uncertainty, e.g. due to missing higher orders, of a few percent only. Such precision predictions were successfully tested against the LHC data. Often the dominant theoretical uncertainty originates from the knowledge of the proton PDFs. Also in this field LHC measurements have already contributed to a refined understanding of the proton structure.

Around the year 2002, two theoretical break-throughs have revolutionised the field of Monte Carlo generators: first, the exact matching of $2 \to 2$ production processes evaluated at NLO accuracy with parton showers, and second, the consistent merging of tree-level matrix elements with varying final-state parton multiplicity with parton showers. These two innovative concepts enable fully exclusive predictions at particle-level with an inclusive production rate accurate to NLO or the first few hardest emissions being modelled through exact real-emission matrix elements. Consequently, Monte-Carlo simulations based on matching or merging of matrix elements and parton showers have become the standard tools in any LHC analysis. Driven by the enormous progress in the evaluation of NLO cross sections (which are largely automated by now), there appeared many multi-leg NLO calculations that can directly be matched with parton showers.

The currently emerging new standard are simulations based on exact NLO calculations with increasing parton multiplicity all matched with parton showers and combined into an inclusive description of the considered production process. However, this is not the end of the development: first prototypes of exact NNLO calculations matched with parton showers have recently been presented. These exciting theoretical developments pave the way to a further scrutiny of our understanding of QCD in the upcoming high-energy and high-luminosity LHC runs. In consequence, they will significantly boost the sensitivity and prospects to find potential new physics at the LHC.

References

1. D.J. Gross, F. Wilczek, Phys. Rev. Lett. **30**, 1343 (1973)
2. H.D. Politzer, Phys. Rev. Lett. **30**, 1346 (1973)
3. T. van Ritbergen, J.A.M. Vermaseren, S.A. Larin, Phys. Lett. B **400**, 379 (1997)
4. Particle Data Group, K. Olive et al., Chin. Phys. C **38**, 090001 (2014)
5. M.L. Mangano et al., JHEP **07**, 001 (2003)
6. F. Krauss, R. Kuhn, G. Soff, JHEP **02**, 044 (2002)
7. T. Gleisberg, S. Höche, JHEP **12**, 039 (2008)
8. A. Cafarella, C.G. Papadopoulos, M. Worek, Comput. Phys. Commun. **180**, 1941 (2009)
9. J. Alwall et al., JHEP **06**, 128 (2011)
10. W. Kilian, T. Ohl, J. Reuter, Eur. Phys. J. C **71**, 1742 (2011)
11. F.A. Berends, W. Giele, Nucl. Phys. B **306**, 759 (1988)
12. C. Duhr, S. Höche, F. Maltoni, JHEP **08**, 062 (2006)
13. M. Dinsdale, M. Ternick, S. Weinzierl, JHEP **03**, 056 (2006)
14. S. Drell, T.M. Yan, Annals Phys. **66**, 578 (1971)
15. G.F. Sterman, S. Weinberg, Phys. Rev. Lett. **39**, 1436 (1977)
16. J.E. Huth et al., in *Proceedings of the Snowmass Summer Study 1990: Research Directions for the Decade*, Snowmass, USA (1990), p. 134
17. G.C. Blazey et al. (2000), arXiv:hep-ex/0005012
18. C. Buttar et al. (2008), arXiv:0803.0678
19. JADE Collaboration, Z. Phys. C **33**, 23 (1986)
20. G.P. Salam, G. Soyez, JHEP **05**, 086 (2007)
21. M. Cacciari, G.P. Salam, G. Soyez, JHEP **04**, 063 (2008)
22. M. Cacciari, G.P. Salam, G. Soyez, Eur. Phys. J. C **72**, 1896 (2012)
23. G.P. Salam, Eur. Phys. J. C **67**, 637 (2010)
24. UA1 Collaboration, Phys. Lett. B **172**, 461 (1986)
25. UA2 Collaboration, Phys. Lett. B **160**, 349 (1985)
26. UA2 Collaboration, Phys. Lett. B **257**, 232 (1991)
27. DØ Collaboration, Phys. Rev. Lett. **101**(062), 001 (2008)
28. CDF Collaboration, Phys. Rev. D **78**, 052006 (2008)
29. ATLAS Collaboration, Phys. Rev. D **86**, 014022 (2012)
30. CMS Collaboration, Phys. Rev. Lett. **107**, 132001 (2011)
31. CMS Collaboration, Phys. Rev. D **87**, 112002 (2013)
32. ATLAS Collaboration (2014), arXiv:1410.8857
33. CMS Collaboration, Phys. Rev. D **90**, 072006 (2014)
34. ATLAS Collaboration, Eur. Phys. J. C **73**, 2509 (2013)
35. V. Gribov, L. Lipatov, Sov. J. Nucl. Phys. **15**, 438 (1972)
36. L. Lipatov, Sov. J. Nucl. Phys. **20**, 94 (1975)
37. Y.L. Dokshitzer, Sov. Phys. JETP **46**, 641 (1977)
38. G. Altarelli, G. Parisi, Nucl. Phys. B **126**, 298 (1977)
39. S. Moch, J. Vermaseren, A. Vogt, Nucl. Phys. B **688**, 101 (2004)
40. A. Vogt, S. Moch, J. Vermaseren, Nucl. Phys. B **691**, 129 (2004)
41. A. Buckley, Phys. Rep. **504**, 145 (2011)

42. G. Marchesini, B. Webber, Nucl. Phys. B **330**, 261 (1990)
43. S. Gieseke, P. Stephens, B. Webber, JHEP **12**, 045 (2003)
44. S. Schumann, F. Krauss, JHEP, **03**, 038 (2008)
45. M. Dinsdale, M. Ternick, S. Weinzierl, Phys. Rev. D **76**, 094003 (2007)
46. S. Plätzer, S. Gieseke, JHEP **01**, 024 (2011)
47. J.C. Winter, F. Krauss, JHEP **07**, 040 (2008)
48. M. Ritzmann, D. Kosower, P. Skands, Phys. Lett. B **718**, 1345 (2013)
49. CMS Collaboration, Eur. Phys. J. C **74**, 2901 (2014)
50. T. Sjöstrand, S. Mrenna, P.Z. Skands, Comput. Phys. Commun. **178**, 852 (2008)
51. CMS Collaboration, Phys. Lett. B **699**, 48 (2011)
52. ATLAS Collaboration, Eur. Phys. J. C **72**, 2211 (2012)
53. CMS Collaboration, JHEP **10**, 87 (2014)
54. ATLAS Collaboration, Phys. Rev. D **86**, 072006 (2012)
55. ATLAS Collaboration, JHEP **05**, 128, (2012)
56. ATLAS Collaboration, Eur. Phys. J. C **73**, 2432 (2013)
57. ATLAS Collaboration, New J. Phys. **16**, 113013 (2014)
58. ATLAS Collaboration, Phys. Rev. Lett. **106**, 172002 (2011)
59. CMS Collaboration, Phys. Rev. Lett. **106**, 122003 (2011)
60. S. Höche, F. Krauss, S. Schumann, F. Siegert, JHEP **05**, 053 (2009)
61. S. Catani, M. Seymour, Nucl. Phys. B **485**, 291 (1997)
62. S. Frixione, Z. Kunszt, A. Signer, Nucl. Phys. B **467**, 399 (1996)
63. D.A. Kosower, Phys. Rev. D **57**, 5410 (1998)
64. Z. Nagy, D.E. Soper, JHEP **09**, 055 (2003)
65. A. Denner, S. Dittmaier, Nucl. Phys. B **734**, 62 (2006)
66. G. Ossola, C.G. Papadopoulos, R. Pittau, Nucl. Phys. B **763**, 147 (2007)
67. C. Anastasiou, R. Britto, B. Feng, Z. Kunszt, P. Mastrolia, Phys. Lett. B **645**, 213 (2007)
68. R.K. Ellis, W. Giele, Z. Kunszt, JHEP **03**, 003 (2008)
69. W.T. Giele, Z. Kunszt, K. Melnikov, JHEP **04**, 049 (2008)
70. C. Berger et al., Phys. Rev. D **78**, 036003 (2008)
71. G. Bevilacqua et al., Comput. Phys. Commun. **184**, 986 (2013)
72. T. Gleisberg et al., JHEP **02**, 056 (2004)
73. T. Gleisberg et al., JHEP **02**, 007 (2009)
74. T. Gleisberg, F. Krauss, Eur. Phys. J. C **53**, 501 (2008)
75. M. Czakon, C. Papadopoulos, M. Worek, JHEP **08**, 085 (2009)
76. K. Hasegawa, S. Moch, P. Uwer, Comput. Phys. Commun. **181**, 1802 (2010)
77. R. Frederix, T. Gehrmann, N. Greiner, JHEP **06**, 086 (2010)
78. G. Bevilacqua, M. Czakon, M. Kubocz, M. Worek, JHEP **10**, 204 (2013)
79. G. Cullen et al., Eur. Phys. J. C **72**, 1889 (2012)
80. A. van Hameren, C. Papadopoulos, R. Pittau, JHEP **09**, 106 (2009)
81. V. Hirschi et al., JHEP **05**, 044 (2011)
82. S. Badger, B. Biedermann, P. Uwer, V. Yundin, Comput. Phys. Commun. **184**, 1981 (2013)
83. F. Cascioli, P. Maierhofer, S. Pozzorini, Phys. Rev. Lett. **108**, 111601 (2012)
84. S. Actis, A. Denner, L. Hofer, A. Scharf, S. Uccirati, JHEP **04**, 037 (2013)
85. T. Binoth et al., Comput. Phys. Commun. **181**, 1612 (2010)
86. S. Alioli et al., Comput. Phys. Commun. **185**, 560 (2014)
87. C. Berger et al., Phys. Rev. Lett. **106**, 092001 (2011)
88. Z. Bern et al., Phys. Rev. D **88**(1), 014025 (2013)
89. H. Ita et al., Phys. Rev. D **85**, 031501 (2012)
90. Z. Bern et al., Phys. Rev. Lett. **109**, 042001 (2012)
91. S. Badger, B. Biedermann, P. Uwer, V. Yundin, Phys. Lett. B **718**, 965 (2013)
92. S. Badger, B. Biedermann, P. Uwer, V. Yundin, Phys. Rev. D **89**, 034019 (2014)
93. G. Bevilacqua, M. Czakon, C. Papadopoulos, M. Worek, Phys. Rev. Lett. **104**, 162002 (2010)
94. S. Badger, A. Guffanti, V. Yundin, JHEP **03**, 122 (2014)
95. ATLAS Collaboration, JHEP **05**, 059 (2014)

96. ATLAS Collaboration (2014), arXiv:1411.1855
97. S. Dittmaier, A. Huss, C. Speckner, JHEP **11**, 095 (2012)
98. S. Alekhin, J. Blümlein, S. Moch, Phys. Rev. D **86**, 054009 (2012)
99. H.L. Lai et al., Phys. Rev. D **82**, 074024 (2010)
100. M. Glück, P. Jimenez-Delgado, E. Reya, Eur. Phys. J. C **53**, 355 (2008)
101. H1 and ZEUS Collaborations, JHEP **01**, 109 (2010)
102. A. Martin, W. Stirling, R. Thorne, G. Watt, Eur. Phys. J. C **63**, 189 (2009)
103. R.D. Ball et al., Nucl. Phys. B **849**, 296 (2011)
104. NNPDF Collaboration, Nucl. Phys. B **849**, 112 (2011)
105. NNPDF Collaboration, Nucl. Phys. B **809**, 1 (2009)
106. CMS Collaboration (2014), arXiv:1410.6765
107. D. Stump et al., Phys. Rev. D **65**, 014012 (2001)
108. G. Watt, R. Thorne, JHEP **08**, 052 (2012)
109. H. Paukkunen, P. Zurita, JHEP **12**, 100 (2014)
110. S. Alekhin et al. (2014), arXiv:1410.4412
111. H1 Collaboration, Eur. Phys. J. C **64**, 561 (2009)
112. R. Thorne, R. Roberts, Phys. Rev. D **57**, 6871 (1998)
113. R. Thorne, Phys. Rev. D **73**, 054019 (2006)
114. NuTeV Collaboration, Phys. Rev. Lett. **99**, 192001 (2007)
115. CMS Collaboration, Eur. Phys. J. C **73**, 2604 (2013)
116. CMS Collaboration (2014), arXiv:1412.1633
117. C. Anastasiou, K. Melnikov, F. Petriello, Phys. Rev. D **69**, 076010 (2004)
118. A. Gehrmann-De Ridder, T. Gehrmann, E.N. Glover, JHEP **09**, 056 (2005)
119. S. Catani, M. Grazzini, Phys. Rev. Lett. **98**, 222002 (2007)
120. M. Czakon, Phys. Lett. B **693**, 259 (2010)
121. C. Anastasiou, L.J. Dixon, K. Melnikov, F. Petriello, Phys. Rev. Lett. **91**, 182002 (2003)
122. K. Melnikov, F. Petriello, Phys. Rev. D **74**, 114017 (2006)
123. S. Catani et al., Phys. Rev. Lett. **103**, 082001 (2009)
124. M. Grazzini, S. Kallweit, D. Rathlev, A. Torre, Phys. Lett. B **731**, 204 (2014)
125. F. Cascioli et al., Phys. Lett. B **735**, 311 (2014)
126. T. Gehrmann et al., Phys. Rev. Lett. **113**, 212001 (2014)
127. S. Catani et al., Phys. Rev. Lett. **108**, 072001 (2012)
128. M. Czakon, P. Fiedler, A. Mitov, Phys. Rev. Lett. **110**(25), 252004 (2013)
129. R.V. Harlander, W.B. Kilgore, Phys. Rev. Lett. **88**, 201801 (2002)
130. C. Anastasiou, K. Melnikov, Nucl. Phys. B **646**, 220 (2002)
131. R.V. Harlander, W.B. Kilgore, Phys. Rev. D **68**, 013001 (2003)
132. O. Brein, A. Djouadi, R. Harlander, Phys. Lett. B **579**, 149 (2004)
133. A. Pak, M. Rogal, M. Steinhauser, JHEP **02**, 025 (2010)
134. P. Bolzoni, F. Maltoni, S. Moch, M. Zaro, Phys. Rev. Lett. **105**, 011801 (2010)
135. G. Ferrera, M. Grazzini, F. Tramontano, Phys. Rev. Lett. **107**, 152003 (2011)
136. J. Currie, A. Gehrmann-De Ridder, E. Glover, J. Pires, JHEP **01**, 110 (2014)
137. J. Currie et al. (2014), arXiv:1407.5558
138. S. Drell, T.M. Yan, Phys. Rev. Lett. **25**, 316 (1970)
139. R. Gavin, Y. Li, F. Petriello, S. Quackenbush, Comput. Phys. Commun. **182**, 2388 (2011)
140. Y. Li, F. Petriello, Phys. Rev. D **86**, 094034 (2012)
141. U. Baur, S. Keller, W. Sakumoto, Phys. Rev. D **57**, 199 (1998)
142. U. Baur et al., Phys. Rev. D **65**, 033007 (2002)
143. S. Dittmaier, M. Kramer, Phys. Rev. D **65**, 073007 (2002)
144. C. Carloni Calame, G. Montagna, O. Nicrosini, A. Vicini, JHEP **10**, 109 (2007)
145. A. Arbuzov et al., Eur. Phys. J. C **54**, 451 (2008)
146. S. Dittmaier, M. Huber, JHEP **01**, 060 (2010)
147. ATLAS Collaboration, Phys. Rev. D **85**, 072004 (2012)
148. CMS Collaboration, JHEP **10**, 132 (2011)
149. CMS Collaboration, Phys. Rev. Lett. **112**, 191802 (2014)

150. CMS Collaboration, JHEP **12**, 030 (2013)
151. ATLAS Collaboration, Phys. Lett. B **725**, 223 (2013)
152. LHCb Collaboration, JHEP **12**, 079 (2014)
153. M. Klasen, M. Brandt, Phys. Rev. D **88**, 054002 (2013)
154. C. Balázs, C. Yuan, Phys. Rev. D **56**, 5558 (1997)
155. M. Guzzi, P.M. Nadolsky, B. Wang, Phys. Rev. D **90**, 014030 (2014)
156. S. Höche, Y. Li, S. Prestel (2014), arXiv:1405.3607
157. A. Karlberg, E. Re, G. Zanderighi, JHEP **09**, 134 (2014)
158. ATLAS Collaboration, JHEP **09**, 145 (2014)
159. ATLAS Collaboration, Phys. Lett. B **705**, 415 (2011)
160. ATLAS Collaboration, Phys. Rev. D **85**, 012005 (2012)
161. CMS Collaboration, Phys. Rev. D **85**, 032002 (2012)
162. J. Pumplin et al., JHEP **07**, 012 (2002)
163. ATLAS Collaboration, Phys. Lett. B **701**, 31 (2011)
164. CMS Collaboration, JHEP **04**, 050 (2011)
165. ATLAS Collaboration, JHEP **12**, 060 (2010)
166. CMS Collaboration, JHEP **01**, 080 (2011)
167. ATLAS Collaboration, Phys. Rev. D **85**, 012003 (2012)
168. ATLAS Collaboration, JHEP **01**, 086 (2013)
169. CMS Collaboration, JHEP **01**, 133 (2012)
170. CMS Collaboration, Eur. Phys. J. C **74**, 3129 (2014)
171. J. Butterworth et al. (2014), arXiv:1405.1067
172. T. Sjöstrand, S. Mrenna, P.Z. Skands, JHEP **05**, 026 (2006)
173. S. Höche, S. Schumann, F. Siegert, Phys. Rev. D **81**, 034026 (2010)
174. T. Binoth, J. Guillet, E. Pilon, M. Werlen, Eur. Phys. J. C **16**, 311 (2000)
175. C. Balázs, E.L. Berger, S. Mrenna, C. Yuan, Phys. Rev. D **57**, 6934 (1998)
176. C. Balázs, E.L. Berger, P.M. Nadolsky, C.P. Yuan, Phys. Rev. D **76**, 013009 (2007)
177. Z. Bern, L.J. Dixon, C. Schmidt, Phys. Rev. D **66**(074), 018 (2002)
178. J.C. Collins, D.E. Soper, Phys. Rev. D **16**, 2219 (1977)
179. S. Catani, F. Krauss, R. Kuhn, B. Webber, JHEP **11**, 063 (2001)
180. L. Lönnblad, JHEP **05**, 046 (2002)
181. F. Krauss, JHEP **08**, 015 (2002)
182. M.L. Mangano, M. Moretti, F. Piccinini, M. Treccani, JHEP **01**, 013 (2007)
183. J. Alwall et al., Eur. Phys. J. C **53**, 473 (2008)
184. L. Lönnblad, S. Prestel, JHEP **03**, 019 (2012)
185. S. Frixione, B.R. Webber, JHEP **06**, 029 (2002)
186. P. Nason, JHEP **11**, 040 (2004)
187. S. Frixione, P. Nason, C. Oleari, JHEP **11**, 070 (2007)
188. S. Alioli, P. Nason, C. Oleari, E. Re, JHEP **06**, 043 (2010)
189. S. Höche, F. Krauss, M. Schönherr, F. Siegert, JHEP **09**, 049 (2012)
190. J. Alwall et al., JHEP **07**, 079 (2014)
191. K. Hamilton, P. Nason, JHEP **06**, 039 (2010)
192. S. Höche, F. Krauss, M. Schönherr, F. Siegert, JHEP **08**, 123 (2011)
193. S. Höche, F. Krauss, M. Schönherr, F. Siegert, JHEP **04**, 027 (2013)
194. L. Lönnblad, S. Prestel, JHEP **03**, 166 (2013)
195. K. Hamilton, P. Nason, C. Oleari, G. Zanderighi, JHEP **05**, 082 (2013)
196. Höche, S. et al. (2014), arXiv:1402.6293
197. CMS Collaboration, Phys. Lett. B **741**, 12 (2015)
198. ATLAS Collaboration, Phys. Rev. D **85**, 092002 (2012)
199. J.M. Campbell, R.K. Ellis, D.L. Rainwater, Phys. Rev. D **68**, 094021 (2003)
200. C. Berger et al., Nucl. Phys. Proc. Suppl. **205–206**, 92 (2010)
201. ATLAS Collaboration, JHEP **07**, 032 (2013)
202. CMS Collaboration, JHEP **01**, 010 (2012)
203. F. Berends et al., Phys. Lett. B **224**, 237 (1989)

204. E. Gerwick, T. Plehn, S. Schumann, P. Schichtel, JHEP **10**, 162 (2012)
205. I.W. Stewart, F.J. Tackmann, Phys. Rev. D **85**, 034011 (2012)
206. BlackHat Collaboration (2014), arXiv:1407.6564
207. ATLAS Collaboration, JHEP **10**, 141 (2014)
208. W.T. Giele, S. Keller, E. Laenen, Phys. Lett. B **372**, 141 (1996)
209. R.K. Ellis, S. Veseli, Phys. Rev. D **60**, 011501 (1999)
210. J.M. Campbell, R.K. Ellis, Phys. Rev. D **62**(114), 012 (2000)
211. J.M. Campbell, R.K. Ellis, F. Maltoni, S. Willenbrock, Phys. Rev. D **69**, 074021 (2004)
212. F. Maltoni, T. McElmurry, S. Willenbrock, Phys. Rev. D **72**, 074024 (2005)
213. J.M. Campbell, R.K. Ellis, F. Maltoni, S. Willenbrock, Phys. Rev. D **73**, 054007 (2006)
214. J.M. Campbell et al., Phys. Rev. D **79**, 034023 (2009)
215. F. Febres Cordero, L. Reina, D. Wackeroth, Phys. Rev. D **80**, 034015 (2009)
216. S. Badger, J.M. Campbell, R. Ellis, JHEP **03**, 027 (2011)
217. C. Oleari, L. Reina, JHEP **08**, 061 (2011)
218. R. Frederix et al., JHEP **09**, 061 (2011)
219. ATLAS Collaboration, Phys. Lett. B **706**, 295 (2012)
220. CMS Collaboration, JHEP **06**, 120 (2014)
221. CMS Collaboration, JHEP **12**, 039 (2013)
222. T. Becher, M. Neubert, D. Wilhelm, JHEP **02**, 124 (2012)
223. A. Banfi, M. Dasgupta, S. Marzani, L. Tomlinson, Phys. Lett. B **715**, 152 (2012)
224. A. Banfi, G.P. Salam, G. Zanderighi, JHEP **08**, 062 (2004)
225. A. Banfi, G.P. Salam, G. Zanderighi, JHEP **06**, 038 (2010)
226. I.W. Stewart, F.J. Tackmann, W.J. Waalewijn, Phys. Rev. Lett. **105**, 092002 (2010)
227. I. Feige, M.D. Schwartz, I.W. Stewart, J. Thaler, Phys. Rev. Lett. **109**, 092001 (2012)
228. E. Gerwick, S. Schumann, B. Gripaios, B. Webber, JHEP **04**, 089 (2013)
229. A. Idilbi, X. Ji, Phys. Rev. D **72**, 054016 (2005)
230. T. Becher, M. Neubert, G. Xu, JHEP **07**, 030 (2008)
231. M. Bonvini, S. Forte, G. Ridolfi, Nucl. Phys. B **847**, 93 (2011)
232. M. Cacciari et al., Phys. Lett. B **710**, 612 (2012)
233. J. Forshaw, J. Keates, S. Marzani, JHEP **07**, 023 (2009)
234. R.M. Duran Delgado, J.R. Forshaw, S. Marzani, M.H. Seymour, JHEP **08**, 157 (2011)
235. A. Banfi, G.P. Salam, G. Zanderighi, JHEP **06**, 159 (2012)
236. A. Banfi, P.F. Monni, G.P. Salam, G. Zanderighi, Phys. Rev. Lett. **109**, 202001 (2012)
237. F.J. Tackmann, J.R. Walsh, S. Zuberi, Phys. Rev. D **86**, 053011 (2012)
238. E. Kuraev, L. Lipatov, V.S. Fadin, Sov. Phys. JETP **45**, 199 (1977)
239. I. Balitsky, L. Lipatov, Sov. J. Nucl. Phys. **28**, 822 (1978)
240. J.R. Forshaw, A. Kyrieleis, M. Seymour, JHEP **06**, 034 (2005)
241. J.R. Andersen, J.M. Smillie, JHEP **01**, 039 (2010)
242. ATLAS Collaboration, JHEP **09**, 053 (2011)
243. ATLAS Collaboration, Eur. Phys. J. C **74**, 3117 (2014)
244. S. Alioli et al., JHEP **04**, 081 (2011)
245. J.R. Andersen, J.M. Smillie, JHEP **06**, 010 (2011)
246. J.R. Andersen, L. Lönnblad, J.M. Smillie, JHEP **07**, 110 (2011)
247. M. Deak, F. Hautmann, H. Jung, K. Kutak, Eur. Phys. J. C **72**, 1982 (2012)
248. S. Alioli et al., Phys. Rev. D **85**, 114034 (2012)
249. S. Höche, M. Schönherr, Phys. Rev. D **86**, 094042 (2012)
250. Y. Hatta et al., Phys. Rev. D **87**(5), 054016 (2013)
251. A. Abdesselam et al., Eur. Phys. J. C **71**, 1661 (2011)
252. A. Altheimer et al., J. Phys. G **39**, 063001 (2012)
253. A. Altheimer et al., Eur. Phys. J. C **74**, 2792 (2014)
254. S.D. Ellis, C.K. Vermilion, J.R. Walsh, Phys. Rev. D **81**, 094023 (2010)
255. CMS Collaboration, JHEP **05**, 090 (2013)
256. CMS Collaboration, JHEP **09**, 029 (2012)
257. CMS Collaboration, JHEP **08**, 173 (2014)

258. CMS Collaboration, JHEP **08**, 174 (2014)
259. D. Krohn, J. Thaler, L.T. Wang, JHEP **02**, 084 (2010)
260. J.M. Butterworth, A.R. Davison, M. Rubin, G.P. Salam, Phys. Rev. Lett. **100**(242), 001 (2008)
261. T. Plehn, G.P. Salam, M. Spannowsky, Phys. Rev. Lett. **104**, 111801 (2010)
262. H.N. Li, Z. Li, C.P. Yuan, Phys. Rev. Lett. **107**, 152001 (2011)
263. H.N. Li, Z. Li, C.P. Yuan, Phys. Rev. D **87**, 074025 (2013)
264. M. Dasgupta, K. Khelifa-Kerfa, S. Marzani, M. Spannowsky, JHEP **10**, 126 (2012)
265. T.T. Jouttenus et al., Phys. Rev. D **88**(5), 054031 (2013)
266. M. Dasgupta, A. Fregoso, S. Marzani, G.P. Salam, JHEP **09**, 029 (2013)
267. M. Dasgupta, A. Fregoso, S. Marzani, A. Powling, Eur. Phys. J. C **73**(11), 2623 (2013)
268. Y.L. Dokshitzer, G. Leder, S. Moretti, B. Webber, JHEP **08**, 001 (1997)
269. M. Wobisch, T. Wengler (1998), arXiv:hep-ph/9907280
270. M. Wobisch, Measurement and QCD analysis of jet cross-sections in deep inelastic positron proton collisions at s**(1/2) = 300-GeV. Ph.D. thesis, RWTH Aachen, DESY-THESIS-2000-049 (2000)
271. M. Dasgupta, L. Magnea, G.P. Salam, JHEP **02**, 055 (2008)
272. S.D. Ellis, Z. Kunszt, D.E. Soper, Phys. Rev. Lett. **69**, 3615 (1992)
273. CDF Collaboration, Phys. Rev. Lett. **70**, 713 (1993)
274. CDF Collaboration, Phys. Rev. D **71**, 112002 (2005)
275. R.K. Ellis, W.J. Stirling, B.R. Webber, *QCD and Collider Physics* (Cambridge University Press, Cambridge, 1996)
276. ATLAS Collaboration, Phys. Rev. D **83**, 052003 (2011)
277. CMS Collaboration, JHEP **06**, 160 (2012)
278. M. Cacciari, J. Rojo, G.P. Salam, G. Soyez, JHEP **12**, 032 (2008)
279. G. Soyez, Phys. Lett. B **698**, 59 (2011)
280. Z. Nagy, Phys. Rev. Lett. **88**, 122003 (2002)
281. Z. Nagy, Phys. Rev. D **68**, 094002 (2003)
282. ZEUS Collaboration, Phys. Lett. B **691**, 127 (2010)
283. ALICE Collaboration, Phys. Lett. B **722**, 262 (2013)
284. H.G. Dosch, P. Gauron, B. Nicolescu, Phys. Rev. D **67**, 077501 (2003)
285. M. Froissart, Phys. Rev. D **123**, 1053 (1961)
286. ATLAS Collaboration, Nucl. Phys. B **489** (2014)
287. H. Yukawa, Proc. Phys. Math. Soc. Jap. **17**, 48 (1935)
288. A. Kaidalov (2001), arXiv:hep-ph/0103011
289. P. Collins, *An Introduction to Regge Theory and High-Energy Physics* (Cambridge University Press, Cambridge, 1977)
290. G.F. Chew, S.C. Frautschi, S. Mandelstam, Phys. Rev. D **126**, 1202 (1961)
291. A. Donnachie, P. Landshoff, Phys. Lett. B **296**, 227 (1992)
292. ATLAS Collaboration, Nat. Commun. **2**, 463 (2011)
293. CMS Collaboration, Phys. Lett. B **722**, 5 (2013)
294. TOTEM Collaboration, Europhys. Lett. **101**(21), 002 (2013)
295. TOTEM Collaboration, Phys. Rev. Lett. **111**(1), 012001 (2013)
296. ARGO-YBJ Collaboration, Phys. Rev. D **80**, 092004 (2009)
297. M. Honda et al., Phys. Rev. Lett. **70**, 525 (1993)
298. R. Baltrusaitis et al., Phys. Rev. Lett. **52**, 1380 (1984)
299. J. Cudell et al., Phys. Rev. Lett. **89**, 201801 (2002)
300. A. Donnachie, P. Landshoff, Phys. Lett. B **727**, 500 (2013)
301. T. Sjöstrand, M. van Zijl, Phys. Rev. D **36**, 2019 (1987)
302. T. Sjöstrand, P.Z. Skands, JHEP **03**, 053 (2004)
303. T. Sjöstrand, P.Z. Skands, Eur. Phys. J. C **39**, 129 (2005)
304. M. Bähr, S. Gieseke, M.H. Seymour, JHEP **07**, 076 (2008)
305. ATLAS Collaboration, Eur. Phys. J. C **74**, 3195 (2014)
306. CMS Collaboration, Eur. Phys. J. C **72**, 2080 (2012)
307. CDF Collaboration, Phys. Rev. D **47**, 4857 (1993)

308. CMS Collaboration, Phys. Rev. D **89**, 092010 (2014)
309. CDF Collaboration, Phys. Rev. D **56**, 3811 (1997)
310. DØ Collaboration, Phys. Rev. D **81**, 052012 (2010)
311. DØ Collaboration, Phys. Rev. D **89**, 072006 (2014)
312. ATLAS Collaboration, New J. Phys. **15**, 033038 (2013)
313. CMS Collaboration, JHEP **03**, 032 (2014)
314. D. Treleani, Phys. Rev. D **76**(076), 006 (2007)
315. M. Bähr, M. Myska, M.H. Seymour, A. Siodmok, JHEP **03**, 129 (2013)
316. M. Diehl, D. Ostermeier, A. Schäfer, JHEP **03**, 089 (2012)

Chapter 6
Higgs-Boson Physics at the LHC

Karl Jakobs, Günter Quast and Georg Weiglein

Abstract The discovery of a Higgs boson or the exclusion of such a particle in a wide mass range from 100 to 1000 GeV was a major design goal of the LHC and the experiments ATLAS and CMS. The discovery of a Higgs-like particle at a mass around 125 GeV marked the beginning of a new era in particle physics. A snapshot of early results at the time of discovery and detailed investigations of the properties of the new particle with the full LHC data set taken in the years 2010–2012 at centre-of-mass energies of 7 and 8 TeV as well as remaining open questions are the main focus of this chapter.

6.1 Electroweak Symmetry Breaking and Higgs Physics

The introduction of spontaneous breaking of the electroweak gauge invariance in the Lagrangian of the electroweak theory was the first proposed theoretically coherent approach to accommodate massive gauge bosons in a gauge-invariant field theory. The theoretical foundations were laid already in the years 1960–1967, in a way today referred to as the "Brout–Englert–Higgs (BEH)"-mechanism [1–7]. The BEH mechanism became a cornerstone of the Standard Model (SM) of particle physics [8–10]. In the SM, the BEH mechanism is realised in a minimal form based on a single complex doublet field, as described in detail in Sect. 4.2.2. The potential formed by this doublet field, consisting of a bilinear and a quartic term, leads to

K. Jakobs (✉)
Albert-Ludwigs-Universität Freiburg,
Hermann-Herder-Str. 3, 79104 Freiburg, Germany
e-mail: karl.jakobs@uni-freiburg.de

G. Quast
Institut für Experimentelle Kernphysik, KIT,
Wolfgang-Gaede-Str. 1, 76131 Karlsruhe, Germany
e-mail: guenter.quast@kit.edu

G. Weiglein
DESY, Notkestr. 85, 22607 Hamburg, Germany
e-mail: georg.weiglein@desy.de

T. Schörner-Sadenius (ed.), *The Large Hadron Collider*,
DOI 10.1007/978-3-319-15001-7_6

a non-vanishing vacuum expectation value. Upon interaction with the gauge-boson fields, this non-trivial structure of the vacuum gives rise to mass terms for the $W^{\pm}$ and Z bosons and also allows fermion mass terms to be generated via mass-dependent Yukawa couplings of fermions to the "Higgs field", see Sect. 4.2.3. The theory does not predict the magnitude of the Yukawa couplings, and hence any fermion masses. The fourth remaining degree of freedom of the complex Higgs doublet of the SM manifests itself as a massive scalar boson, the "Higgs boson", with the same spin and *CP* quantum numbers as the vacuum, i.e. spin zero and *CP*-even. Within the SM it is necessary to invoke the complex doublet field together with its charge conjugate in order to provide mass terms for both up-type and down-type fermions. In many extensions of the SM, for instance its minimal supersymmetric extension (MSSM), see Sect. 10.2, two separate complex Higgs doublets are involved in generating the masses for up-type and down-type fermions. The additional four degrees of freedom arising from the second doublet in such a case lead to the prediction of five physical Higgs bosons (three neutral ones and a pair of charged Higgs bosons) instead of the single Higgs state predicted by the SM.

Within the SM, the mass of the Higgs boson, M_H, is not predicted and corresponds to a free parameter of the theory. In the MSSM, on the other hand, the additional symmetries of the model give rise to an upper bound on the mass of the light *CP*-even Higgs boson. As one of the main characteristics of the BEH-mechanism, the Higgs-boson couplings to massive gauge bosons and fermions are proportional to the masses of these particles. Higgs self-interactions are the crucial ingredient of the symmetry-breaking sector via the BEH mechanism, and their experimental determination provides access to the Higgs potential that endows the vacuum with a non-trivial structure. Higgs-boson self-interactions can be probed in processes with two produced Higgs bosons and, for non-minimal Higgs sectors, also through decays of heavier Higgs bosons into a pair of lighter ones.

After its formulation more than four decades ago, the SM of particle physics passed numerous high-precision experimental tests, in particular by experiments at the e^+e^- collider LEP at CERN, at the Stanford Linear Collider, and at the Tevatron $p\bar{p}$ collider. Predictions of the SM were confirmed to very high precision reaching in some cases the permille level. However, the Higgs boson predicted by the SM remained elusive. Indirect constraints on the mass of the Higgs boson within the SM became possible after the measurement of the top-quark mass by the Tevatron experiments, indicating a preference for a relatively light Higgs boson.

An alternative to the BEH mechanism predicting at least one fundamental Higgs boson could be the presence of a new kind of strong interaction. Similarly to the case of the pion in ordinary QCD, such a new strong interaction could give rise to bound states resulting in a composite Higgs state and possibly further resonances. Models of this kind (see Sect. 11.3.3 for further discussion) may arise as low-energy effective theories of models constructed in higher dimensions of space-time, where the Higgs-boson degree of freedom is identified as the fifth component of a five-dimensional gauge field.

6.1.1 Theoretical Higgs-Boson Mass Bounds

Prior to the start of the LHC there was already strong evidence that signals of the underlying mechanism of electroweak symmetry breaking would have to manifest themselves at the TeV scale. This was due to the fact that the theoretical formulation based on just the known fundamental particles (i.e. without a Higgs boson) and the known interactions was known to break down (i.e. produce unphysical results) at the TeV scale. This problem becomes apparent in particular in the scattering of longitudinal vector bosons, W_L or Z_L. The lowest-order prediction based on just the gauge-boson contributions leads to a result for the scattering amplitude that grows with the square of the centre-of-mass energy of the process and therefore violates probability conservation, i.e. unitarity, at sufficiently high energies. This bad high-energy behaviour would exactly be cancelled by Higgs-boson contributions if the Higgs-boson couplings have the form predicted by the BEH mechanism, i.e. if they are proportional to the gauge coupling times the mass of the involved particle.

Requiring this kind of unitarity cancellations by the exchange of fundamental scalar particles puts an upper bound on the mass of the involved Higgs boson [11, 12], which turns out to be below the scale of 1 TeV. In the case where such a Higgs-boson contribution is absent, the gauge-boson sector would become strongly interacting at the TeV scale. Unitarity would then be restored by this new kind of strong interaction, giving rise to phenomenological consequences at the TeV scale. It should be noted in this context that the presence of a Higgs-like state by itself does not guarantee the appropriate unitarity cancellations in the gauge-boson sector. If the couplings of the Higgs-like state are such that its exchange leads to only an approximate unitarity cancellation, the gauge-boson sector will still become strongly interacting, but the onset of the effects of the strong interaction is shifted to higher energies compared to the case without a Higgs-like particle.

Thus, while within the SM the mass of the Higgs boson is a free parameter, the requirement of appropriate unitarity cancellations puts a theoretical upper bound on this parameter. Furthermore, if one demands that the SM should remain valid all the way up to the Planck scale, where gravity becomes comparable in strength to the other interactions, the precise numerical values of the top-quark mass m_t and the Higgs-boson mass M_H have an important impact on the shape of the Higgs potential extrapolated to these high scales. This sensitivity on m_t and M_H can in turn be used to set constraints on these parameters if one requires the vacuum to be sufficiently well-behaved at high scales while assuming the absence of any contributions of physics beyond the SM, see e.g. [13].

As mentioned above, in supersymmetric extensions of the SM a direct upper bound can be inferred from symmetry relations, implying that at least one of the neutral Higgs bosons in the spectrum has to be light. Within the MSSM, at lowest order the upper bound on the mass of the light *CP*-even Higgs boson is given by the mass of the Z boson, i.e. about 90 GeV. However, this bound is shifted upwards by large higher-order corrections to about 135 GeV for TeV-scale masses of the scalar superpartners of the top quark, see Sect. 10.2. Accordingly, the higher-order effects are crucial for bringing a SM-like Higgs boson predicted within the MSSM in accordance with the existing experimental constraints. It is interesting to note that in the next-to-minimal supersymmetric extension of the SM (NMSSM) there is an additional contribution to the upper bound at lowest order, reducing somewhat the need for relatively large higher-order contributions.

6.1.2 Indirect Experimental Constraints on the Mass of the Higgs Boson in the Standard Model

Another possibility to set constraints on the parameters of a given model arises from confronting high-precision experimental data with accurate theory predictions at the quantum level, where the whole structure of the model enters via loop contributions. Within the SM, constraints on the Higgs-boson mass arise in this way from virtual contributions of the Higgs boson to precision observables. The leading Higgs-boson contributions have a logarithmic dependence on the Higgs-boson mass, while the leading effects of the top-quark mass are quadratic in m_t. Therefore, indirect constraints on M_H from electroweak precision observables require a high experimental precision on m_t, but the two remain strongly correlated.

Precision measurements of the experiments ALEPH, DELPHI, L3 and OPAL at the LEP e^+e^- collider, the experiment SLD at the Stanford linear collider, the experiments CDF and DØ at the Tevatron $p\bar{p}$ collider as well as low-energy electroweak precision measurements such as the muon lifetime provided indirect constraints on M_H, as summarised in [14, 15]. In particular, from muon decay one obtains a relation where the W-boson mass can be predicted in terms of the Z-boson mass, the Fermi constant and the fine structure constant at lowest order, and other parameters enter at the loop level. As discussed in Sect. 4.5.5, these relations can be used to indirectly determine the W-boson and top-quark masses from precision measurements. Within their errors, the predicted masses agree with the direct measurements, as shown in Fig. 4.12. On the contrary, using the direct mass measurements, an indirect prediction for the Higgs-boson mass within the SM can be inferred, leading to a preferred light value of M_H. Considering all precision measurements prior to the LHC, the best-fit value of the Higgs-boson mass within the SM is $M_H = 94^{+29}_{-24}$ GeV, or, expressed as a limit neglecting constraints from direct searches, $M_H < 152$ GeV at the 95 % confidence level [15].

The agreement between the precision observables and the theoretical predictions is even slightly improved in the MSSM, since SUSY contributions tend to give rise to an upward shift in M_W [16].

6.1.3 Higgs-Boson Decay Modes

Within the SM, the Higgs-boson mass, M_H, is the only remaining free parameter of the theory. Accordingly, all Higgs-boson branching ratios and also its total width can be calculated within the SM as a function of M_H, and the whole structure of

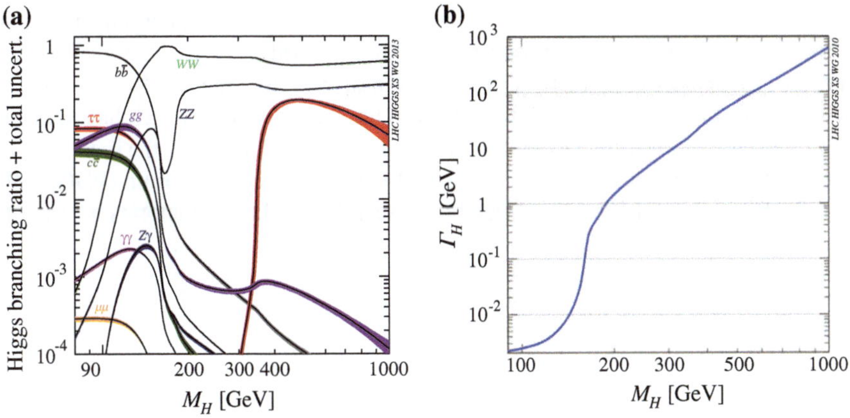

Fig. 6.1 Predictions for **a** the branching ratios and **b** the total width of the Higgs boson within the SM, as given by the LHC Higgs Cross Section Working Group. Adapted from [17, 18]

the model enters via higher-order corrections. In Fig. 6.1 the predictions for the branching ratios and the total width are shown. While the total width for a SM-like Higgs boson is very small in the low-mass region, it grows rapidly above the threshold for on-shell WW and ZZ production. For a Higgs-boson mass of about 125 GeV the Higgs-boson phenomenology within the SM is particularly rich, since many different decay modes are potentially detectable. Since the decays into WW^* and ZZ^* are phase-space suppressed in this mass region, the branching ratios for these decays depend very sensitively on the precise numerical value of the Higgs-boson mass for the decay of an on-shell Higgs boson, and on the other hand off-shell Higgs-boson contributions can be sizeable for these decay modes.

A typical feature of many extended Higgs sectors is a sum rule according to which the squared couplings of all neutral Higgs bosons of the model add up to the SM value. For extended Higgs sectors, where one of the states in the spectrum has SM-like couplings to gauge bosons, this implies that the couplings to gauge bosons of all other Higgs-boson states are strongly suppressed compared to a SM-like Higgs boson of the same mass. This means in particular that the total width of heavy Higgs bosons in extended Higgs sectors is typically much smaller than the one of a heavy SM-like Higgs boson.

6.2 Early Higgs-Boson Searches

The investigation of electroweak symmetry breaking has been a major goal of particle physics experiments over the past decades. Searches for a Higgs boson have been in the focus of experiments since the middle of the 1980s. From the very beginning of their operation in the year 1989, direct searches for a Higgs boson were carried out by the experiments at the LEP e^+e^- collider. After the shutdown of LEP, searches continued at the Tevatron $p\bar{p}$ collider in Run II starting in the year 2001.

Studies of K, B and Υ decays in combination with theoretical arguments led to a lower bound on the Higgs-boson mass within the SM of $M_H > 3.9\,\text{GeV}$ before the start of operation of LEP. A Higgs boson at very low mass, $M_H < 15\,\text{MeV}$, would contribute a long-range component to the nuclear force, which has not been observed. For an overview of search results prior to LEP, we refer the reader to the compilation of results and their discussion by the Particle Data Group (PDG) [19], and for more recent results on Higgs-boson searches at flavour factories we refer to the latest PDG listing [20].

6.2.1 Direct Searches at the LEP e^+e^- Collider

Searches for the Higgs boson were performed during the LEP I phase at centre-of-mass energies around the Z pole, and later at LEP II at steadily increasing centre-of-mass energies ultimately reaching up to $\sqrt{s} = 209\,\text{GeV}$ in the year 2000. The dominant production mechanisms for Higgs bosons in e^+e^- in this energy range is Higgs-strahlung, $e^+e^- \to ZH$.

In e^+e^- collisions with known centre-of-mass energy $e^+e^- \to ZH$ events can be selected from the decay of the Z boson alone, independent of the Higgs-boson decay mode, and the invariant-mass distribution of the system recoiling against the Z boson can be reconstructed. At values of the Higgs-boson mass in reach at LEP II, $M_H \lesssim 115\,\text{GeV}$, the Higgs boson of the SM is expected to be an extremely narrow resonance and decays almost exclusively to pairs of b quarks or τ leptons, as can be seen from Fig. 6.1.

The LEP experiments ALEPH, DELPHI, L3 and OPAL performed searches concentrating on the Higgs-strahlung process in various combinations of Z and Higgs-boson final states, $ZH \to q\bar{q}b\bar{b}$, $\nu\nu b\bar{b}$, $e^+e^-b\bar{b}$, $\mu^+\mu^-b\bar{b}$, $\tau^+\tau^-b\bar{b}$ and $b\bar{b}\tau^+\tau^-$. The analyses made extensive use of the good trigger and identification performance of all LEP detectors for electrons, muons, τ leptons and for b-flavoured jets.

The total integrated luminosity at high energies at LEP II was about $2.5\,\text{fb}^{-1}$ at $\sqrt{s} > 189\,\text{GeV}$, of which $0.5\,\text{fb}^{-1}$ were taken at the highest centre-of-mass energy $\sqrt{s} > 206\,\text{GeV}$. To set the scale, the production cross section of a SM Higgs boson of mass $M_H = 110\,\text{GeV}$ is about 0.3 pb at $\sqrt{s} = 209\,\text{GeV}$; it drops rapidly with increasing Higgs-boson mass and is only 0.01 pb for $M_H = 115\,\text{GeV}$. Typical production rates of backgrounds from pair-produced Z or W bosons are larger by three or four orders of magnitude.

Higgs-boson candidates were identified by constraining the masses of the H-boson candidate and of the Z boson and by exploiting energy and momentum conservation in channels involving $Z \to \nu\nu$ decays. Signal-to-background levels of 1–2 were achieved after the final selection near the sensitivity limit of $M_H = 115\,\text{GeV}$, at a signal efficiency of about 50 %. As an illustration, the mass distribution from the combination of the searches by the four LEP experiments is shown in Fig. 6.2a.

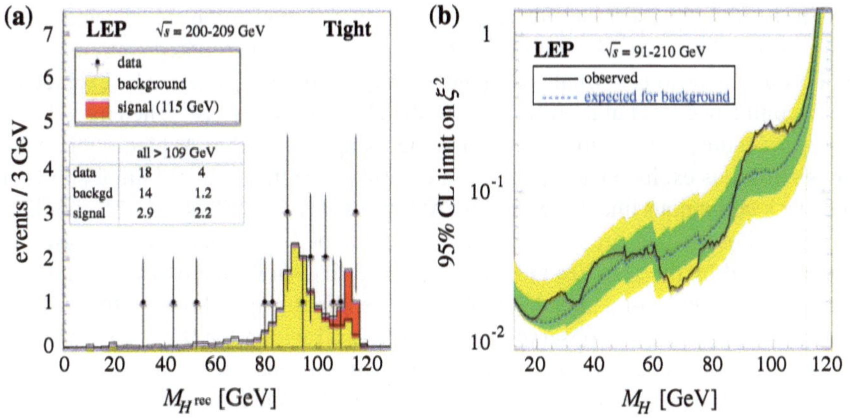

Fig. 6.2 Combined result from searches for the Higgs boson by the LEP experiments ALEPH, DELPHI, L3 and OPAL. **a** Illustrative distribution of the main discriminating variable, the reconstructed Higgs-boson mass, of Higgs-boson candidates after the final selection at LEP II. **b** 95 % upper confidence limit on the existence of a Higgs boson as a function of its mass, at LEP I and LEP II. Adapted from [21]

The LEP experiments also have set the standards for the statistical treatment and the limit calculation [22, 23], which were only slightly modified for applications to searches at Tevatron and LHC. The test-statistic q for the separation between the "background-only" (H_b) and the "signal+background" (H_{s+b}) hypothesis is defined as two times the negative logarithm of the likelihood ratio,

$$q = -2\ln\left(\frac{\mathcal{L}_{s+b}}{\mathcal{L}_b}\right).$$

In fact, q is determined as the sum of contributions to the negative log-likelihood from observed candidate events. The expected probability density functions of q under the hypotheses H_b and H_{s+b} are obtained from a large number of simulated event configurations. In the limit of infinite statistics, q becomes exactly equal to the difference in χ^2 between H_{s+b} and H_b. Integrating the probability density functions for H_b from $-\infty$ to the value q_{obs} determined from the observed data, and from q_{obs} to $+\infty$ for H_{s+b}, one obtains the p-values with respect to the two hypotheses, 1-CL_b and $\text{CL}_{\text{s+b}}$, where the names CL_b and $\text{CL}_{\text{s+b}}$, respectively, were introduced by the LEP collaborations to quantify the confidence level with respect to H_b and H_{s+b}. To obtain the confidence level for the exclusion of a signal, which is robust against setting too low exclusion limits in case of downward-fluctuations of the background, the quantity "CL_s" was introduced, defined as $\text{CL}_\text{s} = \frac{\text{CL}_{\text{s+b}}}{\text{CL}_\text{b}}$. This rescaling of the p-value of H_{s+b} by the probability to observe the expected background is known as the modified frequentist (or CL_s) method. A 95 % exclusion limit is set at the value of the Higgs-boson mass where $\text{CL}_\text{s} = 0.05$.

The results of the searches for a Higgs boson in ZH production at LEP I and LEP II are shown in Fig. 6.2b. The limit is expressed in terms of the squared coupling of the H boson to Z bosons normalised to the SM expectation, ξ^2, that can be excluded at 95 % confidence level at a given value of the Higgs-boson mass.[1] The Higgs-boson decay branching ratios in this plot are assumed to agree with the SM prediction. A SM Higgs boson is excluded at those values of M_H where the observed limit, shown as the black continuous line, is below one. At lower mass, Higgs bosons with couplings to gauge bosons that are suppressed by a factor of ξ^2 compared to the SM case are excluded by the searches at LEP. The dotted line shows the expected exclusion limit in the absence of a signal, and the dark and light bands show the 68 and 95 % confidence regions. The observed exclusion is well within the 95 % (or 2 σ) band. As a result of the Higgs-boson search at LEP, the existence of a SM Higgs boson is excluded in the mass range $10 < M_H < 114.4$ GeV. Due to an excess of observed candidates, this observed limit is slightly smaller than the expected exclusion limit of 115.3 GeV.

The exclusion limits in Fig. 6.2b set important constraints on additional light Higgs bosons of an extended Higgs sector. However, if the coupling of such a state to gauge bosons is below 10 % of the SM coupling (corresponding to $\xi^2 = 0.01$), such a scenario is not constrained by the LEP limits. As discussed above, in an extended Higgs sector where one Higgs state is SM-like, the additional Higgs bosons at higher or lower masses would typically have heavily suppressed couplings to gauge bosons which could be well below the LEP limits.

6.2.2 Searches at the Tevatron Collider

The LEP limit on the mass of the Higgs boson of the SM remained the strongest one for a decade, before Higgs-boson searches at the Tevatron proton-antiproton collider running at a centre-of-mass energy of 1.96 TeV became sensitive to the SM Higgs boson. The Tevatron had continuously been running from the year 2001 until 2011, and the experiments CDF and DØ collected data corresponding to integrated luminosities of up to 9.7 fb^{-1} used for Higgs boson searches.

The dominant production channels for a SM Higgs boson at the Tevatron at masses below 140 GeV, ordered by size, are gluon fusion, Higgs radiation off a W or Z boson followed by the vector-boson-fusion process. The cross section for vector-boson fusion falls less rapidly with rising Higgs-boson mass, and above 180 GeV vector-boson fusion dominates over the WH production mode. Associated production modes with a pair of bottom or top quarks occur at rates more than one order of magnitude lower.

In contrast to LEP, at the Tevatron the Higgs-boson searches at lower masses in the $H \to b\bar{b}$ channels have to deal with a tiny signal on top of an overwhelming

[1] ξ^2 is equivalent to the cross section normalised to the one expected in the SM, commonly denoted as "signal-strength modifier", μ.

background of b-flavoured jets originating from QCD processes. For example, the cross section for pairs of b quarks is more than eight orders of magnitude larger than the production cross section of a 150 GeV Higgs boson. This very unfavourable signal-to-background ratio prohibits exploitation of purely hadronic final states for Higgs-boson searches. Instead, VH modes with leptonically decaying W or Z bosons must be considered. $H \to W^+W^-$ is the dominant decay channel above $M_H = 140$ GeV; it reaches a branching ratio of almost 100 % near the W-pair threshold (see Fig. 6.1). Requiring one W boson to decay leptonically provides good trigger and selection efficiency.

First sensitivity to a SM Higgs boson was reported in the $H \to W^+W^-$ channel by the Tevatron experiments in the year 2010 with an integrated luminosity of about 5 fb^{-1} [24]. Using all production modes in the combined result, the mass range 162–166 GeV was excluded at the 95 % confidence level based on the prediction for the SM rate at that time. With the final data set, the exclusion around the W-pair threshold was extended to cover the range 149–182 GeV [25].

For the searches in the low-mass region, from 90/100–150 GeV, furthermore the decay modes $H \to b\bar{b}$ and $H \to \tau^+\tau^-$ and also $H \to \gamma\gamma$ in the VH production mode were considered. The largest sensitivity arises from the $H \to b\bar{b}$ channel. The size of an expected Higgs-boson signal on top of the backgrounds is shown in Fig. 6.3a. For each of the analysed channels and sub-categories, the events are classified according to their signal-to-background ratio, s/b, thus providing an illustrative representation of the data, which preserves the importance of each event.

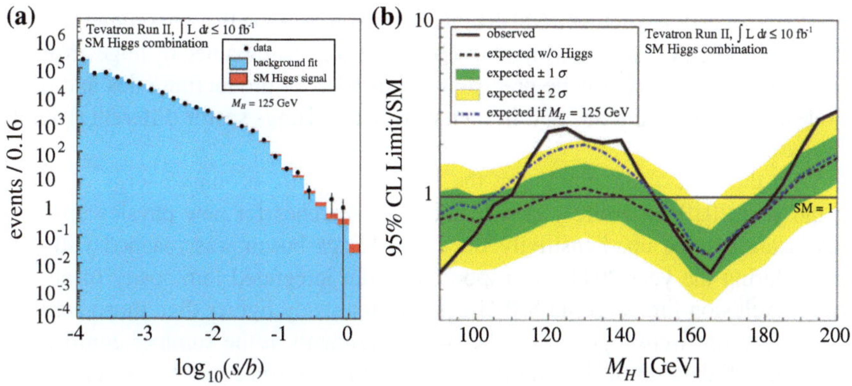

Fig. 6.3 Combination of Higgs-boson searches at the Tevatron. **a** Distribution of the logarithm of the signal-to-background ratio. The expectation for a SM Higgs-boson signal with $M_H = 125$ GeV is also shown. **b** 95 % CL upper limits derived from the full data set of the Tevatron experiments CDF and DØ on the ratio of the observed Higgs-boson production cross section in the decay channels $H \to W^+W^-$ and $H \to b\bar{b}$, normalised to the SM expectation, as a function of the Higgs-boson mass. The *dashed line* represents the expected limit, while the *full black line* shows the observed limit. The *short-dashed line* shows median expected limits assuming that the SM Higgs boson be present at $M_H = 125$ GeV. Adapted from [25]

The final result from a combination of all searches for the SM Higgs boson by the collaborations CDF and DØ with an integrated luminosity of up to 10 fb^{-1} is shown in Fig. 6.3. The statistical treatment for limit setting was based on the CL_s method introduced above and described in more detail in Sect. 6.3.2 below. The likelihood ratio used as the test-statistic accounts for small signals on a large background with a large number of sources of systematic uncertainties.

The excluded mass ranges at 95 % confidence level of a SM Higgs boson are 90–109 and 149–182 GeV, while the expected exclusion in the absence of a signal is 90–120 and 140–184 GeV.

An excess of events exceeding the 2 σ band of the expected exclusion under the assumption of the absence of a signal is seen in the mass region 115–140 GeV. Also shown is the expected limit under the assumption of the presence of a SM Higgs boson at $M_H = 125$ GeV. Expressed in terms of a deviation of the observation from the background-only hypothesis, the p-value (or local significance) of the observed excess corresponds to about 3 standard deviations.

6.3 Higgs-Boson Phenomenology at the LHC

The Higgs-boson searches at LEP and the Tevatron, giving rise to limits on the parameter space of the SM and extensions of it, went on in parallel to the preparation of the LHC experiments. A primary goal in the design of the general-purpose experiments ATLAS and CMS and of the parameters of the accelerator was the capability to clarify the question whether a SM-like Higgs boson exists in the mass range between 100 and 1000 GeV. The experimental activities were accompanied by huge efforts on the theoretical side to provide precision calculations of production rates and decay modes, both within the SM and in scenarios with a Higgs sector different from the SM case.

With the start of data-taking at the LHC at a centre-of-mass energy of 7 TeV in the year 2010, an entirely new window for searches for new physics up to the TeV scale opened up. Sensitivity for a SM Higgs boson was reached with data taken during the year 2011, corresponding to an integrated luminosity of about 5 fb^{-1}. Exclusion limits on the SM Higgs boson already tighter than those achieved at the Tevatron were reported by the two experiments at the summer conferences. With the full data set of the year 2011, a SM Higgs boson could be excluded up to masses of about 600 GeV, leaving only a small window below 127 GeV due to an observed excess of events. In the year 2012 both the centre-of-mass energy and the instantaneous luminosity could be further increased (see Fig. 2.18 in Chap. 2), such that data corresponding to an integrated luminosity of about 20 fb^{-1} could be accumulated at $\sqrt{s} = 8$ TeV. In combination with the excellent performance of the detectors and of the full analysis chains in the ATLAS and CMS experiments, this marked the begin of a new "Higgs Physics" era, heralded by the discovery of "a new boson at a mass around 125 GeV" on 4 July 2012. Based on the analysis of all data of Run 1 of the LHC, a remarkable amount of information about the properties of the new boson has already been accumulated.

6.3.1 Higgs-Boson Production at the LHC

Since the Higgs-boson couplings to all particles are proportional to the masses of these particles, Higgs-boson production processes predominantly arise from diagrams involving heavy particles. At the LHC, the main production processes for a SM-like Higgs boson involve the Higgs-boson couplings to the top quark and to the weak bosons W and Z. Because of the abundance of gluons in the relevant Bjorken-x range the dominant production mechanism is in fact the loop-induced process of gluon fusion involving loops of virtual top or bottom quarks (in extensions of the SM loop contributions of additional particles of new physics could also be sizeable). Vector-boson fusion, Higgs radiation and associated production with top quarks provide characteristic signatures which can be experimentally tagged. Examples of production diagrams are shown in Fig. 6.4.

Precise theoretical predictions for the signal and the background processes in connection with reliable estimates of the remaining theoretical uncertainties are crucial for Higgs physics at the LHC. This applies to the setting of limits obtained from the Higgs-boson searches at the LHC, to the discovery of a signal in the Higgs-boson searches, and also to the interpretation of this signal in view of the predictions of the SM and possible alternatives of it. Enormous progress has been achieved during the last years in the predictions for the observables that are relevant for Higgs physics at the LHC. Higher-order QCD and also electroweak corrections within the SM and other models of new physics have been obtained for the total cross sections, decay rates and also differential cross sections of the relevant signal processes. Improved techniques for obtaining higher-order corrections to multi-leg processes have led to a very significant progress in the theoretical predictions for background processes. New developments on the merging of predictions at NLO and also NNLO with parton showers have greatly improved the accessibility of higher-order corrections in Monte Carlo event generators that can be applied for experimental analyses.

There are two kinds of sources of theoretical uncertainties: (i) parametric uncertainties that are induced by the experimental errors of the input parameters (in the

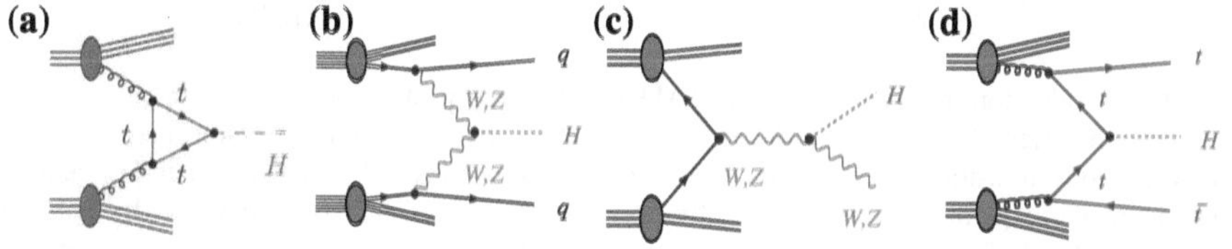

Fig. 6.4 Lowest-order Feynman diagrams for the dominant production mechanisms of a SM-like Higgs boson at hadron colliders: **a** gluon-gluon fusion; **b** vector-boson fusion; **c** Higgs-strahlung; **d** associated $t\bar{t}$ production

case of QCD corrections in the SM this refers to α_s and the heavy-quark masses) and the uncertainties of the PDFs; (ii) theoretical uncertainties from unknown higher-order corrections. The latter uncertainties can be very different for total cross sections and distributions and in general depend on the applied cuts. Reliable estimates of the theoretical uncertainties from unknown higher-order corrections are mandatory for the confrontation of the experimental results with theoretical predictions. They are, however, in general not easy to achieve. Simple scale variations of the renormalisation and/or factorisation scales are not sufficient for this purpose if at higher order new channels open up or new types of corrections enter. Furthermore, scale variations are often not suitable for estimating the effect of unknown electroweak higher-order corrections.

During the last years, the LHC Higgs Cross Section Working Group [17, 18, 26] played an important role in the communication between theory and experiment. It stimulated a very significant effort on the theory side, and at the same time it was instrumental for ensuring that the most advanced theoretical predictions and tools were actually used for the experimental analyses. One of the activities of the LHC Higgs Cross Section Working Group was to compile the state-of-the art predictions for total and differential cross sections and to provide prescriptions for estimating the remaining theoretical uncertainties [17, 18]. The results for the total production cross sections for the Higgs boson of the SM at centre-of-mass energies of 7 and 8 TeV are shown in Fig. 6.5. For each of the production cross sections, the figure also indicates at which level of perturbation theory the prediction has been obtained, and the displayed bands represent the estimate of the remaining theoretical uncertainties (showing the combined effect of parametric uncertainties and theoretical uncertainties from unknown higher-order corrections) according to the recipe of the LHC Higgs Cross Section Working Group.

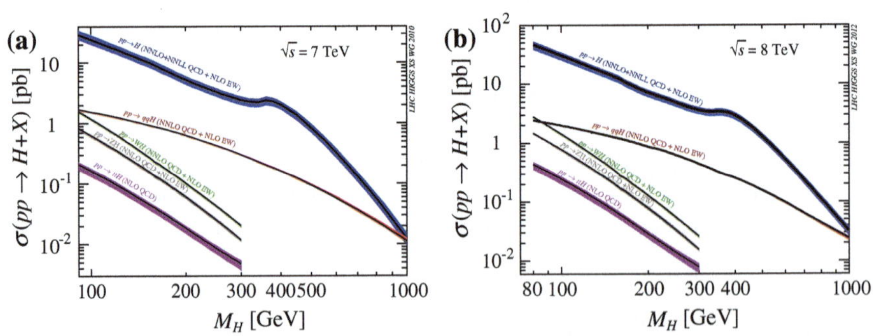

Fig. 6.5 Production cross sections for the SM Higgs boson in proton-proton collisions as published by the LHC Higgs Cross Section Working Group. The predictions for the gluon-fusion process, for vector-boson fusion, for Higgs radiation from W or Z bosons and for the associated production with pairs of top quarks are shown for centre-of-mass energies of 7 TeV (**a**) and 8 TeV (**b**). For each process the level of perturbation theory on which its prediction is based has been indicated, and the displayed bands correspond to the current level of theoretical uncertainties. Adapted from [17, 18]

As can be seen in Fig. 6.5, the dominant production process for both collision energies in the displayed mass range is indeed gluon fusion, owing to the large gluon flux in proton-proton collisions. As mentioned above, this process is loop-induced and, within the SM, dominated by the contribution of the top-quark loop. Bottom-loop contributions can have an important impact for the differential cross section in the region of low p_{T} [27–29]. Loop contributions from non-SM particles can in principle be sizeable for this process, since they enter at the same order as the leading SM contribution, i.e. they do not have to compete with a SM-type lowest-order contribution.

For the total cross section, the higher-order QCD corrections in the SM are very large, more than 100 %, and positive [30–36]. The available phase space plays an important role in this context. Compared to partonic process at leading order consisting of the production of an on-shell Higgs boson from two on-shell gluons, the phase space region for the next-to-leading order (NLO) contribution [30–33], which involves radiation of a gluon, is significantly enlarged. Predictions for the total cross section can to good approximation be obtained in the limit of an infinitely heavy top quark. This fact has facilitated the computation of QCD corrections beyond NLO [37], while at NLO the full quark-mass dependence is known. Next-to-next-to-leading order (NNLO) QCD corrections [34–36, 38] and a next-to-next-to-leading logarithmic (NNLL) resummation [39] have been obtained, while electroweak corrections have been incorporated up to the two-loop (NLO) level [40–42]. The strong theory effort that has been ongoing for many years towards the computation of higher-order contributions to the gluon-fusion process has made it possible to reduce the theoretical uncertainties from unknown higher-order corrections within the SM for this process to a level of below 10 %, which is similar to the current PDF uncertainties.

The fact that the contribution of an infinitely heavy top quark to the gluon-fusion amplitude does not go to zero but rather approaches a constant,[2] has an interesting consequence for a possible extension of the SM by an additional generation of fermions that receive their masses via the BEH mechanism. Each of the two heavy quarks of such a doublet would yield approximately the same contribution as the top-quark loop in the SM, so that the amplitude would be enhanced by about a factor of three and correspondingly the total cross section by about a factor of nine compared to the SM case. Accordingly, this kind of models could be severely constrained already with the early search results from the Tevatron and the LHC.

For the differential $pp \to H + X$ cross section, the approximation of an infinitely heavy top quark is not valid in the high-p_{T} region. The mass dependence needs to be properly taken into account in this case, which makes it challenging to go beyond NLO for the differential cross section [27–29, 43].

The cross section giving rise to the second-highest rate after gluon fusion is (weak) vector-boson fusion (VBF). This is an electroweak process at leading order, and the colour exchange between the two colliding partons is also suppressed in higher orders. Experimentally the signature of this process is characterised by two hard jets in the

[2] This is due to a cancellation between the $1/m_t$ behaviour of the loop contribution and the m_t factor arising from the Yukawa coupling.

far-forward and far-backward directions and relatively little hadronic activity in the central detector ("rapidity gap"). The QCD and electroweak corrections to the total cross section for this process within the SM are moderate and of similar size, about 5 % [44–48]. The remaining theoretical uncertainties from unknown higher-order corrections within the SM are at the level of about 1 %, which is below the current PDF uncertainties. However, the details of the applied cuts, in particular in case of a central jet veto, are important for the assessment of the theoretical uncertainties in the predictions for this channel [18, 49, 50].

The Higgs-strahlung process, i.e. $pp \to HW/HZ+X$, receives QCD corrections to the total cross section that are about 25 % for HW production and about 30 % for HZ production [51, 52], while the electroweak corrections amount to about 7 % and 5 %, respectively [53]. The remaining theoretical uncertainties from unknown higher-order corrections within the SM are at the level of few percent, which is slightly below the current PDF uncertainties. The corrections to differential cross sections can be potentially larger [54, 55].

The process of associated Higgs production with a $t\bar{t}$ pair, $pp \to t\bar{t}H + X$, with $H \to b\bar{b}$, results in a final state comprising four fermions (from the two W bosons produced in the top quark decays) and four b quarks. The QCD corrections for the total and differential cross sections are moderate [56]. For the signal process, the remaining theoretical uncertainties from unknown higher-order corrections within the SM are at the level of about 10 %, which is similar to the current PDF uncertainties. The predictions for the background processes of $t\bar{t}b\bar{b}$ and $t\bar{t}$ + 2jet production are known with NLO QCD accuracy, but the residual scale dependence of the background cross section normalisation is still about 20 % [57, 58].

The associated Higgs production with a $b\bar{b}$ pair plays a minor role in the SM. This production process can be significantly enhanced, however, in extensions of the SM, such as supersymmetric or two-Higgs-doublet models.

The above discussion of Higgs production processes implicitly assumes that the full process involving production and decay can be factorised into the on-shell production of a Higgs boson and the subsequent decay into final-state particles. Despite the small total width of a 125 GeV Higgs boson in the SM of about 4 MeV, off-shell effects beyond the narrow-width approximation can nevertheless be relevant. This is in particular the case for Higgs boson decays into ZZ^* and WW^*, leading to a 4-fermion final state, because of the large phase space suppression below the threshold for on-shell ZZ and WW production. The inclusion of off-shell effects requires a proper treatment of the finite width of the Higgs boson, which takes into account that the pole of the Higgs propagator is located off the real axis in the complex plane. The unitarity cancellations that are induced by Higgs-exchange contributions can give rise to a large destructive interference between signal and background for those processes [59–62]. In order to take these effects into account, accurate predictions for the full partonic process $gg \to 4\ell$ are necessary.

Higgs phenomenology in models of physics beyond the SM can differ from the SM case in various respects. In general, the additional structure of new physics can modify SM-like production and decay processes via different couplings and via new sources for important higher-order contributions, and it can also give rise to new production and decay modes that are not present in the SM. For instance, an enhanced Higgs-boson coupling to bottom quarks would have the effect that the bottom loop in gluon fusion would yield a larger contribution compared to the SM case, which could have an important impact on the p_T distribution. For the loop-induced processes of Higgs production in gluon fusion and Higgs boson decay into a pair of photons or gluons, the impact of additional particles in the loop could be particularly important. Furthermore, off-shell contributions at large invariant masses are sensitive to threshold effects where the decay mode of a Higgs boson at that mass to a pair of new particles would open up.

6.3.2 Statistical Treatment: Exclusion Limits and Significance of a Discovery

Looking for small signals on top of large backgrounds requires careful considerations of the statistical procedures. Systematic errors, entering the statistical interpretation as nuisance parameters, must be properly taken into account. The statistical procedure adopted by the LHC experiments is based on a profile likelihood ratio as the test-statistic and the modified frequentist (CL_s) method [22, 23] to set limits, and is very similar to the procedures adopted at LEP and at the Tevatron. The methodology was developed by the ATLAS and CMS collaborations in a common working group [63]. In the limit of sufficiently high statistics, asymptotic formulae describe the distributions of the test-statistic [64]. Otherwise or as a cross-check, the distributions of the test-statistic were derived by Monte Carlo sampling. Bayesian methods with marginalisation of the nuisance parameters were also studied and found to give compatible results.

Parameters a of the signal model, such as the signal-strength modifier, $\mu = \sigma_{obs}/\sigma_{SM}$, or the signal position resp. mass, are estimated from scans of the profile likelihood ratio $q(a)$, given by

$$q(a) = -2 \ln \left(\frac{\mathcal{L}(\text{data} \,|\, s(a) + b,\ \hat{\theta}_a)}{\mathcal{L}(\text{data} \,|\, s(\hat{a}) + b,\ \hat{\theta})} \right). \tag{6.1}$$

Here, s is the signal model, b stands for the backgrounds, and θ signifies the set of nuisance parameters describing systematic uncertainties. $\hat{\theta}_a$ maximises the likelihood for a given value of a, while $\hat{a}$ and $\hat{\theta}$ represent the parameter values at the overall maximum of the likelihood.

From the expression for the negative log-likelihood ratio q, the test-statistic used to set limits on the signal strength is obtained by taking the signal contribution as $s(\mu) = \mu \cdot s$ and fixing the shape and mass,

$$q_\mu = -2\ln\left(\frac{\mathcal{L}(\text{data}\,|\,\mu\cdot s+b,\,\hat{\theta}_\mu)}{\mathcal{L}(\text{data}\,|\,\hat{\mu}\cdot s+b,\,\hat{\theta})}\right) \quad \text{with } 0\leq\hat{\mu}\leq\mu\,.$$

It is the normalisation to the global maximum of the likelihood that enables one to use the asymptotic formulae in the limit of sufficiently large statistics.[3]

Limits are derived using $\mathrm{CL_s}$, the ratio of a-posteriori probabilities to observe a value of the test-statistic at least as large as the one observed in data under the signal+background and the background-only hypotheses, H_{s+b} and H_b, respectively,

$$\mathrm{CL_s} = \frac{\mathrm{P}(q_\mu \geq q_\mu^{\text{obs}}\,|\,H_{s+b})}{\mathrm{P}(q_\mu \geq q_\mu^{\text{obs}}\,|\,H_b)}\,.$$

If $\mathrm{CL_s}(\mu) < \alpha$, larger values of μ are excluded at a confidence level of $1-\alpha$.

To quantify the presence of an excess of events over what is expected from the background, the likelihood appearing in the numerator of the likelihood ratio is the one for the background-only hypothesis, $\mu = 0$, and the test-statistic then is $q_0 = q(\mu = 0)$. The statistical significance of a signal-like excess, expressed as the "p-value", is computed from the a-posteriori probability

$$p = \mathrm{P}(q_0 \geq q_0^{\text{obs}}\,|\,H_b)\,,$$

which is sometimes also given in terms of an equivalent number of standard deviations of a Gaussian distribution, z, given by inversion of the relation $p = \int_z^{+\infty}\frac{1}{\sqrt{2\pi}}\exp(-x^2/2)\,\mathrm{d}x$.

6.4 Higgs-Boson Searches at the LHC

Searches for the SM Higgs boson at the LHC have been performed in various decay modes, the choice of which is given by the signal rates and the signal-to-background ratios in the different mass regions.

The branching fractions of the Higgs boson of the SM strongly depend on the mass, M_H, as shown in Fig. 6.1. To determine the sensitivity of a search in a given production channel and decay mode, however, other factors like the selection efficiencies, the resolution on the reconstructed mass and—most importantly—the expected background levels as illustrated in Fig. 3.22 in Sect. 3.6 are crucial.

Promising search strategies and background-rejection methods had been established in many studies before the start of data-taking at the LHC [65–67], and the results of such studies are briefly summarised below.

[3] At LEP and Tevatron, the background-only hypothesis was used as the denominator of the likelihood ratio.

6.4.1 Discovery Channels

Among the most important channels are the inclusive $H \to \gamma\gamma$ and $H \to ZZ^{(*)} \to 4\ell$ decay channels. These channels are characterised by a very good mass resolution at the level of $\pm(1\text{–}2)\,\%$. In the low-mass region, the Higgs boson appears as a sharp resonance (the width of which is dominated by the detector resolution) on top of flat backgrounds, which are dominated by $\gamma\gamma$ and $ZZ^{(*)}$ continuum production, respectively. These backgrounds can be estimated from side-band regions using data-driven methods. In the high-mass region, additional decay modes, like $H \to ZZ \to \ell\ell\,\nu\bar{\nu}$, and $H \to ZZ \to \ell\nu q\bar{q}$, can also be studied. In addition, the $H \to WW^{(*)} \to \ell\nu\ell\nu$ decay mode contributes, in particular in the mass region around 160 GeV. However, due to the neutrinos in the final state, no mass peak can be reconstructed. Higgs-boson production manifests itself in this channel as a broad peak in the transverse mass distribution (see below).

From the fermionic decays, only the modes $H \to \tau\tau$ and $H \to b\bar{b}$ have been accessible with the data collected in Run 1. For the $H \to \tau\tau$ decays, the selection of the vector-boson-fusion production mode is important to improve the signal-to-background ratio by exploiting forward-jet tagging [68–72]. Although for a light Higgs boson the dominant decay channel is $H \to b\bar{b}$, there is an overwhelming background from b jets produced in multi-jet events via QCD processes. This already prohibits triggering a $b\bar{b}$ signal if no additional particles are present, such as leptons from a vector-boson decay in the VH production mode, or a pair of top quarks in the $t\bar{t}H$ mode. In addition, the invariant mass of the $b\bar{b}$ system can only be reconstructed with a modest mass resolution of about $\pm 10\,\%$.

The sensitivities of the different search channels studied by the CMS experiment are illustrated in Fig. 6.6, where the expected 95 % confidence level upper limits on the cross section, normalised to the SM expectation σ_{SM}, are shown as a function of M_H. They were evaluated under the assumption that only the expected background is observed in the data recorded in the year 2011, taking into account trigger and selection efficiencies for signal and background events.

The $H \to WW$ channel is the most important one in the intermediate mass range $130 < M_H < 200$ GeV. At very low masses below 125 GeV the $H \to \gamma\gamma$ channel becomes the most sensitive one. In the region 110–130 GeV all three channels, $H \to \gamma\gamma$, $H \to WW^*$ and $H \to ZZ^*$, contribute significantly. For the ATLAS experiment, similar sensitivities were found and the same conclusions on the importance of the various decay channels in different mass regions apply.

6.4.2 Exclusion Limits with Early Data at the LHC

Significant progress in constraining the allowed mass range of a SM Higgs boson was made by the ATLAS and CMS experiments by analysing the data recorded in the year 2011, corresponding to an integrated luminosity of up to 4.9 fb^{-1} per experiment.

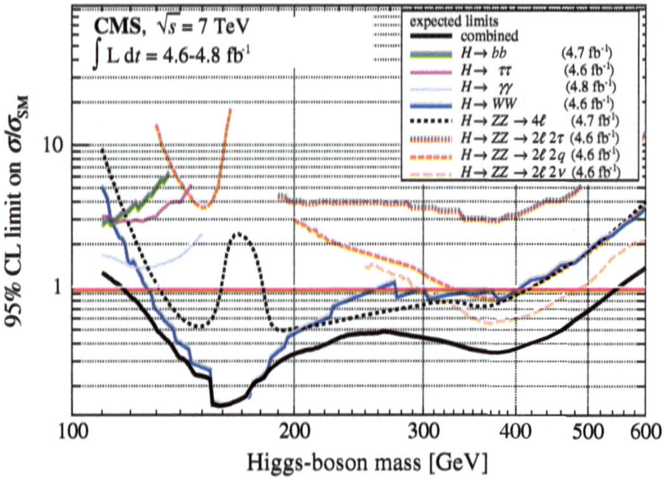

Fig. 6.6 Expected 95 % CL upper limits on the cross section ratio σ/σ_{SM} as a function of the Higgs-boson mass for different Higgs-boson decay channels, corresponding to the integrated luminosity recorded by the CMS experiment in the year 2011. Adapted from [73]

All analyses were performed "blindly", i.e. they were developed on Monte Carlo samples and special data samples, selected such that no signal contribution was expected. The signal regions were inspected only before the major conferences in summer 2011 and for the extraction of the final results.

Searches were performed in numerous channels, but only the combination of the $H \to \gamma\gamma$, $H \to WW^{(*)}$, and $H \to ZZ^{(*)}$ channels reached sensitivity to the presence of a SM Higgs boson in the mass range from 120–500 GeV at that time. The search channels were split into several sub-channels, such as $4e$, $2e2\mu$, or 4μ in case of $H \to ZZ^{(*)}$, and different sub-categories depending on the reconstruction quality of a possible signal. As these categories have varying signal-to-background ratios, this procedure maximises the statistical sensitivity.

The test-statistic used to separate the background-only hypothesis, H_b, from the background + signal hypothesis, H_{s+b}, is based on the profile likelihood ratio $q_\mu(\mu)$, as described in Sect. 6.3.2. The value of the scale factor $\mu = \sigma_{\text{obs}}/\sigma_{\text{SM}}$ that can be excluded at 95 % confidence level is determined from full fits to the combined likelihood of all signal categories, including nuisance parameters describing systematic uncertainties. Exclusion limits were set by both experiments using the CL_s prescription. The results are shown in Fig. 6.7.

In the ATLAS experiment a SM Higgs boson in the mass ranges 112.9–115.5, 131–238 and 251–466 GeV could be excluded at 95 % confidence level. The expected exclusion range, under the assumption that no Higgs boson is present, covers the full mass range from 124–519 GeV. The corresponding range excluded by the CMS experiment was 127–600 GeV, with an expected exclusion range from 118–543 GeV. It is striking that already with this first data set the full mass range from about 127 GeV up to 600 GeV could be excluded, leaving only a small mass window at low mass open for the existence of a SM-like Higgs boson.

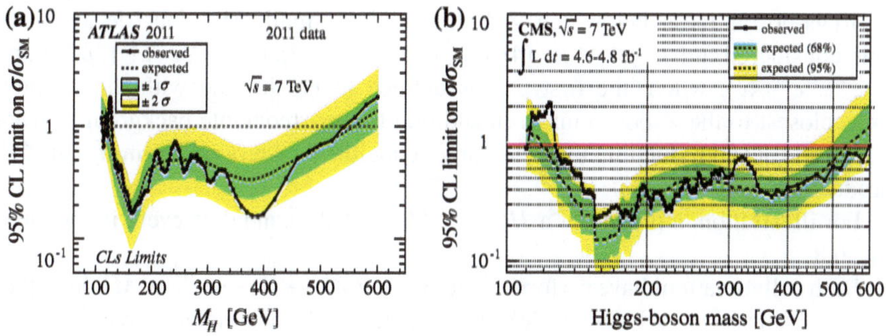

Fig. 6.7 Observed 95 % CL upper limits on the cross section ratio σ/σ_{SM} as a function of the Higgs-boson mass based on the data set taken in the year 2011 by the ATLAS (**a**) and CMS (**b**) experiments. Adapted from [73, 74]

At these low mass values both collaborations consistently observed an excess of events, and therefore the observed exclusion limit was above the expectation, exceeding the $2\,\sigma$ band around 124–126 GeV. The excess with a p-value under the background-only hypothesis corresponded to slightly more than 3 standard deviations in each experiment, not sufficient to claim a discovery at that time.

6.5 Discovery of a Higgs Boson at the LHC

The higher centre-of-mass energy of $\sqrt{s} = 8$ TeV in the year 2012 led to an increase of the production cross section for a light Higgs boson of about 25 %, and already during the first months of running the accumulated integrated luminosity exceeded that of the year 2011. Higher levels of concurrent proton-proton collisions in the same bunch crossing required modifications of the selection procedures in particular in the $H \to \gamma\gamma$ and $H \to WW$ channels to make them more robust against additional activity in the detectors. As already done during the searches in 2011, distributions with an expected contribution from signal events were blinded and not looked at before the analyses were considered final.

In the following the results of analyses of the three bosonic decay channels by the ATLAS and CMS collaborations are discussed [75–77]. All results are based on the data set collected by the experiments until June 2012, corresponding to integrated luminosities of about 5 fb^{-1} at $\sqrt{s} = 7$ TeV and 5–6 fb^{-1} at $\sqrt{s} = 8$ TeV.

6.5.1 The $H \to ZZ^* \to 4\ell$ Signal

The decay channel $H \to ZZ^* \to 4\ell$ provides a rather clean signature in the mass range 115 GeV $< M_H < 2M_Z$. However, due to the small branching fractions of the involved H and Z boson decays, the yield of signal events for a Higgs-boson mass below the ZZ threshold is small. The four decay channels considered are $\mu^+\mu^-\mu^+\mu^-$ (4μ), $e^+e^-\mu^+\mu^-$ ($2e2\mu$), $\mu^+\mu^-e^+e^-$ ($2\mu2e$) and $e^+e^-e^+e^-$ ($4e$), where the first pair is defined to be the one with the dilepton mass closest to the Z-boson mass. In addition to the irreducible backgrounds from $ZZ^{(*)}$ and $Z\gamma^*$ production, there are large reducible backgrounds from $t\bar{t}$ and $Zb\bar{b}$ production.

For illustration, a picture of a $H \rightarrow ZZ^* \rightarrow 2e2\mu$ candidate event is shown in Fig. 6.8.

Both collaborations have performed the $H \rightarrow ZZ^* \rightarrow 4\ell$ search for M_H hypotheses in the mass range 110–600 GeV. Calorimeter and track-isolation requirements together with impact-parameter requirements were used to suppress the reducible background down to the level of the irreducible ZZ^* continuum background. The residual Z+jets and $t\bar{t}$ backgrounds, which have an impact mostly for low invariant four-lepton masses, were estimated from control regions in the data. The irreducible ZZ^* background was predicted using Monte Carlo simulation. The events were categorised according to the lepton-flavour combinations. Mass resolutions of approximately 1.5 % in the four-muon channel and 2 % in the four-electron channel were achieved at $M_H \sim 120$ GeV. After final requirements, the signal-to-background ratio is found to be about two in the signal mass window 120–130 GeV.

The four-lepton invariant mass provides a powerful discriminating variable, which is complemented by kinematic information on the $H \rightarrow 4\ell$ contained in the invariant masses of the two lepton pairs and the decay angles characterising the decay topology in the rest frame of the four-lepton system.

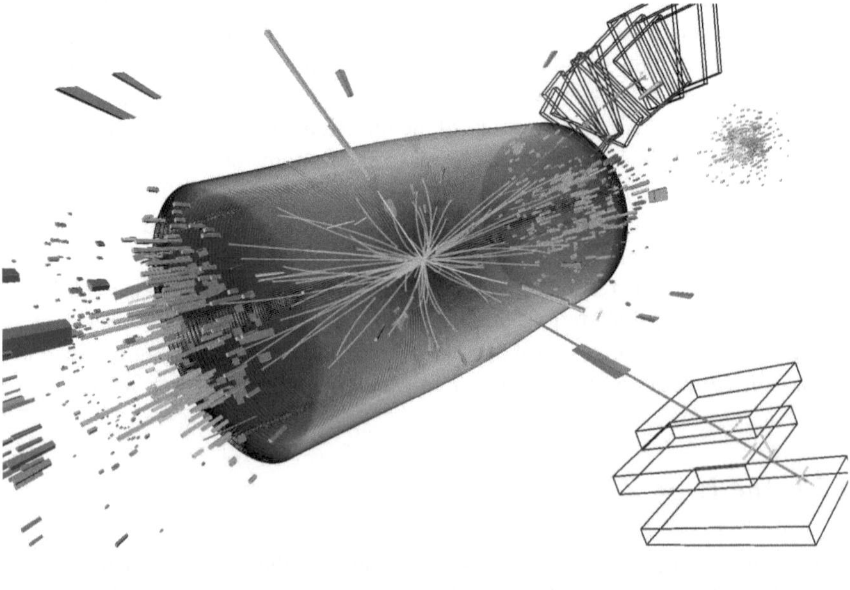

Fig. 6.8 Event display of a $H \rightarrow ZZ$ candidate event in the CMS detector. One of the Z bosons decays to a pair of electrons (tracks with clusters in the electromagnetic calorimeters), the other one to a pair of muons (tracks reaching to the muon chambers). *Source* CERN

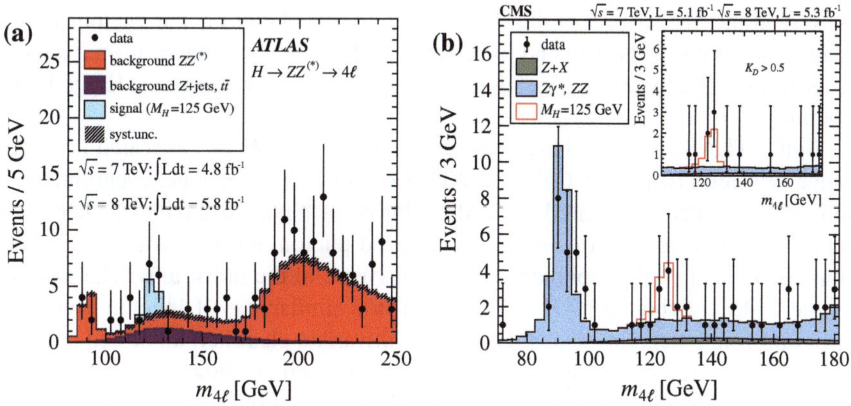

Fig. 6.9 Distributions of the invariant mass of selected events with two pairs of unlike-sign electrons or muons measured by the ATLAS (**a**) and CMS experiments (**b**). Adapted from [75, 76]

The observed and expected mass distributions for events selected after the full selection are displayed in Fig. 6.9.

There are excesses of events in both experiments in the mass region around 125 GeV. In the ATLAS experiment 4.9 events are expected from SM background processes in the mass window from 120–130 GeV, whereas 13 events are observed in the data. The corresponding numbers for the CMS experiment in the mass window 121.5–130.5 GeV are 7.7 events expected from background processes, compared to 18 events observed.

In order to quantify these excesses, the a-posteriori probabilities for the background-only hypotheses were calculated. The minimum local p-values in the data of the ATLAS and CMS experiments occur at 125.0 and 125.6 GeV, respectively, and have an observed (expected) significance of 3.6 (2.7) σ and 3.2 (3.8) σ.

6.5.2 The $H \rightarrow \gamma\gamma$ Signal

$H \rightarrow \gamma\gamma$ is a rare decay mode, which is only detectable in a limited Higgs-mass region between 100 and 150 GeV, where both the production cross section and the decay branching ratio are relatively large. Excellent energy and angular resolution are required to observe the narrow mass peak above the irreducible prompt $\gamma\gamma$ continuum. In addition, there is a reducible background resulting from direct-photon production or from multi-jet production via QCD processes. These processes contribute if one or both jets are misidentified as a photon. The background can be determined from a fit to the data (side-bands) such that uncertainties from Monte Carlo predictions or uncertainties due to normalisations in control regions are avoided.

Within the SM, decays of the Higgs boson to two photons occur via loops dominantly involving W bosons and top quarks, with some contribution from b quarks. This decay mode therefore depends on bosonic as well as on fermionic Higgs-boson couplings.

Both ATLAS and CMS have presented results on the $H \to \gamma\gamma$ search in a mass range between 110 and 150 GeV. The ATLAS analysis [75] separates events into nine independent categories, based on the direction of each photon and on whether it was reconstructed as a converted or an unconverted photon, together with the momentum component of the diphoton system transverse to the thrust axis. The analysis of the CMS collaboration [76, 77] is done in a similar way: Diphoton events are split into four categories depending on a multivariate classifier, a boosted decision tree (BDT), based on the rapidity of the photons and their shower shape characteristics. The mass resolution depends on the classification of the photon and is found to be 1.4 GeV (1.1 GeV) in the best category and 1.7 GeV (1.8 GeV) on average for $m_{\gamma\gamma}$ = 125 GeV in the ATLAS (CMS) experiment. In addition, a further class of events to select the vector-boson fusion (VBF) topology is introduced in both experiments. By requiring two jets with a large separation in pseudo-rapidity, a class of events is defined for which the expected signal-to-background ratio is about an order of magnitude larger than for the events in the other classes defined by photon properties.

The distributions of the invariant mass of the two photons, weighted with category-dependent factors reflecting the signal-to-background ratios, are shown in Fig. 6.10, together with the fit for the background and with the signal expectation for the best-fit value of the Higgs-boson mass in each experiment in the $\gamma\gamma$ channel. The fit to the background was performed separately in each category in the mass range 100–160 (180) GeV in ATLAS (CMS) by using an exponential function with free slope and normalisation parameters or a polynomial of third to fifth order. The statistical analysis employed likelihood functions constructed from the various categories of the 7 and 8 TeV data samples. Clear excesses of events above the background are seen around a mass of ~125 GeV in both experiments.

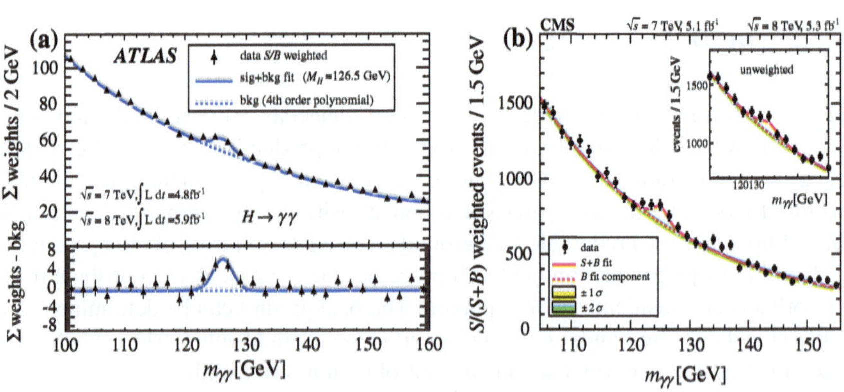

Fig. 6.10 Weighted distributions of the invariant mass of photon pairs measured by the ATLAS (**a**) and CMS (**b**) experiments. The *inset* shows the unweighted distribution. Adapted from [75, 76]

When confronted to the background-only hypothesis, the largest observed (expected) local significances reach $4.5\,\sigma$ ($2.5\,\sigma$) at $M_H = 126.5$ GeV for the ATLAS and 4.1σ (2.8σ) at $M_H = 125$ GeV for the CMS experiment, respectively.

6.5.3 The $H \to WW^* \to \ell\nu\ell\nu$ Signal

The decay mode $H \to WW^* \to \ell\nu\ell\nu$ has the highest sensitivity for a SM-like Higgs boson for Higgs-boson masses around the W^+W^- threshold of 160 GeV. As described above, based on searches in this channel, mass regions could be excluded by both the Tevatron and the LHC experiments already in summer 2011. However, this channel is more challenging in the low-mass region around 125 GeV since the expected signal rates are small due to the reduced $H \to WW$ branching ratio. The presence of neutrinos makes it impossible to reconstruct a narrow mass peak, and evidence for a signal must be extracted from an excess of events above the expected backgrounds. Usually, the WW transverse mass (m_{T}), computed from the lepton and the missing transverse momentum,

$$m_{\mathrm{T}} = \sqrt{(E_{\mathrm{T}}^{\ell\ell} + E_{\mathrm{T}}^{\mathrm{miss}})^2 - |\mathbf{p}_{\mathrm{T}}^{\ell\ell} + \mathbf{p}_{\mathrm{T}}^{\mathrm{miss}}|^2}\,,$$

where $E_{\mathrm{T}}^{\ell\ell} = \sqrt{|\mathbf{p}_{\mathrm{T}}^{\ell\ell}|^2 + m_{\ell\ell}^2}$ and $|\mathbf{p}_{\mathrm{T}}^{\mathrm{miss}}| = E_{\mathrm{T}}^{\mathrm{miss}}$, is used to discriminate between signal and background. In the definition of m_{T}, CMS assumes massless objects to build it, and uses as the discriminant a two-dimensional approach based on m_{T} and the invariant mass, $m_{\ell\ell}$, of the two charged leptons. The WW, $t\bar{t}$ and single-top production processes constitute severe backgrounds, and the signal significance depends critically on their absolute knowledge.

In order to optimise the sensitivity, the analyses of the ATLAS and CMS collaborations are subdivided into categories depending on different lepton final states (ee, $e\mu$ and $\mu\mu$) and jet multiplicities. Typical selection requirements are the presence of two isolated high-p_{T} leptons with significant missing transverse energy and a small azimuthal angular separation. The latter requirement is motivated by the decay characteristics of a spin-0 boson decaying into two W bosons with their subsequent decay $W \to \ell\nu$. The V-A structure of the W-boson decays implies a correlation between the directions of the charged leptons, which can be exploited to reject the WW background. These correlations lead to the use of quantities such as the dilepton invariant mass, $m_{\ell\ell}$, and the angular separation, $\Delta\phi_{\ell\ell}$, in the selection criteria. The various jet categories are sensitive to different Higgs-boson production mechanisms, but also show very different background compositions. The 0-jet category has non-resonant WW production as major background, while the 2-jet category, sensitive to Higgs production via vector-boson fusion, is dominated by $t\bar{t}$ background. The distributions of two powerful discriminating variables, the transverse mass and the invariant mass of the two leptons, as measured in July 2012, after the application of all final selection criteria, are shown in Fig. 6.11.

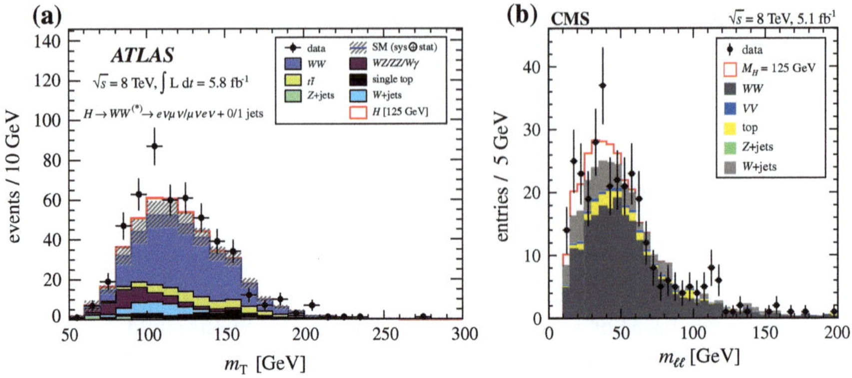

Fig. 6.11 **a** Distribution of the transverse mass, m_T, in the 0- and 1-jet channels as measured by ATLAS. **b** Distribution of the invariant mass of the two leptons in the 0-jet $e\mu$ category as measured by CMS. Adapted from [75, 76]

Excesses of events above the expectations from SM processes excluding Higgs-boson production are visible. A shallow minimum of the local p-value appears around 125.0 GeV, with a local observed (expected for $M_H = 125$ GeV) significance of $2.8\,(2.3)\,\sigma$ in ATLAS and $1.6\,(2.4)\,\sigma$ in CMS.

Other decay channels, in particular $H \to \tau\tau$ and $VH \to Vb\bar{b}$, were also investigated, but were not yet sensitive.

6.5.4 Combined Significances

In summer 2012, the ATLAS and CMS experiments observed an excess of events near $M_H = 125$ GeV in the $H \to ZZ^* \to 4\ell$ and $H \to \gamma\gamma$ channels. These excesses were confirmed by the highly sensitive but low-resolution $H \to WW^* \to \ell\nu\ell\nu$ channel.

The observed local p-value from the combination of all channels in the ATLAS experiment is shown as a function of M_H in Fig. 6.12 for the low mass range. The largest local significance was found for a SM Higgs-boson mass hypothesis of $M_H = 126.5$ GeV, where it reaches $6.0\,\sigma$, with an expected value in the presence of a SM Higgs-boson signal at that mass of $4.9\,\sigma$. A compatible picture was observed in the CMS experiment with an observed (expected) local significance of $5.0\,(5.8)\,\sigma$ at a mass near 125 GeV.

A first test of the compatibility of the observed signal with the SM Higgs boson was provided by the examination of the best-fit value of the common signal strength $\mu = \sigma/\sigma_{SM}$, defined as the ratio between the fitted cross section and the expectation for a SM Higgs boson. It was obtained by a combination of all channels.

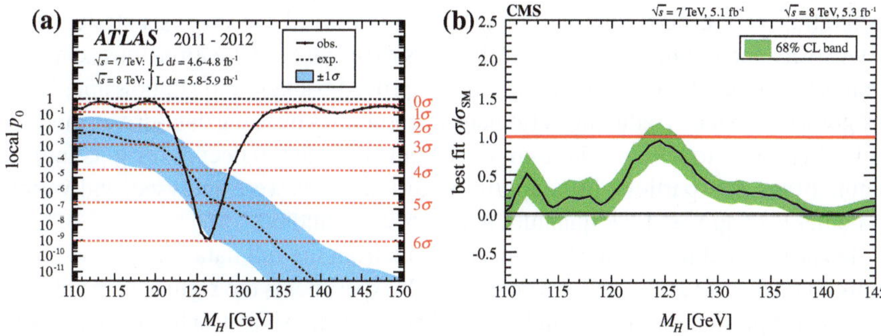

Fig. 6.12 **a** The observed local p-value as a function of M_H (*solid line*) and the expectation with its $\pm 1\,\sigma$ band assuming the presence of a SM Higgs boson at that mass (*dashed line*), from the combination of the ZZ, $\gamma\gamma$ and WW channels by the ATLAS experiment. The *horizontal dashed lines* indicate the p-values corresponding to significances of 1 to 6 σ. **b** Best-fit signal strength as a function of M_H as determined by the CMS experiment from the combination of the ZZ, $\gamma\gamma$, WW, $\tau\tau$ and $VH \to Vb\bar{b}$ channels. Around the fitted signal strength $\mu = \sigma/\sigma_{\rm SM}$ the $\pm 1\,\sigma$ uncertainty band is shown. Adapted from [75, 76]

The result of a scan of the fitted values of μ as a function of M_H for the CMS experiment is shown in Fig. 6.12b. The observed value for the excess at $M_H = 125.5$ GeV is 0.87 ± 0.23. The corresponding value determined by the ATLAS collaboration is 1.40 ± 0.30 at a mass of 126.0 GeV.

6.5.5 Summary: Status in Summer 2012

The results summarised above provide conclusive evidence for the discovery of a new particle with a mass around 125 GeV in the data taken by the experiments ATLAS and CMS in 2011 and during the first half of 2012. The measured signal strength was found to be consistent with the one expected for the production of a SM Higgs boson. The decays to pairs of vector bosons with a net electric charge of zero identified the new particle as a neutral boson. The observation in the diphoton channel excluded the spin-1 hypothesis for the production of an on-shell resonance by virtue of the Landau–Yang theorem [78, 79]. Although these results were compatible with the interpretation in terms of the SM Higgs boson, a wide range of other interpretations was possible as well. It was obvious that more data and refined analyses would be needed to further clarify the nature of the new particle.

6.6 Results from the Full Run 1 Data Set

After the announcement of a discovery of a "new boson at a mass around 125 GeV" on 4 July 2012, each of the two general-purpose LHC experiments collected an additional integrated luminosity of about 15 fb^{-1}. The collaborations finalised the calibration of the detectors and significantly improved the analysis strategies by extensively deploying multivariate techniques for event classification, thus achieving an increased number of signal events and improved background suppression. Much progress was made in reducing systematic uncertainties by using data-driven methods of background modelling. These improvements have led to deeper insights into the nature of the new particle. Its couplings to SM particles were analysed, and various hypotheses of spin and *CP* quantum numbers were studied.

In spring 2013 first, still preliminary results from the full data set provided more hints on the nature of the new particle. In both experiments the signals in the already established production modes and decay channels were shown to be consistent with the expectations for the SM Higgs boson. The CMS collaboration showed first evidence for decays of the new boson into τ leptons. In addition, Higgs-boson searches were extended to higher masses, and the existence of heavy Higgs-like bosons with production rates as large as those expected for the SM Higgs boson was excluded up to a mass of nearly 1 TeV. Consequently, the name of the object changed to "Higgs-like particle at a mass around 125 GeV".

Further improvements to the measurements were made and presented at the 2013 summer conferences, further establishing the Higgs-like nature.

The experimental evidence also convinced the Nobel Prize Committee, and the 2013 Nobel Prize in Physics was awarded to François Englert and Peter W. Higgs "for the theoretical discovery of a mechanism that contributes to our understanding of the origin of mass of subatomic particles, and which recently was confirmed through the discovery of the predicted fundamental particle, by the ATLAS and CMS experiments at CERN's Large Hadron Collider".[4]

The main task remaining at the end of Run 1 was to determine the properties of the new particle as precisely as possible. The final results from studies by the ATLAS and CMS collaborations are presented in the following. They are based on the full Run 1 data set, corresponding to an integrated luminosity of nearly $5\,\mathrm{fb}^{-1}$ recorded at $\sqrt{s} = 7$ TeV and of about $20\,\mathrm{fb}^{-1}$ at $\sqrt{s} = 8$ TeV.

6.6.1 Signals in the Bosonic Decay Modes

The bosonic decay modes $H \to ZZ^* \to 4\ell$, $H \to \gamma\gamma$ and $H \to WW^* \to \ell\nu\ell\nu$ already played a prominent role for the discovery of the Higgs boson. Given the large data sets, the experiments also aimed at separating the various production modes within the channels to get access to the couplings of the Higgs boson to vector bosons and fermions. The gluon-fusion process and the decay to two photons occur via loops, which make them particularly sensitive to physics beyond the SM.

[4]Citation from http://www.nobelprize.org, Oct. 2013.

$H \to ZZ^* \to 4\ell$

Studies of Higgs-boson properties with the full Run 1 data set and improved analyses were performed by both the ATLAS [80] and the CMS [81] experiments. In addition to improvements of lepton identification and calibration, multivariate discriminants were essential in both experiments to achieve optimal separation of the signal from the backgrounds. The discriminants used in the ATLAS and CMS experiments are based on boosted decision trees or multi-dimensional likelihoods and include kinematic information via leading-order matrix elements of signal and background processes. Given the small number of signal events, event-by-event estimates of the mass resolution can be used to achieve a better discrimination.

A two-dimensional distribution of the output of the kinematic discriminant versus the reconstructed four-lepton mass, $m_{4\ell}$, is shown for the candidate events selected by the ATLAS experiment in Fig. 6.13. A clear signal component is visible at high output values of the discriminant and at a mass around 125 GeV. The compatibility of the data with the expectations of a signal and the expected backgrounds was checked using pseudo-experiments generated according to the expected two-dimensional distributions and good agreement was found.

In Fig. 6.14 the $m_{4\ell}$ distributions of the selected candidate events in the ATLAS and CMS experiments are shown. Besides the Higgs-boson signal, the Z peak from $Z \to 4\ell$ decays is clearly visible. For the data from the CMS experiment, the full mass range of the analysis is shown. No events are observed with $m_{4\ell} > 800$ GeV.

The $H \to ZZ^* \to 4\ell$ signal has been observed in the ATLAS experiment with a significance of $8.1\,\sigma$ ($6.2\,\sigma$ expected) at $M_H = 125.36$ GeV, the combined ATLAS measurement of the Higgs-boson mass from the $H \to \gamma\gamma$ and $H \to ZZ^* \to 4\ell$ channels. The signal strength is $\mu = 1.44^{+0.34}_{-0.31}\,(\text{stat})^{+0.21}_{-0.11}\,(\text{syst})$, consistent with

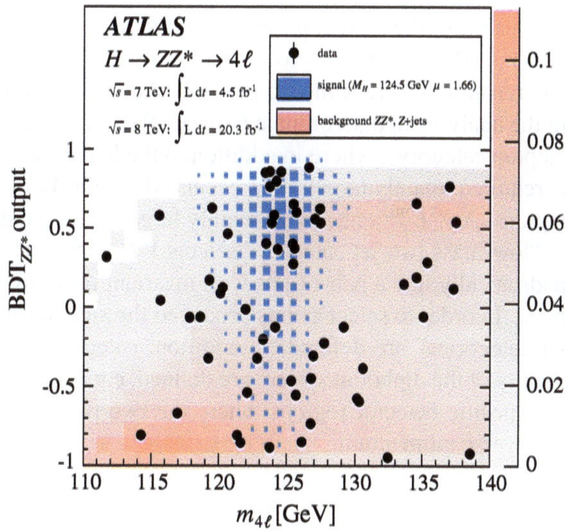

Fig. 6.13 Two-dimensional distribution of the output of the kinematic discriminant (BDT_{ZZ*}) versus the reconstructed mass $m_{4\ell}$ as measured in the ATLAS experiment. The expected distribution for a SM Higgs boson with $M_H = 125$ GeV is indicated by the size of the *square boxes*, and the total background is indicated by the intensity of the *red shading*. Adapted from [80]

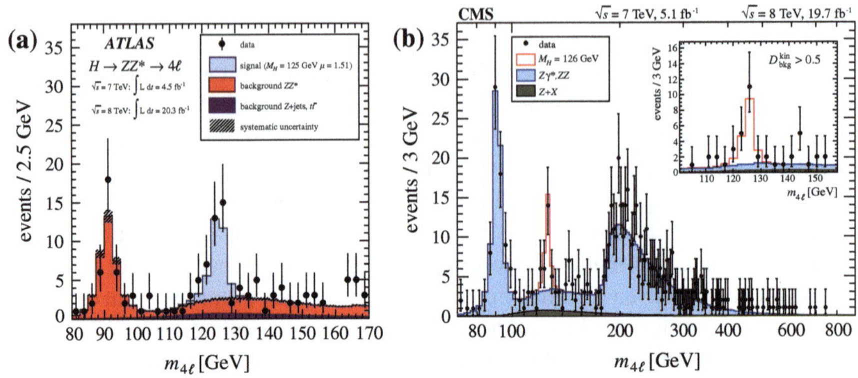

Fig. 6.14 Distributions measured in the ATLAS (**a**) and CMS (**b**) experiments of the reconstructed four-lepton mass for the sum of the $4e$, 4μ, and $2e2\mu$ channels. The expected distributions for a SM Higgs-boson signal for $M_H = 125\,\text{GeV}$ (ATLAS) or 126 GeV (CMS), normalised to the measured signal strength, as well as the background expectations are shown as *histograms*. Adapted from [80, 81]

the SM expectation of one. The measured signal strength in the CMS experiment is $\mu = 0.93^{+0.26}_{-0.23}\,(\text{stat})^{+0.13}_{-0.09}\,(\text{syst})$ and corresponds to a statistical significance of $6.8\,\sigma$ ($6.7\,\sigma$ expected).

Given the full reconstructed final state and the excellent mass resolution, this channel also contributes significantly to the measurement of the Higgs-boson mass and to the determination of its spin and *CP* properties, as discussed in Sects. 6.8.1 and 6.8.2, respectively.

$H \rightarrow \gamma\gamma$

Compared to the results published at the time of discovery (see Sect. 6.5), a better understanding of the detector calibration and in particular of the photon energy scale and resolution were achieved [82, 83]. Additional event categories were introduced in the analyses to discriminate between the main Higgs-boson production modes. A "lepton category", where in addition to the two photons a high-$p_{\rm T}$ electron or muon is required, mainly targets VH events where the W or Z bosons decay into charged leptons. An "$E_{\rm T}^{\rm miss}$ category" mainly targets VH events with $W \rightarrow \ell\nu$ or $Z \rightarrow \nu\nu$. A "low-mass two-jet category" selects VH events where the W or Z bosons decay hadronically into a pair of jets with invariant mass around the W-boson or Z-boson mass. In order to select events produced through the VBF process, "high-mass two-jet categories" are defined. In addition, categories referring to the reconstruction quality of the diphoton system are defined, e.g. taking account of converted photons, the specific detector regions where the two photons are detected, or the diphoton transverse momentum.

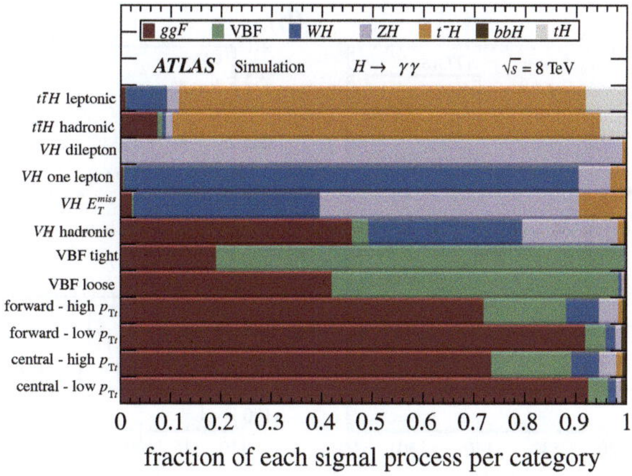

Fig. 6.15 Illustration of event categories used in the $H \to \gamma\gamma$ analysis in the ATLAS experiment. For each category the contributions of the different production modes at $\sqrt{s} = 8\,\text{TeV}$ are given. Adapted from [82]

The statistical benefit from the categorisation of events is large: In ATLAS, the use of these categories improves the sensitivity by about 40 % compared to the inclusive analysis. An overview of the categories is shown in Fig. 6.15, which also shows the contributions of the various production processes to each of the categories, based on Monte Carlo simulation. These contributions are important for the extraction of the signal strengths for the different production mechanisms and hence for the determination of the coupling strengths of the Higgs boson to bosons and fermions, as described in Sect. 6.8.4.

The data are fitted by categories, using parametrised shapes for the background as well as for the signal, specific to each category. The inclusive diphoton invariant-mass distributions, weighted by factors based on the category-dependent signal-to-background ratios, are shown in Fig. 6.16. The results of the fits are superimposed on the data.

At the combined ATLAS Higgs-boson mass of $M_H = 125.36\,\text{GeV}$, the excess corresponds to a significance of $5.2\,\sigma$, while $4.6\,\sigma$ are expected. The signal strength is measured to be $\mu = 1.17 \pm 0.23\,(\text{stat})\,^{+0.10}_{-0.08}\,(\text{syst})\,^{+0.12}_{-0.08}\,(\text{theory})$ [82]. In the CMS experiment, the maximum deviation from the background-only expectation occurs at a mass of $M_H = 124.7\,\text{GeV}$. The measured inclusive signal strength is $\mu = 1.14 \pm 0.21\,(\text{stat})\,^{+0.09}_{-0.05}\,(\text{syst})\,^{+0.13}_{-0.09}\,(\text{theory})$ at a statistical significance of $5.7\,\sigma$ ($5.2\,\sigma$ expected) [83].

Differential Cross-Section Measurements

Based on the data set collected during Run 1, the experiments also presented first measurements of differential cross sections in the $H \to ZZ^* \to 4\ell$ and $H \to \gamma\gamma$

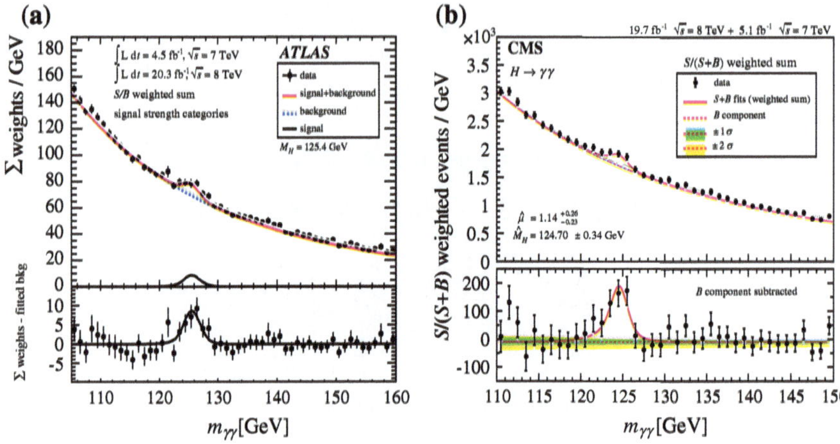

Fig. 6.16 Distributions of the invariant mass of the selected diphoton candidates measured by the ATLAS (**a**) and CMS experiments (**b**). Weights are applied to account for the different signal-to-background ratios in the event categories. The residuals of the data with respect to the fitted background are displayed in the *lower panels*. Adapted from [82, 83]

decay channels. These cross sections have been determined within fiducial phase-space regions and corrected for detection efficiency and resolution effects.

For the $H \to ZZ^* \to 4\ell$ channel the fiducial region is defined for events with four leptons, within $|\eta| < 2.47$ for electrons and $|\eta| < 2.7$ for muons, that pass the following selection criteria: $p_T > 20, 15, 10,$ and 100 GeV, with the mass requirements $50 < m_{12} < 106$ GeV, and $12 < m_{34} < 115$ GeV. For the $H \to \gamma\gamma$ channel two isolated photons are required with transverse momenta larger than 35 % (25 %) of the diphoton invariant mass and within $|\eta| < 2.37$.

As an example, the cross sections measured in the ATLAS experiment in bins of transverse momentum of the four-lepton system for the $H \to ZZ^* \to 4\ell$ channel [84] and and in bins of pseudo-rapidity of the $\gamma\gamma$ system in the $H \to \gamma\gamma$ channel [85] are shown in Fig. 6.17. Good agreement is found between the measured cross sections and theoretical expectations.

Given the larger number of signal events, more detailed measurements have been performed in the $H \to \gamma\gamma$ channel. Additional fiducial regions have been defined, selecting events that contain additional jets, an additional lepton, and events with two jets that have large dijet invariant mass and a large rapidity separation. The latter two fiducial regions enhance the contributions from Higgs-boson production via associated production with a vector boson and vector-boson fusion, respectively. The differential measurements have been performed for several kinematic variables, among them the cosine of the angle between the beam axis and the photons in the Collins–Soper frame of the Higgs boson, $|\cos\theta^*|$, and the azimuthal angle between the two leading jets, $|\Delta\phi_{jj}|$, in events containing two or more jets.

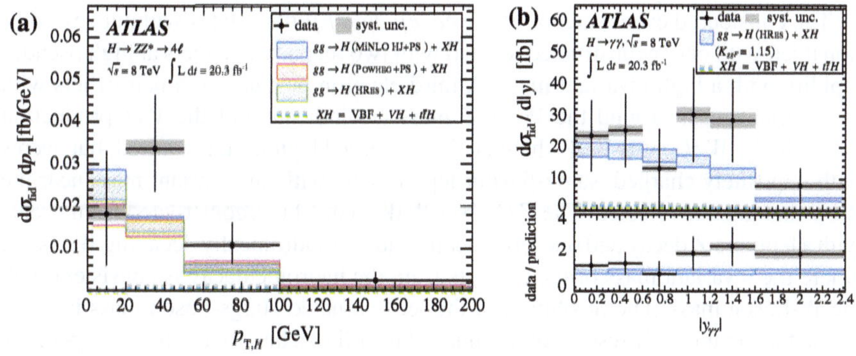

Fig. 6.17 Differential unfolded fiducial cross sections for **a** the transverse momentum $p_{T,H}$ of the Higgs boson in the $H \to ZZ^* \to 4\ell$ decay channel and for **b** the rapidity $y_{\gamma\gamma}$ of the diphoton system in the $H \to \gamma\gamma$ decay channel. For the $H \to ZZ^* \to 4\ell$ channel the measured distributions are compared to different theoretical calculations of the gluon-fusion process. The contributions from vector-boson fusion, ZH/WH and $t\bar{t}H$ are added to the distributions from gluon fusion. All theoretical calculations are normalised to the most precise SM inclusive cross-section predictions currently available. The *error bars* on the data points show the total (stat + syst) uncertainty, while the *grey bands* denote the systematic uncertainties. The bands of the theoretical predictions indicate the total uncertainty. Adapted from [84, 85]

These measurements probe the Higgs-boson kinematics, the jet activity produced in association with the Higgs boson, and the contributions of vector-boson fusion, as well as the spin, charge-conjugation and parity nature of the Higgs boson. In all cases the data are in good agreement with Standard Model expectations.

$H \to WW^* \to \ell\nu\ell\nu$

As discussed above, the dominant SM backgrounds in this channel are $WW^{(*)}$, $t\bar{t}$ and Wt production. The event selection proceeds as described above in Sect. 6.5, i.e. by requiring two isolated high-p_T leptons with a small opening angle accompanied by large missing transverse momentum. The signal sample is sub-divided into categories characterised by the number of reconstructed jets, (N_{jet}), which each receive significantly different contributions from the gluon fusion, vector-boson fusion and associated VH production. Given the relatively large signal sample, this channel provides good sensitivity for the separation of the production modes. The classification by jet multiplicity also allows control of the background from top-quark production, characterised by the presence of b-quark jets. Drell–Yan (DY) events ($pp \to Z/\gamma^* \to \ell\ell$) may be reconstructed with significant missing transverse momentum because of leptonic τ-lepton decays or the degradation of the E_T^{miss} measurement in the high pile-up environment of the 2012 data-taking. Finally, W+jets production in which a jet is reconstructed as a lepton, and the diboson processes $W\gamma$, W^*Z, WZ^* and ZZ^*, are also important backgrounds.

The considered event categories are ee, $e\mu$, and $\mu\mu$ final states without any additional jet, with one reconstructed jet, and with two reconstructed jets at high pseudo-rapidity with a high invariant mass. A third electron or muon or a pair of jets with an invariant mass around the W or Z mass are indicative of the VH production mode. In the $WH \to WWW$ channel, three charged leptons are required, but events with oppositely charged, same-flavour lepton pairs with an invariant mass near the Z-boson mass are rejected. The $ZH \to ZWW$ channel is reconstructed from events with a leptonic Z decay and one leptonically and one hadronically decaying W boson, where the invariant mass of the two jets from the hadronic decay is consistent with the W-boson mass. The resolution of the reconstructed Higgs-boson mass is much worse than in the high-resolution channels, but still reaches about 15 % for specially selected sub-categories.

The distribution of the transverse mass for events passing the full selection of the $H \to WW^* \to \ell\nu\,\ell\nu$ analysis in the ATLAS experiment [86], summed over all lepton flavours for final states with $N_{\text{jet}} \leq 1$, is shown in Fig. 6.18. The signal is clearly visible in the bottom panel that compares the background-subtracted data to the expected m_{T} distribution of a SM Higgs boson.

In the ATLAS analysis an excess of events relative to the background-only expectation is observed in the data with a maximum deviation occurring at $M_H = 130$ GeV. For $M_H = 125.36$ GeV, a significance of 6.1 σ is observed, compared with an

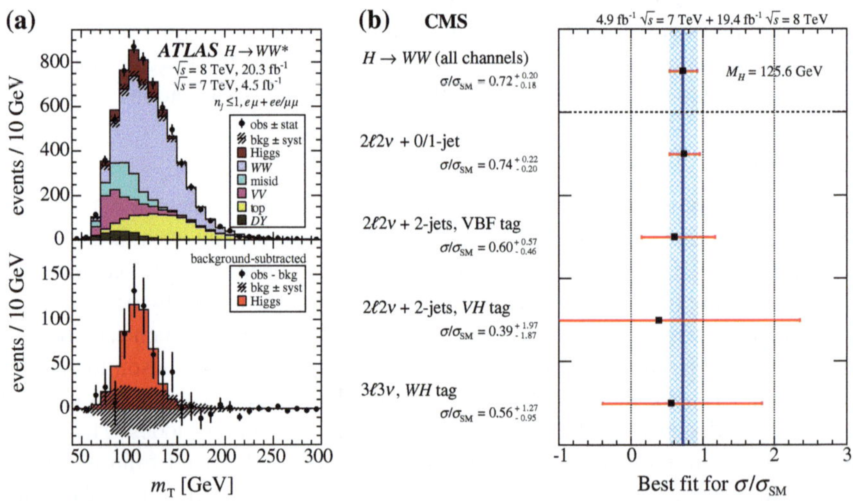

Fig. 6.18 **a** The distribution of the transverse mass for events passing the full selection of the $H \to WW^* \to \ell\nu\,\ell\nu$ analysis, summed over all lepton flavours for final states with $N_{\text{jet}} \leq 1$ in the ATLAS experiment. The *hatched area* represents the total uncertainty on the sum of the signal and background yields from statistical, experimental and theoretical sources. **b** Observed signal strength μ for each category in the $H \to WW$ analysis of the CMS experiment. Adapted from [86, 87]

expected value of $5.8\,\sigma$ for a SM Higgs boson. The corresponding signal strength is $\mu = 1.09^{+0.16}_{-0.15}\,(\text{stat})^{+0.17}_{-0.14}\,(\text{syst})$.

Given the relatively large signal rates in this channel, the signal strength can also be determined for the gluon-fusion and VBF production with reasonable accuracy. The results obtained in a simultaneous fit of μ_{ggF} and μ_{VBF} by the ATLAS collaboration are: $\mu_{\text{ggF}} = 1.02 \pm 0.19\,(\text{stat})^{+0.22}_{-0.18}\,(\text{syst})$ and $\mu_{\text{VBF}} = 1.27^{+0.44}_{-0.40}\,(\text{stat})^{+0.30}_{-0.21}\,(\text{syst})$. These results are consistent with the expectations for the Standard Model Higgs boson and provide evidence for Higgs-boson production via gluon fusion and vector-boson fusion in this channel alone.

The CMS collaboration measures a total signal strength of $\mu = 0.72^{+0.20}_{-0.18}$ at a significance of $4.3\,\sigma$ ($5.8\,\sigma$ expected) for a Higgs-boson mass of $M_H = 125.6\,\text{GeV}$ [87]. The measured signal strengths in the different event categories in the CMS analysis are shown in Fig. 6.18b.

6.6.2 Signals in Fermionic Decay Modes

In order to establish the mass generation for fermions as implemented in the SM, it is of prime importance to demonstrate the direct couplings of the Higgs boson to fermions and their proportionality to mass. The most prominent candidate decay modes are the decays into τ leptons, $H \to \tau\tau$, and into b quarks, $H \to b\bar{b}$. Given the overwhelming background from multi-jet production via QCD processes, the $b\bar{b}$ channel requires the restriction to Higgs bosons produced in association with vector bosons or $t\bar{t}$ pairs. In these cases, either the leptonic decay of the vector boson or of the top quark(s) may provide high-p_{T} leptons, such that the events can be triggered and the signal-to-background ratios improved. More favourable signal-to-background conditions are expected for $H \to \tau\tau$ decays.

The CDF and DØ collaborations have reported a combined excess of events in the search for the $H \to b\bar{b}$ decay mode [88]. This excess has a local significance of $2.8\,\sigma$ at a mass consistent with 125 GeV. Results on the search for both the $H \to \tau\tau$ and $H \to b\bar{b}$ decay modes at the LHC are discussed in the following.

$H \to \tau\tau$

In the search for $H \to \tau\tau$ decays all combinations of leptonic ($\tau \to \ell\nu\bar{\nu}$, with $\ell = e, \mu$) and hadronic ($\tau \to$ hadrons ν) tau decays are considered. The search is designed to be sensitive to the major production processes of a SM Higgs boson. The different production processes lead to different final-state signatures, which have been exploited by defining an event categorisation. Dedicated categories are considered to achieve both a good signal-to-background ratio and a good resolution for the reconstruction of the $\tau\tau$ invariant mass. A "VBF category", enriched in events produced via vector-boson fusion, is defined by the presence of two jets with a large separation in pseudo-rapidity.

The overall sensitivity is further enhanced by subdividing the 0- and 1-jet categories according to the reconstructed transverse momentum, p_{T}, of the Higgs-boson candidate giving rise to a "boosted category" for large transverse momentum. It is dominated by events produced via gluon fusion with additional jets from gluon radiation. Given the challenging signal-to-background conditions, and in order to exploit correlations between final-state observables, multivariate analysis techniques based on BDTs are used. The category without jets and with a low-p_{T} Higgs candidate has only a very low signal contribution and serves to control the background normalisation.

The main and largely irreducible $Z/\gamma^* \rightarrow \tau\tau$ background is modelled using $Z/\gamma^* \rightarrow \mu\mu$ events from data, where the muon tracks and associated energy depositions in the calorimeters are replaced by the corresponding simulated signatures of the final-state particles of the τ-lepton decay. In this approach, essential features such as the modelling of the kinematics of the produced boson, the modelling of the hadronic activity of the event (jets and underlying event) as well as contributions from pile-up are taken from data. Thereby the dependence on Monte Carlo simulation is minimised and only the well-understood τ decays and the detector response of the τ-lepton decay products are based on simulation. This procedure also ensures a very good description of time-dependent effects, arising e.g. from changing beam conditions or detector calibration. Other control samples from data are used to estimate backgrounds resulting from multi-jet or from W+jets production.

The analyses by the ATLAS and CMS collaborations [89, 90] of such mutually exclusive categories search for an excess in the signal bins of the output of a BDT classifier or for a broad excess in the reconstructed $\tau\tau$ mass distribution. The signal contamination from the $H \rightarrow WW^*$ channel has been removed by explicitly treating it as a background contribution.

In the ATLAS experiment the observed signal strength is determined from a global maximum-likelihood fit to the BDT output distributions in data. In Fig. 6.19a the expected and observed event numbers are shown, in bins of $\log_{10}(S/B)$. Here, S/B is the signal-to-background ratio calculated assuming $\mu = 1.4$ (see below) for each BDT bin. A clear excess is visible in the high-sensitivity bins. This excess is also observed in the reconstructed mass distribution, which is shown in Fig. 6.19b. The excess corresponds to a deviation from the background-only hypothesis of 4.5 σ, compared to an expected significance of 3.4 σ. The overall combination of all channels and categories results in a signal strength of $\mu = 1.43^{+0.27}_{-0.26}\,(\text{stat})^{+0.33}_{-0.27}\,(\text{syst})$ at the combined measured mass of $M_H = 125.36\,\text{GeV}$.

Results obtained by the CMS experiment are illustrated in Fig. 6.20. The distribution in Fig. 6.20a gives a visual impression of the observed excess. After combining all decays channels and production modes, clear evidence at the level of 3.2 σ above the background expectation is observed (3.7 σ expected). The signal strength is $\mu = 0.78 \pm 0.27$.

Figure 6.20b illustrates the determination of the Higgs-boson mass from $H \rightarrow \tau\tau$ decays alone in the CMS experiment. The result, $M_H = 122 \pm 7\,\text{GeV}$, is consistent with mass measurements from the high-resolution channels $H \rightarrow ZZ$ and $H \rightarrow \gamma\gamma$.

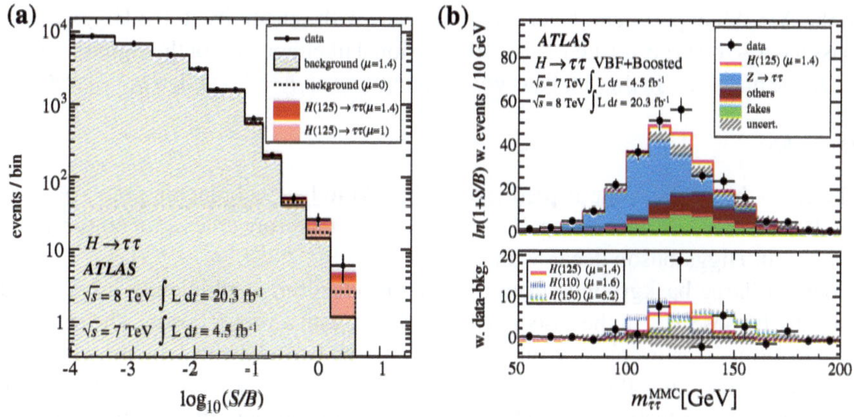

Fig. 6.19 Results on the search for $H \to \tau\tau$ decays in the ATLAS experiment. **a** Event yields as a function of $\log_{10}(S/B)$, where S (signal yield) and B (background yield) are taken from each bin in the BDT. The predicted background is obtained from the global fit (with $\mu = 1.4$), and signal yields are shown for $M_H = 125$ GeV, at $\mu = 1$ and $\mu = 1.4$ (the best-fit value). The background only distribution (*dashed line*), is obtained from the global fit, but fixing $\mu = 0$. **b** Weighted distribution ($w = \ln(1 + S/B)$) of the reconstructed $m_{\tau\tau}$ mass. The *bottom panel* shows the difference between weighted data events and weighted background events (*black points*), compared to the weighted signal yields. The background predictions are obtained from a global fit with the $M_H = 125$ GeV signal hypothesis ($\mu = 1.4$). The $M_H = 125$ GeV signal is plotted as a *solid line*, and, for comparison, the $M_H = 110$ GeV and $M_H = 150$ GeV signals are also shown. All signal strengths are set to their best-fit values. Adapted from [89]

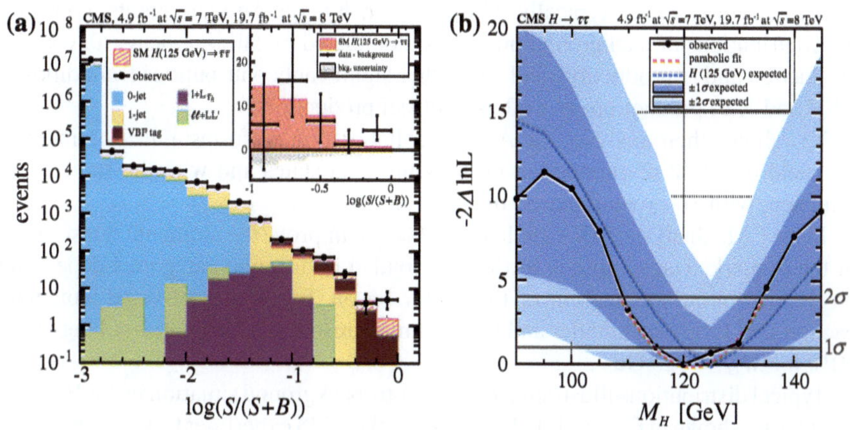

Fig. 6.20 Results on the search for $H \to \tau\tau$ decays in the CMS experiment. **a** Combination of the observed and predicted distributions of $\log_{10}(S/(S+B))$, calculated from the number of signal and background events, S and B, respectively, in each bin of the final discriminant for all decay modes and event categories. The *inset* shows the difference between the signal and the expected background. **b** Scan of the negative log-likelihood difference with profiling of all other parameters as a function of M_H. The background-only hypothesis includes the $H(125\,\text{GeV}) \to WW$ process. Adapted from [90]

In the ATLAS experiment, a cut-based analysis with sensitivity to the Higgs-boson mass has also been performed. A two-dimensional likelihood fit for the signal strength μ and the mass M_H in this analysis gives a minimum at $\mu = 1.4$ and $M_H = 125$ GeV.

$H \to b\bar{b}$

The $H \to b\bar{b}$ decay mode is predicted in the SM to have a branching ratio of 58 % for $M_H = 125$ GeV. Accessing $H \to b\bar{b}$ decays is therefore crucial for constraining the overall Higgs-boson decay width.

Due to large backgrounds from multi-jet production, a sufficiently large signal can only be observed in the associated production with a vector boson. The analyses are performed for events containing zero, one, or two charged leptons (electrons or muons), targeting the $Z \to \nu\nu$, $W \to \ell\nu$ or $Z \to \ell\ell$ decay modes of the vector boson, respectively. In addition to $Z \to \nu\nu$ decays, the 0-lepton channel has a small, but not insignificant, contribution from $W \to \ell\nu$, when the lepton is produced outside of the acceptance or not identified. b-tagging algorithms are used to identify jets consistent with the hypothesis of their originating from b quarks, as e.g. from $H \to b\bar{b}$ decays. To improve the sensitivity, the three channels are each split according to the vector-boson transverse momentum, the number of jets (two or three), and the number of b-tagged jets. Topological and kinematic selection criteria are applied within each of the resulting categories.

The analyses by the ATLAS and CMS collaborations [91, 92] use multivariate methods incorporating various kinematic variables in addition to the dijet mass, as well as b-tagging information, to provide the final discriminating variables. Dedicated control samples, typically with loosened b-tagging requirements, constrain the contributions of background processes, dominated by (W/Z)+heavy-flavour-jet production and $t\bar{t}$ production, and with other significant contributions from diboson (WZ and ZZ) production, as well as multi-jet production.

To validate their analysis, the ATLAS collaboration performs a measurement of the yield of $(W/Z)Z$ production in the same final states and with the same event selection, with $H \to b\bar{b}$ replaced by $Z \to b\bar{b}$.

In the CMS analysis, MVA methods are used to improve the momentum resolution of the b quark. The sample is further subdivided in different categories depending on the transverse momentum of the vector boson. For high transverse momenta, background levels are smaller and the dijet mass resolution is better, reaching about 10 % for $M_H = 125$ GeV.

Typical distributions illustrating the signal-to-background situation in the $H \to b\bar{b}$ channel are shown in Fig. 6.21. In Fig. 6.21a by the CMS experiment, events are sorted in bins of similar expected signal-to-background ratio (S/B), as given by the output of their corresponding BDT discriminant, trained assuming a Higgs-boson mass of 125 GeV. A comparison of the data with the expectation shows evidence for an excess in the signal-enhanced bins.

As an example from the ATLAS experiment, Fig. 6.21b shows the reconstructed $m_{b\bar{b}}$ distribution in data after subtraction of all backgrounds except for the diboson

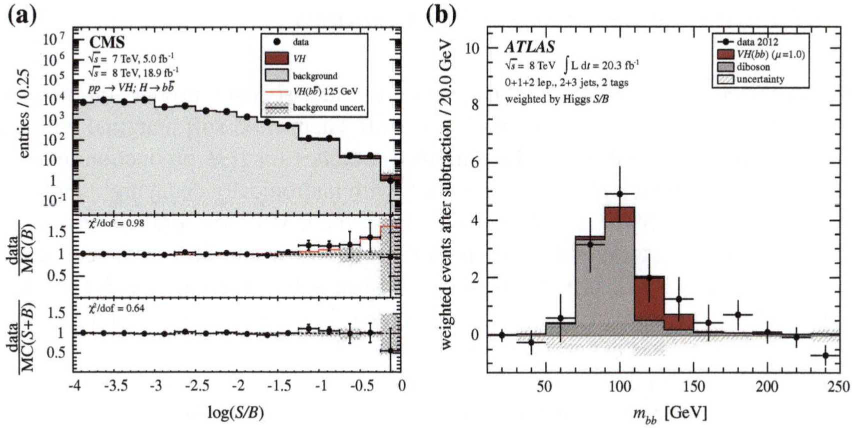

Fig. 6.21 Results on searches for $H \to b\bar{b}$ decays form a study of the associated VH production in the ATLAS and CMS experiments. **a** The distribution of $\log(S/B)$ for the combination of all channels studied in the CMS experiment. The two bottom *insets* show the ratio of the data to the background-only prediction (*above*) and to the predicted sum of background and a SM Higgs boson signal with a mass of 125 GeV (*below*). **b** The $m_{b\bar{b}}$ distribution as measured in the ATLAS experiment for data collected at $\sqrt{s} = 8$ TeV after subtraction of all backgrounds except for the diboson processes and for the associated WH and ZH production of a SM Higgs boson with $M_H = 125$ GeV. The contribution of the associated WH and ZH production is shown as expected for the SM cross section ($\mu = 1.0$). The size of the combined statistical and systematic uncertainty on the fitted background is indicated by the band. Adapted from [91, 92]

processes and for the associated WH and ZH production of a SM Higgs boson with $M_H = 125$ GeV. The contributions from all channels and data-taking periods are summed, weighted by their respective values of Higgs-boson signal-over-background ratio.

Binned maximum-likelihood fits are used to extract the signal yield and the background normalisations from the distributions of the final discriminant. The ATLAS distribution shows a clear signal from the expected diboson $(W/Z)Z$ production with $Z \to b\bar{b}$, from which the ratio of the observed signal yield to the SM expectation is found to be $0.74 \pm 0.09\,(\text{stat}) \pm 0.14\,(\text{syst})$, in agreement with the SM prediction. For a Higgs-boson mass of 125.36 GeV, the observed (expected) deviation from the background-only hypothesis corresponds to a significance of $1.4\,\sigma$ ($2.6\,\sigma$) and the ratio of the measured signal yield to the SM expectation is measured to be $0.52 \pm 0.32\,(\text{stat}) \pm 0.24\,(\text{syst})$ [92].

The CMS collaboration observes an excess in the data corresponding to $2.1\,\sigma$ for a mass of $M_H = 125$ GeV, at an expected sensitivity of also $2.1\,\sigma$. The measured signal strength relative to the one expected for the SM Higgs boson is 1.0 ± 0.5 [91].

The direct measurements of the Higgs-boson decays to fermions in the $\tau\tau$ and $b\bar{b}$ final states by the ATLAS and CMS experiments, as reported in this section, provide very strong evidence for the coupling of the discovered Higgs boson to "down-type" fermions. The combination of the $H \to \tau\tau$ and $H \to b\bar{b}$ final states by the CMS experiment results in a significance of $3.8\,\sigma$ [93].

6.6.3 Rare Production and Decay Channels

The expected sensitivity for the observation of the associated production of a SM Higgs boson with a pair of top quarks, the $t\bar{t}H$-channel, is still marginal with the current expected signal yields. In combined searches for $t\bar{t}H$ production, considering the decay modes $H \rightarrow b\bar{b}$, $H \rightarrow \tau\tau$ (with hadronically decaying τ leptons), $H \rightarrow \gamma\gamma$, $H \rightarrow WW^*$ and $H \rightarrow ZZ^*$ [94], the CMS collaboration observes a significant signal excess with a strength (normalised to the SM cross section) of $\sigma_{t\bar{t}H}/\sigma_{\rm SM} = 2.8 \pm 1.0$. The statistical significance of the observation is $3.4\,\sigma$, while the expected significance is $1.2\,\sigma$. The observed signal would represent an upward fluctuation of the SM expectation by $2.1\,\sigma$ and corresponds to a p-value under the signal hypothesis of 2 %.

In addition to the decay modes discussed above, searches for the decays $H \rightarrow Z\gamma$, with $Z \rightarrow \ell\ell$ [95, 96] and $H \rightarrow \mu\mu$ [97, 98] were performed. These channels have very low branching ratios in the SM, but new physics might lead to a significant enhancement, such that these channels might become visible. In the analyses of the ATLAS and CMS experiments no significant signal excesses were observed. The reported upper limits on the signal strengths by ATLAS (resp. CMS) at $M_H \simeq 125$ GeV are 11 (10) times the SM expectation for the $H \rightarrow Z\gamma$ decay, and 7.0 (7.4) times the SM expectation for the $H \rightarrow \mu\mu$ decay. The latter results correspond to an upper limit of 0.0015 (0.0016) on the branching ratio $\mathrm{B}(H \rightarrow \mu\mu)$, which is consistent with the small expected value of only $\mathrm{B}_{\rm SM}(H \rightarrow \mu\mu) = 2.2 \times 10^{-4}$. The CMS collaboration also measured the limit on $\mathrm{B}(H \rightarrow ee)$ to be smaller than 0.0019 . These results confirm that, indeed, the couplings of the observed state to light fermions are small, and, if compared to the results obtained for decays to pairs of τ leptons, that the leptonic couplings of the new boson are not flavour-universal.

Another important ingredient of our knowledge of the discovered signal is contributed by searches for invisible final states [99, 100], which were performed by both collaborations. Within the SM, decays to neutrinos via the ZZ channel constitute a well-known contribution which can be accounted for. However, searches for invisible decays are also sensitive to additional decay modes into new particles, which are predicted in some theories beyond the SM. A relatively clean signature is provided by the ZH production channel, where the Z boson decays to a pair of electrons or muons, and invisible decays of the Higgs boson lead to large missing transverse energy. The CMS analysis also considers decays of the Z boson to pairs of b quarks and makes use of the VBF production channel, which contributes particularly at high values of M_H. Since no excess above the expected backgrounds from SM processes are observed, 95 % CL upper limits on the cross section times branching ratio for a Higgs boson decaying into invisible particles are set. They are shown in Fig. 6.22. Assuming the SM production cross section for ZH production for a Higgs boson with a mass of 125 GeV, these limits can be translated into limits on the branching fraction to invisible particles of the discovered Higgs boson. They are found to be 75 % in the ATLAS experiment and 58 % in the CMS experiment.

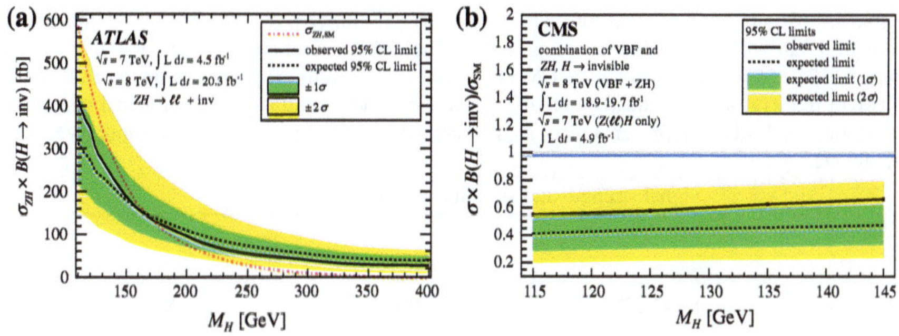

Fig. 6.22 **a** The observed (*solid*) and expected (*dashed*) 95 % CL production cross-section limits times branching ratio on the production of invisibly decaying Higgs bosons from the ATLAS experiment. The *dash-dotted curve* indicates the SM cross section for ZH production, i.e. assuming $\mathrm{B}(H \to \mathrm{inv.}) = 1$. **b** The observed and expected limits in the region around 125 GeV from the CMS experiment, normalised to the total cross section for Higgs-boson production in the SM. Adapted from [99, 100]

6.6.4 Summary of Results on Signal Strengths

The Higgs-boson signal has been clearly established by measurements at the LHC in decays to ZZ^*, $\gamma\gamma$, and WW^* final states, and evidence for decays to pairs of τ leptons has been obtained. The results of the ATLAS and CMS experiments on the signal strength μ and the significance z are summarised in Table 6.1.

There is good agreement with the SM expectation of $\mu = 1$ for all individual channels and also between the measurements performed by the two experiments. It should, however, be noted that for the CMS experiment these results have been extracted in the respective fits of the signal strength in each channel, and the likelihood maxima are reached at different mass values M_H. The numbers for the ATLAS experiment are given assuming a Higgs-boson mass of 125.36 GeV, corresponding to the measured combined mass value. The systematic uncertainties receive similar contributions from theoretical uncertainties on the signal cross section (QCD scale in the gluon-fusion production, PDFs and branching ratios) and all other, mainly experimental, sources.

The measurements of the signal strengths discussed so far do not give direct information on the relative contributions of the different production mechanisms. Therefore, the signal strengths of different production processes contributing to the same decay mode are determined, exploiting the sensitivity offered by the use of event categories in the analyses. The data are fitted separating vector-boson-mediated processes, VBF and VH, from gluon-mediated processes, ggF and $t\bar{t}H$, involving fermion (mainly top-quark) loops or legs. Two dedicated signal-strength modifiers, $\mu^f_{\mathrm{ggF}+t\bar{t}H} = \mu^f_{\mathrm{ggF}} = \mu^f_{t\bar{t}H}$ and $\mu^f_{\mathrm{VBF}+VH} = \mu^f_{\mathrm{VBF}} = \mu^f_{VH}$ are introduced for each final state considered. Contours for the different decay channels are illustrated in Fig. 6.23a.

Table 6.1 Compilation of measured signal strengths, $\mu = \sigma_{obs}/\sigma_{SM}$, and statistical significances, z, for different decay channels of the Higgs boson. The ATLAS numbers are given for a Higgs-boson mass of 125.36 GeV, whereas for the CMS experiment the results are quoted at the best-fit values. If two or more errors are indicated, the first is statistical, the second is the experimental systematic error, and the third one represents theoretical uncertainties. The results are taken from [80–83, 86, 87, 89–92]

Channel	Signal strength and significance values			
	ATLAS		CMS	
	μ	z	μ	z
$H \to ZZ$	$1.44^{+0.34}_{-0.31}{}^{+0.21}_{-0.11}$	8.1	$0.93^{+0.26}_{-0.23}{}^{+0.13}_{-0.09}$	6.8
$H \to \gamma\gamma$	$1.17 \pm 0.23\,^{+0.16}_{-0.11}$	5.2	$1.14 \pm 0.21^{+0.09}_{-0.05}{}^{+0.13}_{-0.09}$	5.7
$H \to W^+W^-$	$1.09^{+0.16}_{-0.15}{}^{+0.17}_{-0.14}$	6.1	$0.72^{+0.20}_{-0.18}$	4.3
$H \to \tau^+\tau^-$	$1.43^{+0.27}_{-0.26}{}^{+0.33}_{-0.27}$	4.5	0.78 ± 0.27	3.2
$VH \to H \to b\bar{b}$	$0.5 \pm 0.3 \pm 0.2$	1.4	1.0 ± 0.5	2.1

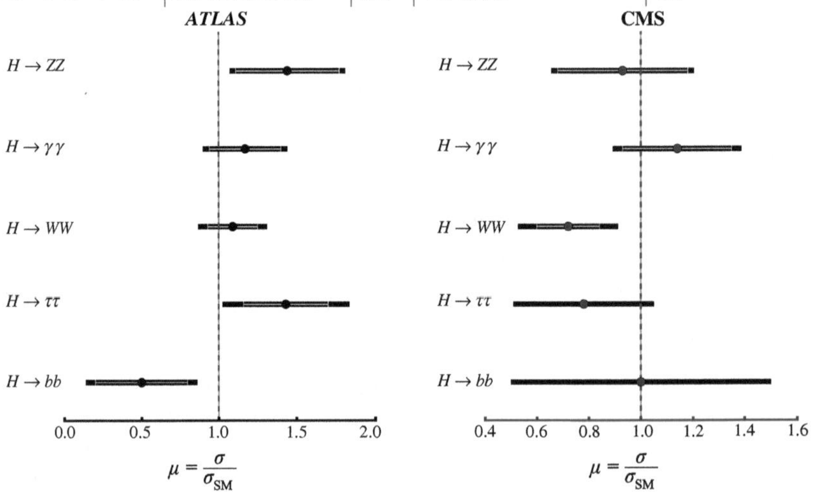

In a more model-independent way, i.e. without assumptions on Higgs-boson branching fractions, ratios of signal strengths were extracted. Results are illustrated in Fig. 6.23b. The numerical values are $\frac{\mu_{VBF+VH}}{\mu_{ggF+t\bar{t}H}} = 0.96^{+0.43}_{-0.31}$ for the ATLAS [115] and $1.25^{+0.62}_{-0.44}$ for the CMS [108] collaboration, consistent with the SM expectation.

Concentrating on the VBF channel alone, the ATLAS collaboration obtains $\frac{\mu_{VBF}}{\mu_{ggF}} = 1.00^{+0.46}_{-0.34}$. This is evidence at the 4.3 σ level that a fraction of Higgs boson production occurs through VBF.

More information on the coupling strengths is gained by a combined analysis of all channels, as discussed below in Sect. 6.8.4.

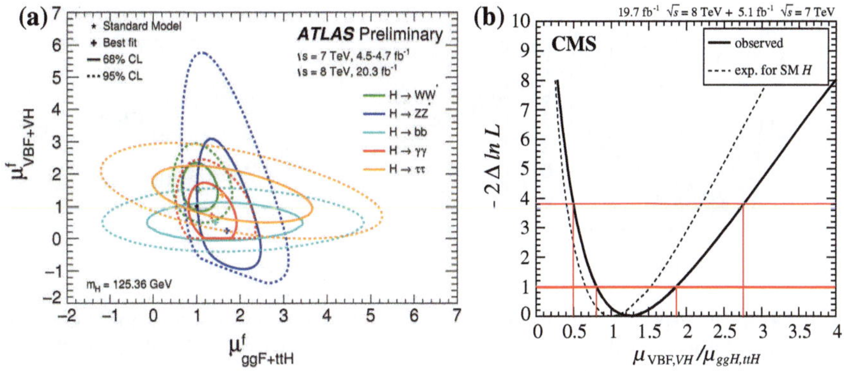

Fig. 6.23 **a** Likelihood contours in the $\mu^f_{\mathrm{ggF}+t\bar{t}H}$–$\mu^f_{\mathrm{VBF}+VH}$ plane for all final states by the ATLAS collaboration. Since at time of publication of the book published results were not yet available, preliminary results are displayed [115], which are expected to be very close to the final ones. **b** Likelihood scan of the ratio $\mu_{\mathrm{VBF}+V\ H}$ / $\mu_{\mathrm{ggF}+t\bar{t}H}$ derived from analogous measurements by the CMS collaboration. Adapted from [108, 115]

6.7 Searches for Additional Higgs Bosons

The particle discovered at a mass around 125 GeV may not be the only one of its kind. If the Higgs sector does not just consist of a single doublet, as in the Standard Model, it will give rise to additional Higgs states. Without any further assumptions, any additional Higgs-like particle has a priori unknown production rates and couplings. As mentioned above, large classes of extended Higgs sectors obey the sum rule that the sum of the squared couplings to gauge bosons of all neutral Higgs bosons add up to the SM value. Accordingly, in such extended Higgs sectors, the closer to the SM values the couplings of the state at 125 GeV to W and Z bosons turn out to be, the more suppressed would the couplings of all other neutral Higgs bosons to gauge bosons. In this context it should be noted that, a priori, it is by no means clear that the observed signal with $M_H = 125$ GeV should be the lightest state of an extended Higgs sector. This means that additional Higgs bosons could be heavier, but also lighter, than the state observed, where in the latter case a significant suppression of the couplings to gauge bosons could make such a light Higgs boson compatible with the existing limits from LEP and the Tevatron.

Several types of searches for additional Higgs bosons have been performed at the LHC. These searches have not shown any evidence for the existence of additional Higgs states so far, and the results of the searches have been reported either as cross-section limits or as limits on the parameter space of particular models.

Searches for decays to WW and ZZ final states are sensitive to heavy Higgs bosons, and the analysis strategies of these channels were refined accordingly. The interest is primarily to search for Higgs-like states with reduced couplings to gauge bosons and, correspondingly, a smaller total width than a SM-like Higgs boson of the same mass.

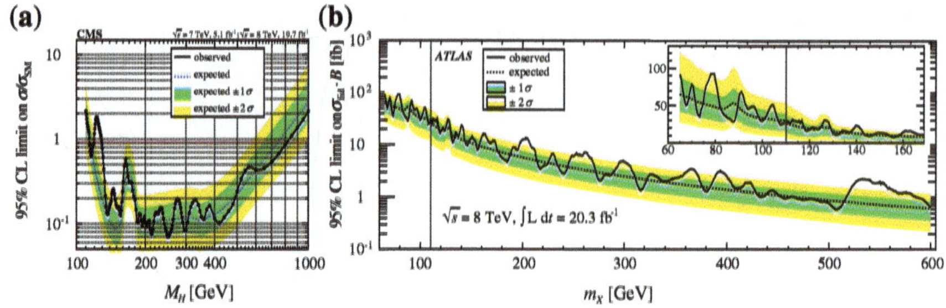

Fig. 6.24 **a** The expected and observed 95 % CL limits on the cross section for Higgs-boson production extracted from the $H \to ZZ \to 4\ell$ channel, normalised to the SM value, from the CMS collaboration. **b** The expected and observed 95 % CL limits on the cross section times branching ratio of additional Higgs-like states decaying to $\gamma\gamma$, from the ATLAS collaboration. The discontinuity in the limit at 110 GeV is due to the transition between a low-mass and a high-mass analysis. Adapted from [81, 101]

As an example, the observed 95 % CL cross-section limit, normalised to the production cross section of a SM Higgs boson, from the CMS experiment in the $H \to ZZ^* \to 4\ell$ channel [81] is shown in Fig. 6.24a. For a SM-like Higgs boson, the limit extends up to a mass of 832 GeV. In the mass range between about 200 and 400 GeV a limit on the cross section of about 10 % of the SM value is obtained. Similar analyses also exist for the WW channel and from the ATLAS experiment.

The ATLAS experiment has published results on searches for narrow resonances in the diphoton decay mode in the mass range 65–600 GeV. In these searches the discovered Higgs boson at about 125 GeV is treated as a background [101]. No evidence for an additional signal is observed, and the results on the excluded values of cross section times branching ratio are shown in Fig. 6.24b.

In supersymmetric models, Higgs-boson couplings to down-type quarks and charged leptons may be increased, while, as mentioned above, the couplings of additional Higgs states to gauge bosons are heavily suppressed if one of the Higgs bosons in the spectrum is SM-like. Therefore, searches for neutral supersymmetric Higgs bosons, generally denoted as Φ, produced in association with b jets in the $\tau\tau$, $\mu\mu$ and also $b\bar{b}$ decay modes are promising. Results obtained by the CMS collaboration [102] for the associated $b\bar{b}H$ production mode are shown in Fig. 6.25 as cross-section limits. Similar, albeit weaker limits have been derived for the production via gluon fusion. No significant deviation from the background expectation is observed in a wide mass range up to 1 TeV. Similar studies are performed by the ATLAS collaboration [103].

Charged Higgs bosons, $H^\pm$, also appear in models with an extended Higgs sector. Depending on the mass, decay modes to $\tau\nu$, cs and tb are dominant. Searches sensitive to the low-mass region were performed in $t\bar{t}$ events in the decay chain $t \to bH^+$, $H^+ \to \tau\nu$ [104, 105]. H^+ production in association with a top quark, $pp \to t(b)H^+ + X$, sensitive in the high-mass region, was also investigated [104]. The final state considered is characterised by the presence of a hadronic τ-lepton decay, $E_{\mathrm{T}}^{\mathrm{miss}}$, b jets, a hadronically decaying W boson, and the absence of any

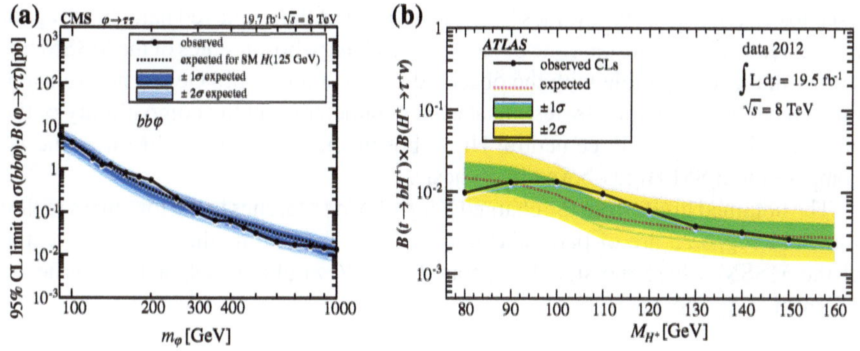

Fig. 6.25 Expected and observed 95 % CL limits on the production cross section times branching ratio for neutral and charged (supersymmetric) Higgs bosons. **a** CMS, neutral Higgs bosons produced in association with b quarks and decaying to τ pairs. **b** ATLAS, light charged Higgs boson produced in decays of top quarks, and decaying to $\tau\nu$. Adapted from [102, 104]

electrons or muons. The analysis by the ATLAS collaboration sets upper limits at the 95 % CL on the branching ratio $\mathrm{B}(t \to bH^+) \times \mathrm{B}(H^+ \to \tau\nu)$ between 0.23 and 1.3 % for a mass range of 80–160 GeV. For the mass range of 180–1000 GeV, upper limits are set on the production cross section times branching ratio between 0.0045 and 0.76 pb. The corresponding results for the low mass region are shown in Fig. 6.25b.

These results have been interpreted in the MSSM, and the corresponding exclusion limits in the M_A–$\tan\beta$ plane from the neutral Higgs-boson searches in the $\tau\tau$ decay mode are shown in Fig. 6.26. In this interpretation, the other MSSM parameters are

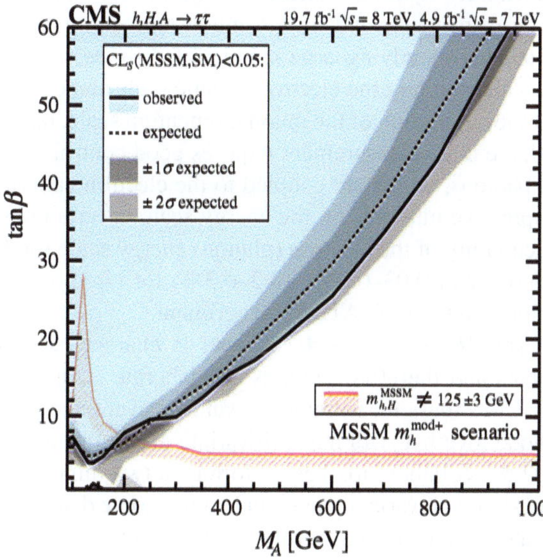

Fig. 6.26 Expected and observed exclusion limits at 95 % CL in the M_A–$\tan\beta$ parameter plane for the MSSM $m_h^{\text{mod}+}$ benchmark scenario [106] from CMS. A test of the compatibility of the data with a signal of the three neutral Higgs bosons h, H and A compared to a SM Higgs-boson hypothesis is performed. The region where the mass of either h or H is compatible with the observed signal at about 125 GeV is indicated. Adapted from [102]

fixed to certain benchmark values [106]. Over most of the displayed parameter region for the neutral Higgs bosons, one of the *CP*-even Higgs-boson states of the MSSM has a mass that is compatible with the observed signal at about 125 GeV. The exclusion limits have been obtained by testing at each parameter point the compatibility of the data to a signal of the three neutral Higgs bosons h, H and A (MSSM hypothesis) compared to a SM Higgs-boson hypothesis.

The obtained limits from the charged Higgs-boson searches in the low-mass region shown in Fig. 6.25 are of particular interest since they probe the parameter region of the MSSM where the signal at about 125 GeV would correspond to the heavy *CP*-even Higgs boson H.

6.8 Properties of the Discovered Higgs Boson

6.8.1 Measurement of the Higgs-Boson Mass

The Higgs-boson mass was the only remaining unknown parameter in the Standard Model prior to the discovery of the 125 GeV Higgs boson. The LHC collaborations have chosen a model-independent approach to measure the Higgs-boson mass based on fitting the spectra of the reconstructed invariant masses of the $H \rightarrow \gamma\gamma$ and $H \rightarrow ZZ^* \rightarrow 4\ell$ decay modes. In these two channels the Higgs boson shows up as a narrow mass peak with a typical experimental resolution of 1.6 to 2 GeV over a smooth background, from which the mass can be extracted without assumptions on the signal production and decay yields. Results on the Higgs-boson mass by the ATLAS [107] and CMS collaborations [81, 83, 108] are based on final calibrations which have reduced systematic uncertainties on the energy scale of the electromagnetic calorimeter and on the muon momentum scale. These improvements were obtained by studying large samples of $Z \rightarrow ee$, $J/\psi \rightarrow ee$ and $Z \rightarrow \ell\ell\gamma$ decays for the calibration of the electromagnetic scale and of $Z \rightarrow \mu\mu$ and $J/\psi \rightarrow \mu\mu$ decays for the calibration of the muon momentum scale, respectively. In the $H \rightarrow \gamma\gamma$ channel, the mass measurement requires good control of the energy scale and resolution of photons, which are coupled to the electron calibration. Based on these analyses impressive numbers on the energy scale uncertainties have been achieved, e.g. the uncertainty of the electron (photon) energy scale for E_T values around 40 GeV is of the order of 0.03–0.05 % (0.2–0.3 %) for rapidity values in the barrel and end-cap calorimeters of the ATLAS experiment.

The $H \rightarrow ZZ^* \rightarrow 4\ell$ channel is characterised by a small number of signal events and therefore requires an elaborate statistical treatment in the analysis. In the CMS mass measurement, event-by-event uncertainties are included in the likelihood together with the multivariate discriminants against background events. The ATLAS collaboration has performed a two-dimensional likelihood fit based on the four-lepton invariant mass and a multivariate discriminant. Care was taken to appropriately model the background in the vicinity of the signal peak.

The combination of the mass measurements in the two channels is done by combining sub-classes with different signal-to-background ratios and resolutions, taking correlated uncertainties into account. For the mass measurement this correlation results mainly from the uncertainty on the energy scale calibration with $Z \to ee$ events, which affects both the electron and photon energy scales. A simultaneous fit of signal strength and Higgs-boson mass to the invariant-mass spectra is performed. Likelihood scans with profiled signal strengths provide the value of the Higgs-boson mass in each channel and the combined mass in each experiment.

An overview of all measurements as well as the result of the combination is given in Table 6.2. The uncertainties are dominated by the statistical component and systematic uncertainties are dominated by experimental effects. Except for common energy-scale uncertainties between decay channels, the errors are therefore largely uncorrelated, and under this assumption the average value of the Higgs-boson mass is $M_H = 125.1 \pm 0.2$ GeV (private average by the authors).

Within the uncertainties the agreement between the ATLAS and CMS measurements is acceptable, with a χ^2 probability of the average of 7.5 %. This value of M_H is well consistent with the indirect determination from electroweak precision measurements, as described in Sect. 4.5.5.

Table 6.2 Compilation of the measured values of the Higgs-boson mass in the high-resolution channels and their combination in the ATLAS and CMS experiments [81, 83, 107]. The combined value in the last line of the table represents a private average by the authors

	Higgs mass (GeV)	
Channel	ATLAS	CMS
$H \to ZZ^* \to 4\ell$	124.51 ± 0.52 (stat) ± 0.06 (syst)	125.6 ± 0.4 (stat) ± 0.2 (syst)
$H \to \gamma\gamma$	125.98 ± 0.42 (stat) ± 0.28 (syst)	124.70 ± 0.31 (stat) ± 0.15 (syst)
$H \to ZZ^*, \gamma\gamma$ combined	125.36 ± 0.37 (stat) ± 0.18 (syst)	$125.02^{+0.26}_{-0.27}$ (stat)$^{+0.14}_{-0.15}$ (syst)
All combined	125.1±0.2 GeV	

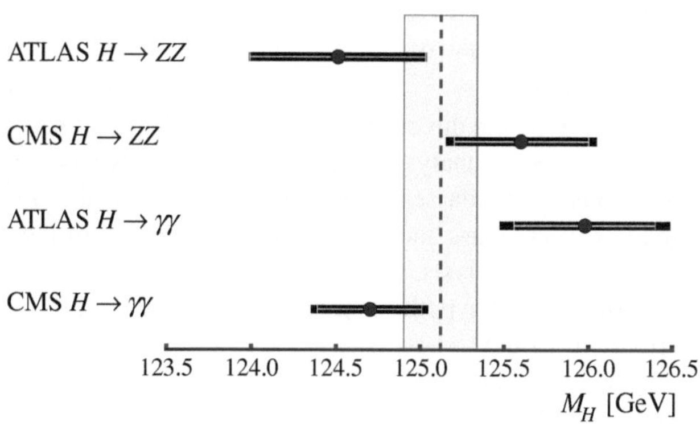

6.8.2 Measurement of Spin and CP Properties

An important property characterising a Higgs boson is its scalar nature. In the Standard Model the Higgs boson is a *CP*-even spin-0 state ($s^{CP} = 0^+$), where *CP* denotes the combined operation of parity (P) and charge conjugation (C). In theories beyond the SM also *CP*-odd states can occur in the Higgs sector, and in general there is also the possibility that a state is not a *CP* eigenstate but rather an admixture of *CP*-even and *CP*-odd components. Experimentally, under the assumption that the observed signal corresponds to a single resonance rather than an overlap of several states with different spins, the spin of the observed particle can be probed by testing distinct hypotheses for spin 0, 1 and 2. Concerning the *CP* properties, the test of distinct hypotheses is only applicable to the case where *CP* conservation is assumed, while in general the analysis needs to be performed for a continuous range of *CP* mixtures.

Information on spin and *CP* properties can be obtained in particular from the analysis of angular relations between either initial-state or final-state objects. The existence of certain production or decay modes already implies spin-*CP* information, as the on-shell production in gluon fusion of a spin-1 particle and its decay to two photons are not possible [78, 79]. In general, branching ratios and relative production rates in gluon fusion and $q\bar{q}$ annihilation are altered under different spin-*CP* assumptions.

There exist a variety of very general spin-2 models consistent with Lorentz invariance, angular-momentum conservation and gauge symmetry, and only a sub-set can reasonably be considered in a given analysis.

The $H \rightarrow ZZ^* \rightarrow 4\ell$ channel offers excellent sensitivity to the spin of the decaying particle by exploiting the angular distributions of the lepton pairs, thus using the decay products of the Z boson as spin analysers. Analyses of the s^{CP} properties [81, 109–111] based on testing distinct hypotheses have been presented by both the ATLAS and CMS experiments using the $H \rightarrow \gamma\gamma$, $H \rightarrow ZZ^* \rightarrow 4\ell$ and $H \rightarrow WW^* \rightarrow \ell\nu\ell\nu$ channels. The hypothesis tests are carried out by constructing a test-statistic, typically a negative log-likelihood ratio of the Standard Model and the alternative hypothesis, $q = -2\ln\frac{\mathcal{L}_{\text{alt}}}{\mathcal{L}_{0^+}}$, from a set of sensitive variables. These variables are chosen to be the direction of flight of the Higgs boson in the Collins–Soper frame in the $H \rightarrow \gamma\gamma$ analysis, five angles characterising the production and decays and the reconstructed masses of the Z and Z^* bosons in the $H \rightarrow ZZ^* \rightarrow 4\ell$ analysis, or the transverse mass and the azimuthal separation of the two leptons in the $H \rightarrow WW^* \rightarrow \ell\nu\ell\nu$ analysis.

The principle of the fixed hypothesis testing is illustrated in Fig. 6.27a, where the expected distributions of the test-statistic are shown for the 0^+ and 2^+ hypotheses for the $H \rightarrow \gamma\gamma$ analysis in the ATLAS experiment. The observed value of q in the data is consistent with the expectations for $s^{CP} = 0^+$ and disfavours the considered $s^{CP} = 2^+$ hypothesis. All hypotheses tested by the ATLAS experiment, namely $s^{CP} = 0^-, 1^+, 1^-$ and 2^+ models, are excluded at confidence levels above 97.8 %. This exclusion holds independently of the assumptions on the coupling strengths to the SM particles and, in the case of the $s^{CP} = 2^+$ model, of the relative fractions of

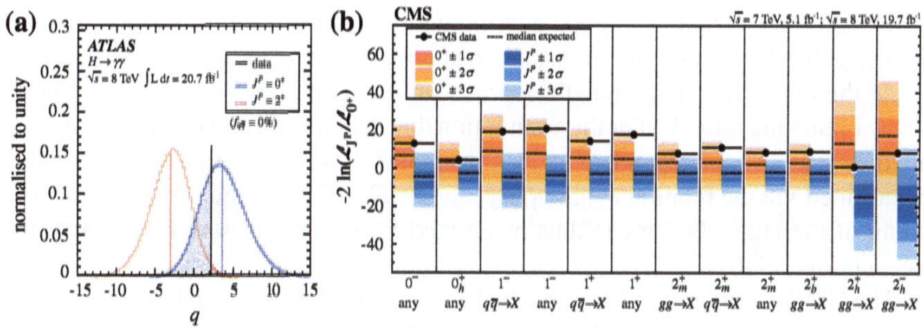

Fig. 6.27 Results on tests of different spin-*CP* hypotheses of the Higgs boson. **a** The expected distributions of the test-statistic for the 0^+ and 2^+ (produced by gluon fusion) hypotheses. The value observed in the data of the ATLAS experiment is indicated by a *continuous vertical line*. **b** Results of tests of different spin-*CP* hypotheses of the Higgs boson based on the $H \to ZZ^* \to 4\ell$ analysis of the CMS experiment. In each of the scenarios indicated, the *left distribution* refers to the 0^+ hypothesis, the *right* one to the tested alternative. Adapted from [81, 110]

gluon-fusion and quark-antiquark production of the spin-2 particle [110]. A summary of results from the CMS experiment based on the $H \to ZZ^* \to 4\ell$ channel is shown in Fig. 6.27b.

The tested spin-1 models assume an exotic vector or pseudo-vector particle decaying to two Z bosons, which must be different from the observed particle decaying to two photons at (almost) the same mass. The $2^+_{\rm m}$ model represents a massive, graviton-like boson suggested, for example, in models with extra dimensions, the other spin-2 models represent extensions or result from higher-dimension operators [81]. The pseudo-scalar and all spin-1 hypotheses tested by the CMS collaboration are excluded at a confidence level of 99.9 % or higher, while the tested spin-2 models are excluded at confidence levels exceeding 95 %.

Going beyond the test of distinct hypotheses, the CMS collaboration has published first results of analyses where *CP*-mixed states are addressed [81]. The presented limits on the fraction of a possible *CP*-odd contribution are still far above theoretical expectations, see Sect. 6.9.

6.8.3 Off-shell Higgs Couplings to the Z Boson

Off-shell production of Higgs bosons plays an important role for the $pp \to H^* \to ZZ^*$ and $pp \to H^* \to WW^*$ processes because of the large phase-space suppression of the branching ratios for the decays of on-shell Higgs boson with a mass of 125 GeV. The analysis of off-shell contributions offers a unique opportunity to measure off-shell couplings of the Higgs boson. In the SM, the ratio of the off-shell cross section, which integrates over the invariant-mass distribution above the threshold $m_{ZZ} = 2M_Z$, to the on-shell signal in the ZZ final state amounts to about 8 %.

The off-shell signal is, in good approximation, independent of the total width of the Higgs boson and provides information about the production and decay couplings, while the on-shell signal strength depends on the total Higgs-boson width via the decay branching ratio. Under the assumption that the on-shell and off-shell couplings are the same, i.e. not dependent on the invariant ZZ mass, the Higgs-boson width is determined via the relation $\mu_{\text{off-shell}}/\mu_{\text{on-shell}} = \Gamma_H/\Gamma_H^{\text{SM}}$. Upper limits on the total width of the Higgs boson can thus be derived from such measurements under this assumption [59–61].

The CMS collaboration has published an analysis of on-shell and off-shell contributions for $H \to ZZ^{(*)} \to 4\ell$ and $H \to ZZ^{(*)} \to 2\ell 2\nu$ [112]. The ATLAS collaboration has published a determination of the off-shell Higgs-boson signal strength in the high mass ZZ and WW final states [113]. The analyses take advantage of improved theoretical calculations of the resonance shape of a light Higgs boson [59] and its implications on the ZZ production rate. In this context also the destructive interference between Z-pair production and the high-mass tail of the Higgs contribution plays an important role, in particular at very high values of the invariant mass of the two Z bosons.

The CMS results are shown in Fig. 6.28. As can be seen in Fig. 6.28a, the effect from the Higgs-boson width is sizeable under this assumption after the suppression of background from $q\bar{q} \to ZZ$ by applying a cut on the "matrix element likelihood" (MELA). Events in which one of the Z bosons decays to neutrinos can be sufficiently well reconstructed at high mass values and also contribute.

Limits obtained on Γ_H in this context are shown in Fig. 6.28b. The combined limit is $\Gamma_H < 22$ MeV at 95 % CL, which corresponds to 5.4 times the SM Higgs-boson width at a value of the Higgs-boson mass of $M_H = 125.6$ GeV (as determined from the 4ℓ channel by the CMS collaboration), while the expected limit is 8.0 times the SM Higgs-boson width. The limit extracted by the ATLAS collaboration is $\Gamma_H < 22.7$ MeV at 95 % CL.

6.8.4 Higgs-Boson Couplings

In general, a modification of a coupling with respect to the SM prediction may give rise to a change of both the coupling strength and the tensor structure of the coupling, where the latter will affect the *CP* properties of the coupling. Accordingly, the tasks of determining the couplings and *CP* properties of the observed signal are closely related to each other. In order to reduce the complexity of the analyses for determining coupling properties during Run 1, a framework has been introduced by the LHC Higgs Cross Section Working Group [26, 114]. It consists of a parametrisation of small deviations from SM-like behaviour of the couplings for various benchmark scenarios, based on several simplifying assumptions, in particular:

- The signals observed in the different search channels are assumed to originate from a single resonance. Thus, the possibility that the signals could result from several overlapping states is not considered.

- The width of the observed particle is neglected, i.e. the zero-width approximation for this state is used. Hence the predicted rate for a given channel can be decomposed as:

$$\sigma \cdot \mathrm{B}(i \to \mathrm{H} \to f) = \frac{\sigma_i \cdot \Gamma_f}{\Gamma_\mathrm{H}} ,$$

  where σ_i is the production cross section through the initial state i, B and Γ_f are the branching ratio and partial decay width into the final state f, respectively, and Γ_H the total width of the Higgs boson.
- Only modifications of coupling strengths are considered, while the tensor structure of the couplings is assumed to be the same as in the Standard Model. This implies in particular that the observed state is assumed to be a *CP*-even scalar.

The coupling scale factors κ_j are defined in such a way that the cross sections σ_j and the partial decay widths Γ_j associated with the SM particle j scale with κ_j^2 compared to the SM prediction [26, 114]. With this notation, and with κ_H^2 being the scale factor for the total Higgs-boson width Γ_H, the cross section for the $gg \to H \to \gamma\gamma$ process, for example, can be expressed as:

$$\frac{\sigma \cdot \mathrm{B}\,(gg \to H \to \gamma\gamma)}{\sigma_{\mathrm{SM}}(gg \to H) \cdot \mathrm{B}_{\mathrm{SM}}(H \to \gamma\gamma)} = \frac{\kappa_g^2 \cdot \kappa_\gamma^2}{\kappa_H^2}.$$

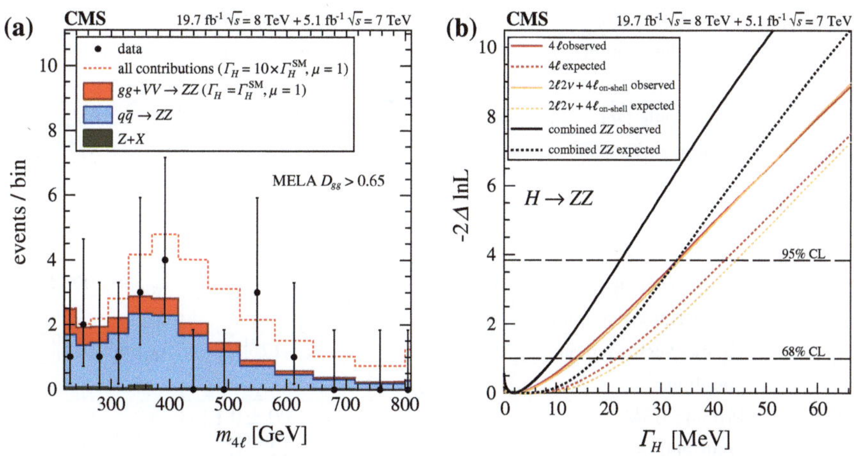

Fig. 6.28 Constraint on the total width of the Higgs boson, Γ_H, by the CMS experiment from a combination of contributions of on-shell and off-shell Higgs-boson production in the $H \to ZZ^* \to 4\ell$ channel under the assumption that the far-off-shell couplings of the Higgs are equal to the on-shell couplings. **a** Distribution of the invariant mass of oppositely charged pairs of leptons originating from the decay of a pair of Z bosons, after applying a cut on the matrix element likelihood to suppress the background from $q\bar{q} \to ZZ$. The expected signal is shown for a Higgs-boson width as expected in the SM and for a value 10 times as large. **b** Likelihood scan versus Γ_H for all measurements of the high-to-low-mass ZZ cross-section ratio. Adapted from [112]

By definition, the currently best available SM predictions for all $\sigma \cdot B$, including higher-order QCD and electroweak corrections, are recovered when all $\kappa_j = 1$.

Since the LHC measurements always involve a combination of Higgs-boson couplings from the production and the decay processes and since there is limited access to the total Higgs-boson width (see above), without further theoretical assumptions only ratios of couplings can be measured at the LHC, rather than absolute values. Results on a set of coupling scale factors of interest, $\{\kappa_j\}$, or, with fewer assumptions, ratios of scale factors are extracted from fits to the data using the profile likelihood ratio $q(\{\kappa_j\})$ according to (6.1). The different benchmark scenarios correspond to different assumptions on which of the couplings are fixed to their SM values and for which of them deviations from their SM strengths are considered. If for the loop-induced processes $H \to \gamma\gamma$ and $gg \to H$ it is assumed that there are no loop contributions from BSM ("beyond SM physics") particles, and if for the total width it is assumed that there are no additional invisible or undetectable decay modes into BSM particles, the scale factors κ_H, κ_g and κ_γ can be expressed in terms of the scale factors entering via the SM-type contributions, namely κ_W, κ_Z, κ_t, κ_b and κ_τ (only the dominant fermion contributions are indicated here for simplicity).

Results obtained by the ATLAS and CMS experiments are described in the following paragraphs and summarised in Table 6.3.

Table 6.3 Fitted values of the scale factors or ratios of scale factors for couplings in the ATLAS [115] and CMS [108] experiments based on the full data set collected during Run 1 corresponding to about 5 fb^{-1} at $\sqrt{s} = 7$ TeV and about 20 fb^{-1} at $\sqrt{s} = 8$ TeV

Scenario	Model description	Fit parameter	Results of the fits	
			ATLAS	CMS
1	Single scale factors for fermion and boson couplings	κ_V, κ_F	$\kappa_V = 1.09 \pm 0.07$ $\kappa_F = 1.11^{+0.17}_{-0.15}$	$\kappa_V = 1.01 \pm 0.07$ $\kappa_F = 0.87^{+0.14}_{-0.13}$
2	Custodial symmetry	λ_{WZ}, (κ_F, κ_Z)	$\lambda_{WZ} = 1.00^{+0.15}_{-0.11}$	$\lambda_{WZ} = 0.92^{+0.14}_{-0.12}$
3	Up-type versus down-type fermions couplings	λ_{du}, (λ_{Vu}, κ_{uu})	$\lambda_{du} = 0.90^{+0.15}_{-0.15}$	$\lambda_{du} = 0.99^{+0.19}_{-0.18}$
4	Quarks versus lepton couplings	$\lambda_{\ell q}$, (λ_{Vq}, κ_{qq})	$\lambda_{\ell q} = 1.12^{+0.22}_{-0.18}$	$\lambda_{\ell q} = 1.03^{+0.23}_{-0.21}$
5	Probing BSM contributions	κ_γ, κ_g, B_{inv}	$\kappa_\gamma = 1.00 \pm 0.12$ $\kappa_g = 1.12^{+0.14}_{-0.11}$ $B_{inv} < 0.27$ (95 % CL)	 $B_{inv} < 0.32$ (95 % CL)

Test of Couplings to Fermions and Vector Bosons

In a first fit scenario a common scale factor is assumed for all couplings to fermions, κ_F ($= \kappa_t = \kappa_b = \kappa_\tau$), and another common scale factor is assumed for the couplings to W and Z, κ_V ($= \kappa_W = \kappa_Z$). It is furthermore assumed that the $H \to \gamma\gamma$ and $gg \to H$ loops and the total Higgs-boson width depend only on κ_F and κ_V, with no contributions from BSM particles.

Figure 6.29 shows the results of the fits to the data for individual channels as well as for their combination for the ATLAS and CMS experiments. Since only the relative sign of κ_F and κ_V is physical, $\kappa_V > 0$ is assumed in the following. Some sensitivity to this relative sign is provided by the negative interference between the W-boson loop and top-quark loop in the $H \to \gamma\gamma$ decay. In both experiments the data prefer the minimum with positive relative sign, which is consistent with the SM prediction. The data are compatible with the expectations from the SM, and the obtained results for both experiments are consistent with the SM values of $\kappa_F = \kappa_V = 1$. The most important direct constraint on the fermion coupling arises from the $H \to \tau\tau$ decay channel. The obtained results of the coupling fits receive similar contributions from statistical and systematic uncertainties.

In this benchmark model, the assumption of no contributions from new particles to the Higgs-boson width provides strong constraints onthe fermion coupling κ_F,

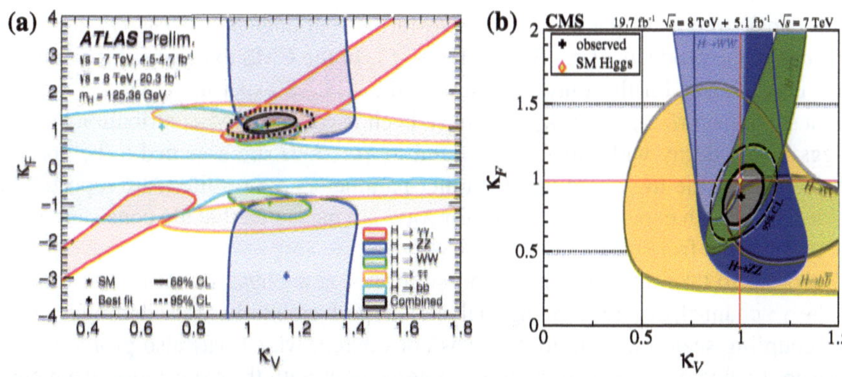

Fig. 6.29 Results of fits for the two-parameter benchmark model that probes different coupling strength scale factors for fermions and vector bosons, assuming only SM contributions to the total width, to $H \to \gamma\gamma$ and $gg \to H$. The values of the coupling scale factors κ_F and κ_V for individual channels as well as for their combination are displayed together with the 68 % CL contours, as measured in the ATLAS (**a**) and CMS (**b**) experiments. The best fit values and the SM expectation are also indicated. Since at time of publication of the book published results from the ATLAS collaboration were not yet available, preliminary results are displayed, which are expected to be very close to the final ones. Adapted from [108, 115]

as about 75 % of the total SM width comes from decays to fermions or involving fermions. If this assumption is relaxed, only the ratio $\lambda_{FV} = \kappa_F/\kappa_V$ can be measured, which still provides useful information on the relationship between Yukawa and gauge couplings.

Custodial Symmetry of the W and Z Couplings

In the SM, custodial symmetry imposes the constraint that the W and Z bosons have related couplings to the Higgs boson at lowest order, $g_{HVV} \sim m_V^2/v$, where v is the vacuum expectation value of the Higgs field, and that $\rho = M_W^2/(M_Z^2 \cdot \cos^2\theta_W)$, where θ_W is the weak mixing angle, is equal to unity. In extensions of the Higgs sector consisting of additional Higgs doublets and/or Higgs singlets, the relation $\rho = 1$ is maintained, and the squared couplings to gauge bosons of the neutral Higgs boson of the model sum up to the SM value (see above). The constraint on the ratio of the Higgs-boson couplings to W and Z is tested by measuring the ratio $\lambda_{WZ} = \kappa_W/\kappa_Z$.

A sensitive measurement can be obtained by considering the observed event yields in the $H \to WW^*$ and $H \to ZZ^*$ channels in combination with information from WH and ZH production, from the VBF process (which in the SM is mediated roughly via 75 % W fusion and 25 % Z fusion) and from the $H \to \gamma\gamma$ decay mode, since within the SM the decay branching ratio receives a dominant contribution from the W loop.

In the fits carried out by the ATLAS and CMS Collaborations the ratio λ_{WZ} has been considered as fit parameter and two further parameters have been taken into account in the products of $\sigma \cdot \mathrm{B}(i \to \mathrm{H} \to f)$. In the CMS experiment κ_F and κ_Z have been chosen for this purpose, while in the ATLAS experiment $\lambda_{FZ} = \kappa_F/\kappa_Z$ and a scale factor $\kappa_{ZZ} = \kappa_Z\kappa_Z/\kappa_H$ have been used. Non-SM contributions to the Higgs-boson decay width to $H \to \gamma\gamma$ and to $gg \to H$ are assumed to be absent. The results of the fit, where the two other parameters are profiled, are reported in Table 6.3. In both experiments values for λ_{WZ} are measured that are consistent with the expectation from custodial symmetry.

In order to take into account the possibility of new physics contributions to the $H \to \gamma\gamma$ channel, a corresponding analysis was performed where an additional effective coupling scale factor ratio $\lambda_{\gamma Z}$ was considered, which was also profiled. The measured value of λ_{WZ} is found to be in agreement with the expectation from custodial symmetry, regardless of the inclusion of the $H \to \gamma\gamma$ channel as an indirect constraint on κ_W.

Relations Within the Fermion-Coupling Sector

Many extensions of the Standard Model contain different couplings of the Higgs boson to up-type and down-type fermions. One example are two-Higgs-doublet models [116, 117], among which the MSSM is the most prominent example. In more general two-Higgs-doublet models the couplings of leptons can be different from those of quarks. Within the MSSM, a non-universality of this kind can occur as a consequence of potentially large higher-order corrections to the relation between the bottom-quark mass and the bottom-Yukawa coupling, while the corresponding relation for the τ lepton is less affected by higher-order contributions. These possible deviations are suppressed in the "decoupling region" of extended Higgs sectors, where the lightest Higgs state in the model behaves SM-like.

In order to test such modifications of fermion couplings the experiments have performed fits where the ratios between coupling scale factors for up-type and down-type quarks, $\lambda_{du} = \kappa_d/\kappa_u$, and for leptons and quarks, $\lambda_{\ell q} = \kappa_\ell/\kappa_q$, are the relevant fit parameters. The other free parameters that are allowed to vary from their SM values in those fits are summarised together with the results of the fits in Table 6.3.

The obtained results are consistent with the SM predictions and do not provide any evidence for asymmetries between up-type and down-type or lepton and quark couplings, respectively.

Probing Contributions Beyond the Standard Model

Many BSM physics scenarios predict the existence of new particles which can contribute in a significant way to loop-induced processes such as $gg \to H$ production and $H \to \gamma\gamma$ decays if they are sufficiently light. In addition, these particles may or may not contribute to the total width of the observed Higgs boson through direct invisible decays or decays into undetectable final states, i.e. final states that cannot be distinguished from the background. In the more general latter case the changes to the total width can be taken into account via both effective scale factors κ_γ and κ_g for the $\gamma\gamma$ decay and gluon-gluon production loops, respectively, and an invisible branching ratio $\mathrm{B}_{\mathrm{inv}}$, assuming that there are no additional undetectable decay modes and that all tree-level couplings are fixed to their SM values. The total Higgs-boson width can in this case be parametrised as

$$\Gamma_H = \frac{\kappa_H^2(\kappa_\gamma, \kappa_g)}{(1 - \mathrm{B}_{\mathrm{inv}})} \Gamma_H^{\mathrm{SM}} .$$

No significant deviations in the effective couplings to photons and gluons have been observed in this kind of fits. The results of the fits are reported in Table 6.3. The data allow upper limits on the branching ratio to invisible particles to be set under these assumptions.

Summary

The ATLAS and CMS Collaborations have used the complete data set collected during Run 1 to test the compatibility of the measured yields in the $H \to \gamma\gamma$, $H \to ZZ^* \to 4\ell$, $H \to WW^* \to \ell\nu\ell\nu$, $H \to \tau\tau$, and $H \to b\bar{b}$ channels with the predictions for the SM Higgs boson under various benchmark assumptions, thereby probing features of the couplings of the Higgs boson. A summary of the measurements of coupling scale factors in the different benchmark models is given in Table 6.3.

In addition to the benchmark models discussed so far, the experiments have considered more generic models where the scale factors to W, Z, t, b and τ are treated independently, and effective coupling scale factors κ_γ and κ_g are introduced that account for particles propagating in the $gg \to H$ production and $H \to \gamma\gamma$ decay loops. Without the assumption on the total width, only ratios of coupling scale factors can be measured. Ratios with respect to κ_Z and a total width scale factor, $\kappa_{gZ} = \kappa_g \cdot \kappa_Z / \kappa_H$, are considered, resulting in seven free parameters in the fit. The results of the fit are graphically displayed in Fig. 6.30. Within the present uncertainties, all results are in good agreement with the expectations for the SM Higgs boson. This fit represents the most model-independent determination of coupling scale factors at the LHC that is currently possible.

Assuming that the Higgs boson does not couple to particles other than the heavy fermions and bosons of the SM, the coupling strengths to W and Z bosons and to t, b, τ and μ can be extracted and their scaling with particle mass can be tested. This is shown in Fig. 6.31 in Sect. 6.10.

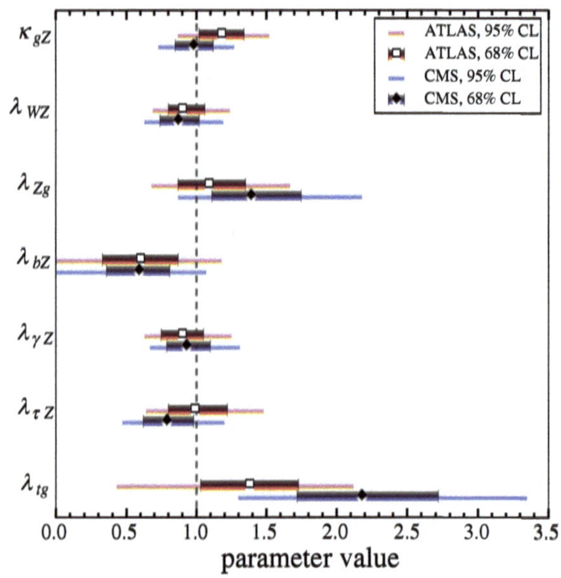

Fig. 6.30 Summary of determinations of ratios of coupling-strength modifiers by the ATLAS and CMS collaborations at their respective values of the Higgs-boson mass ($\kappa_{gZ} = \kappa_g \kappa_Z / \kappa_H$, $\lambda_{ab} = \kappa_a / \kappa_b$). The measured parameter values are taken from [108, 115]

6.9 A Critical Look on the Interpretation of the Observed Higgs-Boson Signal

In the following we discuss how the experimental results obtained so far for the observed signal can be interpreted. We begin with a brief discussion of the different property measurements listed above, where the achieved level of accuracy and the underlying theoretical assumptions of the analyses are addressed. Based on this discussion, we briefly comment on the compatibility of the experimental results with different scenarios of electroweak symmetry breaking.

6.9.1 Mass of the Observed Particle

The mass measurement has already reached the remarkable relative precision of better than 0.2 %. The measured value of about 125 GeV is well consistent with the indirect prediction within the SM based on the other electroweak precision measurements, as described in Sect. 4.5.5.

While a high-precision measurement of this fundamental parameter is of interest in its own right, M_H is also needed as a crucial input parameter for Higgs-boson physics. As can be seen in Fig. 6.1, in particular the branching ratios $\mathrm{B}(H \to ZZ^* \to 4\ell)$ and $\mathrm{B}(H \to WW^* \to \ell\nu\ell\nu)$ depend very sensitively on the precise numerical value of the Higgs-boson mass. This is due to the phase-space suppression far below the threshold for on-shell production of WW and ZZ. Within the SM, an experimental error of 0.2 GeV of the Higgs-boson mass induces parametric uncertainties of 2.5 % for $\mathrm{B}(H \to ZZ^* \to 4\ell)$ and 2.2 % for $\mathrm{B}(H \to WW^* \to \ell\nu\ell\nu)$. Consequently, a further improvement of the accuracy of the mass of the detected signal will be instrumental for reducing parametric theoretical uncertainties in the Higgs sector.

6.9.2 Spin and CP Properties

As discussed in Sect. 6.8.2, most experimental analyses of spin and *CP* properties carried out so far have been concerned with the test of distinct hypotheses for spin and *CP*-states in comparison with the *CP*-even, spin-0 hypothesis corresponding to the Higgs boson of the SM. All tested alternatives have been excluded at high levels of significances, while good compatibility with the data has been found for the $s^{CP} = 0^+$ hypothesis.

The modelling of possible scenarios with spin 2 is non-trivial in this context. As mentioned above, there is in principle a large variety of possible spin-2 models, while at the same time care is necessary to ensure that the discriminating power against the tested hypothesis actually refers to the spin properties rather than to some unphysical behaviour of the spin-2 implementation. As a consequence, it is difficult to draw overall conclusions that apply to all possible manifestations of spin-2 scenarios.

The test of distinct hypotheses is sufficient for determining the spin of the observed particle if the discovered signal corresponds to a single resonance rather than an overlap of several states with different spins. Concerning the *CP* properties, a priori the detected signal could consist of any admixture of *CP*-even and *CP*-odd components (where a pure *CP*-odd state is strongly disfavoured by the analyses discussed above).

The analyses carried out so far are mainly based on observables involving the coupling of the new state to two gauge bosons, HVV, where $V = W, Z$, in particular $H \to ZZ^* \to 4\ell$, $H \to WW^* \to \ell\nu\ell\nu$ and Higgs-boson production in vector-boson fusion. The angular and kinematic distributions in these processes will only provide sensitivity for a discrimination between *CP*-even and *CP*-odd properties if a possible *CP*-odd component of the new state couples with sufficient strength to WW and ZZ. However, in a large variety of models with *CP* violation in the Higgs sector there is no lowest-order coupling between a pseudoscalar A and a pair of gauge bosons, so that the AVV coupling is strongly suppressed compared to the coupling of the *CP*-even component. In this case, the angular and kinematic distributions will show hardly any deviation from the expectation of a pure *CP*-even state, even if the state had a sizeable *CP*-odd component [26]. The difference between a pure *CP*-even state and a state that is a mixture of *CP*-even and *CP*-odd components would rather manifest itself as a reduction of the total rate. However, such a reduction in rate could also be caused by other effects (and there could even be a compensation with other contributions leading to an enhancement of the rate).

As a consequence, only rather weak bounds exist so far on a possible admixture of *CP*-even and *CP*-odd components. Limits have been presented by the CMS collaboration [81] on the fraction of the *CP*-odd contribution, $f_{a3} = |a_3|^2\sigma_3/(\sum |a_i|^2\sigma_i)$, where the summation in the denominator runs over all *CP*-even and *CP*-odd contributions to the production cross section. It should be noted that even a relatively large *CP*-odd component would only result in a small effect in the quantity f_{a3}, if the coefficient $|a_3|$ is loop-suppressed. Accordingly, an extremely high precision in f_{a3} would be needed to probe possible deviations from the SM. While the 95 % CL limit published in [81] is at the level of about 0.5, an accuracy of 10^{-5} is set as a target on f_{a3} in [118].

Channels involving only Higgs-boson couplings to fermions could potentially provide a significantly higher sensitivity for analysing *CP* properties, since in this case the *CP*-even and *CP*-odd components can have the same magnitude.

6.9.3 Constraints on the Total Width

Within the SM, the total width of the Higgs boson with a mass around $M_H = 125$ GeV is expected to be about 4.2 MeV, which is several orders of magnitude below the experimental resolution even in channels with a high resolution on the reconstructed

Higgs-boson mass like $H \to 4\ell$ and $H \to \gamma\gamma$, and only comparatively weak limits can be put on the natural width of the Higgs boson.

While at e^+e^- colliders it is possible to make an absolute measurement of the total production cross section $e^+e^- \to ZH$, making use of Z-boson decays into e^+e^- and $\mu^+\mu^-$ and reconstructing the recoil against the Z boson independently of the decay properties of the Higgs boson, all LHC channels rely on the identification of Higgs-boson decay products. Since the LHC measurements always involve a combination of Higgs-boson couplings from the production and the decay process, the total Higgs-boson width cannot be measured in these analyses without additional assumptions.

It has recently been noted [61] that constraints on the total width can in principle be obtained from the different dependence of on-peak and off-peak contributions on the total width in Higgs-boson decays to ZZ^*. Also interference effects between signal and background in Higgs-boson decays to $\gamma\gamma$ provide some sensitivity to the total width [119–122].

The analyses described in Sect. 6.8.3 make the assumption that the on-shell Higgs-boson couplings are the same as the couplings far off-shell, however, the relation between the couplings can be severely affected by new-physics contributions, in particular via threshold effects. In fact, new-physics effects of this kind may be needed to give rise to a Higgs-boson width that differs from the one of the SM by the currently probed amount, see also the discussion in [123]. The constraints obtained in these analyses therefore do not represent a model-independent bound on the total Higgs-boson width, but they provide an important consistency test of the SM.

It should also be noted that the destructive interference between the signal and background contributions, which is related to the unitarity cancellations caused by the Higgs-boson exchange contribution, limits the sensitivity to the total Higgs-boson width via this method in the region where $\Gamma_H / \Gamma_H^{\mathrm{SM}} \approx 1$. It will therefore be difficult to reach the level of the SM width in this way, even with high energy and integrated luminosity.

6.9.4 Couplings to Gauge Bosons and Fermions

The analysis of the couplings of the observed signal to gauge bosons and fermions has so far been carried out within a framework for coupling scale factors, based on certain theoretical assumptions. In this way ratios of scale factors or, with an additional assumption on the total width of the Higgs boson, the scale factors themselves have been determined in different benchmark scenarios.

The analysis of the signal strengths in the different observed channels (Sect. 6.6.4) and the analysis of the couplings (Sect. 6.8.4) overall show good agreement with the SM predictions at the present level of accuracy. While these results set important constraints on possible realisations of physics beyond the SM, on the other hand many alternatives to the SM predict deviations in the Higgs-boson couplings from the

SM values that are significantly smaller than the currently reached level of precision, see e.g. the discussion in [118].

An important goal for extending the present analyses will be to take into account the possibility that a modification of a coupling can affect the coupling strength and the tensor structure of the coupling, i.e. the *CP* properties, at the same time. This implies that the exploration of the couplings and of the spin and *CP* properties have to be treated together within a coherent framework.

Such a more general framework for parametrising possible deviations from the SM can be provided by an effective Lagrangian [26]. Under the assumption that new physics appears only at a scale Λ that is much larger than M_H, $\Lambda \gg M_H \sim 125$ GeV, deviations from the SM can be parametrised in a systematic way via an expansion in inverse powers of Λ. A suitable effective Lagrangian framework needs to be sufficiently general (in particular, it should not assume a priori that the observed state is a *CP*-even scalar, as it is done in the present framework), while on the other hand the number of involved parameters needs to be practically feasible. Furthermore, it will be crucial that in the limit $\Lambda \to \infty$ the predictions obtained within the effective Lagrangian framework recover the best available SM prediction, including all relevant higher-order QCD and electroweak corrections.

If the predictions of the effective Lagrangian framework are confronted not only with the measurements of total rates but also with differential distributions, it will be important to carefully monitor the range of validity of the effective Lagrangian. The (non-renormalisable) effective operators generated by expanding in inverse powers of Λ will in general have a bad high-energy behaviour if they are probed at p_{T} values that are outside the range of validity of the effective field theory (which is based on the assumption that Λ is much larger than any other involved scale). The effects of an unphysical behaviour of the effective operators therefore need to be clearly distinguished from a true sensitivity to deviations from the SM. Furthermore, the effective Lagrangian framework is not valid in the presence of relatively light particles of physics beyond the SM. It will therefore be important to complement analyses that are carried out in the effective Lagrangian framework with analyses in specific models of BSM physics. In this way it can also be investigated in how far deviations from the SM behaviour that are probed in the effective Lagrangian framework can actually be realised in different models. The impact of relatively light particles of new physics, for instance colour-neutral states for which the direct-production cross sections at the LHC are low, can best be studied in specific models. The conclusions drawn from the analysis of specific models may therefore be significantly different from those obtained from the effective Lagrangian framework.

6.9.5 Higgs-Boson Self-Couplings

The Higgs-boson self-couplings are essentially unconstrained so far. Information on these couplings is crucial for establishing the BEH mechanism and for getting experimental access to the Higgs potential. Access to the trilinear self-coupling of the SM-like state at 125 GeV, HHH, via double Higgs-boson production processes will require an enormous amount of integrated luminosity. Another way of probing triple-Higgs couplings would be the decay of a heavier Higgs boson into a pair of lighter ones, which can occur in extended Higgs sectors.

The quartic Higgs-boson self-coupling currently seems to be out of reach of any foreseeable experimental facility.

6.9.6 Vector-Boson Scattering

Another experimental way to study the physics of electroweak symmetry breaking is the investigation of vector-boson scattering (see Sect. 4.7). If the WW-scattering cross section is not completely unitarised by the exchange of one or more Higgs-like states, this will lead to the onset of strongly interacting effects in the gauge boson sector.

6.9.7 Compatibility of the Experimental Results with Different Scenarios of Electroweak Symmetry Breaking

As discussed above, the properties of the discovered particle, within experimental and theoretical uncertainties, are so far compatible with the interpretation in terms of the SM Higgs boson. However, a wide range of other interpretations of the discovered particle is also possible, corresponding to very different underlying physics (see also Chap. 11 for a discussion of some of these scenarios, especially composite Higgs models). This includes the possibility that the observed state is composite or that it is an admixture or shares properties with other states of new physics. In the context of extended Higgs sectors the signal at about 125 GeV can be interpreted as the lightest state of the Higgs sector, but interpretations involving at least one lighter Higgs state below 125 GeV, having significantly suppressed couplings to gauge bosons as compared to the SM case, are also possible. The sensitivity for discriminating among the different possible interpretations correlates with the achievable precision in confronting the experimental results with the theory predictions.

Many extended Higgs theories have a parameter region corresponding to a "decoupling" limit, where the lightest Higgs scalar becomes SM-like, i.e. in particular it couples essentially with SM strength to gauge bosons. In this parameter region the other Higgs states are relatively heavy and decouple from the gauge bosons. This implies that in the search for BSM Higgs bosons the decay modes into ZZ and WW as well as the weak-boson fusion and VH associated production modes may be heavily suppressed.

While within the SM the mass of the Higgs boson is a free parameter, in many extensions of the SM the mass of the discovered signal can directly be confronted with the model predictions. In the MSSM with TeV-scale masses in the scalar top sector an upper bound on the mass of the lightest Higgs state of about 135 GeV is obtained, which is well compatible with the observed value of the signal. It is interesting to note that interpreting the observed signal as the light *CP*-even Higgs boson of the MSSM leads to a lower bound on the mass of the *CP*-odd Higgs boson, M_A, of about 200 GeV. This allowed range of M_A corresponds to the decoupling region of the MSSM. This implies that the properties of the state at 125 GeV are expected to be SM-like, and that one generically would not have expected any deviations from SM-like properties in the LHC measurements of the new particle carried out so far. In fact, a significant improvement of the accuracy of the Higgs-boson coupling determinations would be needed to probe scenarios of this kind.

6.10 Conclusions and Outlook

The spectacular discovery of a Higgs boson by the ATLAS and CMS experiments constitutes clearly the most significant result of the first data-taking period at the LHC. With this discovery an important milestone in the investigation of electroweak symmetry breaking has been reached, and particle physics has entered a new era.

Based on the analysis of the complete data set collected during Run 1, the bosonic decay modes $H \to ZZ^* \to 4\ell$, $H \to \gamma\gamma$ and $H \to WW^* \to \ell\nu\ell\nu$ have been consolidated and measured with high significance, and strong evidence for decays into fermions via the $H \to \tau\tau$ and $H \to b\bar{b}$ decay modes has meanwhile been established. Within the current uncertainties, all measurements are in good agreement with the predictions for the Higgs boson of the Standard Model.

The mass of the discovered boson is measured, with an excellent precision, to be 125.1 ± 0.2 GeV, which is in agreement with the expectations based on the analysis of electroweak precision observables. The observed spin and *CP* properties are found to be consistent with the Standard Model value of $s^{CP} = 0^+$, while there is limited sensitivity at present to discriminate a pure *CP*-even state from an admixture of *CP*-even and *CP*-odd components.

The couplings to bosons and fermions are found to scale with the particle mass as predicted by the BEH mechanism. This proportionality to the particle mass, as illustrated in Fig. 6.31, is strikingly different from the behaviour of the known gauge interactions, which exhibit universality within the three particle families of quarks and leptons.

With the forthcoming Run 2 of the LHC, both the higher centre-of-mass energy of 13 to 14 TeV and the higher instantaneous luminosity will allow these measurements to be continued with higher precision. Although the discovered signal is so far compatible with a Standard Model-like Higgs boson, a variety of other interpretations, corresponding to very different underlying physics, is still possible. The discovered Higgs boson might be a portal to new physics, and its precise investigation therefore is of prime importance.

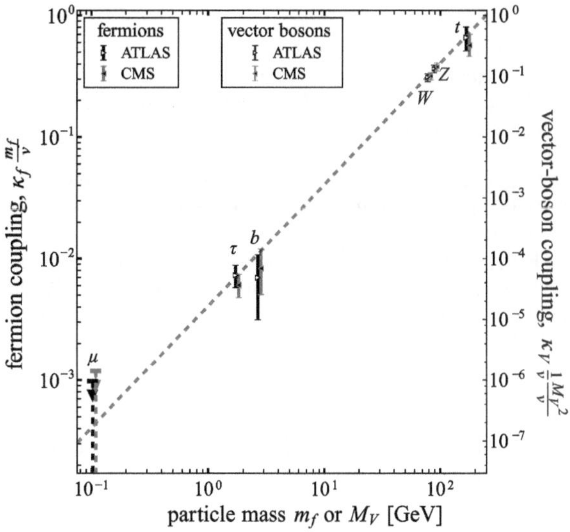

Fig. 6.31 Illustration of the mass dependence of the Higgs-boson couplings, calculated from the result of fits by the ATLAS and CMS collaborations with six free coupling-strength modifiers, see Sect. 6.8.4; particle masses and the value of the vacuum expectation value, $v = 246\,\text{GeV}$, are taken from [124]. For muons, the 95 % CL limits are shown, for all other particles the error bars correspond to 68 % confidence regions. The overall scales of the couplings are different for fermions (*left*) and bosons (*right*) and are fixed by the assumption that the Higgs boson only couples to SM particles

While the minimal Higgs sector of the SM predicts just a single physical Higgs boson, a generic prediction of extended Higgs sectors is the presence of additional Higgs boson states. The search for additional Higgs bosons, which may be heavier but also lighter than the observed state at about 125 GeV, and the investigation of vector-boson scattering are key research areas for the ultimate understanding of electroweak symmetry breaking. The longer-term goals include the measurement of the Higgs-boson self-coupling. To reach the required precision and to interpret the results, large experimental efforts are necessary, which must be accompanied by corresponding theoretical work. The precise exploration of the Higgs sector will constitute a rich physics programme for the high-luminosity upgrade of the LHC and for future e^+e^- and pp colliders.

References

1. Y. Nambu, G. Jona-Lasinio, Phys. Rev. **122**, 345–358 (1961)
2. F. Englert, R. Brout, Phys. Rev. Lett. **13**, 321 (1964)
3. P.W. Higgs, Phys. Lett. **12**, 132 (1964)
4. P.W. Higgs, Phys. Rev. Lett. **13**, 508 (1964)
5. G.S. Guralnik, C.R. Hagen, T.W.B. Kibble, Phys. Rev. Lett. **13**, 585 (1964)
6. P.W. Higgs, Phys. Rev. **145**, 1156 (1966)

7. T.W.B. Kibble, Phys. Rev. **155**, 1554 (1967)
8. S.L. Glashow, Nucl. Phys. **22**, 579 (1961)
9. S. Weinberg, Phys. Rev. Lett. **19**, 1264 (1967)
10. A. Salam, in *Elementary Particle Physics: Relativistic Groups and Analyticity*, ed. by N. Svartholm. Proceedings of the Eighth Nobel Symposium (Almqvist & Wiskell, Stockholm, 1968), p. 367
11. B.W. Lee, C. Quigg, H. Thacker, Phys. Rev. Lett. **38**, 883–885 (1977)
12. B.W. Lee, C. Quigg, H. Thacker, Phys. Rev. D **16**, 1519 (1977)
13. G. Degrassi et al., JHEP **08**, 098 (2012)
14. ALEPH, DELPHI, L3, OPAL and SLD Collaborations, the LEP Electroweak Working Group, the SLD Electroweak and Heavy Flavour Groups, Phys. Rep. **427**, 257 (2006)
15. ALEPH, DELPHI, L3 and OPAL Collaborations, the LEP Electroweak Working Group, Phys. Rep. **532**, 119 (2013)
16. S. Heinemeyer, W. Hollik, G. Weiglein, L. Zeune, JHEP **12**, 084 (2013)
17. LHC Higgs Cross Section Working Group (2011), arXiv:1101.0593, https://twiki.cern.ch/twiki/bin/view/LHCPhysics/CrossSections
18. LHC Higgs Cross Section Working Group (2012), arXiv:1201.3084, https://twiki.cern.ch/twiki/bin/view/LHCPhysics/CrossSections
19. J.J. Hernandez et al., Phys. Lett. B **239**(3–4) (1990)
20. Particle Data Group, Chin. Phys. C **38**, 090001 (2014)
21. LEP Working Group for Higgs boson searches, ALEPH Collaboration, DELPHI Collaboration, L3 Collaboration, OPAL Collaboration, Phys. Lett. B **565**, 61–75 (2003)
22. T. Junk, Nucl. Instrum. Methods A **434**, 435 (1999)
23. A.L. Read, J. Phys. G **28**, 2693 (2002)
24. CDF and DØ Collaborations, Phys. Rev. Lett. **104**, 061802 (2010)
25. CDF and DØ Collaborations, Phys. Rev. D **88**, 052014 (2013)
26. LHC Higgs Cross Section Working Group (2013), arXiv:1307.1347, https://twiki.cern.ch/twiki/bin/view/LHCPhysics/CrossSections
27. E. Bagnaschi, G. Degrassi, P. Slavich, A. Vicini, JHEP **02**, 088 (2012)
28. H. Mantler, M. Wiesemann, Eur. Phys. J. C **73**, 2467 (2013)
29. M. Grazzini, H. Sargsyan, JHEP **09**, 129 (2013)
30. A. Djouadi, M. Spira, P. Zerwas, Phys. Lett. B **264**, 440–446 (1991)
31. S. Dawson, Nucl. Phys. B **359**, 283–300 (1991)
32. D. Graudenz, M. Spira, P. Zerwas, Phys. Rev. Lett. **70**, 1372–1375 (1993)
33. M. Spira, A. Djouadi, D. Graudenz, P. Zerwas, Nucl. Phys. B **453**, 17–82 (1995)
34. R.V. Harlander, W.B. Kilgore, Phys. Rev. Lett. **88**, 201801 (2002)
35. C. Anastasiou, K. Melnikov, Nucl. Phys. B **646**, 220–256 (2002)
36. V. Ravindran, J. Smith, W.L. van Neerven, Nucl. Phys. B **665**, 325–366 (2003)
37. R.V. Harlander, Phys. Lett. B **492**, 74–80 (2000)
38. S. Marzani et al., Nucl. Phys. B **800**, 127–145 (2008)
39. S. Catani, D. de Florian, M. Grazzini, P. Nason, JHEP **07**, 028 (2003)
40. U. Aglietti, R. Bonciani, G. Degrassi, A. Vicini, Phys. Lett. B **595**, 432–441 (2004)
41. G. Degrassi, F. Maltoni, Phys. Lett. B **600**, 255–260 (2004)
42. S. Actis, G. Passarino, C. Sturm, S. Uccirati, Phys. Lett. B **670**, 12–17 (2008)
43. D. de Florian, G. Ferrera, M. Grazzini, D. Tommasini, JHEP **11**, 064 (2011)
44. M. Ciccolini, A. Denner, S. Dittmaier, Phys. Rev. Lett. **99**, 161803 (2007)
45. M. Ciccolini, A. Denner, S. Dittmaier, Phys. Rev. D **77**, 013002 (2008)
46. K. Arnold et al., Comput. Phys. Commun. **180**, 1661–1670 (2009)
47. T. Figy, S. Palmer, G. Weiglein, JHEP **02**, 105 (2012)
48. P. Bolzoni, F. Maltoni, S.O. Moch, M. Zaro, Phys. Rev. Lett. **105**, 011801 (2010)
49. I.W. Stewart, F.J. Tackmann, Phys. Rev. D **85**, 034011 (2012)
50. F.J. Tackmann, J.R. Walsh, S. Zuberi, Phys. Rev. D **86**, 053011 (2012)
51. T. Han, S. Willenbrock, Phys. Lett. B **273**, 167–172 (1991)
52. O. Brein, A. Djouadi, R. Harlander, Phys. Lett. B **579**, 149–156 (2004)

53. M. Ciccolini, S. Dittmaier, M. Kramer, Phys. Rev. D **68**, 073003 (2003)
54. G. Ferrera, M. Grazzini, F. Tramontano, Phys. Rev. Lett. **107**, 152003 (2011)
55. A. Denner, S. Dittmaier, S. Kallweit, A. Muck, JHEP **03**, 075 (2012)
56. W. Beenakker et al., Phys. Rev. Lett. **87**, 201805 (2001)
57. A. Bredenstein, A. Denner, S. Dittmaier, S. Pozzorini, JHEP **08**, 108 (2008)
58. G. Bevilacqua, M. Czakon, C. Papadopoulos, M. Worek, Phys. Rev. Lett. **104**, 162002 (2010)
59. N. Kauer, G. Passarino, JHEP **08**, 116 (2012)
60. G. Passarino, C. Sturm, S. Uccirati, Nucl. Phys. B **834**, 77–115 (2010)
61. F. Caola, K. Melnikov, Phys. Rev. D **88**, 054024 (2013)
62. J.M. Campbell, R.K. Ellis, C. Williams, JHEP **04**, 060 (2014)
63. ATLAS and CMS Collaborations, LHC Higgs Combination Group, in *Procedure for the LHC Higgs Boson Search Combination in Summer 2011*. ATL-PHYS-PUB 2011–11, CMS NOTE 2011/005 (2011), http://cdsweb.cern.ch/record/1379837
64. G. Cowan, K. Cranmer, E. Gross, O. Vitells, Eur. Phys. J. C **71**, 1554 (2011)
65. ATLAS Collaboration, ATLAS detector and physics performance: Technical design report, 2, CERN, Geneva, Switzerland, 1999, https://atlas.web.cern.ch/ATLAS/GROUPS/PHYSICS/TDR/access.html
66. CMS Collaboration, J. Phys. G **34**, 995–1579 (2007)
67. ATLAS Collaboration, Expected Performance of the ATLAS Experiment - Detector, Trigger and Physics, CERN-OPEN-2008-020 (2009)
68. D.L. Rainwater, D. Zeppenfeld, JHEP **12**, 005 (1997)
69. D.L. Rainwater, D. Zeppenfeld, K. Hagiwara, Phys. Rev. D **59**, 014037 (1998)
70. T. Plehn, D.L. Rainwater, D. Zeppenfeld, Phys. Rev. D **61**(093), 005 (2000)
71. D.L. Rainwater, D. Zeppenfeld, Phys. Rev. D **60**, 113004 (1999)
72. N. Kauer, T. Plehn, D.L. Rainwater, D. Zeppenfeld, Phys. Lett. B **503**, 113–120 (2001)
73. CMS Collaboration, Phys. Lett. B **710**, 26 (2012)
74. ATLAS Collaboration, Phys. Lett. B **710**, 49 (2012)
75. ATLAS Collaboration, Phys. Lett. B **716**, 1 (2012)
76. CMS Collaboration, Phys. Lett. B **716**, 30 (2012)
77. CMS Collaboration, JHEP **06**, 081 (2013)
78. L. Landau, Dokl. Akad. Nauk Ser. Fiz. **60**, 207–209 (1948)
79. C.N. Yang, Phys. Rev. **77**, 242–245 (1950)
80. ATLAS Collaboration, Phys. Rev. D **91**, 012006 (2015)
81. CMS Collaboration, Phys. Rev. D **89**, 092007 (2014)
82. ATLAS Collaboration, Phys. Rev. D **90**, 112015 (2014)
83. CMS Collaboration, Eur. Phys. J. C **74**(10), 3076 (2014)
84. ATLAS Collaboration, Phys. Lett. B **738**, 234–253 (2014)
85. ATLAS Collaboration, JHEP **09**, 112 (2014)
86. ATLAS Collaboration (2014), arXiv:1412.2641
87. CMS Collaboration, JHEP **2014**(1), 96 (2014)
88. CDF and DØ Collaborations, Phys. Rev. Lett. **109**, 071804 (2012)
89. ATLAS Collaboration (2015), arXiv:1501.04943
90. CMS Collaboration, JHEP **05**, 104 (2014)
91. CMS Collaboration, Phys. Rev. D **89**, 012003 (2014)
92. ATLAS Collaboration, JHEP **01**, 069 (2015)
93. CMS Collaboration, Nat. Phys. **10** (2014)
94. CMS Collaboration, JHEP **09**, 087 (2014)
95. ATLAS Collaboration, Phys. Lett. B **732**, 8–27 (2014)
96. CMS Collaboration, Phys. Lett. B **726**, 587–609, 29 p (2013)
97. ATLAS Collaboration, Phys. Lett. B **738**, 68–86 (2014)
98. CMS Collaboration (2014), arXiv:1410.6679
99. ATLAS Collaboration, Phys. Rev. Lett. **112**, 201802 (2014)
100. CMS Collaboration, Eur. Phys. J. C **74**(8), 2980 (2014)
101. ATLAS Collaboration, Phys. Rev. Lett. **113**(17), 171801 (2014)

102. CMS Collaboration, JHEP **10**, 160 (2014)
103. ATLAS Collaboration, JHEP **11**, 056 (2014)
104. ATLAS Collaboration, JHEP **03**, 088 (2015)
105. CMS Collaboration, JHEP **07**, 143 (2012)
106. M. Carena et al., Eur. Phys. J. C **73**, 2552 (2013)
107. ATLAS Collaboration, Phys. Rev. D **90**, 052004 (2014)
108. CMS Collaboration (2014), arXiv:1412.8662
109. CMS Collaboration, Phys. Rev. Lett. **110**, 081803 (2013)
110. ATLAS Collaboration, Phys. Lett. B **726**, 120–144 (2013)
111. CMS Collaboration (2014), arXiv:1411.3441
112. CMS Collaboration, Phys. Lett. B **736**, 64 (2014)
113. ATLAS Collaboration, arXiv:1503.01060
114. LHC Higgs Cross Section Working Group (2012), arXiv:1209.0040
115. ATLAS Collaboration, ATLAS-CONF-2015-007
116. H.P. Nilles, Phys. Rep. **110**, 1–162 (1984)
117. H.G. Haber, G.L. Kane, Phys. Rep. **117**, 75 (1985)
118. S. Dawson et al. (2013), arXiv:1310.8361
119. L.J. Dixon, M.S. Siu, Phys. Rev. Lett. **90**, 252001 (2003)
120. S.P. Martin, Phys. Rev. D **86**, 073016 (2012)
121. S.P. Martin, Phys. Rev. D **88**(1), 013004 (2013)
122. L.J. Dixon, Y. Li, Phys. Rev. Lett. **111**, 111802 (2013)
123. C. Englert, M. Spannowsky, Phys. Rev. D **90**, 053003 (2014)
124. Particle Data Group, K. Olive et al., Chin. Phys. C **38**, 090001 (2014)

Chapter 7
Top-Quark Physics at the LHC

Kevin Kröninger, Andreas B. Meyer and Peter Uwer

Abstract The top quark is the heaviest of all known elementary particles. It was discovered in 1995 by the CDF and DØ experiments at the Tevatron. With the start of the LHC in 2009, an unprecedented wealth of measurements of the top-quark's production mechanisms and properties have been performed by the ATLAS and CMS collaborations, most of these resulting in smaller uncertainties than those achieved previously. At the same time, huge progress was made on the theoretical side yielding significantly improved predictions up to next-to-next-to-leading order in perturbative QCD. Due to the vast amount of events containing top quarks, a variety of new measurements became feasible and opened a new window to precisions tests of the Standard Model and to new-physics contributions.

7.1 Introduction

Top quarks have been a subject of scientific research ever since Kobayashi's and Maskawa's speculations about a third family of quarks in the context of solving the problem of weak *CP* violation in the early 1970s [1]. After a two-decade long period of searches at various colliders and experiments, the top quark was finally discovered in 1995 by the CDF and DØ experiments at Fermilab's Tevatron, a proton-antiproton collider, operated at the time at a centre-of-mass energy of $\sqrt{s} = 1.80\,\text{TeV}$ [2, 3] (see Sect. 1.2.4). Since then, and in particular after the upgrade of the Tevatron to a centre-of-mass energy of 1.96 TeV, pioneering precision measurements were performed at this machine. Prime examples are the measurements of the total cross

K. Kröninger (✉)
Technische Universität Dortmund, Physik E IV, 44221 Dortmund, Germany
e-mail: kevin.kroeninger@tu-dortmund.de

A.B. Meyer
DESY, Notkestr. 85, 22607 Hamburg, Germany
e-mail: andreas.meyer@desy.de

P. Uwer
Humboldt-Universität zu Berlin, Newtonstr. 15, 12489 Berlin, Germany
e-mail: peter.uwer@physik.hu-berlin.de

T. Schörner-Sadenius (ed.), *The Large Hadron Collider*,
DOI 10.1007/978-3-319-15001-7_7

section for single-top and top-quark pair ($t\bar{t}$) production and the measurements of the top-quark mass (see [4, 5] and references therein). Currently, both CDF and DØ are in the process of publishing their legacy measurements based on data sets corresponding to an integrated luminosity of about $10\,\mathrm{fb}^{-1}$ collected by each experiment during Tevatron Run 2—and these measurements set a standard for the LHC.

A new era in experimental top-quark physics was marked by the start of the LHC in 2009. At the LHC, top quarks are produced abundantly due to the high centre-of-mass energy, the resulting large rise of the parton luminosities, and the large instantaneous luminosity of the accelerator. During LHC's Run 1 more than 5 million top-quark events were produced at the collision points in ATLAS and CMS each. Based on these huge data samples, most of the measurements performed at the Tevatron have already been improved and/or complementary studies have been performed. This has only been made possible by inheriting a wealth of analysis techniques from the Tevatron experiments where they were pioneered and brought to perfection. At the LHC, top quarks have become tools for searches for new physics, e.g. the search for rare decays, for measurements of top-quark couplings to gauge bosons, or for the investigation of the proton structure.

In this chapter the experimental findings obtained during LHC Run 1 are summarised and put in context with our current understanding of the Standard Model (SM). Emphasis is placed on three aspects: (i) the description of the precision measurements in comparison with results from the Tevatron and with theory predictions, e.g. production cross sections or top-quark properties; (ii) the presentation of new measurements that were not performed at the Tevatron and that either improve the current understanding of already measured quantities or enable measurements of yet unexplored processes and quantities, e.g. small couplings or associated production processes; (iii) the discussion of the physics lessons and insights gained from the LHC Run 1 results. These results do not only impact on our understanding of the Standard Model, but also help us prepare for future investigations at the LHC and beyond. Several important measurements based on the data from LHC Run 1 are still being performed. This chapter can thus not be complete, and only a selection of the results is presented.

7.2 Top-Quark Pair Production

In hadronic collisions, top-quark pairs are dominantly produced through the strong interaction. At the parton level, the production mechanisms are quark-antiquark annihilation and gluon fusion. Figure 7.1 shows the corresponding leading-order (LO) Feynman diagrams. The differential cross section for quark-antiquark annihilation can easily be obtained from $e^+e^- \to \mu^+\mu^-$ by replacing the QED coupling α_{QED} with the QCD coupling α_s and introducing an appropriate colour factor (2/9) (see also Sect. 5.3):

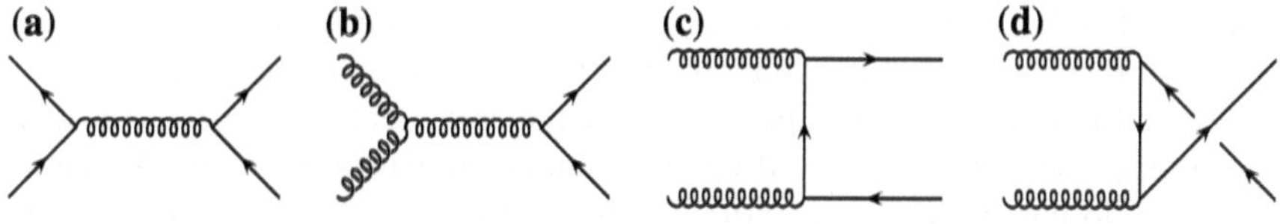

Fig. 7.1 Leading-order Feynman diagrams contributing to top-quark production in hadronic collisions

$$\frac{d\hat{\sigma}_{q\bar{q}\to t\bar{t}}}{dz} = \frac{\pi\alpha_s^2}{9s}\beta\left(2-\left(1-z^2\right)\beta\right).$$

Here s denotes the center-of-mass energy squared, $\beta = \sqrt{1-4m_t^2/s}$ is the velocity of the top quark in the partonic centre-of-mass system, m_t is the top-quark mass, and $z = \cos(\theta)$ is the cosine of the scattering angle defined as the angle between the incoming quark and the outgoing top quark. The corresponding result for gluon fusion reads

$$\frac{d\hat{\sigma}_{gg\to t\bar{t}}}{dz} = \frac{\pi\alpha_s^2}{96s}\beta\frac{7+9z^2\beta^2}{\left(1-z^2\beta^2\right)^2}\left(1+2\beta^2-2z^2\beta^2-2\beta^4+2z^2\beta^4-z^4\beta^4\right).$$

The hadronic cross section is obtained from the partonic cross section through a convolution with the parton distribution functions (PDFs) $f_{i/H}(x,\mu_F)$ which, roughly speaking, describe the probability to find a parton i inside a hadron H with a momentum fraction between x and $x+dx$ of the mother hadron. The factorisation scale μ_F denotes the scale at which, in higher-order calculations, the initial-state singularities are factorised into the parton distribution functions. The final formula for the hadronic cross section is thus given by

$$d\sigma_{H_1H_2\to t\bar{t}+X} = \sum_{i,j}\int dx_1dx_2 f_{i/H_1}(x_1,\mu_F)f_{j/H_2}(x_2,\mu_F) \\ \times\ d\hat{\sigma}_{ij\to t\bar{t}+X}(x_1P_1,x_2P_2,k_t,k_{\bar{t}},\alpha_s(\mu_R),\mu_F), \qquad (7.1)$$

where P_1, P_2 are the momenta of the incoming hadrons, while k_t, $k_{\bar{t}}$ are the momenta of the outgoing top quarks. The sum is over all possible partons. The coupling constant α_s is evaluated at the renormalisation scale μ_R. Note that in higher order, the partonic cross section $d\hat{\sigma}$ also depends on the factorisation scale μ_F such that the factorisation scale dependence of the parton distribution functions is cancelled order by order and the hadronic cross section becomes independent of the non-physical scale μ_F at fixed order in perturbation theory. For more details we refer to Sect. 4.8. Because the QCD coupling constant is not small ($\alpha_s \approx 0.1$), higher-order corrections can give sizeable contributions and need to be taken into account in most cases. Since the top-quark mass sets a large energy scale, non-perturbative effects are essentially cut off and QCD perturbation theory is believed to give reliable predictions. The next-to-leading order (NLO) QCD corrections for $t\bar{t}$ production were calculated a long time ago [6, 7]. Further improvements were obtained by resumming soft-gluon corrections, which lead to a logarithmic enhancement of the cross section, to next-to-leading logarithmic accuracy [8–10]. Soft-gluon resummation has been extended recently to the next-to-next-to-leading logarithmic (NNLL) accuracy [11–15]. Details about the theoretical framework can be found in Sect. 4.8. Despite the fact that top quarks do not form bound states (because of their short lifetime), binding effects still lead to minor corrections of the cross section close to the $t\bar{t}$-pair production threshold. In principle, such a would-be bound state could be observed as a narrow peak in the $t\bar{t}$ invariant-mass spectrum, just below the production threshold. However, the energy resolution of the LHC experiments is not sufficient to resolve this effect.[1] The corresponding corrections to the inclusive cross section are small and have been studied in detail [15–17]. Electroweak corrections have also been investigated [18–24]. For the inclusive cross section of $t\bar{t}$ production at the LHC operating at 14 TeV, they are negative and at the percent level. On the other hand, weak Sudakov logarithms can result in suppression of the cross section at the level of 10–20 % for differential distributions at large momentum transfer (see also Chap. 4). Since this region is precisely the one where new heavy resonances could lead to an increase of the cross section, neglecting weak corrections could potentially hide signs of new physics.

Very recently, full next-to-next-to-leading order (NNLO) QCD corrections have been presented for the inclusive cross section [25]. For a centre-of-mass energy of 8 TeV, the result—including soft-gluon resummation at NNLL accuracy (assuming $m_t = 173.3\,\text{GeV}$ and the MSTW2008nnlo68cl PDF set [26])—reads

$$\sigma_{t\bar{t}}(\sqrt{s} = 8\,\text{TeV}) = 245.8^{+6.2}_{-8.4}\,(\text{scale})^{+6.2}_{-6.4}\,(\text{PDF})\,\text{pb}. \tag{7.2}$$

Note that the result does not include the aforementioned weak corrections. The first uncertainty is due to the residual[2] scale dependence which is used as an estimate of the unknown higher-order contributions. It has been determined by a variation of the renormalisation and factorisation scales in the range $m_t/2 \ldots 2m_t$. The second uncertainty is due to the incomplete knowledge of the parton distribution functions. From (7.2) we conclude that the total cross section is known with a precision better than 5 %. Moving to higher collider energies will slightly improve the PDF uncertainties since less weight is put to the large x region where the PDFs are less precisely known.

So far, only very few NNLO results exist for differential distributions. Most predictions are currently restricted to NLO accuracy (extended in some cases by the resummation of soft-gluon corrections). Fixed-order NLO corrections are available, for example, through the parton-level Monte Carlo (MC) program MCFM [27].Combining

[1] At a future e^+e^- collider, operating at the top-quark production threshold, the effect would be visible thanks to the high energy resolution and could be used for very precise measurements of the top-quark mass.

[2] We call the remaining scale dependence "residual" because it is formally of higher order.

parton-level NLO calculations with parton-shower simulations is in general not straight forward. A naive combination would count real emission processes twice, since real emission is simulated through the parton shower, but, on the other hand, is also explicitly taken into account in the real-emission processes contributing at NLO. A consistent matching that avoids double-counting has been developed in the past [28, 29]. Differential distributions for top-quark pair production at NLO including parton shower effects are given in [30, 31].

In all calculations mentioned before, the production of stable top quarks is assumed, which is then followed by an on-shell decay. This corresponds to the "narrow-width" or "double-pole approximation". The naive expectation is that corrections to this approximation should be suppressed like Γ_t/m_t or even Γ_t^2/m_t^2, where Γ_t denotes the top-quark decay width (see also [32]). Obviously, the naive expectation can only be true if the observable under consideration is not directly related to off-shell effects. A counter-example is the invariant-mass distribution of the top-quark decay products. Off-shell effects for $t\bar{t}$ production have been investigated in detail in [33–35], where the QCD NLO corrections for the process $pp \to W^+W^-b\bar{b}$ have been calculated. Indeed, the effects are typically small, unless observables of the type mentioned before are studied.

As can be seen from the vast amount of different theoretical studies, $t\bar{t}$ production is well understood in the Standard Model. A variety of different corrections have been considered in the past, and precise theoretical predictions for inclusive as well as differential quantities are available. Top-quark pair production is thus an ideal laboratory to test the consistency of the SM and to search for possible deviations. In the following sections we describe the measurements of inclusive and exclusive cross sections for top-quark pair production.

7.2.1 Inclusive $t\bar{t}$ Cross Section

First measurements of the inclusive $t\bar{t}$ cross section were published by the ATLAS and CMS collaborations already in the year 2010 [36, 37]. These are based on data corresponding to an integrated luminosity of about 3 pb^{-1}, a fraction of the data collected in 2010. More precise measurements, based on the full 2010 data sample corresponding to an integrated luminosity of 35 pb^{-1} became available shortly after [38–42].

The amount of data collected in the years 2011 ($\sqrt{s} = 7\,\text{TeV}$) and 2012 ($\sqrt{s} = 8\,\text{TeV}$) correspond to integrated luminosities of about 5 and $20\,\text{fb}^{-1}$, respectively—altogether more than a factor 500 more than that of 2010. With this wealth of data, top-quark physics entered a completely new realm of precision and detail. With the 2011 data samples, measurements of the inclusive top-quark pair cross section were performed in all decay channels of the $t\bar{t}$ system (except the one with two τ leptons in the final state), reaching an unprecedented level of statistical and systematic precision [43–53].

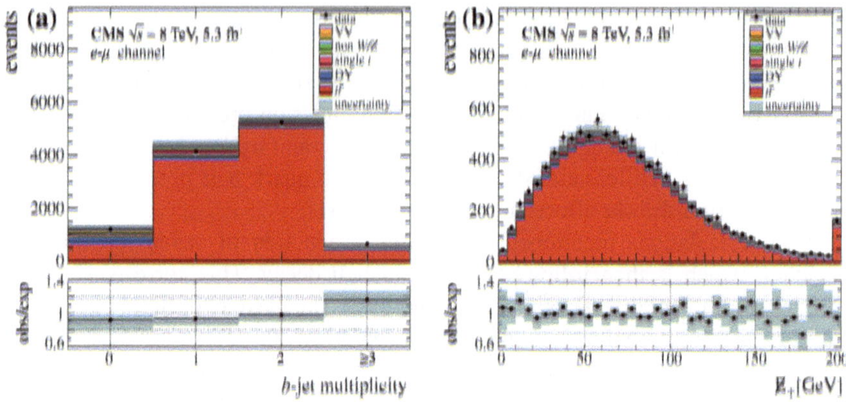

Fig. 7.2 Distributions of the b-tag multiplicity and missing transverse momentum of the CMS event sample used for the measurement of the inclusive $t\bar{t}$ cross section at a centre-of-mass energy of 8 TeV. Adapted from [54]

In the limit of large statistics, the most precise measurements of the inclusive $t\bar{t}$ cross section can be obtained in the e-μ channel, as the backgrounds and associated uncertainties are minimal. The first published measurement of the top-pair cross section at a centre-of-mass energy of 8 TeV was performed by CMS [54]. It makes use of the decay channel of top-quark pairs with two opposite-charge leptons, one electron and one muon, in the final state. In this channel, backgrounds from non-top-quark events are minimal. Dominant contributions arise from Drell–Yan processes with two τ leptons in the final state that both decay into a lepton. These backgrounds are suppressed by requiring at least one of the two jets to be b-tagged. Smaller background contributions come from single top-quark production (see Sect. 7.6) and from top-quark events in other decay channels where one of the jets is misidentified as a lepton. In Fig. 7.2 the distributions of the b-tag multiplicity and the missing transverse momentum after the final event selection are shown. The final result of this measurement is $\sigma_{t\bar{t}}(\sqrt{s} = 8\,\text{TeV}) = 239.0 \pm 2.1\,(\text{stat}) \pm 11.3\,(\text{syst}) \pm 6.2\,(\text{lumi})$ pb, in good agreement with the NNLO prediction quoted in (7.2).

Most recently, the ATLAS collaboration published a measurement yielding the most precise experimental result [53]. In this analysis the numbers of events with exactly one and with exactly two b-tagged jets are counted and used to simultaneously determine $\sigma_{t\bar{t}}$ and the efficiency to reconstruct and b-tag a jet from a top-quark decay, thereby minimising the associated systematic uncertainties. In Fig. 7.3 the distributions of the b-tag multiplicity and the transverse momentum of the b-tagged jets are displayed. The cross sections for centre-of-mass energies $\sqrt{s} = 7$ TeV and 8 TeV are measured to be $\sigma_{t\bar{t}}(\sqrt{s} = 7\,\text{TeV}) = 182.9 \pm 3.1\,(\text{stat}) \pm 4.2\,(\text{syst}) \pm 3.6\,(\text{lumi}) \pm 3.3\,(\text{beam})$ pb, and $\sigma_{t\bar{t}}(\sqrt{s} = 8\,\text{TeV}) = 242.4 \pm 1.7\,(\text{stat}) \pm 5.5\,(\text{syst}) \pm 7.5\,(\text{lumi}) \pm 4.2\,(\text{beam})$ pb, respectively, where the latter uncertainty is due to the beam-energy uncertainty.

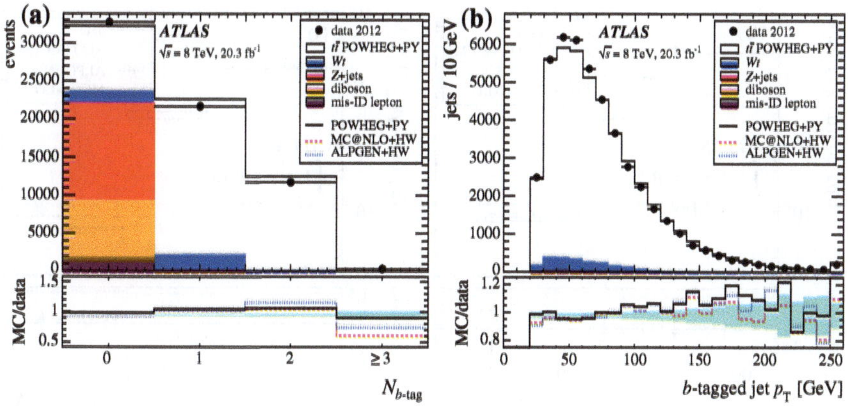

Fig. 7.3 Distributions of the b-tag multiplicity and b-jet transverse momentum of the ATLAS event sample used for the measurement of the inclusive $t\bar{t}$ cross section at a centre-of-mass energy of 8 TeV. Adapted from [53]

7.2.2 Differential $t\bar{t}$ Cross Sections

Additional information about top-quark production and decay can be gained from measurements of differential distributions. These do not only probe QCD predictions and provide input to an improved choice of QCD model and scale parameters, but they also have the potential to constrain the parton distribution functions of gluons at large momentum fractions x. Moreover, the differential distributions are potentially sensitive to new physics, e.g. to decays of massive Z-like bosons into top-quark pairs that would become visible at high $t\bar{t}$ invariant masses (see Chap. 11).

The kinematic properties of a top-quark pair are determined from the four-momenta of all final-state objects by means of reconstruction algorithms. For a general introduction to the different decay channels, see e.g. [55] and references therein. In the single-lepton channels, kinematic-fitting algorithms are applied to obtain the kinematics of both top quarks. In the dilepton channels, due to the presence of two neutrinos, the kinematic reconstruction is under-constrained. Ambiguities between several solutions are resolved by prioritisation, e.g. by the use of the expected neutrino energy distribution.

A large number of distributions of the top quark and the top-quark pair system, as well as their decay products, has been measured at the LHC [56–58]. In contrast to the situation at the Tevatron, the large $t\bar{t}$ production rate at the LHC leads to a substantial reduction of the statistical uncertainties in each bin. The ATLAS and CMS collaborations report normalised differential cross sections, i.e. shape measurements, in which normalisation uncertainties are removed. In Fig. 7.4 the distributions of the invariant mass and of the transverse momentum of the top-quark pair system as measured by ATLAS are displayed [56]. The data are very well described by the various calculations up to an energy scale of about 1 TeV. The results from CMS

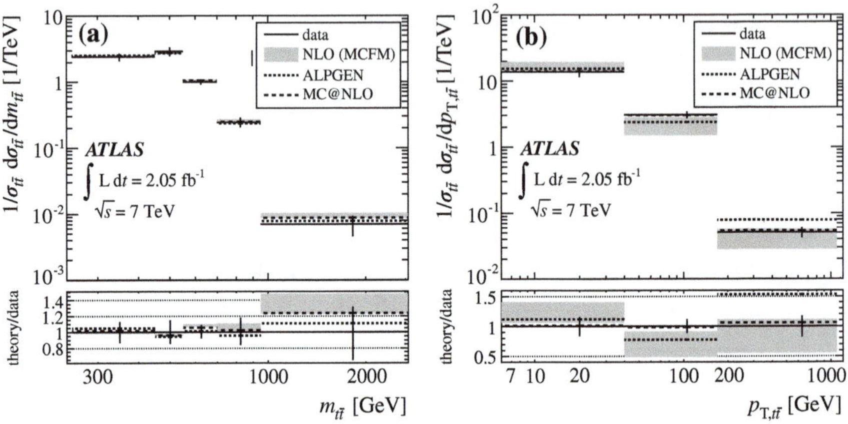

Fig. 7.4 Normalised differential $t\bar{t}$ production cross sections, measured in pp collisions at a centre-of-mass energy of 7 TeV by the ATLAS collaboration, as a function of **a** the invariant mass of the top-quark pair and **b** the transverse momentum of the top-quark pair system. Adapted from [56]

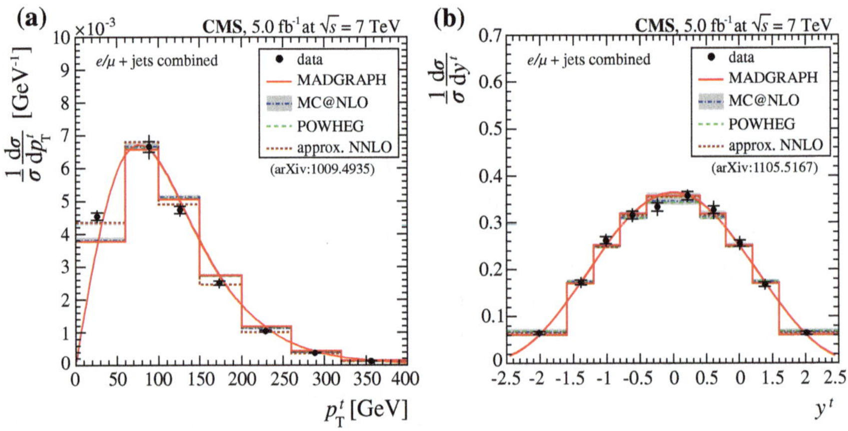

Fig. 7.5 Normalised differential $t\bar{t}$ production cross sections, measured in pp collisions at a centre-of-mass energy of 7 TeV by the CMS collaboration, as a function of **a** the transverse momentum and **b** the rapidity of each of the top quarks. Adapted from [57]

agree with these findings [57]. The transverse momentum and rapidity distribution of each of the top quarks were also measured and the results from CMS are presented in Fig. 7.5. Different theoretical predictions are confronted with the data, and they are generally found to give a good description of the data. However, most Monte Carlo simulations predict the transverse momentum distribution of the top quarks to be somewhat harder than what is seen in the data. This discrepancy between data and simulation is presently under investigation. For the time being it constitutes an important source of uncertainty for many analyses.

7.2.3 Top-Quark Pairs and Additional Jets

At LHC energies, a large fraction of top-quark pairs is accompanied by additional high-p_T jets. Demanding, for example, a minimal transverse momentum of 50 GeV for such additional jets, about 30 % of all top-quark pairs are produced together with at least one further jet [59]. From an experimental point of view, the jet activity needs to be understood since the appearance of additional jets affects the event reconstruction. Owing to the large rate, $t\bar{t}$ production with jets may also lead to sizeable backgrounds for other SM studies or searches for new physics. As an example, $t\bar{t}$ + 1-jet + X production is the dominant background for Higgs production via vector-boson fusion. From a theoretical perspective, the additional jet activity can be used for further tests of the underlying production and decay mechanisms. Anomalous $t\bar{t}g$ couplings can be constrained, for example, through a detailed analysis of the process $pp \to t\bar{t}$ + 1-jet + X. Assuming the validity of our theoretical understanding, $t\bar{t}$ production in association with a jet can also be used to measure the top-quark mass [60]. Since the process $pp \to t\bar{t}$ + 1-jet + X is proportional to α_s^3, NLO contributions can easily give corrections of the order of 30 %. For a precise understanding it is thus mandatory to take these corrections into account. In Born approximation, the partonic processes $gg \to t\bar{t}g$, $q\bar{q} \to t\bar{t}g$, $qg \to t\bar{t}q$ and $g\bar{q} \to t\bar{t}\bar{q}$ contribute to $t\bar{t}$ production in association with a jet. The last three processes are related by crossing. The leading-order partonic matrix elements can be calculated e.g. with the help of MADGRAPH [61]. The hadronic cross sections are then calculable through a numerical Monte Carlo integration, using again (7.1). For the evaluation of the NLO corrections, the one-loop corrections to the aforementioned Born processes, together with real-emission processes, need to be evaluated. Since the two contributions are individually infrared (IR) divergent—the divergences cancel only in the sum—a method to organise this cancellation needs to be applied. In [59, 62], the one-loop amplitudes have been calculated using a traditional tensor reduction for the one-loop integrals. The cancellation of the IR divergences is achieved using the Catani–Seymour subtraction method [63, 64]. In [65], an alternative calculation of the NLO corrections, based on the unitarity method, has been presented. In a subsequent study [66], also the on-shell decay of the top quark has been taken into account. NLO results for $t\bar{t}$ production in association with a photon are also available at NLO QCD [67], since this process is closely related to $t\bar{t}$ production in association with jets. In Fig. 7.6a, the cross section for $t\bar{t}$ + 1-jet + X production is shown as a function of the renormalisation scale μ_R which is set equal to the factorisation scale. For the parton distribution functions the CTEQ6 set [68] has been used.

The transverse momentum of the additional jet is required to be at least 20 GeV. The leading-order result strongly depends on the renormalisation scale μ_R, a fact that directly reflects the running of the coupling constant $\alpha_s(\mu_R)$. The Born approximation can thus at best be considered as a rough estimate of the cross section. The NLO corrections as a function of the renormalisation scale, however, show a flat behaviour around $\mu_R = m_t$. This may be considered as an indication that m_t provides a natural scale for this process. It can also be seen from Fig. 7.6a that the correc-

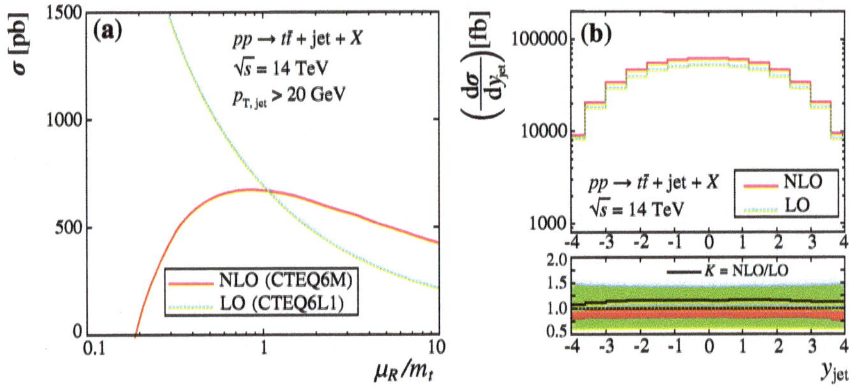

Fig. 7.6 **a** Cross section for $t\bar{t}$ + 1-jet + X production at $\sqrt{s}$ = 14 TeV as a function of the renormalisation scale. **b** Rapidity distribution of the additional jet at NLO. Adapted from [59, 62]

tions are rather small for $\mu_R = m_t$. Similar observations can be made in Fig. 7.6b where the rapidity distribution of the additional jet, calculated at NLO, is shown. In [59, 67] a large variety of differential distributions have been investigated. In particular the transverse momentum distribution of the top quark, the $t\bar{t}$ system and the additional jet have been calculated. It turns out that for large transverse momentum p_{T}, significant QCD corrections together with a large scale uncertainty are observed. In principle, this is not surprising since at a large transverse momentum an additional scale—different from m_t—is introduced. It is conceivable that a phase-space-dependent renormalisation scale could improve the behaviour of the perturbation theory by effectively resumming large logarithmic corrections. In Table 7.1, the dependence of the cross section on the required minimal transverse momentum of the additional jet, $p_{\mathrm{T}}^{\mathrm{cut}}$, is shown. A strong dependence on $p_{\mathrm{T}}^{\mathrm{cut}}$ is found. For $p_{\mathrm{T}}^{\mathrm{cut}} \to 0$ the cross section diverges logarithmically (the divergence cancels a similar divergence in the cross section for inclusive $t\bar{t}$ production at NNLO, when $t\bar{t}$ + 1-jet + X production is combined with the two-loop corrections to inclusive $t\bar{t}$ production).

Table 7.1 Cross section $\sigma_{pp\to t\bar{t}+1\text{-jet}+X}$ at the LHC for different values of $p_{\mathrm{T}}^{\mathrm{cut}}$ for $\mu = \mu_F = \mu_R = m_t$ [59]. In parentheses the uncertainties due to Monte Carlo integrations are quoted. The uncertainties correspond to changes in the scale, namely $\mu = m_t/2$ and $\mu = 2m_t$

$p_{\mathrm{T}}^{\mathrm{cut}}$ (GeV)	$\sigma_{pp\to t\bar{t}+1\text{-jet}+X}$ (pb)	
	LO	NLO
20	$710.8(8)^{+358}_{-221}$	$692(3)3^{-40}_{-62}$
50	$326.6(4)^{+168}_{-103}$	$376.2(6)^{+17}_{-48}$
100	$146.7(2)^{+77}_{-47}$	$175.0(2)^{+10}_{-24}$
200	$46.67(6)^{+26}_{-15}$	$52.81(8)^{+0.8}_{-6.7}$

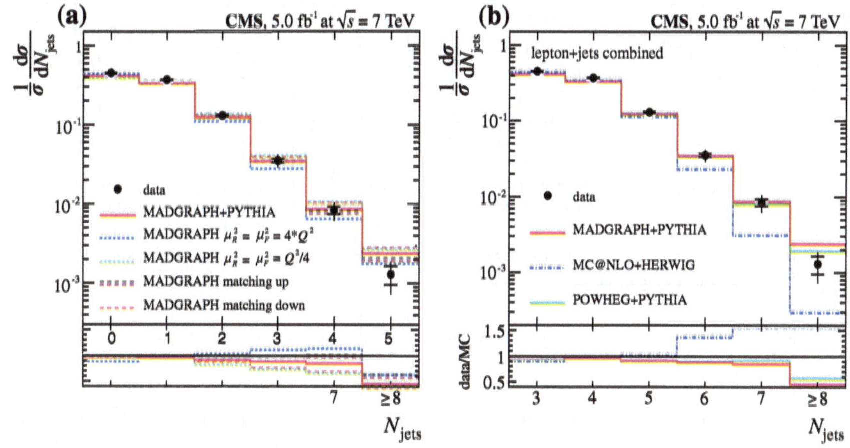

Fig. 7.7 Differential distributions of jet multiplicity in $t\bar{t}$ events with exactly one lepton and at least three jets in the final state. **a** Data compared with the MADGRAPH simulation for different choices of renormalisation and factorisation scales and the matching scale between matrix element and parton shower. **b** Data compared with several Monte Carlo simulations. Adapted from [75]

With the exception of $p_T^{cut} = 20\,\text{GeV}$, where very small and negative corrections are observed, the corrections are positive and typically about 15–20 %.

Recently, NLO corrections for $t\bar{t}$ production in association with two additional jets were studied. For details, we refer to the original work [69–74].

The ATLAS and CMS collaborations studied the distributions of jet multiplicities and additional jets due to QCD radiation in detail [75–77]. The multiplicity distributions of jets for $t\bar{t}$ events in the single-lepton channel as measured by the CMS collaboration is shown in Fig. 7.7. The data are generally well described by the Monte Carlo predictions obtained using MADGRAPH, POWHEG and MC@NLO. Towards very large multiplicities, the MC@NLO generator interfaced with parton showers from HERWIG predicts significantly less events than MADGRAPH or POWHEG, which both use PYTHIA to generate the parton showers.

An alternative way of investigating additional activity in $t\bar{t}$ events is to study "gap-fraction" distributions [75, 76]. In these studies, events are vetoed if they contain an additional jet with transverse momentum above a given threshold in a central rapidity interval. The fraction of events surviving the jet veto, the gap fraction, is presented as a function of the threshold. The gap-fraction distributions for jets as measured by ATLAS are displayed in Fig. 7.8. A qualitatively similar trend is observed as in the multiplicity distribution (Fig. 7.7) in that the MC@NLO generator predicts a larger fraction of events that have no jet activity beyond the jets originating directly from the top-quark decays. However, vetoing jets just in the forward region, at rapidities $|y| > 1.5$, all simulations predict a smaller fraction of events with no additional jet than is seen in the data. These results can be used to improve the choice of models, scale parameters and tunes in Monte Carlo simulations for an optimal description of the data.

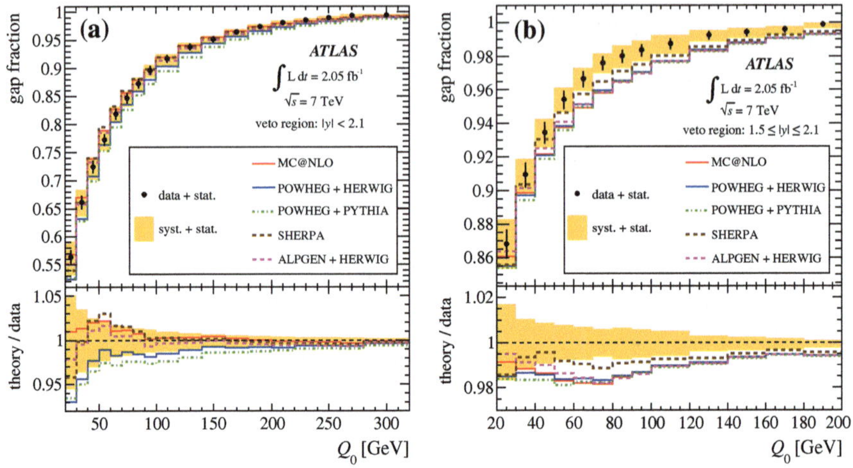

Fig. 7.8 The measured gap fraction as a function of the jet transverse momentum threshold Q_0 above which there are no additional jets. The data are compared with the predictions from NLO and multi-leg LO MC generators, applying jet vetos in two different rapidity regions **a** $|y| < 2.1$ and **b** $1.5 < |y| < 2.1$. Adapted from [76]

Building up on the insights about inclusive distributions of additional jets in top-quark pair events, both ATLAS and CMS have also been studying distributions and rates of events with additional jets originating from heavy quarks. These form an important background to events in which Higgs bosons are produced in association with top-quark pairs, with subsequent decay of the Higgs boson into $b\bar{b}$ or $c\bar{c}$ pairs, see Sect. 6.6.3. The ATLAS experiment published a measurement of top-quark pairs together with heavy-flavour quarks, $t\bar{t} + b + X$ or $t\bar{t} + c + X$ at $\sqrt{s} = 7\,\text{TeV}$ [78]. The separation between heavy-flavour and light-flavour contents of additional jets is achieved by a fit to the vertex-mass distribution of b-tagged jets, shown in Fig. 7.9.

Fig. 7.9 Vertex-mass distribution for high-purity b-tagged jets in top-quark pair events with two leptons in the final state. For this figure, no cut on the b-tagged jet multiplicity is applied. Adapted from [78]

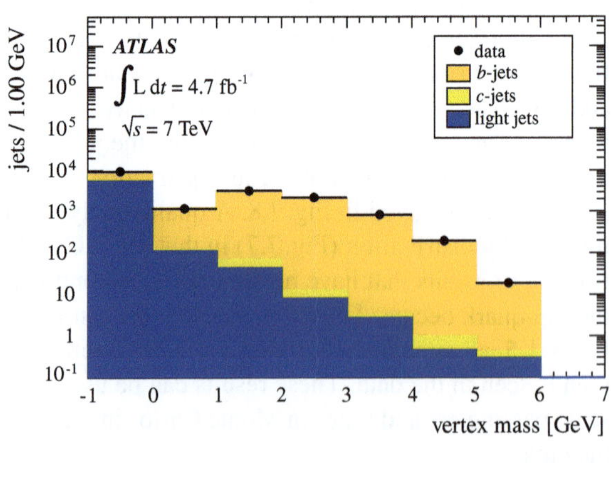

From the fit, the relative contribution of heavy quarks is extracted to be $R_{\text{HF}} = [6.2 \pm 1.1(\text{stat}) \pm 1.8(\text{syst})]\%$, consistent within uncertainties with leading-order expectations, see [78] and references therein.

7.3 Top-Quark Mass

In the SM, the couplings of the top quark to the gauge bosons are governed by the gauge structure (see also Chap. 4). The only free parameters in top-quark physics are thus the three corresponding CKM matrix elements and the top-quark mass, m_t. Instead of the top-quark mass, one may also use the Yukawa coupling λ_t to the Higgs boson as a free parameter since the two are related by $m_t = v/\sqrt{2}\lambda_t$, where v is the vacuum expectation value of the Higgs field. Once the CKM matrix elements and the top-quark mass are known, the SM makes testable predictions for all top-quark properties. Precise measurements of these properties can thus be used to test the consistency of the SM. A prominent example is provided by the simultaneous measurements of the top-quark mass, the W-boson mass and the mass of the Higgs boson since these three masses are related in the SM: The W-boson mass can be calculated as a function of the top-quark mass and the mass of the Higgs boson. A comparison with the measured values thus allows the mechanism of electroweak symmetry breaking predicted by the SM to be indirectly tested. This is demonstrated in Fig. 7.10 where, for comparison, also results within the minimal supersymmetric Standard Model (MSSM, see Chap. 10) are shown [79].

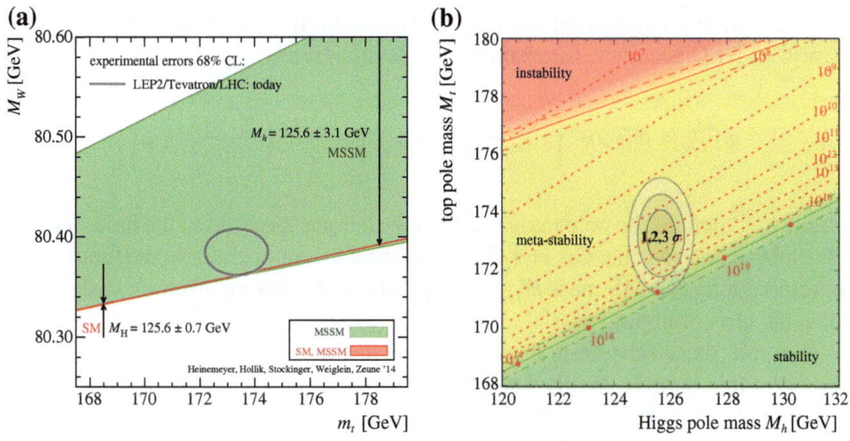

Fig. 7.10 **a** Consistency of the Standard Model: the W-boson mass as function of the top-quark mass for different Higgs-boson masses. **b** Stability of the vacuum for different top-quark masses. Adapted from [79–81]

Recently, also the question of vacuum stability has attracted a lot of attention (see for example [80, 81]). Through quantum corrections, the top quark influences the effective Higgs potential that is responsible for electroweak symmetry breaking. In principle, it is conceivable that these quantum corrections modify the effective potential such that it develops a second minimum or even becomes unbounded from below. As a consequence, the electroweak vacuum might become metastable or even unstable. Calculating the corresponding lifetime of the vacuum and comparing it with the age of the universe provides a further consistency test of the SM.

Experimentally, in most measurements the mass of the top quark is determined through the reconstruction of the top quark's decay products. The top-quark mass can be estimated by comparing the measured values with the value of the mass parameter used in the simulation. Measurements employing this approach generally achieve the most precise results.

In contrast, from a theoretical point of view, a meaningful definition of the top-quark mass requires to specify the renormalisation scheme used to define the parameter in the theoretical predictions. In this respect the top-quark mass should be treated similar to a coupling constant. In the theoretical description of hadronic collisions, the so-called "on-shell" or "pole mass" scheme and the "minimal subtraction scheme" ($\overline{\text{MS}}$) are commonly used. Quantitatively, the pole mass and the mass measured from final state reconstruction are expected to agree within $\mathcal{O}(1\,\text{GeV})$ [82, 83].

In the pole mass scheme, the renormalised mass is defined as the location of the pole of the renormalised quark propagator, including higher-order corrections. In the (modified) minimal subtraction scheme, the renormalised parameters are defined through a minimal subtraction of the ultraviolet (UV) divergences. The renormalisation constants are chosen such that they just cancel the UV divergences encountered in the loop corrections (together with some irrelevant transcendental constants in the modified minimal subtraction scheme). The two definitions are related in perturbation theory. At NLO accuracy, for example, the relation reads

$$m_t^{\text{pole}} = \overline{m_t}(\mu_R)\left\{1 + \frac{\alpha_s(\mu_R)}{\pi} C_F \left[1 - \frac{3}{4}\ln\left(\frac{\overline{m_t}(\mu_R)^2}{\mu_R{}^2}\right)\right]\right\}, \tag{7.3}$$

where m_t^{pole} defines the pole mass and $\overline{m_t}(\mu_R)$ defines the mass in the modified minimal subtraction scheme. $C_F = 4/3$ denotes the Casimir operator in the fundamental representation. The $\overline{\text{MS}}$ mass $\overline{m_t}(\mu_R)$ depends on the renormalisation scale, which coins the term "running mass".

The pole mass scheme is closely related to the intuitive understanding of the mass of a free particle. While for many applications this intuitive picture is a good approximation—for many purposes the top quark behaves like a free quark—it should be clear that this picture is doomed to fail if it comes to ultimate precision. Indeed, it has been shown that the pole mass suffers from the so-called "renormalon ambiguity" that leads to an intrinsic uncertainty of the pole mass of the order of Λ_{QCD} [84, 85]. Despite the fact that the pole mass and the $\overline{\text{MS}}$ mass definitions are related in perturbation theory, it is in practice not straightforward to convert theoretical results

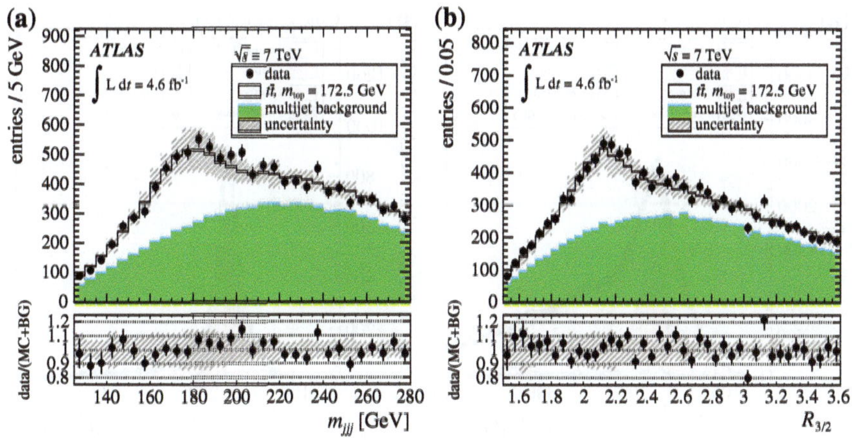

Fig. 7.11 Distributions of **a** the 3-jet invariant mass and **b** the ratio of the three-jet mass to the dijet mass $R_{3/2}$, measured in data and compared to expectations after applying all event selection criteria. The *open histograms* show the expected distributions for $t\bar{t}$ events for a top-quark mass value of 172.5 GeV. The *shaded histograms* indicate the contribution from multi-jet backgrounds. Adapted from [91]

from one scheme to the other. So far, this translation has only been calculated for the inclusive cross section for $t\bar{t}$ production and for a few differential distributions in $t\bar{t}$ production.

Experimentally, the top-quark mass has been measured at the LHC using a large variety of methods and observables, and in different decay channels [40, 86–91]. In addition, the difference of the masses of top quarks and antiquarks have been measured [92, 93].

A recent measurement using $t\bar{t}$ events with fully hadronic final states is presented by the ATLAS collaboration [91]. Events are selected if they contain at least six jets, and exactly two b-tagged jets are required to be among the four leading jets. The top-quark mass is extracted from a binned likelihood fit to the $R_{3/2}$ distribution, shown in Fig. 7.11, where $R_{3/2}$ is the ratio of the reconstructed three-jet and two-jet masses. In this distribution systematic effects that are common to the masses of the reconstructed top quark and the associated W boson cancel. The contribution from multi-jet backgrounds is determined from the data using the event yields in different regions of b-tag jet multiplicity and 6th-jet momentum. The measurement yields a value for the top-quark mass of

$$m_t = 175.1 \pm 1.4\,(\text{stat}) \pm 1.2\,(\text{syst})\,\text{GeV}.$$

The systematic uncertainties are dominated by the residual uncertainties of the jet energy scale, in particular for b-quark jets, and by the uncertainties from hadronization modelling.

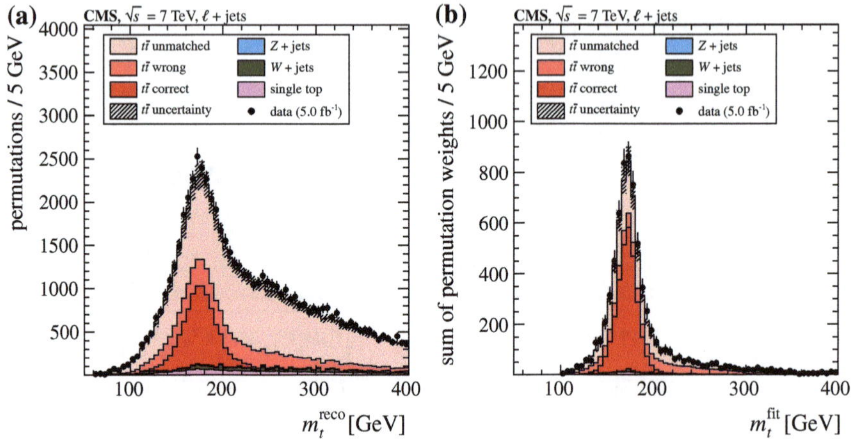

Fig. 7.12 Reconstructed top-quark mass distributions **a** before and **b** fitted top-quark mass distributions after kinematic reconstruction with entries weighted by the reconstructed goodness-of-fit parameter $P_{\rm gof}$. Adapted from [90]

The CMS collaboration presented a measurement of the top-quark mass using the full data set collected at a centre-of-mass energy of $\sqrt{s} = 7$ TeV with one electron or muon and at least four jets in the final state [90]. A kinematic fit to the four leading jets, the lepton and the missing transverse momentum is employed to constrain the selected events to the hypothesis of the production of two heavy particles of equal mass, each one decaying to a W boson and a b quark. The reconstructed masses of the two W bosons are constrained in the fit to 80.4 GeV. The reconstructed invariant-mass distribution is shown in Fig. 7.12. Events can enter the distributions with different parton-jet assignments (permutations). For simulated $t\bar{t}$ events, the parton-jet assignments can be classified as correct, wrong and unmatched permutations where, in the latter, at least one quark from the $t\bar{t}$ decay is not matched to any of the four selected jets. The actual top-quark mass value is determined simultaneously with the jet energy scale using a joint likelihood fit. The joint likelihood is constructed based on the "ideogram method" in which the likelihood for each event is evaluated from analytic expressions obtained from simulated events. Biases arising due to this method are determined using pseudo-experiments and corrections are applied accordingly. The dominant uncertainty of the final result comes from the uncertainty of the difference in the jet energy responses for jets originating from light (u, d, s) or bottom quarks, as well as from statistical uncertainties in the determination of differences between different models for colour-reconnection processes. The final result is

$$m_t = 173.49 \pm 0.43 \text{ (stat+JES)} \pm 0.98 \text{ (syst) GeV},$$

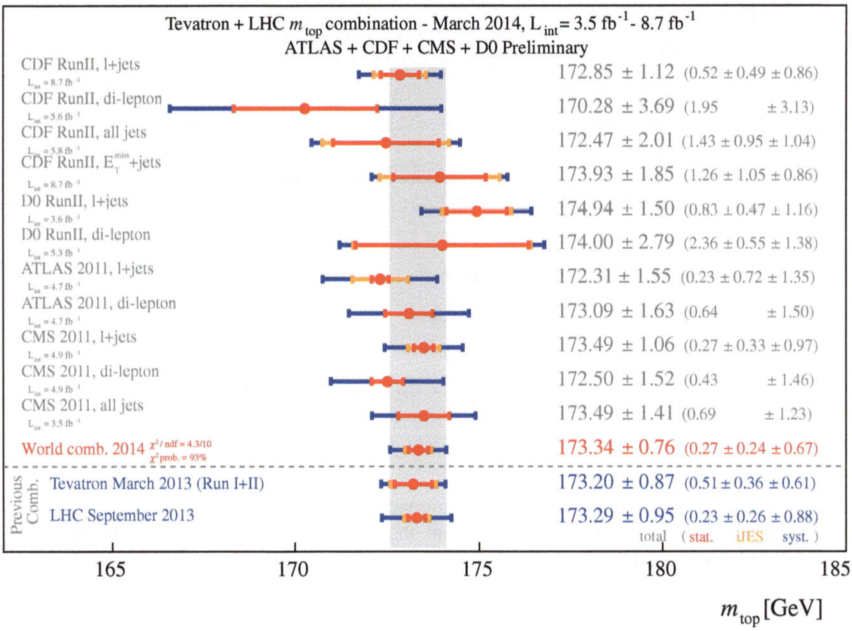

Fig. 7.13 Mass measurements included in the world combination 2014. Also shown are the results from the combinations of LHC-only data and Tevatron-only data. Adapted from [94]

corresponding to an optimal jet energy scale correction of 0.994 ± 0.003 (stat) $\pm$ 0.008 (syst) with respect to the CMS calibration.

In spring 2014, this and other precise results from the LHC and the Tevatron, both preliminary and final, were combined to obtain a first world average of the top-quark mass [94]. A summary is shown in Fig. 7.13.

An alternative approach to the determination of the top-quark mass is to extract its value from the measured inclusive cross section. This approach has the advantage that the cross section and the pole mass are directly related, such that the extraction yields a theoretically well-defined quantity. Both ATLAS and CMS have used their cross-section measurements to extract the top-quark pole mass [53, 95] as defined at NNLO accuracy [25]. The extractions are performed for different parton distribution functions and take into account the experimental dependence of the measured cross section on the assumed top-quark mass. A summary of the results is shown in Fig. 7.14.

Conversely, assuming equality between the pole mass and the directly measured mass within 1 GeV, the cross-section measurements can be used to determine the strong coupling constant α_s [95].

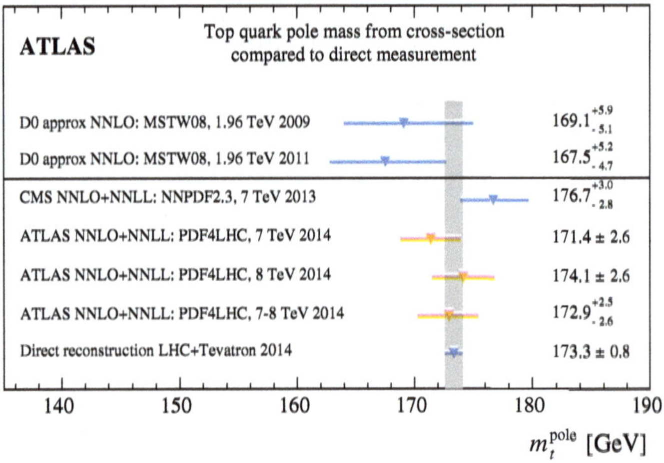

Fig. 7.14 Summary of determinations of the top-quark pole mass from cross-section measurements. Also shown is the top-quark mass world average as obtained from combination of direct experimental measurements [94]. Adapted from [53]

7.4 Tests of QCD Predictions

Perturbative QCD (pQCD) is a fundamental building block for the understanding of physics processes at the LHC. In addition to cross-section predictions, several other aspects of pQCD can be probed in the top-quark sector. Examples are the small but finite charge asymmetry of $t\bar{t}$ pairs and the correlation between the top-quark and antiquark spins in $t\bar{t}$ production; both quantities are sensitive to the relative proportions of the different production processes of $t\bar{t}$ pairs. Another example is the polarisation of top quarks in $t\bar{t}$ events, which tests the C and CP structure of $t\bar{t}$ production.

7.4.1 Charge Asymmetry

At leading order QCD, the reactions $pp \rightarrow t\bar{t}$ and $p\bar{p} \rightarrow t\bar{t}$ are symmetric under the exchange of top quarks and antiquarks. The relevant hard scattering processes are quark-antiquark annihilation, $q\bar{q} \rightarrow t\bar{t}$, and gluon-fusion, $gg \rightarrow t\bar{t}$. At NLO, two additional types of processes have to be considered which break this charge symmetry, namely quark-gluon scattering, $qg \rightarrow t\bar{t}q$, and radiative corrections to quark-antiquark annihilation [96, 97]. In both cases, the charge asymmetry is induced by interference effects, e.g. interference between processes with initial-state and final-state radiation or interference between Born and box diagrams. All gluon-fusion processes are symmetric under the exchange of the top quark and antiquark.

The charge asymmetry results in an asymmetry of the $t\bar{t}$ event kinematics: Top quarks (antiquarks) are preferentially emitted in the direction of the incoming quark (antiquark). The observables with which the charge asymmetry can be measured are chosen depending on the colliding particles and the centre-of-mass energy. Quark-antiquark annihilation dominates the production of $t\bar{t}$ pairs at the Tevatron, and so the top quark and antiquark will preferentially be emitted in the direction of the incoming protons and antiprotons, respectively. The most common observable is thus a forward-backward asymmetry,

$$A^{t\bar{t}} = \frac{N(\Delta y > 0) - N(\Delta y < 0)}{N(\Delta y > 0) + N(\Delta y < 0)},$$

where Δy is the difference between the rapidities of the top quark and antiquark, i.e. $\Delta y = y_t - y_{\bar{t}}$, and N is the number of events. The predictions for this observable depend on several kinematic variables. The inclusive forward-backward asymmetry is predicted to be $A^{t\bar{t}} = 0.088 \pm 0.006$ [98], while for invariant $t\bar{t}$ masses larger than 450 GeV the prediction increases to $A^{t\bar{t}}(m_{t\bar{t}} > 450\,\text{GeV}) = 0.129^{+0.008}_{-0.006}$ [98]. The CDF and DØ experiments have measured the forward-backward asymmetry both inclusively and as a function of several kinematic quantities, e.g. $m_{t\bar{t}}$ [99, 100]. They found an excess compared to the NLO QCD predictions with significances of several standard deviations. These measurements gave rise to speculations about contributions to $t\bar{t}$ production due to physics beyond the Standard Model during the past few years. Although refined theoretical studies and further measurements appear to have resolved the issue, measurements of the charge asymmetry are still in the focus of the LHC top-physics programme.

At the LHC with its symmetric pp initial state, the charge asymmetry can not be measured as a forward-backward asymmetry. Instead, a central-decentral asymmetry is defined. As valence quarks carry a larger average momentum fraction and top (anti)quarks are produced preferentially in the direction of the incoming (anti)quark, the average top-quark rapidity is larger than that of top antiquarks. A useful observable is defined as

$$A_C = \frac{N(\Delta|y| > 0) - N(\Delta|y| < 0)}{N(\Delta|y| > 0) + N(\Delta|y| < 0)},$$

where $\Delta|y|$ is the difference between the absolute values of the top-quark and top-antiquark rapidities, i.e. $\Delta|y| = |y_t| - |y_{\bar{t}}|$. The NLO QCD prediction including electroweak effects for the inclusive asymmetry is $A_C = 0.0123 \pm 0.0005$ [98]. Predictions are also available for different values of the invariant mass, the rapidity and the transverse momentum of the $t\bar{t}$ pair, and they range between -0.6 and 2.8 % (see discussion in [101]). The asymmetry depends on the first two variables because they are correlated to the fraction of quark-antiquark annihilation in $t\bar{t}$ production. It depends on the latter quantity because the amount of initial-state and finale-state radiation changes with increasing transverse momentum.

After an initial measurement using only a subset of the available data recorded at $\sqrt{s} = 7$ TeV [102], the ATLAS collaboration has studied the charge asymmetry based on the full data set, which corresponds to an integrated luminosity of 4.7 fb^{-1} [101]. Events are selected that are consistent with the single-lepton decay mode of $t\bar{t}$ production. For each event, the top-quark pair is reconstructed using a likelihood-based kinematic fit [103], and the rapidities of the top quark and antiquark are reconstructed. The measured distribution of $\Delta|y|$ includes background events and is distorted by detector and acceptance effects. A Bayesian unfolding technique [104] is applied on the background-subtracted spectrum to remove such effects. The measured inclusive asymmetry is

$$A_\mathrm{C} = 0.006 \pm 0.010,$$

where the largest sources of uncertainty are the statistical uncertainty and the uncertainty due to lepton and jet reconstruction. In addition, the asymmetry for invariant $t\bar{t}$ masses greater than 600 GeV is found to be $A_\mathrm{C}(m_{t\bar{t}} > 600\,\mathrm{GeV}) = 0.018 \pm 0.022$, which is in good agreement with the predicted value of $0.0175^{+0.0005}_{-0.0004}$. The asymmetry is also measured as a function of the transverse momentum, the absolute value of the rapidity and the invariant mass of the $t\bar{t}$ pair. The latter measurement is repeated for a subset of the events featuring a high longitudinal $t\bar{t}$ velocity, i.e. requiring $\beta_{z,t\bar{t}} > 0.6$. The asymmetries as a function of the invariant $t\bar{t}$ mass are shown in Fig. 7.15 without (a) and with (b) the additional velocity requirement. All four differential measurements are in agreement with the SM predictions.

The CMS collaboration has measured the charge asymmetry in a data set corresponding to an integrated luminosity of 5.0 fb^{-1} [105, 106] using events with exactly one charged lepton in the final state. After event reconstruction and calculation of the top-quark and top-antiquark rapidities, the $\Delta|y|$ distribution is determined. Subsequently, the estimated background contributions are subtracted from the data and the spectra are corrected for detector and acceptance effects using a regularised unfolding procedure via matrix inversion [107]. The measured inclusive asymmetry is

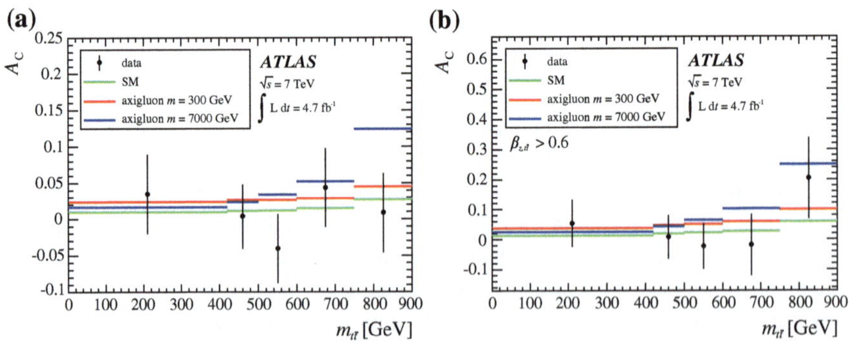

Fig. 7.15 Charge asymmetry measured by the ATLAS collaboration as a function of the invariant $t\bar{t}$ mass **a** without and **b** with the requirement of a large longitudinal $t\bar{t}$ velocity. Adapted from [101]

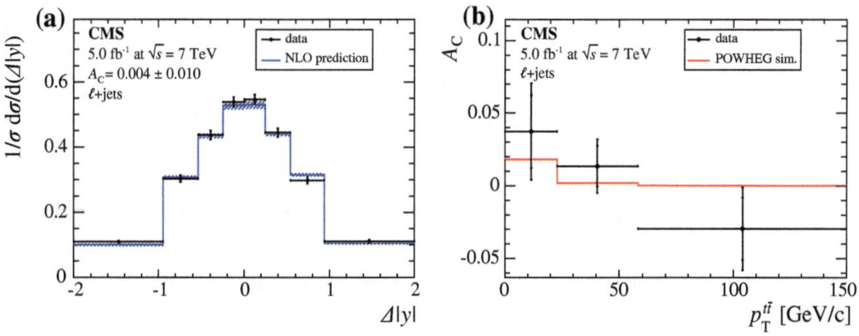

Fig. 7.16 **a** The background-subtracted and unfolded $\Delta|y|$ distribution obtained by CMS for a single-lepton sample. **b** The charge asymmetry as a function of the transverse momentum of the $t\bar{t}$ pair. Adapted from [106]

$$A_C = 0.004 \pm 0.010\,(\text{stat}) \pm 0.011\,(\text{syst}),$$

where the major sources of systematic uncertainty are the residual model dependence of the unfolding procedure and the lepton reconstruction. As in the ATLAS measurement, the asymmetry is also measured as a function of the transverse momentum, the rapidity and the invariant mass of the $t\bar{t}$ system. Figure 7.16a shows the background-subtracted and unfolded $\Delta|y|$ distribution for the inclusive case, and Fig. 7.16b shows the charge asymmetry as a function of the transverse momentum of the $t\bar{t}$ pair. All measurements are consistent with the SM predictions.

In a further study using the full 7 TeV data set, the CMS collaboration measured the charge asymmetry in a sample of dileptonically decaying $t\bar{t}$ pairs [108]. In addition to the observable A_C, a "lepton charge asymmetry" is defined as

$$A_C^{\text{lep}} = \frac{N(\Delta|\eta_\ell| > 0) - N(\Delta|\eta_\ell| < 0)}{N(\Delta|\eta_\ell| > 0) + N(\Delta|\eta_\ell| < 0)},$$

where $\Delta|\eta_\ell| = |\eta_{\ell^+}| - |\eta_{\ell^-}|$ and where $\eta_{\ell^\pm}$ are the pseudo-rapidities of the positively and negatively charged leptons in each event. The SM prediction for this observable is $A_C^{\text{lep}} = 0.0070 \pm 0.0003$ [98]. The event reconstruction is performed using the "analytical matrix weighting technique" (AMWT) [40]. After subtracting all background contributions from the measured $\Delta|y|$ and $\Delta|\eta_\ell|$ distributions, the spectra are unfolded using singular value decomposition [109]. The measured asymmetries are

$$\begin{aligned} A_C &= -0.010 \pm 0.017\,(\text{stat}) \pm 0.008\,(\text{syst}) \text{ and} \\ A_C^{\text{lep}} &= 0.009 \pm 0.010\,(\text{stat}) \pm 0.006\,(\text{syst}), \end{aligned}$$

where the largest sources of systematic uncertainty are residual biases in the unfolding procedure and uncertainties in the $t\bar{t}$ modelling and the jet reconstruction. The lepton charge asymmetry is also calculated as a function of the same three kinematic variables as for the single-lepton analysis. No deviations between the measurements and the predictions are found.

The charge asymmetry in $t\bar{t}$ production is predicted to cause a small effect on a variety of observables at the LHC. None of the measurements performed by the ATLAS and CMS collaborations are in conflict with the SM predictions while the current experimental precision is of the order of the size of the prediction itself. Although a variety of models of physics beyond the Standard Model can be excluded with the set of measurements already performed, the analysis of the 8 TeV data (which was not finished at the time of publication) will provide further sensitivity to the predictions of perturbative QCD. It should be noted, that the interpretation of the leptonic charge asymmetries relies on a solid understanding of top-quark production and decay. In particular, a non-standard top-quark polarisation could affect the leptonic asymmetries. It is thus important to cross check the polarisation through explicit measurements.

7.4.2 Top-Quark Polarisation and Spin Correlation in Top-Quark Pairs

Top-quark polarisation in $t\bar{t}$ events and the correlation between the top-quark and antiquark spins in $t\bar{t}$ production are probes of perturbative QCD and observables that are sensitive to anomalous production mechanisms. Measurements of such quantities are only feasible because of the extremely short lifetime of the top quark of $\tau_t \approx 1.5 \times 10^{-25}$ s. The lifetime is roughly one order of magnitude smaller than the time scale at which hadronization takes place, $\tau_{\text{had}} \approx 1/\Lambda_{\text{QCD}} \approx 3 \times 10^{-24}$ s, and in particular shorter than the time needed to decorrelate the spin configuration of the $t\bar{t}$ pair, $\tau_{\text{decorr}} \approx \frac{\hbar m_t}{\Lambda^2_{\text{QCD}}} \approx 3 \times 10^{-21}$ s [110, 111]. Top quarks will thus decay before they can form bound states. As mentioned before, the large top-quark width cuts off non-perturbative effects. The polarisation of the top-quark is thus not diluted by hadronization effects and can be calculated reliably within perturbation theory. The parity-violating weak decay can then be used to analyse the top-quark polarisation through the angular distribution of the decay products. The correlation of the top-quark and antiquark spins is reflected in the angular correlation of the top-quark and antiquark decay products. This is a unique feature since all lighter quarks form hadrons for which—due to the hadronization process—the initial spin information of the mother particle is diluted or even entirely lost. In contrast, the spin information of the top quark is directly transferred to its decay products.

In top-quark pairs, information about the polarisation of top quarks and the correlation between top-quark and antiquark spins can be obtained from the differential cross section

$$\frac{1}{\sigma}\frac{d\sigma}{d\cos(\theta_1)d\cos(\theta_2)} = \frac{1}{4}\left(1+\alpha_1 P_1\cos(\theta_1)+\alpha_2 P_2\cos(\theta_2)\right.$$
$$\left.-\alpha_1\alpha_2 A\cos(\theta_1)\cos(\theta_2)\right), \qquad (7.4)$$

where $\theta_1(\theta_2)$ are the angles between the momentum direction of a daughter particle of the top (anti)quark and a chosen reference axis. The coefficients $\alpha_1(\alpha_2)$ and $P_1(P_2)$ are the "spin-analysing power" of the daughter particle and the degree of polarisation (with respect to the reference axis) of the top (anti)quark, respectively. The spin-analysing power quantifies the amount of spin information transferred to the daughter particle and depends on the particle type. It is approximately one for charged leptons and down-type quarks from the subsequent decay of the W boson [112]. The coefficient A is a measure of the spin correlation between top quark and antiquark.

As the strong interaction conserves parity, the polarisation of top (anti)quarks in $t\bar{t}$ production within the production plane is expected to be zero. QCD absorptive parts, sometimes also called final-state interactions,[3] introduce a tiny transverse polarisation at the one-loop level [113, 114]. Electroweak corrections lead to a small amount of net polarisation of $\alpha_l P_l = 0.003 \pm 0.001$ [115]. It can be shown that polarisation can be induced by the imaginary part of a chromo-electric dipole moment which in turn can lead to P-odd and CP-odd terms in the matrix elements [115]. Such effects can stem from processes beyond the SM.

The correlation coefficient A can be expressed as an asymmetry variable in the number of events N with parallel and antiparallel spin,

$$A = \frac{N(\uparrow\uparrow)+N(\downarrow\downarrow)-N(\uparrow\downarrow)-N(\downarrow\uparrow)}{N(\uparrow\uparrow)+N(\downarrow\downarrow)+N(\uparrow\downarrow)+N(\downarrow\uparrow)},$$

where $\uparrow$ and $\downarrow$ indicate the spin projections onto the reference axis. The prediction of the correlation coefficient depends on the particular choice of reference axis. While for measurements at the Tevatron the "off-diagonal basis" and the "beam basis" are suitable choices, measurements at the LHC are most sensitive to the correlation coefficient in the "helicity basis". In the beam basis, the direction of the beam is used as reference axis for the top-quark as well as for the antiquark. In the helicity basis, the direction of flight of the top-quark/antiquark is used as respective reference axis. The strength of the correlation predicted by the SM using the helicity basis is $A = 0.031$ [116] with an uncertainty of approximately 1 %.

Close to the threshold, top-quark pairs produced via gluon fusion are in a $^1S_0$ state while top-quark pairs produced via quark-antiquark annihilation are in a $^3S_1$ state. As a consequence, the spins of a top-quark pair produced in quark-antiquark annihilation tend to be parallel while in gluon fusion they tend to be antiparallel. A measurement of the coefficient A is thus a direct probe of the production mechanism. Contributions from additional production mechanisms, e.g. yet unknown intermediate vector

[3]These are due to imaginary parts of the loop integrals caused by the on-shell production of intermediate states.

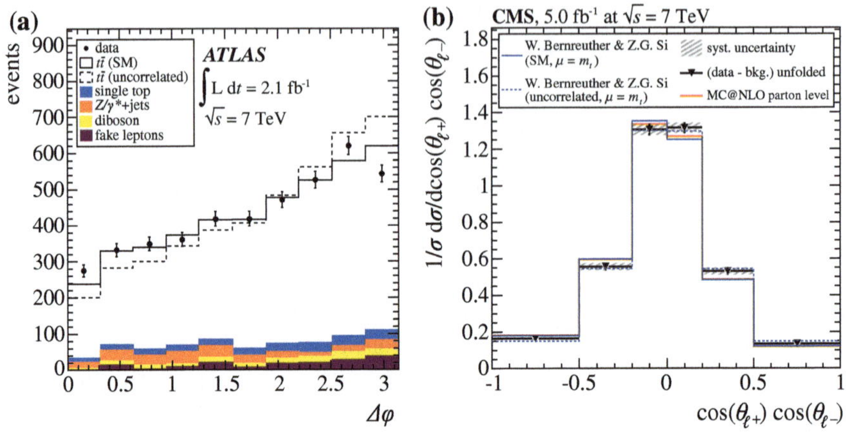

Fig. 7.17 **a** Observed and predicted $\Delta\phi$ spectrum in a sample of dileptonic $t\bar{t}$ events recorded with the ATLAS experiment. **b** Distribution of $\cos(\theta_{\ell+})\cos(\theta_{\ell-})$ in dileptonic $t\bar{t}$ events as observed by CMS together with predictions for two different correlation scenarios. Adapted from [118, 120]

bosons, can lead to altered predictions for A. Note that the correlation coefficient depends both on the centre-of-mass energy and on the initial-state particles, and so the measurements conducted at the Tevatron and the LHC are complementary.

While an evidence for correlated spins in top-quark pairs was already reported by the DØ collaboration [117], the hypothesis that the spins of the top quark and the antiquark are uncorrelated was fully disproved for the first time by a measurement of the ATLAS collaboration [118]. The measurement is based on a data set corresponding to an integrated luminosity of 2.1 fb^{-1} recorded at $\sqrt{s} = 7$ TeV. Events consistent with the signature of $t\bar{t}$ events decaying in the dilepton mode are selected, and the spin correlation is probed using the difference in azimuthal angle between the two charged leptons, $\Delta\phi$, calculated in the laboratory frame. The advantage of this observable compared to those in (7.4) is that no kinematic reconstruction of the top-quark momenta is necessary while the sensitivity to the correlation strength is largely retained [111]. Figure 7.17a shows the observed $\Delta\phi$ spectrum as well as the predictions for the assumption of SM correlations and the absence of correlations. The measured correlation coefficient is not compatible with zero,

$$A = 0.40 \pm 0.04\ \text{(stat)}^{+0.08}_{-0.07}\ \text{(syst)},$$

with a significance of 5.1 standard deviations. The dominating sources of systematic uncertainty are the estimate of events with misidentified leptons and the jet reconstruction. In a second publication based on the full 7 TeV data set [119], the spin correlation is measured in the dilepton channel using a variety of alternative observables, and it also measured in the single-lepton channel.

The CMS collaboration has analysed a data set corresponding to an integrated luminosity of 5.0 fb^{-1} recorded at $\sqrt{s} = 7$ TeV [120]. Events are selected that contain exactly two charged leptons with large transverse momenta. Three angular variables are calculated for each event—the $\Delta\phi$ variable defined earlier as well as the angles θ_{ℓ^+} and θ_{ℓ^-}, which are defined as the angles of the positively and negatively charged lepton in the helicity frame, respectively. From the distributions of these variables, two asymmetries are derived which provide discrimination between the two scenarios of SM correlations and no correlations:

$$A_{\Delta\phi} = \frac{N(\Delta\phi > \pi/2) - N(\Delta\phi < \pi/2)}{N(\Delta\phi > \pi/2) + N(\Delta\phi < \pi/2)},$$

$$A_{\cos\cos} = \frac{N(\cos(\theta_{\ell^+})\cos(\theta_{\ell^-}) > 0) - N(\cos(\theta_{\ell^+})\cos(\theta_{\ell^-}) < 0)}{N(\cos(\theta_{\ell^+})\cos(\theta_{\ell^-}) > 0) + N(\cos(\theta_{\ell^+})\cos(\theta_{\ell^-}) < 0)}.$$

The latter asymmetry is a measure for the correlation coefficient in the helicity basis, i.e. $A = -4 \cdot A_{\cos\cos}$ [115]. The predictions at NLO perturbation theory for the case of SM-like correlations (no correlations) are $A_{\Delta\phi} = 0.115^{+0.014}_{-0.016}$ ($A_{\Delta\phi} = 0.210^{+0.013}_{-0.008}$) and $A_{\cos\cos} = -0.078 \pm 0.006$ ($A_{\cos\cos} = 0$), see [115, 116] and references in [120]. The angles $\theta_{\ell^\pm}$ require the explicit reconstruction of both the top quark and the antiquark, which is done using the AMWT technique. In addition, the relation between the asymmetry variable and the correlation coefficient is valid only if no acceptance cuts and detector effects distort the measurement. The distributions of $\Delta\phi$ and $\cos(\theta_{\ell^+})\cos(\theta_{\ell^-})$ are thus unfolded using singular value decomposition. Figure 7.17b shows the unfolded $\cos(\theta_{\ell^+})\cos(\theta_{\ell^-})$ distribution and the predictions for the two correlation scenarios. The asymmetries are measured to be

$$A_{\Delta\phi} = 0.133 \pm 0.010\,(\text{stat}) \pm 0.007\,(\text{syst}) \pm 0.012\,(\text{top } p_T) \text{ and}$$

$$A_{\cos\cos} = -0.021 \pm 0.023\,(\text{stat}) \pm 0.027\,(\text{syst}) \pm 0.010\,(\text{top } p_T),$$

where the statistical and systematic uncertainties as well as an uncertainty associated with the modelling of the p_T spectrum of the top quark are given. The largest systematic uncertainties come from the unfolding procedure as well as the jet reconstruction, the background estimate and the modelling of $t\bar{t}$ events.

The measurements conducted by ATLAS and CMS show that the spins of the top quark and the antiquark in $t\bar{t}$ events are indeed correlated and that the amount of correlation is as expected from perturbation theory at NLO. The production mechanisms of $t\bar{t}$ events is thus consistent with that predicted by QCD, and no indications for additional production mechanisms are found.

The polarisation of top quarks in $t\bar{t}$ events is studied by CMS using the same data set and event selection as for the measurement of the $t\bar{t}$ spin correlation [120]. The polarisation is estimated from the unfolded distribution of the angle $\theta_{\ell^\pm}$ and the resulting asymmetry,

$$A_P = \frac{N(\cos(\theta_\ell) > 0) - N(\cos(\theta_\ell) < 0)}{N(\cos(\theta_\ell) > 0) + N(\cos(\theta_\ell) < 0)}.$$

The polarisation in the helicity basis is then $P = 2A_P$. The asymmetry is calculated using positively and negatively charged leptons under the assumption of *CP* invariance and is measured to be

$$A_P = 0.005 \pm 0.013\,(\text{stat}) \pm 0.020\,(\text{syst}) \pm 0.008\,(\text{top } p_\text{T}),$$

where the uncertainties are again the statistical and systematic ones and the uncertainty due to the mismodelled p_T spectrum of the top quark. The two largest systematic uncertainties are uncertainties on the top-quark mass and on the jet reconstruction.

An ATLAS measurement of the top-quark polarisation [121] uses $t\bar{t}$ events decaying in the single-lepton and dilepton mode. The data were collected at $\sqrt{s} = 7$ TeV and correspond to an integrated luminosity of $4.66 \pm 0.08\,\text{fb}^{-1}$. The full $t\bar{t}$ system is reconstructed using a kinematic likelihood [103] in the single-lepton channel and the neutrino-weighting method [122] in the dilepton channel. The angle $\theta_{\ell^\pm}$ is calculated in the helicity basis for each lepton, and the corresponding distributions are fitted with templates of partially polarised top quarks. The spin correlation is assumed to be that of the SM. The fits are done separately for each lepton type as well as for positively and negatively charged leptons. The latter is done in order to distinguish scenarios in which the polarisation comes from *CP*-conserving processes and maximally *CP*-violating processes. For such cases, the expressions $\alpha_\ell P$ are either the same for top quark and antiquark, or they differ by a sign. Figure 7.18 shows the distributions of $\cos(\theta_{\ell^\pm})$ measured in single-lepton events for the two scenarios. The combined polarisation measured for the two scenarios is

$$\begin{aligned}
\alpha_\ell P &= -0.035 \pm 0.014\,(\text{stat}) \pm 0.037\,(\text{syst})\ (\mathit{CP}\text{ invariance}),\\
\alpha_\ell P &= 0.020 \pm 0.016\,(\text{stat})^{+0.013}_{-0.017}\,(\text{syst})\ (\mathit{CP}\text{ violation}).
\end{aligned}$$

The major systematic uncertainty stems from the jet reconstruction.

The ATLAS and CMS measurements are both consistent with the SM prediction that top quarks produced in $t\bar{t}$ pairs via the strong interaction are not polarised.

7.5 Tests of Electroweak Predictions

Studies of the electroweak couplings of the top quark comprise measurements of a number of different observables. While the polarisation of W bosons from top-quark decays is a consequence of the $V-A$ structure of the Wtb vertex, the cross sections for single top-quark production depend directly on the strength of the coupling to W bosons (see Sect. 7.6). The coupling strength to Z bosons and photons can be probed by measurements of $t\bar{t}$ production with additional such bosons.

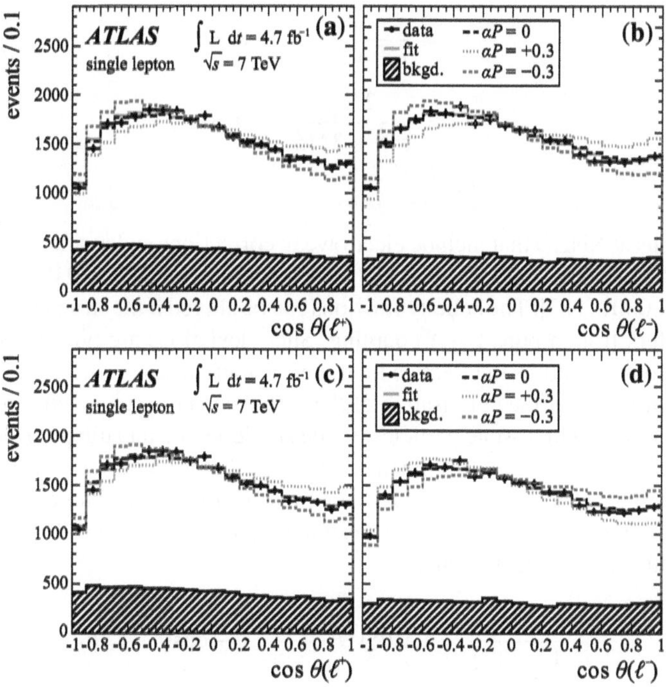

Fig. 7.18 ATLAS polarisation measurements in the single-lepton channel with three different polarisation assumptions. **a** and **c** show the distributions for positively charged leptons, while **b** and **d** show the distributions for negatively charged leptons. (**a, b**) assume *CP* invariance, while for (**c, d**) maximal *CP* violation is assumed. Adapted from [121]

7.5.1 W-Boson Polarisation

The massive W bosons produced in top-quark decays are real spin-1 particles and thus have three possible polarisation states. We will refer to them as longitudinally, left-handedly or right-handedly polarised W bosons. The net amount of polarisation is given by the fractions of the partial decay widths for differently polarised W bosons, Γ_0, Γ_L and Γ_R, respectively. These "helicity fractions" are

$$F_{0/L/R} = \frac{\Gamma_{0/L/R}}{\Gamma_0 + \Gamma_L + \Gamma_R}.$$

In perturbation theory at LO, and neglecting the mass of the bottom quark, these fractions depend solely on the masses of the top quark and the W boson, i.e.

$$F_0 = \frac{m_t^2}{m_t^2 + 2M_W^2} \approx 0.70,$$
$$F_L = \frac{2M_W^2}{m_t^2 + 2M_W^2} \approx 0.30,$$
$$F_R = 0.$$

Calculations at NNLO that include electroweak corrections and assume a finite mass of the bottom quark yield $F_0 = 0.687 \pm 0.005$, $F_L = 0.311 \pm 0.005$ and $F_R = 0.0017 \pm 0.0001$ [123]. The helicity fractions can be altered if the structure of the Wtb vertex differs from a pure $V-A$ coupling. Such deviations are typically described by anomalous couplings in effective field theory approaches [124, 125].

Information about the polarisation of the W boson can be obtained from angular distributions of the final-state particles. In the single-lepton and dilepton channels of $t\bar{t}$ production, the angle θ^* is defined as the angle between the reverse momentum of the leptonically decaying top quark and the direction of the charged lepton, both evaluated in the rest frame of the corresponding W boson [126]. The differential decay width can then be written as

$$\frac{1}{\Gamma}\frac{d\Gamma}{d\cos(\theta^*)} = \sin(\theta^*)^2 F_0 + \frac{3}{8}\left(1 - \cos(\theta^*)\right)^2 F_L + \frac{3}{8}\left(1 + \cos(\theta^*)\right)^2 F_R. \tag{7.5}$$

An angle for the hadronically decaying top quark can be defined analogously. The ATLAS and CMS collaborations have both made use of the angular dependence described by (7.5) to estimate the helicity fractions.

The ATLAS collaboration has analysed a data set corresponding to an integrated luminosity of $1.04\,\mathrm{fb}^{-1}$ taken at a centre-of-mass energy of $\sqrt{s} = 7$ TeV [127]. Two sets of selection criteria are defined so as to enrich samples with events stemming from $t\bar{t}$ production with subsequent decay either in the single-lepton or dilepton decay modes. In both cases, the reconstruction of the top quarks is based on the final-state particles and assumptions on the detector performance. For both event types, an individual analysis strategy is followed. While the first strategy is based on a template fit of the $\cos(\theta^*)$ distributions, the second makes use of the angular asymmetries derived from unfolded $\cos(\theta^*)$ spectra. The individual results are all found to be in agreement and are combined using the BLUE method [128]. Although both analysis methods are based on the same data set, the combined result has a smaller overall uncertainty due to the different sensitivities to sources of systematic uncertainty. The largest systematic uncertainties are due to the signal and background modelling, to the jet reconstruction and to method-specific uncertainties. The combined helicity fractions are

$$F_0 = 0.67 \pm 0.03\ \text{(stat)} \pm 0.06\ \text{(syst)},$$
$$F_L = 0.32 \pm 0.02\ \text{(stat)} \pm 0.03\ \text{(syst)},$$
$$F_R = 0.01 \pm 0.01\ \text{(stat)} \pm 0.04\ \text{(syst)},$$

with a correlation between F_0 and F_L of -0.96.

The CMS collaboration has analysed the full 7 TeV data set corresponding to an integrated luminosity of $5.0\,\text{fb}^{-1}$ [129]. Events are selected that are compatible with the single-lepton decay mode, and top quarks are reconstructed from the final-state particles using a constrained fit. An estimate of the helicity fractions is obtained using a reweighting procedure and a subsequent fit to the $\cos(\theta^*)$ distribution. The largest sources of systematic uncertainty are the background estimate and the jet reconstruction. The helicity fractions are estimated to be

$$F_0 = 0.682 \pm 0.030\ \text{(stat)} \pm 0.033\ \text{(syst)},$$
$$F_L = 0.310 \pm 0.022\ \text{(stat)} \pm 0.022\ \text{(syst)},$$
$$F_R = 0.008 \pm 0.012\ \text{(stat)} \pm 0.014\ \text{(syst)},$$

with a correlation between F_0 and F_L of -0.95. A more recent measurement of the helicity fractions using a data sample collected at $\sqrt{s} = 8\,\text{TeV}$ and enriched with single top-quark events is in good agreement with these values [130].

The results obtained by the ATLAS and CMS collaborations are consistent with one another and with the NNLO predictions. Both measurements are more precise than those published by Tevatron experiments.

The measured helicity fractions are also interpreted in terms of anomalous couplings. Figure 7.19a shows the 68 % and 95 % CL contours in the plane of the (real) tensor couplings g_R and g_L assuming vector couplings of $V_L = 1$ and $V_R = 0$ as

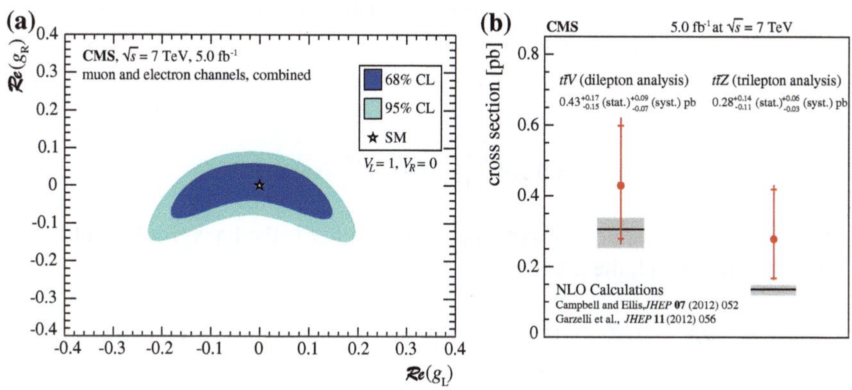

Fig. 7.19 **a** The 68 % and 95 % CL contours in the plane of the (real) tensor couplings g_R and g_L assuming vector couplings of $V_L = 1$ and $V_R = 0$ as obtained by the CMS collaboration. **b** Measured and predicted values of the cross sections for $t\bar{t}V$ and $t\bar{t}Z$ production. The *inner* and *outer error bars* indicate the statistical and total uncertainties, respectively. Adapted from [129, 131]

obtained by the more precise CMS result. A second region of solutions featuring large values of the real part of g_R is excluded from the fit as it is not compatible with the measurement of the single-top t-channel cross section. The results are consistent with the absence of anomalous couplings, i.e. $g_L = g_R = 0$, and are in very good agreement with the predicted $V - A$ structure of the Wtb vertex.

7.5.2 Top-Quark Pairs and Additional Gauge Bosons

While the production of a top-quark pair and an additional photon has only been observed at the Tevatron [132], the CMS collaboration was the first to measure the production of top-quark pairs with additional Z bosons ($t\bar{t}Z$) and W bosons ($t\bar{t}W$) [131, 133]. Such rare processes are expected in the SM. For a centre-of-mass energy of $\sqrt{s} = 7\,\text{TeV}$, calculations at NLO yield predictions for the corresponding cross sections $\sigma_{t\bar{t}Z}$ [134] and $\sigma_{t\bar{t}W}$ [135] of

$$\sigma_{t\bar{t}Z} = 0.137^{+0.012}_{-0.016}\,\text{pb},$$
$$\sigma_{t\bar{t}W} = 0.169^{+0.029}_{-0.051}\,\text{pb}.$$

Here the analysis at $\sqrt{s} = 7$ TeV is briefly described [131]. The data set collected corresponds to an integrated luminosity of $5.0\,\text{fb}^{-1}$. Events are selected according to the number of charged leptons in the final state. The process

$$pp \to t\bar{t}Z \to (t \to b\ell^{\pm}\nu)(t \to bjj)(Z \to \ell^{\pm}\ell^{\mp})$$

is searched for by requiring two leptons with the same flavour but opposite electric charge and with an invariant mass compatible with the mass of the Z boson, and one additional charged lepton. After the event selection, nine events are observed while 3.2 ± 0.8 background events are expected. The resulting cross section is estimated to be

$$\sigma_{t\bar{t}Z} = 0.28^{+0.14}_{-0.11}\,(\text{stat})^{+0.06}_{-0.03}\,(\text{syst})\,\text{pb},$$

where the dominant source of systematic uncertainty is the background yield.

On the other hand, the processes

$$pp \to t\bar{t}Z \to (t \to b\ell^{\pm}\nu)(t \to bjj)(Z \to \ell^{\pm}\ell^{\mp}), \text{ and}$$
$$pp \to t\bar{t}W \to (t \to b\ell^{\pm}\nu)(t \to bjj)(W \to \ell^{\pm}\nu)$$

are searched for by selecting events with exactly two leptons of the same electric charge. After the event selection, 16 events remain, while the background expectation is 9.2 ± 2.6. The combined cross section, $\sigma_{t\bar{t}V}$, where $V = W, Z$, is measured as

$$\sigma_{t\bar{t}V} = 0.43^{+0.17}_{-0.15}\,(\text{stat})^{+0.09}_{-0.07}\,(\text{syst})\,\text{pb}.$$

Both measurements and their NLO predictions are shown in Fig. 7.19b. The measured cross sections are in agreement with the predictions, which indicates that no deviation from the strength of the top-quark coupling to Z and W bosons predicted by the SM is observed. The results from the more recent CMS analysis at $\sqrt{s} = 8$ TeV [133] are also consistent with this conclusion.

7.6 Single Top-Quark Production

Single top-quark production, in contrast to top-quark pair-production which proceeds through the strong interactions, takes place by virtue of charged-current interactions. The production rate for single top quarks is suppressed with respect to top-quark pair production by a factor of 2–3 due to the different couplings strengths α_W and α_s of weak and strong interactions, which is only partially compensated by the larger partonic fluxes due to the lower production threshold. Feynman diagrams of the LO processes contributing to the scattering amplitudes are shown in Fig. 7.20.

Depending on whether the W boson is space-like (Fig. 7.20a), time-like (Fig. 7.20b) or real (Fig. 7.20c, d), one distinguishes between the t-channel, the s-channel and the tW-channel. In the latter case, a single top quark is produced in association with a W boson in the final state. The dominant contribution to single top-quark production at the Tevatron and the LHC is the t-channel. As can be seen from Fig. 7.20a, this channel assumes the existence of a b quark inside the proton and thus requires in the theoretical description the so-called five-flavour scheme, in which u, d, s, c, b are treated as active flavours inside the proton. In the four-flavour scheme, t-channel production occurs formally at higher orders of the QCD coupling, as is illustrated in Fig. 7.21. In general, calculations in these two schemes should give

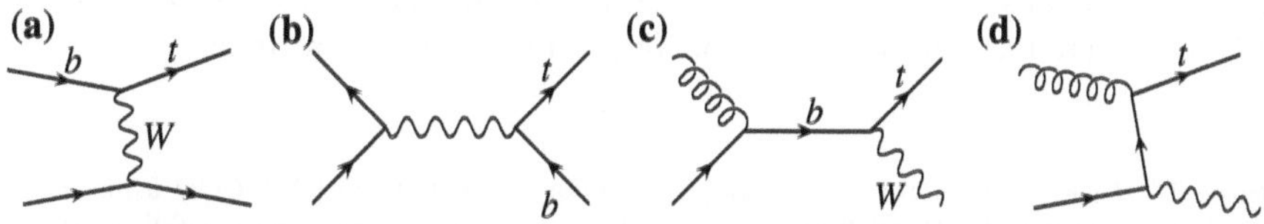

Fig. 7.20 Leading-order Feynman diagrams of processes contributing to single top-quark production: **a** t-channel, **b** s-channel, **c, d** tW-channel

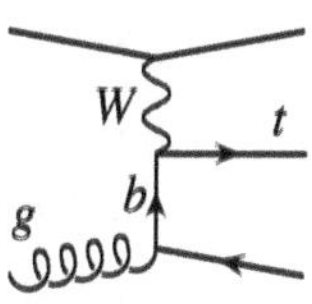

Fig. 7.21 Leading-order Feynman diagram for t-channel single top-quark production in the four-flavour scheme

similar results. Differences occur if logarithmic corrections of the form $\ln({m_b}^2/Q^2)$ become large, where Q denotes a generic energy scale. The five-flavour scheme may partially resume these logarithms through the evolution of the b-quark parton distribution function, while in the four-flavour scheme these logarithms are kept at fixed order only. In [136, 137] it has been checked through an explicit calculation that the two schemes lead indeed to consistent results for the cross section.

At the LHC, the second important production channel for single top quarks is the tW channel. Due to phase-space suppression and the small gluon luminosity, this channel gives only a tiny contribution at the Tevatron. In contrast, s-channel production, which is roughly responsible for one third of the cross section at the Tevatron, leads only to a contribution of a few percent at the LHC. Single top-quark production at the LHC is thus to some extent complementary to that at the Tevatron. A further major difference arises from the fact that the initial state at the Tevatron is a *CP* eigenstate. Since *CP*-violating effects are negligible in single top-quark production within the SM, the numbers of produced single top quarks and antiquarks are identical at the Tevatron. At the LHC, however, the initial state is not a *CP* eigenstate, and more top quarks than antiquarks are produced because in pp collisions the flux of up-type quarks is larger than the flux for down-type quarks.

The NLO QCD corrections for inclusive single top-quark production have been presented in [138–140] for the t-channel, in [141] for the s-channel and in [142, 143] for the tW channel.

These cross sections are shown in Table 7.2, assuming $m_t = 173.3$ GeV and using the MSTW2008lo/nlo68cl pdf set [26]. It turns out that the NLO corrections are only a few percent for the t-channel and slightly larger for the tW channel. In contrast, the NLO corrections to the s-channel contribution are about 30 %. However, as has been pointed out before, this channel gives only a small contribution at the LHC. Since at NLO, no colour transfer between the two quark lines is allowed (the corresponding box contributions vanish after interference with the Born amplitude), it is conceivable that the small corrections are accidental and that the small scale uncertainty observed at NLO underestimates the possible size of the NNLO corrections. Very recently, the vertex corrections for the t-channel have been calculated at NNLO QCD [144]. The corrections are found to be comparable in size to the NLO corrections.

Table 7.2 Cross sections for single top-quark production at LO and NLO QCD using $m_t = 173.3$ GeV and the MSTW2008lo/nlo PDF set. See text for references. The notation is $\sigma(\mu = m_t)^{\sigma(\mu=2m_t)-\sigma(\mu=m_t)+\text{ PDF err. up}}_{\sigma(\mu=\frac{m_t}{2})-\sigma(\mu=m_t)-\text{ PDF err. down}}$

	LHC 8 TeV				Tevatron	
	σ_t^{LO}	$\sigma_{\bar{t}}^{LO}$	σ_t^{NLO}	$\sigma_{\bar{t}}^{NLO}$	$\sigma_{t,\bar{t}}^{LO}$	$\sigma_{t,\bar{t}}^{NLO}$
t	53.8	29.1	$55.2^{+1.6\ +0.35}_{-0.9\ -0.32}$	$30.1^{+0.9\ +0.29}_{-0.5\ -0.32}$	1.03	$0.998^{+0.025\ +0.029}_{-0.022\ -0.032}$
s	2.22	1.24	$3.30^{-0.06\ +0.07}_{+0.08\ -0.05}$	$1.90^{-0.04\ +0.04}_{+0.05\ -0.03}$	0.28	$0.442^{-0.023\ +0.015}_{+0.025\ -0.011}$
tW	8.86	8.85	$9.12^{+0.21\ +0.29}_{-0.38\ -0.36}$	$9.11^{+0.21\ +0.29}_{-0.38\ -0.36}$	0.069	$0.070^{-0.002\ +0.008}_{-0.001\ -0.009}$

Fully differential results at NLO accuracy for single top-quark production have been presented in [145–147]. In [148–151], the analysis has been extended by including the semileptonic decay of the top quark in the narrow-width approximation. In addition, the systematic combination of the NLO corrections with the parton shower has been investigated in MC@NLO [152, 153] as well as in the POWHEG framework [154, 155]. Beyond fixed-order perturbation theory, the impact of logarithmic corrections due to soft-gluon emission has been studied [156–161].

As will be seen below, the experimental study of single top-quark production is challenging—despite the sizeable cross sections—because of the complicated event signature and the large backgrounds. In contrast to $t\bar{t}$ production, singly-produced top quarks are highly polarised—a consequence of their $V-A$ coupling to the W boson. The measurement of the top-quark polarisation thus provides a further test of the $V-A$ structure of the Wtb vertex. Assuming that the production mechanism is well understood, t-channel single top-quark production can also be used to constrain the b parton distribution in the proton. Furthermore, it can be used for a direct measurement of the CKM matrix element V_{tb} which is otherwise only indirectly accessible, via the assumptions of CKM matrix unitarity and of the existence of only three quark families.

7.6.1 *t*-Channel Production

Measurements of the t-channel production cross section have been performed using events with exactly one isolated lepton (electron or muon) originating from the decay of the W boson and two or three jets in the final state. One of the jets has to be identified as a b jet. Additional requirements on kinematic observables, such as the missing transverse momentum or the transverse mass of the W boson, are imposed in order to further remove background. Detailed and precise results are available from ATLAS and CMS [162–166].

An early analysis of t-channel single-top production was performed by the ATLAS collaboration based on an integrated luminosity of about 1 fb^{-1} [162]. Events with two or three jets are selected. To separate t-channel single top-quark signal events from backgrounds, several kinematic variables are combined into one discriminant by employing a neural network that also exploits correlations between the variables. The most discriminating variable for the two-jet sample is the invariant mass of the system formed by the b-tagged jet, the charged lepton and the neutrino, $m_{\ell\nu b}$, see Fig. 7.22a. In the three-jet category, the invariant mass of the two leading jets and the absolute value of the difference in pseudo-rapidity of the leading and the lowest-p_{T} jet are among the most discriminating variables. Multi-jet event yields are determined with data-driven techniques, while contributions from W+jets events are derived from simulation and normalised to data in control regions using a cut-based analysis. All other backgrounds and the t-channel signal expectation are normalised to theoretical cross sections. To extract the signal content of the selected sample, a maximum-likelihood fit is performed to the output distributions of the neural net

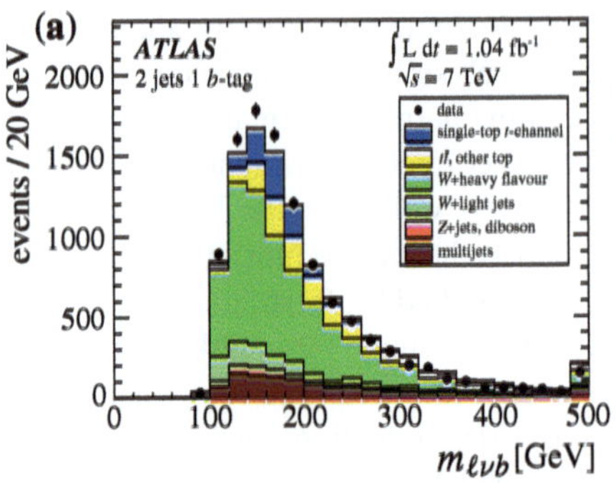

Fig. 7.22 Distribution of the invariant mass of the b-tagged jet, the charged lepton and the neutrino in the b-tagged sample for two-jet events **a** for the inclusive sample and **b** after cuts. The last histogram bin includes overflow. Adapted from [162]

in the two-jet and three-jet data sets. From a simultaneous measurement in the two channels, a cross section of

$$\sigma_{t+\bar{t}} = 83 \pm 4 \text{ (stat)}^{+20}_{-19} \text{ (syst) pb}$$

is measured. This result is confirmed in a cut-based analysis which is illustrated by the distribution in Fig. 7.22b.

More recently, the CMS collaboration performed an analysis based on the full statistics available for $\sqrt{s} = 8$ TeV [165]. The event sample is selected by the application of simple criteria: The events must contain exactly one muon or electron with large transverse momentum. They are categorised according to the numbers of jets and b-tagged jets, and the category enriched with t-channel signal is the one with two jets and one tag. One of the jets, denoted by j', is expected to not originate from b quarks, and its pseudo-rapidity distribution is typical of the t-channel processes where a light parton recoils against a much more massive particle like the top quark. Signal events populate forward regions in the $|\eta_{j'}|$ spectrum, and this feature is used to distinguish the signal from background. Background events from $t\bar{t}$ and from W+jets processes are determined by the use of control categories. In all categories the invariant mass $m_{\ell\nu b}$ is used to define a signal region and a side-band region that contain events inside and outside the reconstructed top-quark mass window of $130 < m_{\ell\nu b} < 220$ GeV, respectively. To determine the contribution from signal events, a binned maximum-likelihood fit is performed to the $|\eta_{j'}|$ distribution of the events in the signal region of the category with two jets and one tag. In Fig. 7.23a the $m_{\ell\nu b}$ distribution is shown for events with forward jets. The figure illustrates that large-purity samples of t-channel single-top quark events can be isolated at the LHC using simple selection criteria. The measured cross section for this process is

$$\sigma_t = 83.6 \pm 2.3 \text{ (stat)} \pm 7.4 \text{ (syst) pb.}$$

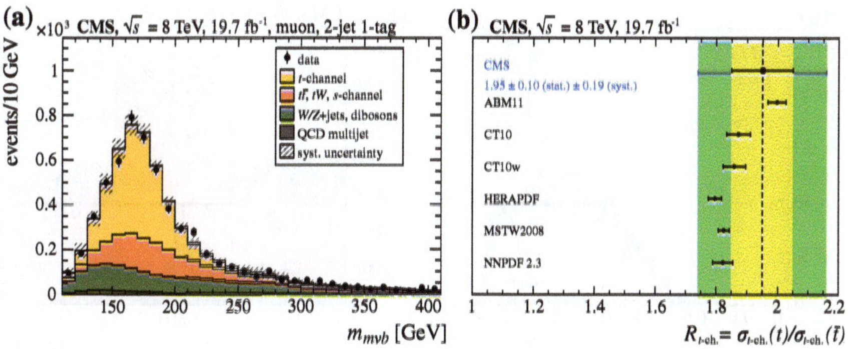

Fig. 7.23 **a** Distribution of the invariant mass $m_{\ell\nu b}$ for the muon decay channel in the region with $|\eta_j| > 2.5$ as measured by CMS. **b** Comparison of the measured t-channel top quark/antiquark production ratio with the predictions obtained using different PDF sets. Adapted from [165]

The largest contributions to the systematic uncertainty come from the choice of the renormalisation and factorisation scales in the simulation of the signal samples and from uncertainties on the jet energy scale and resolutions. From another fit, the cross sections for t quarks and $\bar{t}$ quarks and the corresponding ratio, $R_t = \sigma_t/\sigma_{\bar{t}}$, are obtained:

$$\sigma_t = 53.8 \pm 4.4 \text{ (stat)} \pm 8 \text{(syst) pb},$$
$$\sigma_{\bar{t}} = 27.6 \pm 1.3 \text{ (stat)} \pm 3.7 \text{ (syst) pb},$$
$$R_t = 1.95 \pm 0.10 \text{ (stat)} \pm 0.19 \text{ (syst)}.$$

In Fig. 7.23b the measured ratio is compared with predictions using different PDF sets.

The largeness of the LHC data samples and of the t-channel single top cross section give access to detailed studies of differential distributions [166] and properties, such as the top-quark polarisation, W helicity distributions, and mass measurements. At the time of preparation of this volume, the publication of the latter measurements is still in progress.

7.6.2 Single Top-Quark Production in Association with a W Boson

At the LHC, the production of single top quarks in association with W bosons becomes experimentally accessible for the first time. First evidence was reported by ATLAS using about the first half of the 7 TeV data recorded in 2011 [167] and was confirmed by CMS [168]. The ATLAS analysis makes use of dileptonic final states with events featuring two isolated leptons (electron or muon) with significant

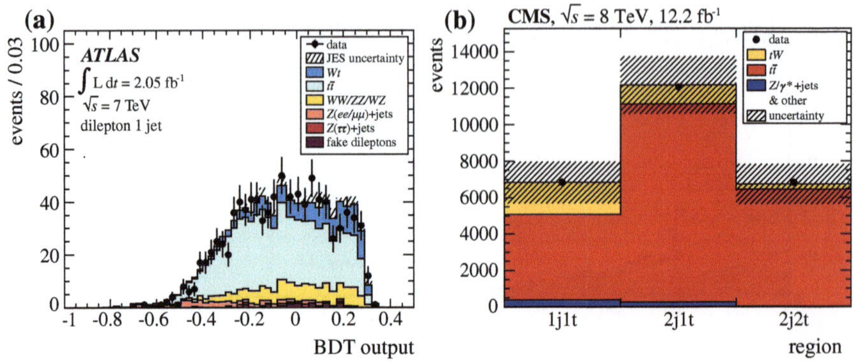

Fig. 7.24 **a** BDT output for selected events in the 1-jet category of the ATLAS analysis. The tW signal is normalised to the theory prediction. **b** Event yields in the signal region with one jet and one b-tagged jet (1j1t) and the control regions (2j1t and 2j2t) for the CMS analysis. The hatched band represents uncertainties of the predicted event yields. Adapted from [167, 169]

transverse missing momentum and at least one jet. A boosted decision tree (BDT) is used to discriminate single top-quark tW events from background events, which mostly arise from top-quark pair production. The result is extracted from a template fit to the BDT output discriminant distribution, which is shown in Fig. 7.24a. It is incompatible with the background-only hypothesis at the 3.3 σ level. The expected sensitivity assuming the Standard Model production rate is 3.4 σ. The measured cross section is

$$\sigma_{tW} = 16.8 \pm 2.9\ (\text{stat}) \pm 4.9\ (\text{syst})\ \text{pb}.$$

The uncertainty of the jet energy scale and of the modelling of the production process are dominant sources of systematic uncertainty.

Most recently, the CMS collaboration reported an observation of the process based on 8 TeV data [169]. Similarly to previous analyses, a multivariate analysis technique makes use of kinematic and topological properties to separate the signal from the dominant $t\bar{t}$ background. An excess consistent with the signal hypothesis is observed, with an observed (expected) significance of 6.1 (5.4) standard deviations above a background-only hypothesis. In Fig. 7.24b the distribution of tW events over different event categories is shown. The measured production cross section is 23.4±5.4 pb, in agreement with the Standard Model prediction.

7.6.3 Determination of V_{tb}

Single top-quark production measurements also give access to the determination of the modulus of the CKM matrix element V_{tb} since in the SM t-channel and s-channel production depend on $|V_{tb}|^2$. The extraction is independent of assumptions about the

Table 7.3 Most precise values and limits for $|V_{tb}|^2$ as extracted from single top-quark cross-section measurements and from R_B

Measurement	$\|V_{tb}\|^2$	$\|V_{tb}\|$ limit at 95 % CL
t-ch., ATLAS [166]	1.02 ± 0.07	>0.88
t-ch., CMS [165]	$0.998 \pm 0.038(\text{exp}) \pm 0.016(\text{theo})$	>0.92
tW-ch., ATLAS [167]	$1.03^{+0.16}_{-0.19}$	N/a
tW-ch., CMS [169]	$1.03 \pm 0.12(\text{exp}) \pm 0.04(\text{theo})$	>0.78
R_B, CMS [170]	$1.014 \pm 0.003(\text{stat}) \pm 0.032(\text{syst})$	>0.975

number of quark generations and about the unitarity of the CKM matrix. The only assumptions required are that $|V_{tb}| \gg |V_{ts}|, |V_{td}|$ and that the Wtb interaction is a SM-like left-handed weak coupling. Using the single-top cross section measurements, $|V_{tb}|^2$ can be extracted from a comparison of the measured and the predicted cross section. Assuming unitarity of the CKM matrix, $|V_{tb}| \leq 1$, a limit can be set. The most precise results are summarised in Table 7.3.

A significantly more precise determination of V_{tb} can be obtained from the measurement of the ratio $R_B = B(t \to Wb)/B(t \to Wq)$. Measurements of R_B have also been performed at the Tevatron [171–173]. CMS presented a measurement of R_B using a binned-likelihood function of the observed b-tagging multiplicity distributions in events with two, three, or four observed jets in the different dilepton

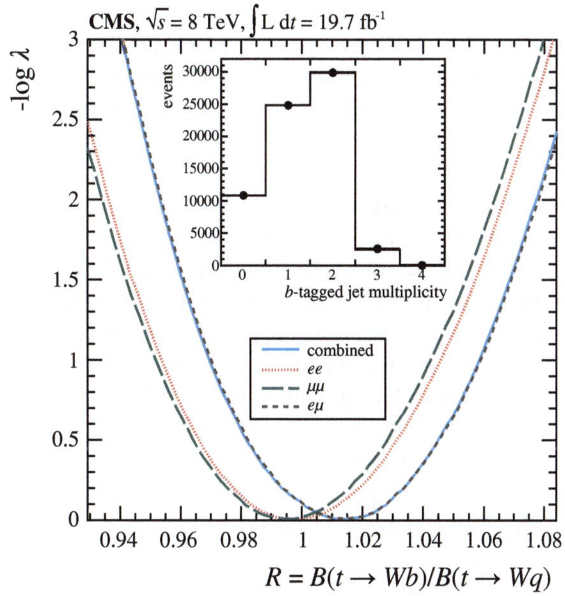

Fig. 7.25 Variations of the likelihood ratio used to extract R_B from the data. The variations observed in the combined fit and in the exclusive ee, $\mu\mu$, and $e\mu$ channels are shown separately. The *inset* shows the inclusive b-tagged jet multiplicity distribution and the fit distribution. Adapted from [170]

channels [170]. In Fig. 7.25 the variation of the profile likelihood ratio is shown. The fit yields a value $R_B = 1.014 \pm 0.003$ (stat) ± 0.032 (syst). Assuming the CKM matrix to be unitary, a lower limit for V_{tb} of 0.975 is set at the 95 % confidence level.

7.7 Conclusions

LHC data have brought a large boost to the field of top-quark physics, and since the LHC start in 2009, tremendous progress has been made.

On the theoretical side, a major breakthrough has been made with the calculation of the NNLO QCD corrections for inclusive production of top-quark pairs. Very recently, the theoretical calculations have been extended to include also predictions for differential distributions. The NNLO QCD corrections have been further improved by including weak corrections, the resummation of logarithmically enhanced contributions due to soft-gluon emission, and bound-state effects. In addition, finite-width effects have been studied in ground-breaking calculations by studying the one-loop QCD corrections including the decay of the top quark. For the experimental analysis, NLO predictions matched and merged to take into account parton-shower corrections and predictions for larger jet multiplicities are available. For single-top-quark production, NLO QCD corrections are known including also the decay of the top quark. For t-channel production, the theoretical predictions have been extended recently to include also partial results at NNLO QCD. In addition, conceptual differences using four or five flavours in the initial state have been studied in detail. For many observables the theoretical uncertainties are at the level of 10 %, and in some cases even the level of a few percent is reached. Beyond the steadily improving quality of theoretical predictions, a variety of new observables that allow precise tests of the underlying theory have been proposed. New methods to measure the top-quark mass are currently under development.

On the experimental side, during LHC Run 1, a rich and diverse program of top-quark measurements and studies has been performed, yielding a comprehensive spectrum of detailed and precise results.

Many of the measurements previously carried out at the Tevatron have been repeated at the LHC, and in many areas the size of the LHC top-quark data samples has facilitated in-depth studies at unprecedented levels of precision. Detailed and precise measurements of differential top-quark cross sections have been performed at the LHC, providing new insights in the top-quark production process and in QCD. Precision measurements of top-quark properties have as yet confirmed the SM expectations. Some of these properties, such as the polarisation of top quarks, the correlation of the top-quark spins in pair production, or the production of top quarks with additional gauge bosons were measured at the LHC for the first time. Stringent limits were set on anomalous couplings, e.g. flavour-changing neutral currents in the top-quark sector.

To date, most LHC top-quark measurements are no longer limited by statistical, but by systematic uncertainties. The latter are related to both the detector and the modelling of top-quark production and decay. Prime examples for the former are the jet energy scale and b-tagging uncertainties, while examples for the latter include the modelling of initial-state and final-state radiation as well as scale and hadronization uncertainties. Substantial effort is being invested to reduce these uncertainties, through auxiliary measurements, through tuning of Monte Carlo generators and through further improvements of the theoretical calculations. The wealth of the LHC Run 1 results is the avant-garde of many more exciting improvements in the future, both in precision measurements as well as in searches for rare processes associated with top quarks.

References

1. M. Kobayashi, T. Maskawa, Prog. Theor. Phys. **49**, 652 (1973)
2. CDF Collaboration, Phys. Rev. Lett. **74**, 2626 (1995)
3. DØ Collaboration, Phys. Rev. Lett. **74**, 2632 (1995)
4. CDF and DØ Collaborations, Phys. Rev. D **89**, 072001 (2014)
5. Tevatron Electroweak Working Group (2014), arXiv:1407.2682
6. P. Nason, S. Dawson, R.K. Ellis, Nucl. Phys. B **303**, 607 (1988)
7. W. Beenakker et al., Phys. Rev. D **40**, 54 (1989)
8. S. Catani et al., Phys. Lett. B **378**, 329 (1996)
9. R. Bonciani et al., Nucl. Phys. B **529**, 424 (1998)
10. N. Kidonakis et al., Phys. Rev. D **64**, 114001 (2001)
11. N. Kidonakis, R. Vogt, Phys. Rev. D **68**, 114014 (2003)
12. S. Moch, P. Uwer, Phys. Rev. D **78**, 034003 (2008)
13. M. Czakon, A. Mitov, Phys. Lett. B **680**, 154 (2009)
14. M. Cacciari et al., Phys. Lett. B **710**, 612 (2012)
15. M. Beneke et al., Nucl. Phys. B **855**, 695 (2012)
16. K. Hagiwara, Y. Sumino, H. Yokoya, Phys. Lett. B **666**, 71 (2008)
17. Y. Kiyo et al., Eur. Phys. J. C **60**, 375 (2009)
18. W. Beenakker et al., Nucl. Phys. B **411**, 343 (1994)
19. J.H. Kuhn, A. Scharf, P. Uwer, Eur. Phys. J. C **45**, 139 (2006)
20. W. Bernreuther, M. Fuecker, Z. Si, Phys. Lett. B **633**, 54 (2006)
21. J.H. Kuhn, A. Scharf, P. Uwer, Eur. Phys. J. C **51**, 37 (2007)
22. W. Bernreuther, M. Fuecker, Z.G. Si, Phys. Rev. D **74**, 113005 (2006)
23. W. Hollik, M. Kollar, Phys. Rev. D **77**, 014008 (2008)
24. W. Bernreuther, M. Fuecker, Z.G. Si, Phys. Rev. D **78**, 017503 (2008)
25. M. Czakon, P. Fiedler, A. Mitov, Phys. Rev. Lett. **110**(25), 252004 (2013)
26. A. Martin et al., Eur. Phys. J. C **63**, 189 (2009)
27. J.M. Campbell, R.K. Ellis, J. Phys. G **42**(1), 015005 (2015)
28. S. Frixione, B.R. Webber, JHEP **06**, 029 (2002)
29. P. Nason, JHEP **11**, 040 (2004)
30. S. Frixione, P. Nason, B.R. Webber, JHEP **08**, 007 (2003)
31. S. Frixione, P. Nason, G. Ridolfi, JHEP **09**, 126 (2007)
32. K. Melnikov, O.I. Yakovlev, Nucl. Phys. B **471**, 90 (1996)
33. A. Denner et al., Phys. Rev. Lett. **106**, 052001 (2011)
34. A. Denner et al., JHEP **10**, 110 (2012)
35. G. Bevilacqua et al., JHEP **02**, 083 (2011)
36. ATLAS Collaboration, Eur. Phys. J. C **71**, 1577 (2011)
37. CMS Collaboration, Phys. Lett. B **695**, 424 (2011)

38. ATLAS Collaboration, Phys. Lett. B **711**, 244 (2012)
39. ATLAS Collaboration, Phys. Lett. B **707**, 459 (2012)
40. CMS Collaboration, JHEP **07**, 049 (2011)
41. CMS Collaboration, Eur. Phys. J. C **71**, 1721 (2011)
42. CMS Collaboration, Phys. Rev. D **84**, 092004 (2011)
43. ATLAS Collaboration, Eur. Phys. J. C **73**, 2328 (2013)
44. ATLAS Collaboration, Phys. Lett. B **717**, 89 (2012)
45. ATLAS Collaboration, JHEP **05**, 059 (2012)
46. CMS Collaboration, Phys. Lett. B **720**, 83 (2013)
47. CMS Collaboration, JHEP **11**, 067 (2012)
48. CMS Collaboration, JHEP **05**, 065 (2013)
49. CMS Collaboration, Eur. Phys. J. C **73**, 2386 (2013)
50. CMS Collaboration, Phys. Rev. D **85**, 112007 (2012)
51. CMS Collaboration, Phys. Lett. B **739**, 23 (2014)
52. ATLAS Collaboration (2014), arXiv:1407.0573
53. ATLAS Collaboration, Eur. Phys. J. C **74**(10), 3109 (2014)
54. CMS Collaboration, JHEP **02**, 024 (2014)
55. P. Uwer, W. Wagner, Top quarks: the peak of the mass hierarchy? in *Physics at the Terascale*, ed. by I. Brock, T. Schörner-Sadenius (Wiley-VCH, Weinheim, 2011), pp. 187–209
56. ATLAS Collaboration, Eur. Phys. J. C **73**, 2261 (2013)
57. CMS Collaboration, Eur. Phys. J. C **73**, 2339 (2013)
58. ATLAS Collaboration, Phys. Rev. D **90**, 072004 (2014)
59. S. Dittmaier, P. Uwer, S. Weinzierl, Eur. Phys. J. C **59**, 625 (2009)
60. S. Alioli et al., Eur. Phys. J. C **73**, 2438 (2013)
61. J. Alwall et al., JHEP **06**, 128 (2011)
62. S. Dittmaier, P. Uwer, S. Weinzierl, Phys. Rev. Lett. **98**, 262002 (2007)
63. S. Catani, M. Seymour, Nucl. Phys. B **485**, 291 (1997)
64. S. Catani et al., Nucl. Phys. B **627**, 189 (2002)
65. K. Melnikov, M. Schulze, Nucl. Phys. B **840**, 129 (2010)
66. K. Melnikov, A. Scharf, M. Schulze, Phys. Rev. D **85**, 054002 (2012)
67. K. Melnikov, M. Schulze, A. Scharf, Phys. Rev. D **83**, 074013 (2011)
68. J. Pumplin et al., JHEP **07**, 012 (2002)
69. A. Bredenstein et al., Phys. Rev. Lett. **103**, 012002 (2009)
70. A. Bredenstein et al., JHEP **08**, 108 (2008)
71. A. Bredenstein et al., JHEP **03**, 021 (2010)
72. G. Bevilacqua et al., Phys. Rev. D **84**, 114017 (2011)
73. G. Bevilacqua, M. Worek, JHEP **07**, 111 (2012)
74. S. Hoeche et al. (2014), arXiv:1402.6293
75. CMS Collaboration, Eur. Phys. J. C **74**(8), 3014 (2014)
76. ATLAS Collaboration, Eur. Phys. J. C **72**, 2043 (2012)
77. ATLAS Collaboration, JHEP **1501**, 020 (2015)
78. ATLAS Collaboration, Phys. Rev. D **89**(7), 072012 (2014)
79. S. Heinemeyer, W. Hollik, G. Weiglein, L. Zeune, JHEP **12**, 084 (2013)
80. G. Degrassi et al., JHEP **08**, 098 (2012)
81. G. Degrassi, Nuovo Cim. C **37**(02), 47–53 (2014)
82. A. Buckley et al., Phys. Rep. **504**, 145–233 (2011)
83. S. Moch et al. (2014), arXiv:1405.4781
84. I.I.Y. Bigi et al., Phys. Rev. D **50**, 2234 (1994)
85. M. Beneke, V.M. Braun, Nucl. Phys. B **426**, 301 (1994)
86. ATLAS Collaboration, Eur. Phys. J. C **72**, 2046 (2012)
87. CMS Collaboration, Eur. Phys. J. C **72**, 2202 (2012)
88. CMS Collaboration, Eur. Phys. J. C **73**, 2494 (2013)
89. CMS Collaboration, Eur. Phys. J. C **74**, 2758 (2014)
90. CMS Collaboration, JHEP **12**, 105 (2012)

91. ATLAS Collaboration (2014), arXiv:1409.0832
92. CMS Collaboration, JHEP **06**, 109 (2012)
93. ATLAS Collaboration, Phys. Lett. B **728**, 363 (2014)
94. ATLAS, CDF, CMS, DØ Collaborations (2014), arXiv:1403.4427
95. CMS Collaboration, Phys. Lett. B **728**, 496 (2014)
96. J.H. Kuhn, G. Rodrigo, Phys. Rev. Lett. **81**, 49 (1998)
97. J.H. Kuhn, G. Rodrigo, Phys. Rev. D **59**, 054017 (1999)
98. W. Bernreuther, Z.G. Si, Phys. Rev. D **86**, 034026 (2012)
99. CDF Collaboration, Phys. Rev. D **87**, 092002 (2013)
100. DØ Collaboration, Phys. Rev. D, **84**, 112005 (2011)
101. ATLAS Collaboration, JHEP **02**, 107 (2014)
102. ATLAS Collaboration, Eur. Phys. J. C **72**, 2039 (2012)
103. J. Erdmann et al., Nucl. Instrum. Methods A **748**, 18 (2014)
104. G. Choudalakis (2012), arXiv:1201.4612
105. CMS Collaboration, Phys. Lett. B **709**, 28 (2012)
106. CMS Collaboration, Phys. Lett. B **717**, 129 (2012)
107. V. Blobel (2002), arXiv:hep-ex/0208022
108. CMS Collaboration, JHEP **04**, 191 (2014)
109. A. Hocker, V. Kartvelishvili, Nucl. Instrum. Methods A **372**, 469 (1996)
110. Y. Grossman, I. Nachshon, JHEP **07**, 016 (2008)
111. G. Mahlon, S.J. Parke, Phys. Rev. D **81**, 074024 (2010)
112. A. Brandenburg, Z. Si, P. Uwer, Phys. Lett. B **539**, 235 (2002)
113. W. Bernreuther, A. Brandenburg, P. Uwer, Phys. Lett. B **368**, 153 (1996)
114. W.G. Dharmaratna, G.R. Goldstein, Phys. Rev. D **41**, 1731 (1990)
115. W. Bernreuther, Z.G. Si, Phys. Lett. B **725**, 115 (2013)
116. W. Bernreuther, Z.G. Si, Nucl. Phys. B **837**, 90 (2010)
117. DØ Collaboration, Phys. Rev. Lett. **108**, 032004 (2012)
118. ATLAS Collaboration, Phys. Rev. Lett. **108**, 212001 (2012)
119. ATLAS Collaboration (2014), arXiv:1407.4314
120. CMS Collaboration, Phys. Rev. Lett. **112**, 182001 (2014)
121. ATLAS Collaboration, Phys. Rev. Lett. **111**, 232002 (2013)
122. DØ Collaboration, Phys. Rev. Lett. **80**, 2063 (1998)
123. A. Czarnecki, J.G. Korner, J.H. Piclum, Phys. Rev. D **81**, 111503 (2010)
124. J. Aguilar-Saavedra et al., Eur. Phys. J. C **50**, 519 (2007)
125. C. Zhang, S. Willenbrock, Phys. Rev. D **83**, 034006 (2011)
126. G.L. Kane, G. Ladinsky, C. Yuan, Phys. Rev. D **45**, 124 (1992)
127. ATLAS Collaboration, JHEP **06**, 088 (2012)
128. A. Valassi, Nucl. Instrum. Methods A **500**, 391 (2003)
129. CMS Collaboration, JHEP **10**, 167 (2013)
130. CMS Collaboration, JHEP **01**, 053 (2015)
131. CMS Collaboration, Phys. Rev. Lett. **110**, 172002 (2013)
132. CDF Collaboration, Phys. Rev. D **84**, 031104 (2011)
133. CMS Collaboration, Eur. Phys. J. C **74**, 3060 (2014)
134. M. Garzelli et al., JHEP **11**, 056 (2012)
135. J.M. Campbell, R.K. Ellis, JHEP **07**, 052 (2012)
136. J.M. Campbell et al., Phys. Rev. Lett. **102**, 182003 (2009)
137. J. Campbell et al., JHEP **10**, 042 (2009)
138. G. Bordes, B. van Eijk, Nucl. Phys. B **435**, 23 (1995)
139. T. Stelzer, Z. Sullivan, S. Willenbrock, Phys. Rev. D **56**, 5919 (1997)
140. T. Stelzer, Z. Sullivan, S. Willenbrock, Phys. Rev. D **58**, 094021 (1998)
141. M.C. Smith, S. Willenbrock, Phys. Rev. D **54**, 6696 (1996)
142. W.T. Giele, S. Keller, E. Laenen, Phys. Lett. B **372**, 141 (1996)
143. S. Zhu, Phys. Lett. B **524**, 283 (2002)
144. M. Brucherseifer, F. Caola, K. Melnikov, Phys. Lett. B **736**, 58 (2014)

145. B. Harris et al., Phys. Rev. D **66**, 054024 (2002)
146. Z. Sullivan, Phys. Rev. D **70**, 114012 (2004)
147. Z. Sullivan, Phys. Rev. D **72**, 094034 (2005)
148. J.M. Campbell, R.K. Ellis, F. Tramontano, Phys. Rev. D **70**, 094012 (2004)
149. Q.H. Cao, C.P. Yuan, Phys. Rev. D **71**, 054022 (2005)
150. Q.H. Cao et al., Phys. Rev. D **72**, 094027 (2005)
151. J.M. Campbell, F. Tramontano, Nucl. Phys. B **726**, 109 (2005)
152. S. Frixione et al., JHEP **03**, 092 (2006)
153. S. Frixione et al., JHEP **07**, 029 (2008)
154. S. Alioli et al., JHEP **09**, 111 (2009)
155. E. Re, Eur. Phys. J. C **71**, 1547 (2011)
156. S. Mrenna, C. Yuan, Phys. Lett. B **416**, 200 (1998)
157. N. Kidonakis, Phys. Rev. D **74**, 114012 (2006)
158. N. Kidonakis, Phys. Rev. D **75**, 071501 (2007)
159. N. Kidonakis, Phys. Rev. D **82**, 054018 (2010)
160. N. Kidonakis, Phys. Rev. D **82**, 114030 (2010)
161. N. Kidonakis, Phys. Rev. D **83**, 091503 (2011)
162. ATLAS Collaboration, Phys. Lett. B **717**, 330 (2012)
163. CMS Collaboration, Phys. Rev. Lett. **107**, 091802 (2011)
164. CMS Collaboration, JHEP **12**, 035 (2012)
165. CMS Collaboration, JHEP **06**, 090 (2014)
166. ATLAS Collaboration, Phys. Rev. D **90**, 112006 (2014)
167. ATLAS Collaboration, Phys. Lett. B **716**, 142 (2012)
168. CMS Collaboration, Phys. Rev. Lett. **110**, 022003 (2013)
169. CMS Collaboration, Phys. Rev. Lett. **112**, 231802 (2014)
170. CMS Collaboration, Phys. Lett. B **736**, 33 (2014)
171. DØ Collaboration, Phys. Rev. Lett. **107**, 121802 (2011)
172. CDF Collaboration, Phys. Rev. D **87**(11), 111101 (2013)
173. CDF Collaboration, Phys. Rev. Lett. **112**, 221801 (2014)

Chapter 8
Quark-Flavour Physics

Stephanie Hansmann-Menzemer and Ulrich Nierste

Abstract Precision measurements in the flavour system are a powerful tool for indirect searches for new physics phenomena. Historically, bottom- and charm-flavour physics is performed in the clean environment of e^+e^- collider experiments. Exploiting the large $b\bar{b}$ and $c\bar{c}$ cross sections in pp collisions, the LHC experiments, however, impressively demonstrate their huge flavour-physics potential and have by now superseded most of the results of previous experiments. In this section we present some of the highlights of the flavour-physics programme of the LHC, give pedagogical introductions to the theoretical concepts, and interpret the experimental results.

8.1 Introduction

The term "flavour" is a synonym for "fermion species". For instance, a charm quark carries the flavour quantum number $C = 1$, the anti-charm has $C = -1$, while all other fermions have $C = 0$. By convention, one assigns a negative flavour quantum number to quarks with negative electric charge, e.g. a bottom quark carries the "beauty" number $B = -1$. This apparently trivial concept is useful, because the five light quarks are confined into hadrons by the strong interaction: Since QCD respects the flavour quantum numbers, we can use them to categorise mesons and baryons. At the same time we can characterise the weak decays, which violate flavour, by the change of flavour quantum numbers occurring in the decay. For example, $B_d^0 \to D^- D_s^+$ comes with $\Delta B = -1$, $\Delta S = +1$, $\Delta C = \Delta U = \Delta D = 0$. From these quantum numbers we can identify the underlying quark transitions as $\bar{b} \to \bar{c}c\bar{s}$ and $\bar{b} \to \bar{u}u\bar{s}$. The former contribution stems from a tree-level W-exchange diagram

S. Hansmann-Menzemer (✉)
Physikalisches Institut, Universität Heidelberg, Im Neuenheimer Feld 226, 69120 Heidelberg, Germany
e-mail: menzemer@physi.uni-heidelberg.de

U. Nierste
Institut für Theoretische Teilchenphysik, Karlsruher Institut für Technologie, Wolfgang-Gaede-Str. 1, 76131 Karlsruhe, Germany
e-mail: ulrich.nierste@kit.edu

T. Schörner-Sadenius (ed.), *The Large Hadron Collider*,
DOI 10.1007/978-3-319-15001-7_8

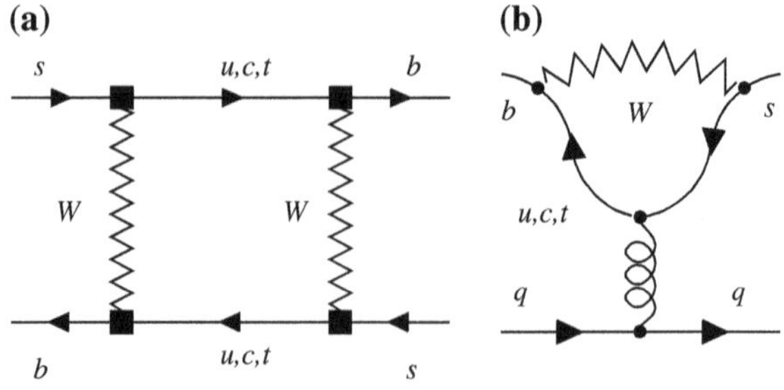

Fig. 8.1 **a** Box diagram for $B_s^0 - \overline{B}_s^0$ mixing. **b** Penguin diagram for the decay $b \to sq\bar{q}$ with the curly line representing a gluon. For $q = u$ or $q = c$ there is also a tree diagram

and is by far dominant, while in the latter process the u quark and the W form a loop as in Fig. 8.1b, with the gluon ending on a charm line. It is advantageous to trade the quantum numbers U and D for the "strong isospin". The QCD Lagrangian is almost invariant with respect to SU(2) rotations of the doublet $(u, d)^T$, because the mass difference between up quark and down quark is so tiny. The strong-isospin quantum numbers (I, I_3) have the same pattern as those of particle spin, the fundamental doublet $(u, d)^T$ having $I = 1/2$ and $I_3 = \pm 1/2$. A prominent example of a doublet of hadrons is $(p, n)^T$, which gave birth the concept of strong isospin three decades before quarks were conceived! The s, c, b, t quarks carry no strong isospin, and therefore $(B^+, B_d^0)^T$, $(D^+, D^0)^T$, and $(K^+, K^0)^T$ are doublets as well. The pions form a strong-isospin triplet $(\pi^+, \pi^0, \pi^-)^T$.

In the Standard Model (SM), the only flavour-violating interaction vertices are the couplings of the W boson to fermions, which connect up-type to down-type fermions of all three generations. This important feature of flavour violation, however, does not originate in the electroweak interaction. Instead, these inter-generational transitions are rooted in the Yukawa sector of the SM (see Sect. 4.2.3) and thus have the same origin as the fermion masses. The main focus of flavour physics is the study of flavour-changing transitions, with two major goals: On one hand, one aims at a determination of the parameters of the Cabibbo–Kobayashi–Maskawa (CKM) matrix, which describes the strength of flavour violation, and its counterpart in the lepton sector, the Pontecorvo–Maki–Nakagawa–Sakata (PMNS) matrix. The CKM matrix involves four parameters, one of which is the Kobayashi–Maskawa phase δ_{KM}, which quantifies the violation of the charge-parity (*CP*) symmetry in the Yukawa sector. To date, *CP* violation has only been observed in quark-flavour physics, in several K and B decays. On the other hand, one wants to probe new physics: To this end, flavour-changing neutral current (FCNC) processes play an important role, because they are highly suppressed in the SM. Two prominent examples of FCNC processes are shown in Fig. 8.1.

Depending on the model of new physics, FCNC processes probe virtual effects of new particles with masses of 100 TeV and beyond. Currently FCNC processes give powerful constraints on the parameter spaces of models of new physics. Another branch of quark-flavour physics deals with the strong interaction that binds the quarks into hadrons. Recent years have brought new insights into hadron spectroscopy, with the discovery of new hadrons and the determination of their properties like mass and spin.

The observation of flavour oscillations of neutrinos has opened the field of experimental lepton-flavour physics. Neutrino data have permitted the three angles of the PMNS matrix to be determined with good accuracy, while no *CP* asymmetry has been measured yet. The established violation of lepton flavour permits FCNC decays of charged leptons, such as $\tau^- \to \mu^-\mu^+\mu^-$, which can be studied at the LHC. In the SM, the branching fractions of such decays are too small to be ever measured. However, in models of new physics charged-lepton FCNC processes can be sizeable. For pedagogical introductions to flavour physics in the Standard Model and beyond see [1–8].

8.1.1 Theoretical Concepts in Flavour Physics

The Yukawa interaction of quarks couples the Higgs doublet Φ (see (4.4)) to the left-chiral quark doublets Q_j and the right-handed quark singlet fields u'_{Rk}, d'_{Rk}. Here $j, k = 1, 2, 3$ labels the generation and

$$\Phi = \begin{pmatrix} \phi^+ \\ \frac{v+H+i\chi}{\sqrt{2}} \end{pmatrix}, \qquad Q_j = \begin{pmatrix} u'_{Lj} \\ d'_{Lj} \end{pmatrix}. \tag{8.1}$$

The quark Yukawa Lagrangian is

$$\mathcal{L}^q_{\text{Yuk}} = -\,\overline{Q}_j\, Y^d_{jk}\, \Phi\, d'_{Rk} \;-\; \overline{Q}_j\, Y^u_{jk}\, \epsilon\, \Phi^* u'_{Rk} + \text{h.c.}\,,$$

where a sum over repeated indices is understood and $\epsilon = i\sigma^2$. The 3×3 Yukawa matrices $Y^{u,d}$ are arbitrary complex matrices. The vacuum expectation value v in (8.1) gives rise to the mass matrices $M^{u,d} = Y^{u,d} v/\sqrt{2}$. To find the physical quark fields, we must diagonalise these matrices by unitary rotations of the fields $u'_{L,Rk}$, $d'_{L,Rk}$ (called "weak eigenstates") to a new basis of "mass eigenstates": $u'_{L,Rj} = S^u_{L,Rjk}\, u_{L,Rk}$, $d'_{L,Rj} = S^d_{L,Rjk}\, d_{L,Rk}$.[1] The unprimed fields correspond to the physical particles, and it is often more convenient to write $u_R = u_{R1}$, $c_R = u_{R2}$ and $t_R = u_{R3}$ with an analogous notation for the left-chiral and down-type quark fields. The piece of $\mathcal{L}^q_{\text{Yuk}}$ involving the—now diagonal—mass matrices describes the quark masses m_j:

$$\begin{aligned} \mathcal{L}_m &= -\, m_{u_j} \left[\bar{u}_{Lj} u_{Rj} + \bar{u}_{Rj} u_{Lj}\right] - m_{d_j} \left[\bar{d}_{Lj} d_{Rj} + \bar{d}_{Rj} d_{Lj}\right] \\ &\equiv -\sum_{q=u,d,s,c,b,t} m_q\, \bar{q} q\,. \end{aligned}$$

[1] The relation to the notation in (4.2.3)–(4.5) is as follows: $Q_L = \Psi_L$, $q'_{\tau j} = \psi_{q_\tau,j}$, $q_{\tau j} = \hat{\psi}_{q_\tau,j}$, $Y^q = G_q$, and $S^q_\tau = U^{q_\tau \dagger}$ for $q = u, d$ and $\tau = L, R$.

Here, we have introduced the usual four-component Dirac field $q \equiv q_L + q_R$ and remind the reader of $\bar{q}_R q_R = \bar{q}_L q_L = 0$.

The four unitary matrices $S^{u,d}_{L,R}$ drop out everywhere in the SM Lagrangian with one important exception: The CKM matrix

$$V = S_L^{u\dagger} S_L^d$$

appears in the quark couplings of the W boson:

$$\mathcal{L}_W^q = \frac{g}{\sqrt{2}} \left[V_{jk}\, \bar{u}_{Lj}\, \gamma^\mu d_{Lk}\, W_\mu^+ + V_{jk}^*\, \bar{d}_{Lk}\, \gamma^\mu u_{Lj}\, W_\mu^- \right]. \tag{8.2}$$

$\mathcal{L}_W^q$ violates the discrete symmetries parity (P), time reversal (T), and charge conjugation (C). The parity transformation $\boldsymbol{x} \to -\boldsymbol{x}$ maps the left-chiral quark fields in the Lagrangian (8.2) onto their right-handed counterparts. Since the W boson does not couple at all to right-handed quarks, P violation in the SM is maximal. The same is true for C violation, because C maps left-chiral fermion fields onto right-chiral antifermion fields. However, the combination of the two transformations, CP, does not change the chirality of the fermion fields and deserves deeper consideration. We may write the effect of the CP transformation on the field monomials in (8.2) as

$$CP: \qquad \bar{u}_{Lj}\, \gamma^\mu d_{Lk}\, W_\mu^+ \quad \longleftrightarrow \quad \bar{d}_{Lk}\, \gamma^\mu u_{Lj}\, W_\mu^- \,. \tag{8.3}$$

Apparently $\mathcal{L}_W^q$ conserves CP if V_{jk} is real. In contrast, $\operatorname{Im} V_{jk} \neq 0$ does not imply that CP is violated: We may always rephase the quark fields, $d_k \to d_k \exp(i\phi_{d_k})$ and $u_j \to u_j \exp(i\phi_{u_j})$, with arbitrary phases ϕ_{d_k} and ϕ_{u_j} without changing the physics predictions of the SM. So we are entitled to combine the CP operation in (8.3) with such arbitrary phase rotations, or, equivalently, to put an arbitrary phase factor into the right-hand side of the definition of CP in (8.3). The rephasing of the fields changes V_{jk} in (8.2) to $V_{jk} \exp(i\phi_{d_k} - i\phi_{u_j})$. In a SM with only two fermion generations $\mathcal{L}_W^q$ conserves CP, because we can always pick the three phases $\phi_{d_1} - \phi_{u_1}$, $\phi_{d_1} - \phi_{u_2}$ and $\phi_{d_2} - \phi_{u_2}$ in a way that the unitary 2×2 matrix V is real. (The fourth phase $\phi_{d_2} - \phi_{u_1}$ is the sum of the other three phases and therefore not an independent quantity). The resulting matrix is conventionally parametrised with the Cabibbo angle θ_C as $V_{11} = V_{22} = \cos\theta_C$, $V_{12} = -V_{21} = \sin\theta_C$. Like CP, the time reversal transformation $t \to -t$ also leaves the chirality of the quark fields unchanged. T, however, is an anti-unitary transformation which involves a complex conjugation $V_{jk} \to V_{jk}^*$, and T and CP have the same effect on $\mathcal{L}_W^q$. One can show that any quantum field theory (if a set of very broadly defined conditions is fulfilled) conserves the combined symmetry CPT [9, 10] (CPT theorem), and in phenomenological discussions often CP violation and T violation are not distinguished.

Prior to the discovery of the third fermion generation, Kobayashi and Maskawa had realised that one cannot rotate all phases in V away if there are more than two fermion generations [11]. They proposed the existence of a third generation as a

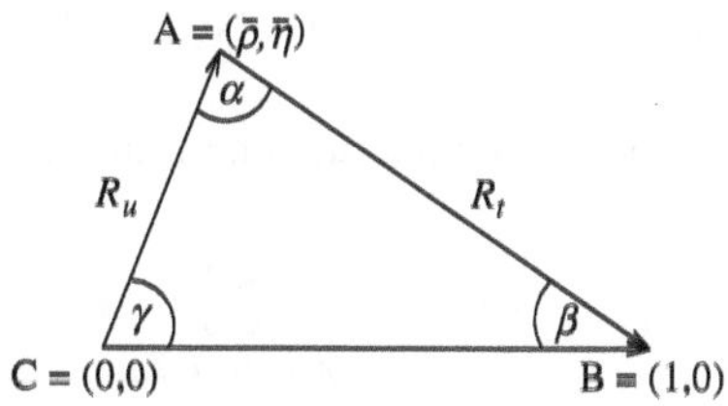

Fig. 8.2 The (standard) unitarity triangle

possibility to explain the *CP* violation found in $K^0 \to \pi^+\pi^-$ decays by Christenson, Cronin, Fitch, and Turlay [12, 13]. Meanwhile, the discovery of a Higgs boson and the analyses of its decay modes have permitted to rule out the possibility that there is a fourth generation in the SM [14].

The *CP* violation encoded in the CKM matrix is depicted in terms of the unitarity triangle: The unitarity of the CKM matrix entails that the three complex numbers $V_{ub}^* V_{ud}$, $V_{cb}^* V_{cd}$, and $V_{tb}^* V_{td}$ sum to zero. This relation can be shown as a triangle in the complex plane. In order to eliminate dependences on arbitrary phase conventions, one divides these three numbers by $V_{cb}^* V_{cd}$ and defines

$$\overline{\rho} + i\overline{\eta} \equiv -\frac{V_{ub}^* V_{ud}}{V_{cb}^* V_{cd}} = R_u e^{i\gamma} . \qquad (8.4)$$

The resulting triangle with unit base line is shown in Fig. 8.2.

The two non-trivial sides of the triangle are

$$R_u \equiv \sqrt{\overline{\rho}^2 + \overline{\eta}^2}, \qquad R_t \equiv \sqrt{(1-\overline{\rho})^2 + \overline{\eta}^2} . \qquad (8.5)$$

CP-violating quantities are associated with the three angles

$$\alpha = \arg\left[-\frac{V_{td} V_{tb}^*}{V_{ud} V_{ub}^*}\right], \quad \beta = \arg\left[-\frac{V_{cd} V_{cb}^*}{V_{td} V_{tb}^*}\right], \quad \gamma = \arg\left[-\frac{V_{ud} V_{ub}^*}{V_{cd} V_{cb}^*}\right] . \qquad (8.6)$$

In the commonly used phase convention of the Particle Data Group [15], V_{ud}, V_{us}, V_{cb}, and V_{tb} are real and positive. The deviations of the other phases from 0 to π are negligibly small (with V_{cd} almost negative), with the important exceptions

$$V_{ub} \simeq |V_{ub}| e^{-i\gamma}, \qquad V_{td} \simeq |V_{td}| e^{-i\beta}, \qquad V_{ts} \simeq |V_{ts}| e^{i\beta_s} . \qquad (8.7)$$

The first two relations can be read off from (8.6), the third phase $\beta_s = 0.02 = 1°$ governs *CP* violation in $B_s^0 \to J/\psi\,\phi$. Measurements define constraints on the $\overline{\rho}$–$\overline{\eta}$ plane; the consistency of the regions allowed by different measurements provides a test of the SM. $\overline{\rho} = 0.15 \pm 0.01$ and $\overline{\eta} = 0.34 \pm 0.01$ are two of the four "improved Wolfenstein parameters", the other two are $\lambda \simeq |V_{us}| = 0.226$ and $A = |V_{cb}|/\lambda^2 = 0.81 \pm 0.03$. The quoted numerical values are obtained from a global fit assuming that there is no physics beyond the SM [16]. One often expands CKM elements in terms of λ.

The strong interaction of quarks is encoded in the QCD Lagrangian

$$\mathcal{L}_{\text{QCD}} = \sum_{q=u,d,s,c,b,t} \bar{q}_L \left[i\not{\partial} - g_s T^a_c \not{G}^a \right] q_R + \mathcal{L}_m - \frac{1}{4} G^a_{\mu\nu} G^{\mu\nu\, a} \tag{8.8}$$

with the notation introduced in Chap. 4, and $\mathcal{L}_m$ of Sect. 8.1.1. $\mathcal{L}_{\text{QCD}}$ respects the discrete symmetries C, P and T. This feature is used to further characterise the hadron states, e.g. $|\pi^0\rangle$ has the quantum numbers $C = 1$, $P = -1$. A challenge to calculations of flavour-changing amplitudes is the presence of widely separated mass scales, ranging from $\Lambda_{\text{QCD}} \sim 350\,\text{MeV}$ over $\overline{m}_c \sim 1.25\,\text{GeV}$ and $\overline{m}_b \sim 4.3\,\text{GeV}$ to $M_W = 80.4\,\text{GeV}$ and $\overline{m}_t \sim 165\,\text{GeV}$. The QCD coupling $\alpha_s = g_s^2/(4\pi)$ changes dramatically over this range of energies: While we can do perturbative calculations (i.e. calculations of Feynman diagrams with quarks and gluons) of the QCD effects associated with scales of $\overline{m}_b$ and above, the fundamental scale of QCD, Λ_{QCD}, is associated with the genuine non-perturbative effects confining the quarks and gluons into colourless hadrons. In a given calculation, we must first separate the physics of the different scales and apply different calculational methods to the different energy regimes. The confinement of quarks also implies that there is no unique natural definition of quark masses, we use masses defined in the $\overline{\text{MS}}$ renormalisation scheme (described in Sect. 7.3) as indicated by the bar. Note that $\overline{m}_t$ is smaller by roughly 7 GeV than the top pole mass used in collider physics (see (7.3)). In order to disentangle high-energy ("short-distance") from low-energy ("long-distance") physics, one constructs an effective Hamiltonian. To explain this concept, we describe the effective Hamiltonian (density) $H^{|\Delta B|=1}$ for the weak decays of beauty-flavoured hadrons, with relevant low-energy scales m_b and Λ_{QCD}. The effective theory defined by $H^{|\Delta B|=1}$ is designed to reproduce the SM amplitudes for these decays with an accuracy of m_b^2/M_W^2:

$$\begin{aligned} &\langle f | \mathbf{T} \exp\left[-i \int \mathrm{d}^4 x H^{\text{SM}}(x) \right] | H_b \rangle \\ &= \langle f | \mathbf{T} \exp\left[-i \int \mathrm{d}^4 x H^{|\Delta B|=1}(x) \right] | H_b \rangle \left[1 + \mathcal{O}\left(\frac{m_b^2}{M_W^2} \right) \right]. \end{aligned} \tag{8.9}$$

Here $H_b = \overline{B}{}^0_d,\ B^-,\ \overline{B}{}^0_s,\ \bar{\Lambda}_b \ldots$ is the decaying hadron, f denotes the final state, and $H^{\text{SM}} = -\mathcal{L}^{\text{SM}}_{\text{int}}$ is the part of the SM Hamiltonian comprising the interactions among the fields. For our case, the most important pieces of H^{SM} are $-\mathcal{L}^q_W$ in (8.2) and the parts of $-\mathcal{L}_{\text{QCD}}$ in (8.8) involving quark-gluon couplings and gluon self-couplings (bilinear terms like $\bar{q}_L i \not{\partial} q_R$ describe the propagation of the particles and do not belong to $\mathcal{L}^{\text{SM}}_{\text{int}}$.). $H^{|\Delta B|=1}$ is further subdivided into the sum of terms which are characterised by the change of the other flavour quantum numbers S, D, C, and U. To keep the notation readable, we write $H^{b\to s}$ for the effective Hamiltonian with

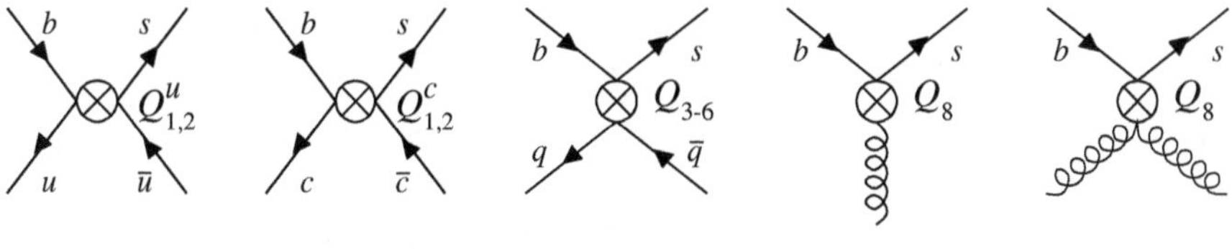

Fig. 8.3 Effective operators of (8.10). There are two types of fermion-gluon couplings associated with the chromomagnetic penguin operator Q_8

$|\Delta B| = 1$, $\Delta B = -\Delta S$, $\Delta C = \Delta U = \Delta D = 0$. We will need $H^{b\to s}$ later to discuss a highlight of LHCb in Run 1, the milestone result on *CP* violation found from the study of $B_s^0 \to J/\psi K^+K^-$. The tree-level transition $b \to sc\bar{c}$ involves the exchange of a *W* boson between the currents $\bar{c}\gamma^\mu b$ and $\bar{s}\gamma^\nu c$. Since the momentum *p* flowing through the *W* propagator satisfies $|p^2| \leq m_b^2$, replacing this propagator as $ig_{\mu\nu}/(M_W^2 - p^2) \to ig_{\mu\nu}/M_W^2$ leads to an error of order m_b^2/M_W^2 or less. This replacement results in an amplitude of a point-like interaction with the four quark-field operators involved in the interaction. This four-quark operator Q_2^c is pictorially obtained by contracting the *W* line to a point and is depicted in Fig. 8.3. In the same way the tree-level transition $b \to su\bar{u}$ gives rise to the operator Q_2^u. Once we include QCD effects, by e.g. adding a gluon connecting the quark lines, more operators emerge. An important feature of the effective theory is the absence of *W* and top-quark fields. The effects of these heavy particles are entirely absorbed into the coupling constants (called Wilson coefficients) of the effective operators. As long as only leading processes in the electroweak interaction (with a single *W* exchange and no additional *W*, *Z* or photon) are considered, $H^{b\to s}$ involves the following list of operators:

$$
\begin{aligned}
Q_1^c &= \bar{s}_L^\alpha \gamma_\mu c_L^\beta \, \bar{c}_L^\beta \gamma^\mu b_L^\alpha\,, & Q_1^u &= \bar{s}_L^\alpha \gamma_\mu u_L^\beta \, \bar{u}_L^\beta \gamma^\mu b_L^\alpha\,, \\
Q_2^c &= \bar{s}_L^\alpha \gamma_\mu c_L^\alpha \, \bar{c}_L^\beta \gamma^\mu b_L^\beta\,, & Q_2^u &= \bar{s}_L^\alpha \gamma_\mu u_L^\alpha \, \bar{u}_L^\beta \gamma^\mu b_L^\beta\,, \\
Q_3 &= \sum_{q=u,d,s,c,b} \bar{s}_L^\alpha \gamma_\mu b_L^\alpha \, \bar{q}_L^\beta \gamma^\mu q_L^\beta\,, & Q_4 &= \sum_{q=u,d,s,c,b} \bar{s}_L^\alpha \gamma_\mu b_L^\beta \, \bar{q}_L^\beta \gamma^\mu q_L^\alpha\,, \\
Q_5 &= \sum_{q=u,d,s,c,b} \bar{s}_L^\alpha \gamma_\mu b_L^\alpha \, \bar{q}_R^\beta \gamma^\mu q_R^\beta\,, & Q_6 &= \sum_{q=u,d,s,c,b} \bar{s}_L^\alpha \gamma_\mu b_L^\beta \, \bar{q}_R^\beta \gamma^\mu q_R^\alpha\,, \\
Q_8 &= \frac{g_s}{16\pi^2}\, m_b\, \bar{s}_L \sigma^{\mu\nu} G^a_{\mu\nu} T^a_c b_R\,.
\end{aligned}
\tag{8.10}
$$

Here α and β are colour indices and $Q_{3-6,8}$ are called "QCD penguin operators". These operators arise from the removal of the top quark from the effective theory, e.g. contracting the right diagram of Fig. 8.1 to a point results in Q_{3-6}. The unitarity of the CKM matrix allows one of the three CKM structures to be eliminated from the problem, thanks to $V_{tb}V_{ts}^* = -V_{cb}V_{cs}^* - V_{ub}V_{us}^*$. Therefore we may write:

$$
\begin{aligned}
H^{b\to s} = & \frac{G_F}{\sqrt{2}} V_{cb} V_{cs}^* \left[\sum_{j=1,2} C_j Q_j^c - \sum_{j=1,\dots 6,8} C_j Q_j \right] \\
& + \frac{G_F}{\sqrt{2}} V_{ub} V_{us}^* \left[\sum_{j=1,2} C_j Q_j^u - \sum_{j=1,\dots 6,8} C_j Q_j \right] .
\end{aligned}
\tag{8.11}
$$

Here, the Fermi constant $G_F \propto 1/M_W^2$ contains the dependence on M_W and the Wilson coefficients $C_{1-6,8}$ are dimensionless. $C_{1-6,8}$ are functions of μ/M_W and m_t/M_W, where μ is the renormalisation scale. These coefficients are determined from (8.9) in a procedure called "matching calculation". The Wilson coefficients have the important feature that they are both independent of the infrared structure of the considered decay and of the external states $|H_b\rangle$ and $|f\rangle$. These properties have two important consequences: Firstly, one can therefore calculate $C_{1-6,8}$ in perturbation theory from amplitudes with external quark states, and secondly, these coefficients are universal coupling constants applicable to *all* $b \to s$ transitions, without dependence on the hadronisation process of the quarks. The decay amplitude for $H_b \to f$ is then given as

$$
\begin{aligned}
A_f &= \langle f | H^{b\to s} | H_b \rangle \\
&= \frac{G_F}{\sqrt{2}} V_{cb} V_{cs}^* \left[\sum_{j=1,2} C_j \langle f | Q_j^c | H_b \rangle - \sum_{j=1,\dots 6,8} C_j \langle f | Q_j | H_b \rangle \right] \\
&+ \frac{G_F}{\sqrt{2}} V_{ub} V_{us}^* \left[\sum_{j=1,2} C_j \langle f | Q_j^u | H_b \rangle - \sum_{j=1,\dots 6,8} C_j \langle f | Q_j | H_b \rangle \right] .
\end{aligned}
\tag{8.12}
$$

In this expression all QCD effects are factorised such that the ultraviolet (UV) physics resides in $C_{1-6,8}$ and the infrared (IR) physics is contained in the hadronic matrix elements $\langle f | Q_j^{u,c} | H_b \rangle$, $\langle f | Q_j | H_b \rangle$. Equation (8.12) is a special case of an operator product expansion (OPE), a "large-mass expansion". OPE are used in many applications of QCD described in this book, for example in the DGLAP equation in which the hard scale is the centre-of-mass energy of the LHC rather than a heavy mass as in our case. The perturbative element of the DGLAP equation, which corresponds to our Wilson coefficients, is the partonic cross section $\hat{\sigma}$ in (5.4). The hadronic matrix elements contain non-perturbative QCD dynamics and are difficult to calculate from first principles. If the momentum transfer between the hadronic states in the initial state and in final state is small (such as in $B \to \pi \ell \nu$ at the kinematic endpoint with a slow π), one can use lattice QCD to calculate the hadronic matrix element involved. In many cases one can relate a given matrix element to another one with the help of a symmetry respected by QCD. Through a combination of different measurements one can then eliminate the hadronic matrix elements and solve for the physical quantity of interest (e.g. a *CP* phase). The most prominent symmetry of QCD used in this context is CP, which ensures e.g. $\langle f|Q_j^c|\overline{B}_d^0\rangle = -\langle \overline{f}|Q_j^{c\dagger}|B_d^0\rangle$. Here the definition $CP|B_d^0\rangle = -|\overline{B}_d^0\rangle$ is used and $\overline{f}$ denotes the state which is CP-conjugate to f:

$$|\overline{f}\rangle = CP\,|f\rangle\,.$$

The coefficients C_j and the (renormalised) operators Q_j depend on the unphysical renormalisation scale μ, but this μ dependence cancels from $H^{b\to s}$. Physical amplitudes (like the ones depicted in Fig. 8.1) contain large logarithms $L = \ln \overline{m_b}/M_W$, and higher orders in the perturbative expansion involve enhancement factors of $\alpha_s^n L^n$, $n = 1, 2, 3\ldots$ with respect to the leading order, which puts ordinary perturbation theory into doubt. The OPE in (8.12) splits this logarithms as $L = \ln(\mu/M_W)+\ln(\overline{m_b}/\mu)$ with the two terms residing in $C_j(\mu)$ and the matrix element, respectively. With the help of renormalisation group equations (RGE) explained around (5.1) one can sum $\alpha_s^n \ln^n(\mu/M_W)$ to all orders $n = 0, 1, 2, \ldots$ in perturbation theory. The matching of full and effective theory is calculated at the scale $\mu = M_W$ and yields $C_j(M_W)$, which is free of large logarithms. With the RGE, $C_j(M_W)$ can be evolved to the low scale $\mu = \overline{m}_b$, and in $C_j(\overline{m}_b)$ the large logarithm L is properly resummed. Matrix elements like $\langle f|Q_j^{u,c,.}(\mu)|H_b\rangle$ in (8.12) contain no large logarithms for the choice $\mu = \overline{m}_b$, so that the products $C_j(\overline{m}_b)\langle f|Q_j^{u,c,.}(\overline{m}_b)|H_b\rangle$ are safe from excessively large α_s corrections. The Wilson coefficients C_j can be systematically improved by calculating higher-order corrections and summing sub-leading large logarithms like $\alpha_s^{n+1}L^n$ with the RGE. Writing $\boldsymbol{C} = (C_1, C_2, \ldots C_6, C_8)$ the numerical values in the leading-log (LL) approximation are $\boldsymbol{C}(\mu = M_W) = (0, 1, 0, 0, 0, 0, 0)$ and $\boldsymbol{C}(\mu = \overline{m}_b) = (-0.27, 1.12, 0.012, -0.027, 0.008, -0.034, -0.151)$, showing the impact of the RG evolution.

For the calculation of CP asymmetries we often do not need the numerical values of the C_j. It is sufficient to write $H^{b\to s}$ in (8.11) as $H^{b\to s} = \xi_c h_c + \xi_u h_u + \text{h.c.}$ with $\xi_q = V_{qb}V_{qs}^*$: We can simply use the CP invariance of the strong interaction as $\langle \overline{f}|h_c|\overline{B}\rangle = -\langle f|h_c^\dagger|B\rangle$ when considering a CP asymmetry in the decay $B \to f$ with $B = B^-, B_d^0, B_s^0$. The asymmetry

$$A_{CP}^{\text{dir}}(B \to f) = \frac{\Gamma(B \to f) - \Gamma(\overline{B} \to \overline{f})}{\Gamma(B \to f) + \Gamma(\overline{B} \to \overline{f})} \tag{8.13}$$

quantifies "direct" CP violation or CP violation "in decay". Expressing the decay rates $\Gamma(B \to f)$, $\Gamma(\overline{B} \to \overline{f})$ in terms of $H^{b\to s}$ and defining the amplitudes $A_q = \langle \overline{f}|h_q|\overline{B}\rangle = -\langle f|h_q^\dagger|B\rangle$ for $q = u, c$ yields

$$
\begin{aligned}
A_{CP}^{\text{dir}}(B \to f) &= \frac{|\langle f|H^{b\to s}|B\rangle|^2 - |\langle \overline{f}|H^{b\to s}|\overline{B}\rangle|^2}{|\langle f|H^{b\to s}|B\rangle|^2 + |\langle \overline{f}|H^{b\to s}|\overline{B}\rangle|^2} \\
&= \frac{|\xi_c^* A_c + \xi_u^* A_u|^2 - |\xi_c A_c + \xi_u A_u|^2}{|\xi_c^* A_c + \xi_u^* A_u|^2 + |\xi_c A_c + \xi_u A_u|^2} \\
&= \frac{2\text{Im}(\xi_u/\xi_c)\text{Im}(A_u/A_c)}{1 + 2\text{Re}(\xi_u/\xi_c)\text{Re}(A_u/A_c) + |\xi_u/\xi_c|^2 |A_u/A_c|^2} \,. \qquad (8.14)
\end{aligned}
$$

In the second line of this equation one easily spots that the *CP*-violating phase (often called "weak phase") changes sign ($\xi_c^*, \xi_u^* \to \xi_c, \xi_u$) when going from $B \to f$ to the *CP*-conjugate decay. A_{CP}^{dir} involves the weak-phase difference $\arg(\xi_u/\xi_c) = -\gamma$ (see (8.7)). By contrast, the phases of A_u and A_c are *CP*-conserving—they are not flipped by the *CP* transformation. These "strong phases" stem from a QCD effect, the strong rescattering $B \to f' \to f$ of the final state in the hadronisation process. Writing $\delta = \arg(A_u/A_c)$ for the strong-phase difference, we observe in (8.14) that $A_{CP}^{\text{dir}}(B \to f) \propto \sin\gamma \sin\delta$, so that a non-vanishing direct *CP* asymmetry requires both non-zero weak-phase and strong-phase differences between the interfering amplitudes. In general, it is impossible to calculate strong phases, making direct *CP* asymmetries less useful for the determination of fundamental *CP* phases than the mixing-induced *CP* phases discussed below in Sect. 8.2. However, a few direct *CP* asymmetries are theoretically tractable by combining several measurements to eliminate uncalculable hadronic quantities: In $B \to \pi\pi$ one can use the isospin symmetry and measure α; in $B^{\pm} \to D^0 K^{\pm}$ one can exploit different D^0 decay modes for this purpose and determine γ.

8.1.2 The LHC—a True B Factory

There are several key-quantities of an experiment which are crucial to perform high precision measurements in the *B* and charm system. Among them are the number of reconstructed signal candidates, which is related to the size of the data samples and the $b\bar{b}$ and $c\bar{c}$ cross section. Precise vertex and thus decay time resolution are a prerequisite to separate the production and decay vertex of the long-lived *B* and *D* mesons and to resolve time-dependent effects. Many *CP*-violation measurements in the *B* system require to identify the production flavour, which is done by so-called flavour-tagging algorithms. Therefore the tagging performance is also a critical parameter in the discussion of the flavour physics potential of an experiment.

In this section, we give an overview of the advantages and disadvantages of the different types of flavour-physics experiments at e^+e^- and at hadron colliders.

The name "*B* factories" was originally introduced for the e^+e^- colliders PEP-II at SLAC with the BaBar experiment [17] and KEKB at KEK with the Belle experiment [18]. Both colliders were operating at a centre-of-mass energy of $\sqrt{s} = 10.58\,\text{GeV}$, corresponding to the mass of the excited $b\bar{b}$ resonance $\Upsilon(4S)$. BaBar was running from 1999 to 2008 and collected a data set with an integrated luminosity of about 550 fb^{-1}. The Belle experiment also started in 1999 and was operational until the end of 2009, collecting about 1 ab^{-1}. A small fraction of the data was taken at a higher centre-of-mass energy corresponding to the $\Upsilon(5S)$ resonance, so that also the production of heavier B hadrons such as the B_s^0 was possible.

The cross section for the production of a $\Upsilon(4S)$ resonance is 1.1 nb^{-1}, which is about a quarter of the total hadronic e^+e^- cross section under BABAR and Belle conditions. The $\Upsilon(4S)$ decays predominantly and in similar fractions into B^+B^- pairs or into $B^0\overline{B}^0$ pairs. Due to the high fraction of B events among all events, no dedicated triggers are needed. There is—contrary to the situation at hadron colliders—no loss in efficiency at this stage of the experiment.

Due to flavour conservation, the two neutral B mesons are produced in a coherent state. As a consequence, the flavours of the two mesons are fully correlated up to the point at which the first of the two decays. The B factories exploit the use of electron and positron beams with slightly different energies ("asymmetric beams")—a setup which results in a moving or "boosted" centre-of-mass system. In this system, the two B mesons are produced almost at rest,[2] and thus the flight direction of the two B mesons in the experiment is entirely given by the boost along the z axis. The measured difference in z between the two B decay vertices is given by the B-meson lifetime of $\tau_B \sim 1.5$ ps and the according Lorentz boost. It is, on average, 250 μm. The decay time resolution is inversely proportional to the average flight distance in the laboratory system. At the BABAR and Belle experiments, it is about 1 ps. In typical $\Upsilon(4S)$ decays at BABAR and Belle, the final state contains only the decay products of the two B hadrons and thus allows both to be reconstructed very efficiently.

Many CP-violation and mixing analysis, require the knowledge of the production flavour of the signal B, which can be extracted from the reconstruction of the second B in the event. The signal B is also referred to as "same-side" B and the second B as "opposite-side" B. Both the coherent production and the very high reconstruction efficiency allow for a high probability to identify the signal B production flavour correctly.

The multi-purpose experiments CDF and DØ at the $p\bar{p}$ collider Tevatron at Fermilab, USA, established that—despite the very busy environment—precision measurements competitive with those obtained at the e^+e^- B factories are possible at hadron colliders. The LHCb experiment follows this path (as do, to a lesser extent, ATLAS and CMS).

In pp collisions at LHC energies, only small fractions of the proton energies are needed to produce a $b\bar{b}$ pair. Thus, despite the symmetric pp collisions, a produced $b\bar{b}$ pair will move in the laboratory frame in either the forward or the backward direction. It is for this reason that the LHCb experiment has been designed as a single-arm forward spectrometer covering the pseudo-rapidity range $1.9 < \eta < 4.9$ As a consequence, about 35 % of all produced $b\bar{b}$ pairs are in the acceptance of the LHCb detector. The total integrated luminosity of the data taken by LHCb in 2011 at $\sqrt{s} =$ 7 TeV corresponds to 1 fb^{-1}. In 2012, at a centre-of-mass energy of $\sqrt{s} = 8$ TeV, additional 2 fb^{-1} data have been recorded. These integrated luminosities correspond

[2] $m(\Upsilon(4S)) \approx 2m(B)$.

to 1×10^{11} and 2×10^{11} $b\bar{b}$ pairs in the LHCb detector acceptance, respectively. Due to the large boost the average flight distance of a B hadron in the LHCb experiment is about 1 cm. With a decay-time resolution of about 50 fs, the experiment is in an excellent position for precision time-dependent analysis, e.g. the measurement of the very high B_s^0 meson oscillation frequency. Since in contrast to the case of the B factories both b quarks hadronise and oscillate separately, in pp collisions it is harder to conclude from the opposite-side B hadron in the event on the production flavour of the signal B hadron.

The $b\bar{b}$ cross section at the LHC is huge—$\sigma_{pp \to b\bar{b}} \sim 0.3$ mb at $\sqrt{s} = 7$ TeV. However, the rate of inelastic pp collisions is even a factor 200 higher. In addition, the average momentum of B daughter tracks is not significantly larger than the momentum of tracks from minimum-bias events. Therefore, the selection of interesting B decays in the LHC pp environment is one of the challenging tasks of the LHCb experiment, requiring dedicated hardware and software triggers and a complete event reconstruction already at an early stage of the (software-based) high-level trigger. Typical trigger and reconstruction efficiencies for B mesons at LHCb vary between 0.1 and 10 % depending on the decay channel.

The large displacement of the B-decay vertex is an essential ingredient of the identification of B events. However, due to the large boost in the forward direction, the association of B hadrons to their production vertex is challenging. Therefore—in order to reduce the number of pp interactions per bunch crossing to an average number of visible interactions of $\nu = 1.5$ that is kept constant over the entire run ("luminosity levelling", see Sect. 2.1.4)—the two proton beams, before colliding in the LHCb experiment, are separated with the help of dedicated magnets.

The clear advantage of B physics experiments at hadron colliders is the huge cross section and the excellent decay-time resolution compared to e^+e^- facilities. The huge available statistics can even compensate a number of drawbacks that come with the pp environment: the inefficiencies caused by the trigger, the reconstruction and the selection that are necessary to reject the background from inelastic events; the approximately 10 times worse performance in identifying the production flavour; and combinatorial background from the underlying event. In fact, time-dependent analyses in the B_s^0 system are only possible at hadron colliders. However, due to their clean environment, the e^+e^- experiments are significantly better in reconstructing decays involving neutral pions, photons or missing energy from B-meson decays with neutrinos in the final state.

For the same reasons which make the LHCb experiment ideally suited for B physics, the experiment is also an excellent charm facility. The $c\bar{c}$ production cross section is about 20 times higher than that of $b\bar{b}$ production. However, the charm-decay vertices are less displaced ($\tau_{D^0} \sim 0.5 - 1.0$ ps) and thus harder to select. LHCb has collected the by far largest samples of charm events in the world. 3×10^{14} $c\bar{c}$ pairs are produced in the LHCb acceptance.[3]

Also ATLAS and CMS profit from the high $b\bar{b}$ cross section at the LHC. The two experiments are, however, limited to the observation of b quarks in the central

[3]This results in 10^9 fully reconstructed events of the most abundant decay channel $D^0 \to K^-\pi^+$.

rapidity region. They are good in detecting modes with leptons in the final state, such as $B_q^0 \to \mu^+\mu^-$ and $B_d^0 \to K^*\mu^+\mu^-$, because these are easier to trigger than hadronic modes, especially in an environment with, on average, up to 20 visible interaction. The decay-time resolution of ATLAS and CMS is, however, lower than that of LHCb due to less boosted B hadrons ($\sigma_t \sim 100\,\text{fs}$). In addition, reconstruction in ATLAS and CMS suffers from more combinatorial background and a momentum and mass resolutions that are about a factor two to four worse than for LHCb, due to more material in the tracking system. For decays such as $B_q^0 \to \mu^+\mu^-$, the higher luminosity of the ATLAS and CMS experiments compensates these disadvantages and results in competitive or even better branching ratio measurements than achievable at LHCb. However, for all time-dependent measurements and for analyses requiring the identification of the production flavour of the B hadrons, LHCb—as a dedicated flavour-physics experiment—is clearly superior to the multi-purpose experiments ATLAS and CMS.

8.2 Theory of Neutral *B* Mixing and *CP* Violation

Meson-antimeson mixing plays a pivotal role in flavour physics. On one hand the mixing frequency of the oscillations between a meson and its antiparticle is highly sensitive to new physics associated with hypothetical very massive particles. On the other hand meson-antimeson mixing permits measurements of *CP* phases which are free from the uncertainties associated with non-perturbative QCD effects. This feature allows us not only to probe new-physics contributions adding to the mixing box diagrams (see Fig. 8.1a), but also to study the *CP* phases of the decay amplitudes whenever both meson and antimeson can decay into the same final state. The ubiquity of the subject calls for a separate theory section, and we exemplify the basic concepts of meson-antimeson mixing here with $B^0-\overline{B}^0$ mixing, covering both B_d^0 and B_s^0. $D-\overline{D}^0$ mixing, another important topic for LHCb, is relegated to Sect. 8.7.1.

The box diagram in Fig. 8.1a describes the transition between a B_s^0 and its antiparticle $\overline{B}_s^0$. This is called a $\Delta B = 2$ process, because the beauty quantum number changes by 2 units. One obtains the $\Delta B = 2$ diagram for $B_d^0-\overline{B}_d^0$ mixing by replacing the external s quark lines by d quark lines in Fig. 8.1a. We discuss B^0–$\overline{B}^0$ mixing here simultaneously for the two cases of interest, B_d^0 and B_s^0. Most of the presented formalism also applies to $K^0-\overline{K}^0$ or $D^0-\overline{D}^0$ mixing.

$B_q^0-\overline{B}_q^0$ mixing, $q = d$ or s, has two striking consequences: First, the flavour eigenstates B_q^0 and $\overline{B}_q^0$ do not coincide with the mass eigenstates B_{qL} and B_{qH} ("L" and "H" standing for "light" and "heavy"). While the *CPT* theorem encountered in Sect. 8.1.1 enforces equal masses and equal total decay widths for B_q^0 and $\overline{B}_q^0$, the states B_{qL} and B_{qH} differ in their masses and widths. Second, a meson tagged (i.e. identified) as a B_q^0 at time $t = 0$ oscillates into a quantum-mechanical superposition of B_q^0 and $\overline{B}_q^0$ for $t > 0$. Denoting the state of such a meson as $|B_q^0(t)\rangle$ one may write

$$|B_q^0(t)\rangle = \langle B_q^0|B_q^0(t)\rangle|B_q^0\rangle + \langle\overline{B}_q^0|B_q^0(t)\rangle|\overline{B}_q^0\rangle + \cdots, \qquad (8.15)$$

with "..." denoting the states into which $B_q^0(t)$ can decay. An analogous formula holds for $|\overline{B}_q^0(t)\rangle$. By studying the time dependence of some decay $B_q^0(t) \to f$, we can probe the interference of the decays $B_q^0 \to f$ and $\overline{B}_q^0 \to f$, provided both of these two decays are allowed. This interference effect gives rise to the most powerful means to study *CP* violation in *B* physics—the comparison of the time-dependent decay rates of the two processes $B_q^0(t) \to f$ and $\overline{B}_q^0(t) \to f$. The studied phenomenon is called "mixing-induced *CP* violation" or "*CP* violation in the interference of mixing and decay" and is explained below.

The time dependence of the coefficients in (8.15) is governed by a Schrödinger equation,

$$i\frac{\mathrm{d}}{\mathrm{d}t}\begin{pmatrix}\langle B_q^0|B_q^0(t)\rangle\\ \langle\overline{B}_q^0|B_q^0(t)\rangle\end{pmatrix} = \left(M^q - i\frac{\Gamma^q}{2}\right)\begin{pmatrix}\langle B_q^0|B_q^0(t)\rangle\\ \langle\overline{B}_q^0|B_q^0(t)\rangle\end{pmatrix}. \qquad (8.16)$$

The coefficients $\langle B_q^0|\overline{B}_q^0(t)\rangle$ and $\langle\overline{B}_q^0|\overline{B}_q^0(t)\rangle$ obey the same equation. The hermitian 2×2 matrices M^q and Γ^q are called mass and decay matrix, respectively. The mass eigenstates $|B_{qL}\rangle$ and $|B_{qH}\rangle$ are found by diagonalising $M^q - i\Gamma^q/2$. These obey an exponential-decay law with the eigenvalues $M_{L,H}^q - i\Gamma_{L,H}^q/2$ of $M^q - i\Gamma^q/2$:

$$\begin{aligned}|B_{qL}(t)\rangle &= |B_{qL}\rangle\exp\left[-iM_L^q t - \frac{\Gamma_L^q}{2}t\right],\\ |B_{qH}(t)\rangle &= |B_{qH}\rangle\exp\left[-iM_H^q t - \frac{\Gamma_H^q}{2}t\right].\end{aligned} \qquad (8.17)$$

The basis transform between $(|B_q^0\rangle, |\overline{B}_q^0\rangle)$ and $(|B_{qL}\rangle, |B_{qH}\rangle)$ reads

$$\begin{aligned}|B_{qL}(t)\rangle &= p_q|B_q^0(t)\rangle + q_q|\overline{B}_q^0(t)\rangle\\ |B_{qH}(t)\rangle &= p_q|B_q^0(t)\rangle - q_q|\overline{B}_q^0(t)\rangle\end{aligned} \qquad (8.18)$$

with complex coefficients q_q, p_q satisfying $|q_q|^2 + |p_q|^2 = 1$. The *CPT* theorem implies that the diagonal elements of M^q and Γ^q are the same, $M_{11}^q = M_{22}^q$ and $\Gamma_{11}^q = \Gamma_{22}^q$. Without this property, the ansatz in (8.18) with just two complex numbers p_q and q_q would not work.

The individual phases of M_{12}^q and Γ_{12}^q are unphysical, because they change under rephasings of $|B_q^0\rangle$ or $|\overline{B}_q^0\rangle$. This leaves five physical quantities in (8.16). The first two are the average mass and the average width of the neutral B^0 meson:

$$M_{B_q^0} = M_{11}^q = M_{22}^q = \frac{M_H^q + M_L^q}{2}, \; \Gamma_q = \Gamma_{11}^q = \Gamma_{22}^q = \frac{\Gamma_H^q + \Gamma_L^q}{2}. \qquad (8.19)$$

The remaining three physical quantities are related to $B_q^0 - \overline{B}_q^0$ mixing:

$$\left|M_{12}^q\right|, \quad \left|\Gamma_{12}^q\right|, \quad \phi_{12}^q \equiv \arg\left(-\frac{M_{12}^q}{\Gamma_{12}^q}\right). \tag{8.20}$$

The CP transformation exchanges B_q^0 and $\overline{B}_q^0$, so that $M_{12}^q/\Gamma_{12}^q \rightarrow M_{21}^q/\Gamma_{21}^q = M_{12}^{q*}/\Gamma_{12}^{q*}$ under CP. Thus in this step ϕ_{12}^q flips sign and we recognise it as a CP-violating phase, while the first two quantities in (8.20) are CP-even. By solving the eigenvalue problem of $M^q - i\Gamma^q/2$, we can relate M_{12}^q and Γ_{12}^q to the mass difference ΔM_q and the width difference $\Delta\Gamma_q$, defined as

$$\Delta M_q = M_H^q - M_L^q, \qquad \Delta\Gamma_q = \Gamma_L^q - \Gamma_H^q. \tag{8.21}$$

While $\Delta M_q > 0$ by definition, the sign of $\Delta\Gamma_q$ must be determined by experiment. The definition in (8.21) is chosen such that $\Delta\Gamma_q > 0$ in the SM.

To find the formula describing the time evolution of $\left|B_q^0(t)\right\rangle$ and $\left|\overline{B}_q^0(t)\right\rangle$ (revealing the oscillatory behaviour advertised above), we first invert (8.18) (to find $\left|B_q^0(t)\right\rangle = (\left|B_{qL}(t)\right\rangle + \left|B_{qH}(t)\right\rangle)/(2p)$ and $\left|\overline{B}_q^0(t)\right\rangle = (\left|B_{qL}(t)\right\rangle - \left|B_{qH}(t)\right\rangle)/(2q)$) and insert the time evolution of (8.17). With (8.18), we then trade $\left|B_{qL}\right\rangle$ and $\left|B_{qH}\right\rangle$ for $\left|B_q^0\right\rangle$ and $\left|\overline{B}_q^0\right\rangle$, finding an expression for $\left|B_q^0(t)\right\rangle$ and $\left|\overline{B}_q^0(t)\right\rangle$ in terms of $M_{L,H}^q$ and $\Gamma_{L,H}^q$. The result is conventionally written as

$$\begin{aligned} \left|B_q^0(t)\right\rangle &= g_+(t)\left|B_q^0\right\rangle + \frac{q_q}{p_q}\, g_-(t)\left|\overline{B}_q^0\right\rangle, \\ \left|\overline{B}_q^0(t)\right\rangle &= \frac{p_q}{q_q}\, g_-(t)\left|B_q^0\right\rangle + g_+(t)\left|\overline{B}_q^0\right\rangle. \end{aligned} \tag{8.22}$$

To get nice expressions for $g_\pm(t)$, we finally use (8.19) and (8.21) to trade $M_{L,H}$ and $\Gamma_{L,H}$ for ΔM_q, $\Delta\Gamma_q$, and the average mass $M_{B_q^0}$ and the average width Γ_q defined in (8.19):

$$\begin{aligned} g_+(t) &= e^{-iM_{B_q^0}t}\, e^{-\Gamma_q t/2}\left[\cosh\frac{\Delta\Gamma_q t}{4}\cos\frac{\Delta M_q t}{2} - i\sinh\frac{\Delta\Gamma_q t}{4}\sin\frac{\Delta M_q t}{2}\right], \\ g_-(t) &= e^{-iM_{B_q^0}t}\, e^{-\Gamma_q t/2}\left[-\sinh\frac{\Delta\Gamma_q t}{4}\cos\frac{\Delta M_q t}{2} + i\cosh\frac{\Delta\Gamma_q t}{4}\sin\frac{\Delta M_q t}{2}\right]. \end{aligned}$$

Setting $t = 0$ in these formulae we verify $g_+(0) = 1$ and $g_-(0) = 0$ and find that $g_\pm(t)$ has no zeros for $t > 0$ if $\Delta\Gamma_q \neq 0$. Hence, an initially produced B_q^0 will never turn into a pure $\overline{B}_q^0$ or back into a pure B_q^0. We will frequently encounter the combinations

$$
\begin{aligned}
|g_\pm(t)|^2 &= \frac{e^{-\Gamma_q t}}{2}\left[\quad \cosh\frac{\Delta\Gamma_q t}{2} \pm \cos\left(\Delta M_q t\right)\right], \\
g_+^*(t)\, g_-(t) &= \frac{e^{-\Gamma_q t}}{2}\left[-\sinh\frac{\Delta\Gamma_q t}{2} + i\sin\left(\Delta M_q t\right)\right].
\end{aligned}
\tag{8.23}
$$

As a first application of (8.22) and (8.23), we look at the survival probability of a meson tagged as B_q^0 at time $t = 0$:

$$
\begin{aligned}
P_{\text{no mixing}}(t) &= |\langle B_q^0|B_q^0(t)\rangle|^2 = |g_+(t)|^2 \\
&= \frac{e^{-\Gamma_q t}}{2}\left[\cosh\frac{\Delta\Gamma_q t}{2} + \cos\left(\Delta M_q t\right)\right] \\
&= |\langle \overline{B}_q^0|\overline{B}_q^0(t)\rangle|^2 .
\end{aligned}
\tag{8.24}
$$

We observe that the usual exponential decay law is modified by the oscillatory term $\cos\left(\Delta M_q t\right)$, and we identify the oscillation frequency with ΔM_q. While $\Delta\Gamma_d/\Gamma_d$ is tiny, one has $\Delta\Gamma_s \sim 0.1\,\Gamma_s$ so that we cannot set $\Delta\Gamma_s$ to zero in studies of B_s^0 decays. Note that $\exp[-\Gamma_q t]\cosh(\Delta\Gamma_q t/2) = (\exp[-\Gamma_L t] + \exp[-\Gamma_H t])/2$, so the decay involves two exponentials with both decay rates of the two mass eigenstates. The mixing probabilities are different for $B_q^0 \to \overline{B}_q^0$ and $\overline{B}_q^0 \to B_q^0$:

$$
\begin{aligned}
P_{B_q^0\to\overline{B}_q^0}(t) &= |\langle \overline{B}_q^0|B_q^0(t)\rangle|^2 = \left|\frac{q_q}{p_q}\right|^2 |g_-(t)|^2 \\
&= \frac{e^{-\Gamma_q t}}{2}\left|\frac{q_q}{p_q}\right|^2\left[\cosh\frac{\Delta\Gamma_q t}{2} - \cos\left(\Delta M_q t\right)\right],
\end{aligned}
\tag{8.25}
$$

while

$$
P_{\overline{B}_q^0\to B_q^0}(t) = \frac{e^{-\Gamma_q t}}{2}\left|\frac{p_q}{q_q}\right|^2\left[\cosh\frac{\Delta\Gamma_q t}{2} - \cos\left(\Delta M_q t\right)\right]. \tag{8.26}
$$

So far we have only used the *CPT* theorem in the derivation of the time-evolution formulae, but no specific property of the theory (SM or some extension) to be probed. The quantities ΔM_q, $\Delta\Gamma_q$, and q_q/p_q, however, depend on the theory and will be calculated for the SM below.

We do not quote the exact solutions for ΔM_q, $\Delta\Gamma^q$, q_q/p_q in terms of M_{12}^q and Γ_{12}^q here. These formulae simplify if one uses $|\Gamma_{12}^q| \ll |M_{12}^q|$, which holds for B^0–$\overline{B}^0$ mixing (but not for $K^0-\overline{K}^0$ or $D^0-\overline{D}^0$ mixing):

$$
\begin{aligned}
\Delta M_q &\simeq 2\,|M_{12}^q|, \\
\Delta\Gamma_q &\simeq 2\,|\Gamma_{12}^q|\cos\phi_{12}^q .
\end{aligned}
$$

The corrections to these formulae are of order $|\Gamma_{12}^q/M_{12}^q|^2 < 10^{-4}$ and negligible in view of the theoretical uncertainties of M_{12}^q and Γ_{12}^q. We see that $|\Gamma_{12}^q| \ll |M_{12}^q|$

implies $|\Delta\Gamma_q| \ll \Delta M_q$, in agreement with experimental findings. The hierarchy between $|\Gamma_{12}^q|$ and $|M_{12}^q|$ originates from the heaviness of the top quark that enhances $|M_{12}^q|$ through the box diagram in Fig. 8.1a. However, $|\Gamma_{12}^q| \ll |M_{12}^q|$ is firmly established experimentally, without the need to resort to theory. (i.e. the possibility to explain the observed $|\Delta\Gamma_q| \ll \Delta M_q$ through $\phi_{12}^q \sim \pi/2$ is ruled out by bounds on $|\phi_{12}^q|$ inferred from *CP* asymmetries measured in semileptonic decays, which are discussed below in detail.) Defining the small quantity

$$a_{sl}^q = \mathrm{Im}\frac{\Gamma_{12}^q}{M_{12}^q} = \left|\frac{\Gamma_{12}^q}{M_{12}^q}\right| \sin\phi_{12}^q\,, \tag{8.27}$$

we can write

$$\frac{q_q}{p_q} = -\frac{M_{12}^{q*}}{|M_{12}^q|}\left[1 - \frac{a_{sl}^q}{2}\right] \tag{8.28}$$

up to corrections of $(\Gamma_{12}^q/M_{12}^q)^2$. For $\phi_{12}^q = 0$, (8.28) entails $a_{sl}^q = 0$, and one finds $|q_q/p_q| = 1$. *CP* violation in mixing is characterised by $|q_q/p_q| \neq 1$ and quantified by a_{sl}^q. This type of *CP* violation affects all $B_q^0 \to f$ decays in the same way. It is usually measured in flavour-specific $B_q^0(t) \to f$ decays, which means that $B_q^0 \to f$ is allowed while $\overline{B}_q^0 \to f$ is forbidden.

To link the theoretical quantities describing time-dependent phenomena to experiment, we need the time-dependent decay rate:

$$\Gamma(B_q^0(t) \to f) = \frac{1}{N_{B_q^0}} \frac{\mathrm{d}N(B_q^0(t) \to f)}{\mathrm{d}t}\,,$$

where $\mathrm{d}N(B_q^0(t) \to f)$ denotes the number of $B_q^0(t) \to f$ decays occurring within the time interval between t and $t+\mathrm{d}t$. $N_{B_q^0}$ is the total number of B_q^0 mesons produced at time $t = 0$. An analogous definition holds for $\Gamma(\overline{B}_q^0(t) \to f)$. With the shorthand notation for decay amplitudes,

$$A_f = A(B_q^0 \to f) = \langle f|H^{|\Delta B|=1}|B_q^0\rangle\,,$$
$$\overline{A}_f = A(\overline{B}_q^0 \to f) = \langle f|H^{|\Delta B|=1}|\overline{B}_q^0\rangle\,,$$

the theoretical expressions for the time-dependent decay rates are

$$
\begin{aligned}
\Gamma(B_q^0(t) \to f) &= \mathcal{N}_f \left| \langle f | H^{|\Delta B|=1} | B_q^0(t) \rangle \right|^2 = \mathcal{N}_f \left| g_+(t) A_f + \frac{q_q}{p_q} g_-(t) \overline{A}_f \right|^2 , \\
\Gamma(\overline{B}_q^0(t) \to f) &= \mathcal{N}_f \left| \langle f | H^{|\Delta B|=1} | \overline{B}_q^0(t) \rangle \right|^2 = \mathcal{N}_f \left| \frac{p_q}{q_q} g_-(t) A_f + g_+(t) \overline{A}_f \right|^2 ,
\end{aligned}
\tag{8.29}
$$

with the time-independent normalisation factor $\mathcal{N}_f$ comprising the result of the phase-space integration. A key quantity to study mixing-induced *CP* violation is the combination

$$
\lambda_f = \frac{q}{p} \frac{\overline{A}_f}{A_f} .
$$

Using (8.23) and (8.28) in (8.29), we readily find

$$
\begin{aligned}
\Gamma(B_q^0(t) \to f) = \mathcal{N}_f \, |A_f|^2 \, e^{-\Gamma_q t} \Bigg\{ & \frac{1 + |\lambda_f|^2}{2} \cosh \frac{\Delta\Gamma_q \, t}{2} \\
& + \frac{1 - |\lambda_f|^2}{2} \cos(\Delta M_q \, t) - \mathrm{Re}\lambda_f \, \sinh \frac{\Delta\Gamma_q \, t}{2} - \mathrm{Im}\lambda_f \, \sin\left(\Delta M_q \, t\right) \Bigg\} ,
\end{aligned}
\tag{8.30}
$$

$$
\begin{aligned}
\Gamma(\overline{B}_q^0(t) \to f) = \mathcal{N}_f \, |A_f|^2 \, (1 + a_{sl}^q) \, e^{-\Gamma_q t} \Bigg\{ & \frac{1 + |\lambda_f|^2}{2} \cosh \frac{\Delta\Gamma_q \, t}{2} \\
& - \frac{1 - |\lambda_f|^2}{2} \cos(\Delta M_q \, t) - \mathrm{Re}\lambda_f \, \sinh \frac{\Delta\Gamma_q \, t}{2} + \mathrm{Im}\lambda_f \, \sin(\Delta M_q \, t) \Bigg\} .
\end{aligned}
\tag{8.31}
$$

Here and everywhere else we neglect terms of order ${a_{sl}^q}^2$ and higher. In the $B_q^0(t) \to \overline{f}$ decay rates, it is advantageous to keep $\overline{A}_{\overline{f}}$ (and trade $A_{\overline{f}}$ for $\lambda_{\overline{f}}^{-1}$), because *CP* relates $\overline{A}_{\overline{f}}$ to A_f (while A_f and $\overline{A}_f$ are usually unrelated):

$$
\begin{aligned}
\Gamma(B_q^0(t) \to \overline{f}) = \mathcal{N}_f \left| \overline{A}_{\overline{f}} \right|^2 e^{-\Gamma_q t} \, (1 - a_{sl}^q) \Bigg\{ & \frac{1 + |\lambda_{\overline{f}}|^{-2}}{2} \cosh \frac{\Delta\Gamma_q \, t}{2} \\
& - \frac{1 - |\lambda_{\overline{f}}|^{-2}}{2} \cos(\Delta M_q \, t) - \mathrm{Re}\frac{1}{\lambda_{\overline{f}}} \sinh \frac{\Delta\Gamma_q \, t}{2} + \mathrm{Im}\frac{1}{\lambda_{\overline{f}}} \sin(\Delta M_q \, t) \Bigg\} ,
\end{aligned}
\tag{8.32}
$$

$$
\begin{aligned}
\Gamma(\overline{B}_q^0(t) \to \overline{f}) = \mathcal{N}_f \left| \overline{A}_{\overline{f}} \right|^2 e^{-\Gamma_q t} \Bigg\{ & \frac{1 + |\lambda_{\overline{f}}|^{-2}}{2} \cosh \frac{\Delta\Gamma_q \, t}{2} \\
& + \frac{1 - |\lambda_{\overline{f}}|^{-2}}{2} \cos(\Delta M_q \, t) - \mathrm{Re}\frac{1}{\lambda_{\overline{f}}} \sinh \frac{\Delta\Gamma_q \, t}{2} - \mathrm{Im}\frac{1}{\lambda_{\overline{f}}} \sin(\Delta M_q \, t) \Bigg\} .
\end{aligned}
\tag{8.33}
$$

Equations (8.30) and (8.33) are the master formulae capable to describe the time evolution of *any* B_q^0 or $\overline{B}_q^0$ decay. One can fit the data directly to these formulae.

We now discuss two important special cases: flavour-specific decays and decays to *CP* eigenstates. The former decays are characterised by $\overline{A}_f = A_{\overline{f}} = 0$, and prototypes are the semileptonic decays $B_q^0 \to X^- \ell^+ \nu_\ell$ and $\overline{B}_q^0 \to X^+ \ell^- \overline{\nu}_\ell$ with $\ell = e$ or $\ell = \mu$ and X denoting an unspecified hadronic final state.[4] (The case $\ell = \tau$ is challenging at hadron colliders.) A flavour-specific decay $B_q^0(t) \to f_{\rm fs}$ projects out the $|B_q^0\rangle$ component of $|B_q^0(t)\rangle$, while the decay into the *CP*-conjugate state $\overline{f}_{\rm fs}$ projects out the $|\overline{B}_q^0\rangle$ component. That is, we expect $\Gamma(B_q^0(t) \to f_{\rm fs}) \propto P_{\rm no\ mixing}(t)$ and $\Gamma(B_q^0(t) \to \overline{f}_{\rm fs}) \propto P_{B_q^0 \to \overline{B}_q^0}(t)$, and with $\lambda_{f_{\rm fs}} = 0$ and $\lambda_{\overline{f}_{\rm fs}}^{-1} = 0$ we indeed recover the time-dependence of (8.24), (8.26) from (8.30) to (8.33). In the following discussion it is further assumed that the considered $B_q^0 \to f_{\rm fs}$ decay has no direct *CP* violation, i.e. that $\overline{A}_{\overline{f}_{\rm fs}} = A_f$. (This is certainly fulfilled for $f_{\rm fs} = X^- \ell^+ \nu_\ell$.) From (8.30) and (8.32) we obtain the mixing asymmetry,

$$\mathcal{A}_0(t) = \frac{\Gamma(B_q^0(t) \to f_{\rm fs}) - \Gamma(\overline{B}_q^0(t) \to f_{\rm fs})}{\Gamma(B_q^0(t) \to f_{\rm fs}) + \Gamma(\overline{B}_q^0(t) \to f_{\rm fs})},$$

as

$$\mathcal{A}_0(t) = \frac{\cos(\Delta M_q\, t)}{\cosh(\Delta\Gamma_q\, t/2)} + \frac{a_{sl}^q}{2}\left[1 - \frac{\cos^2(\Delta M_q\, t)}{\cosh^2(\Delta\Gamma_q\, t/2)}\right].$$

Note that $\mathcal{A}_0(t)$ is not a *CP* asymmetry—this quantity is used for the measurements of the oscillation frequency ΔM_q.

Next we consider the "*CP* asymmetry in flavour-specific decays" (called "semileptonic *CP* asymmetry" when measured in $B_q^0 \to X^- \ell^+ \nu$). We have defined a_{sl}^q in (8.27) in such a way that a_{sl}^q coincides with this *CP* asymmetry:

$$a_{sl}^q = \frac{\Gamma(\overline{B}_q^0(t) \to f_{\rm fs}) - \Gamma(B_q^0(t) \to \overline{f}_{\rm fs})}{\Gamma(\overline{B}_q^0(t) \to f_{\rm fs}) + \Gamma(B_q^0(t) \to \overline{f}_{\rm fs})}.$$

Importantly, no flavour tagging is needed to measure a_{sl}. With the untagged rate

$$\Gamma[f, t] = \Gamma(B_q^0(t) \to f) + \Gamma(\overline{B}_q^0(t) \to f),$$

one finds

$$a_{sl,unt}^q(t) \equiv \frac{\Gamma[f_{\rm fs}, t] - \Gamma[\overline{f}_{\rm fs}, t]}{\Gamma[f_{\rm fs}, t] + \Gamma[\overline{f}_{\rm fs}, t]} = \frac{a_{sl}^q}{2} - \frac{a_{sl}^q}{2}\frac{\cos(\Delta M_q\, t)}{\cosh(\Delta\Gamma_q t/2)}. \qquad (8.34)$$

The second application of the master formulae are decays into *CP* eigenstates, $B_q^0 \to f_{\rm CP}$. *CP* eigenstates satisfy $|\overline{f}_{\rm CP}\rangle = \eta_{f_{\rm CP}} |f_{\rm CP}\rangle$ with $\eta_{f_{\rm CP}} = 1$ ($\eta_{f_{\rm CP}} = -1$) for

[4] All formulae presented for flavour-specific decays equally apply to the inclusive semileptonic decay or to exclusive decays like $B_q^0 \to D_s^{*-} \ell^+ \nu_\ell$.

CP-even (*CP*-odd) states. The time-dependent *CP* asymmetry is

$$a_{f_{\rm CP}}(t) = \frac{\Gamma(\overline{B}_q^0(t) \to f_{\rm CP}) - \Gamma(B_q^0(t) \to f_{\rm CP})}{\Gamma(\overline{B}_q^0(t) \to f_{\rm CP}) + \Gamma(B_q^0(t) \to f_{\rm CP})} .$$

Using (8.30) and (8.31) one finds

$$a_{f_{\rm CP}}(t) = -\frac{A_{CP}^{\rm dir}\cos(\Delta M_q\, t) + A_{CP}^{\rm mix}\sin(\Delta M_q\, t)}{\cosh(\Delta\Gamma_q\, t/2) + A_{\Delta\Gamma}\sinh(\Delta\Gamma_q\, t/2)} + \mathcal{O}(a_{sl}^q) , \qquad (8.35)$$

with (for $f = f_{\rm CP}$)

$$A_{CP}^{\rm dir} = \frac{1-|\lambda_f|^2}{1+|\lambda_f|^2}, \qquad A_{CP}^{\rm mix} = -\frac{2\,{\rm Im}\lambda_f}{1+|\lambda_f|^2}, \qquad A_{\Delta\Gamma} = -\frac{2\,{\rm Re}\lambda_f}{1+|\lambda_f|^2} .$$

If we neglect the small deviation of $|q/p|$ from unity, the definition of $A_{CP}^{\rm dir}$ coincides with the one in (8.13) for the considered case $f = f_{\rm CP}$. The other two quantities $A_{CP}^{\rm mix}$ and $A_{\Delta\Gamma}$ encode properties of $B_q^0 - \overline{B}_q^0$ mixing and have no counterparts in $B^\pm$ decays. Note that $|A_{CP}^{\rm dir}|^2 + |A_{CP}^{\rm mix}|^2 + |A_{\Delta\Gamma}|^2 = 1$.

Experimentally one can track the time-dependence of $a_f(t)$ and read off the coefficients of $\cos(\Delta M\, t)$ and $\sin(\Delta M\, t)$, so that one can determine $|\lambda_f|$ and ${\rm Im}\lambda_f$. In general it is not possible to relate this information to the *CP* phase in the Lagrangian, due to the untractable non-perturbative QCD in $A_{f_{\rm CP}}$. However, this obstacle is not there for so-called "golden modes", which have the property that the decay amplitude is dominated by a single combination $V_{\rm CKM}$ of CKM elements. An example for a golden mode is $B_s^0 \to J/\psi\,\phi$ with $V_{\rm CKM} = V_{cb}^* V_{cs}$—the competing "penguin" amplitude involves $V_{ub}^* V_{us}$ which is ~40 times smaller. Since QCD respects *CP*, golden modes simply fulfill

$$\frac{\overline{A}_{f_{\rm CP}}}{A_{f_{\rm CP}}} = -\,\eta_{f_{CP}} \frac{V_{\rm CKM}^*}{V_{\rm CKM}} . \qquad (8.36)$$

(The minus sign stems from the convention $CP|B_q^0\rangle = -|\overline{B}_q^0\rangle$. λ_f is independent of such conventions.) From (8.36) one reads off $|\overline{A}_{f_{\rm CP}}| = |A_{f_{\rm CP}}|$ and, neglecting a_{sl}^q, finds $|\lambda_{f_{\rm CP}}| = 1$. Thus golden modes satisfy

$$A_{CP}^{\rm dir} = 0, \qquad A_{CP}^{\rm mix} = -{\rm Im}\lambda_{f_{\rm CP}} .$$

Next we discuss the SM predictions for ΔM_q, $\Delta\Gamma_q$, and a_{sl}^q: The quantities M_{12}^q and Γ_{12}^q are calculated from the box diagram in Fig. 8.1a. M_{12}^q is dominated by the internal top contribution, so that $\Delta M_s = 2|M_{12}^s|$ reads [19]

$$\Delta M_s = \left(18.8 \pm 0.6_{V_{cb}} \pm 0.3_{m_t} \pm 0.1_{\alpha_s} \right) {\rm ps}^{-1} \frac{f_{B_s}^2\, B_{B_s}}{(220\,{\rm MeV})^2} .$$

This number is found from the next-to-leading order (NLO) QCD calculation of [20] with up-to-date values of the input parameters. The uncertainties stemming from the errors of $|V_{cb}|$, α_s, and m_t are indicated. The by far largest source of uncertainty is the parameter $f_{B_s}\sqrt{B_{B_s}}$ that quantifies a hadronic matrix element. Reference [21] quotes the lattice QCD result $f_{B_s}\sqrt{B_{B_s}} = (216 \pm 15)$ MeV, giving the SM prediction

$$\Delta M_s = (18.2 \pm 2.5)\,\mathrm{ps}^{-1}\,.$$

For the mass difference in the $B_d^0-\overline{B}_d^0$ system one finds

$$\Delta M_d = (0.49 \pm 0.02)\,\mathrm{ps}^{-1}\left(\frac{|V_{td}|}{0.0082}\right)^2 \frac{f_{B_d}^2 B_{B_d}}{(175\,\mathrm{MeV})^2}\,.$$

Reference [21] recommends $f_{B_d}\sqrt{B_{B_d}} = (176 \pm 12)$ MeV. ΔM_d determines $|V_{td}|$ and is an important ingredient in global fits of the apex $(\overline{\rho}, \overline{\eta})$ of the unitarity triangle in Fig. 8.2, fixing one side through $|V_{td}| \propto R_t$ (see (8.5)). In order to predict $\Delta\Gamma_q$ and a_{sl}^q, one must calculate Γ_{12}^q. Unlike ΔM_q, this calculation requires an expansion in powers of $\Lambda_{\mathrm{QCD}}/m_b$, which involves several hadronic matrix elements. Using NLO calculations [22–25] one finds [26]

$$\left.\frac{\Delta\Gamma_d}{\Gamma_d}\right|_{\mathrm{SM}} = (42 \pm 8)\cdot 10^{-4}, \qquad \left.\frac{\Delta\Gamma_s}{\Gamma_s}\right|_{\mathrm{SM}} = 0.137 \pm 0.027\,. \tag{8.37}$$

These predictions use the experimental information on ΔM_d and ΔM_s, respectively, to reduce hadronic uncertainties and are therefore sensitive to new physics in these quantities. The *CP* asymmetries in flavour-specific decays (semileptonic *CP* asymmetries) are predicted as [26]

$$a_{sl}^{s,\mathrm{SM}} = (1.9 \pm 0.3)\cdot 10^{-5}, \qquad a_{sl}^{d,\mathrm{SM}} = -(4.1 \pm 0.6)\cdot 10^{-4} \tag{8.38}$$

for $|V_{ub}| = (3.56^{+0.15}_{-0.20})\cdot 10^{-3}$.

The semileptonic *CP* asymmetries are very sensitive to new physics [23, 25–29].

8.3 Measurements of *B* Oscillations and *CP* Asymmetries

8.3.1 *Measurements of the $B_d^0-\overline{B}_d^0$ and $B_s^0-\overline{B}_s^0$ Oscillation Frequencies*

The oscillation frequency of neutral *B* mesons is measured in flavour-specific final states, which allow directly to determine *B* flavour at the decay out of the *B* decay products. The most abundant fully reconstructed flavour-specific final states are those of the $B_d^0 \to J/\psi K^*$, $B_d^0 \to D^-\pi^+$ and $B_s^0 \to D_s^-\pi^+$ decays. In order to determine whether a B meson has the same or opposite flavour at production and decay, its production flavour also needs to be determined. Therefore so-called "flavour-tagging algorithms" are used. There are two conceptual different types of tagging algorithms, so-called "opposite-side" and "same-side" flavour-tagging algorithms. The opposite-side flavour-tagging algorithms exploit the fact that b quarks are always produced in $b\bar{b}$ quark pairs. By partially reconstructing the second (opposite-side) B hadron in the event, one can determine its decay flavour. Assuming no oscillation of the opposite-side B[5] one can thus conclude on the signal-production flavour. Alternatively, the same-side tagging algorithm exploits the signal B fragmentation properties to determine signal-production flavour. The physical[6] and the actually measured time-dependent asymmetries $\mathcal{A}_{\text{phys}}(t)$ and $\mathcal{A}_{\text{meas}}(t)$ are defined as

$$\mathcal{A}_{\text{meas}}(t) = \frac{N_{\text{unmixed}}(t) - N_{\text{mixed}}(t)}{N_{\text{unmixed}}(t) + N_{\text{mixed}}(t)} \sim D_{\text{tag}} D_t \mathcal{A}_{\text{phys}}(t)\,,$$
$$\mathcal{A}_{\text{phys}}(t) = \frac{\cos(\Delta M_q t)}{\cosh(\Delta \Gamma_q t/2)}\,,$$

respectively. Here, N_{unmixed} refers to the number of B decays which have been tagged to have the same production and decay flavour, while N_{mixed} refers to those which have been tagged to have different production and decay flavours. The physical asymmetry is damped by two dilution factors: One results from the imperfect tagging performance and is about $D_{\text{tag}} \sim 0.3$ at LHCb. The damping due to the decay-time resolution is about $D_t \sim 0.73$, corresponding to $\sigma_t \sim 50$ fs for the fast $B_s^0-\overline{B}_s^0$ oscillation and negligible ($D_t \sim 1$) for the slow $B_d^0-\overline{B}_d^0$ oscillation. Using 1 fb^{-1} of data, the following oscillation frequencies have been measured at the LHCb experiment [30, 31] in the decays $B_s^0 \to D_s^-\pi^+$ (see Fig. 8.4), $B_d^0 \to J/\psi K^*$and $B_d^0 \to D^-\pi^+$, respectively:

$$\Delta M_d = 0.5178 \pm 0.0061(\text{stat}) \pm 0.0037(\text{sys})\ \text{ps}^{-1},$$
$$\Delta M_s = 17.768 \pm 0.023(\text{stat}) \pm 0.006(\text{sys})\ \text{ps}^{-1}.$$

Additionally, the measurement of the $B_d^0-\overline{B}_d^0$ oscillation frequency in semileptonic decays is currently under investigation by the LHCb experiment. This analysis has—due to the large number of signal events—the potential to result in the globally most precise measurement. However, the correction for the missing momentum caused by the final-state neutrino might result in a limiting systematic uncertainty.

[5] 17 % of all B_d^0 mesons and 50 % of all B_s^0 mesons oscillate before they decay.

[6] The definition of $\mathcal{A}_{\text{phys}}(t)$ follows directly from 8.24 and 8.25.

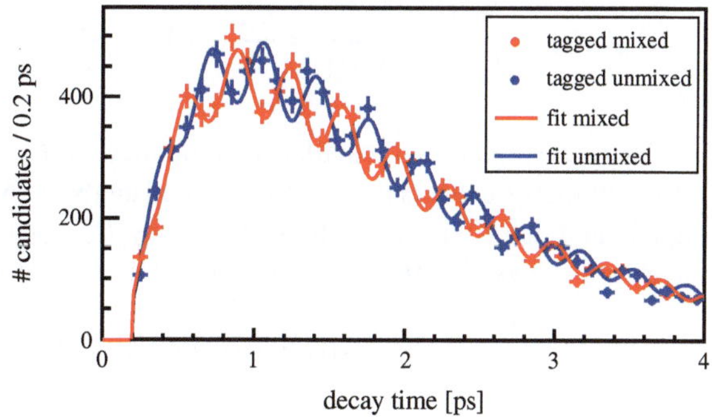

Fig. 8.4 Time-dependent decay rates of B_s^0 mesons with same (unmixed, *blue symbols*) and different (mixed, *red symbols*) flavour at production and decay time. Adapted from [30]

8.3.2 CP Violation in Neutral B-Meson Mixing

CP violation in neutral *B* meson mixing is predicted to be tiny in the Standard Model (see (8.38)). However, there is a measurement of the dimuon asymmetry from the DØ collaboration which indicates a deviation from this prediction [32]. This result still needs to be confirmed or falsified by LHC experiments. *B* hadrons preferentially undergo decays with positively charged leptons ($\bar{b} \rightarrow \bar{c}\ell^+\nu_\ell$), while anti-*B* decay into negatively charged leptons ($b \rightarrow c\ell^-\overline{\nu}_\ell$). Thus a pair of negatively charged muons[7] indicate the transformation of a *B* hadron into an anti-*B* hadron, while a pair of positively charged muons indicate the transformation of an anti-*B* hadron into a *B* hadron. The dimuon asymmetry is defined as:

$$\mathcal{A} = \frac{N(\mu^+\mu^+) - N(\mu^-\mu^-)}{N(\mu^+\mu^+) + N(\mu^-\mu^-)},$$

where $N(\mu^+\mu^+)$ and $N(\mu^-\mu^-)$ are the numbers of events with two positively or two negatively charged muons, respectively. This asymmetry is a combination of the B_d^0 and B_s^0 mixing asymmetries a_{sl}^d and a_{sl}^s. Other analysis from BABAR, Belle, CLEO, DØ and LHCb focus on separate measurements of a_{sl}^d or a_{sl}^s by reconstructing all charged stable particles of the semileptonic decays.

Measuring tiny physics asymmetries requires a precise understanding of various experimental or instrumental effects, e.g. of detection (A_D) and material interaction (A_IA) asymmetries. For example the reaction of a negatively charged kaon with a proton from the detector material is possible ($K^- + p \rightarrow \lambda + \pi$), while the charge-conjugated process does not exist in the experiment. The LHCb experiment also has to take production asymmetries (A_P) into account: Due to the (particle-antiparticle) asymmetric initial state in *pp* collisions, the number of produced *B* and anti-*B* particles is not the same. The measured asymmetry of *B* decays into a final state *f* or its conjugated state $\bar{f}$ is given by (see (8.34))

[7] Experimentally, muons are the lepton species most easily detected.

$$\mathcal{A}_{\text{meas}}(t) = \frac{N_f(t) - N_{\bar{f}}(t)}{N_f(t) + N_{\bar{f}}(t)} = \frac{a_{sl}^q}{2} + (A_{\text{P}} - \frac{a_{sl}^q}{2})\frac{\cos \Delta M_q t}{\cosh(\Delta\Gamma_q t/2)} + A_{\text{D}} + A_{\text{IA}} \, .$$

Due to the high B_s^0–$\overline{B}_s^0$ oscillation frequency, ΔM_s, the term involving the production asymmetry, cancels in a time-integrated analysis. About 180,000 $B_s^0 \to D_s^- \mu^+ X$ signal candidates with the subsequent decay $D_s^- \to \phi(K^+K^-)\pi^-$ found in 1 fb^{-1} have been analysed by the LHCb collaboration, and the asymmetry is found to be [34]:

$$a_{sl}^s = -0.06 \pm 0.50(\text{stat}) \pm 0.36(\text{syst})\,\% \, .$$

This measurement is a pure counting analysis. Fitting the D_s mass distribution of $B_s^0 \to D_s^- \ell^+ \nu X$ and $\overline{B}_s^0 \to D_s^- \ell^+ \nu X$ signal candidates gives a handle on the time-integrated rates N_f and $N_{\bar{f}}$, respectively.[8] Contrary to the mixing analysis, no flavour tagging is needed. Due to the similar momentum of the two oppositely charged kaons in the final state,[9] interaction asymmetries have been a minor concern for this analysis. Interaction and detector effects affect B and $\overline{B}$ decays in the same way and thus cause no residual asymmetries. Further improvements of this analysis include a larger data set, additional decay modes (e.g. the entire $D_s^- \to K^+K^-\pi^-$ Dalitz space) and the thus required detailed treatment of the according material interaction asymmetries. In order to measure a_{sl}^d a time-dependent analysis is needed that separates the constant and the time-dependent contributions and thus to eliminate the production asymmetry. LHCb recently published this complex analysis [33] with a final result of

$$a_{sl}^d = -0.02 \pm 0.19(\text{stat}) \pm 0.30(\text{syst})\,\% \, .$$

The measured size of *CP* violation in neutral *B* mixing is consistent with the Standard Model expectation and does not confirm the tension seen by the DØ experiment in earlier measurements. An overview of the status of semileptonic asymmetries as of spring 2014 is displayed in Fig. 8.5.

8.3.3 CP Violation in the Interference of Mixing and Decay

The *B* factories BABAR and Belle had originally been built to discover *CP* violation in the interference of mixing and decay in the decay to the *CP* eigenstate $B_d^0 \to J/\psi K_S^0$.[10] This goal was achieved in the year 2002, establishing for the first time *CP* violation outside of the kaon system [35, 36]. In the B_d^0 system, the decay-width difference $\Delta\Gamma_d$ is negligible. Therefore, the time-dependent decay-rate asymmetry

[8] Due to the missing neutrino energy, the reconstruction of the B^0 mass peak is not possible.

[9] We explicitly ask the two kaons to come from an intermediate ϕ resonance.

[10] The tiny contribution of *CP* violation in the subsequent K_S decay is ignored.

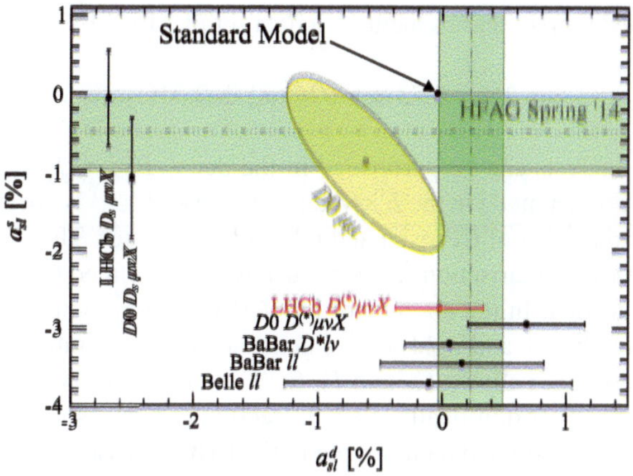

Fig. 8.5 Status of semileptonic asymmetries. The *horizontal band* represents combined LHCb and DØ results from analysis of B_s^0 events. The *vertical band* corresponds to the combination of B_d^0 results from BABAR, Belle, CLEO and DØ. The *ellipse* represents the DØ dimuon asymmetry measurement. The measurement points with *error bars* indicate which measurements entered in the combination displayed in the *horizontal* and *vertical bands*. The latest result, which is not included in the combination, is shown in *red*. Adapted from auxiliary material provided with [33]

can be written as (see (8.35)[11]):

$$\mathcal{A}_{J/\psi K_S^0}(t) = \frac{N(B_d^0 \to J/\psi K_S^0)(t) - N(\overline{B_d^0} \to J/\psi K_S^0)(t)}{N(B_d^0 \to J/\psi K_S^0)(t) + N(\overline{B_d^0} \to J/\psi K_S^0)(t)}$$
$$= S_{J/\psi K_S^0} \sin(\Delta M_d\, t) - C_{J/\psi K_S^0} \cos(\Delta M_d\, t)\,.$$

The decay $B_d^0 \to J/\psi K_S^0$ is an almost pure tree-level decay, therefore *CP* violation in decay can be ignored. Furthermore, *CP* violation in *B* mixing is negligible, and therefore $C_{J/\psi K_S^0} \approx 0$ and $S_{J/\psi K_S^0} \sim \sin 2\beta$, where β can be expressed in terms of the CKM matrix elements involved in the B_d^0 mixing diagram (Fig. 8.1a) $\beta = \arg[-(V_{cd}V_{cb}^*)/(V_{td}V_{tb}^*)]$. The theoretical interpretation of the asymmetry in terms of $\sin 2\beta$ is very clean and therefore often referred to as "golden mode". Experimentally, the decay $B_d^0 \to J/\psi K_S^0$ is also a golden mode. It has a large branching ratio, and the decay of the J/ψ into two leptons is easily triggered and selected. Additional requirements on the displacement of the *B* vertex from the primary vertex allow an almost background-free reconstruction even in an hadronic environment. The current world average from the analysis of this mode is [37]:

$$\sin 2\beta = 0.665 \pm 0.024\,.$$

[11] $S_{J/\psi K_S^0} \equiv A_{CP}^{\mathrm{m}/x}$ and $C_{J/\psi K_S^0} \equiv -A_{CP}^{\mathrm{d}/r}$ in this notation. Additionally $\sinh(\Delta\Gamma_d t/2) = 0$ and $\cosh(\Delta\Gamma_d t/2) = 1$ due to $\Delta\Gamma_d = 0$.

LHCb performed a first measurement on a data set of 1 fb^{-1} [38], resulting in

$$\sin 2\beta = 0.73 \pm 0.07(\mathrm{stat}) \pm 0.04(\mathrm{syst})\,.$$

An analogous analysis can be performed in the decay $B_s^0 \to J/\psi\,\phi$, which is the corresponding golden mode in the B_s^0 system. However $J/\psi\,\phi$ is not a CP eigenstate but a linear combination of CP-even and CP-odd eigenstates. Therefore, in order to disentangle the three polarisation states[12] that have well-defined CP values, an angular analysis is needed. In the absence of any CP violation, the two CP eigenstates correspond to the heavy and light mass eigenstate. Thus separating statistically the polarisation amplitudes allows Γ_H and Γ_L (or Γ_s and $\Delta\Gamma_s$, respectively) to be measured simultaneously. Additionally, a small fraction of non-resonant $B_s^0 \to J/\psi KK$ decays are part of the studied data sample in the LHCb measurement.

The main challenge of this analysis is to resolve the high B_s^0 mixing frequency and to tag the B_s^0 production flavour. Consequently, the knowledge of the tagging performance is the source for the largest (albeit still small!) systematic uncertainty on the measured CP phase ϕ_s. The uncertainty related to the tagging is proportional to the absolute value of ϕ_s, so that for a small central value, the uncertainties are also small.[13] In absence of penguin contributions to the decay, the measured phase ϕ_s corresponds to $-2\beta_s$, where $\beta_s = \arg[(V_{ts}V_{tb}^*)/(V_{cs}V_{cb}^*)]$. The analysis of about 27,500 $B_s \to J/\psi KK$ decays at LHCb found [39]

$$\begin{aligned}\phi_s &= 0.07 \pm 0.09(\mathrm{stat}) \pm 0.01(\mathrm{syst}),\\ \Gamma_s &= 0.663 \pm 0.005(\mathrm{stat}) \pm 0.006(\mathrm{syst})\mathrm{ps}^{-1},\\ \Delta\Gamma_s &= 0.100 \pm 0.016(\mathrm{stat}) \pm 0.003(\mathrm{syst})\mathrm{ps}^{-1}.\end{aligned}$$

Possible extensions of this default analysis allow a non-zero direct CP-violation component and different CP phases for all three polarisation amplitudes to be determined. All these results are, however, still statistically limited.

The phase ϕ_s can also be measured in a second mode, namely in $B_s^0 \to J/\psi\,\pi^+\pi^-$ [40, 41]. This has been shown to be an almost pure CP-odd final state [42], where the phase can be extracted without an angular analysis. Therefore, the analysis of a sample with about a factor five less signal candidates results in a similar sensitivity to ϕ_s. However, external input is needed because Γ_s and $\Delta\Gamma_s$ cannot be measured simultaneously in this mode.

Another decay to a CP eigenstate—which therefore also requires no angular analysis to extract ϕ_s—is the decay $B_s^0 \to D_s^+D_s^-$. The corresponding LHCb analysis based on about 3300 events in 3 fb^{-1} is, however, still statistical limited [43].

A simultaneous fit of both $B_s^0 \to J/\psi\,K^+K^-$and $B_s^0 \to J/\psi\,\pi^+\pi^-$ decays on a sample of 1 fb^{-1} of LHCb data resulted in the world's best measurement of ϕ_s, Γ_s

[12]The two vector mesons in the final state can have a relative angular momentum of $L = 0, 1, 2$. This results in two transverse and one longitudinal polarisation states contributing to this decay.

[13]The current contribution from flavour tagging to the systematic uncertainties on ϕ_s is $\sigma_{\phi_s}/\phi_s \approx 10\,\%$.

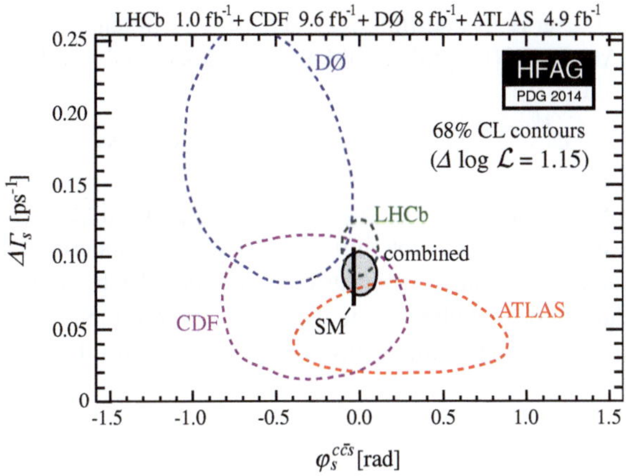

Fig. 8.6 Different 68 % CL contours in the ϕ_s–$\Delta\Gamma_s$) plane. The *gray array* reflects the combined experimental result. The *thin black rectangle* is the SM prediction [26]. The combination does not yet include the latest update of the ATLAS analysis. Adapted from [37]

and $\Delta\Gamma_s$ [39][14]:

$$\phi_s = 0.01 \pm 0.07(\text{stat}) \pm 0.01(\text{syst}),$$
$$\Gamma_s = 0.661 \pm 0.004(\text{stat}) \pm 0.006(\text{syst})\text{ps}^{-1},$$
$$\Delta\Gamma_s = 0.106 \pm 0.011(\text{stat}) \pm 0.007(\text{syst})\text{ps}^{-1}.$$

The ATLAS collaboration also performed an angular analysis of the decay $B_s^0 \to J/\psi\,\phi$. Due to the experiment's worse decay-time resolution and tagging performance, the sensitivity to ϕ_s is significantly smaller (for the same amount of reconstructed signal decays); however, the measurement of Γ_s and $\Delta\Gamma_s$ gives comparable precision [44]:

$$\phi_s = 0.12 \pm 0.25(\text{stat}) \pm 0.05(\text{syst}),$$
$$\Gamma_s = 0.677 \pm 0.007(\text{stat}) \pm 0.004(\text{syst})\text{ps}^{-1},$$
$$\Delta\Gamma_s = 0.053 \pm 0.021(\text{stat}) \pm 0.010(\text{syst})\text{ps}^{-1}.$$

A summary of all ϕ_s and $\Delta\Gamma_s$ measurements is shown in Fig. 8.6.

A conceptually very similar analysis is the analysis of the penguin decay $B_s^0 \to \phi\,\phi$. The phase $\phi_s^{B_s^0\to\phi\,\phi}$ measured in the interference between the amplitudes of the mixing and the decay, is expected to be zero in the absence of higher-order decay diagrams in the Standard Model. The CKM-related phases of the mixing box diagram and the penguin decay diagram cancel exactly. Any new physics entering the mixing box diagram should affect both the measurement of $\phi_s^{B_s^0\to J/\psi\,\phi}$ and of $\phi_s^{B_s^0\to\phi\,\phi}$, while

[14] Note that in case of sizeable *CP* violation, the two decays lead to different numerical values for ϕ_s.

the decay $B_s^0 \to \phi\phi$ is in addition very sensitive to potential additional penguin contributions. About 4000 signal candidates have been analysed [45]:

$$\phi_s^{B_s^0 \to \phi\phi} = -0.17 \pm 0.15(\text{stat}) \pm 0.03(\text{syst})\,.$$

So far, all results are consistent with Standard Model predictions. The measurements of the *CP* phase ϕ_s are by far not limited by statistical uncertainties, so that they will greatly benefit from the coming upgrade of the LHCb experiment. Once the experimental data start to challenge the theoretical predictions, we however have to go carefully through all the assumptions and approximations that have been made to derive these predictions (e.g. neglecting penguin contributions to the decay).

8.4 *B* Hadron Lifetimes

There are two main motivations to measure lifetimes: the test for contribution of new physics in the B_s^0 system and the test of heavy-quark expansion (HQE).

The HQE is an operator product expansion of the total decay rate $\Gamma(H_b)$ of a b-flavoured hadron H_b. The presented formalism applies to a hadron composed of a single b (or $\bar{b}$) quark and not containing c or $\bar{c}$ quarks [46–49]. The HQE amounts to an expansion in powers of $\Lambda_{\text{QCD}}/m_b \sim 0.1$:

$$\Gamma(H_b) \equiv \frac{1}{\tau_{H_b}} = \Gamma_0 + \frac{1}{m_b^2}\Gamma_2 + \frac{1}{m_b^2}\Gamma_3 + \cdots\,, \tag{8.39}$$

with the lifetime τ_{H_b}. There are three important features of the HQE: (i) The leading contribution Γ_0 equals the QCD-corrected decay rate of a free b quark, and thus Γ_0 involves no hadronic uncertainties; (ii) there are no $1/m_b$ corrections; (iii) Γ_3 is enhanced by a phase-space factor of $16\pi^2$, because the dominant contributions to Γ_0 are the three-body $b \to c\bar{u}d$ and $b \to c\bar{c}s$ decays, while Γ_3 contains contributions with a two-particle phase space. Examples of decay modes with only two quarks in the final state are weak-annihilation (WA) or weak-scattering (WS) processes, such as $\bar{b}d \to \bar{c}u$, which contributes to $\Gamma(B_d^0)$. The numerically dominant contribution to Γ_3, however, is the Pauli interference (PI) diagram which arises e.g. in B^+ decays: The final state has two u quarks, one of which is the "spectator quark", while the other is produced in the weak decay, $\bar{b}u \to \bar{c}\bar{d}uu$. The anti-symmetrisation of the final state in the two u-quark momenta (Pauli interference) also gives rise to a phase-space enhanced contribution. Thus, to leading order in Λ_{QCD}/m_b, all b-flavoured hadrons have the same lifetime. The dominant source of lifetime splittings is Γ_3, and one expects effects of order $16\pi^2\Lambda_{\text{QCD}}^3/m_b^3 \approx 15\,\%$. A generic feature of the HQE is that the lifetime splitting among U-spin partners is small: $\tau_{B_s^0} = \tau_{B_d^0}$ up to corrections of at most $1\,\%$, and similarly $\tau_{\Lambda^0} \simeq \tau_{\Xi^0}$, both in excellent agreement with experiment. U-spin is a symmetry that is similar to the strong-isospin symmetry introduced in Sect. 8.1: In the limit $m_s = m_d$ the QCD lagrangian in (8.8) is invariant under SU(2) rotations of the doublet $(d, s)^T$. $(B_s^0, B_d^0)^T$ and $(\Lambda_b, \Xi_b^0)^T$ are U-spin doublets. A precise prediction of lifetime differences requires the calculation of hadronic matrix element by non-perturbative methods such as lattice QCD. The only available calculations are quite old and do not include dynamical sea quarks. An update of the NLO QCD predictions for $\tau_{B^+}/\tau_{B_d^0}$ of [50] can be found in [26]. Reference [50] also addresses τ_{Ξ^-} versus $\tau_{\Xi^0} \simeq \tau_{\Lambda^0}$, but for $\tau_{\Xi^-} - \tau_{\Xi^0}$ no lattice calculation is currently available.

The HQE is also applied to Γ_{12}^q for the calculation of the width difference $\Delta\Gamma_q$ between the B_q^0 mass eigenstates B_{qL} and B_{qH} and the semileptonic CP asymmetries leading to the predictions in (8.37) and (8.38). In the calculation of Γ_{12}, the terms Γ_0 and Γ_2 in (8.39) are absent, because a $|\Delta B| = 2$ process necessarily involves the valence quark q (see Fig. 8.1). Therefore the measurement of $\Delta\Gamma_s$ or $\Delta\Gamma_d$ directly probes the HQE at order $1/m_b^3$.

For all B hadrons except the B_s^0, the time-dependent decay rate can be described by a single exponential distribution. In the B_s^0 system, however, due to the non-vanishing decay-width difference $\Delta\Gamma_s$, one has to deal with two different lifetimes. If the CP violation in the decay can be neglected, the lifetimes of the CP-even and CP-odd eigenstates are give by

$$e^{-\Gamma_{\text{even/odd}}t} = (1 \pm \cos\phi_{12}^s)e^{-\Gamma_L t} + (1 \mp \cos\phi_{12}^s)e^{-\Gamma_H t},$$

where ϕ_{12}^s is the CP phase of (8.20), which is measured with the semileptonic CP asymmetries (see (8.27)). This phase is small, so that $\Gamma_{\text{even}} = \Gamma_L$ and $\Gamma_{\text{odd}} = \Gamma_H$ to a very good approximation. Lifetime measurements in CP-even and in CP-odd final states give input to $\Delta\Gamma_s$.

If non-Standard Model amplitudes exist, which give rise to additional CP violation in any of the studied decays, this can be potentially revealed by comparing the lifetimes measured in different CP-even and CP-odd eigenstates.

Due to the dependence of the lifetime on the CP content of the final state, measurements in the B_s^0 system that describe the decay-time distribution with a single exponential function are generally referred to as "effective-lifetime" measurements. The measurement of the lifetime in a flavour-specific eigenstate gives rise to the average decay width

$$\Gamma_s = \frac{\Gamma_L + \Gamma_H}{2}.$$

The current status of measurements of B_s^0 effective lifetimes is summarised in Table 8.1.

To measure the decay-width difference $\Delta\Gamma_d$ is beyond any experimental precision. However, a limit can be placed by comparing the lifetime in the CP-odd eigenstate $B_d^0 \to J/\psi K_S^0$ to the one in the flavour-specific eigenstate $B_d^0 \to J/\psi K^*$ [51]:

$$\Delta\Gamma_d/\Gamma_d = -0.044 \pm 0.025(\text{stat}) \pm 0.011(\text{syst}). \tag{8.40}$$

Table 8.1 B_s^0 effective-lifetime measurements of the LHCb experiment. The first quoted number always refers to the statistical and the second to the systematic uncertainties

Decay	Γ_L (ps^{-1})	τ_L (ps)	Size of data set (fb^{-1})	References
Mainly *CP*-even states				
$B_s^0 \to K^+ K^-$	Not quoted	$1.407 \pm 0.016 \pm 0.007$	1.0	[52]
$B_s^0 \to D_s^+ D_s^-$	$0.725 \pm 0.014 \pm 0.009$	Not quoted	3.0	[53]
Mainly *CP*-odd states				
Decay	Γ_H (ps^{-1})	τ_H (ps)	Size of data set (fb^{-1})	References
$B_s^0 \to J/\psi f_0$	$0.588 \pm 0.014 \pm 0.009$	$1.700 \pm 0.040 \pm 0.026$	1.0	[54]
$B_s^0 \to J/\psi K_S^0$	Not quoted	$1.75 \pm 0.12 \pm 0.07$	1.0	[55]
Γ_S in flavour-specific states				
Decay	Γ_s (ps^{-1})	τ_s (ps)	Size of data set (fb^{-1})	References
$B_s^0 \to D_s^- D_s^+$	Not quoted	$1.52 \pm 0.15 \pm 0.01$	3.0	[53]
$B_s^0 \to \pi^+ K^-$	Not quoted	$1.60 \pm 0.06 \pm 0.01$	1.0	[52]
$B_s^0 \to D_s^- \pi^+$	Not quoted	$1.535 \pm 0.015 \pm 0.014$	1.0	[56]

A detailed study of all average B hadron lifetimes and their ratios is of great interest to test HQE. Especially for the Λ_b^0 lifetime there were, for a long time, significant discrepancies between measurements and HQE predictions. In 2003, a widely quoted average of all data gave $\tau_{\Lambda_b^0}/\tau_{B^0} = 0.798 \pm 0.052$ [57]. The HQE predictions were >0.9. This discrepancy has been entirely resolved with the recent LHCb measurements. All measured ratios are nowadays in good agreement with predictions.

Experimentally, the main challenge in measuring a lifetime is to properly determine the detector acceptance. To effectively trigger and select the signal B candidates, their long decay length is exploited, which results in a non trivial decay-time acceptance. Due to differences in the resolution in data and simulation, the acceptances cannot simply be taken from simulation. One way out is to measure lifetime ratios relative to a kinematically similar decay of a B hadron with well-known lifetime. In such a quantity, the acceptance effects are expected to cancel to a large extent. Deviations from this cancellation are studied on simulated data and are then assigned as systematic uncertainties. One example for this method is the measurement of the average B_s^0 lifetime in the decay $B_s^0 \to J/\psi f_0$, which is measured with respect to the B_d^0 lifetime in the decay $B_d^0 \to J/\psi K^*$. This approach works well for all B_s^0 and Λ_b^0 lifetime measurements, which have for the moment still large statistical uncertain-

Table 8.2 *B*-hadron lifetimes measured by the LHCb experiment. The first uncertainty is statistical, the second systematic

Lifetime	Value (ps)	Size of data set (fb^{-1})	References
$\tau_{B^0 \to K^+\pi^-}$	1.524 ± 0.011 ± 0.004	1.0	[52]
$\tau_{B^+ \to J/\psi K^+}$	1.637 ± 0.004 ± 0.003	1.0	[51]
$\tau_{B_d^0 \to J/\psi K^*}$	1.524 ± 0.006 ± 0.004	1.0	[51]
$\tau_{B_d^0 \to J/\psi K_S^0}$	1.499 ± 0.013 ± 0.005	1.0	[51]
$\tau_{\Lambda_b^0 \to J/\psi \Lambda}$	1.415 ± 0.027 ± 0.006	1.0	[51]
$\tau_{\Lambda_b^0 \to J/\psi p K^-}$	1.479 ± 0.009 ± 0.010	3.0	[58]
$\tau_{B_c^+ \to J/\psi \mu\nu, X}$	0.509 ± 0.008 ± 0.012	2.0	[59]
$\tau_{\Xi_b^- \to J/\psi \Xi^-}$	1.599 ± 0.041 ± 0.022	3.0	[60]

ties compared to the world average result on the B_d^0 and B^+ lifetimes. To perform world-best absolute lifetime measurements also for B_d^0 and B^+ decays is, however, significantly more complicated. Trigger and offline selections need to be optimised in order to avoid biases, and the impact of unavoidable biases needs to be measured in data. LHCb has recently published world-record absolute measurements of all B-hadron lifetimes in $B \to J/\psi X$ modes. These are listed in Table 8.2, together with LHCb results for the Λ_b^0, B_c^+, and Ξ_b^- lifetimes.

8.5 Direct *CP* Violation and the CKM Angle γ

Numerous searches for direct *CP* violation have been performed at LHCb. Among others, the first observation of *CP* violation in the B_s^0 system has been established in the decay $B_s^0 \to K^-\pi^+$ [61]. However, quantitative SM predictions for direct *CP* violation are challenging because of the presence of hadronic factors in the decay amplitudes that cannot be accurately calculated from quantum chromodynamics at present. One solution to this problem is to simultaneously determine the physics parameters related to the CKM matrix elements and the hadronic nuisance parameters in the data. This approach is followed in the time-integrated analysis of the CKM angle γ from tree-mediated decays. An alternatively approach to the determination of γ exploiting symmetry arguments in order to get a handle on the unknown hadronic parameters is pursued in the time-integrated analysis of loop-induced decays.

8.5.1 Time-Integrated Analysis of γ from Tree Decays

The angle γ, defined as $\gamma = \arg[-V_{ud}V_{ub}^*/V_{cd}V_{cb}^*]$ is the only *CP*-violating phase which is measurable in tree-mediated decays. The relevant theoretical predictions are extremely precise ($\Delta\gamma/\gamma \sim \mathcal{O}(10^{-6})$). Experimentally, however, the angle is still quite poorly known—the precision measurement of the CKM angle γ is the last missing piece in the Standard Model flavour sector. The most powerful modes to measure γ are $B^+ \to D^0 K^+$ and $B^+ \to \overline{D}^0 K^+$ decays, with the quark-level transitions $\bar{b} \to \bar{c}u\bar{s}$ and $\bar{b} \to \bar{s}u\bar{c}$, respectively. In Sect. 8.1.1 we have learned that a non-vanishing direct CP asymmetry requires two interfering decay amplitudes (denoted A_u and A_c in (8.14)) with different strong and weak phases. In order to observe the interference of the $\bar{b} \to \bar{c}u\bar{s}$ and $\bar{b} \to \bar{s}u\bar{c}$ amplitudes, the D^0 and $\overline{D}^0$ mesons emerging from the two decay channels are required to decay into a common final state. Different possibilities of common final states are discussed below, when the individual methods are presented. Both B decays can be described with three parameters, namely $r_b(\mathcal{O}(0.1))$, the absolute value of their amplitude ratio, δ_b, the connected strong phase difference, and the CKM angle γ. The size of the interference effect and thus the sensitivity to γ is directly proportional to $1/r_b$.

The analysis exploiting the decay $D^0 \to K^+\pi^-$ and the doubly Cabibbo-suppressed partner $D^0 \to K^-\pi^+$ is called Gronau–London–Wyler (GLW) method [62, 63]. Additional unknown parameters entering the analysis are r_D, the absolute value of the ratio of amplitudes of the Cabibbo-favoured and Cabibbo-suppressed D decays, and δ_d, their relative strong-phase difference. These two parameters have been measured precisely at the CLEO experiment and are here used as external input.

Alternatively one can study D^0 decays to CP eigenstates (e.g. K^+K^- and $\pi^+\pi^-$). The relevant parameters describing the D decay are $r_D = 1$ and the CP eigenvalue δ_d, which is $\pm$ 1 depending on the the final state. This type of analysis is called Atwood–Dunietz–Soni (ADS) method [64, 65].

Additional multi-body D^0 decays, such as $D^0 \to \pi^\pm K^\mp \pi^+\pi^-$ can be exploited (multi-body ADS method). A common final state is, however, not the only prerequisite for interference effects. The decays must, in addition, have the same phase space. Consequently, the magnitudes of the D^0 decay amplitudes and the strong-phase difference both become functions of the phase space. This might also introduce too many additional parameters given the limited available statistics. One way out is to introduce effective quantities averaged over the phase space, namely an effective strong phase $\delta_{K3\pi}$, an effective amplitude ratio $r_{K3\pi}$ and a so-called coherence factor $\kappa_{K3\pi}$, which modifies all interference terms. Exploiting the full statistical power of the data requires a Dalitz-plot analysis. This is the so-called Giri–Grossman–Soffer–Zupan (GGSZ) method [66], which uses self-conjugated three-body final states such as $D^0 \to K_S^0\pi^+\pi^-$ or $D^0 \to K_S^0 K^+K^-$. The Dalitz plot can be fitted with model-dependent assumptions on the decay amplitudes. A model-independent approach is to use external input on the strong-phase difference and D amplitude ratios in bins of the Dalitz space.

A further method to extract γ, which is used for the first time by LHCb, is the Grossmann–Ligeti–Soffer (GLS) method. It is similarly to the GLW method, but exploits singly Cabibbo-suppressed D^0 decays such as $D^0 \to K^\mp K^{*\mp}$. Although the total rate of this decay is quite low, r_D is of $\mathcal{O}(1)$. For a model-independent analysis, external input on the strong-phase variation over the D^0 Dalitz plot is required.

The observables of all the above discussed analyses are ratios and asymmetries of B^+ and B^- decays to final states with either a D^0 or $\overline{D}^0$ involved. The measured observables are connected to the physics parameters related to the CKM matrix elements and the hadronic nuisance parameters via trigonometrical functions. This leads to e.g. ambiguous solutions and makes the statistical treatment with the size of the current data set very challenging. For this reason, the LHCb experiment performs its own γ combination based on a frequentist approach.

As an alternative to the main decay $B^\pm \to DK^\pm$, similar decays with excited D or K mesons can be studied [67]. The K can also be replaced by a pion. These decays are, however, significantly rarer and thus more sensitive to uncertainties from D^0 mixing and CP violation in D^0 decays.

The combination of the two-body and four-body ADS [68, 69], the GLW [68] and the model-independent GGSZ analysis [70] exploiting both $B \to DK$ and $B \to D\pi$ modes based on $1\,\mathrm{fb}^{-1}$ of data[15] results in a best-fit value of $\gamma = 72.6°$ modulo 180° [72]. The following confidence intervals are set:

$$\gamma \in [55.4, 82.3]^\circ \text{ at } 68\,\%\text{ CL},$$
$$\gamma \in [40.2, 92.7]^\circ \text{ at } 95\,\%\text{ CL}.$$

The results of the model-dependent GGSZ analysis [73] and the GLS analysis [74] are not included in this combination.

This combination is the world's best single-experiment result and in good agreement with the result of the B factories [75, 76]:

$$\gamma = (69^{+17}_{-16})^\circ \quad \text{at } 68\,\%\text{ CL} \quad (\text{BaBar}),$$
$$\gamma = (68^{+15}_{-14})^\circ \quad \text{at } 68\,\%\text{ CL} \quad (\textit{Belle}).$$

Note that the three experiments use partially common external input for their determination of γ, so that the uncertainties are not uncorrelated.

8.5.2 Time-Integrated Analysis of γ from Loop Decays

An alternative strategy to determine γ using loop-mediated two-body B decays—namely $B^0_d \to \pi^+\pi^-$ and $B^0_s \to K^+K^-$—has also been proposed [77–79]. The hadronic parameters entering these decays are related by the U-spin symmetry explained in Sect. 8.4, i.e. by the exchange of d and s quarks in the decay diagrams, and this relation can be exploited to cancel the unknown hadronic factors. There are, however, non-factorisable corrections breaking U-spin symmetry of up to 20–30 %.

Assuming corrections of 50 %, the result of a fit to $B^0_d \to \pi^+\pi^-$ and $B^0_s \to K^+K^-$ in a data set of $1\,\mathrm{fb}^{-1}$ of data exploiting Bayesian fitting techniques is found to be [80]

[15] The updated result on $3\,\mathrm{fb}^{-1}$ [71] was not yet published at the time of writing of this volume.

$$\gamma \in [56.8, 70.7]^\circ \text{ at } 68\,\% \text{ CL},$$

expressed modulo 180°. It is interesting to compare the γ values of the analyses of tree-mediate decays, which are almost insensitive to contributions from new physics, and the one from loop-mediated decays. With the current precision, the LHCb results are in good agreement.

8.5.3 Time-Dependent Measurement of γ

A time-dependent analysis of $B_s^0 \to D_s^\pm K^\mp$ gives another handle on the CKM phase γ. The direct decay $B_s^0 \to D_s^+ K^-$ interferes with a mixing transition $B_s^0 \to \overline{B}_s^0$ and the subsequent decay $\overline{B}_s^0 \to D_s^+ K^-$, so that here a mixing-induced *CP* asymmetry can be exploited while no direct *CP* asymmetry is present. The weak-phase difference observed in this decay is $\gamma - 2\beta_s$. Assuming that higher-order penguin processes do not contribute, the phase introduced by the mixing is $-2\beta_s = \phi_s$, where ϕ_s is taken as external input from the analysis of the decay $B_s^0 \to J/\psi h^+ h^-$ described above. As the final state is not a *CP* eigenstate, the strong-phase difference is unknown and is treated as a nuisance parameter to be simultaneously determined in the analysis. Contrary to the time-integrated analysis of charged B decays (which is mainly a counting experiment), the time-dependent analysis requires to resolve the high B_s^0–$\overline{B}_s^0$ oscillation frequency and relies on flavour tagging and its correct calibration. This complex analysis results in $\gamma = (122^{+25}_{-39})^\circ$ at 68 % CL, expressed modulo 180° [81].

8.6 Rare Decays

The suppression of loop diagrams makes FCNC meson decays powerful probes of new physics, giving complementary information to the observables related to particle-antiparticle mixing. In this section we discuss the analysis of angular distributions of rare B decays such as $B_d^0 \to K^{*0}\mu^+\mu^-$, $B_d^0 \to K^{*0}e^+e^-$ $B_s^0 \to \phi\mu^+\mu^-$, $B^+ \to K^+\mu^+\mu^-$, $B_d^0 \to K_S^0\mu^+\mu^-$. Furthermore, we present searches for very rare decays such as $B_{d,s}^0 \to \mu^+\mu^-$, $D^0 \to \mu^+\mu^-$, $K_S^0 \to \mu^+\mu^-$, and the flavour-violating decays $\tau^- \to \mu^-\mu^+\mu^-$, and $B_{d,s}^0 \to \mu^\pm e^\mp$.

8.6.1 Theory of Rare Decays and Electroweak Penguins

In order to describe (semi-)leptonic rare decays we must supplement the operator basis of our effective Hamiltonian in (8.10) by operators describing the couplings of quarks to photons or leptons. Specifying to muons, we need the operators

$$Q_7 = \frac{e}{16\pi^2} m_b \bar{s}_L \sigma^{\mu\nu} F_{\mu\nu} b_R \,, \tag{8.41}$$

$$Q_9 = \frac{e^2}{16\pi^2} \bar{s}_L \gamma_\mu b_L \, \bar{\mu} \gamma^\mu \mu \,, \tag{8.42}$$

$$Q_{10} = \frac{e^2}{16\pi^2} \bar{s}_L \gamma_\mu b_L \, \bar{\mu} \gamma^\mu \gamma_5 \mu \tag{8.43}$$

to describe $B_d^0 \to K^0\mu^+\mu^-$, $B_d^0 \to K^{*0}\mu^+\mu^-$, their charged counterparts, and $B_s^0 \to \phi\mu^+\mu^-$ in the SM. Here, e is the electromagnetic coupling constant. Q_7 is the magnetic penguin operator describing the effective coupling of $\bar{s}$ and b to a photon. Z-penguin and box diagrams generate both Q_9 and Q_{10}, while the photon penguin diagram contributes only to the coefficient of Q_9. The SM coefficients of these operators equal $C_7 = -0.29$, $C_9 = 4.07$, and $C_{10} = -4.31$ at the scale $\mu = \overline{m}_b$. One can fit $C_{7,9,10}$ to the experimental data on branching fractions and decay distributions to quantify the possible size of new physics. These fits usually further include $C'_{7,9,10}$, which are the coefficients of the operators Q'_7, Q'_9, Q'_{10} obtained from their unprimed counterparts by flipping the chiralities of the quark fields. Within the SM, these coefficients are very small. However, in many scenarios of new physics they might be sizeable. Furthermore, new physics can render all six coefficients complex. The decay modes permit several kinematic distributions to be measured, which are described in the experimental section below. In the decay $B_d^0 \to K^{*0}\mu^+\mu^-$, the K^{*0} meson is detected through $K^{*0} \to K^+\pi^-$, so that this decay is flavour-specific. By contrast, the decays $B_d^0 \to K_S^0\mu^+\mu^-$ and $B_s^0 \to \phi\mu^+\mu^-$ interfere with their charge-conjugate modes $\overline{B}_d^0 \to K_S^0\mu^+\mu^-$ and $\overline{B}_s^0 \to \phi\mu^+\mu^-$ and are thereby sensitive to mixing-induced *CP* violation. The decay amplitudes involve a charm-quark loop, which is connected to the muon pair via a photon. When studying $B \to K^{(*)}\mu^+\mu^-$, one must exclude the region of lepton-invariant mass $q^2 \approx M_\psi^2$, where ψ represents any charmonium (meaning a c–$\bar{c}$ bound state). For low q^2, one can calculate the decay amplitude with the help of QCD factorisation [82–84]. For high $q^2 \gg 4m_c^2$ above the charmonium resonances, one can expand in inverse powers of q^2 by using an operator product expansion [85–87].

The SM contribution to the leptonic rare decay $B_s^0 \to \mu^+\mu^-$ only involves Q_{10}. The other two operators in (8.41) involve a vector coupling to the muons that produces a muon pair in a $C = -1$ state. A muon pair with zero total angular momentum, however, is always in a $C = 1$ state, so that Q_{10} cannot contribute to $B_s^0 \to \mu^+\mu^-$. The only hadronic quantity entering $B_s^0 \to \mu^+\mu^-$ is the decay constant, which is calculated with high precision to $f_{B_s} = (227.7 \pm 4.5)$ MeV with lattice QCD [21]. The branching ratio of $B_d^0 \to \mu^+\mu^-$ is smaller than $B(B_s^0 \to \mu^+\mu^-)$ by a factor of $|V_{td}/V_{ts}|^2 f_{B_d}^2 / f_{B_s}^2 M_{B_d^0}/M_{B_s^0}$ (neglecting a tiny phase space factor). The SM predicts the branching ratios

$$B(B_s^0 \to \mu^+\mu^-) = (3.65 \pm 0.23) \times 10^{-9} \,,$$
$$B(B_d^0 \to \mu^+\mu^-) = (1.06 \pm 0.09) \times 10^{-10} \,.$$

These numbers are taken from the calculation of [88], which use the quoted value for f_{B_s} and f_{B_d} = (190.5 ± 4.2) MeV. The value for $B(B_s^0 \to \mu^+\mu^-)$ takes the effect of $B_s^0-\overline{B}_s^0$ mixing into account [89].

The decay of a scalar particle into a muon pair needs a helicity flip of one of the muons, which suppresses the decay amplitude by a factor of $m_\mu/M_{B_s^0}$. Therefore, $B_s^0 \to \mu^+\mu^-$ is sensitive to new physics coming with a scalar or pseudoscalar coupling to muons. The SM Higgs contribution is in this category but is negligibly small because of the smallness of the muon Yukawa coupling. In models with more than one neutral Higgs boson, however, it is possible to have Higgs contributions to $B_s^0 \to \mu^+\mu^-$ which compete with or even exceed the SM amplitude [90–93]. Most importantly, $B_s^0 \to \mu^+\mu^-$ puts powerful constraints on the parameter space of the minimal supersymmetric Standard Model (MSSM), even if minimal flavour violation (MFV[16]) is employed to minimise the impact on FCNC observables [96–99]. In the MFV-MSSM prediction, the most important parameters governing $B_s^0 \to \mu^+\mu^-$ are $\tan\beta$ (the ratio of the vacuum expectation value of the two Higgs doublets) and the mass M_A^0 of the pseudoscalar Higgs boson. One then has

$$B(B_s^0 \to \mu^+\mu^-) \propto \frac{\tan^6\beta}{(M_A^0)^4},$$

so that $B(B_s^0 \to \mu^+\mu^-)$ probes large $\tan\beta$ in conjunction with small M_A^0. Interestingly, ATLAS and CMS searches for $A^0 \to \tau\tau$ probe the same region of the parameter space, and (depending on the other MSSM parameters, notably the stop mixing angle) currently both processes are competitive [99, 100].

The decay $D^0 \to \mu^+\mu^-$ is a $c \to u$ FCNC decay. The amplitude suffers from severe cancellations between loops involving an s and those with a d quark. FCNC amplitudes vanish in the limit in which the quark masses in the loop are degenerate (this feature is called Glashow–Iliopoulos–Maiani (GIM) mechanism [101]): Any $\Delta C = -1$ FCNC amplitude can be written as $\mathcal{A} = V_{cb}^* V_{ub} f(m_b^2) + V_{cs}^* V_{us} f(m_s^2) + V_{cd}^* V_{ud} f(m_d^2)$ with $V_{cq}^* V_{uq} f(m_q^2)$ denoting the sum of the loop diagrams with internal quark $q = d, s$ or b. We next use CKM unitarity to eliminate $V_{cd}^* V_{ud} = -V_{cb}^* V_{ub} - V_{cs}^* V_{us}$ and find $\mathcal{A} = V_{cb}^* V_{ub}(f(m_b^2) - f(m_d^2)) + V_{cs}^* V_{us}(f(m_s^2) - f(m_d^2))$. The dominant CKM structure $V_{cs}^* V_{us}$ comes with a suppression factor of $(m_s^2 - m_d^2)/M_W^2$ found by expanding $f(m_s^2) - f(m_d^2)$ to first order. This cancellation precludes robust theory predictions, because loop calculations sensitive to scales as low as m_s are not reliable because of quark confinement. With current techniques, FCNC processes of charmed hadrons are theoretically not tractable, and SM predictions are mainly educated guesses and—prior to experimental determinations—often vary over a few orders of magnitude. However, some new-physics interaction coupling differently to d, s, b quarks may lift the GIM cancellation and enhance $B(D^0 \to \mu^+\mu^-)$ by oders

[16] MFV is a theoretical concept implying that all contributions from new physics to a flavour-violating amplitude are governed by the same CKM elements as the SM contribution [94, 95].

of magnitude as well. Thus a reliable measurement of this branching ratio—or other charm FCNC processes—will trigger the question whether the measured value is due to new physics or just to long-distance QCD effects.

The rare decays $K^0_{S,L} \to \mu^+\mu^-$ are dominated by loops with internal charm and up quarks, which largely cancel due to the GIM mechanism. Historically, the observed smallness of $B(K^0_L \to \mu^+\mu^-)$ has lead to the postulation of the charm quark to remove FCNC couplings of the Z and to tame the contribution from the up-quark loop [101]. Neglecting the small CP violation in $K^0 - \overline{K}^0$ mixing, K^0_L is CP-odd and decays into a $\mu^+\mu^-$ pair in an S-wave. By contrast, $K^0_S \to \mu^+\mu^-$ produces a muon pair in a P-wave. Both decays receive large long-distance contributions from an intermediate two-photon state, which rescatters into two muons. The sensitivity of the K^0_L and K^0_S decays to new physics is different, so there is a benefit in the search for $K^0_S \to \mu^+\mu^-$. The SM predicts $B(K^0_S \to \mu^+\mu^-) = (5.0 \pm 1.5) \times 10^{-10}$ [102, 103].

In the SM, the decays $\tau^- \to \mu^-\mu^+\mu^-$ and $B^0_d \to \mu^\pm e^\mp$ vanish for equal neutrino masses because of the GIM mechanism in the lepton sector. The branching fractions of these decays are unmeasurably small within the SM, thus any observation of a lepton-flavour-violating decay would be an unambiguous proof of new physics.

8.6.2 Angular Distribution of Electroweak Penguin Decays

The observation of a decay with a branching ratio significantly larger than the SM expectation would be an unambiguous sign for new physics. At the LHC, searches for very rare decays with predicted branching ratios ranging from 10^{-8} to 10^{-18} for dimuon decays of neutral B and D mesons and kaons as well as searches for lepton-flavour-violating processes with predicted branching ratios of smaller than 10^{-40} are performed. Experimentally, these searches are straightforward, as would be the interpretation of any significant deviation. However, the theoretical uncertainties of absolute branching-ratio predictions are typically larger than 20 %. To exclude or discover effects which result in small modifications of the observed amplitudes, relative quantities such as angular distributions are better suited as in these a large part of the theoretical uncertainties potentially cancels out.

The angular distribution of the final-state particles of the decay $B^0_d \to K^{*0}\mu^+\mu^-$, where K^{*0} indicates the $K^{*0} \to K^+\pi^-$ decay, is described by three angles (θ_l, θ_K, and ϕ) and the dimuon invariant mass squared q^2. θ_l is the angle between the flight direction of the μ^+ (μ^-) and the B^0_d ($\overline{B}^0_d$) meson in the dimuon rest frame, θ_K is the angel between the flight direction of the charged kaon and the B^0_d ($\overline{B}^0_d$) meson in the K^{*0} (K^{*0}) rest frame, and ϕ is the angle between the decay planes of the K^{*0} ($\overline{K}^{*0}$) and the dimuon system in the B^0_d ($\overline{B}^0_d$) meson rest frame. A formal definition of the angles can be found in [104]. Using the notation of [105], the decay distribution of the B^0_d corresponds to

$$\frac{d^4\Gamma}{dq^2\, d\cos\theta_l\, d\cos\theta_K\, d\phi} = \frac{9}{32\pi}\Big[I_1^s \sin^2\theta_l + I_1^c \cos^2\theta_K$$
$$+ I_2^s \sin^2\theta_K \cos 2\theta_l + I_2^c \cos^2\theta_K \cos 2\theta_l$$
$$+ I_3 \sin^2\theta_K \sin^2\theta_l \cos 2\phi + I_4 \sin 2\theta_K \sin 2\theta_l cos\phi$$
$$+ I_5 \sin 2\theta_K \sin\theta_l \cos\phi + I_6 \sin^2\theta_K \cos\theta_l$$
$$+ I_7 \sin 2\theta_K \sin\theta_l \sin\phi + I_8 \sin 2\theta_K \sin 2\theta_l \sin\phi$$
$$+ I_9 \sin^2\theta_K \sin^2\theta_l \sin 2\phi \Big], \qquad (8.44)$$

where the 11 coefficients I_j are bilinear combinations of K^{*0} decay amplitudes and vary with q^2. The decay amplitudes are functions of the Wilson coefficients of the operators introduced in (8.41) and below, which contain information about short-distance effects and are sensitive to physics beyond the SM, and of hadronic form factors, which depend on long-distance effects.

Combining B_d^0 and $\overline{B}_d^0$ decays, and assuming equal numbers of both, the following normalised *CP* averages and *CP* asymmetries can be introduced, respectively:

$$S_j = (I_j + \bar{I}_j)/\frac{d\Gamma}{dq^2} \quad \text{or} \quad A_j = (I_j - \bar{I}_j)/\frac{d\Gamma}{dq^2}\,.$$

In the limit of the dimuon mass getting large compared to the muon mass, $q^2 \gg 4m_\mu^2$, some of the coefficients can be related to the longitudinal polarisation of the K^{*0} meson, F_L ($S_1^c = -S_2^c = F_L$ and $\frac{4}{3}S_1^s = 4S_2^s = 1 - F_L$). The angular term $\sin^2\theta_K \cos\theta_l$ generates a forward-backward asymmetry, A_{FB}, of the dimuon system ($A_{\mathrm{FB}} = \frac{3}{4}S_6$) [106].

In an ideal world with infinite statistics, one would like to measure 2×11 angular coefficients in a fine binning of q^2. The largest currently analysed data set of 1 fb^{-1} of data from the LHCb experiment contains, however, only about 900 reconstructed signal candidates. Thus we can only extract information about a subset of these parameters simultaneously. This can either be done by integrating over two of the three angles, as was done in a previous analysis based on 340 pb^{-1} [107]. Alternatively, one can introduce angular transformations such as

$$\hat{\phi} = \begin{cases} \phi + \pi & \text{if}\, \phi < 0\,, \\ \phi & \text{otherwise}\,. \end{cases}$$

Applying this transformation results in

$$
\begin{aligned}
\frac{1}{\mathrm{d}\Gamma/\mathrm{d}q^2}\frac{1}{\mathrm{d}q^2\,\mathrm{d}\cos\theta_l\,\mathrm{d}\cos\theta_K\,\mathrm{d}\hat{\phi}} = \frac{9}{16\pi}\Big[& F_L\cos^2\theta_K + \frac{3}{4}(1-F_L)(1-\cos^2\theta_K^2) \\
& - F_L\cos^2\theta_K(2\cos^2\theta_l - 1) \\
& + \frac{1}{4}(1-F_L)(1-\cos^2\theta_K)(2\cos^2\theta_l - 1) \\
& + S_3(1-\cos^2\theta_K)(1-\cos^2\theta_l)\cos 2\hat{\phi} \\
& + \frac{4}{3}A_{FB}(1-\cos^2\theta_K)\cos\theta_l \\
& A_9(1-\cos\theta_K)(1-\cos^2\theta_l)\sin 2\hat{\phi}\Big] .
\end{aligned}
$$

All the angular terms in (8.44) are orthogonal,[17] so that in the limit of infinite statistics the measurements of the coefficients are uncorrelated and the central values and the uncertainties should be identical, irrespective of such a transformation having been applied or not. The main challenge of this analysis with limited statistics is to guarantee the stability of the fit and to make sure to not end up in local minima or at phase-space boundaries.

The LHCb experiment performed a simultaneous measurement of the differential branching fraction ($\mathrm{d}B/\mathrm{d}q^2$), A_{FB}, F_L, S_3 and A_9 of the $B_d^0 \to K^{*0}\mu^+\mu^-$ decay in six bins of q^2 [104] (Fig. 8.7). These results are in good agreement with the SM predictions as well as with statistically less precise individual measurements of ($\mathrm{d}B/\mathrm{d}q^2$), A_{FB} and F_L of the B factories and the CMS and ATLAS experiments.

Combinations of F_L and S_i have been proposed independently by several authors[18] as alternative fit variables. In particular in the large recoil limit (low q^2), the observables denoted as $P_4^{'}$, $P_5^{'}$, $P_6^{'}$ and $P_8^{'}$ [109] are largely free from form-factor uncertainties. These observables are defined as

$$
P_{4,5,6,8}^{'} = \frac{S_{j=4,5,7,8}}{\sqrt{F_L(1-F_L)}} .
$$

In a second, complementary analysis of the $B_d^0 \to K^{*0}\mu^+\mu^-$ candidates in the first $1\,\mathrm{fb}^{-1}$ of data of the LHCb experiment, a suitable set of angular transformation has been chosen to determine these observables of interest in six q^2 bins each. While the results $P_4^{'}$, $P_6^{'}$ and $P_8^{'}$ are in good agreement with the Standard Model expectations, some tension is present in the variable $P_5^{'}$ [108] (Fig. 8.8).

The observed deviation corresponds to 3.7 Gaussian standard deviations (σ). If the 24 measurements (six q^2 bins $\times$ four observables) are considered as independent, the probability that at least one varies from the expected value by $3.7\,\sigma$ or more is approximately 0.5 %.

In order to interpret this result in the context of new physics, one must respect all other experimental results involving the same Wilson coefficients. Reference [110]

[17] The product of two terms integrated over phase space is equal to zero.

[18] For a complete list see [108] and references therein.

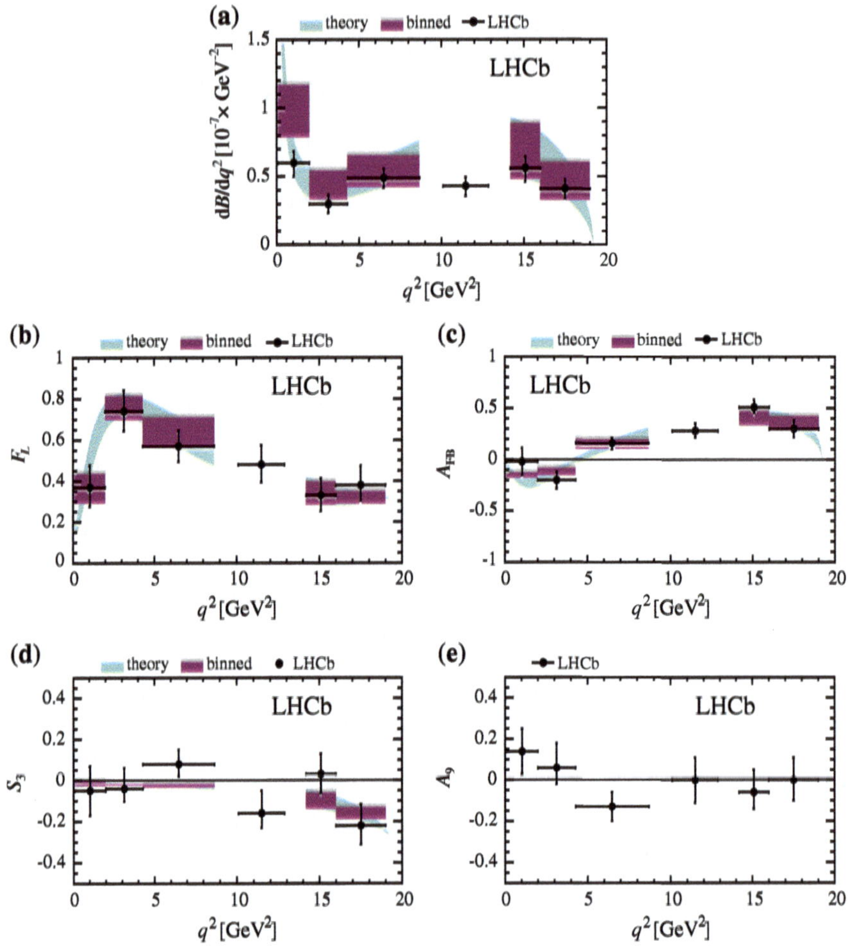

Fig. 8.7 Differential branching fraction, fraction of longitudinal polarisation of the K^{*0} F_L, dimuon system forward-backward asymmetry $A_{\rm FB}$, and the angular observables S_3 and A_9 from the $B_d^0 \to K^{*0}\mu^+\mu^-$ decays as a function of the dimuon invariant mass squared, q^2. The experimental data points are compared to the SM predictions (*green smooth band*), and a rate average of the SM prediction across each q^2 bin is indicated by the *purple rectangular* regions. No theory prediction is included for A_9, which is vanishingly small in the SM. Adapted from [104]

has performed such an analysis, combining several measured observables in $B \to X_s\gamma$, $B \to K^*\gamma$, $B \to K\mu\mu$ and $B \to X_s\mu\mu$ decays with the $B \to K^*\mu\mu$ data described above. The P_5' data can be explained with large negative contributions to C_9 and/or C_7. The global analysis of [110] finds a 68 % CL interval $2.2 \le C_9 \le 2.8$, which is to be compared with the SM value $C_9 = 4.07$. A negative contribution from new physics to C_7 is also favoured, but the 68 % CL region also covers the SM

Fig. 8.8 Measured values of $P_5^{'}$ (*points*) compared with SM predictions (*shaded bands*). The *error bars* indicate the 68.3 % CL. Adapted from [108]

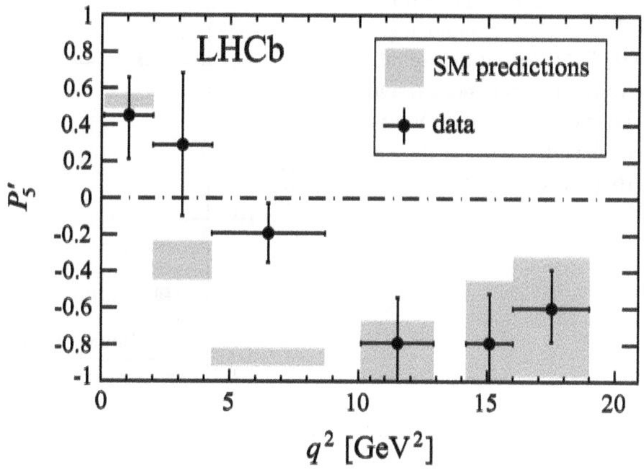

value. There is no evidence for non-zero chirality-flipped coefficients $C'_{7,9,10}$, which essentially vanish in the SM.

The power of the effective Hamiltonian ansatz is that the same Wilson coefficients are used to describe all (electroweak penguin) decays. The aim is to measure them in as many as possible decays and thus to get rid of the uncertainties from the hadronic form factors. In this spirit, further similar but still statistically limited modes are investigated at LHCb, like $B^+ \to K^+\mu^+\mu^-$, $B_d^0 \to K_S^0\mu^+\mu^-$ [111], $B^+ \to K^+e^+e^-$ [112] and $B_s^0 \to \phi\mu^+\mu^-$ [113].

8.6.3 Searches for Very Rare Decays

The analysis strategy for searches for rare decay modes is very similar for all searches performed by the LHCb collaboration: A boosted decision tree (BDT) is trained on simulated events to separate signal from physical background. Only variables which are uncorrelated to the invariant mass are explored in the BDT. Those are, for example, impact-parameter values of decay daughter tracks, the decay-length significance of the signal candidate, and momentum variables. The performance of the BDT is then evaluated for the signal process using kinematically similar reference modes in data. In case of $B_{d,s}^0 \to \mu^+\mu^-$, for example, $B^0 \to K^+\pi^-$, $J/\psi \to \mu^+\mu^-$ and $B^+ \to J/\psi K^+$ candidates are exploited. The performance of the BDT on the combinatorial background is determined on mass side-bands. Simulation is only used to estimate the size of potential differences in performance between the reference modes and the signal. An example of the performance of the BDT for $B_{d,s}^0 \to \mu^+\mu^-$ events exploited by the LHCb collaboration is shown in Fig. 8.9.

Next, the invariant-mass distribution is fitted in bins of the BDT output variable. In case of no observation, a modified frequentist approach, the $\mathrm{CL_s}$ method [115], is used to evaluate a limit on the number of observed signal candidates. To translate this

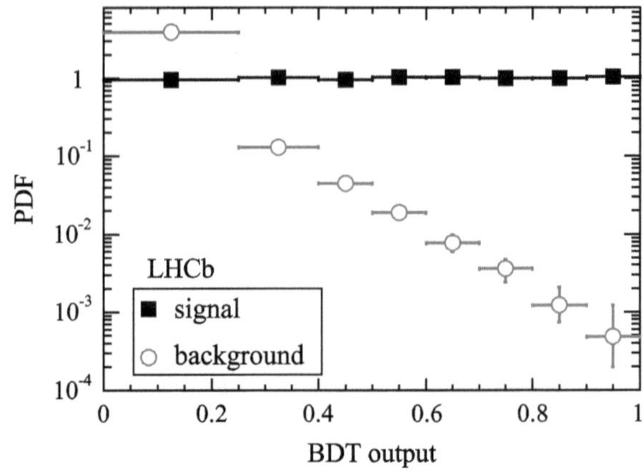

Fig. 8.9 BDT trained for $B^0_{d,s} \to \mu^+\mu^-$ candidates. The BDT output for signal (from reference modes) and combinatorial background (from side-bands) is shown. Adapted from [114]

number into a limit on the branching fraction, efficiencies relative to a reference mode with well-known branching ratio are determined. The more similar the reference is to the signal mode, the smaller are the related systematic uncertainties. The measured limits of the lepton-violating decays $\tau^- \to \mu^-\mu^+\mu^-$, $B^0_{d,s} \to e^\pm\mu^\mp$ and the rare decays $D^0 \to \mu^+\mu^-$ and $K^0_S \to \mu^+\mu^-$ are listed in Table 8.3.

The most prominent search for rare decays at the LHC is that for $B^0_{d,s} \to \mu^+\mu^-$ events. A first evidence of the B^0_s decay has been established at the end of 2012 by the LHCb experiment [125]. In summer 2013, both the LHCb and the CMS collaborations presented a clear signal peak in the $\mu^+\mu^-$ invariant-mass distribution at the B^0_s mass and a small excess at the B^0_d mass (Fig. 8.10). The following branching ratios and limits have been determined by the LHCb collaboration, based on 3.0 fb^{-1} of data [114]:

Table 8.3 Search for rare decays by the LHCb collaboration. Quoted limits are at 90 % CL

Channel	SM predictions	World/previous best limit	LHCb limit	LHCb data set (fb^{-1})
$\tau^- \to \mu^-\mu^+\mu^-$	$\mathcal{O}(10^{-40})$	Belle: 2.1×10^{-8} [116]	4.6×10^{-8} [117]	3.0
$B^0_d \to e^\pm\mu^\mp$	–	CDF: 6.4×10^{-8} [118]	3.7×10^{-9} [119]	1.0
$B^0_s \to e^\pm\mu^\mp$	–	CDF: 2.0×10^{-7} [118]	1.1×10^{-8} [119]	1.0
$D^0 \to \mu^+\mu^-$	$<6.0 \times 10^{-11}$	Belle: 1.4×10^{-7} [120]	6.2×10^{-9} [121]	0.9
$K^0_S \to \mu^+\mu^-$	$(5.0 \pm 1.5) \times 10^{-12}$ [122]	3.2×10^{-7} [123]	9.0×10^{-9} [124]	1.0

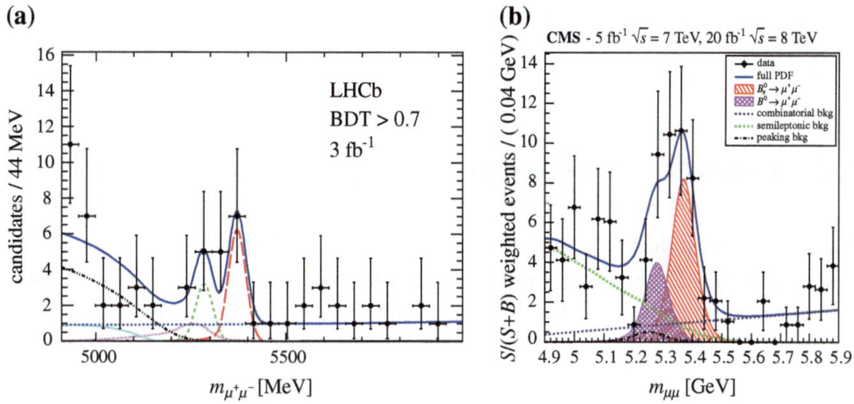

Fig. 8.10 **a** Dimuon invariant mass for $B^0_{d,s} \to \mu^+\mu^-$ signal candidates with a high BDT output value from the LHCb experiment. **b** Weighted diumon invariant-mass distribution for $B^0_{d,s} \to \mu^+\mu^-$ signal candidates from the CMS experiment. Both plots are for illustration only and do not represent the full power of the analysed data sets. Adapted from [114, 126]

$$\begin{aligned} \mathrm{B}(B^0_s \to \mu^+\mu^-) &= (2.9^{+1.1}_{-1.0}) \times 10^{-9}\,, \\ \mathrm{B}(B^0_d \to \mu^+\mu^-) &< 7.4 \times 10^{-10}\ 95\,\%\,\mathrm{CL}\,. \end{aligned}$$

The CMS collaboration, based on $25.0\,\mathrm{fb}^{-1}$ of data, obtained the following results [126]:

$$\begin{aligned} \mathrm{B}(B^0_s \to \mu^+\mu^-) &= (3.0^{+1.0}_{-0.9}) \times 10^{-9}\,, \\ \mathrm{B}(B^0_d \to \mu^+\mu^-) &< 1.1 \times 10^{-9}\ 95\,\%\,\mathrm{CL}\,. \end{aligned}$$

Due to the additional helicity suppression, the Standard Model predictions for branching ratios for the modes with electrons in the final state are even smaller than for muonic events. The treatment of electrons in the final state is, however, technically significantly more challenging because of the energy loss related to bremsstrahlung. Furthermore, the LHCb calorimeter (see Sect. 3.3) at LHCb is mainly designed for usage in the trigger rather than for clean electron or photon reconstruction.

8.7 Charm Physics

For the same reasons that make LHCb a great place to study B physics, it is also an excellent place to study charm physics. Both on the experimental and on the theoretical side, charm physics has significantly gained interest in recent years. This was mainly caused by the first single-experiment observation of neutral D meson mixing and a potential hint for a sizeable and thus eventually non-SM compatible

CP violation in $D^0 \to h^+h^-$ decays. The rare decay $D^0 \to \mu^+\mu^-$ is discussed above in Sect. 8.6.

8.7.1 Theoretical Background

The general formulae derived for the time evolution of B_d^0 and B_s^0 mesons in Sect. 8.2 also apply to D–$\overline{D}^0$ mixing. However, in charm physics one uses different notations, which we summarise here. The mass eigenstates are typically denoted as

$$|D_1\rangle = p|D^0\rangle + q|\overline{D}^0\rangle, \qquad |D_2\rangle = p|D^0\rangle - q|\overline{D}^0\rangle, \tag{8.45}$$

with masses and widths $m_{1,2}$ and $\Gamma_{1,2}$, respectively. $|D_1\rangle$ is defined as the state "which is dominantly *CP*-even", meaning that $B(D_1 \to K^+K^-) \gg B(D_2 \to K^+K^-)$. (The same holds for all other tree-level decays to *CP*-even final states like $\pi^+\pi^-$.) Since—unlike in K or B mixing—the definition in (8.45) makes no reference to masses or lifetimes, the dimensionless quantities

$$x = \frac{m_1 - m_2}{\Gamma}, \qquad y = \frac{\Gamma_1 - \Gamma_2}{2\Gamma} \tag{8.46}$$

can both have either sign, to be determined by experiment. *CP* violation in the neutral D system is tiny and still undiscovered. Contrary to the situation with B^0–$\overline{B}^0$ mixing, one cannot expand in $|\Delta\Gamma/\Delta M| = |y/x|$, since x and y are similar in magnitude.

The SM predictions for x and y are calculated from box diagrams, which are obtained from those in Fig. 8.1a by replacing the external quark lines by c and u and the internal lines by d, s, b. The contribution with the dominant CKM factors involves internal d and s quarks. These diagrams cancel up to terms of order $(m_s^4 - m_d^4)/(m_c^2 M_W^2)$, so that the GIM cancellation is less than in $B(D^0 \to \mu^+\mu^-)$ discussed in Sect. 8.6.1, and theory predictions are even less reliable. As a consequence of the smallness of box diagrams, the time evolution is slow, and one can expand the (8.30) and (8.33) in t, keeping only linear and quadratic terms. In the subsequent sections, we give these expanded time-evolution formulae in the contexts of the experimental discussions.

In the SM, *CP* violation in the charm system is small because *CP* asymmetries vanish for two fermion generations and the effect of the third generation on charm FCNC amplitudes is suppressed by the small CKM combination $V_{cb}V_{ub}^*$. While *CP* violation in mixing and mixing-induced *CP* violation is expected to be of order 10^{-3}, [127] has shown that *CP* asymmetries of order 1 % are possible within the SM. For the same reason, direct *CP* asymmetries are small in the SM. Non-zero direct *CP* asymmetries occur in $|\Delta C| = 1$ decays like $D^0 \to K^+K^-$ or $D^0 \to \pi^+\pi^-$, but the mismatch of the weak phases of the two interfering amplitudes (see (8.14)) is again governed by $V_{cb}V_{ub}^* \approx 1.5 \times 10^{-4} \cdot \exp(i\gamma)$. Adapting (8.14) to D decays, the first factor in the numerator reads $\mathrm{Im}(V_{cb}V_{ub}^*/V_{cs}V_{us}^*) \approx 6 \times 10^{-4}$. If hadronic effects enhance the ratio of the amplitudes in a favourable way ("penguin enhancement"), we can hope for *CP* asymmetries of several tenths of a percent [128]. New physics can lift the *CP* asymmetries easily to much larger values.

8.7.2 Measurement of Charm Mixing and CP Asymmetries

8.7.2.1 Mixing

In order to detect mixing, the production and the decay flavour of the D meson needs to be known.

Typically, D mesons in the flavour-conserving strong-interaction decay $D^{*+} \to D^0\pi_s^+$ are used for this analysis. The charge of the soft pion, π_s, indicates the production flavour of the D. The mass difference $m(D^{*+}) - m(D^0)$ is only slightly larger than the mass of a pion. Thus a requirement on the upper limit of the mass difference $m(D^{*+}) - m(D^0) - m(\pi)$ is a very powerful tool to reject background from combinations of D^0 candidates with random pions. The $D^{*+} \to D^0(\to K^-\pi^+)\pi_s^+$ process is denoted as "right sign" (RS), and $D^{*+} \to D^0(\to K^+\pi^-)\pi_s^+$ as "wrong sign" (WS). The RS decay rate is dominated by a Cabibbo-favoured amplitude. The WS rate arises from the interfering amplitudes of the doubly Cabibbo-suppressed $D^0 \to K^+\pi^-$ decay and the Cabibbo-favoured $\overline{D}^0 \to K^+\pi^-$ decay following oscillation, each of similar magnitude. About 2.3×10^5 signal WS D^0 and $\overline{D}^0$ decays and approximately 230 times more RS decays are reconstructed at the LHCb experiment in 1 fb^{-1} of data. In the limit of $|x|$, $|y| \ll 1$, and assuming negligible *CP* violation, the time-dependent ratio $R(t)$ of the WS-to-RS decay rates is given by

$$R(t) \approx R_D + \sqrt{R_D}\, y' \frac{t}{\tau} + \frac{x'^2 + y'^2}{4} (\frac{t}{\tau})^2 , \tag{8.47}$$

where t is the decay time, τ is the average D^0 lifetime and R_D the ratio of suppressed-to-favoured decay rates. The parameters x' and y' depend linearly on the mixing parameters as $x' \equiv x \cos\delta + y \sin\delta$, where δ is the strong-phase difference between the suppressed and favoured amplitudes. This relation gets modified if *CP* violation is permitted. In the absence of D^0 meson mixing $R(t)$ would be identical to R_D. The pure mixing contribution enters quadratically in the ratio $R(t)$ in (8.47). The linear term in the ratio describes the interference of the Cabibbo-suppressed and the mixing-related WS decays. Any time-dependent behaviour of the ratio is related to the mixing and therefore an unambiguous sign of neutral D-meson mixing. To test for *CP* violation, the analysis is separately performed for initial D^0 and $\overline{D}^0$ signal candidates and thus two sets of parameters (R_D^+, x'^{2+}, y'^+) and (R_D^-, x'^{2-}, y'^-) are measured, respectively. $R_D^+ \neq R_D^-$ would be a sign for direct *CP* violation, and any difference between (x'^{2+}, y'^+) and (x'^{2-}, y'^-) would indicate *CP* violation in mixing or mixing-induced *CP* violation. Figure 8.11 shows the mixing parameters

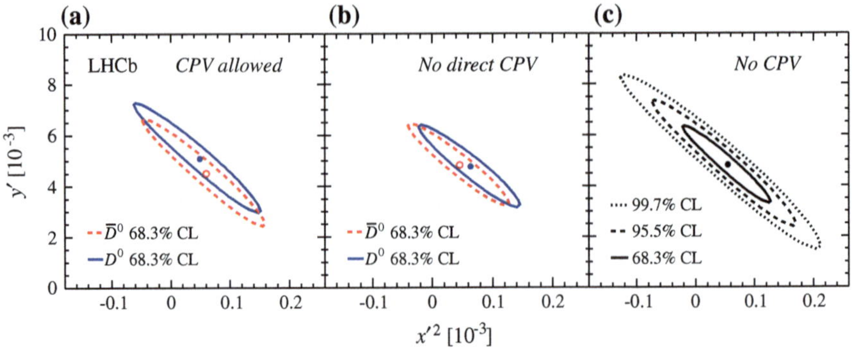

Fig. 8.11 Two-dimensional confidence regions in the (x'^2–y') plane obtained **a** without any restrictions on *CP* violation, **b** assuming no direct *CP* violation, and **c** assuming *CP* conservation. The *dashed* (*solid*) *curves* in (**a**) and (**b**) indicate the contours of the mixing parameters associated with $\overline{D}^0$ (D^0) decays. Adapted from [129]

for different fitting scenarios. With the current precision, no sign for *CP* violation has been revealed, and a mixing signal with a significance of more than 10σ has been established. The measured parameters, assuming no *CP* violation, are [129]:

$$\begin{aligned} R_D &= (3.568 \pm 0.058 \pm 0.033) \times 10^{-3}, \\ y' &= (4.8 \pm 0.8(\text{stat}) \pm 0.5(\text{syst})) \times 10^{-3}, \\ x'^2 &= (5.5 \pm 4.2(\text{stat}) \pm 2.6(\text{syst})) \times 10^{-5}. \end{aligned}$$

These values are consistent with the SM expectations—the GIM mechanism indeed suppresses $|x'|$ and $|y'|$ well below 1. The quoted central values are roughly larger by a factor of 1000 than the results of a naive calculation of quark diagrams neglecting QCD. While new physics can also enhance x and y, the impact on x is typically much larger than the effect on y, and one would then expect $|x| \gg |y|$.

8.7.2.2 *CP* Violation in *D* Decays

The time-dependent *CP* asymmetry $A_{CP}(f;t)$ for D^0 decays into a *CP* eigenstate f (with $f = \overline{f}$) is defined as

$$A_{CP(f;t)} \equiv \frac{\Gamma(D^0(t) \to f) - \Gamma(\overline{D}^0(t) \to f)}{\Gamma(D^0(t) \to f) + \Gamma(\overline{D}^0(t) \to f)},$$

where Γ is the decay rate for the process indicated. In general $A_{CP}(f;t)$ depends on f. For $f = K^-K^+$ and $f = \pi^-\pi^+$, $A_{CP}(f;t)$ can be expressed in terms of two contributions: a direct component associated with *CP* violation in the decay amplitudes, and an indirect component associated with *CP* violation in mixing or mixing-induced *CP* violation. In the limit of U-spin symmetry, the direct component is equal in magnitude and opposite in sign for K^-K^+ and $\pi^-\pi^+$, though the size of U-spin breaking effects remains to be quantified precisely. The asymmetry $A_{CP}(f;t)$ may thus be written to first order in t as

$$A_{CP}(f;t) = a_{CP}^{\rm dir} + \frac{t}{\tau} a_{CP}^{\rm ind} ,$$

where $a_{CP}^{\rm dir}(f)$ is the direct *CP* asymmetry, τ is the D^0 lifetime, and $a_{CP}^{\rm ind}$ is the indirect *CP* asymmetry. To a good approximation the latter quantity is universal, while the first one depends on the final state f. The time-integrated asymmetry measured by an experiment, $A_{CP}(f)$, depends upon the time acceptance of that experiment. It can be written as

$$A_{CP}(f) = a_{CP}^{\rm dir}(f) + \frac{\langle t \rangle}{\tau} a_{CP}^{\rm ind} ,$$

where $\langle t \rangle$ is the average decay time in the reconstructed sample. Denoting by Δ the differences between quantities for $D^0 \to K^-K^+$ and $D^0 \to \pi^-\pi^+$, it is then possible to write

$$\begin{aligned}\Delta A_{CP} &\equiv A_{CP}(K^-K^+) - A_{CP}(\pi^-\pi^+)\\ &= a_{CP}^{\rm dir}(K^-K^+) - A_{CP}^{\rm dir}(\pi^-\pi^+) + \frac{\Delta\langle t \rangle}{\tau} a_{CP}^{\rm ind} .\end{aligned}$$

In the limit that $\Delta\langle t \rangle$ vanishes, ΔA_{CP} is equal to the difference in the direct *CP* asymmetry between the two decays.

The measurement of ΔA_{CP} has been performed in two different samples. Firstly $D^{*+} \to D^0\pi_s^+$ decays are studied, where the D^{*+} is directly produced in the pp collision and the soft pion, π_s^+, is used to determine the D^0 production flavour ("tag"). A complementary measurement is performed using $\overline{B} \to D^0\mu^-\overline{\nu}_\mu X$ decays, where $\overline{B}$ include $\overline{B}^0$ and B^- mesons. In this case, the muon is used to tag the D^0 production flavour. Experimentally, ΔA_{CP} is a very robust quantity. Nuisance asymmetries, such as D production asymmetries or detection asymmetries of the tag particle, π_s and μ, respectively, effect the asymmetries in the K^-K^+ and the $\pi^-\pi^+$ channels in a very similarly way. Minor corrections need to be considered to compensate the slightly different phase spaces of the two final states. A first asymmetry measurement exploiting promptly produced D^0 mesons was published in January 2012 [130], based on a data set of 0.67 fb^{-1}. The measurement indicated a 3.5 σ evidence for a non-zero value of ΔA_{CP} [130]:

$$\Delta A_{CP} = \left[-0.82 \pm 0.21(\text{stat}) \pm 0.11(\text{syst})\right]\% .$$

This result caused quite some excitement and activity both on the experimental and the theoretical side. The discrepancy is, however, not confirmed by the analysis exploiting charm decays from semileptonic B decays [131], which is based on a data set of 3 fb^{-1}:

$$\Delta A_{CP} = [+0.14 \pm 0.16(\text{stat}) \pm 0.08(\text{syst})]\,\%\,.$$

Also a preliminary update of the analysis using promptly produced D^0 mesons indicates a value well consistent with zero.

The analysis exploiting charm decays from semileptonic B decays also included a—much more complex—measurement of the individual asymmetries, for which all nuisance asymmetries have to be measured to permille-level precision in data. Assuming negligible CP violation in charm mixing and in Cabibbo-favoured D decays, the individual asymmetry is found to be [131]

$$A_{CP}(K^-K^+) = [-0.06 \pm 0.15(\text{stat}) \pm 0.10(\text{syst})]\,\%\,,$$

where the correlation coefficient between ΔA_{CP} and $A_{CP}(K^-K^+)$ is $\rho = 0.28$.

To measure the indirect CP violation in the charm system, the following observable is introduced:

$$A_\Gamma(f) \equiv \frac{\hat{\tau}(f) - \hat{\bar{\tau}}(f)}{\hat{\tau}(f) + \hat{\bar{\tau}}(f)}\,.$$

Here, the effective lifetimes $\hat{\tau}$ ($\hat{\bar{\tau}}$) refer to D^0 ($\overline{D}^0$) lifetimes measured using a single-exponential model in a specific decay mode f. Assuming small asymmetries, in very good approximation $A_\Gamma = -a_{CP}^{\text{ind}}$ (8.7.2.2). Measuring the effective lifetimes rather than determining a_{CP}^{ind} from the slope of the time-dependent ratio of decay rates requires the precise knowledge of the decay-time-dependent acceptances. A dedicated entirely data-driven method called "swimming" is applied to determine event-by-event decay-time acceptances. Exploiting a sample of 1.0 fb^{-1} of promptly produced $D^{*+} \to D^0\pi_s^+$ decays, the following asymmetries have been measured [132]:

$$A_\Gamma(K^-K^+) = (-0.35 \pm 0.62(\text{stat}) \pm 0.12(\text{syst})) \times 10^{-3}\,,$$
$$A_\Gamma(\pi^-\pi^+) = (0.33 \pm 1.06(\text{stat}) \pm 0.14(\text{syst})) \times 10^{-3}\,.$$

In the measurement of effective lifetimes, another parameter can be determined:

$$y_{CP} = \frac{\tau(D^0 \to K^-\pi^+)}{\tau(D^0 \to K^-K^+)} - 1\,.$$

In the absence of CP violation, this parameters is identical to the mixing parameter y. In combination with an earlier measurement of A_Γ based on a data set of 29 pb^{-1} y_{CP} is found to be [133]

$$y_{CP} = (5.5 \pm 6.3(\text{stat}) \pm 4.1(\text{syst})) \times 10^{-3}\,.$$

There are still more interesting searches for *CP* violation in *D* decays into multi-body final states. As has been stated above, only final states that cover the same phase space can interfere. Therefore, the size of *CP* asymmetries might vary as a function of the phase space, and local effects may be larger than phase-space-integrated ones. Model-dependent and model-independent time-integrated analyses of multi-body *D* decays in the modes $D^+ \to \pi^-\pi^+\pi^+$ [134], $D^0 \to K^-K^+\pi^-\pi^-$ [135], $D^+ \to \phi\pi^+$ and $D_s^+ \to K_S^0\pi^+$ [136] have been performed. The advantage of this type of analysis is that an overall nuisance asymmetry between D^0 and $\overline{D}^0$ or $D^+_{(s)}$ and $D^-_{(s)}$ decays has no impact on the relative variations of the asymmetries as function of phase space. No sign for *CP* violation in the charm system has been revealed in any measurement up to now.

8.8 Spectroscopy of Exotic Resonances

Hadron spectroscopy is one of the areas at LHCb that recently started to attract more attention and is foreseen to be explored more extensively in the future. One of the highlights of the spectroscopy results from LHCb is the observation of the resonant character of the $Z(4430)^-$ state in the decay $Z(4430)^- \to \psi'\pi^-$ [137]. Observed charmonium resonances below the treshold of $2m(D)$ are in agreement with predictions. There are, however, several excited resonances which are predicted, but not yet experimentally confirmed—and vice versa (Fig. 8.12a). The resonances above the threshold are generally rather broad, which might explain their having escaped observation so far. More puzzling are the observed resonances which have not been expected to be there. Efforts are ongoing to understand their nature. Two frequently quoted potential explanations are $D\overline{D}^0$ molecules or tetra-quarks. LHCb has now confirmed the $Z(4430)^-$ state, which had already been seen by the Belle collaboration [138] before. Furthermore, LHCb has proven the resonance character of the state by measuring a phase shift of π in the amplitude, scanning over the mass range of the resonance (Fig. 8.12b). The measured quantum numbers exclude the $\overline{D}^*(2010)D_1^*(2420)$ hypothesis, which was before considered as the most likely molecule candidate, and support the tetra-quark hypothesis.

8.9 Summary and Outlook

The LHC experiments and especially LHCb have demonstrated impressively their huge physics potential in the last years. The LHCb experiment has operated successfully far beyond its original design, and a wealth of analyses (>200 papers) have been published. Many parameters have been determined with a significantly better precision than ever expected. Systematic uncertainties are well under control and are for most analyses significantly smaller than the statistical ones. However, no sign of new physics has been revealed up to know in indirect searches in the flavour system. Increasing the centre-of-mass energy from 8 to 13 TeV in Run 2 will potentially boost the physics reach of the direct searches performed at the general-purpose experiments significantly. For flavour physics, however, the energy increase only corresponds to a gain of about a factor of two in the $b\bar{b}$ and $c\bar{c}$ cross sections. A major step forward in available statistics and thus in the physics reach will be done with the upgraded LHCb experiment, which is planned to be ready for Run 3. LHCb plans to read out the entire detector at a rate of 40 MHz and perform a full reconstruction of all events in real-time at the high-level software trigger. The current bottleneck of limited readout rate of 1 MHz will thus be resolved. About a factor 100 more signal events will become available for the LHCb experiment within the next 10 years. Once the measurements start to challenge the theoretical uncertainties, a close cooperation of experimental and theoretical physicists will become even more important to make sure that the precision of the theoretical predictions will match those of the experiment. Already now e.g. the experimental precision of $\Delta\Gamma_s$ reported in Sect. 8.3.3 exceeds the accuracy of the prediction in (8.37).

Complementary input in the B and charm systems is expected from Belle II that will start data-taking in 2017. While Belle II has less statistics than LHCb, it can also study decay modes which LHCb cannot reconstruct, such as decays with missing energy.

While waiting for significantly more data to update the key measurements of the experiment, the LHCb collaboration will use Run 2 to test its ambitious trigger plans

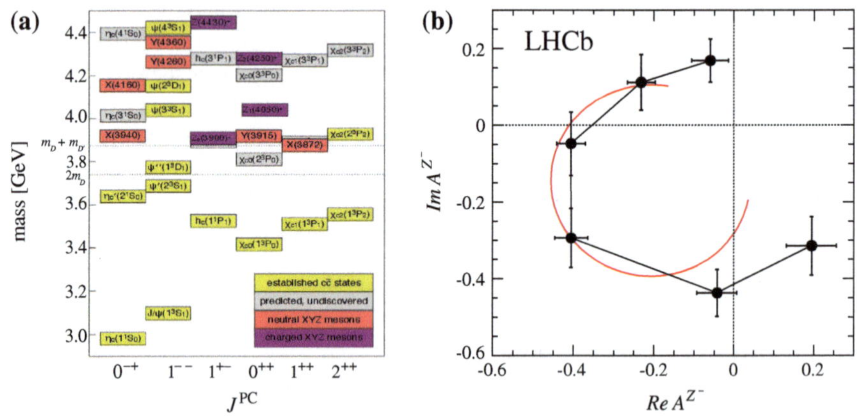

Fig. 8.12 **a** The spectra of charmonium-like mesons. **b** Fitted values of the Z^- amplitude in six $m^2_{\psi'\pi^-}$ bins, shown in an *Argand diagram*. The *red curve* is the prediction from the Breit–Wigner formula with a resonance mass (width) of 4475(172) MeV and magnitude scaled to intersect the bin with the largest magnitude centred at (4477 MeV)^2. Units are arbitrary. This plot confirms the resonance character of the observed broad peak in the $m_{\psi'\pi^-}$ distribution. Adapted from [137, 139]

for Run 3. Additionally the Run 1 and Run 2 data sets will be used to broaden the LHCb physics programme. Potentially interesting areas are spectroscopy measurements and Dalitz analyses of electroweak penguin decays. Furthermore LHCb covers a unique rapidity range, which is valuable for measurements of electromagnetic and diffractive cross sections, which have only been little explored up to know.

References

1. A.J. Buras (1998), arXiv:hep-ph/9806471
2. K. Anikeev et al. (2001), arXiv:hep-ph/0201071
3. R. Fleischer (2004), arXiv:hep-ph/0405091
4. Y. Nir (2005), arXiv:hep-ph/0510413
5. G. Hiller, U. Uwer (2011) Quark flavour physics. in *Physics at the Terascale*, ed. by I. Brock, T. Schörner-Sadenius, Wiley-VCH, Weinheim, Germany, pp. 163–187.doi:10.1002/9783527634965.ch8
6. U. Nierste (2009), arXiv:0904.1869
7. G.C. Branco, L. Lavoura, J.P. Silva, *CP Violation (International Series of Monographs on Physics)* (Oxford University Press, Oxford, 1999)
8. I.I. Bigi, A. Sanda, *CP violation (Cambridge Monographs on Particle Physics, Nuclear Physics and Cosmology)* (Cambridge University Press, Cambridge, 2008)
9. Luders, G. (1954) Kong. Dan. Vid. Sel. Mat. Fys. Med. 28N5, 1–17
10. W. Pauli, Exclusion principle, Lorentz group and reflection of space-time and charge, in *Niels Bohr and the Development of Physics*, ed. by W. Pauli, L. Rosenfeld, V. Weisskopf (MacGraw-Hill, New York, 1955), pp. 30–51
11. M. Kobayashi, T. Maskawa, Prog. Theor. Phys. **49**, 652–657 (1973)
12. J. Christenson, J. Cronin, V. Fitch, R. Turlay, Phys. Rev. Lett. **13**, 138 (1964)
13. J. Christenson, J. Cronin, V. Fitch, R. Turlay, Phys. Rev. **140B**, 74 (1965)
14. O. Eberhardt et al., Phys. Rev. Lett. **109**, 241802 (2012)
15. Particle Data Group, Chin. Phys. C **38**, 090001 (2014)
16. CKMfitter Group (2005) Eur. Phys. J. C **41**, 1–131. Regular updates at http://ckmfitter.in2p3.fr. See also http://www.utfit.org
17. BaBar Collaboration, Nucl. Instrum. Meth. A **729**, 615–701 (2013)
18. Belle Collaboration, Nucl. Instrum. Meth. A **479**, 117–232 (2002)
19. U. Nierste, J. Phys. Conf. Ser. **447**, 012017 (2013)
20. A.J. Buras, M. Jamin, P.H. Weisz, Nucl. Phys. B **347**, 491–536 (1990)
21. S. Aoki et al. (2013), arXiv:1310.8555
22. M. Beneke et al., Phys. Lett. B **459**, 631–640 (1999)
23. M. Beneke, G. Buchalla, A. Lenz, U. Nierste, Phys. Lett. B **576**, 173–183 (2003)
24. M. Ciuchini et al., JHEP **08**, 031 (2003)
25. A. Lenz, U. Nierste, JHEP **06**, 072 (2007)
26. A. Lenz, U. Nierste (2011), arXiv:1102.4274
27. S. Laplace, Z. Ligeti, Y. Nir, G. Perez, Phys. Rev. D **65**, 094040 (2002)
28. A. Lenz at al., Phys. Rev. D **83**, 036004 (2011)
29. A. Lenz et al., Phys. Rev. D **86**, 033008 (2012)
30. LHCb Collaboration, New J. Phys. **15**, 053021 (2013)
31. LHCb Collaboration, Phys. Lett. B **719**, 215–318 (2013)
32. DØ Collaboration, Phys. Rev. D **89**, 012002 (2014)
33. LHCb Collaboration (2014), Phys. Rev. Lett. **114**, 041601 (2015)
34. LHCb Collaboration, Phys. Lett. B, **728**, 607–615 (2014)
35. B.A.B.A.R. Collaboration, Phys. Rev. D **79**, 072009 (2009)

36. Belle Collaboration, Phys. Rev. Lett. **108**, 171802 (2012)
37. Heavy Flavour Averaging Group (2012), arXiv:1207.1158v2
38. LHCb Collaboration, Phys. Lett. B, **721**, 24–31 (2013)
39. LHCb Collaboration, Phys. Rev. D **87**, 112010 (2013)
40. LHCb Collaboration, Phys. Rev. D **87**, 052001 (2013)
41. LHCb Collaboration, Phys. Lett. B **736**, 186 (2014)
42. LHCb Collaboration, Phys. Lett. B **713**, 378–386 (2012)
43. LHCb Collaboration, Phys. Rev. Lett. **113**, 211801 (2014)
44. ATLAS Collaboration, Phys. Rev. D **90**, 052007 (2014)
45. LHCb Collaboration, Phys. Rev. D **90**, 052011 (2014)
46. V.A. Khoze, M.A. Shifman, Sov. Phys. Usp. **26**, 387 (1983)
47. M.A. Shifman, M. Voloshin, Sov. J. Nucl. Phys. **41**, 120 (1985)
48. M.A. Shifman, M. Voloshin, Sov. Phys. JETP **64**, 698 (1986)
49. I.I. Bigi, N. Uraltsev, A. Vainshtein, Phys. Lett. B **293**, 430–436 (1992)
50. M. Beneke et al., Nucl. Phys. B **639**, 389–407 (2002)
51. LHCb Collaboration, JHEP **04**, 114 (2014)
52. LHCb Collaboration, Phys. Lett. B **736**, 446–454 (2014)
53. LHCb Collaboration, Phys. Rev. Lett. **112**, 111802 (2014)
54. LHCb Collaboration, Phys. Rev. Lett. **108**, 241801 (2012)
55. LHCb Collaboration, Nucl. Phys. B **873**, 275–292 (2013)
56. LHCb Collaboration, Phys. Rev. Lett. **113**, 172001 (2014)
57. M. Battaglia et al. (2003), arXiv:hep-ph/0304132
58. LHCb Collaboration, Phys. Lett. B **734**, 122 (2014)
59. LHCb Collaboration, Eur. Phys. J. C **74**, 2839 (2014)
60. LHCb Collaboration, Phys. Rev. Lett. **113**, 172001 (2014)
61. LHCb Collaboration, Phys. Rev. Lett. **110**, 221601 (2013)
62. M. Gronau, D. Wyler, Phys. Lett. B **265**, 172 (1991)
63. M. Gronau, D. London, Phys. Lett. B **253**, 483 (1991)
64. D. Atwood, I. Dunietz, A. Soni, Phys. Rev. Lett. **78**, 3257 (1997)
65. A. Atwood, I. Dunietz, A. Soni, Phys. Rev. D **63**, 036005 (2001)
66. A. Giri, Y. Grossman, A. Soffer, J. Zupan, Phys. Rev. D **68**, 054018 (2003)
67. LHCb Collaboration, Phys. Rev. D **90**, 112002 (2014)
68. LHCb Collaboration, Phys. Lett. B **712**, 203 (2012)
69. LHCb Collaboration, Phys. Lett. B **723**, (2013)
70. LHCb Collaboration, Phys. Lett. B **718**, 43 (2012)
71. LHCb Collaboration, JHEP **10**, 97 (2014)
72. LHCb Collaboration, Phys. Lett. B **726**, 151–163 (2013)
73. LHCb Collaboration, Nucl. Phys. B **888**, 169–193 (2014)
74. LHCb Collaboration, Phys. Lett. B **733**, 36 (2014)
75. Babar Collaboration, Phys. Rev. D **87**, 052015 (2013)
76. Belle Collaboration (2013), arXiv:1301.2033
77. R. Fleischer, Eur. Phys. J. C **52**, 267–281 (2007)
78. R. Fleischer, Phys. Lett. B **459**, 306–320 (1999)
79. R. Fleischer, R. Knegjens, Eur. Phys. J. C **71**, 1532 (2011)
80. LHCb Collaboration, Phys. Lett. B **741**, (2015)
81. LHCb Collaboration, JHEP **11**, 060 (2014)
82. T. Feldmann, J. Matias, JHEP **01**, 074 (2003)
83. S. Jaeger, J. Martin Camalich, JHEP **05**, 043 (2013)
84. S. Descotes-Genon, L. Hofer, J. Matias, J. Virto, JHEP **1412**, 125 (2014)
85. B. Grinstein, D. Pirjol, Phys. Rev. D **70**, 114005 (2004)
86. C. Bobeth, G. Hiller, D. van Dyk, JHEP **07**, 098 (2010)
87. M. Beylich, G. Buchalla, T. Feldmann, Eur. Phys. J. C **71**, 1635 (2011)
88. C. Bobeth et al., Phys. Rev. Lett. **112**, 101801 (2014)
89. K. De Bruyn et al., Phys. Rev. Lett. **109**(041), 801 (2012)

90. Heather E. Logan, Ulrich Nierste, Nucl. Phys. B **586**, 39–55 (2000)
91. A. Crivellin, A. Kokulu, C. Greub, Phys. Rev. D **87**(9), 094031 (2013)
92. X.Q. Li, J. Lu, A. Pich, JHEP **06**, 022 (2014)
93. W. Altmannshofer, D.M. Straub, JHEP **08**, 121 (2012)
94. R. Sekhar Chivukula, Howard Georgi, Phys. Lett. B **188**, 99 (1987)
95. G. D'Ambrosio, G.F. Giudice, G. Isidori, A. Strumia, Nucl. Phys. B **645**, 155–187 (2002)
96. K. Babu, C.F. Kolda, Phys. Rev. Lett. **84**, 228–231 (2000)
97. A. Dedes, H.K. Dreiner, U. Nierste, Phys. Rev. Lett. **87**, 251804 (2001)
98. A.J. Buras, P.H. Chankowski, J. Rosiek, L. Slawianowska, Phys. Lett. B **546**, 96–107 (2002)
99. W. Altmannshofer, M. Carena, N.R. Shah, F. Yu, JHEP **01**, 160 (2013)
100. W. Altmannshofer, PoS Beauty **2013**, 024 (2013)
101. S.L. Glashow, J. Iliopoulos, L. Maiani, Phys. Rev. D **2**, 1285–1292 (1970)
102. G. Ecker, A. Pich, Nucl. Phys. B **366**, 189–208 (1991)
103. G. Isidori, R. Unterdorfer, JHEP **01**, 009 (2004)
104. LHCb Collaboration, JHEP **08**, 131 (2013)
105. W. Altmannshofer et al., JHEP **01**, 019 (2009)
106. C. Bobeth, G. Hiller, G. Piranishvili, JHEP **07**, 106 (2008)
107. LHCb Collaboration, Phys. Rev. Lett. **108** (2012)
108. LHCb Collaboration, Phys. Rev. Lett. **111**, 191801 (2013)
109. S. Descotes-Genon, T. Hurth, J. Mathias, J. Virto, JHEP **05**, 137 (2013)
110. S, Descotes-Genon, J. Matias, J. Virto, Phys. Rev. D **D88**(7), 074002 (2013)
111. LHCb Collaboration, JHEP **05**, 082 (2014)
112. LHCb Collaboration, JHEP **05**, 159 (2013)
113. LHCb Collaboration, JHEP **07**, 084 (2013)
114. LHCb Collaboration, Phys. Rev. Lett. **111**, 101805 (2013)
115. A.L. Read, J. Phys. G **28**, 2693 (2002)
116. Belle Collaboration, Phys. Lett. B **687**, 139 (2010)
117. LHCb Collaboration (2014), arXiv:1409.8548
118. CDF Collaboration, Phys. Rev. Lett. **102**, 201801 (2009)
119. LHCb Collaboration, Phys. Rev. Lett **111**, 141801 (2013)
120. Belle Collaboration, Phys. Rev. D **81**, 091102 (2010)
121. LHCb Collaboration, Phys. Lett. B **725**, 15–24 (2013)
122. G. Isidori, R. Unterdorfer, JHEP **01**, 009 (2004)
123. S. Gjesdal et al., Phys. Lett. B **44**, 217 (1973)
124. LHCb Collaboration, JHEP **01**, 090 (2013)
125. LHCb Collaboration, Phys. Rev. Lett. **110**, 021801 (2013)
126. CMS Collaboration, Phys. Rev. Lett **111**, 101804 (2013)
127. M. Bobrowski, A. Lenz, J. Riedl, J. Rohrwild, JHEP **03**, 009 (2010)
128. J. Brod, Y. Grossman, A.L. Kagan, J. Zupan, JHEP **10**, 161 (2012)
129. LHCb Collaboration, Phys. Rev. Lett. **111**, 251801 (2013)
130. LHCb Collaboration, Phys. Rev. Lett. **108**, 111602 (2012)
131. LHCb Collaboration, JHEP **07**, 041 (2014)
132. LHCb Collaboration, Phys. Rev. Lett. **112**, 041801 (2014)
133. LHCb Collaboration, JHEP **04**, 129 (2012)
134. LHCb Collaboration, Phys. Lett. B **728**, 585–595 (2014)
135. LHCb Collaboration, Phys. Lett. B **726**, 623–633 (2013)
136. LHCb Collaboration, JHEP **06**, 112 (2013)
137. LHCb Collaboration, Phys. Rev. Lett. **112**, 222002 (2014)
138. Belle Collaboration, Phys. Rev. Lett **100**, 142001 (2008)
139. S. Olsen (2014), arXiv:1403.1254

Chapter 9
Heavy-Ion Physics at the LHC

Ralf Averbeck, John W. Harris and Björn Schenke

Abstract The first collisions of lead nuclei delivered by the Large Hadron Collider in late 2010 mark the beginning of a new chapter in the study of relativistic heavy-ion collisions. With $\sqrt{s_{NN}} = 2.76$ TeV, the energy record so far held by BNL's Relativistic Heavy-Ion Collider was exceeded by more than one order of magnitude. Substantial data samples of Pb–Pb collisions were recorded during the first 3 years of LHC operation, furthering the understanding of the properties of strongly interacting matter under extreme conditions. Not only did the LHC programme confirm the formation of a hot and dense medium with a ratio of shear viscosity to entropy density close to the predicted lower bound for quantum fluids it also opens up new approaches to determine in detail the properties of this elusive medium. In this chapter, an overview is given of the results from the heavy-ion programme published so far, with emphasis on measurements of correlation observables and, in particular, hard probes of the medium, which are produced abundantly for the first time in heavy-ion collisions at LHC energies.

R. Averbeck (✉)
EMMI, GSI Helmholtzzentrum für Schwerionenforschung,
Planckstr. 1, 64291 Darmstadt, Germany
e-mail: r.averbeck@gsi.de

J.W. Harris
Department of Physics, Yale University,
268 Whitney Avenue, PO Box 208120, New Haven, CT 06520-8120, USA
e-mail: john.harris@yale.edu

B. Schenke
Physics Department, Brookhaven National Laboratory, Upton, NY 11973, USA
e-mail: bschenke@bnl.gov

T. Schörner-Sadenius (ed.), *The Large Hadron Collider*,
DOI 10.1007/978-3-319-15001-7_9

9.1 Introduction

9.1.1 Heavy-Ion Collisions: From the Bevalac, AGS and SPS via RHIC to the LHC

Fundamental properties of strongly interacting nuclear matter can be explored by studying the interior of the atomic nucleus as well as the nucleon, the interior of neutron stars, and the matter created in heavy-ion collisions.

The goal of heavy-ion physics research is to advance our understanding of the strong force and of extended systems governed by QCD (see Chap. 5.1).

This involves in particular the determination of the phase structure of nuclear matter, the understanding of colour confinement and the dynamic properties of (deconfined) nuclear matter under the extreme conditions of high temperature and density.

First heavy-ion collisions were conducted at the Bevalac, a combination of the Bevatron (Billions of eVSynchrotron) and SuperHILAC linear accelerator at Lawrence Berkeley National Laboratory in 1971 [1]. Early on fluid-dynamic behaviour of the produced matter was predicted [2, 3] and observed [4]. These studies were followed by experiments at the Super Proton Synchrotron (SPS) at CERN and at the Alternating Gradient Synchrotron (AGS) at Brookhaven National Laboratory (BNL). AGS experiments studied the high baryon density region of relativistic heavy-ion collisions (for a review see [5]). The SPS programme was a massive effort, with involvement of experimental groups from all over the world. In 2000, an assessment of the results of the SPS led to the statement that "[...] we have compelling evidence that a new state of matter has indeed been created, at energy densities which have never been reached over appreciable volumes in laboratory experiments before and which exceed by more than a factor of 20 that of normal nuclear matter. The new state of matter found in heavy-ion collisions at the SPS features many of the characteristics of the theoretically predicted Quark-Gluon Plasma." (key references can be found in [6]). One main evidence for this was the "J/ψ suppression", i.e. the fact that J/ψ mesons were produced much less abundantly than expected. However, in light of the following RHIC experiments this suppression has proven more difficult to interpret [7–10].

With the advent of the high-energy colliders RHIC (the Relativistic Heavy Ion Collider operating at BNL since 2000) and the LHC, the physics of relativistic heavy-ion collisions has entered a new era: The energies available for the collisions are sufficiently high to allow creation of new forms of QCD matter that are characterised by high parton densities. These new forms of matter include both the wavefunctions of the incoming nuclei, which develop high gluon densities leading to a saturated gluon state, and the partonic matter produced in the intermediate stages of the collision, which is expected to form a deconfined quark-gluon plasma (QGP).

Indeed, the measurements by the four original experimental collaborations at RHIC [7–10] have established—and recent data from the LHC have (as will be discussed in this review) confirmed—a novel quantitative framework for the theoretical description of nuclear matter at high energy densities: QCD matter at energy densities greater than 1 GeV/fm^3 (more than six times the normal nuclear energy density) acts like a strongly interacting plasma of quarks and gluons, which behaves like a nearly inviscid fluid and is highly opaque to energetic coloured probes.

9.1.2 QCD at High Density and Temperature

As alluded to above, heavy-ion collisions allow us to explore the phase structure of nuclear matter. In particular a confinement-deconfinement phase transformation between hadronic matter and the QGP is predicted from QCD calculations on the lattice. Over the past few years, increasingly precise lattice calculations of thermal QCD, extrapolated to the continuum and thermodynamic limits and to small quark masses and even non-zero baryon chemical potential, have become available (see [11] and references therein).

These calculations show a rapid rise of the energy density as a function of temperature in the temperature range 150–170 MeV. This and various other quantities, such as the strange quark number susceptibility, the Polyakov loop, the chiral condensate and the trace anomaly, indicate a rapid cross-over to a deconfined (and chiral-symmetry-restored) quark-gluon phase [11].

Because of the complexity and the non-perturbative nature of hot and dense QCD, calculations require approximations. One possibility to make theoretical progress is the above-mentioned lattice implementation, where space-time is discretised and finite temperature equilibrium systems can be studied.

Alternatively, perturbative methods can be used in some situations, in particular when dealing with large momentum scales, i.e. small coupling. In heavy-ion collisions, thermal perturbation theory, which uses resummations of certain classes of loop corrections, has made significant progress recently in describing the pressure, energy density, and entropy density of the quark-gluon plasma [12–15].

Perturbative calculations can also be applied to the study of interactions of high-momentum partons passing through the created bulk medium. This can be used to study the phenomenon of "jet quenching", the suppression of high-momentum probes in the presence of a hot and dense medium. For recent comparative studies of various calculations of jet quenching see [16, 17].

An effective theory of QCD at asymptotically high energies is the so called colour-glass condensate (CGC) [18] that describes the state of a highly energetic nucleus as a dense and saturated system of gluons. Within this framework, a dynamically generated "saturation scale" Q_s characterises the system. Its size typically guarantees a small coupling $\alpha_s(Q_s) \ll 1$, which allows for a perturbative treatment. The system, however, is strongly interacting, because of the high occupation number ($\sim 1/\alpha_s$) of gluons with transverse momenta smaller than Q_s. The early stages of a heavy-ion collision are then described by the collision of two such colour-glass condensates, resulting in the so called "glasma" [19].

To describe the later stages of a heavy-ion collision, when the system is strongly interacting and not necessarily perturbative, the effective theory of hydrodynamics supplemented with the QCD equation of state from lattice QCD can be employed. Hydrodynamics uses coarse graining to correctly represent the long-distance dynamics of a strongly interacting system [20, 21].

We will describe the various theoretical approaches in more detail below as we discuss the experimental observables they are used to describe.

9.1.3 Geometry of Heavy-Ion Collisions

9.1.3.1 Collision Centrality

In collisions of extended nuclei, the volume of the interaction region depends on the impact parameter, b, which is the distance between the centres of the two nuclei in a plane transverse to the beam direction. The "centrality" of the collision reflects the initial overlap area of the colliding nuclei and is related to the impact parameter. It is determined by a comparison of measured data with simulations of the collision geometry.

For this purpose, the purely geometric Glauber model [22–24] is commonly used [25], in which the collision of two nuclei with mass number A is treated as a superposition of binary nucleon-nucleon collisions. Nucleons experiencing at least one binary collision with a nucleon from the other nucleus are called "participant" nucleons. "Spectator" nucleons are those which do not undergo such a binary collision. The number of participants, N_{part}, is a measure for the initial overlap area of the colliding nuclei. N_{part}, the corresponding number of spectators, $N_{\text{spec}} = 2 \times A - N_{\text{part}}$, and the total number of binary nucleon-nucleon collisions, N_{coll}, are calculated for a given impact parameter using a Monte Carlo approach, in which the nuclear collision process is simulated event-by-event. In the first step of a Glauber Monte Carlo [26], the positions of nucleons in the two nuclei are defined stochastically following the nuclear density function typically modelled via a modified Woods–Saxon function.

In the second step, the nucleus-nucleus collision is simulated for an impact parameter randomly chosen according to the geometrical distribution $\mathrm{d}P/\mathrm{d}b \sim b$. The nuclear collision is simulated as a sequence of binary nucleon-nucleon collisions with the nucleons moving along straight-line trajectories and a constant inelastic nucleon-nucleon cross section $\sigma_{NN}^{\text{INEL}}$ for all binary collisions. If the transverse distance d between two nucleons is small enough $\left(d < \sqrt{\sigma_{NN}^{\text{INEL}}/\pi}\right)$, they are assumed to collide. N_{coll} and N_{part} are determined by counting the binary collisions and the nucleons that experience at least one collision, respectively. The effective nucleon luminosity in the nuclear collision is reflected by the nuclear overlap function $T_{AA} = N_{\text{coll}}/\sigma_{NN}^{\text{INEL}}$. Geometrical quantities such as the mean values $\langle N_{\text{part}}\rangle$, $\langle N_{\text{coll}}\rangle$ and $\langle T_{AA}\rangle$ are calculated as a function of the impact parameter b by averaging over many simulated nucleus-nucleus collisions. ALICE [27], ATLAS [28], and CMS [29] employ Glauber

Monte Carlo implementations for a consistent description of *pA* and *AA* collisions such that data from the different experiments and from different collision systems can be compared with each other and with theoretical calculations.

However, neither the impact parameter b itself nor the geometrical quantities N_{part} or N_{coll} can be measured directly. Instead, experimental observables reflecting the collision geometry are used to classify nucleus-nucleus collisions according to their centrality. Most relevant is the average multiplicity of produced particles, which increases monotonically with decreasing impact parameter, i.e. from peripheral to central collisions. In ALICE the amplitude of the signal registered in the V0 detectors ($2.8 < \eta < 5.1$ and $-3.7 < \eta < -1.7$, see Sect. 3.7) is commonly used [27], which is proportional to the charged particle multiplicity N_{ch}. In ATLAS the total transverse energy $\sum E_{\text{T}}$ measured at the electromagnetic energy scale in the FCAL calorimeters ($3.2 < |\eta| < 4.9$, Sect. 3.3) is the basis for the centrality measurement [28], and in CMS the energy deposit E^{HF} in the HF calorimeters ($3 < |\eta| < 5.2$, Sect. 3.3) is employed [29].

In practice, the collision centrality is quoted as a percentage of the total nucleus-nucleus interaction cross section. The centrality percentile c of a collision with impact parameter b is then

$$c = \frac{1}{\sigma_{AA}} \int_0^b \frac{\mathrm{d}\sigma}{\mathrm{d}b'}\,\mathrm{d}b' \approx \frac{1}{\sigma_{AA}} \int_x^\infty \frac{\mathrm{d}\sigma}{\mathrm{d}x'}\,\mathrm{d}x',$$

where x is the experimental observable used to determine the centrality in the various experiments.

9.1.3.2 Event Plane and Eccentricity

The beam direction and the impact-parameter vector define a reaction plane with azimuth Ψ_{RP} for each non-central collision. A hallmark feature of the quark-gluon plasma produced in heavy-ion collisions is the azimuthal anisotropy of particle emission with respect to the orientation of the reaction plane.

In a simplified picture of continuous nucleon density within an initial, spatially asymmetric nuclear overlap region with an almond-like shape, the initial collision geometry is determined solely by the impact parameter. If, in a more realistic picture, the positions of the individual nucleons participating in the collision are taken into account, the overlap region has an irregular shape which fluctuates on an event-by-event basis around the reaction plane. This shape can be characterised by the eccentricities ε_n that, for instance in a Glauber model calculation, can be estimated from the positions (r, ϕ) of the participating nucleons in the transverse plane relative to their centre-of-mass [30]:

$$\varepsilon_n = \frac{\sqrt{\langle r^n \cos n\phi \rangle^2 + \langle r^n \sin n\phi \rangle^2}}{\langle r^n \rangle}$$

Partons interact strongly with each other in the spatially asymmetric initial state allowing local thermal equilibrium to be reached and pressure gradients to build up that lead to collective expansion of the collision region [31]. The pressure gradients are anisotropic, providing the largest acceleration in the direction of the short axis of the collision almond. This gives rise to an anisotropic momentum distribution of the produced particles which is typically quantified in terms of a Fourier expansion of the final-state particle yields [32, 33]:

$$E\frac{\mathrm{d}^3N}{\mathrm{d}^3p} = \frac{1}{2\pi}\frac{\mathrm{d}^2N}{p_\mathrm{T}\,\mathrm{d}p_\mathrm{T}\,\mathrm{d}y}\left(1+\sum_{n=1}^{\infty}2v_n(p_\mathrm{T},y)\cos\left[n\left(\phi-\Psi_\mathrm{n}\right)\right]\right).$$

Here E, p_T, y, and ϕ are the particle's energy, transverse momentum, rapidity, and azimuthal angle, respectively. Ψ_n is the azimuthal angle of the initial spatial plane of symmetry of the n-th harmonic. In the simplified case of continuous nucleon density in the colliding nuclei Ψ_n and Ψ_RP are identical and, because of symmetry arguments, the odd Fourier coefficients $v_{2\mathrm{n}+1}$ are zero. Only via event-by-event fluctuations of the initial spatial asymmetry odd harmonic symmetry planes are generated, which then give rise to non-zero odd harmonics $v_{2\mathrm{n}+1}$.

9.1.3.3 The Case of Proton-Nucleus Collisions

In order to have an understanding of nucleus–nucleus collisions, it is essential to measure and understand proton-proton (*pp*), proton-nucleus (pA) and nucleus-nucleus (AA) collisions in a consistent framework. Initial-state and final-state effects must be identified and separated both experimentally and theoretically, if possible. Proton-lead (*p*-Pb) measurements at the LHC will contribute substantially to our knowledge of the initial state by establishing the extent of gluon saturation in the nucleus and investigating the existence of a colour-glass condensate at low-x in nuclei. Investigation of cold nuclear matter effects and their impact on the interpretation of the measurements of observables in the final state, such as colour screening and jet quenching, are critical to our understanding of the lead-lead (Pb–Pb) data. Common aspects of the data, such as the multiplicity and overall centre-of-mass energy dependence of *p*-Pb and Pb–Pb collisions, allow comparisons of measurements across experiments and energies to be made, from RHIC to the LHC.

9.2 Characterisation of the Final State at Freeze-Out

9.2.1 Introduction

A nucleus-nucleus collision is a very violent, highly dynamic event. The energy deposited in the collision volume depends on the density of virtual partons (mostly gluons) confined at low Bjorken-x in the colliding nuclei. With increasing collision energy, the relevant x range at the LHC extends to lower values, down to $x \approx 10^{-3}$ or even lower. Upon the impact of the colliding nuclei, i.e. during the passage time $t_{pass} = 2R/\gamma_{cm}$ of two nuclei with radius R moving with γ_{cm} in their centre-of-mass system, these dense, virtual parton fields turn real via initial collision processes, some of which occur with large momentum transfer ("hard processes"). The created primordial, strongly interacting medium thermalises quickly, i.e. within not more than 1 fm, and expands rapidly as a quark-gluon plasma. The QGP cools down while it continues to expand, mainly longitudinally, until it reaches the critical temperature at which the phase transition from a deconfined QGP to a gas of hadrons occurs. In principle, this hadron gas can expand and cool down further until it becomes so dilute that inelastic collisions cease to happen. Since from this moment on the relative yields of the various hadron species do not change, it is called "chemical freeze-out". At the LHC, and also at RHIC, it seems that the chemical freeze-out coincides with hadronisation, i.e. the phase transition from a QGP to a hadron gas. After the chemical freeze-out, the hadron gas continues to expand and cool until elastic collisions effectively cease because the hadron gas becomes too dilute. At this moment, called "kinetic freeze-out", the hadron momentum distributions and their correlations are frozen and the hadrons or their decay products stream to the detectors.

The main challenge of heavy-ion physics at the LHC is to characterise the properties of the deconfined medium in the early stage of the collision with measurements that are performed via hadrons or their decay products, which decouple from the system only at freeze-out. Direct information from the QGP phase can only be obtained via probes that do not interact strongly, such as electrons and photons. While exploratory photon measurements have been conducted [34, 35] at low p_T in Pb–Pb collisions at the LHC, final results from Run 1 are not yet published. Hadronic observables also carry precious information from the QGP phase of Pb–Pb collisions at the LHC. The interpretation of the corresponding measurements, however, must rely on model comparisons where hydrodynamic calculations set the standard at high collision energies, sometimes in combination with hybrid approaches using hadronic transport calculations.

Before addressing observables that can be connected to the early phases of nuclear collisions, the following measurements characterising the system in its final state at freeze-out will be discussed.

9.2.2 *Particle Multiplicity*

The LHC experiments have measured the multiplicity density of charged particles in order to estimate the energy density created in Pb–Pb collisions at the LHC. In addition, comparisons with theoretical model calculations and measurements at lower energies can be used to isolate particle-production mechanisms and potentially identify the presence of new features such as a colour-glass condensate. The requirement to reproduce multiplicity measurements both at RHIC and the much larger LHC

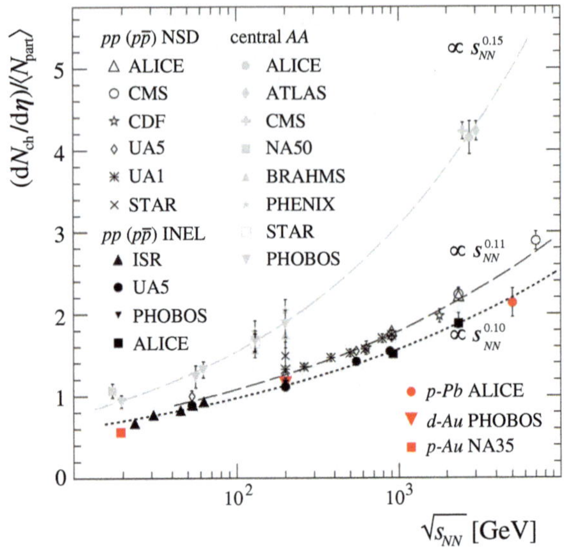

Fig. 9.1 Charged-particle pseudo-rapidity density per participant measured at mid-rapidity as a function of $\sqrt{s_{NN}}$ for non-single diffractive *pp*, *pA* (or *dA*) and central *AA* collisions. *Curves* representing the $\sqrt{s_{NN}}$ dependences are superimposed individually on the *pp*, *pA* (or *dA*) and *AA* data for reference. Compiled from [36]

collision energy places significant constraints on models and predictions. Measurements of the charged-particle density scaled by the average number of participant nucleons, derived from the Glauber Model [25], are presented in Fig. 9.1 for *pp*, *pA* (or *dA*), and central *AA* collisions as a function of $\sqrt{s_{NN}}$ [36].

9.2.3 Spectra of Charged Particles and Identified Hadrons

The transverse-momentum distributions of charged particles are sensitive to properties of different phases of the nuclear collisions at low and high p_T. Above momenta of a few GeV charged particles originate mainly from the fragmentation of jets produced in hard scattering processes in the first phase of the collision. This regime will be discussed in detail in Sect. 9.4.1. At low p_T charged-particle production is governed by soft processes and consequently the momentum distributions reflect the properties of the bulk medium at kinetic freeze-out. Figure 9.2a shows transverse-momentum distributions of identified pions, kaons, and protons measured with ALICE [37] in the 5 % most central Pb–Pb collisions at $\sqrt{s_{NN}} = 2.76$ TeV in comparison with corresponding results from Au–Au collisions at $\sqrt{s_{NN}} = 0.2$ TeV [38, 39]. At both energies the spectra reflect a clear mass hierarchy. At any given p_T the local inverse slope of the p_T distributions increases with increasing hadron mass. This can be interpreted in a scenario in which the hadrons are emitted from a thermalised medium expanding radially with a velocity profile common to all hadrons. Heavier hadrons, such as protons, obtain a larger momentum than the lighter ones, such as kaons and pions, if they participate in the collective radial flow with the same velocity.

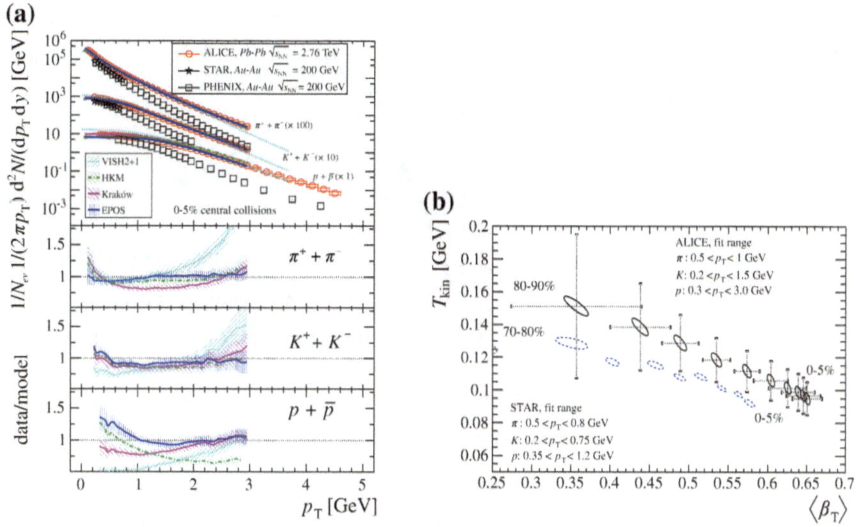

Fig. 9.2 **a** Transverse-momentum spectra of pions, kaons, and protons (sums of particles and antiparticles) produced in the 5 % most central Pb–Pb collisions at $\sqrt{s_{NN}} = 2.76$ TeV [37] compared to corresponding measurements from RHIC at 0.2 TeV [38, 39] and to hydrodynamical model calculations. **b** Results from blast-wave fits at the LHC and at RHIC. Adapted from [37]

This qualitative picture can be quantified via so-called "blast-wave fits" [41], where the p_T distributions of protons, kaons, and pions (excluding pions with $p_T < 0.5$ GeV and kaons with $p_T < 0.35$ GeV [37] for which resonance decays contribute significantly to the yield) are fit simultaneously with a parametrisation inspired by hydrodynamic principles, which has only two parameters describing the spectral shapes: the temperature T_{kin} at kinetic freeze-out and the average radial-flow velocity $\langle\beta_T\rangle$ over an assumed velocity profile. Figure 9.2b shows these two freeze-out parameters as obtained from blast-wave fits for different centrality selections of heavy-ion collisions at the LHC and at RHIC [37]. In general, for a given collision centrality both T_{kin} and $\langle\beta_T\rangle$ are larger for the higher-energy collisions at the LHC compared to RHIC, reaching $T_{kin} \approx 95$ MeV and $\langle\beta_T\rangle \approx 0.65$ for the 5 % most central Pb–Pb collisions at the LHC. Towards more peripheral collisions the expansion velocity becomes smaller and the freeze-out temperature grows. It is interesting to note that the transverse-momentum distributions of light hadrons measured in p-Pb collisions at $\sqrt{s_{NN}} = 5.02$ TeV show an evolution with the charged-particle multiplicity [42] similar to the pattern observed in Pb–Pb collisions at the LHC. Models which include final-state effects in p-Pb collisions describe the data better than models without such effects [42].

The collision-energy dependence of the kinetic freeze-out temperature and the average expansion velocity in central heavy-ion collisions are shown in Fig. 9.3 with results from the experiments EOS [43], FOPI [44], NA49 [45], STAR [39, 46], and ALICE [47]. At low energies, both T_{kin} and $\langle\beta_T\rangle$ grow rapidly with increasing $\sqrt{s_{NN}}$.

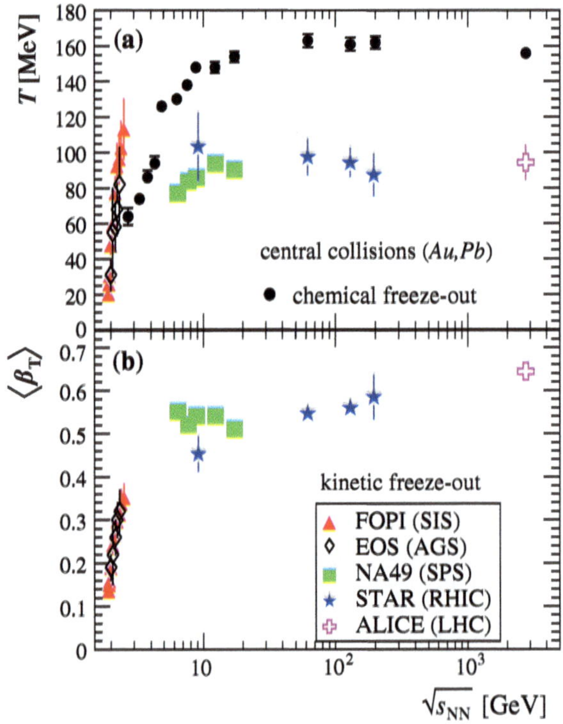

Fig. 9.3 Collision-energy dependence of **a** the kinetic and chemical freeze-out temperatures and **b** the average radial-flow velocity at kinetic freeze-out. Adapted from [40]

For $\sqrt{s_{NN}} > 5$ GeV T_{kin} remains approximately constant at $\approx$100 MeV and $\langle \beta_T \rangle$ continues a gradual rise until a maximum radial-flow velocity is reached at the LHC. For comparison Fig. 9.3a also displays the temperatures at chemical freeze-out, that will be discussed in Sect. 9.2.4. The temperature at chemical freeze-out is typically larger than that at kinetic freeze-out by 50–60 MeV. An exception are collisions at the lowest energies up to $\sqrt{s_{NN}}$ values of a few GeV where the chemical freeze-out temperature is smaller than T_{kin}, which seems unphysical and is not yet understood.

Blast-wave fits can be extended to include further hadron species, and the assumption of common freeze-out parameters for all particles can be varied. Further insight can be gained from the comparison of the measured distributions with full hydrodynamical calculations [48–51] as displayed in Fig. 9.2a. Such models are applicable in the p_T range up to 2–3 GeV, and they describe hadron production reasonably well for central collisions. Towards more peripheral collisions and higher p_T, the hydrodynamical description of the data worsens, potentially indicating the growing contribution from hard processes to particle production.

Measurements in Au–Au collisions at RHIC [55, 56] reveal a pronounced enhancement of baryon yields relative to mesons in the region of intermediate p_T ($2 < p_T < 8$ GeV) that is not observed in pp collisions, the so-called "baryon anomaly". As demonstrated in Fig. 9.4a also in central Pb–Pb collisions at the LHC, the proton-to-pion ratio [53] exhibits a peak around p_T values of 3 GeV where it is more than a

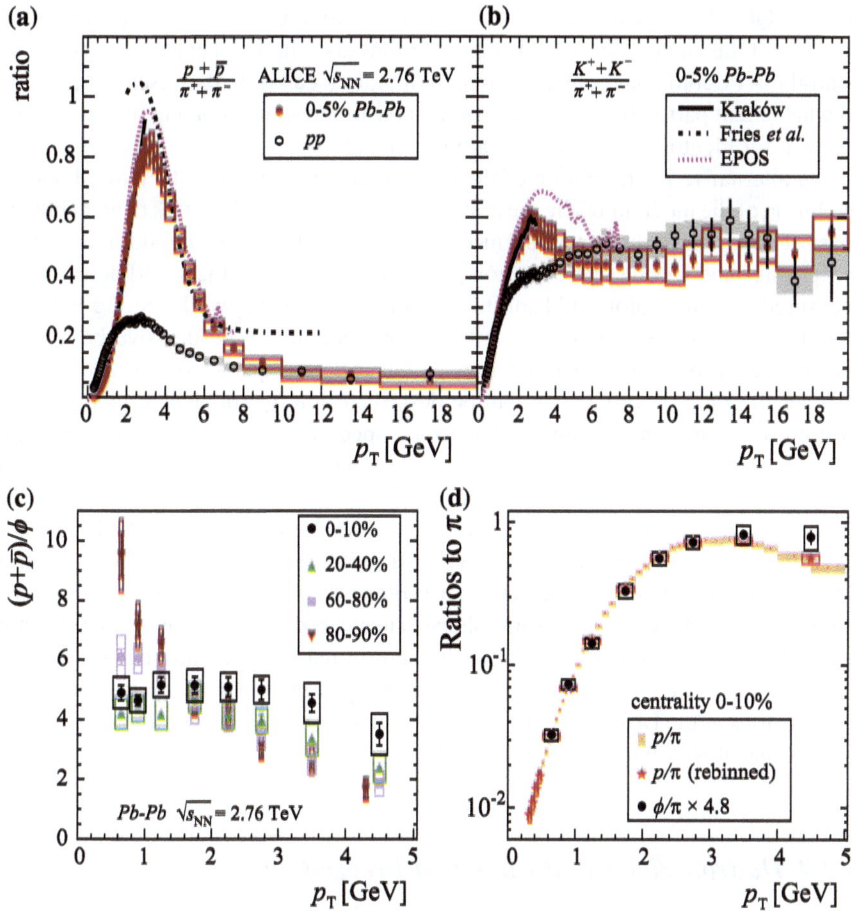

Fig. 9.4 Transverse-momentum dependence of proton (**a**) and kaon yields (**b**) relative to the pion yield in the 5 % most central Pb–Pb collisions at the LHC compared to theoretical predictions [49, 50, 52]. Ratios of proton to ϕ-meson yields (**c**) as a function of p_T for four centrality classes of Pb–Pb collisions at the LHC, and proton-to-pion and ϕ-meson-to-pion ratios versus p_T (**d**), where the ϕ-meson yield was scaled up to match the proton yield (see legend). Adapted from [53, 54]

factor three larger than that measured in *pp* collisions. A similar observation was made in the strangeness sector via the measurement of the p_T dependence of the ratio Λ/K^0_s [57]. It was suggested that this baryon anomaly could reflect new hadronisation mechanisms such as baryon string-junction transfer along the axis of a fragmenting jet [58] or parton recombination from the QGP [52], which would give baryons on average a larger p_T than mesons. Models in which the recombination of shower partons from neighbouring jets is considered [59] predict proton-to-pion ratios that are up to an order of magnitude larger than those measured. Calculations which consider only the recombination of soft thermalised partons from the radially flowing QGP [52] are in reasonable agreement with the data. The observation that the ratio of the yields of charged kaons and pions also shows an enhancement [53] in central Pb–Pb collisions compared to pp collisions (see Fig. 9.4b) raises the question to what extent parton recombination or the constituent-quark content of hadrons are relevant for identified hadron ratios at intermediate p_T.

An alternative interpretation of the baryon anomaly involves the hydrodynamic evolution of the medium where the relevant parameter is not the quark content but the mass of a hadron. Hydrodynamical models without [50] and with the addition of high-p_T physics and special hadronisation processes for quenched jets [49] describe the measured proton-to-pion and kaon-to-pion ratios reasonably well (see Fig. 9.4a, b). For this issue, measurements of ϕ-meson production can be decisive because ϕ mesons and protons have similar masses but different quark content and baryon number. If ϕ-meson production can be described in a hydrodynamic picture, then protons and ϕ mesons should exhibit similar spectral shapes despite their different quark content. The yield ratio $(p + \bar{p})/\phi$ is shown as a function of p_T for various centrality classes of Pb–Pb collisions at $\sqrt{s_{NN}} = 2.76$ TeV [54] in Fig. 9.4c. While in non-central collisions $(p+\bar{p})/\phi$ decreases with increasing p_T, the ratio is flat over the full measured p_T range in the most central collisions. Consequently, in central Pb–Pb collisions also the yield ratios of protons to pions and ϕ mesons to pions, shown in Fig. 9.4d, exhibit the same shape. These observations indicate that it is not the number of constituent quarks but the hadron mass that mainly determines the shapes of hadron-momentum distributions at low and intermediate p_T in central Pb–Pb collisions at the LHC, contrary to the expectation from a recombination scenario but consistent with hydrodynamical models.

9.2.4 Hadron Yields and Chemical Freeze-Out

The hadron yields at mid-rapidity can be obtained by integration of the transverse-momentum spectra extrapolated beyond the lowest measured p_T values down to zero p_T. Thermal models have been very successful in describing hadron yields in heavy-ion collisions [60–62] (see [63] for a review). Such models describe the particle abundances at the chemical freeze-out of a heavy-ion collision statistically via a hadron gas in thermal and chemical equilibrium. Three parameters are sufficient to calculate the abundances of the various hadron species in a grand-canonical ensemble approach, i.e. the temperature T, the baryochemical potential μ_B, and the volume V of the system at chemical freeze-out. Feed-down from all resonances and other unstable states must be taken into account before results from the thermal-model calculation can be compared with measured hadron yields. A good fit of the hadron yields measured in central Pb–Pb collisions at the LHC (summarised in [64]) is obtained with the parameters $T = 156$ MeV, $\mu_B = 0$, and $V = 5330\,\text{fm}^3$ as

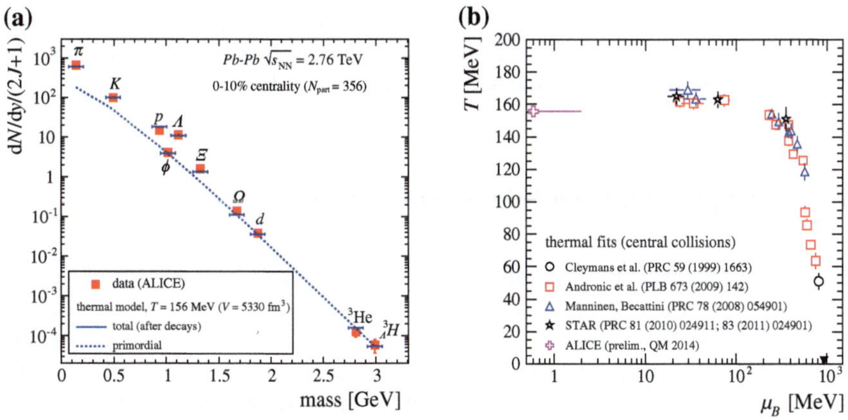

Fig. 9.5 **a** Hadron yields per unit rapidity measured at mid-rapidity in the 10 % most central Pb–Pb collisions at the LHC divided by their corresponding spin-degeneracy factors and compared with a thermal-model fit. **b** Chemical freeze-out temperature versus baryochemical potential obtained from thermal-model fits of hadron yields at various collision energies. Ground-state nuclear matter is indicated by the *triangle* close to zero temperature. Adapted from [40]

shown in Fig. 9.5a, which also demonstrates the role of decays after freeze-out. At zero baryochemical potential the thermal model predicts that matter and antimatter are produced with equal yields, which is consistent with the experimental observation [47]. The measured proton yield, however, is slightly smaller than one would expect from the thermal-model fit, an observation which is not yet fully understood.

It is important to point out that in the thermal model also the measured yields of multi-strange hyperons [65] are well described. An enhancement of strangeness production per participant nucleon in heavy-ion collisions compared to *pp* collisions was among the earliest signatures proposed for the formation of a QGP [66, 67]. This expectation was based on the argument that in a deconfined medium the relevant quark masses should not be the constituent but the bare quark masses and, consequently, the abundances of light parton species in equilibrium should become similar. Indeed strangeness enhancement was observed at the SPS [68–70], at RHIC [71] and also at the LHC [65]. However, it is now understood that for strangeness production in *pp* collisions the grand-canonical ensemble approach is not appropriate [72], and a proper canonical description leads to reduced strangeness yields. This accounts for the observed differences between hyperon yields per participant nucleon in heavy-ion and *pp* collisions, i.e. strangeness production is not enhanced in the former but suppressed in the latter.

The relationship between the temperature T and the baryochemical potential μ_B at chemical freeze-out in heavy-ion collisions is depicted in Fig. 9.5b. Each point in this diagram was obtained via a fit of the measured hadron yields in central Au–Au or Pb–Pb collisions at a given collision energy $\sqrt{s_{NN}}$. With increasing $\sqrt{s_{NN}}$, μ_B becomes smaller and T grows until it reaches a saturation value close to 160 MeV for $\mu_B \lesssim 300\,\text{MeV}$. This saturation of T can provide a connection to the QCD phase boundary between a hadron gas and a QGP, where the link is the conjecture that chemical freeze-out can coincide with hadronisation [73].

9.2.5 The Quest for the Initial Temperature

As it was discussed in the preceding Sects. 9.2.3 and 9.2.4, the final state of a central Pb–Pb collision can be characterised as a hot hadron gas with a temperature at chemical freeze-out close to the expected hadronisation temperature, i.e. the temperature at the QCD phase boundary between a QGP and a hadron gas. One of the "holy grails" of heavy-ion physics is the measurement of the initial temperature of the QGP. This phase of the collision is not directly accessible via strongly interacting probes, but only with electroweak probes which can leave the hot and dense medium without strong final-state interactions. The measurement of thermal "black body" radiation emitted from the hot fireball might, therefore, be the only experimental approach to determine the initial temperature.

In central Au–Au collisions at RHIC, a signal consistent with thermal photon production has been extracted for $p_T \lesssim 4$ GeV from the photon p_T distribution, measured via electron-positron pairs, i.e. virtual photons [74]. An inverse slope of the thermal photon p_T distribution above 200 MeV was reported. Since photons are emitted over the whole evolution of the system, the measured spectrum is a convolution of photon spectra from the initial, highest-temperature phase to the late hadron-gas phase, such that the initial temperature can only be deduced via models implementing the full system evolution. Further complications arise from the radial expansion of the fireball, which results in blue (and red) shifts in the photon-momentum distribution. Hydrodynamic-model calculations employing initial temperatures between 300 and 600 MeV are qualitatively consistent with the measured thermal-photon spectrum at RHIC [74]. First low-p_T photon measurements have been conducted at the LHC with ALICE [34], where an indication for a thermal-photon component with an inverse slope close to 300 MeV was observed in the real-photon p_T distribution in central Pb–Pb collisions.

A promising alternative approach which, however, requires larger data sets of Pb–Pb collisions than could be recorded during Run 1 at the LHC is the measurement of the mass distribution of virtual photons, i.e. electron or muon pairs, at masses above $\approx$1 GeV. Mass spectra are not affected by blue shifts or other effects that have an impact on the momentum distributions of photons. In measurements of muon-pair production in In–In collisions at the SPS($\sqrt{s_{NN}} = 17.3\,\text{GeV}$) a strong excess of muons beyond known sources could be isolated and interpreted as thermal radiation from the hot medium [75, 76]. Above a mass of 1 GeV an inverse slope of the muon-pair mass distribution of about 200 MeV, considerably larger than the QCD phase-transition temperature, was measured implying that the muon pairs where emitted from a partonic source [75, 76]. Similar measurements at the LHC will be promising in the future.

9.3 Correlations in Heavy-Ion Collisions

In heavy-ion collisions, multi-particle correlations as a function of the particles' azimuthal angles with respect to the reaction plane reveal a clear signal of collective flow of the produced matter. This was demonstrated early on at the SPS [77] and at RHIC [78]. The observed azimuthal multi-particle distributions are best described when treating the system like a fluid with a very small ratio of the shear viscosity η to the entropy density s, η/s, i.e. as an almost perfect fluid. The fluid picture gives a simple explanation for the measured anisotropies: Matter is pushed outwards according to the pressure gradients of the initial collision geometry in the transverse plane (see Sect. 9.1.3). An anisotropic initial energy-density distribution in coordinate space thus leads to an anisotropic momentum distribution of the observed particles. This conversion is more efficient when the system has a smaller viscosity. Explicit simulations (see [21] for a recent review) indicate that the ratio of shear viscosity to entropy density needs to be close to the conjectured minimum value for an infinitely strongly-coupled theory [79, 80].

At the LHC, similar measurements at 14 times higher centre-of-mass energy than the top RHIC energy have found qualitatively similar results [81–83]. In this section, we review the LHC measurements and recent theoretical developments on azimuthal correlations and the physical insight they provide.

9.3.1 Azimuthal Distributions in Heavy-Ion Physics

To quantify the anisotropy of the azimuthal particle distribution, it is useful to expand the distribution in a Fourier series:

$$\frac{\mathrm{d}N}{\mathrm{d}\phi_i} \propto 1 + 2\sum_{n=1}^{\infty} v_n \cos\left[n\left(\phi_i - \psi_n\right)\right].$$

Here $(\phi_i - \psi_n)$ is the angle of particle i with respect to an event plane ψ_n that is common to all particles and depends on the harmonic number n. Experimentally, this angle can be obtained from the so-called Q-vector, defined via

$$Q_n = \sum_{l=1}^{M} e^{in\phi_l} = |Q_n| e^{in\psi_n} = Q_{n,x} + i\,Q_{n,y},$$

where the sum is over all M particles in an event. The angle $\tilde{\psi}_n = \frac{1}{n}\tan^{-1}\left(\frac{Q_{n,y}}{Q_{n,x}}\right)$ can be used to determine the observed

$$v_n^{\mathrm{obs}} = \langle\langle \cos[n(\phi - \tilde{\psi}_n)] \rangle\rangle,$$

where the $\langle\langle \ldots \rangle\rangle$ stands for an average over all particles in an event followed by an average over events. The finite number of particles leads to a limited event-plane resolution of $R = \langle \cos[n(\tilde{\psi}_n - \psi_n)] \rangle$. Using this, the flow coefficients v_n can be determined according to (see [33] and references therein)

$$v_n = \langle\langle \cos[n(\phi - \psi_n)] \rangle\rangle = v_n^{\text{obs}}/R.$$

This "event plane method" contains complex multi-particle correlations. A simpler way to determine the v_n is via two-particle correlations, quantified by

$$c_n\{2\} = \langle\langle \cos[n(\phi_i - \phi_j)] \rangle\rangle = \langle v_n^2 \rangle + \delta_{2,n} = \langle v_n \rangle^2 + \sigma_n^2 + \delta_{2,n},$$

where σ_n^2 characterises event-by-event fluctuations and $\delta_{2,n}$ represents non-flow two-particle correlations from e.g. resonance decays or jets. For example, two particles will be correlated if they emerge from the decay of the same resonance—this correlation is short ranged (in η) and is unrelated to the correlation all particles have because they emerge from a flowing fluid. Non-flow correlations can be suppressed less by requiring a minimum difference in pseudo-rapidity η between the two particles.

Another way to suppress non-flow correlations is to study correlations of more than two particles. In particular, the four-particle cumulant, defined as

$$\begin{aligned} c_n\{4\} &= \langle e^{in(\phi_1+\phi_2-\phi_3-\phi_4)} \rangle - 2\langle e^{in(\phi_1-\phi_2)} \rangle^2 \\ &= \langle v_n^4 + \delta_{4,n} + 4v_n^2\delta_{2,n} + \delta_{2,n}^2 - 2(v_n^2 + \delta_{2,n})^2 \rangle = \langle -v_n^4 + \delta_{4,n} \rangle, \end{aligned}$$

has, by construction, no contribution from 2-particle non-flow correlations [84].

Higher cumulants as well as other methods like Lee–Yang zeroes [85] and the scalar-product method [86] have also been used to accurately determine the flow coefficients v_n.

9.3.2 Long-Range Correlations in Rapidity

The azimuthal structure of multi-particle correlations discussed above is accompanied by long-range correlations in pseudo-rapidity η. This leads to the so-called "ridge" structure in the $\Delta\eta$–$\Delta\phi$ correlation strength, where $\Delta\eta$ is the difference in pseudo-rapidity between the two particles and $\Delta\phi$ the difference in azimuthal angle. Two-particle correlations can be quantified by measuring the associated particle yield per trigger particle,

$$\frac{1}{N_{\text{trig}}} \frac{d^2 N^{\text{pair}}}{d\Delta\eta d\Delta\phi} = B(0,0) \times \frac{S(\Delta\eta, \Delta\phi)}{B(\Delta\eta, \Delta\phi)}. \tag{9.1}$$

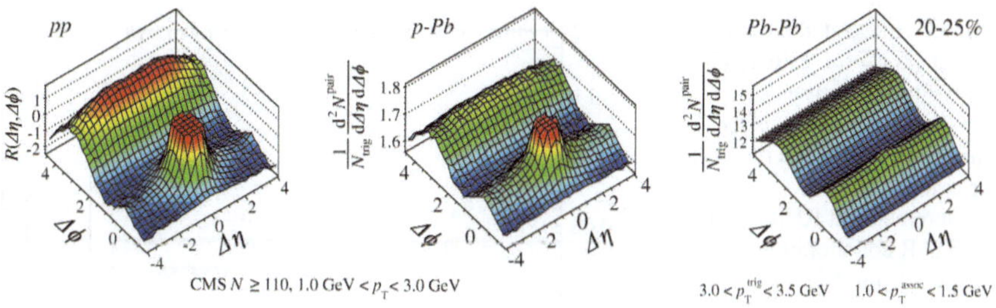

Fig. 9.6 The ridge in high-multiplicity pp, p-Pb and Pb–Pb collisions. Adapted from [87, 90, 91, 95]

The signal S is the per-trigger-particle yield of pairs from the same event, and the background B is obtained by pairing a trigger particle from one event with all the associated particles from a different event. $B(0, 0)$ is a normalisation factor, and N_{trig} is the number of trigger particles.

At the LHC, two-particle correlations have been measured in Pb–Pb collisions [87–89], high-multiplicity pp collisions [90], and p-Pb collisions [91–94].

Figure 9.6 shows the differential per-trigger-particle yield (9.1) in heavy-ion collisions and proton-lead collisions. For high multiplicity pp collisions, the correlation R is shown, which is equivalent to the per-trigger-particle yield but differs in normalisation and is centered around zero [90]. In pp and p-Pb collisions, a typical jet-like correlation (from the fragmentation of the jet) can be seen on the "near-side" around $\Delta\eta = 0$ and $\Delta\phi = 0$. The "away-side" jet (around $\Delta\phi = \pi$) also contributes to short-range correlations in $\Delta\eta$. These correlations are also present in Pb–Pb collisions. For the centrality and p_{T} ranges shown in the example, they are however mostly overwhelmed by correlations of the bulk medium.

All two-particle correlation measurements shown exhibit similar angular correlations at large $\Delta\eta$ (long range). This modulation can be decomposed into harmonics v_n as discussed above, with v_2 being prominent in all these examples. The presence of long-range correlations in $\Delta\eta$ means that particles with a large spatial separation are correlated. Causality demands that these correlations emerge at early times when the particles were still causally connected. As noted above, the azimuthal structure is very well described by viscous fluid dynamics using fluctuating energy-density distributions in the transverse plane of the collision (see Sect. 9.1.3). The correlation in $\Delta\eta$, however, must emerge at very early times, before fluid dynamics becomes valid. In all fluid-dynamic simulations, this correlation is incorporated into the initial condition by the assumption that the transverse geometry does not vary significantly as a function of rapidity.

Recent theoretical focus has been on the interpretation of the observed correlations in high-multiplicity pp and p-Pb collisions. On the one hand, there is the possibility that fluid dynamics can be used even in these comparatively small systems [96–99]. On the other hand, the initial state in the colour-glass condensate framework has been shown to produce correlations compatible with the data in these smaller systems without the need for additional final-state collective effects [100–104].

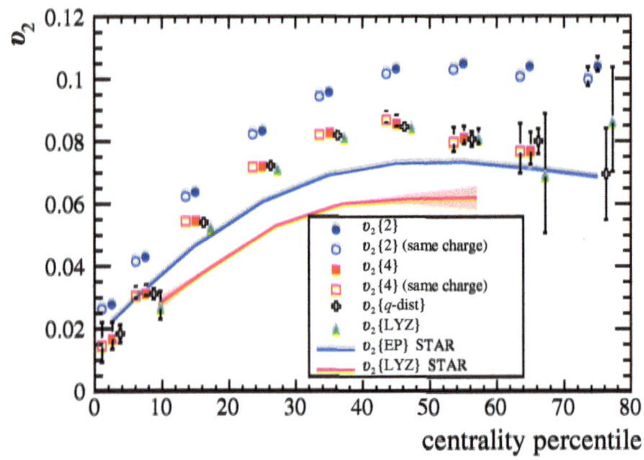

Fig. 9.7 Integrated elliptic flow at $\sqrt{s_{NN}} = 2.76$ TeV in Pb–Pb collisions as a function of centrality using various methods. The results are compared to measurements at RHIC from the STAR collaboration. Adapted from [81]

9.3.3 *Elliptic Flow: v_2*

With the exception of the most central events, the most prominent flow harmonic is the elliptic flow v_2. This can be explained by the fact that the interaction region in a non-central heavy-ion collision is approximately elliptical in the plane transverse to the beamline. The larger the distance between the centres of the nuclei (impact parameter) gets, the larger the initial anisotropy will be. This causes a correlation between the total number of produced particles and the initial ellipticity of an event. If the produced matter interacts strongly, this initial anisotropy is transformed into an anisotropy of the particle-momentum distribution. In a fluid-dynamic framework, this process can be understood as the system reacting to pressure gradients that vary in strength depending on the azimuthal direction.

Indeed, the resulting correlation between the multiplicity and the size of v_2 has been observed in heavy-ion collisions at various energies [81]. Figure 9.7 shows that this is also observed at the LHC, the difference being that v_2 is generally larger at the higher collision energy.

Figure 9.8 shows the energy dependence of v_2 in a given centrality bin, for measurements from a wide range of experiments [81, 105, 106]. There is a continuous increase in the magnitude of the elliptic flow for this centrality region from RHIC to LHC energies. In comparison to the elliptic-flow measurements in Au–Au collisions at $\sqrt{s_{NN}} = 0.2$ TeV, one observes an increase of approximately 30 % in the magnitude of v_2 at $\sqrt{s_{NN}} = 2.76$ TeV.

As can be seen from the comparison of the transverse-momentum dependent elliptic flow at various energies [81, 82, 107, 108] in Fig. 9.9, the difference of the integrated elliptic flow stems mostly from the harder spectra at LHC energies, which shift the weight in the integrated v_2 towards higher-momentum particles.

The pseudo-rapidity dependence of the elliptic flow in central and mid-central collisions is very weak [82, 83], indicating boost-invariance of the system. This is

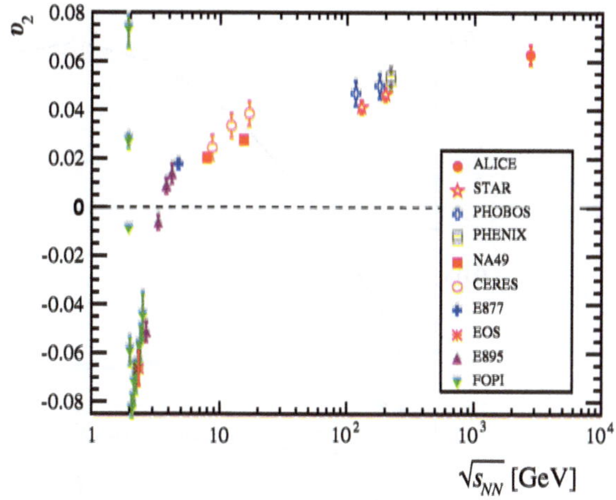

Fig. 9.8 Integrated elliptic flow at $\sqrt{s_{NN}} = 2.76$ TeV in Pb–Pb in the 20–30 % centrality class compared with results from lower energies taken at similar centralities. Adapted from [81]

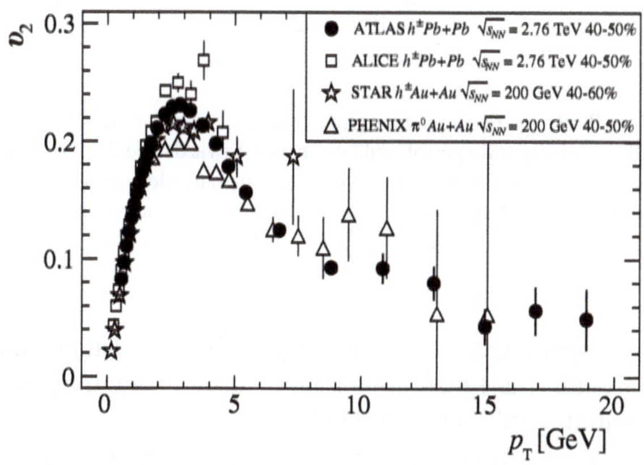

Fig. 9.9 v_2 versus hadrons' transverse momentum p_T at $|\eta| < 1$ in the 40–50 % centrality interval from the ATLAS collaboration, compared to previous experimental data: ALICE $v_2\{2\}$ for inclusive charged particles, PHENIX v_2 for identified π^0, and STAR data on $v_2\{2\}$ for inclusive charged particles for the 40–60 % interval. Adapted from [82]

in contrast to the strong variation in $v_2(\eta)$ observed by the PHOBOS experiment at $\sqrt{s_{NN}} = 0.2$ TeV [109], which decreases by approximately 30 % between $\eta = 0$ and $\eta = 2.5$.

9.3.4 Higher Moments

Measurements of higher flow harmonics provide deeper insight into the initial state of heavy-ion collisions. A combined analysis of different anisotropic-flow coefficients v_n allows the theoretical description of the initial state to be constrained and thereby provides a better starting point to determine properties of the created hot and dense nuclear matter. This will be discussed in more detail in Sect. 9.3.5.

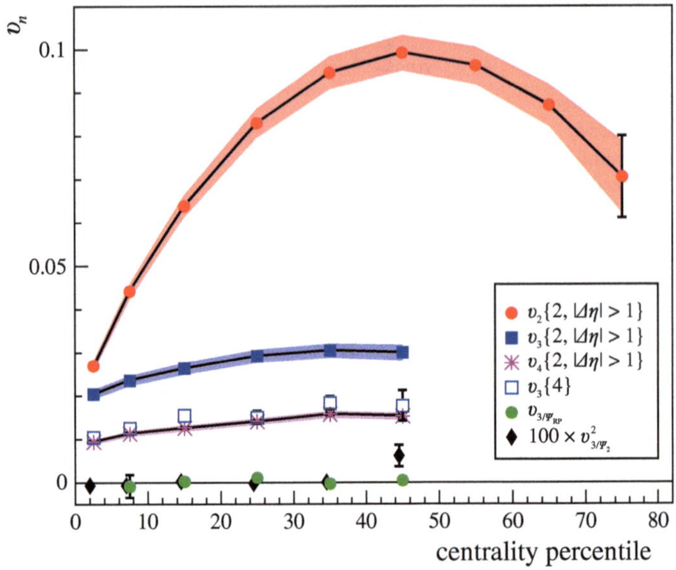

Fig. 9.10 v_2, v_3 and v_4 integrated over $0.2 < p_T < 5.0$ GeV as a function of event centrality, with the more central (peripheral) collisions towards the *left* (*right*). *Full* and *open squares* show the v_3 from two-particle correlations and four-particle correlations, respectively. Also shown are v^2_{3/Ψ_2} and $v_{3/\Psi_{RP}}$, which represent the triangular flow measured relative to the second-order event plane and the reaction plane, respectively. Adapted from [88]

Experiments at the LHC have produced detailed analyses of flow harmonics v_1 through v_6 as functions of transverse momentum p_T, pseudo-rapidity η, and collision centrality [88, 89, 110–113].

Odd harmonics are entirely driven by fluctuations in the initial state, most prominently by geometric fluctuations of nucleon positions. If the interaction region had an idealised smooth almond shape, odd moments would be zero due to symmetry. In Fig. 9.10, results from the ALICE collaboration [88] are presented for p_T-integrated v_2, v_3, and v_4 as a function of centrality. The gap in $\Delta\eta$ excludes short range-correlations that emerge from jets or resonance decays, reducing non-flow effects. Higher harmonics v_3 and v_4 are smaller than v_2, as is expected from the average shape of the interaction region. The third harmonic measured with respect to the impact-parameter direction, $v_{3/\Psi_{RP}}$, or with respect to the spatial plane of elliptic symmetry in each event, Ψ_2, is consistent with zero. This indicates that v_3 is entirely due to fluctuations and not correlated with the average geometry or with v_2.

In ultra-central collisions, where the average interaction region is almost round in the transverse plane, v_2 is also mostly fluctuation-driven, and v_3 can become larger than v_2. Results for v_n as a function of transverse momentum p_T in 0–0.2 % central collisions by the CMS collaboration are shown in Fig. 9.11. A characteristic crossing of $v_3(p_T)$ and $v_4(p_T)$ above $v_2(p_T)$ can be observed.

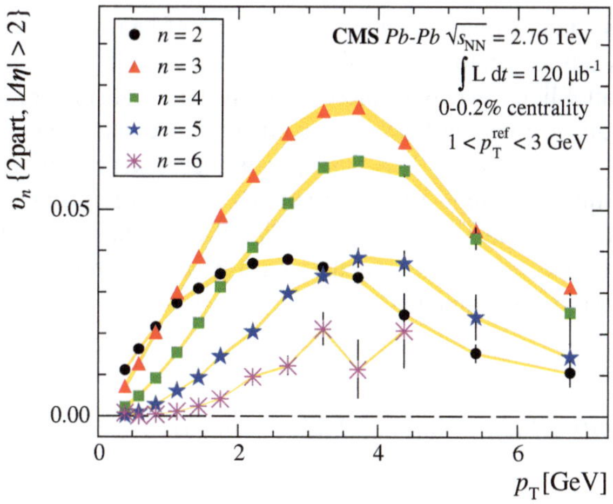

Fig. 9.11 Values of v_2 to v_6 as a function of p_T in 0–0.2 % central Pb–Pb collisions at $\sqrt{s_{NN}}$ = 2.76 TeV. Adapted from [112]

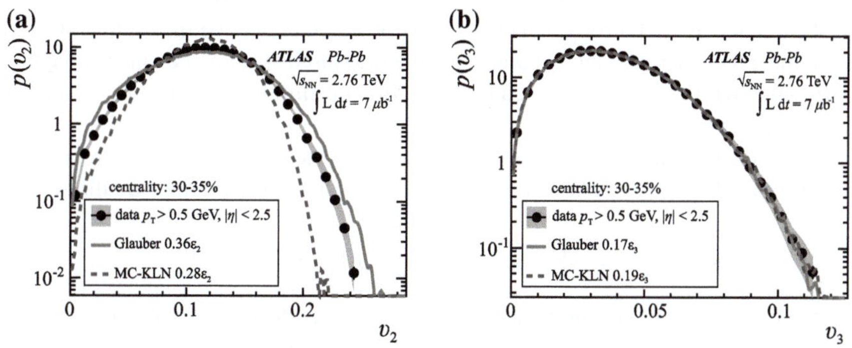

Fig. 9.12 The event-by-event v_2 (**a**) and v_3 (**b**) distributions compared with the eccentricity ε_n distributions from two initial-geometry models: a Monte-Carlo Glauber model (*solid lines*) and the MC-KLN model (*dashed lines*). The ε_n distributions were rescaled to have the same mean values as the v_n distributions. The scale factors are indicated in the legends. Adapted from [111]

The ATLAS collaboration has measured the event-by-event distributions of v_2, v_3, and v_4 [111]. Results for v_2 and v_3 for one given centrality are shown in Fig. 9.12. These distributions can be compared to distributions of the initial eccentricities ε_n of the interaction region. There is a direct correspondence between the two distributions for events that are not too peripheral [111, 114]. For more peripheral events, non-linear effects in the final-state dynamics of the system introduce differences between the ε_n and v_n distributions [115].

A comparison of the measured v_n distributions to theoretical calculations allows information on the initial state to be extracted, in particular because they are largely independent of the properties of the flowing medium. An example is the value of the ratio of shear viscosity to entropy density, η/s [114].

Another very interesting direction that the field of heavy-ion physics has taken is the recent study of proton-lead collisions and their relation to heavy-ion collisions. High-multiplicity events of these small-system collisions produce v_n values strikingly similar to those in heavy-ion collisions of comparable final-state particle multiplicity [92–94, 116]. This is surprising because the initial geometries are expected to be quite different. The observation has triggered strong interest in determining whether the same mechanism, namely collective flow, is responsible for the measured v_n in p-Pb collisions, or whether multi-particle correlations from the initial state can explain the phenomenon.

9.3.5 *Viscosity of the Produced Medium: η/s*

Since the discovery that the medium produced in heavy-ion collisions behaves like a strongly interacting fluid, both experimental and theoretical research has been focused on extracting quantitative properties of this fluid. The first question to ask is how perfect this fluid is. Quantitatively this can be expressed by the value of its ratio of shear viscosity to entropy density, η/s. This ratio can only be indirectly extracted from the experimental data by comparing the measured spectra and anisotropic flow harmonics with results from viscous fluid-dynamic simulations (see [21] and references therein).

In the ideal (non-viscous) case, the evolution of the system created in relativistic heavy-ion collisions is described by the following five conservation equations:

$$\partial_\mu T^{\mu\nu}_{\text{id}} = 0\,, \quad \partial_\mu J^\mu_B = 0,$$

where $T^{\mu\nu}_{\text{id}}$ is the energy-momentum tensor and J^μ_B is the net baryon current. These are usually re-expressed using the time-like flow four-vector u^μ as

$$T^{\mu\nu}_{\text{id}} = (\varepsilon + \mathcal{P})u^\mu u^\nu - \mathcal{P}g^{\mu\nu}\,, \quad J^\mu_B = \rho_B u^\mu,$$

where ε is the energy density, $\mathcal{P}$ is the pressure, ρ_B is the baryon density and $g^{\mu\nu} = \text{diag}(1, -1, -1, -1)$ is the metric tensor. The equations are then closed by adding the equilibrium equation of state

$$\mathcal{P} = \mathcal{P}(\varepsilon, \rho_B)$$

as a local constraint on the variables.

The appropriate equation of state can be determined from lattice QCD calculations [117–121] combined with hadron resonance gas models.

In the first-order or Navier–Stokes formalism for viscous hydrodynamics, the stress-energy tensor is decomposed into

$$T^{\mu\nu}_{\text{1st}} = T^{\mu\nu}_{\text{id}} + S^{\mu\nu},$$

where $T_{\mathrm{id}}^{\mu\nu}$ is given above and the viscous part of the stress-energy tensor is

$$S^{\mu\nu} = \eta\left(\nabla^{\mu}u^{\nu} + \nabla^{\nu}u^{\mu} - \frac{2}{3}\Delta^{\mu\nu}\nabla_{\alpha}u^{\alpha}\right),$$

where $\Delta^{\mu\nu} = g^{\mu\nu} - u^{\mu}u^{\nu}$ is the local 3-metric and $\nabla^{\mu} = \Delta^{\mu\nu}\partial_{\nu}$ is the local spatial derivative. Note that $S^{\mu\nu}$ is transverse with respect to the flow velocity since $\Delta^{\mu\nu}u_{\nu} = 0$ and $u^{\nu}u_{\nu} = 1$. Hence, u^{μ} is a (timelike) eigenvector of the whole stress-energy tensor with the same eigenvalue ϵ. η is the shear viscosity of the medium. The Navier–Stokes form is conceptually simple but introduces unphysical super-luminal signals that lead to numerical instabilities. The second-order Israel–Stewart formalism [122–124] avoids this super-luminal propagation, as do more recent approaches [125, 126].

In the Israel–Stewart formalism for a conformal fluid [127], the stress-energy tensor is decomposed as

$$T^{\mu\nu} = T_{\mathrm{id}}^{\mu\nu} + \pi^{\mu\nu}.$$

The evolution equations are

$$\begin{aligned}\partial_{\mu}T^{\mu\nu} &= 0,\\ \Delta_{\alpha}^{\mu}\Delta_{\beta}^{\nu}u^{\sigma}\partial_{\sigma}\pi^{\alpha\beta} &= -\frac{1}{\tau_{\pi}}\left(\pi^{\mu\nu} - S^{\mu\nu}\right) - \frac{4}{3}\pi^{\mu\nu}(\partial_{\alpha}u^{\alpha}),\end{aligned}$$

where for brevity we do not show bulk viscosity ζ, heat-flow, vorticity and terms that turn out to be numerically irrelevant.

We note that the Israel–Stewart formalism is based on two main choices: The 14-moment approximation to truncate the single-particle distribution function, and the choice of the equations of motion for the dissipative currents [126], which is not unique [128, 129]. Alternative derivations of second-order dissipative hydrodynamics [129] have been shown to achieve better agreement with microscopic kinetic-theory calculations [130], as do approaches allowing for an arbitrarily large momentum anisotropy [130–132].

Fluid dynamics does not in itself have a way to know when the system stops interacting strongly, or "freezes out". Hence, this process is usually implemented by introducing a freeze-out temperature below which particles are assumed to stream freely towards the detector. Alternatively, once the system becomes too dilute to be properly described by viscous-fluid dynamics, one can switch to a microscopic transport description. This is done by sampling the nearly thermal particle distributions in the fluid-dynamic simulation and then propagating the individual particles including their mutual interactions (see [21] and references therein).

9.3.5.1 Initial Conditions for Fluid-Dynamic Calculations

Fluid-dynamic simulations determine the evolution of a strongly interacting system given an initial condition for the energy momentum tensor. The accurate determination of the initial state of a heavy-ion collision is highly relevant for the determination of fluid properties from comparison with experimental data.

The initial geometry of the interaction region and its fluctuations strongly affect the generation of flow and the values of all v_n. It is therefore very important to have a reliable theoretical model. Important constraints on the initial-state model can be obtained by comparing all computed v_n coefficients to experimental data simultaneously and studying the event-by-event fluctuations of the flow harmonics. The latter are largely independent of the details of the fluid-dynamic evolution and thus reflect features of the initial state rather directly [114].

Different models exist that can generate fluctuating initial conditions for hydrodynamic simulations. One of the most commonly used models is the Monte-Carlo (MC) Glauber model [25]. In its simplest implementation, uncorrelated nucleons are sampled from measured density distributions. Then the two nuclei that are brought into collision are arranged according to a random impact parameter b and projected onto the transverse x–y plane, assuming straight-line trajectories for all nucleons. Interaction probabilities are then computed using the relative distance between two nuclei and the measured nucleon-nucleon inelastic cross section. Every wounded nucleon (sometimes also a fraction of binary-collision points) is then assigned an energy or entropy density, parametrised e.g. as a two-dimensional Gaussian in the transverse plane. The model can be improved by the inclusion of many-body correlations between the nucleons [133].

Another commonly used initial state model is the MC-KLN (Monte-Carlo Kharzeev–Levin–Nardi) model [134–138], which is based on the colour-glass condensate framework. Within this framework, one takes into account the well-established feature of QCD that at small Bjorken-x, a novel regime governed by large gluon densities and non-linear coherence phenomena takes over [139]. These high gluon densities correspond to strong classical fields, permitting calculations of the wave function using classical techniques. Quantum corrections are then incorporated via non-linear renormalisation group equations such as the Jalilian-Marian–Iancu–McLerran–Weigert–Leonidov–Kovner (JIMWLK) equations [140–145] or, in the large N_c limit, the Balitsky-Kovchegov (BK) equations [146] that describe the evolution of the wavefunction towards smaller x.

The MC-KLN model computes the initial energy-density distribution from a convolution of the unintegrated gluon distributions of the two colliding nuclei, after sampling nucleon positions in a way similar to the MC Glauber model. The model has been extended to include running-coupling (rc) BK evolution of the gluon distributions in the rcBK model [147]. MC-KLN-models, as the MC Glauber model, need additional negative binomial fluctuations of multiplicities for a given number of participants in order to be able to describe measured multiplicity distributions [148].

Another approach based on the colour-glass condensate is the IP-Glasma model [149, 150], which combines the impact-parameter dependent saturation (IP-Sat) model [151, 152] of high-energy nucleon (and nuclear) wavefunctions with the classical Yang–Mills (CYM) dynamics of the glasma fields (see Sect. 9.1) produced in a heavy-ion collision [153–158]. After fixing the free parameters of the IP-Sat model by fits to small-x HERA deep-inelastic scattering (DIS) data from electron-proton collisions and fixed-target nuclear DIS data [159, 160], the IP-Sat model provides an excellent description of these data. The IP-Glasma model includes fluctuations of nucleon positions as well as sub-nucleonic fluctuations of colour charges. It further includes the non-linear pre-equilibrium evolution of the initial gluon fields.

Starting the dynamical evolution at time zero after the collision leads to the build up of early-time flow and an independence of the exact time when one switches to a fluid-dynamic description [115]. However, not all features of the early-stage dynamics are included in the IP-Glasma model. The energy-momentum tensor of the glasma fields is very anisotropic so that when switching to a fluid-dynamic description, one implicitly assumes that some process causes instant isotropisation and thermalisation of the matter. Instabilities triggered by quantum fluctuations, and subsequent strong scattering of over-occupied fields, can lead to rapid isotropisation, but their inclusion requires a full $3+1$-dimensional simulation with a realistic description of quantum fluctuations. The IP-Glasma model only includes the $2+1$-dimensional transverse dynamics of the classical fields without quantum fluctuations. However, significant progress is being made [161–168], and the IP-Glasma model may be extended to include these important effects. Matching of the full stress-energy tensor, including viscous corrections and flow, to the hydrodynamic simulation will then be possible. Another possibility is to couple the IP-Glasma initial condition to anisotropic hydrodynamic simulations [131, 169–171]. The additional colour-charge fluctuations in the IP-Glasma model naturally lead to negative binomial fluctuations in the event-by-event multiplicity.

Apart from MC Glauber and CGC-based frameworks, several other Monte Carlo models are used to determine fluctuating initial conditions. These are, for example, UrQMD [172], EPOS [49], and AMPT [173, 174], which all incorporate Monte Carlo techniques to compute initial particle production and to then convert the soft part of the spectrum into the bulk energy-density distribution used in hydrodynamic simulations. They also provide initial flow and, in principle, the full stress-energy tensor including viscous corrections.

9.3.5.2 Fluid-Dynamic Description of LHC Data

Here, various calculations within the fluid-dynamic framework that have produced pre- (and post-) dictions for flow results at LHC energies will be presented. Depending on the initial conditions and details of the implementation, different values for η/s have been extracted.

Fluid-dynamic simulations that were fit to RHIC data have provided predictions for the spectra and anisotropic flow harmonics at LHC energies [175, 176]. These calculations were performed using $2+1$-dimensional viscous-fluid dynamics with $\eta/s = 0.08$ for smooth Glauber initial conditions and $\eta/s = 0.16$ for KLN initial conditions. A comparison of this calculation with experimental data from LHC indicates that v_2 can be described using the same η/s value as used at RHIC energies.

Ideal hydrodynamic-calculations coupled to a hadronic-cascade model also produce predictions for the mean transverse momentum and elliptic flow in heavy-ion collisions at LHC energies [177, 178]. These calculations show that the transverse-momentum spectra of hadrons become harder but that the p_T dependence of v_2 does not change appreciably between RHIC and LHC energies. This explains the previously-discussed increase in the experimentally observed integrated v_2 by the increase of the mean p_T.

In contrast to the viscous-fluid-dynamic calculations in [175, 176], simulations using similar 2+1-dimensional fluid dynamics, coupled to a hadronic cascade simulation [179], find best agreement with experimental v_2 data at RHIC for $(\eta/s)_{QGP} = 0.16$ and at LHC for $(\eta/s)_{QGP} = 0.2{-}0.24$. They also investigate a temperature-dependent η/s, but results for spectra and v_2 are insensitive to the exact shape of that temperature dependence.

Detailed calculations were also performed omitting the hadronic cascade [48]. Here the value of $\eta/s = 0.2$, which was the best fit for v_2 when using the MC-KLN model at RHIC energies, was used to predict v_2 at LHC energies. The experimental data are slightly over-predicted using this value.

In [180] identified-particle v_2 values are compared to ALICE data, and good agreement is found for $(\eta/s)_{QGP} = 0.16$. The authors note that the difference in η/s to the previous result of $(\eta/s)_{QGP} = 0.2{-}0.24$ [179] within the same simulation framework results from trying to match a calculation that does not include fluctuations to different measurements of v_2 that differ from one another due to the effect of fluctuations, as discussed above.

Including event-by-event fluctuations using an MC Glauber model leads to a reasonable agreement with experimental LHC data on v_2 for $\eta/s = 0.08$ in a 3+1-dimensional fluid-dynamic simulation [181].

Another 3+1-dimensional viscous-fluid-dynamic simulation without fluctuations [50] also finds reasonable agreement with experimental data for $p_T \leq 1$ GeV when using $\eta/s = 0.08$ and non-zero but small bulk viscosity $\zeta/s = 0.08$.

A range of possible effective values for η/s was estimated in [182], where v_n in ultra-central collisions were studied. This range is $0.07 \leq \eta/s \leq 0.43$. All values quoted in the literature (apart from ideal calculations where $\eta/s = 0$) fall within this range.

Employing fluctuating IP-Glasma initial conditions, v_1 through v_5 at both RHIC and LHC energies were studied in [115]. Best agreement with experimental data is achieved for $\eta/s = 0.12$ at RHIC and 0.2 at LHC, the striking feature being the excellent simultaneous description of v_2 through v_5. Comparisons of results from this model and experimental data from LHC [88, 89] are shown in Fig. 9.13.

Bulk viscosity has thus far been rarely included in fluid-dynamic simulations. Apart from the simulation [50] mentioned above that uses a constant ζ/s, bulk viscosity was studied in detail in [183], where it was found that its inclusion increases all v_n coefficients, opposite to the effect of shear viscosity.

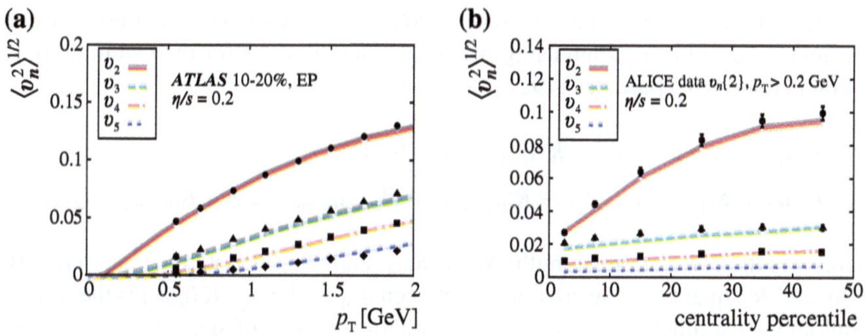

Fig. 9.13 **a** The transverse-momentum-dependent $v_n(p_{\rm T})$ coefficients in 10–20 % central events measured using the event-plane method by the ATLAS collaboration compared to calculations in the IP-Glasma+fluid-dynamics model. **b** The $p_{\rm T}$-integrated $v_n\{2\}$ coefficients as a function of centrality measured by the ALICE collaboration, compared to the calculated root-mean-square v_n from the IP-Glasma+fluid-dynamics model. Adapted from [115]

9.3.6 Directed Flow: v_1

The dipolar anisotropy quantified by v_1 has a contribution that is even in rapidity and generated by fluctuations, and a contribution that is odd in rapidity. The first contribution can be explained by collective flow much like the other v_n discussed above [184, 185].

The observed negative slope in rapidity of the second contribution has about three times smaller magnitude than that observed at the highest RHIC energy [186]. This suggests a smaller longitudinal tilt of the initial system and disfavours a picture of strong fireball rotation predicted at LHC energies [187, 188].

9.3.7 Femtoscopy

Hanbury–Brown–Twiss (HBT) interferometry, used to measure e.g. the size of a stellar emission source of photons [189], has been widely used in heavy-ion collisions to determine the space-time extent of the created fireball [190, 191]. In the heavy-ion case hadrons, mostly pions, are used. One measures a two-particle correlation function $C(\mathbf{q}) = A(\mathbf{q})/B(\mathbf{q})$, where $A(\mathbf{q})$ is the measured distribution of the difference $\mathbf{q} = \mathbf{p}_2 - \mathbf{p}_1$ between the three-momenta of two particles from the same event and $B(\mathbf{q})$ the corresponding distribution obtained using particles from different events. The momentum difference is computed in the longitudinally co-moving system, where the longitudinal pair momentum vanishes. $\mathbf{q}$ is decomposed into $(q_{\rm out}, q_{\rm side}, q_{\rm long})$, with the "out" axis pointing along the pair transverse momentum, the "side" axis perpendicular to it in the transverse plane, and the "long" axis along the beam.

The correlation function can then be fitted by an expression accounting for Bose–Einstein enhancement (when using boson correlations) and for the Coulomb interaction between two particles [192]:

$$C(\mathbf{q}) = \mathcal{N}[(1-\lambda) + \lambda K(q_{\mathrm{inv}})(1 + G(\mathbf{q}))]$$
$$G(\mathbf{q}) = \exp(-(R^2_{\mathrm{out}}q^2_{\mathrm{out}} + R^2_{\mathrm{side}}q^2_{\mathrm{side}} + R^2_{\mathrm{long}}q^2_{\mathrm{long}} + 2|R_{\mathrm{ol}}|R_{\mathrm{ol}}q_{\mathrm{out}}q_{\mathrm{long}})).$$

Here, λ is the correlation strength, R_{long}, R_{out}, and R_{side} denote the Gaussian HBT radii, and R_{ol} quantifies the cross term between q_{out} and q_{long}. $K(q_{\mathrm{inv}})$ is the squared Coulomb wave function averaged over a spherical source of size equal to the mean of R_{long}, R_{out} and R_{side}. q_{inv} is equal to $|\mathbf{q}|$ calculated in the pair rest frame.

First measurements of Pb–Pb collisions at $\sqrt{s_{NN}} = 2.76$ TeV at the LHC by the ALICE collaboration [193] found 10–35 % larger radii for the 5 % most central collisions compared to those measured by STAR in central Au–Au collisions at $\sqrt{s_{NN}} = 200$ GeV [194]. The measurement at LHC shows a continued trend of R_{long} growing with the cube root of the charged particle pseudo-rapidity density that was established previously at lower collision energies. R_{out} follows a similar trend with a smaller slope, while R_{side}, which is most closely related to the transverse size of the emission source, showed a flat behaviour with a shallow minimum between AGS and SPS energies, but is found to be larger at LHC energies. This behaviour can be interpreted as being due to the transition from baryon to meson dominance at freeze-out [195].

Various hydrodynamic model calculations [119, 196–199] as well as a hadronic kinematic-base model [200] are able to describe various trends of the experimental data. Some even reproduce all observed trends [193]. A comparison of ALICE results with HBT radii computed from the UrQMD transport model [201] indicates that the emission in the model is less explosive than that observed in the data.

The systematic behaviour of the product of the three radii, which is connected to the volume of the homogeneity region, is shown in Fig. 9.14 [193]. It shows a linear

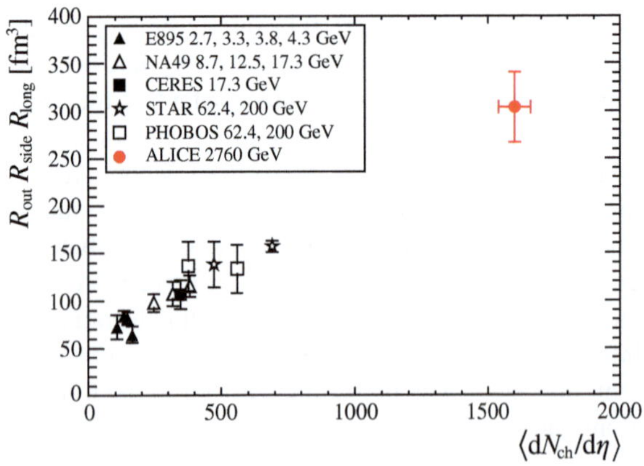

Fig. 9.14 Product of the HBT radii at pion transverse momenta $k_{\mathrm{T}} = 0.3$ GeV, obtained from two-pion correlations. The ALICE result (*filled circle*) is compared to results obtained for central Au–Au and Pb–Pb collisions at lower energies at the AGS, SPS, and RHIC. Adapted from [193]

dependence on the charged-particle pseudo-rapidity density and is two times larger at the LHC than at RHIC [39, 45, 193, 194, 202–208]. Not shown is the decreasing trend of HBT radii with increasing k_T, which is a characteristic feature of expanding particle sources since the HBT radii describe the homogeneity length rather than the overall size of the particle-emitting system [193].

Within hydrodynamic scenarios, the decoupling time for hadrons at mid-rapidity can be estimated in the following way: The size of the homogeneity region is inversely proportional to the velocity gradient of the expanding system. The longitudinal velocity gradient in a high-energy nuclear collision decreases with time as $1/\tau$ [209]. Therefore, the magnitude of R_{long} is proportional to the total duration of the longitudinal expansion, i.e. to the decoupling time of the system [210].

The decoupling time τ_f, extracted from R_{long} measurements at various collision energies is plotted as a function of the cube root of the charged-particle pseudo-rapidity density in Fig. 9.15 [39, 45, 193, 194, 202–208]. One finds a linear dependence and a maximum value of $\tau_f = 10$–$11\ \text{fm}/c$ in central Pb–Pb collisions at $\sqrt{s_{NN}} = 2.76$ TeV. Uncertainties due to transverse expansion and a finite pion chemical potential may lead to an underestimation of the actual decoupling time by ~25 % [211].

The ALICE collaboration has also studied three-pion correlations and finds a suppression of three-pion compared to two-pion Bose–Einstein correlations [212]. This seems to suggest a non-zero coherent component to pion production in heavy-ion collisions. It is significant at low momentum of the pion triplet while vanishing for high triplet momentum. This observation is qualitatively consistent with the formation of a Bose–Einstein condensate, which is expected to radiate coherently at low momentum. More experimental and theoretical work is needed to rule out alternative explanations.

In addition to the radius, more detailed information on the shape of the emission source can be obtained by performing azimuthally sensitive HBT measurements [213–216]. So far, no results on this have been published by the LHC experiments.

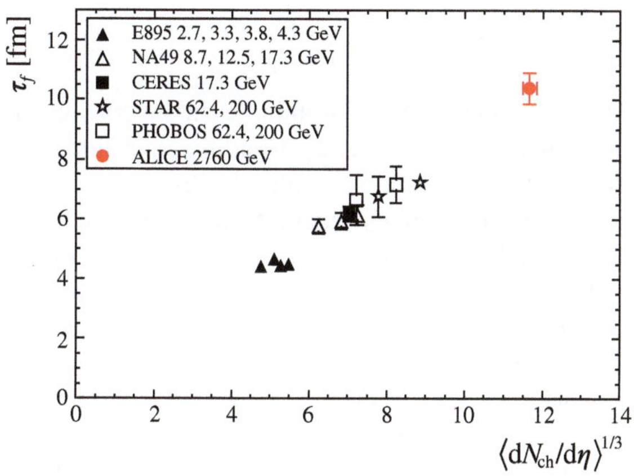

Fig. 9.15 The decoupling time extracted from $R_{\text{long}}(k_T)$. The ALICE result (*filled circle*) is compared to results obtained for central Au–Au and Pb–Pb collisions at lower energies at the AGS, SPS, and RHIC. Adapted from [193]

A comparison of the HBT radii in Pb–Pb, *p*-Pb, and *pp* collisions indicates similar radii in *pp* and *p*-Pb systems, while Pb–Pb collisions are typically larger at the same multiplicity [217].

9.4 Hard Probes

9.4.1 High Transverse-Momentum Processes

9.4.1.1 Introduction

The study of the products of hard scattering processes (large transverse-momentum transfer) between constituent partons (quarks and gluons) is another means of investigating the hot QCD medium that is formed in relativistic heavy ion collisions (see [218, 219]). The production rates of these hard processes are calculable in the Standard Model assuming QCD factorisation. Initial-state effects in heavy-ion collisions, e.g. gluon saturation and possible colour-glass effects, can also be investigated and compared to assumptions of nuclear parton distribution functions. After the initial parton-parton collision, scattered partons must traverse the coloured medium and hadronise (fragment). The interactions of partons with the medium and momentum exchange lead to parton-energy loss altering the final distributions of the fragmentation products. Understanding parton-energy loss and its consequences in a hot QCD medium requires detailed experimental and theoretical investigation. Modifications of the fragmentation of partons can be studied by comparing hard scattering results with those in elementary *pp* collisions. Since hard processes at the LHC occur at higher rates than at lower energies, jets, high momentum particles, particles containing heavy quarks, and quarkonia are attractive probes to study properties of the medium in heavy-ion collisions.

9.4.1.2 Techniques

Jets play a key role in determining the kinematics of the hard scattering of partons in particle physics. Typical jet-finding algorithms are utilised to find and determine the energy of jets in an event. However, the extraction of the parton kinematics from jet measurements becomes complicated in heavy-ion collisions, as there exists an underlying background of soft hadrons that fluctuates from event to event. This hadronic background depends on the impact parameter of the nuclear collisions, i.e. the nuclear overlap, which determines on average the number of binary nucleon-nucleon collisions and the particle multiplicity in a nucleus-nucleus event. Thus, precise determination of the energy of jets at lower transverse momenta $\sim$10–100 GeV becomes more complicated and must incorporate techniques for determining contributions from the soft background. Jet finding and jet measurements will be described in more detail in Sect. 9.4.1.4 below.

Leading particles and correlations (see Sect. 9.3) are often utilised to provide information on hard scattering in addition to, or as a substitute for, jets. The spectra of various types of particles produced in heavy-ion collisions can be compared to those in pp collisions, scaled by the number of binary nucleon-nucleon collisions in the nucleus-nucleus geometry. This is called the nuclear modification factor and is represented by

$$R^{i}_{AA}(p_{\mathrm{T}}) = \frac{\mathrm{d}^2 N^{i}_{AA}/\mathrm{d}p_{\mathrm{T}}\,\mathrm{d}\eta}{\langle T_{AA}\rangle \mathrm{d}^2\sigma^{i}_{NN}/\mathrm{d}p_{\mathrm{T}}\,\mathrm{d}\eta},$$

where N^{i}_{AA} is the yield for particle (or jet) type i in nucleus–nucleus collisions and σ^{i}_{NN} is the cross section for particle type i in nucleon-nucleon collisions. $\langle T_{AA}\rangle$, also known as the nuclear overlap function, is the ratio of the number of binary nucleon–nucleon collisions (typically calculated in the Glauber model) to the inelastic nucleon–nucleon cross section. R_{AA} is unity for the case where nucleus-nucleus collisions are a superposition of nucleon–nucleon collisions, i.e. in the absence of nuclear effects such as parton-energy loss or colour screening. A generalised version of this ratio has also been used where instead of comparison to nucleon–nucleon collisions the ratio is taken for central to peripheral nucleus–nucleus collisions and labelled R_{CP}.

In this section, results in heavy-ion collisions will be presented for leading particles (R_{AA}) and azimuthal correlations, jet spectra (jet R_{AA} and R_{CP}), jet-shape modifications, heavy-quark jets, dijet correlations and energy imbalance, and photon-tagged jet correlations.

9.4.1.3 Particle Yields at Large Transverse Momenta

In order to understand parton-energy loss, the nuclear modification factor R_{AA} is investigated in heavy-ion collisions as a function of p_{T} for particles as a function of collision energy $\sqrt{s}$ and centrality, and for different types of particles. Shown separately in Fig. 9.16 is R_{AA} for central collisions of heavy ions for charged hadrons and pions at SPS, RHIC and LHC energies. The three regimes are labelled and easily visible in the figure, with details in the legend. At SPS energies there is an enhancement, commonly attributed to the Cronin effect [221]. A large suppression is observed for the higher energies of RHIC and the LHC at intermediate p_{T} (2–20 GeV) with a gradual rise as p_{T} increases beyond 20 GeV. Various model calculations are also shown, with widely diverging predictions. Parameters used in the models are derived by fitting to the RHIC data, then extrapolated to the LHC energy. The predictions depend on ingredients in each of the models, in particular the parton density and parton energy-loss mechanisms such as scattering and radiation (see [222–226] for details). In all cases the suppression is a result of parton-energy loss in the dense medium.

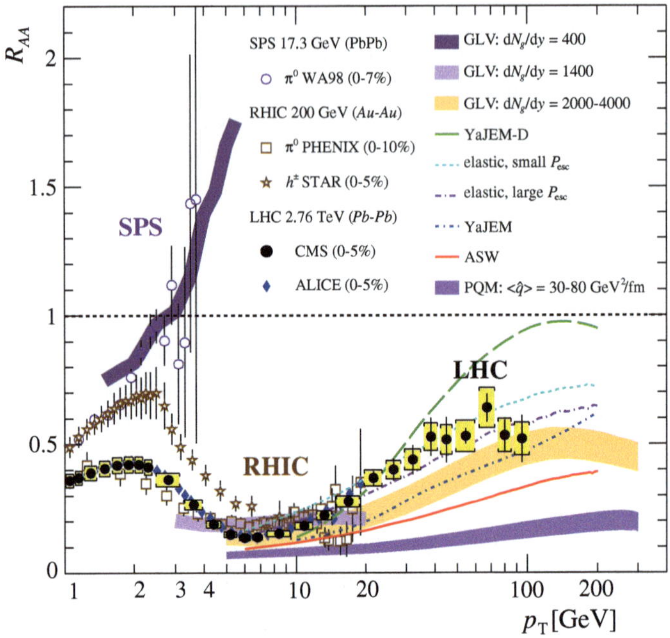

Fig. 9.16 R_{AA} for central collisions of heavy ions for charged hadrons and pions at SPS, RHIC and LHC energies as described in the legend. Also shown are various model predictions (see text for details). Adapted from [220]

In order to investigate potential cold nuclear matter effects in the initial state, p-Pb collisions were measured. Shown in Fig. 9.17 are $R_{p\text{-Pb}}$ for charged particles and $R_{\text{Pb-Pb}}$ for photons and for $W^{\pm}$ and Z^0 bosons. Note that, as expected, $R_{p\text{-Pb}}$ is consistent with a value of one for the non-strongly interacting gauge bosons. This is also the case for charged particles in p-Pb reactions, indicating no observation of initial state nuclear effects in the production of charged particles at mid-rapidity. Also shown in Fig. 9.17 is $R_{\text{Pb-Pb}}$ for charged particles for central collisions at the LHC, as presented in Fig. 9.16, which exhibits a strong suppression.

Looking further into the centrality dependence in Pb–Pb collisions, and comparing to p-Pb, one sees in Fig. 9.18a that the suppression in Pb–Pb is maximal for the most central collisions and decreases for peripheral (i.e. less central) collisions. Also shown, for reference, are the p-Pb data that exhibit no suppression. Investigating the particle-type dependence for hadrons, one sees in Fig. 9.18b that pions, kaons and protons behave differently, with a hierarchy of increased suppression for the lighter hadrons over $1 < p_T < 8$ GeV. The pions and kaons have a similar suppression above $p_T \sim 4$ GeV, whereas protons are less suppressed over the entire range $1 < p_T < 8$ GeV. It appears that in this p_T range there is a combination of overall strong suppression and a mass-dependent radial flow or various recombination effects in Pb–Pb that orchestrates the mass hierarchy observed in $R_{\text{Pb-Pb}}$.

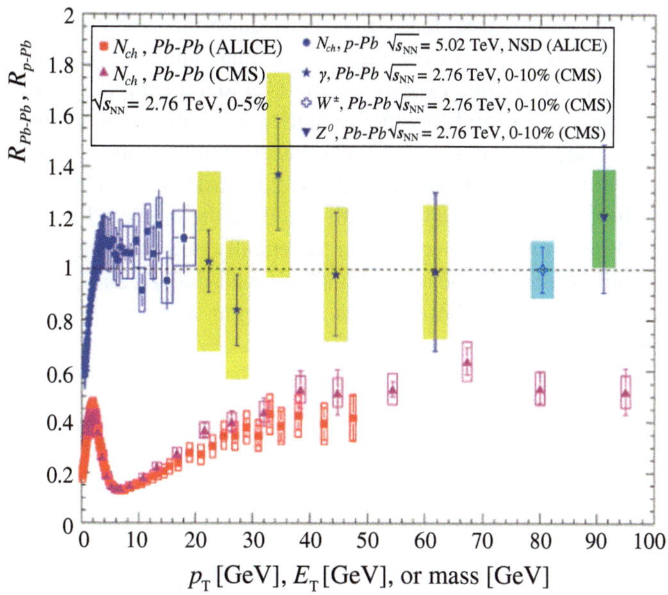

Fig. 9.17 R_{Pb-Pb} versus p_T for charged particles and for gauge bosons (γ, $W^{\pm}$and Z^0) in central collisions as given in the legend. Also shown are $R_{p\text{-}Pb}$ data for charged particles [227]. Compiled from [220, 227–231]

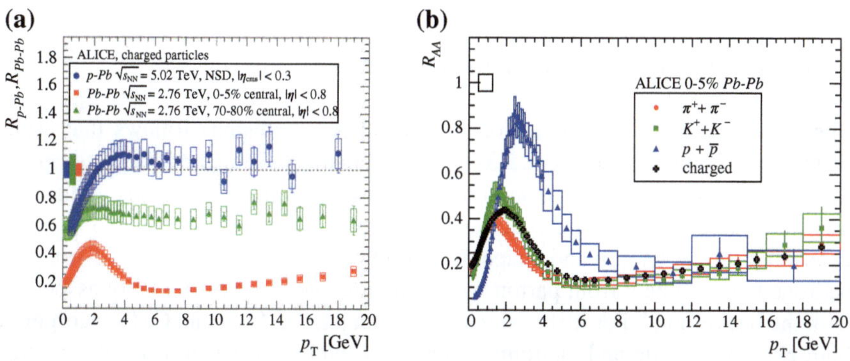

Fig. 9.18 **a** R_{Pb-Pb} for central and peripheral collisions and $R_{p\text{-}Pb}$. **b** R_{Pb-Pb} for π, K, p. In (**b**), the *box* at the *top left* represents the total normalisation uncertainty for *pp* and Pb–Pb collisions. Adapted from [53, 227]

In summary, the suppression observed for more central collisions in Pb–Pb and the R_{Pb-Pb} mass hierarchy in the range $1 < p_T < 8$ GeV coupled with the absence of suppression in *p*-Pb suggests the presence of a final-state effect, as expected from the parton-energy loss in the medium.

In order to understand the process of parton-energy loss in QCD at high temperatures in a fundamental way, its dependence on the properties of the parton and of the medium need to be addressed, leading to many questions. What is the dependence of parton-energy loss on properties (e.g. flavour, mass and energy) of the traversing parton that may be accessible experimentally (via identification of particles) in the final state? What is the extent to which radiative and collisional energy-loss mechanisms play a role? How does the energy loss depend on other properties of the medium, such as temperature? What happens to the lost energy? How is it dependent on the virtuality of the parton and its path length in the medium? Answers to these questions will need to be pursued both theoretically and experimentally, and they may come from differential measurements with well-defined and controlled initial- and final-state geometries. These studies have begun and will be described in the next section.

9.4.1.4 Jets

Measuring jets in heavy-ion collisions and comparing the results to pp and pA collisions allows more detailed and differential measurements at the parton level to be performed. Thus, the kinematics of the initial binary parton-parton scattering and the impact of the final state on partons traversing the medium can be measured via jets. Furthermore, dijet and γ-jet asymmetries, especially relative to initial geometries in terms of centrality and reaction plane, can provide more information about parton-energy loss.

Jet Reconstruction in the Heavy-Ion Environment

The technique of jet reconstruction in pA and AA collisions follows that established in pp collisions (see Sect. 5.3). However, there are some additional complications associated with the large hadronic background from soft particles that are not associated with the jets. This requires background subtraction in order to correctly determine the jet energy and shape. Clustering algorithms are used to distinguish hadrons that originate from parton fragmentation and those presumably associated with the background. For identifying products in jets, ATLAS and CMS incorporate both electromagnetic and hadronic calorimetry, while ALICE utilises electromagnetic calorimetry and charged-particle tracking to identify the hadronic component of jets. Clustering algorithms that are both infrared and collinear safe have been developed and are used to identify jets and to compare directly with theory. For jet finding the anti-k_t algorithm [232] as implemented in the FASTJET package [233, 234] has been used primarily. The soft background also fluctuates from event to event and is identified from the underlying event typically through the use of the k_t [235] or iterative-cone algorithms. Finally, the background is subtracted, detector efficiency and resolution corrections implemented, and the jet energy scale adjusted via data-driven and/or Monte Carlo techniques in order to derive a final jet energy spectrum [236]. Details of the implementation of these various techniques for jet finding in heavy-ion collisions can be found in [237].

Jet Spectra and Jet Quenching

The first jet measurements in Pb–Pb collisions at the LHC were unique for the field of heavy-ion collisions. Initially ATLAS [238] and then also CMS [29] presented results on the azimuthal angular correlation of dijets along with comparisons of the energy imbalance of the dijet pair. The ATLAS results are shown in Fig. 9.19. The energy asymmetry A_J, where $A_J = (E_{T1} - E_{T2})/(E_{T1} + E_{T2})$, is presented in Fig. 9.19a for $\sqrt{s_{NN}} = 2.76$ TeV Pb–Pb collisions and compared to $\sqrt{s} = 7$ TeV *pp* data and Monte Carlo simulations using PYTHIA jets embedded in HIJING events. T1 is defined as the highest-energy jet with transverse energy $E_{T1} > 100$ GeV and T2 the highest transverse energy jet in the opposite hemisphere with $E_{T2} > 25$ GeV. The Pb–Pb events exhibit a large asymmetry A_J with the away-side jet having lost considerable energy compared to that observed in the *pp* data and HIJING + PYTHIA simulations. This is a clear indication of jet quenching in Pb–Pb collisions at the LHC. In Fig. 9.19b, the azimuthal angular correlation $\Delta\phi$ between the leading jet and the opposite-side jet indicates that the jet angles, however, are not appreciably altered and that they exhibit the expected peak at $\Delta\phi = \pi$ as seen in the *pp* data and HIJING + PYTHIA simulations.

Jet spectra have been measured in $\sqrt{s} = 0.9, 2.76, 7$ and 8 TeV *pp*, $\sqrt{s_{NN}} = 5.02$ TeV *p*-Pb and in $\sqrt{s_{NN}} = 2.76$ TeV Pb–Pb collisions at the LHC. Displayed in Fig. 9.20 is the ratio R_{CP} of production in central compared to peripheral Pb–Pb collisions at $\sqrt{s_{NN}} = 2.76$ TeV for charged jets in ALICE [239] and calorimeter jets in ATLAS [240]. Jets are reconstructed with a value of the resolution parameter $R = 0.3$ that is utilized in the anti-k_t algorithm of FASTJET. Also displayed are the R_{CP} for charged particles [220] measured by ALICE and CMS. The R_{CP} for jets is observed to be significantly below one, and appears to increase slightly between the

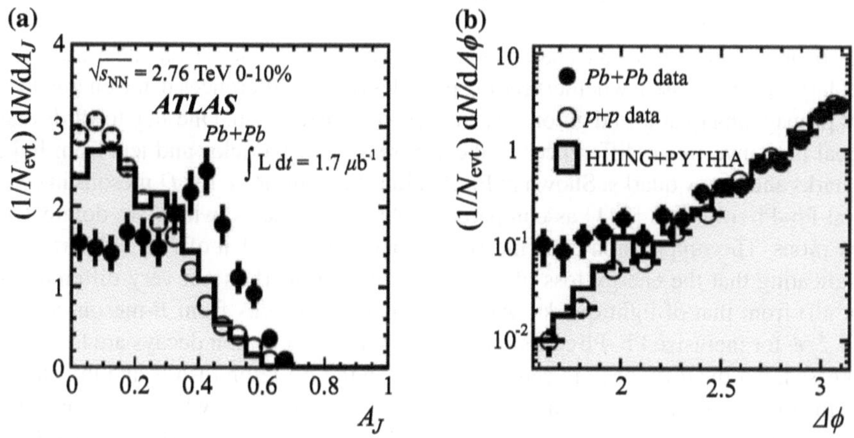

Fig. 9.19 **a** Dijet energy asymmetry A_J (defined in text) measured in $\sqrt{s_{NN}} = 2.76$ TeV central Pb–Pb events and in $\sqrt{s} = 7$ TeV *pp* events compared to simulations using HIJING + PYTHIA dijet events. **b** Corresponding azimuthal angular correlation between leading and next-to-leading jet on opposite-side. Adapted from [238]

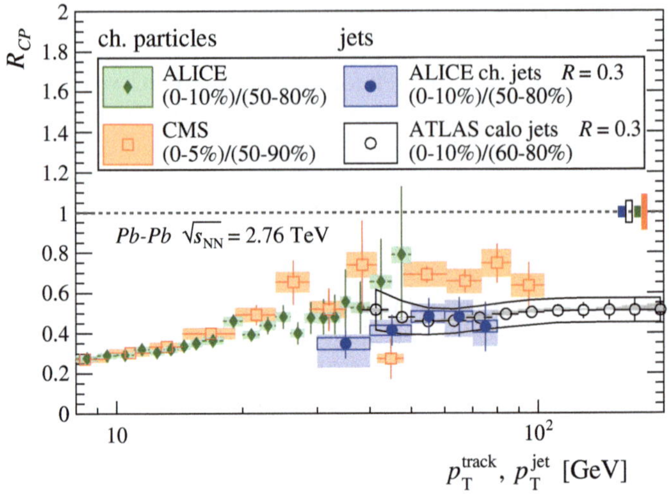

Fig. 9.20 R_{CP} for charged jets in ALICE [239] and calorimeter jets in ATLAS [240] in $\sqrt{s_{NN}}$ = 2.76 TeV Pb–Pb collisions compared to that for charged particles [220] from ALICE and CMS for the centralities given in the legend. Adapted from [239]

lowest measured jet p_T (30 GeV) up to jet transverse momentum $p_T = 200$ GeV. The jet R_{CP} has values similar to those of charged particles, indicating that the partons have interacted with the medium and that the resultant jet energy is redistributed outside the jet cone of $R = 0.3$ utilized by the anti-k_t algorithm in the jet-finding. Analyses of jet nuclear modification factors in p-Pb collisions at $\sqrt{s_{NN}} = 5.02$ TeV have found $R_{p\text{-Pb}}$ consistent with one over the entire range of measured jet momenta $25 < p_T < 800$ GeV. The $R_{p\text{-Pb}} \approx 1$ indicates no large initial state effects and suggests that the suppression in central Pb–Pb collisions is a final state effect, namely a result of jet quenching.

In order to determine the characteristics of parton-energy loss from measurements of large-p_T processes, whether leading particles or jets, differences in the energy loss of propagating quarks and gluons are important to distinguish. One step towards that goal is to investigate differences in the suppression of particles and jets from light quarks and heavy quarks. Shown in Fig. 9.21a is the ratio R_{AA} for D mesons in central Pb–Pb collisions [241] as compared to charged particles, which are dominated by pions. The suppression of D mesons is consistent with that of charged particles, indicating that the energy loss of the heavier charm quark is not very different (if at all) from that of light quarks. Also shown are the results from B-meson decays to J/ψ for inclusive Pb–Pb collisions [242, 243]. The B-meson decays are less suppressed, but the difference in centrality between the B-meson results and the others in the figure must be noted. Thus, it is not clear from this figure whether b quarks are less suppressed for a given collision centrality (which, to some extent, may reflect path length). Shown in Fig. 9.21b are the R_{AA} from b-quark jets and inclusive jets measured in central Pb–Pb collisions [244, 245]. This result clearly indicates that the

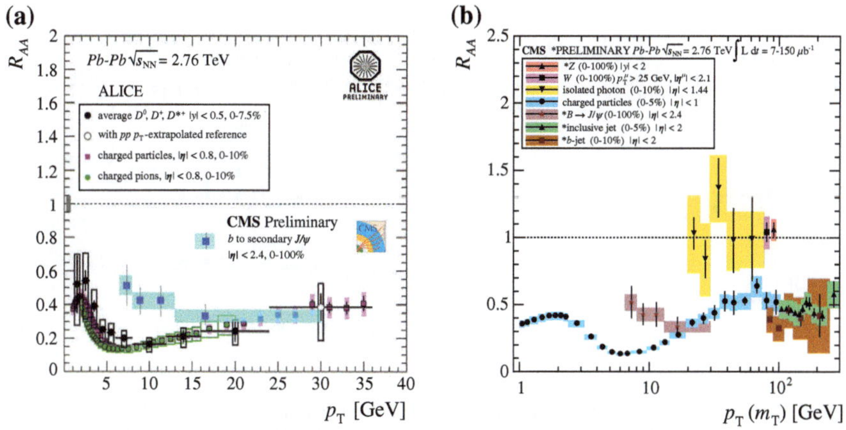

Fig. 9.21 **a** R_{AA} for D mesons and charged particles in central Pb–Pb collisions from ALICE, and B-meson decays to J/ψ for inclusive Pb–Pb collisions from CMS. **b** R_{AA} for various particles and jets (see legend) measured in central Pb–Pb collisions from CMS. Adapted from [220, 228–230, 241–246]

b-jet suppression is of similar magnitude to that of inclusive jets and suggests a lack of flavour dependence in the quenching mechanism at these energies.

Parton-energy loss in the QCD medium is expected to modify jet fragmentation in Pb–Pb collisions as compared to that in pp collisions. The patterns of jet fragmentation into charged particles can be investigated and compared to pp collisions to determine potential modifications. The fragmentation functions are measured as a function of the variable z or $\xi = ln(1/z)$, where $z = p/p_{\rm jet}$ is the momentum fraction of the jet of momentum $p_{\rm jet}$, that is carried by the detected charged particle of momentum p. The ratio of this fragmentation function in Pb–Pb collisions relative to pp collisions is shown in Fig. 9.22a for central Pb–Pb collisions. An enhancement is observed at large ξ. This represents more particles with low momentum fraction z in Pb–Pb as compared to pp collisions and indicates that jets are significantly modified in Pb–Pb compared to pp collisions, suggesting parton-energy loss in the medium.

Jet modification can also be investigated by studying jet shapes or the charged-particle density in jets in Pb–Pb collisions relative to that in pp collisions. The ratio of jet R_{CP} values for anti-$k_t$jet radii $R = 0.3$, 0.4 and 0.5 compared to those of $R = 0.2$ jets is presented in Fig. 9.22b as a function of $p_{\rm T}$ for 0–10 % central Pb–Pb collisions. These studies of the particle densities as a function of cone size for central Pb–Pb collisions exhibit enhanced particle densities for larger jet cones relative to pp [240]. Smaller jet cone sizes in Pb–Pb appear mostly like pp jets. These results exhibit definite broadening of the jet in central Pb–Pb collisions, as expected from interactions (including gluon bremsstrahlung) of the fast parton with the medium.

ATLAS, ALICE and CMS have measured positive values of v_2 for particles and jets at large transverse momenta in Pb–Pb collisions at $\sqrt{s_{NN}} = 2.76$ TeV. All

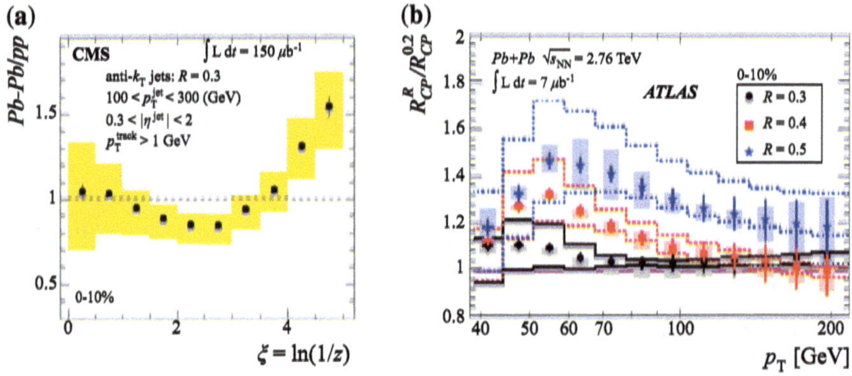

Fig. 9.22 **a** Fragmentation function ratios from CMS as a function of $\xi = ln(1/z)$ of jets as defined in the legend. **b** Ratios of jet R_{CP} values measured by ATLAS for jet radii $R = 0.3$, 0.4 and 0.5 compared to $R = 0.2$ jets as a function of p_T for 0–10 % central Pb–Pb collisions. Adapted from [240, 247]

three experiments agree well in the measurements of $v_2(4)$ for $p_T < 20$ GeV over the entire range of centralities [248, 249]. At larger momenta, CMS has measured a value of $v_2 \simeq 0.02$ over the range $20 < p_T < 50$ GeV for particle tracks in 0–5 % central collisions [249]. ATLAS finds $v_2^{\rm jet} \simeq 0.02$ over the range $50 < p_T < 200$ GeV for jets in 5–10 % central collisions [248]. ALICE measures a somewhat larger $v_2^{\rm jet}$ than ATLAS but with large uncertainties for charged jets in the range $50 < p_T < 100$ GeV [250]. Since hydrodynamics is not expected to play any direct role at these larger transverse momenta, a significant value of v_2 is highly indicative of a path-length dependence of parton-energy loss. The partons lose more energy in the out-of-plane direction than in-plane due to the elongation out-of-plane in the overlap region. This results in a positive value of v_2 as a result of the different path lengths and total energy loss of partons in- and out-of-plane.

Direct photons are not expected to interact strongly with the medium in Pb–Pb collisions. Thus, they are unmodified by the medium and can be used to determine the kinematics of the initial hard scattering. This is observed in Pb–Pb collisions where $R_{AA}^{\gamma} = 1.0$ within uncertainties signifying that the yields of isolated photons in Pb–Pb are a superposition of those measured in *pp* data. Photons at leading order are produced back-to-back with an associated parton with nearly identical transverse momenta. Thus, measurements of back-to-back photons and jets can be used to test jet quenching and the energy lost by the parton by comparison with the photon. Shown in Fig. 9.23 are $R_{J\gamma}$ of isolated prompt photons back-to-back with jets from CMS [251]. The data are for isolated photons with transverse momentum $p_T^{\gamma} > 60$ GeV and an associated jet with $p_T^{\rm jet} > 30$ GeV within a back-to-back azimuthal angular range $\Delta\phi_{J\gamma} > 7\pi/8$. Also shown are Monte Carlo simulations of hard scattering events from PYTHIA alone and using PYTHIA +HYDJET. *pp* data are shown as reference. The ratio $R_{J\gamma}$ is observed to decrease with increasing centrality ($N_{\rm part}$) and remains

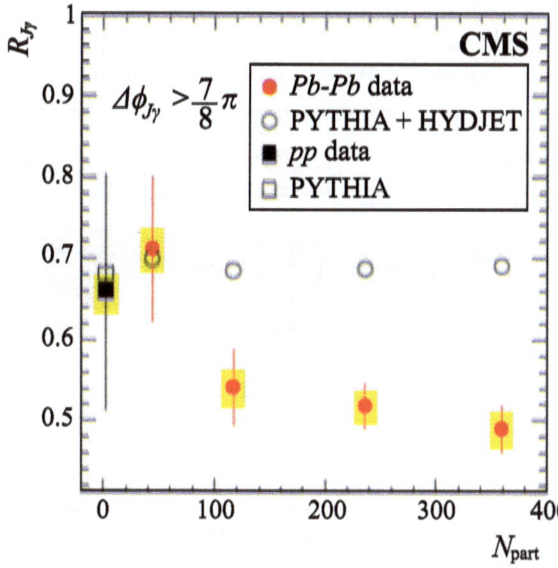

Fig. 9.23 $R_{J\gamma}$ of isolated prompt photons with back-to-back jets from CMS with $R = 0.4$. The data are shown as a function of centrality as represented by N_{part} for $\sqrt{s_{NN}} = 2.76$ TeV Pb–Pb collisions. See the text for details. Adapted from [251]

below the values from PYTHIA + HYDJET and the *pp* reference data, except in the least central Pb–Pb case. In addition, no angular broadening is observed beyond that seen in the *pp* data and the MC reference at all centralities (not shown). ATLAS has presented similar results for Z-jet back-to-back coincidence measurements [252]. Furthermore, the average ratio of jet transverse momentum to photon transverse momentum for the most central events was found by CMS to be $0.73 \pm 0.02(\text{stat.}) \pm 0.04(\text{syst.})$, which is lower than the value of 0.86 seen in the *pp* data and that predicted by PYTHIA + HYDJET at the same centrality [251].

It is important to understand where the energy ends up that is lost by partons as they traverse the medium. A novel approach is to measure and correlate charged particles with the momentum imbalance of dijets and centrality [29]. In a corresponding measurement, CMS found the dijet imbalance, which was presented above in Fig. 9.19 for ATLAS, to increase with centrality. It was observed that, as the dijet momentum imbalance increases, the fragmentation pattern on the opposite-side (in azimuth) of the leading jet becomes softer. This can be seen in Fig. 9.24. Plotted is the missing p_{T} of charged tracks projected onto the axis of the leading jet (jet with the largest p_{T} in the event) as given by

$$\not{p}_{\text{T}}^{\parallel} = \sum_i -p_{\text{T}}^i \cos(\phi_i - \phi_{\text{leading-jet}})$$

for tracks with $p_{\text{T}}^{\text{jet}} > 0.5\,\text{GeV}$ at mid-rapidity ($|\ \eta\ | < 2.4$) and jets with $p_{\text{T}}^{\text{jet}} > 120$ GeV at $|\ \eta\ | < 1.6$. In Fig. 9.24a, the momentum is balanced between the leading-jet hemisphere (negative values of $\not{p}_{\text{T}}^{\parallel}$) and opposite-side hemisphere (positive $\not{p}_{\text{T}}^{\parallel}$). Note that the leading-jet hemisphere primarily contains tracks with

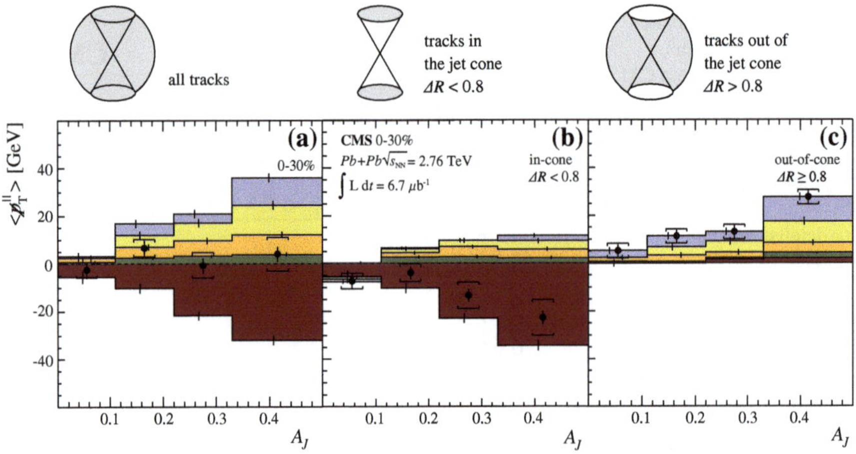

Fig. 9.24 Missing transverse momentum $\not{p}_{\mathrm{T}}^{\parallel}$ of charged tracks projected onto the leading-jet axis as a function of dijet momentum asymmetry A_J for central-collision event class 0–30 %. Average values are plotted as data points (>0.5 GeV), with shaded (colour where available) regions for contributions for each track p_{T} range (in GeV) listed from *top* to *bottom*: 0.5–1.0 (*light purple*), 1.0–2.0 (*yellow*), 2.0–4.0 (*orange*), 4.0–8.0 (*green*), and >8.0 (*brown*). **a** All charged tracks; **b** only charged tracks inside the jet cones of $\Delta R < 0.8$; **c** charged tracks outside the jet cones. Track selections are depicted relative to the jet cones indicated in the diagram above each panel. Adapted from [29]

$p_{\mathrm{T}} > 8$ GeV, while the opposite-side hemisphere primarily contains tracks with $p_{\mathrm{T}} < 8$ GeV. When only tracks inside the jet cones are included in the calculation of $\not{p}_{\mathrm{T}}^{\parallel}$ (Fig. 9.24b), there is an excess of track momentum carried by high-p_{T} tracks in the leading-jet direction with the excess increasing with A_J. For tracks outside the jet cones (Fig. 9.24c) a large excess of momentum appears at low p_{T} (<8 GeV) that increases with A_J.

In summary, results on dijets from ATLAS and CMS show that there is an increase in the dijet energy and momentum asymmetry as the centrality of the collision increases that is consistent with energy loss of the parton as it traverses a larger distance in the medium. CMS results indicate that there is a dijet momentum imbalance that increases with centrality of the collision and that this imbalance is recovered only when lower-p_{T} tracks down to $p_{\mathrm{T}} = 0.5$ GeV are included in the calculation of $\not{p}_{\mathrm{T}}^{\parallel}$.

9.4.2 Heavy-Flavour Production

9.4.2.1 Introduction

Heavy-flavour hadrons that carry charm or beauty quarks are unique probes for the QGP produced in heavy-ion collisions at the LHC [253]. Owing to the large quark masses ($m_c = 1.275 \pm 0.025$ GeV, $m_b = 4.18 \pm 0.03$ GeV [254]), heavy quark-antiquark pairs are produced predominantly in hard scattering processes of partons in the earliest phase of a heavy-ion collision. Their formation time ($\tau_{\text{form}} \simeq 1/2m_q < 0.1$ fm) is small compared to the timescale of QGP formation/thermalisation. Once produced, a small fraction ($\simeq$1–2 % in pp collisions, almost independent of $\sqrt{s}$) of the heavy quark-antiquark pairs hadronises into a bound colour-neutral quarkonium state, i.e. charmonium ($c\bar{c}$) or bottomonium ($b\bar{b}$). The fate of these states in heavy-ion collisions is discussed in Sect. 9.4.3. The vast majority of charm and beauty quark-antiquark pairs, however, fragment individually in two heavy-flavour jets that contain two open heavy-flavour hadrons composed of a heavy quark and a light antiquark (or vice versa). In-medium heavy-flavour production during later stages of the system evolution does not play a substantial role. Again the reason is the mass of $c\bar{c}$ and $b\bar{b}$ pairs, which is significantly larger than even the initial temperature of the QGP produced at the LHC so that thermal production rates are small. However, a measurement of thermal heavy-flavour production on top of the yields from initial production via hard scattering could provide a sensitive measure of the medium's initial temperature.

Heavy quarks experience the full evolution of the system while they propagate through the medium and interact with its constituents. Since the strong interaction preserves the quark flavour, the charm and beauty yields remain unaffected, i.e. they should scale with the number of binary collisions N_{coll} or with the nuclear overlap function in nuclear collisions, and only their phase-space distributions are modified. Two experimental observables related to the interaction of heavy quarks with the medium are investigated for heavy-flavour hadrons or their decay products as a function of p_{T}, the rapidity y, and the collision centrality: the nuclear modification factor R_{AA} and the azimuthal anisotropy quantified via the elliptic-flow coefficient v_2.

These two quantities are intimately connected with each other. Sufficient scattering of heavy quarks in the medium can reduce their momenta such that they eventually thermalise and participate in the collective expansion of the system. Therefore, the simultaneous measurement of R_{AA} and v_2 may provide access to the heavy-quark transport coefficients in the QGP.

In general, heavy quarks lose energy via radiative [255, 256] and collisional processes [257, 258] while traversing the dense medium, leading to a reduction of heavy-flavour yields at high p_{T}. However, differences between the energy loss of heavy quarks and light quarks are expected. Due to the so-called dead-cone effect [259, 260], the parton-energy loss via medium-induced bremsstrahlung, i.e. the emission of gluon radiation, depends on the parton mass. Due to a destructive interference, gluon radiation is suppressed in a cone around the direction of the parton's direction of motion. The opening angle of this cone increases with the ratio m/E of the parton mass and energy such that at a given parton momentum the radiative energy loss of heavy quarks is smaller than that for light quarks [259, 260]. The resulting mass hierarchy is expected to be most pronounced for momenta close to the quark masses and to disappear in the limit $m \ll E$. This mass ordering of heavy- and light-flavour energy loss is reinforced by the fact that, at LHC energies, light-flavour hadrons mostly originate from gluon fragmentation while heavy-flavour hadrons are produced via the fragmentation of heavy quarks. Since gluons carry a larger colour charge than quarks—an effect that is quantified by a Casimir coupling factor of 3 for gluons and 4/3 for quarks—gluons suffer a correspondingly larger radiative energy loss. Even though additional collisional energy loss, which results in a larger suppression of heavy quarks, should reduce the difference between heavy-quark and light-quark suppression, a mass hierarchy is expected for the parton-energy loss, ΔE_{parton}, in the kinematic regimes where the mass cannot be neglected with respect to the parton momentum: $\Delta E_{\text{gluon}} > \Delta E_{\text{LF}} > \Delta E_{\text{charm}} > \Delta E_{\text{beauty}}$ ("LF" stands for light quarks). When confronting the measured nuclear modification factors R_{AA} of light-flavour and heavy-flavour hadrons with this expected hierarchy, a number of additional caveats need to be considered, including the initial spectral shapes of the produced partons and their fragmentation functions, which are harder for heavy quarks compared to light quarks. Furthermore, at low p_{T} light-flavour hadrons are mostly produced via soft processes—which is not the case for heavy-flavour hadrons, which originate almost exclusively from hard partonic scattering processes.

The energy loss of heavy quarks contributes to the azimuthal anisotropy of their emission in a twofold way: Since the distance that heavy quarks travel through the medium while losing energy is smaller in the direction of the reaction plane than perpendicular to it, v_2 reflects the path-length dependence of heavy-quark energy loss [261, 262]. At low p_{T}, v_2 in addition is sensitive to the rescattering of heavy quarks in the medium. A positive heavy-flavour v_2 would indicate that heavy quarks participate in the collective motion of the medium, which is a prerequisite for thermalisation with the latter [31].

In order to fully exploit the potential of heavy-flavour measurements in heavy-ion collisions, it is mandatory to conduct additional reference measurements in elementary *pp* collisions and in proton(deuteron)-nucleus collisions, as it was done for the first time at RHIC at $\sqrt{s_{NN}} = 0.2$ TeV (see [253] for a review). Compared to these pioneering studies, the measurements conducted during Run 1 at the LHC could take advantage of the much larger heavy-flavour production cross sections at the LHC energy compared to RHIC. Furthermore, the LHC experiments profited from the availability of high-resolution vertex spectrometers, which allow the decay topology of heavy-flavour hadrons to be exploited, from the first day of running—an experimental feature not available to the RHIC experiments during their first decade of operations.

Proton-proton collisions not only provide a reference for heavy-flavour measurements in nuclear collision systems, they are also interesting in their own right. Since a hard scale is introduced by the quark mass, perturbative QCD methods are applicable even for the calculation of low-momentum heavy-quark production via hard scattering processes. Therefore, such measurements offer a critical testing ground for QCD in the new energy regime of the LHC.

In *p*-Pb collisions, effects of the presence of cold nuclear matter (CNM) on heavy-flavour observables can be investigated. Such studies are crucial for the heavy-ion programme because in Pb–Pb collisions CNM effects are present in the initial state. Most important are modifications of the parton distribution functions in nuclei with respect to those in nucleons such as shadowing or gluon saturation (see [263] for a review). In general, these modifications lead to a depletion of nuclear parton densities at the Bjorken-x values relevant for heavy-flavour production at the LHC. As a consequence, heavy-flavour production cross sections as a function of p_T may be reduced compared to pp cross sections (scaled with the appropriate number of binary collisions). In addition, the Cronin effect plays a role [264, 265]. Commonly, this is attributed to the so-called k_T broadening, i.e. multiple soft parton scatterings in the initial state before the hard scattering process leading to heavy-flavour production takes place. Finally, partonic energy loss both in the initial state and in the final state cannot be neglected completely [266, 267].

9.4.2.2 Experimental Approach

At the LHC, the only experiment capable of performing exclusive heavy-flavour measurements at mid-rapidity down to low values of p_T is ALICE. As this kinematic region is the most relevant for QGP studies in heavy-ion collisions, the following discussion is focused on ALICE, and unique contributions from the other LHC experiments are indicated when data are available.

The most direct access to heavy-flavour production is the exclusive reconstruction of heavy-flavour hadron decays, which is only possible through hadronic decay channels that, however, have typical branching ratios of only a few percent. ALICE has measured the production of various D-meson species down to low p_T at mid-rapidity in pp [268–270], p-Pb [270, 271], and Pb–Pb collisions [241, 272, 273] via the decays [254] $D^0 \rightarrow K^-\pi^+$, $D^+ \rightarrow K^-\pi^+\pi^+$, $D^{*+} \rightarrow D^0\pi^+$ (strong decay), and $D_s^+ \rightarrow \phi\pi^+ \rightarrow K^+K^-\pi^+$ and their charge conjugates. These measurements exploit the decay topologies and take advantage of the lifetimes of the weakly decaying heavy-flavour hadrons. The measured inclusive D-meson yields include contributions from the fragmentation of charm quarks ("prompt" D mesons) and feed down from B-meson decays ("non-prompt" D mesons). The latter contribution is usually estimated and subtracted from the inclusive D-meson yields. Mid-rapidity exclusive heavy-flavour hadron measurements with ATLAS [274, 275] and CMS [276–278] are restricted to higher p_T values in pp and p-Pb collisions. At forward rapidity, LHCb has measured the production of various heavy-flavour meson species in pp collisions [279–281].

An alternative approach is the measurement of electrons or muons from semi-leptonic decays of heavy-flavour hadrons. The relatively large branching ratios of typically 10 % and the missing combinatorial background in single-particle measurements is advantageous. However, because of the incomplete reconstruction of the decaying heavy-flavour hadron, the kinematics of the decay lepton is well correlated with the kinematics of the heavy quark only at high momenta. Furthermore, background from other lepton sources has to be measured independently. For electrons, Dalitz decays of light neutral mesons, mainly the π^0 and η, and conversions of photons in the detector material are the most important background sources. For muons, weak decays of pions and kaons and, at high momenta, decays of W and Z bosons are relevant. ALICE has measured electron production from heavy-flavour decays down to low p_{T} at mid-rapidity and the corresponding muon production at forward rapidity in pp [282–287], p-Pb [288], and Pb–Pb collisions [287, 289]. Up to now, ATLAS [290] and CMS measurements [291, 292] of leptons from heavy-flavour decays at mid-rapidity are published only for pp collisions, and they are mostly restricted to higher values of p_{T} ($p_{\mathrm{T}} \gtrsim 5\,\mathrm{GeV}$).

Finally, beauty-hadron production can be investigated via the measurement of non-prompt J/ψ mesons, i.e. J/ψ mesons not originating from the primary collision vertex but from the displaced decay vertex of B hadrons. Such measurements of beauty production have been conducted by ALICE [293], CMS [242, 294] and LHCb [295].

9.4.2.3 Heavy-Flavour Production in *pp* Collisions

Heavy-flavour production has been investigated in pp collisions at $\sqrt{s} = 2.76$ and 7 TeV in all channels mentioned above as a function of p_{T} and y. Figure 9.25 summarizes four results from pp collisions at 7 TeV most relevant to heavy-ion collisions. Shown as a function of p_{T} are the mid-rapidity production cross sections of D^+ mesons measured by ALICE in the channel $D^+ \rightarrow K^-\pi^+\pi^+$ [268] (Fig. 9.25a) and of electrons from heavy-flavour hadron decays measured by ALICE [282] and ATLAS [290] (Fig. 9.25c), as well as the production cross section of muons from heavy-flavour hadron decays measured by ALICE at forward rapidity [286] (Fig. 9.25d). The cross section of non-prompt J/ψ mesons originating from B-hadron decays as measured by ALICE [293], CMS [294], and LHCb [295] as a function of rapidity is shown in Fig. 9.25b). The data are described by perturbative QCD calculations going beyond leading order within the combined experimental and theoretical uncertainties, in particular by calculations within the fixed-order next-to-leading-log (FONLL) [296–298], the general-mass variable flavour numbering scheme (GM-VFNS) [299, 300], and the k_{T} factorisation [301–304] pQCD implementations (the latter one is not shown).

The main uncertainties limiting the precision of theoretical calculations arise from assumptions on the factorisation and renormalisation scales, μ_F and μ_R, as well as the quark masses and the parton distribution functions used in the calculations. While the theoretical uncertainties are small at high p_{T}, they grow substantially towards low p_{T} where precision data have the potential to constrain model parameters already now or in the near future. With the availability of heavy-flavour measurements in pp collisions and with their successful theoretical description, it is the first time that heavy quarks can be used to their full potential as calibrated probes in heavy-ion collisions in all channels. Cross sections measured in pp collisions at $\sqrt{s} = 7$ TeV have been extrapolated [305] to the Pb–Pb and p-Pb collision energies of $\sqrt{s_{NN}} =$ 2.76 and 5.02 TeV, respectively, employing the $\sqrt{s}$ evolution of corresponding cross sections in pQCD calculations in order to construct the necessary pp references for the measurement of heavy-flavour nuclear modification factors. This procedure was validated via the comparison of the extrapolated references with measurements in pp collisions at $\sqrt{s} = 2.76$ TeV [269, 283, 287].

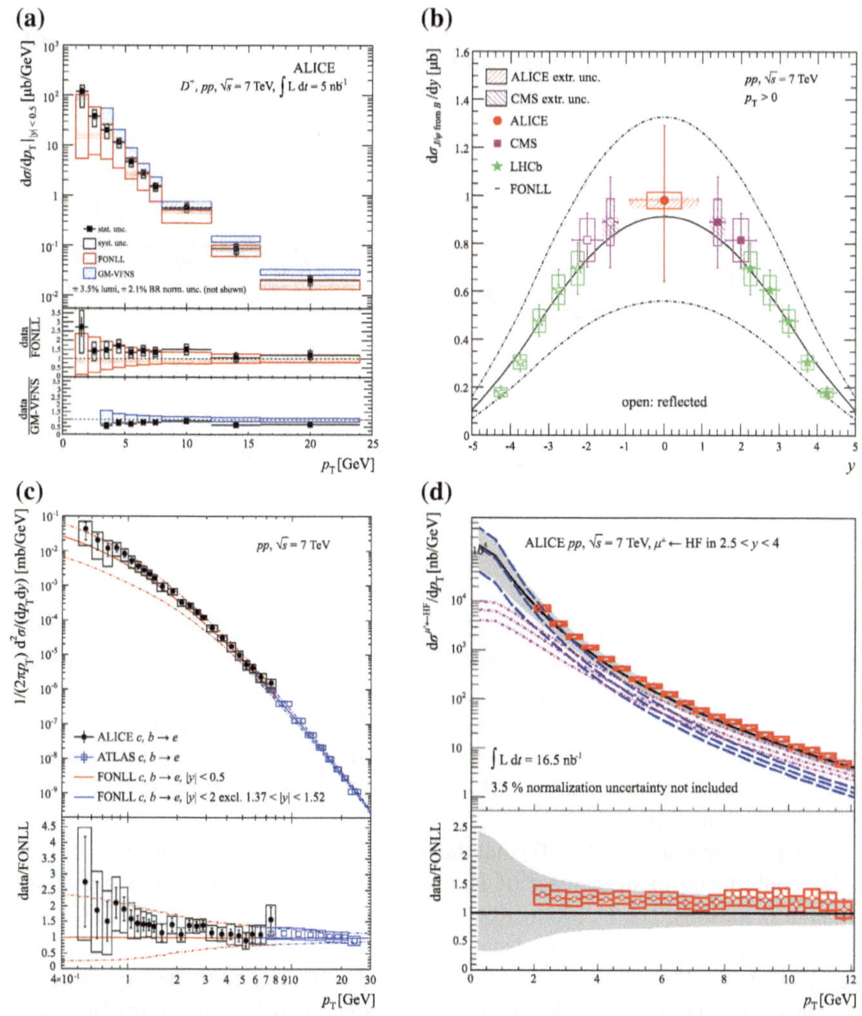

Fig. 9.25 Heavy-flavour production cross sections measured in pp collisions at $\sqrt{s}$ = 7 TeV as a function of p_T or rapidity in various channels: **a** cross section of D^+production at mid-rapidity; **b** cross section of non-prompt J/ψ mesons from B-hadron decays; **c** cross section of electrons from heavy-flavour hadron decays at mid-rapidity; **d** cross section of muons from heavy-flavour hadron decays at forward rapidity. The data are compared with results from FONLL [296–298] and GM-VFNS [299, 300] pQCD calculations. Adapted from [268, 282, 286, 293]

9.4.2.4 Nuclear Modification Factors in Pb–Pb and p-Pb Collisions

The nuclear modification factor R_{AA} of D mesons fully reconstructed via their hadronic decays has been measured for the first time in heavy-ion collisions with ALICE in Pb–Pb collisions at $\sqrt{s_{NN}}$ = 2.76 TeV as a function of transverse

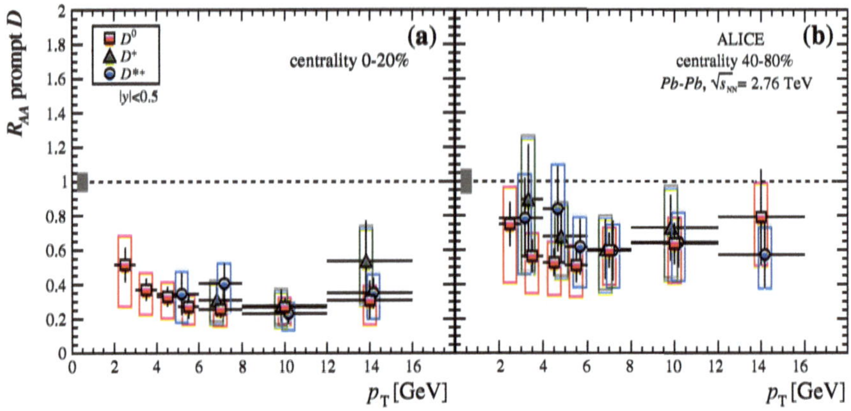

Fig. 9.26 R_{AA} of prompt D^0, D^+ and D^{*+} mesons in **a** 0–20 % and **b** 40–80 % central Pb–Pb collisions measured with ALICE at $\sqrt{s_{NN}} = 2.76$ TeV [241]. The feed down from B-hadron decays has been subtracted. Error bars (*boxes*) represent the statistical (systematic) uncertainties. Adapted from [241]

momentum and collision centrality [241]. As shown in Fig. 9.26, R_{AA} is similar for the D-meson species D^0, D^+ and D^{*+}, and a strong suppression by a factor 3–4 is observed at high p_T in central collisions. Towards lower p_T, the D^0 nuclear modification factor shows a tendency towards less suppression. Also prompt D mesons are less suppressed for more peripheral collisions. These observations are qualitatively consistent with a scenario in which charm quarks suffer energy loss in the QGP before fragmentation.

It was suggested in [306, 307] that the presence of abundant strange quarks in the QGP could result in an increase of D_s production with respect to the yield of non-strange D mesons. The D_s production cross section was measured with ALICE in pp collisions at $\sqrt{s} = 7$ TeV in the decay channel $D_s \rightarrow K^+K^-\pi^+$ [270]. Preliminary results from Pb–Pb collisions [308] suggest that the D_s R_{AA} could be larger than that of non-strange D mesons, which would be consistent with the conjecture of enhanced D_s production. The experimental uncertainties, however, are still too large to draw a firm conclusion.

In order to exclude the possibility that the observed charm suppression in Pb–Pb collisions is due to the presence of the Pb nuclei in the initial-state, charm production was measured in p-Pb collisions at $\sqrt{s} = 5.02$ TeV [271]. The resulting nuclear modification factor $R_{p\text{-Pb}}$ in p-Pb collisions is compared with R_{AA} of prompt D mesons [241] as a function of p_T in central and semi-peripheral Pb–Pb collisions in Fig. 9.27a. The suppression observed in Pb–Pb collisions is not present in the p-Pb system, and cold nuclear matter effects are small. This confirms that charm suppression in Pb–Pb collisions is a consequence of the interaction of charm quarks with the hot and dense QGP.

$R_{p\text{-Pb}}$ is consistent with unity for $p_T > 2$ GeV. In Fig. 9.27b, predictions from models which include cold nuclear matter effects are compared with the

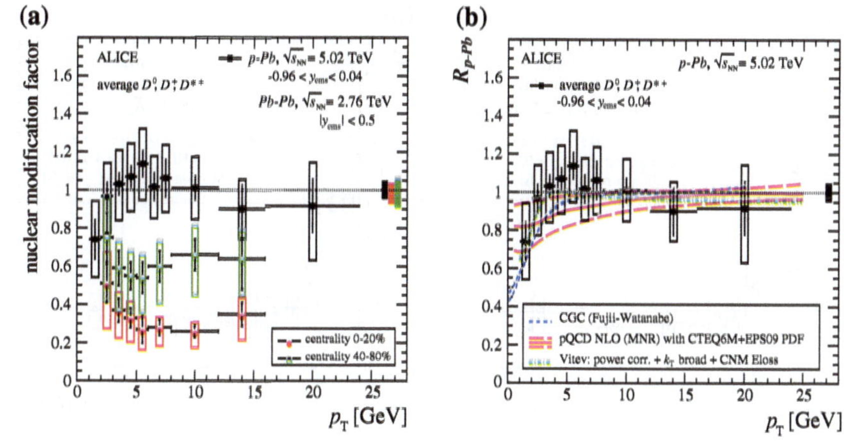

Fig. 9.27 **a** Comparison of $R_{p\text{-Pb}}$ with corresponding nuclear modification factors measured in central and semi-peripheral Pb–Pb collisions [241]. **b** Average $R_{p\text{-Pb}}$ of prompt D mesons in minimum-bias p-Pb collisions [271] as a function of p_T compared to predictions from models [309–312] including cold nuclear matter effects. Adapted from [271]

measurement. A model implementing charm production via NLO pQCD processes [309] and including the EPS09 prescription [310] of the nuclear modification of parton distribution functions in cold nuclear matter describes the measurement within uncertainties. So does another calculation that considers initial-state effects only, which is based on the colour-glass condensate [311]. A third model [312] including final-state energy loss in cold nuclear matter in addition to nuclear shadowing and k_T broadening in the initial state is also in good agreement with the data.

Qualitatively the same observations—namely a strong suppression of yields at high p_T in central Pb–Pb collisions with respect to cross sections in pp collisions scaled with the nuclear overlap function and the absence of large cold nuclear matter effects—have also been made in other heavy-flavour channels. Examples are electrons [289] (muons [287, 288]) from heavy-flavour decays at mid- (forward) rapidity. Immediately, the question arises whether the expected energy-loss hierarchy of partons in the QGP is reflected in the measured nuclear modification factors of the various probes mentioned above. This issue is addressed in Fig. 9.28, which shows as a function of p_T in central Pb–Pb collisions a comparison of the relevant nuclear modification factors. In Fig. 9.28a the R_{AA} of prompt D mesons [241] is compared with that of charged particles [231] (mostly pions that in the p_T range shown are dominantly produced via gluon fragmentation at the LHC). The fact that the radiative energy loss of gluons and light quarks should be larger than that of charm quarks is not obvious from R_{AA} which is quite similar for the two particle species with a hint of a slightly larger R_{AA} of D mesons compared to that of charged particles. When the high-p_T suppression of heavy flavour was first discovered via the measurement of electrons from heavy-flavour decays in Au–Au collisions at $\sqrt{s_{NN}} = 0.2$ TeV at RHIC [313], the quantitative similarity with light-flavour hadron suppression

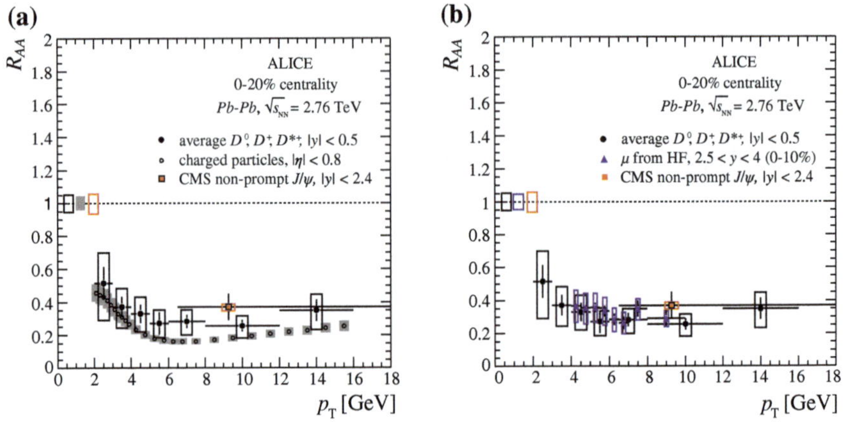

Fig. 9.28 Average nuclear modification factor of prompt D mesons [241] as a function of p_T in central Pb–Pb collisions compared with **a** R_{AA} of charged particles [231] and of non-prompt J/ψ mesons from B-hadron decays [242] and **b** with muons from heavy-flavour decays measured at forward rapidity [287]. Adapted from [241]

could not be explained theoretically by the models available then, which included in-medium gluon radiation as the only energy-loss mechanism of partons in the QGP. Many recent models include also collisional energy loss, which has in general improved the simultaneous description of the charm and charged-particle R_{AA}. A comparison of the D-meson data with current models will be discussed in more detail in Sect. 9.4.2.6.

Of particular importance is also the comparison of charm and beauty probes because a mass hierarchy is expected, i.e., the heavier beauty quarks should suffer less energy loss in the QGP because of the dead-cone effect. B-meson yields have not yet been measured in exclusive channels in Pb–Pb collisions and, consequently, other approaches are needed. The first indication for beauty suppression at the LHC came from muons [287] (and also electrons [289]) from heavy-flavour decays, a dominant fraction of which at high p_T ($p_T \gtrsim 5$ GeV) originates from beauty-hadron decays. Since then, the measurement of non-prompt J/ψ mesons, which are almost exclusively produced via B-hadron decays at all p_T, gave more direct access to beauty suppression [242]. The comparison of the R_{AA} of D mesons [241], heavy-flavour decay muons [287] and non-prompt J/ψ mesons [242] shown in Fig. 9.28b, however, has to be interpreted carefully. Since both the muons and the non-prompt J/ψ mesons with a given p_T are decay products from beauty hadrons at higher p_T, their R_{AA} cannot be directly compared with that of D mesons at the same p_T. While in Fig. 9.28 only a hint for a larger R_{AA} for beauty compared to charm is visible, recent results based on higher statistics data [243], taking properly into account the decay kinematics of beauty hadrons to J/ψ mesons, demonstrate that indeed the R_{AA} of beauty is larger compared to that of charm, consistent with the expected mass hierarchy of heavy-quark energy loss in the QGP.

9.4.2.5 Azimuthal Anisotropy of Heavy-Flavour Production in Pb–Pb Collisions

As discussed above, in-medium energy loss and the azimuthal anisotropy of particle emission are intimately related. Sufficient re-scattering of heavy quarks in the QGP inevitably will result in a participation of the latter in the collective expansion of the medium, and eventually thermalisation may be reached. The observable that reflects the collective expansion is the elliptic-flow coefficient v_2. The elliptic flow of charm was investigated with ALICE via the measurement of v_2 of D^0, D^+ and D^{*+} mesons [272, 273] using the event-plane [33], scalar product [86], and two-particle cumulant methods [315]. The values of v_2 of the different meson species were found to be consistent with each other within the uncertainties. As an example, $v_2(p_T)$ of D^0 mesons is compared with the corresponding charged-particle v_2 [314] for 0–10, 10–20 and 30–50 % central Pb–Pb collisions in Fig. 9.29. A significant, positive v_2 is observed, which is quantitatively consistent with the charged-particle v_2. An increase of v_2 towards more peripheral collisions is observed both for charged particles and for charm, which is consistent with the on-average larger initial eccentricity of the collision zone in peripheral collisions as compared to central collisions. These data constitute the first direct measurement of elliptic flow of charm and demonstrate that charm quarks participate in the collective dynamics of the QGP. The D-meson v_2 measurements are complemented by corresponding measurements of electrons [316] and muons [317] from heavy-flavour decays at mid and forward rapidity, respectively, which show quantitatively similar effects as previous measurements of semileptonic heavy-flavour decays at RHIC [318].

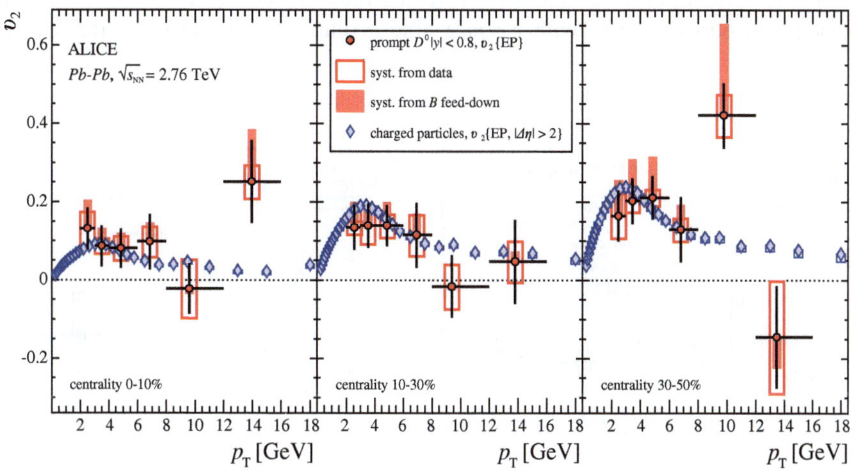

Fig. 9.29 Prompt D^0-meson [273] and charged-particle v_2 [314] as a function of p_T in three Pb–Pb centrality classes as measured with ALICE. Adapted from [273]

9.4.2.6 Heavy-Flavour R_{AA} and v_2 Versus Models

Heavy-flavour measurements have reached a precision with which they start to constrain models that aim at the description of transport properties in the QGP. In Fig. 9.30 the average D-meson nuclear modification factor and elliptic-flow coefficient measured with ALICE [241, 273] are compared with predictions from models calculating simultaneously both R_{AA} and v_2. Similar comparisons have been made for electrons and muons from heavy-flavour decays, with similar conclusions.

The Wicks, Horowitz, Djordjevic and Gyulassy (WHDG) model [319, 320] is a perturbative QCD calculation including both radiative and collisional energy loss on the parton level and a realistic collision geometry. The hydrodynamical expansion is omitted so that the D-meson v_2 is underestimated while R_{AA} is described reasonably well. Another pQCD-based model, which gives a v_2 close to the data but overestimates R_{AA} at low p_T, is the hybrid MC@sHQ+EPOS approach [321], which combines heavy-quark energy loss via collisional and radiative mechanisms with a fluid-dynamical expansion of the medium based on the EPOS model [49, 322]. Another class of models are transport models. POWLANG [323, 324] is based on collisional processes implemented using Langevin dynamics. Relativistic viscous hydrodynamics is used to describe the expanding medium. This model predicts too large high-p_T suppression and too small v_2. Also in UrQMD [325–328] heavy-quark transport is treated using the Langevin approach, and the medium evolution is modelled by a combination of ideal hydrodynamics and hadronic transport. Model parameters were tuned to heavy-flavour measurements at RHIC and were kept for the calculations at the LHC, which fail to reproduce the suppression at low p_T but are otherwise in good agreement with the data. The Cao, Qin and Bass model [329] uses quasi-elastic scattering and radiative energy loss in a Langevin-based transport approach with a space-time evolution of the medium treated via

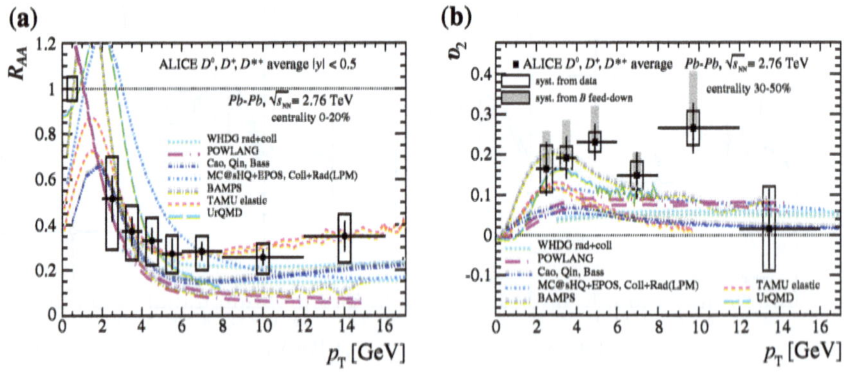

Fig. 9.30 Comparison of model calculations with **a** the average D-meson R_{AA} in 0–20 % central Pb–Pb collisions [241] and **b** with the corresponding v_2 in the 30–50 % centrality class [273] as measured with ALICE (see text for details on the models). Adapted from [273]

viscous hydrodynamics. The measured D-meson R_{AA} is well reproduced in central collisions, but v_2 is significantly underestimated. In the TAMU elastic model [330], which is exclusively based on collisional, elastic processes, the interactions proceed via the formation of resonances in the medium. The hydrodynamic evolution of the medium is taken into account. Overall, this model provides a reasonable description of the data. Finally, the BAMPS partonic transport model [331, 332] follows the Boltzmann approach. Only collisional processes are considered for the interaction of heavy quarks with the medium, and the impact of radiative processes is mimicked by an additional scaling factor tuned to describe the heavy-flavour R_{AA} and v_2 at RHIC. The D-meson v_2 at the LHC is described well, while R_{AA} is underestimated at high p_T.

In general, the comparisons demonstrate that it is challenging for models to describe the suppression of heavy flavour in central collisions and, simultaneously, reproduce their azimuthal anisotropy in non-central collisions. Consequently, heavy-flavour measurements provide a crucial tool to come to a quantitative understanding of the properties of strongly interacting matter at the extreme temperature and density reached in Pb–Pb collisions at the LHC.

9.4.3 Quarkonium Production

9.4.3.1 Introduction

Measurements of quarkonia, i.e. bound charmonium ($c\bar{c}$) or bottomonium ($b\bar{b}$) states, are among the most promising tools to study the properties of the hot and dense medium produced in high-energy heavy-ion collisions. Quarkonia were suggested [333] as the prime probes sensitive to deconfinement. In fact, the in-medium dissociation or "melting" of the J/ψ meson via the colour analogue of Debye screening in the medium was originally proposed as the tell-tale signature of the formation of a deconfined QGP [333]. An extension of the original idea considering more quarkonium species led to the picture of "sequential suppression" [334, 335], a characteristic suppression pattern of the various quarkonium states reflecting the different binding energies. Since the binding energies are related to the sizes of the quarkonium states, and because the screening length in a deconfined medium depends on the temperature of the latter [334], the observed suppression pattern gives information about the medium's temperature. The screening can be investigated further theoretically via lattice QCD calculations, even though the interpretation remains challenging (see [336] for a review).

Charmonium production has been studied experimentally at the CERN SPS and at RHIC for $\sqrt{s_{NN}} \approx$ 20–200 GeV. As predicted in the screening scenario (see [337] for a review), a strong suppression of the J/ψ yields has been observed in central heavy-ion collisions compared to pp collisions that are scaled to the nuclear collision geometry using the nuclear overlap function. The interpretation, however, is complicated by at least three issues, which need to be taken into account in addition to in-medium dissociation of quarkonia. First, quarkonia production is affected by cold nuclear matter effects, such as the nuclear modification of parton distribution functions or the nuclear absorption of quarkonia or their $q\bar{q}$ precursor states. Second, feed down from excited quarkonium states needs to be taken into account because such decays contribute significantly to the inclusive J/ψ yield in *pp* collisions. Third, and most relevant for quarkonium physics at LHC energies, the large abundance of charm quarks in the deconfined medium can lead to new production mechanisms that become important in addition to direct $q\bar{q}$ production via hard scattering processes and that mask the suppression of the latter component. Such new quarkonium production mechanisms involve the formation of bound $q\bar{q}$ state from charm or beauty quarks and antiquarks present in the deconfined QGP. Two distinctly different scenarios have been proposed.

In the statistical hadronisation model [338, 339], the heavy quarks and antiquarks are produced in initial hard scattering processes. Full screening is assumed in the deconfined medium, such that no quarkonium states are formed and the suppression is maximal. Instead the heavy quarks thermalise in the QGP and hadronise at chemical freeze-out statistically together with all other quarks [338, 340]. The total yields of quarkonia predicted in this model depend on the chemical freeze-out parameters and the charm and beauty cross sections in *pp* collisions used as input for the calculation. The centrality dependence of the quarkonia yields is related to the issue of canonical suppression of hadrons carrying open charm [341, 342].

In contrast to the quarkonia production mechanism considered in the statistical hadronisation model discussed above, the kinetic recombination model for heavy quarks and antiquarks in the QGP [343] follows a different approach. In this transport model, charmonium states are continuously dissociated and subsequently regenerated over the entire lifetime of the QGP. The model includes a dynamic expansion of the deconfined system, interaction cross sections at the quark level and initial direct charm and beauty production cross sections as a starting point for the heavy-quark transport. Recent results from this approach are summarised in [344–346].

J/ψ and Υ measurements at the LHC were expected to settle the question on the role of (re)generation mechanisms and to provide a view on quarkonia dissociation from an important new angle. The LHC and its experiments have met this challenge, as will be discussed in the following.

9.4.3.2 Charmonium Production at the LHC

To quantify the J/ψ yields in Pb–Pb or *p*-Pb collisions relative to the binary-collision-scaled *pp* reference, the previously-defined nuclear modification factors R_{AA} and $R_{p\text{-Pb}}$ are useful. Measurements of inclusive J/ψ production conducted by ALICE [347], ATLAS [348] and CMS [242] in the dimuon decay channel at high p_T demonstrate a strong suppression of J/ψ yields in central Pb–Pb collisions at $\sqrt{s_{NN}} = 2.76$ TeV that extend beyond expected nuclear shadowing effects in the initial state (see below) and are similar in magnitude with the suppression of open-charm hadrons.

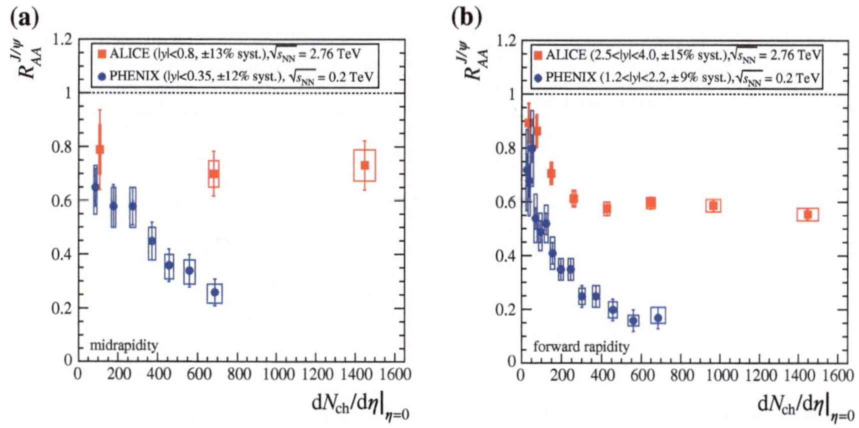

Fig. 9.31 Nuclear modification factor R_{AA} of inclusive J/ψ production at **a** mid-rapidity and **b** forward rapidity as a function of the multiplicity density, measured at $\eta = 0$. ALICE data [349] from Pb–Pb collisions at the LHC are compared with PHENIX data [350] from lower energy Au–Au collisions at RHIC. Adapted from [40]

Since models based on sequential dissociation typically predict the largest suppression at low p_T, while this is where the (re)generation models feature their largest enhancement (reflecting the phase-space distribution of charm quarks in the deconfined medium), it is crucial to turn attention away from the high-p_T region. In this context, the measurement of R_{AA} of inclusive J/ψ, i.e. J/ψ production measured down to zero p_T, is of relevance. This quantity is shown in Fig. 9.31 as a function of the charged-particle multiplicity density (as a proxy for the energy density) at mid-rapidity and in the forward direction.

In this figure, J/ψ measurements with ALICE at the LHC [349] are compared with corresponding PHENIX results from lower-energy collisions at RHIC [350]. The inclusive J/ψ R_{AA} is significantly larger at the LHC than at RHIC. Qualitatively, this observation is in line with the idea that the larger charm-quark densities at the LHC compared to RHIC lead to more (re)generation of J/ψ in the deconfined medium. The question whether (re)generation models can describe the inclusive J/ψ R_{AA} is addressed in Fig. 9.32a, which confronts predictions from the statistical hadronisation model [351] and from transport models [344–346] with ALICE data [349]. While in the first model, all J/ψ mesons are produced at hadronisation and no J/ψ mesons remain from the direct production in the initial stage of the collision, in the latter class of models primordial J/ψ mesons contribute about one third to the final-state yield. The models describe the data well, including the centrality dependence, even though the theoretical uncertainties (indicated by the bands in Fig. 9.32a) are substantial. The precision of the models can be improved by reduced uncertainties on the total charm production cross section which is needed as model input.

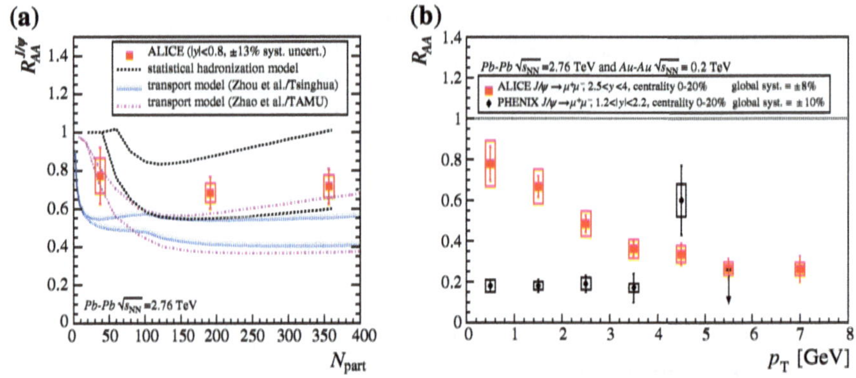

Fig. 9.32 **a** Nuclear modification factor R_{AA} of inclusive J/ψ production at mid-rapidity at the LHC as a function of the number of participants $N_{\rm part}$. Calculations from the statistical hadronisation model [351] and from transport models using the TAMU [344, 345] and Tsinghua implementations [346] are compared to the ALICE data [349]. **b** Nuclear modification factor R_{AA} of J/ψ production at forward rapidity as a function of $p_{\rm T}$ in the 20 % most central collisions. ALICE data [349] are compared with PHENIX data [350] from lower energy collisions at RHIC. Adapted from [40, 349]

An even more striking demonstration of the $p_{\rm T}$ dependence of J/ψ (re)generation is shown in Fig. 9.32b, where the J/ψ R_{AA} at forward rapidity, given as a function of $p_{\rm T}$ for the 20 % most central Pb–Pb collisions measured at the LHC, is compared with ALICE data [349] and the corresponding measurement in Au–Au collisions with PHENIX at RHIC [350]. At low $p_{\rm T}$, J/ψ production at the LHC is suppressed less by about a factor of four (compared to RHIC), while at the same time the energy density is larger at the LHC by a similar factor (see Fig. 9.31). Consequently, the initial temperature of the QGP should be larger at the LHC than at RHIC.

If indeed a significant fraction of the J/ψ mesons observed in Pb–Pb collisions at the LHC is produced via (re)generation processes, this can have an impact on the J/ψ azimuthal distribution with respect to the orientation of the reaction plane as quantified via the elliptic-flow coefficient v_2. For primordial J/ψ, only the fact that they travel a shorter path through the medium—and, therefore, are less likely to dissociate when they are emitted in the reaction plane compared to the emission perpendicular to the reaction plane—can lead to an azimuthal asymmetry. A positive J/ψ v_2, observed in Pb–Pb collisions at the SPS [352], was interpreted in terms of this path-length dependence because at $\sqrt{s_{\rm NN}} = 17.2$ GeV the charm-production cross section is insufficient for (re)generation processes to play a significant role compared to primordial J/ψ production. If charm quarks participate in the collective expansion of the medium or, potentially, even thermalise, J/ψ mesons from (re)generation will reflect the elliptic flow of the charm quarks from which they are formed. Since (re)generation becomes more important with increasing charm-production cross section and therefore with increasing collision energy, it is interesting to study the elliptic flow of J/ψ mesons as a function of $\sqrt{s_{\rm NN}}$.

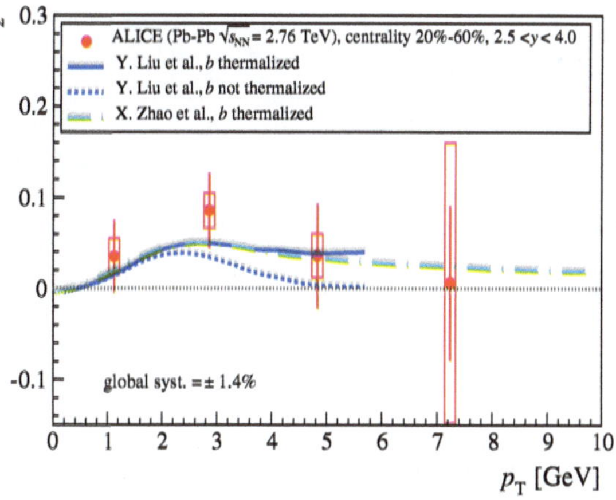

Fig. 9.33 Elliptive-flow coefficient v_2 of J/ψ measured with ALICE in 20–60 % central Pb–Pb collisions at forward rapidity as a function of p_T in comparison with transport model calculations [345, 355]. Adapted from [354]

In Au–Au collisions at RHIC ($\sqrt{s_{NN}} = 200$ GeV), where models suggest that up to about one third of the J/ψ yield stems from (re)generation processes, the J/ψ v_2 is consistent with zero, albeit with uncertainties too large to draw definite conclusions [353]. The situation changes at the LHC where a positive v_2 is observed in semi-central Pb–Pb collisions [354], however with limited statistical significance. Transport model calculations incorporating (re)generation of J/ψ [345, 355] describe the data reasonably well, as demonstrated in Fig. 9.33, which compares the model results with ALICE data [354].

The features exhibited by the J/ψ data from Run 1 at the LHC clearly show that (re)generation mechanisms play an important role for charmonium production at such high energies. The idea that the charmonium yields may serve as a "thermometer" of the medium seems not to hold. However, J/ψ production can still be considered as a probe for deconfinement, not only via the dissociation mechanism proposed originally [333], but specifically through the (re)generation mechanisms [338, 343], which, by definition, play a role only in a deconfined system. The question of which (re)generation scenario provides a better description of nature, be it statistical hadronisation or kinetic recombination, cannot be answered on the basis of the J/ψ data available at this time. Precision measurements, also addressing $\psi(2S)$ production, at the LHC's design energy will help to clarify this important issue.

A quantitative interpretation of the heavy-ion data also requires a detailed understanding of cold nuclear matter (CNM) effects. J/ψ production was measured in p-Pb collisions at the LHC with ALICE [356] and LHCb [360]. While the nuclear modification factors are under-predicted by a model based on saturation [361], shadowing calculations [358, 362] yield a better description of the data. These can still be improved further by also considering energy loss in cold nuclear matter [359]. Intriguing indications for possible final-state effects were revealed in the

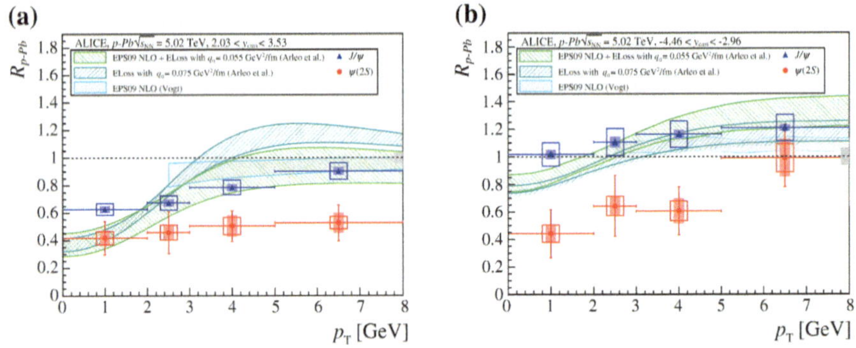

Fig. 9.34 Nuclear modification factor $R_{p\text{-Pb}}$ of J/ψ [356] and $\psi(2S)$ [357] measured with ALICE as a function of p_T in **a** p-Pb and **b** Pb-p collisions. Error boxes (*shaded areas*) denote uncorrelated (partially correlated) systematic uncertainties. Fully correlated uncertainties are indicated by the *filled boxes* at $R_{p\text{-Pb}} = 1$. Model calculations of the J/ψ $R_{p\text{-Pb}}$ are shown including shadowing [358] and coherent energy loss [359]. Adapted from [357]

measurement of $\psi(2S)$ production in p-Pb collisions at the LHC [357]. The nuclear modification factors $R_{p\text{-Pb}}$ of J/ψ [356] and ψ(2S) [357] mesons have been measured with ALICE in two rapidity windows, at forward rapidity in the centre-of-mass system (where "forward" is defined by the direction of the proton beam) and at backward rapidity. In Fig. 9.34, these two configurations are denoted as p-Pb and Pb-p collisions, respectively. The figure confronts model calculations of the J/ψ $R_{p\text{-Pb}}$ including shadowing [358] and coherent energy loss [359] with the data. Corresponding calculations of the $\psi(2S)$ $R_{p\text{-Pb}}$ do not deviate significantly from the J/ψ case and, therefore, are not shown separately. The models are in fair agreement with the measured J/ψ $R_{p\text{-Pb}}$ in both rapidity regions, but they fail to reproduce the $R_{p\text{-Pb}}$ of ψ(2S), which are more suppressed than J/ψ in a systematical way and without a significant p_T dependence. This implies that the established initial-state nuclear effects are not the only driving force behind the suppression of the $\psi(2S)$ yields in p-Pb collisions at the LHC and that final-state effects might play a role as well.

9.4.3.3 Bottomonium

Measurements of $b\bar{b}$ production allow an investigation of quarkonium dissociation from a different angle. The three members of the Υ family, i.e. the $\Upsilon(1S)$, $\Upsilon(2S)$ and $\Upsilon(3S)$, can be measured in their dimuon decay channels. They feature very different binding energies, a fact that makes them good candidates to test sequential dissociation. Compared to charmonia, bottomonium production via (re)generation in the deconfined medium should be less important because of the much smaller cross section for beauty production.

Bottomonium production has been studied with ALICE at forward rapidity [364] and with CMS at midrapidity [363] in Pb–Pb collisions at the LHC. Dimuon-mass

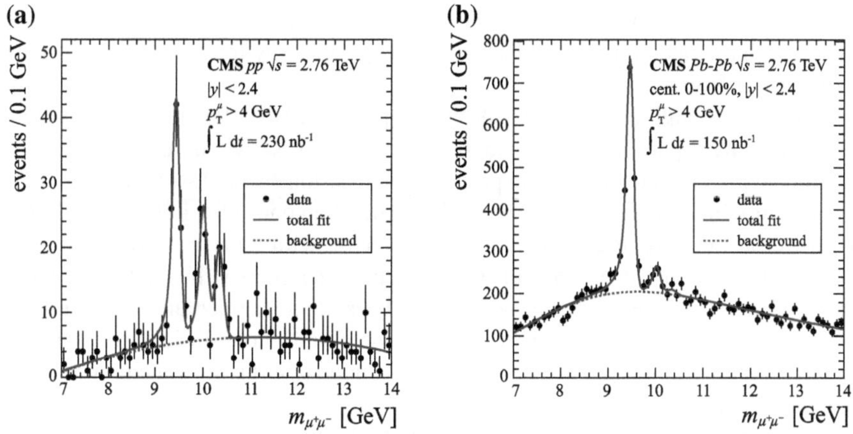

Fig. 9.35 Dimuon invariant-mass distribution measured with CMS in **a** *pp* and **b** Pb–Pb collisions at $\sqrt{s_{NN}} = 2.76$ TeV. The *solid* (*dashed*) *lines* indicate simultaneous fits to the data for the sum of signal and background (background only). Adapted from [363])

spectra obtained with CMS are shown for *pp* and Pb–Pb collisions in Fig. 9.35. The excellent resolution of the CMS muon system allows the three Υ states to be separated from each other in *pp* collisions. In Pb–Pb collisions, where the mass resolution is similar, the $\Upsilon(2S)$ is strongly suppressed and the $\Upsilon(3S)$ is no longer visible as a peak in the mass distribution, a pattern reminiscent of the expectation from sequential dissociation. At forward rapidity, $\Upsilon(1S)$ production is suppressed more than at mid-rapidity. A model which implements screening in a hydrodynamic calculation [365] does not describe the data. A better description of the LHC Υ data is achieved with transport model calculations [366] in which only a small (re)generation component is added to the majority of the bottomonia, which are mainly produced directly in hard scattering processes and which are subject to screening in the hot and dense medium.

9.5 Conclusions

9.5.1 Lessons Learned

The heavy-ion programme at the LHC has confirmed the discovery of almost perfect fluid behavior of strongly interacting matter created in collisions of highly energetic nuclei. Detailed measurements of the azimuthal momentum distribution of low transverse momentum hadrons have yielded valuable insight into the medium properties and initial-state fluctuations. First calculations indicate a larger (by ~60 %) effective shear viscosity to entropy density ratio at LHC compared to the medium created in heavy-ion collisions at RHIC energies. This is an important first step towards understanding the temperature dependent properties of strongly interacting matter under extreme conditions.

The study of high-momentum hadrons and jets at the LHC, where the cross section to produce hard partons is significantly larger than at RHIC energies, can provide additional insight into the medium properties and parton energy loss mechanism. The nuclear modification factor R_{AA} is much smaller than one at the LHC like at RHIC. However at the LHC it exhibits a gradual increase with transverse momentum for $p_{\mathrm{T}} \gtrsim 20\,\mathrm{GeV}$, transverse momenta that were not accessible with sufficient statistics at RHIC energies. This, together with the analysis of jet fragmentation functions, jet shapes, jet v_2, and dijet asymmetries, all of which are more difficult to determine at RHIC, allows us to advance our theoretical understanding of in-medium showers and constrain energy-loss calculations.

Another hard probe that can be exploited to its full potential for the first time in heavy-ion collisions at the LHC is heavy flavour. The large masses of charm and beauty quarks introduce a hard scale even at zero quark momentum. Heavy quarks, therefore, experience the full evolution of the quark-gluon plasma while interacting with it. The resulting modifications of the phase-space distributions of heavy-flavour hadrons in Pb–Pb collisions with respect to *pp* collisions carry unique information on the in-medium energy-loss mechanisms of partons and on the transport properties of the QGP.

Charmonium has been verified to be an important probe for deconfinement. However, dissociation is accompanied by significant regeneration due to the large charm quark densities produced at the LHC energies. Measurements of bottomonium have shown with unprecedented precision the dissociation pattern of the various Υ states in heavy-ion collisions, consistent with the formation of a quark-gluon plasma at high temperatures.

9.5.2 Open Questions

The creation of an almost perfect fluid of quarks and gluons in heavy-ion collisions at both RHIC and LHC has been confirmed. However, the precise nature of the initial state and pre-equilibrium dynamics and the properties of the fluid across the phase diagram of temperature and net-baryon density are still largely unknown. Furthermore, an understanding of how the fluid property emerges microscopically has not been achieved. The higher luminosities of the next LHC heavy-ion run should allow answers to many questions by providing more detailed measurements over an unprecedented range of temperatures and over a broader range of scales with different probes of the medium. These include more differential measurements of the particle production and flow, jets, electromagnetic probes, various particle and jet correlations, quarkonium dissociation and heavy-quark diffusion in the medium.

Thus, together with a beam-energy scan at RHIC that will shed light on the phase structure of nuclear matter and potentially discover a critical point that marks the change from a cross over to a first-order phase transition, the future heavy-ion programme at the LHC has great potential to advance our understanding of strongly interacting matter at high temperatures and densities, as well as the complex many-body physics governed by quantum chromodynamics.

In addition, detailed investigations of collisions of protons and heavy ions at both RHIC and LHC energies are necessary to determine the origin of the long-range correlations observed in high-multiplicity events. This can lead to new insights into the mechanism of particle production from colliding highly energetic nuclei, which are dominated by a large number of gluons, and also yield a limit on how small the smallest droplet of fluid ever observed in nature can be.

References

1. R. Stock, J. Phys. G **30**, S633–S648 (2004)
2. H. Stöcker, J. Maruhn, W. Greiner, Z. Phys, A **293**, 173 (1979)
3. L.P. Csernai, H. Stöcker, Phys. Rev. C **25**, 3208 (1981)
4. W. Reisdorf, H. Ritter, Ann. Rev. Nucl. Part. Sci. **47**, 663–709 (1997)
5. M. Tannenbaum, Rept. Prog. Phys. **69**, 2005–2060 (2006)
6. U.W. Heinz, M.Jacob (2000), arXiv:nucl-th/0002042
7. BRAHMS Collaboration, Nucl. Phys. A **757**, 1–27 (2005)
8. PHOBOS Collaboration, Nucl. Phys. A **757**, 28–101 (2005)
9. STAR Collaboration, Nucl. Phys. A **757**, 102–183 (2005)
10. PHENIX Collaboration, Nucl. Phys. A **757**, 184–283 (2005)
11. H.T. Ding, Nucl. Phys. A **931**, 52–62 (2014)
12. J.O. Andersen, M. Strickland, N. Su, JHEP **08**, 113 (2010)
13. J.O. Andersen, L.E. Leganger, M. Strickland, N. Su, Phys. Lett. B **696**, 468–472 (2011)
14. N. Haque, M.G. Mustafa, M. Strickland, Phys. Rev. D **87**, 105007 (2013)
15. N. Haque et al., JHEP **05**, 027 (2014)
16. N. Armesto et al., Phys. Rev. C **86**, 064904 (2012)
17. K.M. Burke et al., 2013. Phys. Rev. C **90**, 014909 (2014)
18. F. Gelis, E. Iancu, J. Jalilian-Marian, R. Venugopalan, Ann. Rev. Nucl. Part. Sci. **60**, 463–489 (2010)
19. F. Gelis, Int. J. Mod. Phys. A **28**, 1330001 (2013)
20. U. Heinz, R. Snellings, Ann. Rev. Nucl. Part. Sci. **63**, 123–151 (2013)
21. C. Gale, S. Jeon, B. Schenke, Int. J. Mod. Phys. A **28**, 1340011 (2013)
22. R. Glauber, Phys. Rev. **100**, 242–248 (1955)
23. R. Glauber (1959) in Lectures in Theoretical Physics, vol. 1, ed. by W.E. Brittin, L.G. Dunham (Interscience Publishers, New York, 1959), p. 315
24. R.J. Glauber, Nucl. Phys. A **774**, 3–13 (2006)
25. M.L. Miller, K. Reygers, S.J. Sanders, P. Steinberg, Ann. Rev. Nucl. Part. Sci. **57**, 205–243 (2007)
26. B. Alver, M. Baker, C. Loizides, P. Steinberg (2008), arXiv:0805.4411
27. ALICE Collaboration, Phys. Rev. C **88**(4), 044909 (2013)
28. ATLAS Collaboration, Phys. Lett. B 710, 363–382 (2012)
29. CMS Collaboration, Phys. Rev. C **84**, 024906 (2011)
30. G.Y. Qin, H. Petersen, S.A. Bass, B. Muller, Phys. Rev. C **82**, 064903 (2010)
31. J.Y. Ollitrault, Phys. Rev. D **46**, 229–245 (1992)
32. S. Voloshin, Y. Zhang, Z. Phys, C **70**, 665–672 (1996)
33. A.M. Poskanzer, S. Voloshin, Phys. Rev. C **58**, 1671–1678 (1998)

34. ALICE Collaboration, Nucl. Phys. A **904–905**, 573c–576c (2013)
35. ALICE Collaboration, J. Phys. Conf. Ser. **446**, 012028, (2013)
36. ALICE Collaboration, Phys. Rev. Lett. **110**, 032301 (2013)
37. ALICE Collaboration, Phys. Rev. C **88**(4), 044910 (2013)
38. PHENIX Collaboration, Phys. Rev. C **69**, 034909 (2004)
39. STAR Collaboration, Phys. Rev. C **79**, 034909 (2009)
40. A. Andronic, Int. J. Mod. Phys. A **29**, 1430047 (2014)
41. E. Schnedermann, J. Sollfrank, U.W. Heinz, Phys. Rev. C **48**, 2462–2475 (1993)
42. ALICE Collaboration, Phys. Lett. B **728**, 25–38 (2014)
43. E.O.S. Collaboration, Phys. Rev. Lett. **75**, 2662–2665 (1995)
44. FOPI Collaboration, Nucl. Phys. A **848**, 366–427 (2010)
45. NA49 Collaboration, Phys. Rev. C **77**, 064908 (2008)
46. STAR Collaboration, Phys. Rev. C **81**(024), 911 (2010)
47. ALICE Collaboration, Phys. Rev. Lett. **024**, 252301 (2012)
48. C. Shen, U. Heinz, P. Huovinen, H. Song, Phys. Rev. C **84**, 044903 (2011)
49. K. Werner et al., Phys. Rev. C **85**, 064907 (2012)
50. P. Bozek, I. Wyskiel-Piekarska, Phys. Rev. C **85**, 064915 (2012)
51. I. Karpenko, Y. Sinyukov, K. Werner, Phys. Rev. C **87**, 024914 (2013)
52. R. Fries, B. Muller, C. Nonaka, S. Bass, Phys. Rev. C **68**, 044902 (2003)
53. ALICE Collaboration, Phys. Lett. B **736**, 196–207 (2014)
54. ALICE Collaboration, Phys. Rev. C **91**, 024609 (2015)
55. PHENIX Collaboration, Phys. Rev. Lett. **91**, 172301 (2003)
56. STAR Collaboration, Phys. Rev. Lett. **92**, 052302 (2004)
57. ALICE Collaboration, Phys. Rev. Lett. **111**(22), 222301 (2013)
58. P. Aurenche, B. Zakharov, Eur. Phys. J. C **71**, 1829 (2011)
59. R.C. Hwa, C. Yang, Phys. Rev. Lett. **97**, 042301 (2006)
60. J. Cleymans, K. Redlich, Phys. Rev. Lett. **81**, 5284–5286 (1998)
61. P. Braun-Munzinger, K. Redlich, J. Stachel (2003), arXiv:nucl-th/0304013
62. M. Petráň, J. Letessier, V. Petráček, J. Rafelski, Phys. Rev. C **88**(3), 034907 (2013)
63. F. Becattini, R. Fries, The QCD confinement transition: hadron formation, in *Springer Materials-The Landolt-Börnstein Database*, ed. by R. Stock, Springer, Berlin-Heidelberg, Germany, pp. 208–239 (2010)
64. M. Floris, Nucl. Phys. A **931**, 103–112 (2014)
65. ALICE Collaboration, Phys. Lett. B **728**, 216–227 (2014)
66. J. Rafelski, B. Muller, Phys. Rev. Lett. **48**, 1066 (1982)
67. P. Koch, B. Muller, J. Rafelski, Phys. Rept. **142**, 167–262 (1986)
68. WA97 Collaboration, Phys. Lett. B **449**, 401–406 (1999)
69. NA57 Collaboration, J. Phys. G **32**, 427–442 (2006)
70. NA49 Collaboration, Phys. Rev. C **78**, 034918 (2008)
71. STAR Collaboration, Phys. Rev. C **77**, 044908 (2008)
72. S. Hamieh, K. Redlich, A. Tounsi, Phys. Lett. B **486**, 61–66 (2000)
73. A. Andronic, P. Braun-Munzinger, J. Stachel, Phys. Lett. B **673**, 142–145 (2009)
74. PHENIX Collaboration, Phys. Rev. Lett. **104**, 132301 (2010)
75. NA60 Collaboration, Phys. Rev. Lett. **100**, 022302 (2008)
76. NA60 Collaboration, AIP Conf. Proc. **1322**, 1–10 (2010)
77. NA49 Collaboration, Phys. Rev. Lett. **80**, 4136–4140 (1998)
78. STAR Collaboration, Phys. Rev. Lett. **86**, 402–407 (2001)
79. G. Policastro, D.T. Son, A.O. Starinets, Phys. Rev. Lett. **87**, 081601 (2001)
80. P. Kovtun et al., Phys. Rev. Lett. **94**, 111601 (2005)
81. ALICE Collaboration, Phys. Rev. Lett. **105**, 252302 (2010)
82. ATLAS Collaboration, Phys. Lett. B **707**, 330–348 (2012)
83. CMS Collaboration, Phys. Rev. C **87**, 014902 (2013)
84. N. Borghini, P.M. Dinh, J.Y. Ollitrault, Phys. Rev. C **63**, 054906 (2001)
85. R. Bhalerao, N. Borghini, J. Ollitrault, Nucl. Phys. A **727**, 373–426 (2003)

86. STAR Collaboration, Phys. Rev. C **66**, 034904 (2002)
87. CMS Collaboration, JHEP **07**, 076 (2011)
88. ALICE Collaboration, Phys. Rev. Lett. **107**, 032301 (2011)
89. ATLAS Collaboration, Phys. Rev. C **86**, 014907 (2012)
90. CMS Collaboration, JHEP **09**, 091 (2010)
91. CMS Collaboration, Phys. Lett. B **718**, 795–814 (2013). Supplemental materials: https://twiki.cern.ch/twiki/bin/view/CMSPublic/ PhysicsResultsHIN12015
92. ATLAS Collaboration, Phys. Rev. Lett. **110**, 182302 (2013)
93. ATLAS Collaboration, Phys. Lett. B **725**, 60–78 (2013)
94. ALICE Collaboration, Phys. Lett. B **726**, 164–177 (2013)
95. CMS Collaboration, Eur. Phys. J. C **72**, 2012 (2012)
96. P. Bozek, W. Broniowski, Phys. Rev. C **88**(1), 014903 (2013)
97. P. Bozek, W. Broniowski, G. Torrieri, Phys. Rev. Lett. **111**, 172303 (2013)
98. A. Bzdak, B. Schenke, P. Tribedy, R. Venugopalan, Phys. Rev. C **87**, 064906 (2013)
99. G.Y. Qin, B. Mller, Phys. Rev. C **89**, 044902 (2014)
100. A. Dumitru et al., Phys. Lett. B **697**, 21–25 (2011)
101. K. Dusling, R. Venugopalan, Phys. Rev. Lett. **108**, 262001 (2012)
102. K. Dusling, R. Venugopalan, Phys. Rev. D **87**, 051502(R) (2013)
103. K. Dusling, R. Venugopolan, Phys. Rev. D **87**, 054014 (2013)
104. K. Dusling, R. Venugopalan, Phys. Rev. D **87**, 094034 (2013)
105. FOPI Collaboration, Phys. Lett. B **612**, 173–180 (2005)
106. S. Voloshin, M. Poskanzer, R. Snellings, Relativistic heavy ion physics, in by R, ed. by *Springer Materials-The Landolt-Börnstein Database* (Berlin, Stock (Springer, 2010), pp. 293–333
107. STAR Collaboration, Phys. Rev. C **72**, 014904 (2005)
108. PHENIX Collaboration, Phys. Rev. Lett. **105**, 142301 (2010)
109. PHOBOS Collaboration, Phys. Rev. C 72(051), 901 (2005)
110. CMS Collaboration, Phys. Rev. Lett. **110**, 042301 (2013)
111. ATLAS Collaboration, JHEP **11**, 183 (2013)
112. CMS Collaboration, JHEP **02**, 088 (2014)
113. CMS Collaboration, Phys. Rev. C **89**, 044906 (2014)
114. H. Niemi, G. Denicol, H. Holopainen, P. Huovinen, Phys. Rev. C **87**, 054901 (2013)
115. C. Gale et al., Phys. Rev. Lett. **110**, 012302 (2013)
116. CMS Collaboration, Phys. Lett. B **724**, 213–240 (2013)
117. M. Laine, Y. Schroder, Phys. Rev. D **73**, 085009 (2006)
118. M. Chojnacki, W. Florkowski, Acta Phys. Polon. B **38**, 3249–3262 (2007)
119. M. Chojnacki, W. Florkowski, W. Broniowski, A. Kisiel, Phys. Rev. C **78**, 014905 (2008)
120. A. Bazavov et al., Phys. Rev. D **80**, 014504 (2009)
121. H. Song, U.W. Heinz, Phys. Rev. C **78**, 024902 (2008)
122. W. Israel, Ann. Phys. **100**, 310–331 (1976)
123. J. Stewart, Proc. Roy. Soc. Lond. A **357**, 59 (1977)
124. W. Israel, J.M. Stewart, Ann. Phys. **118**, 341–372 (1979)
125. A. Muronga, Phys. Rev. Lett. **88**, 062302 (2002)
126. G. Denicol, E. Molnr, H. Niemi, D. Rischke, Eur. Phys. J. A **48**, 170 (2012)
127. R. Baier et al., JHEP **04**, 100 (2008)
128. G. Denicol, T. Koide, D. Rischke, Phys. Rev. Lett. **105**, 162501 (2010)
129. G. Denicol, H. Niemi, E. Molnar, D. Rischke, Phys. Rev. D **85**, 114047 (2012)
130. W. Florkowski, R. Ryblewski, M. Strickland, Nucl. Phys. A **916**, 249–259 (2013)
131. M. Martinez, M. Strickland, Nucl. Phys. A **848**, 183–197 (2010)
132. R. Ryblewski, W. Florkowski, Phys. Rev. C **85**, 064901 (2012)
133. M. Alvioli, H. Holopainen, K. Eskola, M. Strikman, Phys. Rev. C **85**, 034902 (2012)
134. D. Kharzeev, M. Nardi, Phys. Lett. B **507**, 121–128 (2001)
135. D. Kharzeev, E. Levin, Phys. Lett. B **523**, 79–87 (2001)
136. H.J. Drescher, Y. Nara, Phys. Rev. C **75**, 034905 (2007)

137. H.J. Drescher, A. Dumitru, A. Hayashigaki, Y. Nara, Phys. Rev. C **74**, 044905 (2006)
138. H.J. Drescher, Y. Nara, Phys. Rev. C **76**, 041903 (2007)
139. E. Iancu, R. Venugopalan (2003), arXiv:hep-ph/0303204
140. I. Balitsky, Nucl. Phys. B **463**, 99–160 (1996)
141. J. Jalilian-Marian, A. Kovner, A. Leonidov, H. Weigert, Nucl. Phys. B **504**, 415–431 (1997)
142. J. Jalilian-Marian, A. Kovner, A. Leonidov, H. Weigert, Phys. Rev. D **59**, 014014 (1999)
143. E. Iancu, A. Leonidov, L.D. McLerran, Nucl. Phys. A **692**, 583–645 (2001)
144. E. Ferreiro, E. Iancu, A. Leonidov, L. McLerran, Nucl. Phys. A **703**, 489–538 (2002)
145. A.H. Mueller, Phys. Lett. B **523**, 243–248 (2001)
146. Y.V. Kovchegov, Phys. Rev. D **60**, 034008 (1999)
147. J.L. Albacete, A. Dumitru (2010),
148. A. Dumitru, Y. Nara, Phys. Rev. C **85**, 034907 (2012)
149. B. Schenke, P. Tribedy, R. Venugopalan, Phys. Rev. Lett. **108**, 252301 (2012)
150. B. Schenke, P. Tribedy, R. Venugopalan, Phys. Rev. C **86**, 034908 (2012)
151. J. Bartels, K.J. Golec-Biernat, H. Kowalski, Phys. Rev. D **66**, 014001 (2002)
152. H. Kowalski, D. Teaney, Phys. Rev. D **68**, 114005 (2003)
153. A. Kovner, L.D. McLerran, H. Weigert, Phys. Rev. D **52**, 6231–6237 (1995)
154. Y.V. Kovchegov, D.H. Rischke, Phys. Rev. C **56**, 1084–1094 (1997)
155. A. Krasnitz, R. Venugopalan, Nucl. Phys. B **557**, 237 (1999)
156. A. Krasnitz, R. Venugopalan, Phys. Rev. Lett. **84**, 4309–4312 (2000)
157. A. Krasnitz, R. Venugopalan, Phys. Rev. Lett. **86**, 1717–1720 (2001)
158. T. Lappi, Phys. Rev. C **67**, 054903 (2003)
159. H. Kowalski, L. Motyka, G. Watt, Phys. Rev. D **74**, 074016 (2006)
160. H. Kowalski, T. Lappi, R. Venugopalan, Phys. Rev. Lett. **100**, 022303 (2008)
161. K. Dusling, F. Gelis, R. Venugopalan, Nucl. Phys. A **872**, 161–195 (2011)
162. K. Dusling, T. Epelbaum, F. Gelis, R. Venugopalan, Nucl. Phys. A **850**, 69–109 (2011)
163. T. Epelbaum, F. Gelis, Nucl. Phys. A **872**, 210–244 (2011)
164. K. Dusling, T. Epelbaum, F. Gelis, R. Venugopalan, Phys. Rev. D **86**, 085040 (2012)
165. J. Berges, S. Schlichting, Phys. Rev. D **87**, 014026 (2013)
166. T. Epelbaum, F. Gelis, Phys. Rev. D **88**, 085015 (2013)
167. T. Epelbaum, F. Gelis, Phys. Rev. Lett. **111**, 232301 (2013)
168. J. Berges, K. Boguslavski, S. Schlichting, R. Venugopalan, Phys. Rev. D **89**, 074011 (2014)
169. M. Martinez, M. Strickland, Nucl. Phys. A **856**, 68–87 (2011)
170. W. Florkowski, R. Ryblewski, Phys. Rev. C **83**, 034907 (2011)
171. R. Ryblewski, W. Florkowski (2011), arXiv:1103.1260
172. H. Petersen, G.Y. Qin, S.A. Bass, B. Muller, Phys. Rev. C **82**, 041901 (2010)
173. L. Pang, Q. Wang, X.N. Wang, Phys. Rev. C **86**, 024911 (2012)
174. L. Pang, Q. Wang, X.N. Wang, Nucl. Phys. A **904–905**, 811c–814c (2013)
175. M. Luzum, P. Romatschke, Phys. Rev. Lett. **103**, 262302 (2009)
176. M. Luzum, Phys. Rev. C **83**, 044911 (2011)
177. T. Hirano, P. Huovinen, Y. Nara, Phys. Rev. C **83**, 021902 (2011)
178. T. Hirano, P. Huovinen, Y. Nara, Phys. Rev. C **84**, 011901 (2011)
179. H. Song, S.A. Bass, U. Heinz, Phys. Rev. C **83**, 054912 (2011)
180. H. Song, S. Bass, U.W. Heinz, Phys. Rev. C **89**, 034919 (2014)
181. B. Schenke, S. Jeon, C. Gale, Phys. Lett. B **702**, 59–63 (2011)
182. M. Luzum, J.Y. Ollitrault, Nucl. Phys. A **904–905**, 377c–380c (2013)
183. J. Noronha-Hostler et al., Phys. Rev. C **88**, 044916 (2013)
184. E. Retinskaya, M. Luzum, J.Y. Ollitrault, Phys. Rev. Lett. **108**, 252302 (2012)
185. ATLAS Collaboration, J. Phys. Conf. Ser. **389**, 012013 (2012)
186. ALICE Collaboration, Phys. Rev. Lett. **111**, 232302 (2013)
187. L. Csernai, V. Magas, H. Stöcker, D. Strottman, Phys. Rev. C **84**, 024914 (2011)
188. J. Bleibel, G. Burau, C. Fuchs, Phys. Lett. B **659**, 520–524 (2008)
189. R. Hanbury Brown, R.Q. Twiss, Nature **4541**, 1046–1048 (1956)
190. S. Pratt, Phys. Rev. D **33**, 1314–1327 (1986)

191. M.A. Lisa, S. Pratt, R. Soltz, U. Wiedemann, Ann. Rev. Nucl. Part. Sci. **55**, 357–402 (2005)
192. Y. Sinyukov et al., Phys. Lett. B **432**, 248–257 (1998)
193. ALICE Collaboration, Phys. Lett. B **696**, 328–337 (2011)
194. STAR Collaboration, Phys. Rev. C **71**, 044906 (2005)
195. CERES Collaboration, Phys. Rev. Lett. **90**, 022301 (2003)
196. E. Frodermann, R. Chatterjee, U. Heinz, J. Phys. G **34**, 2249–2254 (2007)
197. P. Bozek, M. Chojnacki, W. Florkowski, B. Tomasik, Phys. Lett. B **694**, 238–241 (2010)
198. I. Karpenko, Y. Sinyukov, Phys. Lett. B **688**, 50–54 (2010)
199. N. Armesto et al., J. Phys. G **35**, 054001 (2008)
200. T.J. Humanic, Phys. Rev. C **79**, 044902 (2009)
201. Q. Li, G. Graf, M. Bleicher, Phys. Rev. C **85**, 034908 (2012)
202. E895 Collaboration, Phys. Rev. Lett. **84**, 2798–2802 (2000)
203. NA49 Collaboration, Phys. Rev. C **66**, 054902 (2002)
204. CERES Collaboration, Nucl. Phys. A **714**, 124–144 (2003)
205. STAR Collaboration, Phys. Rev. C **80**, 024905 (2009)
206. PHOBOS Collaboration, Phys. Rev. C **73**, 031901 (2006)
207. PHOBOS Collaboration, Phys. Rev. C **74**, 021901 (2006)
208. B. Back et al., Phys. Rev. Lett. **91**, 052303 (2003)
209. J. Bjorken, Phys. Rev. D **27**, 140–151 (1983)
210. A. Makhlin, Y. Sinyukov, Z. Phys, C **39**, 69 (1988)
211. Y. Sinyukov, S. Akkelin, A.Y. Tolstykh, Nucl. Phys. A **610**, 278C–285C (1996)
212. ALICE Collaboration, Phys. Rev. C **89**, 024911 (2014)
213. U.A. Wiedemann, Phys. Rev. C **57**, 266–279 (1998)
214. E895 Collaboration, Phys. Lett. B **496**, 1–8 (2000)
215. S. Voloshin, W. Cleland, Phys. Rev. C **53**, 896–900 (1996)
216. STAR Collaboration, Phys. Rev. Lett. **93**, 012301 (2004)
217. ALICE Collaboration, Phys. Lett. B **739**, 139–151 (2014)
218. D. d'Enterria, Jet quenching, in SpringerMaterials–The Landolt-Börnstein Database, ed. by R. Stock, Springer, Berlin-Heidelberg, Germany, pp. 471–520 (2010)
219. U. Wiedemann, Jet quenching in heavy-ion collisions, in *SpringerMaterials–The Landolt-Börnstein Database*, ed. by R. Stock, Springer, Berlin-Heidelberg, Germany, pp. 521–562 (2010)
220. CMS Collaboration, Eur. Phys. J. C **72**, 1945 (2012)
221. J. Cronin et al., Phys. Rev. D **11**, 3105 (1975)
222. I. Vitev, M. Gyulassy, Phys. Rev. Lett. **89**, 252301 (2002)
223. T. Renk, H. Holopainen, R. Paatelainen, K.J. Eskola, Phys. Rev. C **84**, 014906 (2011)
224. C.A. Salgado, U.A. Wiedemann, Phys. Rev. D **68**, 014008 (2003)
225. N. Armesto, A. Dainese, C.A. Salgado, U.A. Wiedemann, Phys. Rev. D **71**, 054027 (2005)
226. A. Dainese, C. Loizides, G. Paic, Eur. Phys. J. C **38**, 461–474 (2005)
227. ALICE Collaboration, Phys. Rev. Lett. **110**, 082302 (2013)
228. CMS Collaboration, Phys. Lett. B **715**, 66–87 (2012)
229. CMS Collaboration, Phys. Lett. B **710**, 256–277 (2012)
230. CMS Collaboration, Phys. Rev. Lett. **106**, 212301 (2011)
231. ALICE Collaboration, Phys. Lett. B **720**, 52–62 (2013)
232. M. Cacciari, G.P. Salam, G. Soyez, JHEP **04**, 063 (2008)
233. M. Cacciari, G.P. Salam, Phys. Lett. B **641**, 57–61 (2006)
234. M. Cacciari, G.P. Salam, G. Soyez, Eur. Phys. J. C **72**, 1896 (2012)
235. S. Catani, Y.L. Dokshitzer, M. Seymour, B. Webber, Nucl. Phys. B **406**, 187–224 (1993)
236. G. Cowan, in *Proceedings of the Workshop on Advanced Statistical Techniques in Particle Physics*, ed. by M. Whalley, L. Lyons, Durham, UK, 2002, pp. 248–257
237. D.G. d'Enterria, J. Phys. G **34**, S53–S82 (2007)
238. ATLAS Collaboration, Phys. Rev. Lett. **105**, 252303 (2010)
239. ALICE Collaboration, JHEP **03**, 013 (2014)
240. ATLAS Collaboration, Phys. Lett. B **719**, 220–241 (2013)

241. ALICE Collaboration, JHEP **09**, 112 (2012)
242. CMS Collaboration, JHEP **05**, 063 (2012)
243. CMS Collaboration, Nucl. Phys. A **904–905**, 194c–201c (2013)
244. CMS Collaboration, Phys. Rev. Lett. **113**, 132301 (2014)
245. CMS Collaboration, Nucl. Phys. A **904–905**, 146c–153c (2013)
246. CMS Collaboration, Nucl. Phys. A **932**, 17–24 (2014)
247. CMS Collaboration, Phys. Rev. C **90**(024), 908 (2014)
248. ATLAS Collaboration, Nucl. Phys. A **931**, 1002–1006 (2014)
249. CMS Collaboration, Phys. Rev. Lett. **109**, 022301 (2012)
250. ALICE Collaboration, Nucl. Phys. A **931**, 382–387 (2014)
251. CMS Collaboration, Phys. Lett. B **718**, 773–794 (2013)
252. ATLAS Collaboration, Nucl. Phys. A **904–905**, 233c–240c (2013)
253. R. Averbeck, Prog. Part. Nucl. Phys. **70**, 159–209 (2013)
254. Particle Data Group, Phys. Rev. D **86**, 010001 (2012)
255. M. Gyulassy, M. Plumer, Phys. Lett. B **243**, 432–438 (1990)
256. R. Baier et al., Nucl. Phys. B **484**, 265–282 (1997)
257. M.H. Thoma, M. Gyulassy, Nucl. Phys. B **351**, 491–506 (1991)
258. E. Braaten, M.H. Thoma, Phys. Rev. D **44**, 2625–2630 (1991)
259. Y.L. Dokshitzer, V.A. Khoze, S. Troian, J. Phys. G **17**, 1585–1587 (1991)
260. Y.L. Dokshitzer, D. Kharzeev, Phys. Lett. B **519**, 199–206 (2001)
261. M. Gyulassy, I. Vitev, X. Wang, Phys. Rev. Lett. **86**, 2537–2540 (2001)
262. E. Shuryak, Phys. Rev. C **66**, 027902 (2002)
263. C. Salgado et al., J. Phys. G **39**, 015010 (2012)
264. M. Lev, B. Petersson, Z. Phys, C **21**, 155 (1983)
265. B. Kopeliovich, J. Nemchik, A. Schafer, A. Tarasov, Phys. Rev. Lett. **88**, 232303 (2002)
266. I. Vitev, Phys. Rev. C **75**, 064906 (2007)
267. F. Arleo, S. Peigne, T. Sami, Phys. Rev. D **83**, 114036 (2011)
268. ALICE Collaboration, JHEP **01**, 128 (2012)
269. ALICE Collaboration, JHEP **07**, 191 (2012)
270. ALICE Collaboration, Phys. Lett. B **718**, 279–294 (2012)
271. ALICE Collaboration. Phys. Rev. Lett. **113**, 232301 (2014)
272. ALICE Collaboration, Phys. Rev. Lett. **111**, 102301 (2013)
273. ALICE Collaboration, Phys. Rev. C **90**, 034904 (2014)
274. ATLAS Collaboration, JHEP **10**, 042 (2013)
275. ATLAS Collaboration, Phys. Rev. D **85**(052), 005 (2012)
276. CMS Collaboration, Phys. Rev. Lett. **106**, 112001 (2011)
277. CMS Collaboration, Phys. Rev. Lett. **106**(252), 001 (2011)
278. CMS Collaboration, Phys. Rev. C **84**, 052008 (2011)
279. LHCb Collaboration, JHEP **04**, 093 (2012)
280. LHCb collaboration, JHEP **08**, 117 (2013)
281. LHCb collaboration, Nucl. Phys. B **871**, 1–20 (2013)
282. ALICE Collaboration, Phys. Rev. D **86**(112), 007 (2012)
283. ALICE Collaboration, Phys. Rev. D **91**, 012001 (2015)
284. ALICE Collaboration, Phys. Lett. B **721**, 13–23 (2013)
285. ALICE Collaboration, Phys. Lett. B **738**, 97–108 (2014)
286. ALICE Collaboration, Phys. Lett. B **708**, 265–275 (2012)
287. ALICE Collaboration, Phys. Rev. Lett. **109**(112), 301 (2012)
288. ALICE Collaboration, Nucl. Phys. A **931**, 546–551 (2014)
289. ALICE Collaboration, Nucl. Phys. A **931**, 514–519 (2014)
290. ATLAS Collaboration, Phys. Lett. B **707**, 438–458 (2012)
291. CMS Collaboration, JHEP **03**, 090 (2011)
292. CMS Collaboration, JHEP **06**, 110 (2012)
293. ALICE Collaboration, JHEP **11**, 065 (2012)
294. CMS Collaboration, Eur. Phys. J. C **71**, 1575 (2011)

295. LHCb Collaboration, Eur. Phys. J. C **71**, 1645 (2011)
296. M. Cacciari, M. Greco, P. Nason, JHEP **05**, 007 (1998)
297. M. Cacciari et al., JHEP **10**, 137 (2012)
298. M. Cacciari, S. Frixione, P. Nason, JHEP **03**, 006 (2001)
299. B. Kniehl, G. Kramer, I. Schienbein, H. Spiesberger, Eur. Phys. J. C **72**, 2082 (2012)
300. P. Bolzoni, G. Kramer, Nucl. Phys. B **872**, 253–264 (2013)
301. P. Hagler et al., Phys. Rev. D **62**, 071502 (2000)
302. S. Baranov, M. Smizanska, Phys. Rev. D **62**, 014012 (2000)
303. H. Jung, M. Kraemer, A. Lipatov, N. Zotov, Phys. Rev. D **85**, 034035 (2012)
304. R. Maciula, A. Szczurek, Phys. Rev. D **87**(9), 094022 (2013)
305. R. Averbeck et al. (2011), arXiv:1107.3243
306. I. Kuznetsova, J. Rafelski, Eur. Phys. J. C **C51**, 113–133 (2007)
307. M. He, R.J. Fries, R. Rapp, Phys. Rev. Lett. **110**(11), 112301 (2013)
308. ALICE Collaboration, Nucl. Phys. A **904–905**, 635c–638c (2013)
309. M.L. Mangano, P. Nason, G. Ridolfi, Nucl. Phys. B **373**, 295–345 (1992)
310. K.J. Eskola, H. Paukkunen, C.A. Salgado, JHEP **04**, 065 (2009)
311. H. Fujii, K. Watanabe, Nucl. Phys. A **920**, 78–93 (2013)
312. R. Sharma, I. Vitev, B.W. Zhang, Phys. Rev. C **80**, 054902 (2009)
313. PHENIX Collaboration, Phys. Rev. Lett. **96**, 032301 (2006)
314. ALICE Collaboration, Phys. Lett. B **719**, 18–28 (2013)
315. A. Bilandzic, R. Snellings, S. Voloshin, Phys. Rev. C **83**, 044913 (2011)
316. ALICE Collaboration, Nucl. Phys. A **904–905**, 661c–664c (2013)
317. ALICE Collaboration, Nucl. Phys. A **904–905**, 677c–680c (2013)
318. PHENIX Collaboration, Phys. Rev. C **84**(044), 905 (2011)
319. S. Wicks, W. Horowitz, M. Djordjevic, M. Gyulassy, Nucl. Phys. A **784**, 426–442 (2007)
320. W. Horowitz, M. Gyulassy, Nucl. Phys. A **872**, 265–285 (2011)
321. M. Nahrgang, J. Aichelin, P.B. Gossiaux, K. Werner, Phys. Rev. C **89**(1), 014905 (2014)
322. K. Werner, I. Karpenko, T. Pierog, M. Bleicher, K. Mikhailov, Phys. Rev. C **82**, 044904 (2010)
323. W. Alberico et al., Eur. Phys. J. C **71**, 1666 (2011)
324. M. Monteno et al., J. Phys. G **38**, 124144 (2011)
325. S. Bass et al., Prog. Part. Nucl. Phys. **41**, 255–369 (1998)
326. M. Bleicher et al., J. Phys. G **25**, 1859–1896 (1999)
327. T. Lang, H. van Hees, J. Steinheimer, M. Bleicher (2012), arXiv:1211.6912
328. T. Lang et al., J. Phys. Conf. Ser. **426**, 012032 (2013)
329. S. Cao, G.Y. Qin, S.A. Bass, Phys. Rev. C **88**(4), 044907 (2013)
330. M. He, R.J. Fries, R. Rapp, Phys. Lett. B **735**, 445–450 (2014)
331. J. Uphoff, O. Fochler, Z. Xu, C. Greiner, Phys. Rev. C **84**, 024908 (2011)
332. J. Uphoff, O. Fochler, Z. Xu, C. Greiner, Phys. Lett. B **717**, 430–435 (2012)
333. T. Matsui, H. Satz, Phys. Lett. B **178**, 416 (1986)
334. S. Digal, P. Petreczky, H. Satz, Phys. Rev. D **64**, 094015 (2001)
335. F. Karsch, D. Kharzeev, H. Satz, Phys. Lett. B **637**, 75–80 (2006)
336. A. Mocsy, P. Petreczky, M. Strickland, Int. J. Mod. Phys. A **28**, 1340012 (2013)
337. L. Kluberg, H. Satz, Color Deconfinement and Charmonium Production in Nuclear Collisions, in *SpringerMaterials-The Landolt-Börnstein Database*, ed. by R. Stock (arXiv:0901.3831), Springer, Berlin-Heidelberg, Germany, pp. 373–423 (2010)
338. P. Braun-Munzinger, J. Stachel, Phys. Lett. B **490**, 196–202 (2000)
339. P. Braun-Munzinger, J. Stachel, Charmonium from Statistical Hadronization of Heavy Quarks – a Probe for Deconfinement in the Quark-Gluon Plasma, in *SpringerMaterials – The Landolt-Börnstein Database*, ed. by R. Stock (arXiv:0901.2500), Springer, Berlin-Heidelberg, Germany, pp. 424–444 (2010)
340. A. Andronic, P. Braun-Munzinger, K. Redlich, J. Stachel, Nucl. Phys. A **789**, 334–356 (2007)
341. P. Braun-Munzinger, J. Stachel, Nucl. Phys. A **690**, 119–126 (2001)
342. M.I. Gorenstein, A. Kostyuk, H. Stöcker, W. Greiner, Phys. Lett. B **524**, 265–272 (2002)
343. R.L. Thews, M. Schroedter, J. Rafelski, Phys. Rev. C **63**, 054905 (2001)

344. X. Zhao, R. Rapp, Nucl. Phys. A **859**, 114–125 (2011)
345. X. Zhao, A. Emerick, R. Rapp, Nucl. Phys. A **904–905**, 611c–614c (2013)
346. K. Zhou, N. Xu, Z. Xu, P. Zhuang, Phys. Rev. C **89**, 054911 (2014)
347. ALICE Collaboration, Phys. Rev. Lett. **109**, 072301 (2012)
348. ATLAS Collaboration, Phys. Lett. B **697**, 294–312 (2011)
349. ALICE Collaboration, Phys. Lett. B **734**, 314–327 (2014)
350. PHENIX Collaboration, Phys. Rev. Lett. **98**, 232301 (2007)
351. A. Andronic, P. Braun-Munzinger, K. Redlich, J. Stachel, J. Phys. G **38**, 124081 (2011)
352. NA50 Collaboration, Eur. Phys. J. C **61**, 853–858 (2009)
353. STAR Collaboration (2013) Phys. Rev. Lett. **111** (5), 052 301
354. ALICE Collaboration, Phys. Rev. Lett. **111**, 162301 (2013)
355. Y. Liu, N. Xu, P. Zhuang, Nucl. Phys. A **834**, 317C–319C (2010)
356. ALICE Collaboration, JHEP **02**, 073 (2014)
357. ALICE Collaboration, JHEP **12**, 073 (2014)
358. J. Albacete et al., Int. J. Mod. Phys. E **22**, 1330007 (2013)
359. F. Arleo, S. Peigne, JHEP **03**, 122 (2013)
360. LHCb Collaboration, JHEP **02**, 072 (2014)
361. H. Fujii, K. Watanabe, Nucl. Phys. A **915**, 1–23 (2013)
362. E. Ferreiro, F. Fleuret, J. Lansberg, A. Rakotozafindrabe, Phys. Rev. C **88**(4), 047901 (2013)
363. CMS Collaboration, Phys. Rev. Lett. **109**, 222301 (2012)
364. ALICE Collaboration, Phys. Lett. B **738**, 361–372 (2014)
365. M. Strickland, A.I.P. Conf, Proc. **1520**, 179–184 (2013)
366. A. Emerick, X. Zhao, R. Rapp, Eur. Phys. J. A **48**, 72 (2012)

Chapter 10
Supersymmetry

Philip Bechtle, Tilman Plehn and Christian Sander

Abstract In this chapter, supersymmetry (SUSY) as a complete and renormalisable candidate for an extension of the Standard Model is described. SUSY at an energy scale not too far above the electroweak scale would solve the hierarchy problem of the SM Higgs boson, dynamically explain electroweak symmetry breaking, and provide a dark-matter candidate. Since it doubles the Standard Model degrees of freedom, SUSY predicts a large number of additional particles, whose properties and effects on precision measurements can be explicitly predicted in a given SUSY model. The chapter motivates SUSY, describes the searches for SUSY particles at the LHC, and describes the status of SUSY in global analyses after the LHC Run 1.

10.1 A Short Motivation

In many ways, the Standard Model (SM) including a Higgs boson [1–5] is a complete theory, both from an experimental and from a theoretical point of view. It explains all observations at the LHC and at preceding high-energy experiments and shows no conceptual inconsistencies. Even better, it is a renormalisable field theory with all gauge couplings remaining perturbative to the Planck scale $M_{\text{Pl}} \sim 10^{19}$ GeV. Aside from gravity—which is missing in the Standard Model—we could be happy with this

P. Bechtle (✉)
Physikalisches Institut, Universität Bonn, Nussallee 12, 53115 Bonn, Germany
e-mail: bechtle@physik.uni-bonn.de

T. Plehn
Institut für Theoretische Physik, Universität Heidelberg, Philosophenweg 16, 69120 Heidelberg, Germany
e-mail: plehn@uni-heidelberg.de

C. Sander
Institut für Experimentalphysik, Universität Hamburg, Luruper Chaussee 149, 22761 Hamburg, Germany
e-mail: christian.sander@physik.uni-hamburg.de

T. Schörner-Sadenius (ed.), *The Large Hadron Collider*,
DOI 10.1007/978-3-319-15001-7_10

rather complete picture after the Higgs discovery [6]. However, if the perturbatively renormalisable Standard Model is supposed to describe all measurements in particle physics over such a wide range of energies, it looks much less complete [7, 8].

If we widen our experimental horizon beyond LHC energies, several shortcomings arise [9, 10]; it does not accommodate the matter-antimatter asymmetry of the universe [11]; and it gives no reason why the three gauge couplings of the Standard Model gauge group almost meet at a large unification scale, but not quite [12].

Gauge coupling unification can be viewed as an experimental motivation for physics beyond the Standard Model, but it is closely tied to a list of open theoretical questions: Where does the flavour structure in the quark and lepton sectors come from? Is there a high-scale simplification of the Standard Model gauge structure? Why is the Higgs mass so much smaller than the Planck scale, which we consider the ultimate physical cut-off scale of the theory [13–15]?

As the only fundamental scalar particles in the Standard Model, the newly discovered Higgs boson causes peculiar problems in its perturbative field theory description. In the presence of a physical cut-off scale, its mass turns out to be extremely sensitive to quantum corrections. This means that we would expect such a scalar to acquire a mass close to the cut-off scale. On the other hand, electroweak symmetry breaking and the unitarity of WW scattering require the Higgs mass to stay sufficiently below the TeV scale. This tension is called the "gauge hierarchy problem" or "hierarchy problem".

Supersymmetry is a leading candidate for a theory that can solve many of these problems [16–19]. It is based on an extension of the symmetry structure of the Standard Model by a symmetry linking fermions and bosons, i.e. matter fields and force carriers [20]. In that sense, supersymmetry completes the concept of symmetries as the guiding principle of elementary particle physics. While models with more than one supersymmetry generator are useful tools to study the structure of gauge theories, we will focus on the phenomenologically most interesting case of $\mathcal{N} = 1$ supersymmetry [21, 22].

This supersymmetric version of the Standard Model [7, 23, 24] protects the Higgs-boson mass from quadratic divergences. It improves the unification of the three gauge couplings and moves the gauge-unification scale to values around $M_{\text{GUT}} \simeq 10^{16}$ GeV, large enough to avoid predicting unobserved proton decays [25–28]. It can provide a weakly interacting dark-matter candidate, and it might help explaining the matter-antimatter asymmetry. And what makes it most relevant to LHC physics: it predicts new fundamental particles at the TeV scale.

10.2 Theoretical Introduction

The minimal supersymmetric Standard Model (MSSM) is the simplest supersymmetric extension of the Standard Model. Through the additional symmetry requirement it first of all doubles the number of fundamental particle fields. Since its underlying symmetry respects the number of degrees of freedom, it predicts two complex scalar fields for each Standard Model fermion, which can be arranged to match the two helicities of the Dirac field. For each Standard Model fermion f we denote these two scalar "sfermions" as $\tilde{f}_{L,R}$. The gauge bosons get mapped onto fermionic fields, where the massless photon and gluon fields only provide enough degrees of freedom for Majorana partners. This means that the discovery e.g. of a strongly interacting Majorana gluino $\tilde{g}$ would be a clear sign for an underlying supersymmetry.

10.2.1 Minimal Supersymmetric Standard Model

Two additional aspects of the MSSM that are of crucial relevance to LHC searches are also directly linked to the properties of supersymmetric gauge theories.

First, the so-called superpotential, written in terms of superfields with their fermionic and bosonic constituent fields, predicts interactions where (scalar) supersymmetric partner states of quarks (squarks) break lepton number and baryon number and mediate rapid proton decay. The easiest way to remove them is by postulating a new symmetry or multiplicative quantum number, called R-parity. All Standard Model states are assigned $R = +1$, which means they transform even under an R-transformation. In contrast, supersymmetric partner states with $R = -1$ are odd. R-parity thus requires that any interaction vertex include an even number of supersymmetric partner states [29, 30]. If, for example, we consider supersymmetric QCD with R-even quark and gluon fields and R-odd (scalar) quark partners $\tilde{q}$ and (fermionic) gluon partners $\tilde{g}$, the allowed supersymmetric 3-point gauge interactions are

$$qqg, \; ggg \;\; \stackrel{\text{SUSY}}{\longrightarrow} \;\; \tilde{q}\tilde{q}g, \; q\tilde{q}\tilde{g}, \; \tilde{g}\tilde{g}g.$$

Three-squark interactions do not exist because they would lead to proton decay and need to be forbidden. In addition, R-parity has a crucial effect on the "lightest supersymmetric partner" (LSP): as an R-odd state, it cannot decay to two Standard Model particles. Since, however, there is also no lighter supersymmetric partner to which it could decay through an allowed interaction, it has to be stable and can therefore serve as a dark-matter candidate.

Second, because of the underlying supersymmetry, the superpotential has to be holomorphic, i.e. it cannot include a superfield and its complex conjugate at the same time. This property directly translates into the Higgs scalar as a constituent of the superfield. As described in Chap. 6, the Higgs sector in the Standard Model gives mass to up-type and down-type fermions through the Higgs field and its conjugate. In the MSSM this is not possible, which means that the Standard Model ($R = +1$) sector of the MSSM has to be modified [31, 32]: instead of one Higgs doublet we need at least two doublets to provide masses to all fermion fields. In the minimal setup of the MSSM, these doublets are arranged such that at tree level one of them couples to up-type fermions and the other to down-type fermions. In combination, their two vacuum expectation values have to provide the weak gauge boson masses, $v_u^2+v_d^2=v^2=(246\text{ GeV})^2$. Two complex Higgs doublets are built of eight degrees of freedom. Just as in the Standard Model, one pseudo-scalar degree of freedom and one set of charged degrees of freedom become the Goldstone modes of the massive Z and $W^\pm$ bosons. The remaining five degrees of freedom, including a pseudoscalar and a charged scalar, turn into physical Higgs fields: two scalar Higgs fields H_1^0 and H_2^0, one pseudo-scalar Higgs A^0, and a pair of charged Higgs fields $H^\pm$. The two scalar states mix to mass eigenstates, h^0 and H^0. As will be discussed below, the mass of the lightest supersymmetric Higgs boson is bounded from above.

In particular the structure of scalar interactions in the MSSM is not obvious from the Standard Model Lagrangian. The reason is that supersymmetry respects the number of degrees of freedom not only of the on-shell particles, but also of the off-shell fields. Already in the simple Wess–Zumino model [18] with its chiral (matter) superfield, we observe that the conservation of off-shell degrees of freedom as well as the consistency of supersymmetry transformations staying within each superfield require an auxiliary field F. It does not feature a kinetic term and can therefore be removed through its equation of motion. After integrating out the F field, this structure ensures that the Standard Model matter fermions and their scalar partners have the same masses. If we include gauge interactions via a vector superfield this links the Standard Model gauge bosons to their fermionic gaugino partners and leads to another auxiliary field, D. This field has gauge couplings to scalars of the kind $D\phi^\dagger\phi$, which after exploiting the equation of motion introduces gauge couplings of the type ϕ^4. In the Higgs sector they replace the free quartic coupling $\lambda\phi^4$ and hence determine the masses of the supersymmetric Higgs bosons. After diagonalising the Higgs eigenstates, we can express the mass of the lightest Higgs boson in terms of the ratio of the two vacuum expectation values $\tan\beta = v_u/v_d$, namely [33–36]

$$M_h = M_Z\cos(2\beta) + \text{radiative corrections} < 135\text{ GeV}, \tag{10.1}$$

where the limit assumes squarks in the TeV range. This is one of the key predictions of the MSSM and is based on the structure of its scalar potential. In the light of the experimental constraints discussed in Sect. 10.6.1, the additional MSSM Higgs states should be significantly heavier.

If we add supersymmetric partners for the entire Standard Model spectrum, keeping in mind that the two Higgs doublets are R-even, we find the MSSM particle content shown in Table 10.1. As mentioned before, every Standard Model fermion acquires a scalar partner, called squark, slepton or sneutrino. The first-generation and second-generation sfermions with their negligible fermion-Higgs couplings can, to a good approximation, be represented by their interaction states, while the heavy-flavour sfermions have to be rotated into their mass eigenstates. This follows from the structure of their masses and mixings, which is determined by supersymmetry breaking.

Every Standard Model boson matches onto a fermionic partner; one example is the gluino as the partner of the gluons. Because supersymmetry respects the number of degrees of freedom, an on-shell gluino without counting the colour structure can only be made out of two degrees of freedom. This does not allow us to define a Dirac

Table 10.1 MSSM mass eigenstates after electroweak symmetry breaking. We assume that left-right scalar mixing is only relevant for the massive third generation

State	Symbol	R	SU(3)$_c$	$3Q_{EM}3$
CP-even Higgs	h^0, H^0	+1	1	0
CP-odd Higgs	A^0	+1	0	0
Charged Higgs	$H^\pm$	+1	1	±1
Gluino	$\tilde{g}$	−1	8	0
Neutralinos	$\tilde{\chi}_1^0, \tilde{\chi}_2^0, \tilde{\chi}_3^0, \tilde{\chi}_4^0$	−1	1	0
Charginos	$\tilde{\chi}_1^\pm, \tilde{\chi}_2^\pm$	−1	1	±1
Up squarks	$\tilde{u}_{L,R}, \tilde{c}_{L,R}, \tilde{t}_{1,2}$	−1	3	2/3
Down squarks	$\tilde{d}_{L,R}, \tilde{s}_{L,R}, \tilde{b}_{1,2}$	−1	3	$-1/3$
Sleptons	$\tilde{e}_{L,R}, \tilde{\mu}_{L,R}, \tilde{\tau}_{1,2}$	−1	1	±1
Sneutrinos	$\tilde{\nu}_e, \tilde{\nu}_\mu, \tilde{\nu}_\tau$	−1	1	0
Gravitino	$\tilde{G}$	−1	1	0

fermion and leaves us with a Majorana fermion, which forms its own anti-particle. In the interaction basis the partner states of the electroweak gauge bosons and the extended Higgs sector occur as one neutral bino, charged and neutral winos, and charged and neutral Higgsinos. These states can mix into mass eigenstates, called neutralinos and charginos and ordered by mass. The neutralinos, just like the gluino, are Majorana fermions, as suggested by the available two on-shell degrees of freedom. The charginos with their electric charge cannot be their own anti-particles, so they instead form Dirac fermions with interactions that violate particle number. Finally, the gravitino as the spin-3/2 partner of the graviton can play a role at the LHC in specific supersymmetric models.

10.2.2 Supersymmetry Breaking

Obviously, supersymmetry is not an exact symmetry of nature, because we have not observed any scalar partners of the Standard Model fermions with the same masses. Without affecting the stabilisation of the quadratically divergent Higgs mass and the improved gauge coupling unification, we can introduce masses for all supersymmetric partners through so-called soft breaking [37]. In this way, the supersymmetric partners receive two contributions to their masses from electroweak symmetry breaking as well as from soft supersymmetry breaking. Such soft-breaking masses remove the quadratic divergence of the Higgs mass, but leave a logarithmic dependence. The largest contribution to the light Higgs mass is proportional to the logarithm of the top-stop mass ratio, $\log(m_{\tilde{t}}/m_t)$. This logarithmic mass dependence leads to the dominant radiative corrections in (10.1) and is referred to as the "little hierarchy problem" of supersymmetry. It is a key ingredient to so-called naturalness arguments. In general, naturalness describes to what degree we have to tune ratios of model parameters in order to accommodate experimental constraints on a model. If the light Higgs mass given in (10.1) is considered a prediction over the MSSM parameter space, a large stop mass moves us away from the supersymmetric limit $m_t = m_{\tilde{t}}$ and is hence fine-tuned or un-natural. While a discussion of these effects will be part of a theoretical analysis of a given supersymmetric model after its discovery, its benefit in guiding LHC analyses is not at all obvious.

Including the soft breaking terms, the relevant Lagrangian of the MSSM reads [7]

$$\begin{aligned}-\mathcal{L}_{\text{soft}} &= m_{H_d}^2 |H_d|^2 + m_{H_u}^2 \, |H_u|^2 - (B\mu \, H_u \cdot H_d + \text{h.c.}) \\ &\quad + m_{\tilde{Q}}^2 \, |\tilde{Q}|^2 + m_{\tilde{U}}^2 \, |\tilde{U}^c|^2 + m_{\tilde{D}}^2 \, |\tilde{D}^c|^2 + m_{\tilde{L}}^2 \, |\tilde{L}|^2 + m_{\tilde{E}}^2 \, |\tilde{E}^c|^2 \\ &\quad + \left(y_u A_u \, \tilde{Q} \cdot H_u \, \tilde{U}^c - y_d A_d \, \tilde{Q} \cdot H_d \tilde{U}^c - y_e A_e \, \tilde{L} \cdot H_d \tilde{E}^c + \text{h.c.} \right) \\ &\quad + \left(\frac{M_3}{2} \, \tilde{g}^a \tilde{g}^a + \frac{M_2}{2} \, \tilde{W}^d \tilde{W}^d + \frac{M_1}{2} \, \tilde{B}^0 \tilde{B}^0 + \text{h.c.} \right), \end{aligned} \tag{10.2}$$

where we omit flavour indices. The first line contributes to the Higgs potential, as do the supersymmetric F and D terms described above. The second and third lines generate masses and mixings for the sfermions. The couplings of the kind $y_j A_j$ describe left-right mixing in the Higgs interaction and turn into mixing terms in the squark-mass matrix after electroweak symmetry breaking. Assuming that they are proportional to the fermion Yukawa interaction ensures that the one-loop contributions to the fermion masses are protected by the chiral symmetry, i.e. they scale with thc fermion masses themselves. In Table 10.1 we neglect them for the first two generations. The final line in the Lagrangian (10.2) contains three gaugino Majorana mass terms.

If we expand the Lagrangian (10.2) in its most general form, the number of free parameters exceeds 100, which makes the MSSM scenario somewhat unpredictive. This parameter set includes complex gaugino masses and trilinear mixing parameters, which are strongly constrained by electric dipole moments, or a general flavour sector with 6×6 matrices for the squark masses, clearly not reflecting the observed flavour symmetries. For direct LHC searches with their given energy range and their limited precision, we can reduce this set to the LHC-relevant parameters, usually referred to as the phenomenological MSSM (pMSSM). Its specific assumptions are: no R-parity violation, no new sources of CP violation, mass-degenerate first and second generation scalars, and no flavour-changing neutral currents. These assumed symmetry properties automatically avoid a large number of indirect constraints on the supersymmetric Lagrangian and allow for a dark-matter candidate. They leave us with 19 supersymmetric model parameters at the TeV scale:

Higgs potential parameters: $\tan\beta$, μ, and M_A;
gaugino masses: M_1, M_2, and M_3;
squark mass parameters: $m_{\tilde{t}_L}$, $m_{\tilde{t}_R}$, $m_{\tilde{b}_R}$, $m_{\tilde{u}_L}$, $m_{\tilde{u}_R}$, and $m_{\tilde{d}_R}$;
slepton and sneutrino mass parameters: $m_{\tilde{\tau}_L}$, $m_{\tilde{\tau}_R}$, $m_{\tilde{e}_L}$, and $m_{\tilde{e}_R}$;
trilinear couplings: A_t, A_b, and A_τ.

In the pMSSM, the lightest neutralino is the dark-matter candidate [38, 39], because the only other weakly interacting candidate, sneutrino dark matter, is ruled out experimentally. The gravitino only couples gravitationally to the rest of the spectrum. As long as it is not the LSP, it will therefore not play a role in LHC phenomenology. However, if it should be the LSP and lead to decays of long-lived particles, it can easily be added to the pMSSM.

At the TeV scale, all pMSSM parameter can be chosen independently. However, if we attempt to link these model parameters to the structure of supersymmetry breaking, we will most likely evolve them to higher energy scales, above which supersymmetry will remain unbroken. Under renormalisation group evolution the squark and gluino masses behave differently: the gluino mass only receives multiplicative corrections, while the squark mass gets corrected by terms proportional to the gluino mass. In this way a heavy gluino will induce large squark masses, which means that we cannot link models with $m_{\tilde{g}} \gg m_{\tilde{q}}$ to high-scale theories [40]. To a lesser extent the same is true for sleptons and electroweak gauginos. From a phenomenological perspective this does not cause any problems, because indirect constraints mostly require heavy scalars. The extreme case of decoupled sfermions is referred to as "split supersymmetry" [41, 42], and will be discussed in Sect. 10.5.1.

Historically, many searches for supersymmetry were not only interpreted, but also set up with a specific SUSY-breaking mechanism in mind. Actual supersymmetry breaking can be described by the auxiliary F and D fields introduced in Sect. 10.2.1. However, for LHC phenomenology supersymmetry breaking is simply an experimental fact; what is more relevant is the way supersymmetry breaking is mediated to the TeV-scale Lagrangian. The most popular mechanism is gravity mediation, either referred to as a supergravity-inspired model (mSUGRA) or as a strongly constrained minimal supersymmetric model (CMSSM). In these models the sector communicating SUSY breaking does not couple to the gauge structure of the Standard Model [43–45]. This requires all gaugino masses, all scalar mass parameters (of Higgs bosons, squarks and sleptons) and all trilinear couplings to unify at the same scale. The CMSSM model parameters are $m_{1/2}$, m_0, and A_0, all defined at $M_{\rm GUT}$. In addition, we specify $\tan\beta$ and the sign of μ in the Higgs sector. Strictly speaking, in the corresponding mSUGRA model the free model parameter is $B = A_0 - m_0$. After renormalisation group evolution, the gaugino masses in the CMSSM scale with constant ratios M_i^2/g_i^2, or $M_1 : M_2 : M_3 \sim 1 : 2 : 6$. The lightest squark will be the lighter stop, its TeV-scale mass mostly depending on $m_{1/2}$. The gravitino is similar in mass to the other superpartners and usually does not affect the LHC phenomenology significantly because it is not the LSP. A variant of the CMSSM is the non-universal Higgs mass ansatz (NUHM), which separates the scalar supersymmetric partner masses from the scalar Higgs sector.

An alternative breaking mechanism is gauge mediation (GMSB). If we assume that some kind of messenger sector communicates SUSY breaking to the MSSM spectrum, gravity mediation assumes that this messenger sector is not charged under any of the Standard Model charges. In gauge mediation we allow for non-gravitational interactions between the SUSY breaking sector and the MSSM particles [46, 47]. The corresponding messenger particles can have masses in the 10^4 GeV range. Again, the gaugino masses scale like constant ratios M_i/g_i^2. Gaugino masses and scalar squared masses are generated at the one-loop and two-loop levels such that eventually they are of similar size. The gravitino can be the LSP, which for the LHC means that the second-lightest supersymmetric partner (NLSP) can be long-lived before it eventually decays e.g. to a gravitino and a photon or an electroweak gauge boson.

In such a scenario where the NLSP is linked to the rest of the supersymmetric spectrum and the LSP is very weakly coupled, the lightest supersymmetric partner relevant for LHC analyses is different from the actual LSP at cosmological time scales. Now, the NLSP does not have to be electrically neutral, so the non-gravitational decay chains at the LHC can as well end with the lightest slepton. The question if we can observe and measure such late decays at the LHC depends on the lifetime of the NLSP.

10.2.3 Signatures of SUSY

While our theoretical understanding of supersymmetry breaking might make it appealing to search for CMSSM and GMSB signatures right away, we should instead identify the most generic features of supersymmetric models and search for those. In the spirit of the pMSSM, we can start with the straight-forward assumption that the LSP is the lightest neutralino and that it is responsible for (part of) the dark matter in the universe.

In that case we expect to see strongly and weakly interacting supersymmetric states at the LHC. Unless they are too heavy, the strongly interacting particles with production cross sections proportional to α_s^2 will be produced much more frequently than the weak states with cross sections proportional to $\alpha_W^2 \sim \alpha_s^2/250$. Some production cross sections are shown in Fig. 10.1 as a function of the mass of the produced supersymmetric particles. Because of the Majorana nature of the t-channel gluino exchange, squarks $\tilde{q}$ and anti-squarks $\tilde{q}^*$ can be produced in pairs as $pp \to \tilde{q}\tilde{q}^*$ and $pp \to \tilde{q}\tilde{q}$. Slepton-pair production is further suppressed because it can only proceed through s-channel diagrams. On the other hand, the LSP cannot be strongly interacting, which means that the directly produced coloured states first need to shed their colour charge by radiating jets, thus decaying to the weakly interacting sector, and then in turn decay to two LSPs. The resulting LHC signature consists of jets plus missing energy. A distinct advantage of this signature is that irreducible background cross sections like W+jets and Z+jets are not huge and produce much softer jets compared to the signal. Top-quark pairs can be a dangerous background and will then have to be suppressed, as described below.

Following the above argument of strongly interacting superpartner production with a cascade decay to the weak sector and the LSP at the end, such a decay chain starting from each gluino or squark could read

$$(\tilde{g} \to)\ \tilde{q} \to \tilde{\chi}_2^0 \to \tilde{\ell} \to \tilde{\chi}_1^0. \qquad (10.3)$$

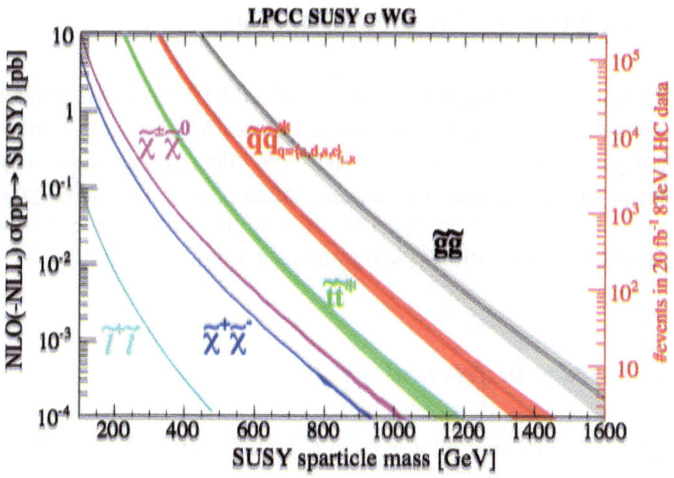

Fig. 10.1 The cross sections expected for different production modes of SUSY particles as a function of their mass. At the same mass scale, gluino production dominates over the production of first-generation squarks. Third-generation squark production is suppressed, since these particles cannot be produced in t-channel processes. Particles not participating in the strong interaction, like gauginos and sleptons, have the smallest production cross section. Adapted from [48]

The corresponding classic kinematic signature for cascade decays arises from the two leptons: two same-flavour opposite-sign leptons with an invariant mass below an edge given by

$$0 < m_{\ell\ell}^2 < \frac{(m_{\tilde{\chi}_2^0}^2 - m_{\tilde{\ell}}^2)(m_{\tilde{\ell}}^2 - m_{\tilde{\chi}_1^0}^2)}{m_{\tilde{\ell}}^2}. \tag{10.4}$$

Typical backgrounds are $\gamma^*/Z^0 \to \ell^+\ell^-$ events as well as leptonically decaying top-quark pairs. The latter can feature both same-flavour and opposite-flavour leptons, which means that they can be controlled by looking at the difference of the same-flavour and opposite-flavour events. The dominant Drell–Yan background for same-flavour leptons should be well understood experimentally and theoretically.

Changing the production process from light-flavour squarks or sbottoms to stop pairs leads to a very different decay signature, for example $\tilde{t} \to t\tilde{\chi}_1^0$ or $\tilde{t} \to b\tilde{\chi}_1^+$. The kinematic edge typical for cascade decays does not feature in this process, and the final state can be indistinguishable from top-quark pair production. Hence, in particular for small stop masses this signal is challenging to extract from the much larger top-pair background. Other ways of searching for supersymmetry, based on electroweak production processes, long-lived particles or R-parity violating decays will be discussed below.

10.3 Generic Searches for Supersymmetry

Given the large number of free parameters (see Sect. 10.2.2), the MSSM in particular and SUSY in general have a plethora of realisations. As discussed before, there are various possible signatures and some of these provide a generic sensitivity to many SUSY models. In this section we discuss various classes of such signatures and describe a few analysis methods used to estimate the remaining background from Standard Model processes after typical event selections.

10.3.1 Searches with Jets and E_T^{miss}

In proton-proton collisions, coloured sparticles are dominantly produced via the strong interaction—at least if they are not too heavy.

Although leptons can also be produced (see Sect. 10.3.2), fully hadronic decay chains have a large probability to be realised. The LSP at the end of the sparticle decay chain is only weakly interacting and therefore escapes the detector unseen, leaving a momentum imbalance in the transverse plane of the event. This imbalance manifests itself as the missing transverse energy E_T^{miss} of an event. Thus, searches in final states with jets and E_T^{miss} but without leptons are expected to have a good sensitivity to a large fraction of supersymmetric models.

In many cases, SUSY events are expected to produce on average more E_T^{miss}, or higher multiplicities of energetic jets. Consequently, searches for new physics are typically selecting events in the tails of distributions of E_T^{miss} and/or other observables, and these quantities have to be precisely understood. However, detector effects, misreconstruction or unknown higher-order corrections to the production of many additional partons in the final state are contributing to potentially large systematic uncertainties. To minimise this uncertainty, most of the searches are using data as much as possible to estimate the SM backgrounds from events in phase-space regions which are dominated by SM processes. An example for this approach is [49] which describes the search for new physics in final states with many jets and missing transverse energy. The background to this search has four major contributions:

Firstly, the irreducible background from $Z(\rightarrow \nu\nu)$+jets events. This background can be estimated from selected $Z(\rightarrow l^+l^-)$+jets events, by removing the two leptons from the event (as if they would be neutrinos) and correcting for lepton efficiencies and acceptance effects as well as for the ratio of the branching fractions of Z into neutrinos and Z into charged leptons. As this ratio is approximately six, this correction leads to large statistical uncertainty. Alternatively, γ+jets events can be used by exploiting their similarity to Z+jets events at high boson momenta where mass effects can be neglected. The statistical uncertainty for this method is significantly smaller, but the systematic uncertainties from unknown higher-order electroweak and QCD corrections on the cross section ratio of Z+jets and γ+jets are particularly large for high jet multiplicities.

Secondly, there is a background originating from W+jets and $t\bar{t}$ events with leptonically decaying W bosons, i.e. $W \rightarrow \ell\nu$ with $\ell = e, \mu$. The neutrino leads to genuine $E_{\mathrm{T}}^{\mathrm{miss}}$, and if the lepton is lost (i.e. not reconstructed, not isolated, or out of acceptance), the event can be selected. The background can be estimated from a μ+jets control sample by reweighting the events according to the muon inefficiencies, which are measured from data.

Thirdly, if a W in W+jets or $t\bar{t}$ events decays into a τ lepton and a neutrino, the τ can decay hadronically and leave a jet-like signature. This contribution can also be estimated from a μ+jets control sample by replacing the muon by a jet with a momentum expected from a hadronic tau decay. Corrections have to be applied for the muon reconstruction and isolation efficiencies, as well as for the branching ratio of the τ lepton into hadronic final states.

A final significant background is QCD multi-jet production with $E_{\mathrm{T}}^{\mathrm{miss}}$ originating from jet energy mismeasurements. This background is estimated from an inclusive jet sample. Each event is, in a first step, rebalanced in the transverse plane, i.e. the jet momenta are adjusted within their uncertainties such that the event has no $E_{\mathrm{T}}^{\mathrm{miss}}$. In a second step, each jet is smeared according to measured jet energy response distributions. It is particularly important to include the non-Gaussian tails in the jet energy response, as they can lead to particularly large values of $E_{\mathrm{T}}^{\mathrm{miss}}$.

Finally, 36 exclusive search regions are defined in $H_{\mathrm{T}} = \sum_{\mathrm{jets}} p_{\mathrm{T}}$, $H_{\mathrm{T}}^{\mathrm{miss}} = |\sum_{\mathrm{jets}} \boldsymbol{p}_{\mathrm{T}}|$, and N_{jets}.

Alternative methods for the background estimation are, for example, the usage of simulated events, which are then reweighted according to the ratio to data in background-dominated control regions.

A comparison of background prediction and data shows a good agreement (see Fig. 10.2), and since no signal is observed upper limits on sparticle masses are set in various supersymmetric models. If the results are interpreted in the CMSSM, as shown in Fig. 10.3a, the fully hadronic searches are sensitive to squark masses up to 1.3 TeV and gluino masses up to 1.8 TeV. Mass limits in less constrained models can be significantly weaker, as will be discussed in Sect. 10.3.4.

Many other analyses with different observables and analysis techniques have been carried out in the fully hadronic final state. One often used observable is $m_{\mathrm{eff}} = H_{\mathrm{T}} + E_{\mathrm{T}}^{\mathrm{miss}}$, which is an estimator for the total energy scale of the event [50]. Other variables are motivated by the process $\tilde{q}\tilde{q} \rightarrow q\tilde{\chi}_1^0 q\tilde{\chi}_1^0$ for which the two jets originating from the emitted quarks are possibly enclosing a small azimuthal angle ($<\pi$), while for QCD dijet events the two leading jets are mostly back-to-back in the transverse plane. Moreover, the momenta of the jets are carrying some information on the mass difference between the initially produced SUSY particles and the final-state LSPs. Therefore, the jet kinematics can be used to construct observables which are having a high potential to separate background from signal, and which are further sensitive to the scale of a given scenario of new physics. In the case of longer decay chains, the jets are clustered in two "hemispheres", and for each of them the four momenta of the jets are added to one "pseudo-jet". The resulting two pseudo-jets are then used for further kinematic investigations. Three of the variables which have

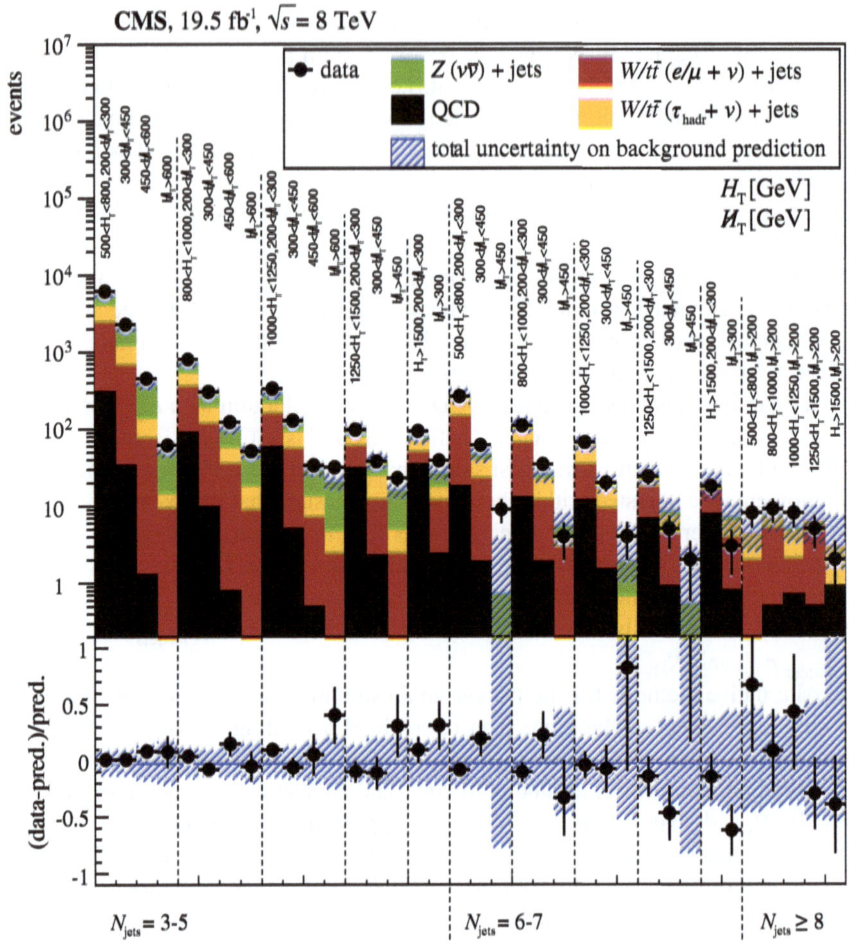

Fig. 10.2 Comparison of background prediction and data for various search regions defined in H_{T}, $H_{\mathrm{T}}^{\mathrm{miss}}$, and N_{jets}. Adapted from [49]

been used in the fully hadronic channel are α_{T} [52, 53], the razor [54], and the "stransverse mass" m_{T2} [55–57], which is further explained in Sect. 10.4. All the different approaches show a comparable sensitivity to new physics.

10.3.2 Final States with Leptons and $E_{\mathrm{T}}^{\mathrm{miss}}$

Leptons can be produced in supersymmetric decay chains either if sleptons are present in the sparticle decay chain or if heavier neutralinos or charginos occur. In the latter case, W and Z bosons can be produced, which can decay further into leptons. If

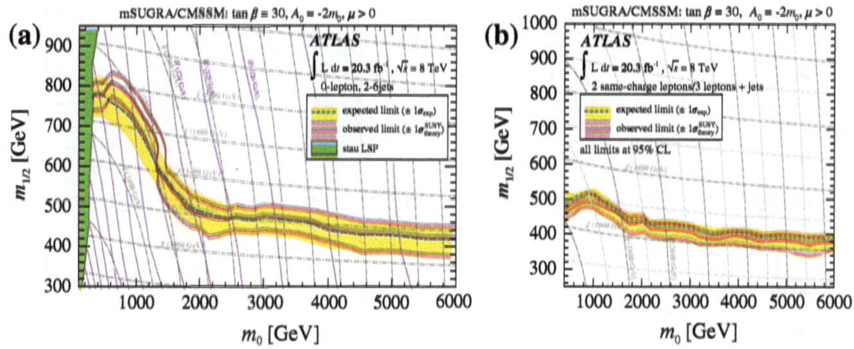

Fig. 10.3 Interpretation of ATLAS searches in final states with jets+$E_{\mathrm{T}}^{\mathrm{miss}}$ or jets+same-sign leptons+$E_{\mathrm{T}}^{\mathrm{miss}}$ in the CMSSM. Adapted from [50, 51]

the initially produced supersymmetric particle is a squark or a gluon, at least one jet per decay chain is produced, and thus another typical final-state signature for many supersymmetric models consists of jets+isolated leptons+$E_{\mathrm{T}}^{\mathrm{miss}}$, where the isolation observable is calculated from further energy deposits in a cone around the lepton. We distinguish here between leptons of the first two generations (electrons and muons) and of the third generation (τ leptons) since the reconstruction of hadronically decaying τ leptons is more complex. Different lepton multiplicities and charge combinations are sensitive to different models: single-lepton searches are quite generic, but often suffer from large backgrounds from Standard Model processes [58, 59]. Final states with three or more leptons are particularly sensitive to the direct production of electroweak superpartners, as will be discussed later. Final states with two leptons can be separated into different classes according to the lepton charges and flavours: same sign or opposite sign with same flavour or opposite flavour. In particular the searches in final states with same-sign leptons provide good sensitivity to many models, because in signal events the two leptons with the same charge can be produced in different decay chains, e.g. $\tilde{q}\tilde{q} \to \bar{q}\bar{q}\tilde{\chi}_1^{\pm}\tilde{\chi}_1^{\pm} \to \bar{q}\bar{q}l^{\pm}\nu l^{\pm}\nu\tilde{\chi}_1^0\tilde{\chi}_1^0$, while Standard Model same-sign processes are very rare and mostly originate from events where a non-prompt lepton, e.g. from a heavy-flavour jet, accidentally fulfills the isolation requirement [51, 60]. This background is usually estimated by selecting leptons with a looser isolation criterion and reweighting the events with a tight-to-loose ratio that has been measured in a background-enriched control region.

So far, no excess beyond the Standard Model expectation has been established at the LHC for searches in final states with leptons, jets and $E_{\mathrm{T}}^{\mathrm{miss}}$, and the results are interpreted in various models of supersymmetry. Figure 10.3b shows the interpretation in the CMSSM. While the jets+$E_{\mathrm{T}}^{\mathrm{miss}}$ searches provide the best sensitivity at small values of m_0 (see Fig. 10.3a), the sensitivity of searches with same-sign leptons are almost independent of m_0, and both searches exclude at large values of m_0 comparable regions in the m_0–$m_{1/2}$ plane.

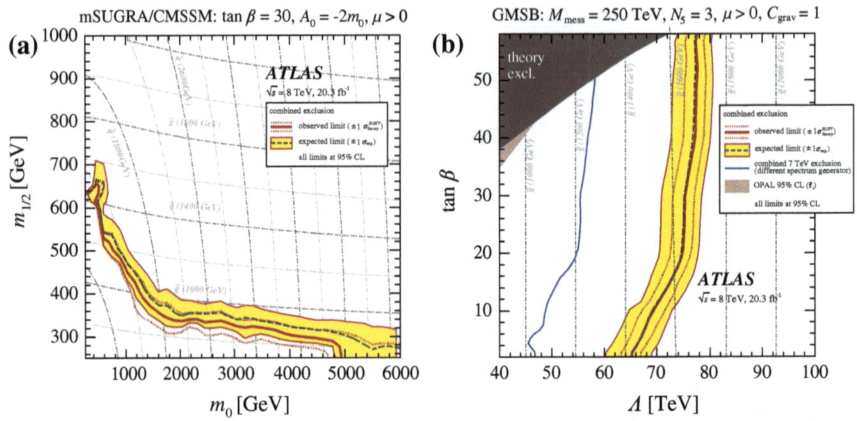

Fig. 10.4 **a** CMSSM interpretation of the ATLAS search with at least one τ lepton, $E_{\mathrm{T}}^{\mathrm{miss}}$ and jets. The search performs best for small m_0, i.e. small $\tilde{\tau}$ masses. **b** Interpretation of the same analysis in the GMSB; the limits are shown in the plane defined by the parameter $\tan\beta$ and the effective SUSY breaking scale Λ. Adapted from [63]

In many supersymmetric models the partner of the τ lepton, the stau ($\tilde{\tau}$) is the lightest slepton. Light staus that are almost mass-degenerate with the LSP are very interesting from a cosmological point of view as will be further explained in Sect. 10.6.1. Thus, if staus are light, it is expected that they occur frequently in decay chains, and that consequently numerous τ leptons are produced. Signatures in which the τ lepton decays further into a light lepton and two neutrinos are covered by the searches with electrons and/or muons as described above. However, in ~65 % of the cases, the τ decays into an odd number of charged hadrons accompanied by neutral hadrons. If the mass splitting between the stau and the LSP is small, the τ leptons can have rather low transverse momenta, and not all might be reconstructed. Therefore, searches with only one or with more τ leptons in final states with additional jets from the decay of the initially produced squarks or gluons and $E_{\mathrm{T}}^{\mathrm{miss}}$ are motivated [61–63]. The dominant background for the search with one τ lepton are W+jets events and, to a smaller extent, QCD multi-jet events with one jet being misreconstructed as a hadronically decaying τ. For the final state with two τ leptons, the dominant backgrounds are again W+jets events with one real τ and one misidentified jet, as well as $t\bar{t}$ events with two real τ leptons. No evidence for new physics has been observed by ATLAS and CMS so far, and limits on supersymmetric parameters have been set. Two interpretations within the CMSSM and within the GMSB model are displayed in Fig. 10.4. Especially for the CMSSM interpretation it can be seen that the best sensitivity is observed for models where m_0 and thus the mass of the stau is small and close to the mass of the LSP.

Furthermore, a very generic search for anomalous production of three or four leptons in various flavour combinations and kinematic search regions, e.g. with low or high accompanying jet activity, has been performed [64]. Once more, no significant excess beyond the expected statistical fluctuations has been observed.

10.3.3 Final States with E_T^{miss} and Photons

In the searches discussed above, it was assumed that the LSP is the lightest neutralino. However, in GMSB models the gravitino is the LSP [65]. As—due to the weakness of gravitation—the couplings of the gravitino to other particles are heavily suppressed, all decay chains are reaching the next-to-lightest supersymmetric particle (NLSP), which then decays further into the gravitino and a SM particle. The type of this SM particle is related to the type of the NLSP: if the NLSP is "bino-like", the SM particle will most likely be a photon, while for "wino-like" NLSPs it will in most cases be a Z boson. This fact motivates searches for supersymmetry in final states with one or more photons in association with jets and E_T^{miss}. Important backgrounds for such searches are originating from photon+jet processes with jet energy mismeasurement leading to artificial E_T^{miss}, from QCD multi-jet production where one of the jets is identified as a photon, and lastly from $W(\rightarrow e\nu)$+jets production with the electron being misreconstructed as a photon. As no ATLAS or CMS search shows a significant excess beyond the Standard Model expectation, mass limits have been set in various models. Figure 10.5 shows two interpretations within general gauge mediation models (GGM) models [66], with a bino-like or wino-like NLSP [67, 68] (GGM is a generalisation of GMSB models).

10.3.4 Simplified Models: Virtues and Challenges

In Sects. 10.3.1 to 10.3.3, various results from SUSY searches have been interpreted in the context of so called "full models". These are characterised by a relatively small number of parameters that fully specify the sparticle mass spectrum and thereby fix

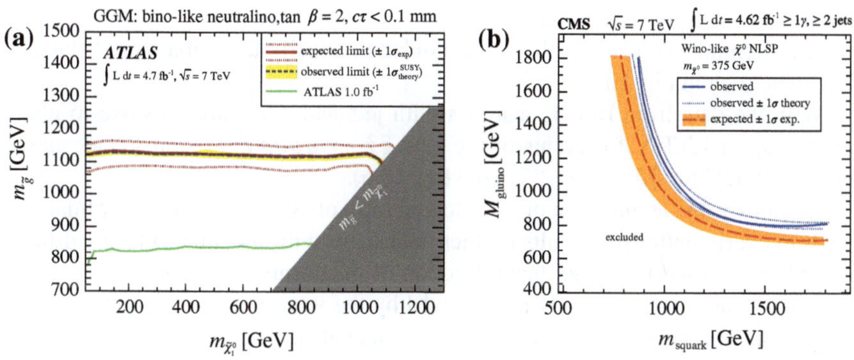

Fig. 10.5 Interpretation of the searches with photons, jets and E_T^{miss} in the GGM. **a** Exclusion limit from a search with two photons in the plane spanned by the NLSP mass and the gluino mass for a bino-like NLSP. **b** Exclusion limit from a search with at least one photon in the plane spanned by the squark mass and the gluino mass for a fixed wino-like NLSP mass. Adapted from [67, 68]

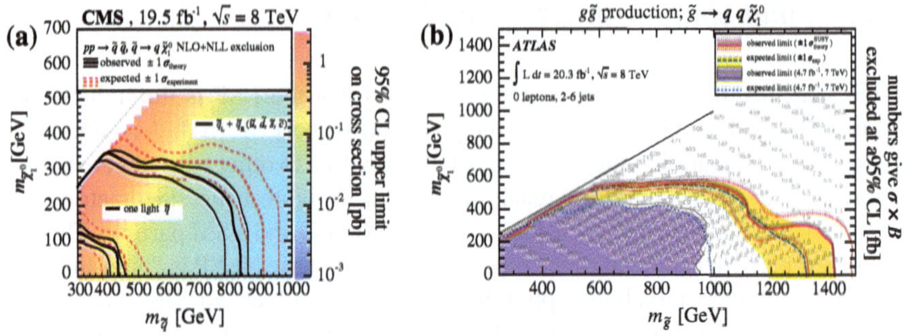

Fig. 10.6 **a** Interpretation of a CMS search with jets+$E_{\mathrm{T}}^{\mathrm{miss}}$ in a simplified model with 100 % branching ratio of $pp \to \tilde{q}\tilde{q} \to q\tilde{\chi}_1^0 q\tilde{\chi}_1^0$, with only one or four squark flavours kinematically accessible, shown in the $m(\tilde{q})$–$m(\tilde{\chi}_1^0)$ plane. **b** A similar search by ATLAS interpreted in a simplified model with the decay topology $pp \to \tilde{g}\tilde{g} \to q\bar{q}q\bar{q}\tilde{\chi}_1^0\tilde{\chi}_1^0$ shown in the $m(\tilde{g})$–$m(\tilde{\chi}_1^0)$ plane. Adapted from [49, 50]

the production cross sections and branching ratios of every sparticle. In such models events can be in general realised in many different ways, e.g. by different primarily produced sparticles and by different combinations of the two decay chains. Although this provides realistic signatures for any assumed SUSY model with any set of specific parameters, the results are difficult to generalise to other models or event topologies. Another limitation is that in a given full model, the masses of different sparticles are often related in a way that can not be generalised. In the CMSSM, for example, there is the well known ∼1 : 2 : 6 relation for $M_1 : M_2 : M_3$.

An alternative approach is the concept of simplified model searches (SMS) [69]: the basic assumption here is that only one particular event topology is realised, e.g. $pp \to \tilde{q}\tilde{q} \to q\tilde{\chi}_1^0 q\tilde{\chi}_1^0$. The only free parameters are then the masses of the two involved sparticles, namely $m_{\tilde{q}}$ and $m_{\tilde{\chi}_1^0}$, which make these "models" rather simple and easy to interpret. The production cross section is fixed by the squark mass [48], which potentially depends on the number of kinematically accessible squark flavours, as can be seen in Fig. 10.6.

With the data from Run 1, searches with jets and $E_{\mathrm{T}}^{\mathrm{miss}}$ are sensitive to gluino masses up to 1.3 TeV for events of the type $\tilde{g}\tilde{g} \to q\bar{q}q\bar{q}\tilde{\chi}_1^0\tilde{\chi}_1^0$, and squark masses up to 880 GeV for $\tilde{q}\tilde{q} \to qq\tilde{\chi}_1^0\tilde{\chi}_1^0$ (see Fig. 10.6).

Obviously, the assumption of a branching ratio of 100 % in a particular final state leads to overly optimistic results if other final states with signatures that are difficult to find are realised for a significant fraction of the events. Therefore, the exclusion limit for a full model can only be achieved by the interpretation of the SMS result within this full model. An example for such an analysis is presented in Sect. 10.6.1.

10.4 The Rest of the Spectrum: SUSY Searches for Electroweak and Third-Generation Production

In the previous section, searches for SUSY were discussed that aim at the most abundant signatures in generic implementations of SUSY. These generic signatures are also featured in the most common highly constrained models such as the CMSSM, where the mass scales of all sfermions are coupled to each other and all gauginos are coupled to each other. Thus, excluding high first-generation squark (and gluino) masses means that in models like the CMSSM also light sleptons and third-generation squarks and gauginos are indirectly excluded. In this section, searches are covered for which the simple assumption of maximally constrained SUSY shall not be a motivation.

The generic searches are motivated by the relatively high cross section of gluino production and first-generation and second-generation squark production, as shown in Fig. 10.1. However, as shown e.g. in Fig. 10.3, the limits obtained in this type of search in highly constrained SUSY models such as the CMSSM correspond to SUSY mass scales of gluinos and squarks of more than $M_{\tilde{g}} = 1.2$ TeV and $M_{\tilde{q}} = 1.8$ TeV, respectively. If all SUSY masses were that heavy, SUSY would cease to help the hierarchy problem, as discussed in Sect. 10.3.

Therefore, in order to retain the feature of solving the hierarchy problem, it is theoretically attractive to think of more general SUSY models in which the mass scale of the sparticles that only interact weakly (the sleptons and gauginos) and the mass scale of the third-generation squarks are decoupled from the scale of the first-generation and second-generation squarks, and from that of the gluinos. It is easily possible to realise such a model in the framework of the MSSM or in even more general models. On the other hand, there is no obvious choice for a benchmark model for such a scenario, defined at the GUT scale with only a very small number of parameters. This is in contrast to the generic searches. Thus, the results of the specific searches for SUSY in electroweak production processes or in the production of third-generation squarks ask for a more model-independent parametrisation, to allow for the later interpretation of the search results in any specific SUSY model. This is mostly realised in the form of the so-called simplified model searches already introduced in the previous section. While this section covers the searches and the translation of their results into SMS limits, the application of these limits is discussed further in Sect. 10.6.1.

10.4.1 Electroweak Production of SUSY Particles

Searches for SUSY events with electroweak interactions at the production vertex can be classified in two different ways: first, by the type of SUSY particles occurring at the production vertex, and second, by the visible final-state particles. Examples for production and decay processes are shown in Fig. 10.7.

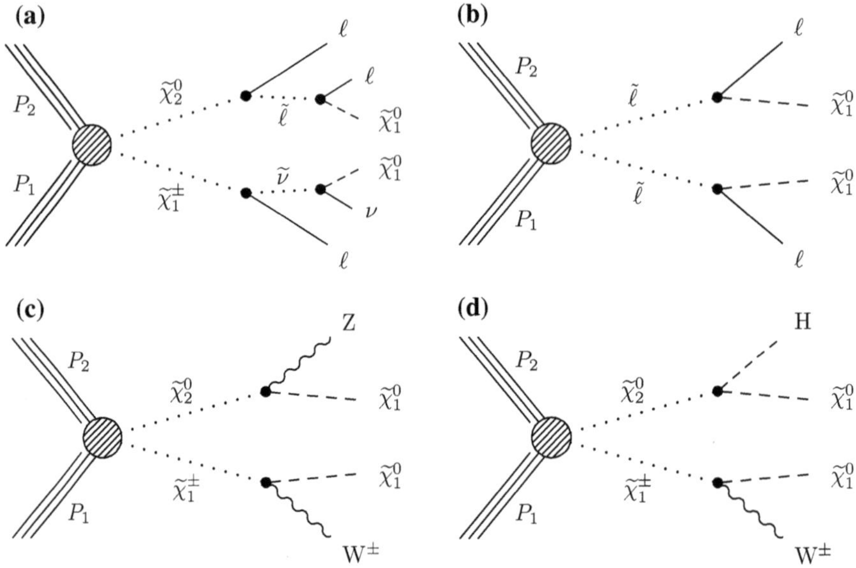

Fig. 10.7 Example diagrams for the production of gauginos or sleptons without the production of strongly interacting sparticles. These production processes involve electroweak interactions

It can be seen that some of the fundamental features of SUSY events already discussed in Sect. 10.3 are retained: leptons in the final state allow events to be distinguished from the QCD background, and in R-parity conserving models missing transverse energy E_T^{miss} is produced by the escaping LSPs, here assumed to be the lightest neutralinos $\tilde{\chi}_1^0$. However, there are also significant differences with respect to generic searches for SUSY events involving the strong interaction at the production vertex: there, the high production cross section of up to $\mathcal{O}(10 \text{ pb})$ (depending on the mass scale, see Fig. 10.1) made possible by the strong interaction allows rather heavy SUSY objects to be produced at the 8 TeV LHC Run 1. However, in electroweak production processes for SUSY particles not interacting strongly, there is a reduction in the production cross section at the same mass scale of more than a factor of 100. Thus, at the same centre-of-mass energy, the accessible mass scales for the direct production of sleptons and gauginos is significantly lower than for first-generation squarks and gluinos. Consequently, also the transverse momenta of the final-state particles and the expected E_T^{miss} are significantly lower than for strong-production processes of heavy squarks and gluinos—a fact that can limit the selective power of the trigger and the offline selection.

While gauginos couple directly to the quarks in the protons via gauge interactions, the sleptons need an intermediate SM gauge boson to couple to the partons in the beam protons. This explains why the direct production of sleptons again is suppressed with respect to the gaugino production by about a factor of 100.

As shown in Fig. 10.7, different production modes can yield very similar final states. Leptonic decays of W and Z bosons are employed, which means that the first three diagrams yield near-identical final-state particle compositions of two or more leptons and missing energy. Depending on the intermediate particles, the expected kinematics may vary. On the other hand, the usable decays of Higgs bosons, e.g. into b quarks, τ leptons or two photons, yield final states with two hadronic heavy quark jets or two photons. Therefore, a strategy is employed where the searches are classified by final-state particles, and each search can then be interpreted in simplified models describing different production mechanisms, which can be optimally constrained by an individual final state. In the following, out of a very large number of available searches [60, 70–75], the following examples are discussed in detail: the production of 2-lepton and 3-lepton final states in conjunction with $E_{\mathrm{T}}^{\mathrm{miss}}$, where the leptons can be electrons or muons; the production of τ-lepton final states with $E_{\mathrm{T}}^{\mathrm{miss}}$; and the specific search for SUSY decays with electroweak production of SUSY particles with a Higgs boson in the decay chain.

The first example discussed here stems from ATLAS and CMS searches for 2 or 3 leptons and $E_{\mathrm{T}}^{\mathrm{miss}}$, where the leptons can come in same-sign or opposite-sign configurations [71, 74]. It can be beneficial to also search for same-sign lepton configurations in cases where one out of three leptons is outside of the detector acceptance in production processes as shown in Fig. 10.7a. The dominant background to these searches are diboson events (W^+W^-, $W^{\pm}Z$ and ZZ), $W^{\pm}$+jets and Z+jets events, and $t\bar{t}$ production with a small contamination of leptons from secondary decays, e.g. from heavy quarks.

Examples for discriminating variables employed in these searches are given in Fig. 10.8. An example for a signal region aiming at SUSY decays involving two $W^{\pm}$ bosons is shown in Fig. 10.8a from the ATLAS experiment. The plot shows the distribution of the invariant mass of two opposite-sign same-flavour (SF) leptons $m_{\ell\ell}$, where $\ell = e, \mu$.

An example of another signal region is shown in Fig. 10.8b. It exploits the "stransverse" mass m_{T2} for two leptons of different flavours (DF) [55, 56], defined as

$$m_{\mathrm{T2}} = \min_{\mathbf{q}_{\mathrm{T}}} \left[\max \left(m_{\mathrm{T}}(p_{\mathrm{T},\ell}1, \mathbf{q}_{\mathrm{T}}), m_{\mathrm{T}}(p_{\mathrm{T},\ell}2, p_{\mathrm{T}}^{\mathrm{miss}} - \mathbf{q}_{\mathrm{T}}) \right) \right],$$

where $p_{\mathrm{T},\ell}1$ and $p_{\mathrm{T},\ell}2$ are the transverse momenta of the two leptons, and $\mathbf{q}_{\mathrm{T}}$ is a transverse vector that minimises the larger of the two transverse masses m_{T}. The latter is defined by

$$m_{\mathrm{T}}(\mathbf{p}_{\mathrm{T}}, \mathbf{q}_{\mathrm{T}}) = \sqrt{2(p_{\mathrm{T}} q_{\mathrm{T}} - \mathbf{p}_{\mathrm{T}} \cdot \mathbf{q}_{\mathrm{T}})}.$$

For SM $t\bar{t}$ and W^+W^- events, in which two $W^{\pm}$ bosons decay leptonically and $p_{\mathrm{T}}^{\mathrm{miss}}$ originates from the two neutrinos, the m_{T2} distribution has an upper end-point at the $W^{\pm}$ mass. For signal events, the undetected LSP contributes to $p_{\mathrm{T}}^{\mathrm{miss}}$, and the m_{T2} end-point is correlated to the mass difference between the slepton or chargino and the lightest neutralino. For large values of this difference, the m_{T2} distribution for signal events extends significantly beyond the distributions obtained from $t\bar{t}$ and W^+W^-

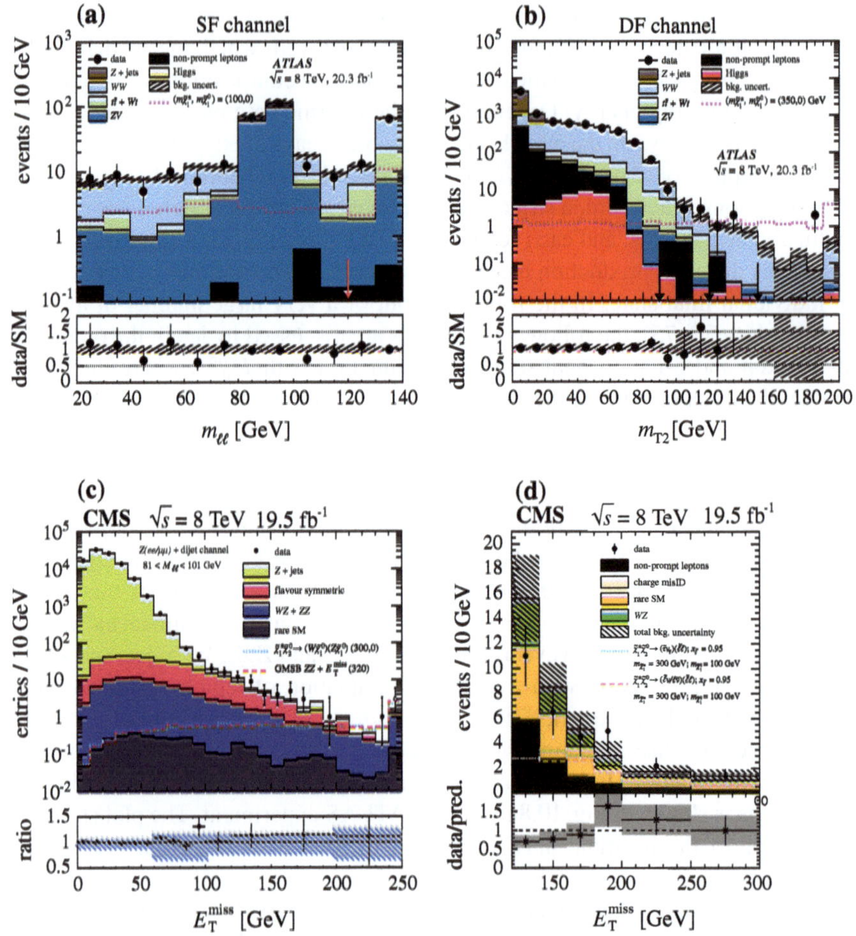

Fig. 10.8 Selected observables from the searches for the production of gauginos and sleptons via the electroweak interaction from ATLAS (**a**, **b**) and CMS (**c**, **d**). See the text for a discussion of the observables, signal properties and backgrounds. Adapted from [71, 74]

events. This is clearly visible in Fig. 10.8b, where the background drops strongly on the logarithmic scale at about $m_{\mathrm{T2}} \approx M_W$, while the signal stays flat.

Other typical observables besides the leptonic invariant mass, m_{T2} and m_{T}, are H_{T} (see Sect. 10.3.1), $E_{\mathrm{T}}^{\mathrm{miss}}$ and $m_{\mathrm{eff}} = H_{\mathrm{T}} + E_{\mathrm{T}}^{\mathrm{miss}}$. Examples for $E_{\mathrm{T}}^{\mathrm{miss}}$ are given in Fig. 10.8c, d, for the signal regions aimed at decays via $W^{\pm}$ and Z (c) and for decays of the gauginos via staus (d). The latter is favourable in many models motivated by the stau co-annihilation mechanism of reducing the predicted amount of dark matter in the universe down to the measured values (see Sect. 10.6.1). There, the stau is the NLSP and thus is kinematically favoured in the decay of heavier SUSY particles. It can be seen that the SM backgrounds drop strongly for growing $E_{\mathrm{T}}^{\mathrm{miss}}$, and only SM processes with true E_T^{miss} from e.g. $Z \to \nu\bar{\nu}$ decays remain at large values. On the other hand, due to the energy carried by the LSP, SUSY processes on average show much higher E_T^{miss} values compared to the background. Thus, high signal-to-background ratios can be reached at very high E_T^{miss}.

Neither ATLAS nor CMS observed a significant SUSY signal in any of these signal regions in Run 1. Thus, the results of the searches are converted into limits on SMS decay chains, as already discussed in Sect. 10.3.4. Since different signal regions can be sensitive to different signals depending on the kinematical configuration, expected limits from all signal regions are derived for all model points and decay chains. The observed limit for a given decay chain and model point is then calculated only for the signal region with the strongest expected limit.

Examples for the limits derived for the searches described above are given in Fig. 10.9. The top row shows examples from the ATLAS experiment, the bottom row from the CMS experiment. The considered SMS decay chains are $\tilde{\chi}_1^+\tilde{\chi}_1^- \to \tilde{\ell}\nu/\tilde{\nu}\ell \to 2\ell\nu\tilde{\chi}_1^0$ with $m_{\tilde{\ell},\tilde{\nu}} = (m_{\tilde{\chi}_1^\pm} - m_{\tilde{\chi}_1^0})/2$ (Fig. 10.9a), $\tilde{\ell}^+\tilde{\ell}^- \to 2\ell^\pm\tilde{\chi}_1^0$ (b),

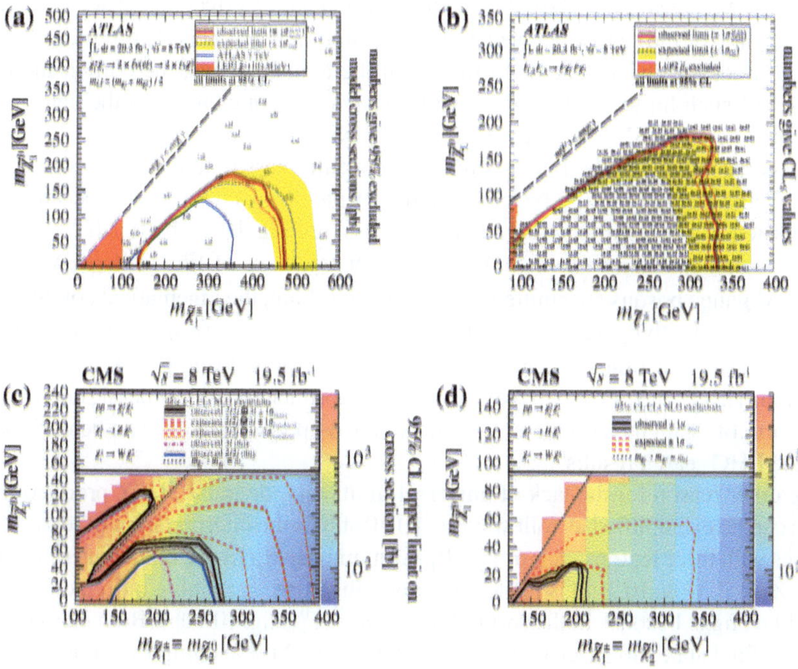

Fig. 10.9 The resulting limits on selected SMS decay chains from ATLAS (**a**, **b**) and CMS (**c**, **d**). The excluded production cross section of the decay chain given in the figures is plotted against the masses of the physical SUSY particles appearing in the decay chain. The expected exclusion lines in the plots are given by the assumption of the full electroweak coupling at the production and of 100 % branching fractions into the studied final state. Individual kinematical assumptions on the particles involved in the decay are also given in the plots. Adapted from [71, 74]

$\tilde{\chi}_2^0\tilde{\chi}_1^\pm \to Z\tilde{\chi}_1^0 W\tilde{\chi}_1^0$ (c), and $\tilde{\chi}_2^0\tilde{\chi}_1^\pm \to h\tilde{\chi}_1^0 W\tilde{\chi}_1^0$ (d). The physical masses of the SUSY particles involved are given on the axes of the plots, while the numbers (ATLAS) respectively color scale (CMS) give the cross section of the exclusive production of the decay chain under investigation that can be excluded at the 95 % CL. If only two SUSY masses are involved, this limit on the cross section of individual decay chains can be interpreted directly in every SUSY model which involves the decay chain, as long as there is no strong dependence on the production process. It is possible that—depending on the nature of the gauginos in a given physical model—the relative strengths of the predicted s-channel and t-channel production processes do vary, which could influence the selection efficiency and thus the limit on the cross section. Therefore, in some cases t-channel production is removed from the simulated process by assuming the particle in the t-channel, e.g. a squark in the SMS implementation, to be very heavy. However, influences from differences in the production process for the same decay chains are typically weak.

A more challenging situation arises for decay chains with more than two different SUSY particles that are not degenerate in mass. One example is the decay chain $\tilde{\chi}_1^+\tilde{\chi}_1^- \to \tilde{\ell}\nu/\tilde{\nu}\ell \to 2\ell\nu\tilde{\chi}_1^0$ decay chain in Fig. 10.9a with the explicit assumption of $m_{\tilde{\ell},\tilde{\nu}} = (m_{\tilde{\chi}_1^\pm} - m_{\tilde{\chi}_1^0})/2$. Also this latter condition has an effect on the acceptance and efficiency of the search. However, this assumption is not generically true in all physical models to which the SMS limit shall be applied. Thus, the correct interpretation of such limits might involve additional studies of changes in the detector acceptance between SMS implementation and physical model.

A general feature of the limits is that, for specific mass relations, the momenta of the visible decay products become small. For the decay chains that proceed via the emission of SM fermions with negligible mass (see Fig. 10.9a, b), this is the case for $m_{\tilde{\chi}_1^\pm} = m_{\tilde{\chi}_1^0}$ or $m_{\tilde{\ell}} = m_{\tilde{\chi}_1^0}$, respectively. For the cases of Fig. 10.9c, d, where heavy SM gauge bosons are emitted from the decay chain, the kinematical condition in which the sensitivity vanishes changes to $m_{\tilde{\chi}_1^\pm} - m_{\tilde{\chi}_1^0} = m_{V,H}$. In these cases, the trigger and selection efficiencies degrade considerably, leading to a very weak exclusion. This is one of the two reasons why—unfortunately—no general statement in the form of "gauginos (or sleptons) are excluded up to a mass of" can be derived from the LHC search results.

The other reason for the lack of universal limits on individual SUSY production modes can be exemplified as follows: Fig. 10.10 shows the interpretation of leptonic searches with three leptons (e, μ or τ) [70]. Assuming an exclusive branching ratio of $\tilde{\chi}_2^0 \to Z + \tilde{\chi}_1^0$ (Fig. 10.10a) is neglecting a possibly large branching ratio including a SM-like Higgs boson h in the form of $\tilde{\chi}_2^0 \to h + \tilde{\chi}_1^0$ (Fig. 10.10b). Both can yield the same final state, however with strongly different SM branching fractions to the visible final-state particles and with different yields of τ leptons versus light leptons. In a physical model, the real limit would have to be derived for a combination of the two decay channels given above. Since the limits derived for the two channels differ significantly due to the much weaker branching fraction of the Higgs boson into light leptons, the limit has to be evaluated for each model point depending on its individual prediction for the branching fractions.

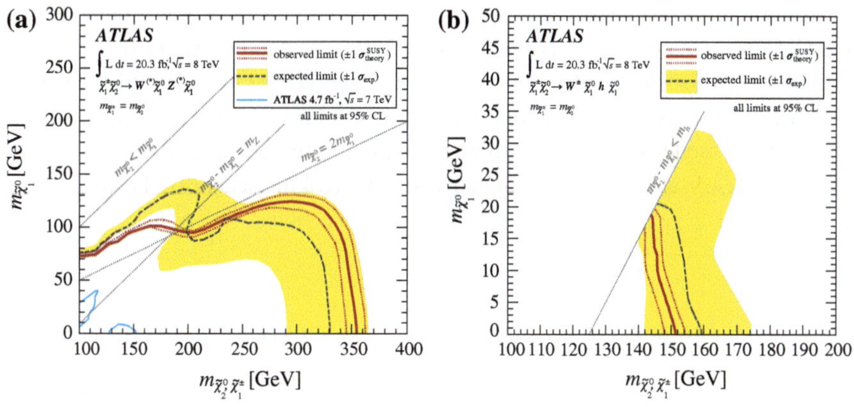

Fig. 10.10 Interpretation of ATLAS searches in final states with three leptons (e, μ or τ) for direct chargino and neutralino production. As expected, the limits on the sparticle masses for an assumed simplified model with $\tilde{\chi}^0_2 \to Z + \tilde{\chi}^0_1$ (**a**) are much stronger than those for the case of decay chains involving the SM-like Higgs boson h: $\tilde{\chi}^0_2 \to h + \tilde{\chi}^0_1$ (**b**). Adapted from [70]

Fig. 10.11 Overview plot of a selection of published limits on gaugino production from the CMS collaboration. All limits are given in the form of SMS limits on individual production processes. Similar results are available from the ATLAS collaboration. Adapted from [74]

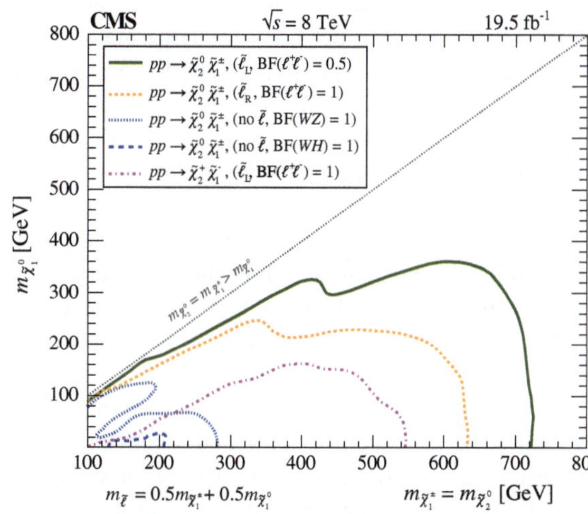

Many other interesting search channels are covered along the lines described above by the ATLAS and CMS experiments. Examples are the search for electroweak SUSY production with τ leptons without accompanying jets [73, 74] or more specific searches with Higgs bosons in the decay chain [76].

A summary of CMS searches for electroweak production of gauginos is given in Fig. 10.11 [74, 75]. Similar results are available from ATLAS [70, 71, 73]. Depending on which particles are involved, different kinematical bounds exist. The strongest limits in terms of the accessible mass range are obtained for $\tilde{\chi}^\pm_1 \tilde{\chi}^0_2 \to \tilde{\ell}\nu/\tilde{\ell}\ell \to 3\ell\nu 2\tilde{\chi}^0_1$ in specific 3-lepton searches [70, 74] that show a very striking signature of 3 light leptons with high transverse momenta and large $E_{\mathrm{T}}^{\mathrm{miss}}$. For both ATLAS and CMS, masses of up to $m_{\tilde{\chi}_1^\pm} = m_{\tilde{\chi}_2^0} \approx 700$ GeV can be excluded at the 95 % CL. However, for small values of $m_{\tilde{\chi}_1^\pm} - m_{\tilde{\chi}_1^0}$, the visible momenta of the leptons disappear. Thus, even for an assumption of an exclusive branching fraction in the given decay mode and full electroweak coupling strength at the vertex, $m_{\tilde{\chi}_1^\pm} \approx m_{\tilde{\chi}_2^0} \approx 200$ GeV is still allowed. In many SUSY breaking scenarios, however, some fine-tuning of the parameters is needed in order to achieve $m_{\tilde{\chi}_1^\pm} \approx m_{\tilde{\chi}_1^0}$.

10.4.2 Searches for SUSY Particles of the Third Generation

As already motivated in Sect. 10.2, there is a strong interest in the third-generation squarks. The main reason for this is twofold: first, third-generation squarks play a key role in determining the mass of the SM-like Higgs boson h in a SUSY model. If their mixing is large, i.e. if large mass splittings between the different mass eigenstates $\tilde{t}_1$ and $\tilde{t}_2$ (or between $\tilde{b}_1$ and $\tilde{b}_2$) occur, then $m_h \approx 126$ GeV can be achieved without excessively unnatural mass hierarchies between the SUSY particles and their SM partners. Furthermore, the large mass splitting yields a situation where $\tilde{t}_1$ and $\tilde{b}_1$ are the lightest predicted squarks. Second, the production cross section of third-generation squarks at the LHC is about one order of magnitude lower than the corresponding cross sections of the SUSY partners of light quarks at the same mass (see Fig. 10.1), despite sharing the same couplings. This is a consequence of the negligible heavy-quark content of the proton. Therefore, $\tilde{t}_{1,2}$ and $\tilde{b}_{1,2}$ can only be produced in s-channel processes via a gluon, and not in the s-channel via a quark or in t-channel processes. While the discovery of $\tilde{t}_{1,2}$ and $\tilde{b}_{1,2}$ thus would be of highest importance in the context of the connection between SUSY and the Higgs sector, their lower cross section makes them more difficult to pin down than the more abundantly produced first-generation squarks.

Figure 10.12 shows the typical production and decay processes of stops. They can either be produced directly as the heavier mass eigenstate with a decay via a Higgs boson (a) and via Z bosons (b), or via the decay of gluinos (an example for a three-body-decay of a gluino with a virtual stop is given in (c)). Further, if the lighter mass eigenstate is produced, its decay could proceed via neutralinos (d) or charginos (e, f). Therefore, a complex collection of different final states can be searched for, the realisation of which depends both on the couplings of the model and on the kinematical configuration.

These kinematical configurations are shown in Fig. 10.13 in the $m_{\tilde{t}_1}$–$m_{\tilde{\chi}_1^0}$ plane. For large $m_{\tilde{t}_1}$, the direct two-body decay $\tilde{t}_1 \to t\tilde{\chi}_1^0$ is open. For very large $m_{\tilde{t}}$, the top quark from the decay is boosted strongly in the transverse direction, which might require specific reconstruction techniques, as outlined below. As soon as $m_{\tilde{t}_1} - m_{\tilde{\chi}_1^0} = m_t$ is reached, the direct two-body decay is closed and the sensitivity drops to zero. For smaller $m_{\tilde{t}_1}$, the decay via a (virtual) chargino into the same final state can be used: $\tilde{t}_1 \to b\tilde{\chi}_1^\pm \to bW^\pm\tilde{\chi}_1^1$. For $m_{\tilde{t}_1} - m_{\tilde{\chi}_1^0} = m_b + M_{W^\pm}$, also this channel

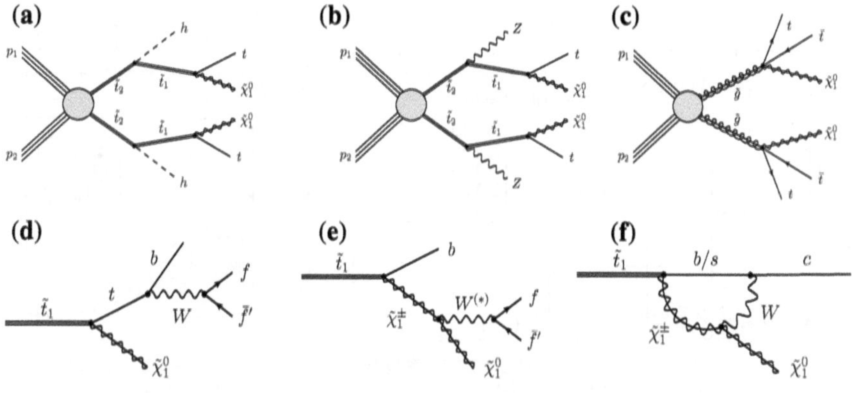

Fig. 10.12 Example diagrams for the production and decay of top quarks in conjunction with the production of the lighter or heavier stops or gluinos

Fig. 10.13 A schematic illustration of the different kinematical areas which determine the selection topology of stop searches. Adapted from [77]

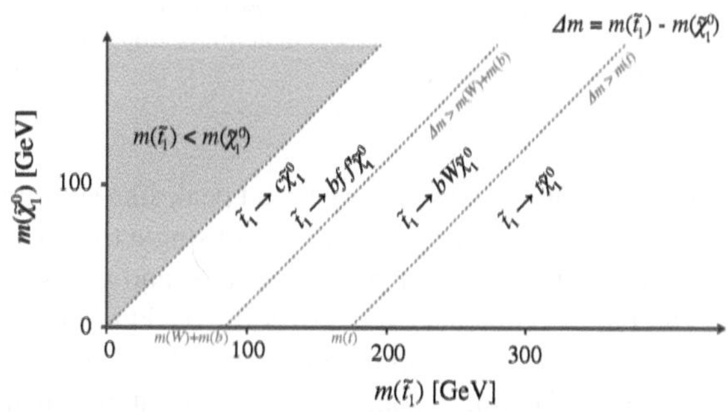

closes and the sensitivity vanishes again. Then, the decay via virtual $W \to ff'$ or the decay $\tilde{t}_1 \to c\tilde{\chi}_1^0$ remains. Finally, the region $m_{\tilde{t}_1} < m_{\tilde{\chi}_1^0}$ is uninteresting from a cosmological point of view.

Out of a very large set of available searches [76–85], a few examples shall be explained in the following. Figure 10.14 shows examples from the event selections of such searches. For the ATLAS experiment, an example from a search for heavy stops is shown [77]. For large $m_{\tilde{t}_1} \gg m_t + m_{\tilde{\chi}_1^0}$, the top quark from the decay and the hadronic decay products of the $t \to bW \to bff'$ decay are strongly boosted, and the top quark cannot easily be reconstructed from the individual jets of its decay products. In this case, the search for so-called "fat jets" or "large-R jets" is employed. First, an anti-k_t jet with a very wide radius parameter $R = 1.0$, $p_{\mathrm{T}} > 150$ GeV and $|\eta| < 2.0$ is reconstructed [86, 87]. Within this jet, jet trimming [88] is applied with an anti-k_t jet finder and $R = 0.3$. From all objects within the large-R jets that carry a fraction of more than 0.05 of the transverse momentum of the large-R jet, an invariant mass is reconstructed. The distribution of this reconstructed mass is shown in Fig. 10.14a after preselection cuts. Flavour tagging is then applied to the sub-jets, which should contain a b-quark jet. It can be seen that for signal events, the jet-

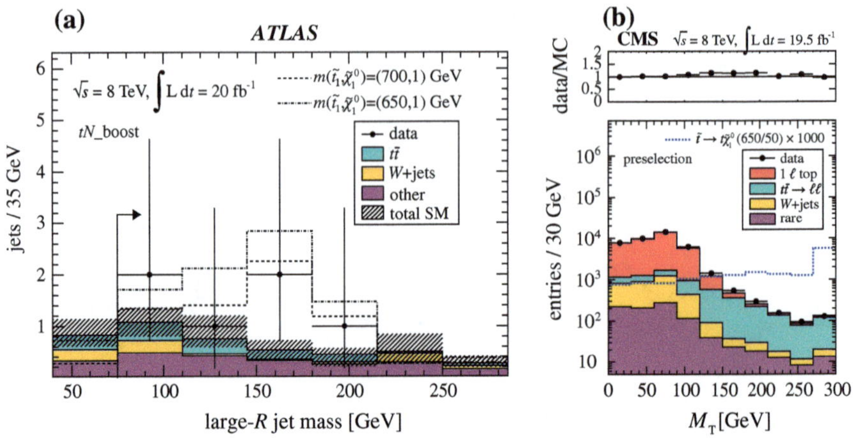

Fig. 10.14 Illustrations of typical selection variables for searches for $\tilde{t}_1$ decays. **a** The ATLAS search for heavy stops yielding fat jets shown in the form of the jet mass of the fat jet. **b** An example from CMS with a search for intermediate-mass stops shown in the form of the transverse-mass distribution. Adapted from [77, 78]

mass distribution shows a very broad peak-like structure around the top-quark mass. The remaining backgrounds are $t\bar{t}$ and single top quark production, $t\bar{t}$ production in association with a vector boson, Z+jets, and diboson production. The data show a slight, but insignificant excess over the background.

An example is a search from the CMS experiment, which is both sensitive to $\tilde{t}_1 \rightarrow t\tilde{\chi}_1^0$ and to non-resonant $\tilde{t}_1 \rightarrow bW\tilde{\chi}_1^0$ at intermediate values of $m_{\tilde{t}_1}$ [78]. There, the variable m_T already introduced in the previous section is used as a discriminator against events where the real missing transverse energy is exclusively stemming from a $W \rightarrow \ell\nu$ decay, such as semileptonic $t\bar{t}$ events. No excess over the background is observed, and the variable is used as an input to a multivariate selection, from which limits are derived.

An overview of the currently published search results and limits is given in Fig. 10.15 using examples from the ATLAS collaboration [77, 80, 82, 85, 89–91]. Similar results are available from CMS [78]. As expected, the observed sensitivity is governed by the kinematic regions defined in Fig. 10.13. The strongest limits reach up to $m_{\tilde{t}_1} > 700$ GeV for the assumption of the full strong-production cross section and, more importantly, of 100 % branching ratio into the given decay. This limit by itself is already touching the areas which could be considered theoretically interesting for an elaboration on the natural ability of SUSY to explain the hierarchy problem of the SM, without unduly fine-tuning the SUSY-parameters themselves. For $m_{\tilde{1}_1} \approx \mathcal{O}(1\ \mathrm{TeV})$ and higher, the difference between the SUSY scale and the electroweak scale becomes too large to explain the hierarchy problem of the SM Higgs mass without additional assumptions. However, as explained in Sect. 10.4.1, the kinematics of the decays close to the kinematic transitions leads to vanishing momenta of the visible final-state SM particles, which in turn yields vanishing exclusions.

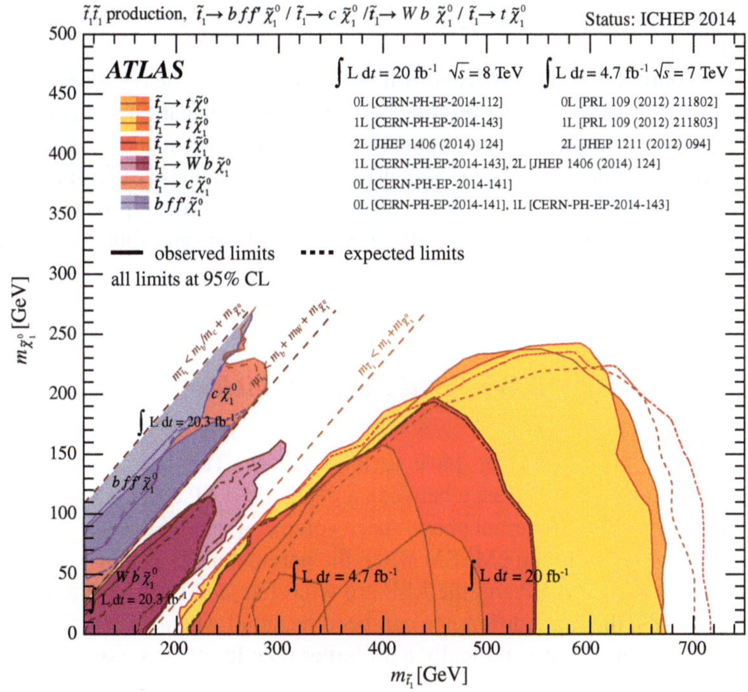

Fig. 10.15 A selection of published limits on the production of third-generation squarks from the ATLAS experiment. All limits are given in the form of SMS limits on individual production processes (see references in the plot). Similar results are obtained from the CMS experiment (see e.g. [78]). Adapted from [77, 80, 82, 85, 89–91]

Therefore, it remains for a global analysis in a given model to look for candidates of SUSY implementations which are still in agreement with the allowed low stop and low neutralino masses. An example of such an analysis is shown in Sect. 10.6.1.

10.5 Exotic SUSY Scenarios

The searches for SUSY described in the previous sections are motivated by models with a wide range of assumptions. These range from cosmological constraints (like a neutral stable LSP) over flavour physics precision measurements (no additional tree-level FCNC) to aesthetic arguments (fine-tuning). However, none of these are strictly required to be applicable for a realistic SUSY scenario. For example, the dark matter in the universe could be explained by something entirely unrelated to SUSY, while SUSY still could explain electroweak symmetry breaking and the mass scale of the Higgs boson. It is the aim of this section to introduce a brief selection of searches in which the motivation goes beyond the most constraint and most beautiful SUSY models.

The first part of this section deals with charged long-lived heavy SUSY particles. The focus of the second part lies on what happens if the explanation of the dark matter in the universe is omitted from the motivation of SUSY. The last part covers scenarios with compressed mass spectra, i.e. scenarios in which the mass difference between the initially produced SUSY particle and the sum of the decay products is small. Such scenarios lead to final states with only soft particles and thus to challenging signatures.

10.5.1 Searches for Long-Lived SUSY Particles

Long-lived SUSY particles can have various origins. They can occur in models with a very small mass difference between a charged NLSP and the neutral LSP, in which case the Lorentz-invariant phase-space volume is small. Such scenarios can be realised, for example, in the pMSSM when the parameter M_2 is much smaller than M_1 and μ. Then, the NLSP and the LSP are wino-like and both have a mass close to M_2. Similarly, the mass difference can become small when μ is small compared to M_1 and M_2, but then it is typically a bit larger than in the first case. Searches for heavy stable charged particles, which are causing large energy deposits in the tracking detectors, are described in Chap. 11 and can be reinterpreted in SUSY models like the ones mentioned.

Another potential origin for long-lived SUSY particles are scenarios with a small coupling of the NLSP to the LSP. Such a situation can come about because of a small coupling constant itself, as in the case of GMSB models where the LSP is the gravitino which couples very weakly to the NLSP. Alternatively, the coupling can be small because of heavy virtual particles in the decay process. One scenario for this is "split SUSY" [41, 42] (see Sect. 10.2.2), where the gluino and the other gauginos might be accessible at the LHC, but all scalar SUSY particles are very heavy. Since the decay of the gluino can only happen via squarks, it is much suppressed, and the gluino lifetime is long. The gluino is then forming a colour-singlet state, a so-called R-hadron, which can pass through the detector and might be stopped by hadronic interactions in one of the detector components, most likely in the dense calorimeter material. Then, after some time, the stopped gluino will decay, providing a distinct signature of hadronic activity in the calorimeter without associated tracks pointing to the interaction point. This also happens in time intervals with no bunch crossings which allows such events to be triggered with low thresholds on the energy deposit in the hadronic calorimeter [92, 93]. Using this technique, gluino masses up to 830 GeV can be excluded for a generic R-hadron model with gluino lifetimes between 10^{-6} and 10^4 s.

For models with intermediate lifetimes of the NLSP, it may not reach the outer parts of the detector, but decay inside it. A possible scenario for such meta-stable particles would be a chargino with the decay $\tilde{\chi}_1^{\pm} \rightarrow W^*(\rightarrow ff')\tilde{\chi}_1^0$. Since here the mass difference between LSP and NLSP is small, the momenta of the visible decay products are very soft and might not be detectable. However, if the lifetime is long enough for the chargino to travel at least through some fraction of the inner tracking detectors, the signature of a disappearing track provides sensitivity to such models. So far, no excess of events with disappearing tracks has been observed at ATLAS or CMS, and a lower limit on the mass of meta-stable charginos has been set at 270 GeV in one particular SUSY model [94].

10.5.2 Searches for R-Parity-Violating Models

In generic SUSY models with minimal particle content, the SUSY Lagrangian includes terms that violate conservation of lepton (L) and baryon (B) number [95, 96], e.g.

$$\frac{1}{2}\lambda_{ijk}L_iL_j\bar{E}_k + \lambda'_{ijk}L_iQ_j\bar{D}_k + \frac{1}{2}\lambda''_{ijk}\bar{U}_i\bar{D}_j\bar{D}_k,$$

where L_i and Q_i indicate the lepton and quark SU(2)-doublet superfields, respectively, while $\bar{E}_i$, $\bar{U}_i$ and $\bar{D}_i$ are the corresponding singlet superfields. The indices i, j and k refer to quark and lepton generations. The Higgs SU(2)-doublet superfield H_2 is the Higgs field that couples to up-type quarks. The λ_{ijk}, λ'_{ijk} and λ''_{ijk} parameters are new Yukawa couplings.

Since the superfields contain both SM particles and SUSY particles, they e.g. couple two SM particles to only one SUSY particle. This means that, generally, single SUSY particles can be produced, and they can decay into final states where only SM particles are present.

In the absence of a protective symmetry, L- and B-violating terms may allow proton decay at a rate that is in conflict with the tight experimental constraints on the proton's lifetime [97]. This difficulty can be avoided by imposing the conservation of R-parity defined as $P_R = (-1)^{3(B-L)+2S}$ as introduced in Sect. 10.2.1. However, experimental bounds on proton decay can also be evaded in R-parity-violating (RPV) scenarios, as long as the Lagrangian conserves either L or B, because then one of the two steps of the proton decay remains forbidden: either the production of a single SUSY particle from an up- and a down-quark in the proton, or the decay of the SUSY particle into light SM particles, such as leptons and a pion or kaon. Since in RPV models, single SUSY particles can decay, it means that in such a model there is no stable LSP, and thus no SUSY candidate for dark matter.

Therefore, an obvious way to generate an interesting search scenario at the LHC that does not contradict proton-lifetime bounds is to just allow non-zero values of λ_{ijk}. Then, B remains conserved and only L can be violated. The SUSY LSP decays into leptons, generating spectacular signatures consisting of a large number of isolated charged leptons and $E_{\mathrm{T}}^{\mathrm{miss}}$ only from the neutrinos. A search along these lines for final states with at least four isolated leptons, including electrons, muons and τ leptons,

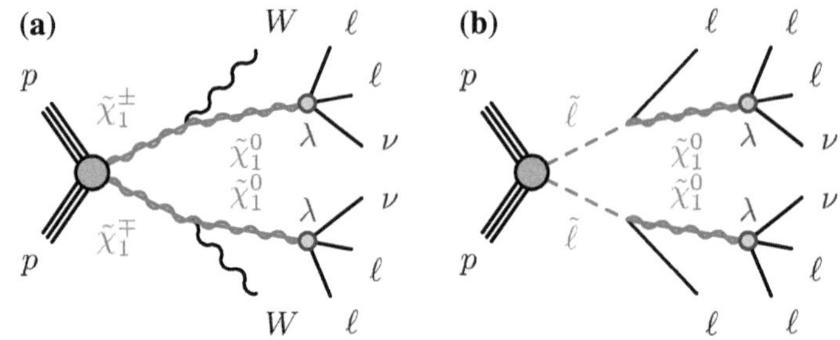

Fig. 10.16 Representative diagrams of SUSY final states in models with R-parity-violating SUSY and leptonic R-parity-violating couplings. **a** Chargino NLSP; **b** slepton NLSP

was performed by the ATLAS experiment [72]. Examples for processes leading to such final states are shown in Fig. 10.16.

The signal regions selected for such a search are defined by only 3 conditions: first, at least four light leptons or τ leptons are required; second, a moderate requirement on $E_{\mathrm{T}}^{\mathrm{miss}} > 50$ GeV is imposed; and third a veto against the Z resonance on same-flavour opposite-sign leptons is applied. The veto is applied because the typical SM backgrounds are rich in Z bosons, such as WZ, ZZ, Z+jets and rare processes such as ZWW, ZZW and $t\bar{t}Z$. No significant excess of the data over the background is observed, and thus limits are set on the allowed cross section at the 95 % CL in SMS scenarios, e.g. on topologies defined by the processes depicted in Fig. 10.16. Assuming the couplings in the SMS scenario and 100 % branching ratio in the final state under study (which, again, is typically not true in any physical model), mass limits on the NLSP and the LSP can be set. As an example, for a chargino NLSP up to $m_{\tilde{\chi}_1^\pm} > 750$ GeV can be excluded for exclusive decays into light leptons, and $m_{\tilde{\chi}_1^\pm} > 550$ GeV for decays involving τ leptons, which allow significantly higher backgrounds from QCD jets than light leptons. In these searches, the limit does not vanish for degenerate masses (as in the searches presented in Sect. 10.4), because for RPV decays the search does not depend heavily on the particles emitted between the NLSP and the LSP. Instead, the sensitivity vanishes for $m_{\mathrm{LSP}} \to 0$ GeV, because then the LSP is highly boosted and its decay products are collinear and cannot be individually reconstructed anymore.

10.5.3 Compressed Spectra

In scenarios where the mass difference of the initially produced SUSY particle and the sum of the decay products is small, like e.g. in the case of compressed SUSY [98], the transverse momentum of the final-state objects is in general small. It is, consequently, very challenging to trigger and select signal events from such models, although—e.g. in the case of gluino or squark production—the production cross section might be very large. One possibility to get some sensitivity to such models is described in Chap. 11, where the search for directly produced dark matter particles will be discussed: initial-state or final-state radiation with high transverse momentum is leading to mono-jet or mono-photon signatures, or to signatures with only one W/Z boson. A reinterpretation of these searches in SUSY scenarios provides some sensitivity to compressed mass spectra.

One example is a search for light $\tilde{t}_1$ which are almost mass-degenerate with the $\tilde{\chi}_1^0$ [85]. For such a scenario, the decay via virtual $W^* \to ff'$ or the CKM-suppressed decay $\tilde{t}_1 \to c\tilde{\chi}_1^0$ are possible (see Sect. 10.4.2 for details). In the latter case the sensitivity can be further improved by distinguishing c jets from jets with different quark flavour. The mass ranges in the $m_{\tilde{t}_1}$–$m_{\tilde{\chi}_1^0}$ plane that are excluded by a search with mono-jet events is shown in Fig. 10.15.

10.6 Current Status

In the previous sections, the experimental status of the SUSY searches after LHC Run 1 were discussed in detail. No statistically significant sign for the existence of supersymmetric particles has been found. Neither did the searches for extended Higgs sectors show any sign of new physics (Sect. 6.7), nor did the precision measurements in the B-meson system (Chap. 8). Precision measurements generally show good agreement with the SM prediction (Sect. 10.4), albeit with one potentially interesting development in the cross-section measurement of W^+W^-production (Sect. 10.6.2). Thus, limits on SUSY have been derived from the direct SUSY searches, either in the form of excluded 2-dimensional hyper-surfaces in a model-parameter space, or in the form of simplified models. But what do these limits mean in terms of the understanding of full physical models of SUSY and *all* their free parameters? How can we apply these limits, and: taking all available information from all experiments together, can we exclude models of SUSY?

It is the aim of this section to explain the status of SUSY. In Sect. 10.3.4 we have already detailed how to use the simplified model search limits. Section 10.6.1 then explains the current status of the global interpretation of SUSY searches and discusses their remaining viability as an explanation of the hierarchy problem, dark matter and the unification of forces. Section 10.6.2 discusses the interpretation of the interesting (albeit statistically not significant) anomaly of the excess in the W^+W^-cross-section measurement in Run 1 in a SUSY model, and the resulting consequences for future direct searches for SUSY at the LHC. The section closes with a summary of the prospects for the LHC Run 2 in Sect. 10.6.3.

10.6.1 Global Fits

If we assume a supersymmetric model with a manageable number of free parameters, we can estimate which regions of parameter space are preferred and which are excluded given our current experimental knowledge. The key observables in such a global parameter fit are the measured Higgs mass and the relic density attributed to the weakly interacting dark-matter agent. If we further assume thermal dark-matter production, the measured relic density requires a sizeable annihilation cross section of the dark-matter states [9, 10]. Given the precision of the Higgs-mass measurement, we also need to include the top mass with appropriate uncertainties as a model parameter. Additional constraints are usually included in our definition of the supersymmetric model: custodial symmetry will ensure that we are not in conflict with electroweak precision data; a flavour symmetry like minimal flavour violation will keep us away from the rigid flavour constraints in the K and B sectors; leaving out sneutrino dark matter accommodates most of the dark-matter direct-detection constraints.

In the MSSM, the measured Higgs mass gives valuable information on the radiative corrections mentioned in (10.1). The tree-level value as well as the radiative corrections depend on $\tan\beta$. The leading radiative correction arises from the top-stop sector and involves the same logarithm $\log m_{\tilde{t}}/m_t$ that governs the little hierarchy problem. A relatively heavy Higgs mass requires large stop masses or at least large stop mixing driven by the trilinear coupling A_t introduced in Sect. 10.2.2. Finally, the plateau value for M_h mentioned in (10.1) will only be reached for heavy additional Higgs states, $M_A > \mathcal{O}(300\ \text{GeV})$.

When we rely on gravity mediation or gauge mediation, we tend to assume that the lightest neutralino is mostly a bino, as discussed in Sect. 10.2.2. Compared to other compositions of the lightest neutralino the typical bino annihilation cross sections are small, which means they predict too large a relic density. Therefore, scenarios with dominantly bino dark matter have to rely on a mechanism which enhances their annihilation rate. One way would be a very light t-channel state, like for example the $\tilde{\tau}$, mediating the annihilation $\tilde{\chi}_1^0\tilde{\chi}_1^0 \to \tau^+\tau^-$; alternatively, an LSP mass matching the mass of a light Higgs h^0 or heavy Higgs A^0 state combined with enough of a wino component to couple to the Higgs sector will allow for a resonant s-channel annihilation; we can also postulate an additional supersymmetric particle roughly within 10 % of the LSP mass to open co-annihilation processes like, e.g., $\tilde{\chi}_1^0\tilde{f} \to Zf$ for any Standard Model fermion f [99–102], $\tilde{\chi}_1^0\tilde{\chi}^+ \to W^+Z$ [100, 103, 104], $\tilde{\chi}_1^0\tilde{\chi}_2^0 \to W^+W^-$, or $\tilde{\chi}_1^0\tilde{\chi}^+ \to t\bar{b}$. Finally, in the so-called focus point region, or, more generally, the so-called well-tempered neutralino regime of the CMSSM the higgsino mass parameter μ resides close to the point where it crosses zero and changes sign. This way the light LSP receives a sizeable higgsino fraction, which is one way to avoid such co-annihilation channels and generate the observed relic density with the annihilation process $\tilde{\chi}_1^0\tilde{\chi}_1^0 \to W^+W^-$. Similarly, in general models like the pMSSM, the LSP can have a significant wino or higgsino content, enhancing the pair-annihilation rate sufficiently.

A third class of experimental constraints come from indirect searches like the branching ratios for $B_s^0 \to \mu^+\mu^-$ and $b \to s\gamma$, and the anomalous magnetic moment of the muon. The bottom-flavour sector most effectively constrains parameter regions with small heavy Higgs masses and large $\tan\beta > \mathcal{O}(30)$. The measured anomalous magnetic moment of the muon requires light sleptons and light electroweak gauginos. In combination with other measurements this last requirement can be hard to fulfill, which means that the measured anomalous magnetic moment often adds an offset to the quality of the global fits.

Direct LHC searches enter the global fits as limits on the masses (and couplings) of the new states. The LHC limits on the weakly interacting sector are relatively mild. In contrast, the weakly interacting gaugino and slepton sectors are highly sensitive to dark matter observables and, if light, to limits from direct LEP searches. The latter essentially require all new states coupling to the weak gauge sector to have masses above half of the LEP centre-of-mass energy. In the strongly interacting sector the observation of a Higgs boson at a mass far beyond its tree-level bound $M_h < M_Z$ but in agreement with the loop-corrected value of (10.1) points to relatively large squark and gluino masses. This is in general agreement with the non-observation of SUSY in direct searches, so the direct LHC searches will not add much information to a global analysis.

Other constraints, e.g. from possible signals in the search for emissions of neutralino annihilation in the universe, do not enter the global fits. The reason is that possible signals extracted either from γ-ray data [105, 106] or from positron data [107] collected with satellite experiments are subject to significant systematic uncertainties on the astrophysical background, and thus is not necessarily a sign of physics beyond the SM.

Without discussing the technical details, e.g. on the differences in the Bayesian and frequentist approaches to supersymmetric parameter analyses, we will start from the relatively simple CMSSM case and then move towards more general models described in Sect. 10.2.2. As described above, the main constraints on the supersymmetric parameter space will come from a combination of the Higgs-mass prediction with and the properties of the weakly interacting dark-matter agent. The more constrained the model is, the more clearly we can then identify preferred parameter regions. In particular the dark matter constraint strongly reduces the number of allowed parameter regions, for example in the CMSSM [108–113]: In Fig. 10.17a we show the profile likelihood in the two mass parameters m_0 and $m_{1/2}$, where the latter largely determines the mass of the LSP. The LHC limits are not included in this

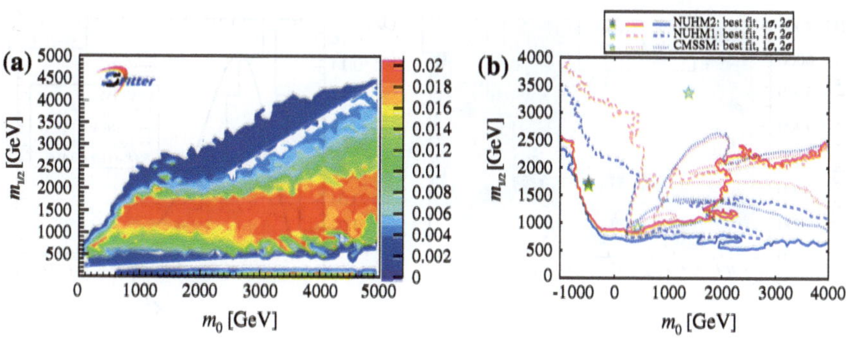

Fig. 10.17 Fit results in the CMSSM. **a** Frequentist profile likelihood in the m_0–$m_{1/2}$ plane. **b** Comparison of the CMSSM and the NUHM. Adapted from [108, 109]

figure, so a narrow strip at low but constant $m_{1/2}$ values indicates h^0-pole annihilation. Such low universal gaugino masses are ruled out by gluino searches at the LHC. The strongly correlated region for $m_{1/2} < 1$ TeV corresponds to stau co-annihilation with a light bino. The absence of the corresponding narrow line-shaped region we see that the focus point region with a light higgsino LSP is essentially ruled out. The bulk region around $m_{1/2} = 1.5$ TeV relies on the heavy Higgs-pole annihilation [108]. It is merged with the so-called well-tempered neutralino scenario, both dominating the allowed CMSSM parameter space.

In this CMSSM parameter study, large values $\tan\beta > 40$ are preferred in order to accommodate the relatively large Higgs-mass value. As expected, the allowed parameter regions by the Higgs mass and the relic density touch the LHC exclusion for low $m_{1/2}$, but the preferred regions are beyond the sensitivity of direct LHC searches (see Fig. 10.3). In Fig. 10.17b, a similar result is shown, now based on the NUHM setup introduced in Sect. 10.2.2. While the best-fit value moves around in the m_0 versus $m_{1/2}$ plane, owing to a relatively flat profile likelihood profile, the qualitative result hardly changes [109].

An interesting question is how well the CMSSM—as one of the most constrained models for the supersymmetric mass spectrum—agrees with all available data in a comprehensive statistical analysis. In Fig. 10.18a the mass spectrum of the global best-fit parameter point of the MSSM is shown [113]. As expected, gluinos and light-flavour squarks are pushed to high masses. The allowed heavy Higgs states cover a wide range, but also prefer to be heavy. Most of the weakly interacting new states prefer to be light, but are not necessarily accessible in direct production in Run 1. The dark-matter annihilation in this best-fit point proceeds through stau co-annihilation. In Fig. 10.18b the statistical analysis of the CMSSM and the corresponding log-likelihoods are shown. The smooth curve shows the naive expectation of the distribution of log-likelihoods, while the histogram shows the actual result from toy experiments. The problem with all CMSSM fits is that slepton and squark masses are linked through the universal scalar mass assumption. Experimentally, the large measured value of the anomalous magnetic moment of the muon prefers

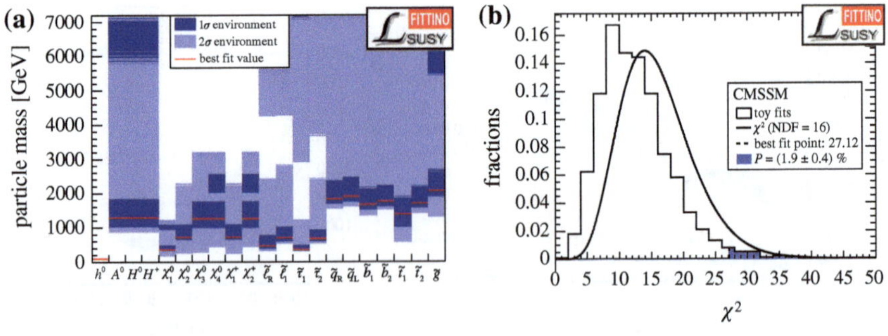

Fig. 10.18 **a** Mass spectrum for the best-fit point in the CMSSM including *error bars*. **b** Calculation of the CMSSM p-value below 5 % CL. Adapted from [113]

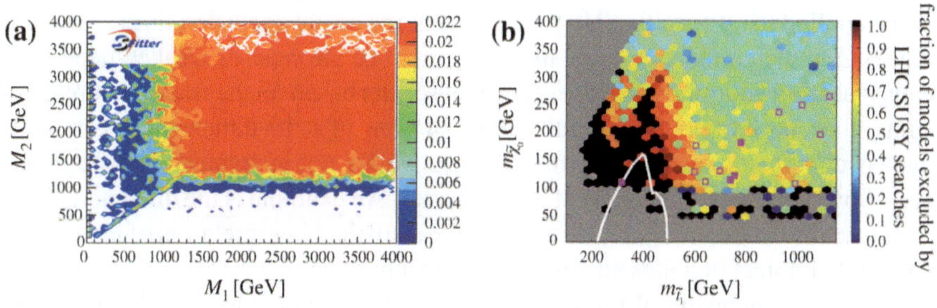

Fig. 10.19 Fit results in the pMSSM. **a** Frequentist profile likelihood in the bino-wino mass plane before applying LHC results. **b** Fraction of excluded pMSSM models in the LSP-$\tilde{t}_1$ mass plane. Overlaid is a limit from a simplified-model interpretation. Adapted from [108, 114]

light sleptons, while the stronger Higgs-mass constraint prefers heavy squarks. Correspondingly, the overall χ^2 equivalent shows that the quality of the fit is poor and the corresponding p-value is low. In essence, the CMSSM is ruled out at more than 95 % CL [113].

This tension in the CMSSM suggests to look at significantly more general, but still predictive parametrisations of SUSY. The biggest challenge in pMSSM parameter studies is to effectively scan the 19-dimensional parameter space [108, 114, 115]. In Fig. 10.19 we show the 2-dimensional profile likelihood in the bino–wino mass plane [108]. As for the CMSSM, the dark matter constraint determines the main features of the allowed parameter range in the weakly interacting sector. The strongly correlated diagonal strip for $M_1 \sim M_2 \sim 1$ TeV corresponds to stau co-annihilation; the individual points around $M_1 \sim 63$ GeV will at sufficiently large statistics form a line resulting from h^0-pole annihilation; the bulk region for large M_1 and M_2 receives contributions from heavy Higgs-pole annihilation, or, more generally the well-tempered neutralino scenario including chargino co-annihilation. There exists allowed parameter space for stop co-annihilation, not covered by this specific parameter analysis.

Based the Higgs-mass constraints a pMSSM analysis can for example map out the preferred parameter space of the extended Higgs sector [116]. One of the most interesting parameter correlation in combination with dark matter then becomes the bino mass versus the lightest stop mass. The former represents the dark matter constraints while the latter probes the Higgs-mass constraints. In Fig. 10.19 the fraction of pMSSM parameter points excluded by the LHC searches is shown. Obviously, parameter points with low masses are more often excluded by the searches discussed in Sect. 10.4. Comparing the contour of constant fraction of rejected points to the contour from a simplified-model interpretation, we observe differences. The reason lies in the assumed decays. On the one hand, actual branching ratios smaller than 100 % result in a weaker limit or fewer excluded points than for the simplified model. On the other hand, decay modes not governed by the simplified model can have an enhanced sensitivity and consequently lead to stronger limits or more excluded points. This difference illustrates the problem with simplified model interpretations. Their limits can be applied to all specific SUSY models, but the actual exclusion lines are not the same in complete models. There, the limit might be stronger or weaker, depending on its mass and coupling structure. Note that results based on the fraction of rejected points must be interpreted carefully. First, a scan of a 19-dimensional parameter space is computationally very demanding, so we are not guaranteed that we do not miss allowed points for example at lower masses. An example is the disrupted line of h^0-pole annihilation in Fig. 10.19. Second, the statistical interpretation of the fraction of excluded points includes an ad-hoc Bayesian measure.

The main conclusion of the supersymmetric parameter studies is that including the Higgs measurements of Run 1, the available data on dark matter and indirect constraints, as well as the LHC exclusion bounds the MSSM is not under significant pressure—in strong contrast to the more popular constraint models as the CMSSM. It still offers an explanation for dark matter, for the hierarchy problem, and for approximate gauge coupling unification.

10.6.2 Experimental Anomalies

From any experiment based on statistical analyses and including significant systematic complexity we should always expect experimental anomalies which lend themselves to new physics interpretations. The crucial test of any anomaly pointing, for example, to physics beyond the Standard Model will be if it becomes more significant when more data are included. A few such anomalies exist at the end of Run 1 at the LHC. They hardly contribute to the global fits, but they maybe interesting to follow up when LHC resumes data taking in 2015.

Already the 7 TeV analyses of W^+W^- production in ATLAS [117] and CMS [118] showed only modest agreement between the data and Standard Model simulations. General problems with this channel, both in the total rate and in some of the distributions, have been known for a long time [119]. The phenomenological analyses of [120, 121] also includes some CMS 8 TeV results in the same channel [122]. The experimental signature that can be interpreted as Standard Model background plus supersymmetry signal, is dileptons plus missing transverse momentum. The signature includes a jet veto to reject the top-quark pair production background. The assumed signal process is

$$pp \to \tilde{t}_1\tilde{t}_1^* \to (\tilde{\chi}_1^+ b)\,(\tilde{\chi}_1^- \bar{b}) \to (\tilde{\chi}_1^0 \ell^+ \nu b)\,(\tilde{\chi}_1^0 \ell^- \bar{\nu} \bar{b}). \tag{10.5}$$

This production and decay process is similar to the cascade decay with its dilepton edge discussed in Sect. 10.2.3. However, rather than through a slepton, the chargino decay proceeds through an off-shell W boson. The final state differs from the W^+W^- background only in the the additional existence of two bottom-quark jets. However, as long as the masses of all new particles involved are small and the mass difference does not exceed $m_{\tilde{t}} - m_{\tilde{\chi}^\pm} \sim 7$ GeV, these jets will be too soft to be observable in the general QCD activity of the events. If the mass difference

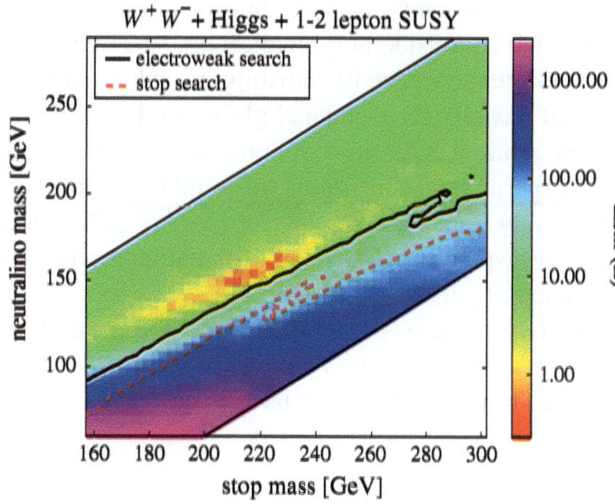

Fig. 10.20 Two-dimensional likelihood for the relevant mass parameters in the decay chain shown in (10.5). The analysis was performed on ATLAS and CMS W^+W^-measurements. Adapted from [120]

between the stop and the chargino is fixed, the two free parameters in this analysis are the stop mass and the LSP mass. Figure 10.20 shows the likelihood of a scan of the two-dimensional mass plane. It is dominated by the event count in the signal regions of the three LHC analyses mentioned above. The stop branching ratio to the final state of (10.5) is assumed to be 100 %. The strong correlation follows constant mass differences between the stop, the chargino, and the LSP. The best-fit point is

$$m_{\tilde{t}} = 212^{+35}_{-40}\,\text{GeV} \qquad \text{and} \qquad m_{\tilde{\chi}_1^0} = 150^{+30}_{-20}\,\text{GeV}.$$

The estimated significance of this single anomaly is around 2.6 standard deviations and can be increased by including tri-lepton signals [120]. The final individual and combined significances will have to be determined in a full experimental analysis.

In a situation like this it is crucial to ensure that the potential signal for new physics is not ruled out by any other LHC searches. For example, searches for direct electroweak gaugino production in the 2-lepton plus missing energy channel contribute to the more constraining of the two exclusion limits shown in Fig. 10.20. The second exclusion limit arises from direct stop searches including bottom-quark jets.

As discussed in Sect. 10.2.3 the invariant mass distribution of two leptons, $m_{\ell\ell}$, is a standard signature for (supersymmetric) cascade decays. There exist indications for such an anomaly in two opposite-sign same-flavor leptons combined with missing transverse momentum in CMS [123]. As discussed in Sect. 10.2.3 the main backgrounds can be controlled by opposite-sign opposite-flavor subtraction. The edge of the corresponding distribution appears around 80 GeV. From this information we can compute the hypersphere of supersymmetric mass values following (10.4).

A sizeable number of events under such an edge points to a strongly interacting hard scattering process. To relate a possible LHC signal to the direct production of light-flavor squarks and gluinos, we have to take into account the limits from generic searches for jets plus missing energy, which only allow heavy light-flavour squarks and gluinos. In comparison, the production rates of third-generation squark pairs are expected to be considerably smaller: first, the light-flavour rates include an additional factor 8 for experimentally indistinguishable squarks, as listed in Table 10.1; second, for light flavours the t-channel gluino exchange contributes up to half of the cross section, while it is strongly suppressed for sbottoms and non-existent for stops. To avoid experimental constraints we could link an observed $m_{\ell\ell}$ edge to the production of relatively light sbottom pairs, $pp \to \tilde{b}\tilde{b}^*$. The typical supersymmetric parameter range corresponding to such an explanation is similar to the masses required for the WW anomaly discussed above.

Of course, these anomalies could have a number of explanations that are not related to supersymmetry. However, it shows that at the end of Run 1, there are open questions in LHC data which have to be watched in the upcoming runs.

10.6.3 Prospects for LHC Run 2

As was shown in Sect. 10.6.1, the CMSSM as the prime example for a constraint SUSY model is at the verge of being completely excluded. However, the exclusion of a single highly constraint model in a global analysis must not be taken as a hint against the realisation of a more general SUSY model. Therefore, also in LHC Run 2, SUSY searches will be of high importance. As motivated in the pMSSM studies in the previous section, special attention will be given to searches for electroweak SUSY partners, sleptons and third-generation squarks. If these remain light (and thus more easily accessible at the LHC), a naturalness problem in the SUSY sector stemming from the difference between the electroweak scale and the mass scale of the SUSY particles can be avoided. Thus gauginos, sleptons and third-generation squarks are prime objectives for the LHC.

As an example of studies for gaugino and top squark searches at higher luminosity, results from ATLAS [124, 125] are shown in Fig. 10.21. The existing observed limits derived from the non-observation of SUSY [71, 77] are compared to the expected reach at the LHC with an ultimate integrated luminosity of 300 fb^{-1} and at the HL-LHC 2.5 with an ultimate integrated luminosity of 3 ab^{-1}. Similar projections exist for CMS [126]. The plots show both the reach for exclusion (dashed) at 95 % CL and for a 5 σ discovery. It is obvious that already the discovery reach at $\mathcal{L} = 300$ fb^{-1} extends significantly beyond the current exclusion, and thus great discoveries might be expected. However, if the mass scales of SUSY particles in all viable models could be pushed beyond the $\mathcal{O}$(TeV) region, SUSY would completely loose its attractiveness as an explanation of the hierarchy problem.

The LHC prospects for searches after gluino pair production are also shown in Fig. 10.21, both for scenarios with 100 % branching ratio of $\tilde{g} \to q\bar{q}\tilde{\chi}_1^0$ and $\tilde{g} \to t\bar{t}\tilde{\chi}_1^0$ [126]. In both scenarios the expected discovery reach is extended to 2 TeV and beyond.

On the other hand, the exclusion and discovery potential is still expected to be small for specific mass cases, e.g. for $m_{\tilde{\chi}_2^0} - m_{\tilde{\chi}_1^0} \approx M_Z$ or $m_{\tilde{t}_1} - m_t \approx m_{\tilde{\chi}_1^0}$. In such

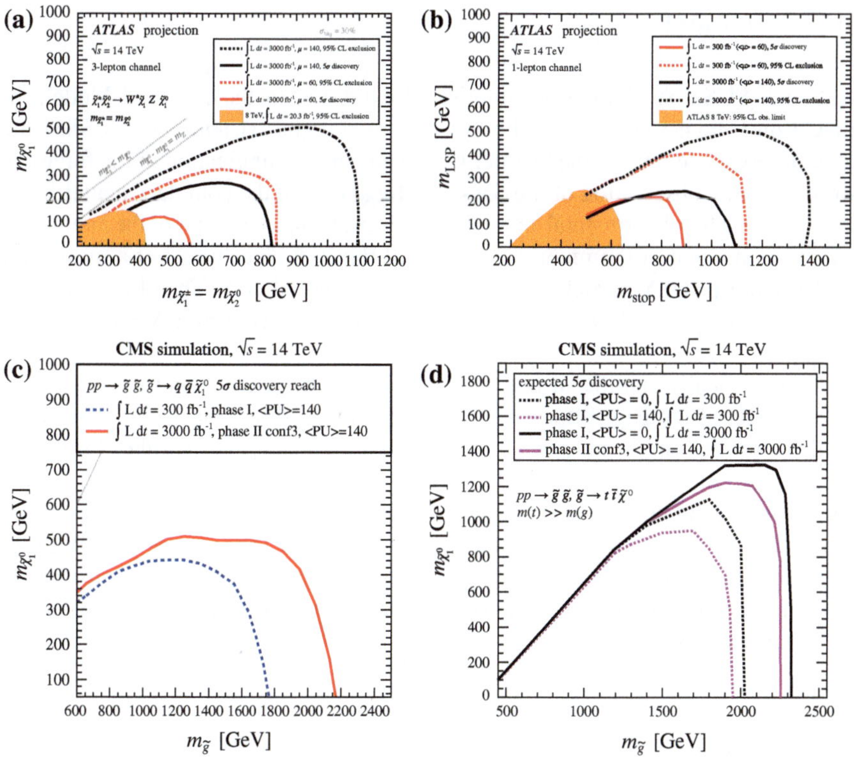

Fig. 10.21 **a** The expected 95 % CL exclusion (*dashed lines*) and discovery contours (*solid lines*) for LHC luminosity scenarios with 300 fb^{-1} (red) and 3000 fb^{-1} (*black*) in the $\tilde{\chi}_1^\pm/\tilde{\chi}_2^0 - \tilde{\chi}_1^0$ mass plane for a simplified model with 100 % branching ratio of $\tilde{\chi}_1^\pm/\tilde{\chi}_2^0 \to (W/Z)\tilde{\chi}_1^0$ [124]. The measured 8 TeV exclusion contour is displayed as the *orange area* [71]. **b** Same as (**a**) but the $\tilde{t}_1$–$\tilde{\chi}_1^0$ mass plane assuming the decay $\tilde{t}_1 \to t\tilde{\chi}_1^0$ with a branching ratio of 100 % in the 1-lepton decay channel [125]. In addition, the observed excluded region from [77] is shown. **c** The 5 σ discovery reach for LHC luminosity scenarios with 300 fb^{-1} (*blue dashed line*) and 3000 fb^{-1} (*red solid line*) in the $\tilde{g}$–$\tilde{\chi}_1^0$ plane for an assumed 100 % branching ratio of $\tilde{g} \to q\bar{q}\tilde{\chi}_1^0$ [126]. **d** Same as (**c**) but for a 100 % branching ratio of $\tilde{g} \to q\bar{t}\tilde{\chi}_1^0$, shown for no (*black*) and realistic (*purple*) pile-up conditions [126]. Adapted from [77, 125, 126]

a case, the momentum of the SM decay particles approaches 0. Therefore, the trigger and selection efficiencies become small and the discovery or exclusion potential can be weakened. The same mechanism applies also to other searches. Thus, it can be expected that also in the unfortunate case of a non-discovery of SUSY in LHC Run 2 and beyond, the fate of SUSY will not be decided by the achievable exclusions directly, but, if at all, by a global analysis looking for SUSY models which fulfill at the same time all constraints from direct searches and all indirect constraints from other sensitive measurements, as outlined in Sect. 10.6.1.

10.7 Summary

Although SUSY still is the best-motivated extension of the Standard Model, the LHC Run 1 did not provide us with a clear verdict on its existence. The MSSM still provides explanations for the hierarchy problem and the unification of the forces, as well as for the observation of the SM-like Higgs boson, the cold dark matter in the universe, and for experimental anomalies such as the measurement of the anomalous magnetic moment of the muon $(g-2)_\mu$. All that can be achieved at a mass scale of the lightest SUSY particles of at most a few hundred GeV. The observed Higgs mass, for example, or dark-matter searches could have ruled out the MSSM, but they did not.

No statistically significant hint for new physics has been found in Run 1. The combination of the Higgs-mass measurement and the direct mass limits on SUSY particles on the one hand, and observables like $(g-2)_\mu$ on the other hand have excluded strongly constrained models like the CMSSM at the 95 % CL level. This is a great success of combined experimental studies at the LHC, at B factories, at the Tevatron, LEP, from low-energy precision physics, and from cosmology. With the increased energy and luminosity of LHC Run 2, we can expect definite answers on new physics models which include strongly interacting states at the few-TeV scale, including the MSSM.

References

1. P.W. Higgs, Phys. Lett. **12**, 132–133 (1964)
2. P.W. Higgs, Phys. Rev. Lett. **13**, 508–509 (1964)
3. F. Englert, R. Brout, Phys. Rev. Lett. **13**, 321–323 (1964)
4. CMS Collaboration, Phys. Lett. B **716**, 30–61 (2012)
5. ATLAS Collaboration, Phys. Lett. B **716**, 1–29 (2012)
6. T. Plehn, Lect. Notes Phys. **886** (2015)
7. S.P. Martin, Adv. Ser. Direct. High Energy Phys. **21**, 1–153 (2010)
8. D.E. Morrissey, T. Plehn, T.M. Tait, Phys. Rept. **515**, 1–113 (2012)
9. G. Jungman, M. Kamionkowski, K. Griest, Phys. Rept. **267**, 195–373 (1996)
10. G. Bertone, D. Hooper, J. Silk, Phys. Rept. **405**, 279–390 (2005)
11. W. Bernreuther, Lect. Notes Phys. **591**, 237–293 (2002)
12. U. Amaldi, W. de Boer, H. Furstenau, Phys. Lett. B **260**, 447–455 (1991)
13. E. Gildener, Phys. Rev. D **14**, 1667 (1976)
14. S. Weinberg, Phys. Rev. D **13**, 974–996 (1976)
15. B.W. Lee, C. Quigg, H. Thacker, Phys. Rev. Lett. **38**, 883–885 (1977)
16. Y. Golfand, E. Likhtman, JETP Lett. **13**, 323–326 (1971)
17. D. Volkov, V. Akulov, JETP Lett. **16**, 438–440 (1972)
18. J. Wess, B. Zumino, Nucl. Phys. B **70**, 39–50 (1974)
19. A. Salam, J. Strathdee, Nucl. Phys. B **76**, 477–482 (1974)
20. R. Haag, J.T. Lopuszanski, M. Sohnius, Nucl. Phys. B **88**, 257 (1975)
21. J. Wess, J. Bagger, *Supersymmetry and Supergravity* (Princeton University Press, Princeton, 1992)
22. J. Terning, *Modern Supersymmetry: Dynamics and Duality* (Oxford University Press, Oxford, 2006). doi:10.1093/acprof:oso/9780198567639.001.0001

23. M. Drees, R. Godbole, P. Roy, *Theory and Phenomenology of Sparticles: An Account of Four-Dimensional N = 1 Supersymmetry in High Energy Physics* (World Scientific, Hackensack, 2004)
24. H. Baer, X. Tata, *Weak Scale Supersymmetry: From Superfields to Scattering Events* (Cambridge University Press, Cambridge, 2006)
25. S. Dawson, H. Georgi, Phys. Rev. Lett. **43**, 821 (1979)
26. M. Einhorn, D. Jones, Nucl. Phys. B **196**, 475 (1982)
27. S. Dimopoulos, H. Georgi, Nucl. Phys. B **193**, 150 (1981)
28. M. Dine, W. Fischler, Phys. Lett. B **110**, 227 (1982)
29. G.R. Farrar, P. Fayet, Phys. Lett. B **76**, 575–579 (1978)
30. H.K. Dreiner, Adv. Ser. Direct. High Energy Phys. **21**, 565–583 (2010)
31. J.F. Gunion, H.E. Haber, G.L. Kane, S. Dawson, Front. Phys. **80**, 1–448 (2000)
32. A. Djouadi, Phys. Rept. **459**, 1–241 (2008)
33. P.H. Chankowski, S. Pokorski, J. Rosiek, Nucl. Phys. B **423**, 437–496 (1994)
34. H.E. Haber, R. Hempfling, Phys. Rev. D **48**, 4280–4309 (1993)
35. R. Hempfling, A.H. Hoang, Phys. Lett. B **331**, 99–106 (1994)
36. M. Carena et al., Nucl. Phys. B **580**, 29–57 (2000)
37. L. Girardello, M.T. Grisaru, Nucl. Phys. B **194**, 65 (1982)
38. H. Goldberg, Phys. Rev. Lett. **50**, 1419 (1983)
39. J.R. Ellis et al., Nucl. Phys. B **238**, 453–476 (1984)
40. J. Jaeckel, V.V. Khoze, T. Plehn, P. Richardson, Phys. Rev. D **85**(015), 015 (2012)
41. N. Arkani-Hamed, S. Dimopoulos, JHEP **06**, 073 (2005)
42. G. Giudice, A. Romanino, Nucl. Phys. B **699**, 65–89 (2004)
43. L.E. Ibanez, G.G. Ross, Phys. Lett. B **105**, 439 (1981)
44. A.H. Chamseddine, R.L. Arnowitt, P. Nath, Phys. Rev. Lett. **49**, 970 (1982)
45. R. Barbieri, S. Ferrara, C.A. Savoy, Phys. Lett. B **119**, 343 (1982)
46. M. Dine, W. Fischler, M. Srednicki, Nucl. Phys. B **189**, 575–593 (1981)
47. S. Dimopoulos, S. Raby, Nucl. Phys. B **192**, 353 (1981)
48. M. Kramer et al., arXiv:1206.2892 (2012)
49. CMS Collaboration, JHEP **06**, 055 (2014)
50. ATLAS Collaboration, JHEP **09**, 176 (2014)
51. ATLAS Collaboration, JHEP **06**, 035 (2014)
52. L. Randall, D. Tucker-Smith, Phys. Rev. Lett. **101**(221), 803 (2008)
53. CMS Collaboration, Eur. Phys. J. C **73**, 2568 (2013)
54. CMS Collaboration, Phys. Rev. Lett. **111**(8), 081 802 (2013)
55. C. Lester, D. Summers, Phys. Lett. B **463**, 99–103 (1999)
56. A. Barr, C. Lester, P. Stephens, J. Phys. G **29**, 2343–2363 (2003)
57. CMS Collaboration, JHEP **10**, 018 (2012); CMS Collaboration, arXiv:1502.04358
58. ATLAS Collaboration, Phys. Rev. D **86**, 092 002 (2012)
59. CMS Collaboration, Eur. Phys. J. C **73**, 2404 (2013)
60. CMS Collaboration, JHEP **01**, 163 (2014)
61. CMS Collaboration, Eur. Phys. J. C **73**, 2493 (2013)
62. ATLAS Collaboration, Eur. Phys. J. C **72**, 2215 (2012)
63. ATLAS Collaboration, JHEP **09**, 103 (2014)
64. CMS Collaboration, Phys. Rev. D **90**(032), 006 (2014)
65. S. Dimopoulos, S.D. Thomas, J.D. Wells, Nucl. Phys. B **488**, 39–91 (1997)
66. P. Meade, N. Seiberg, D. Shih, Prog. Theor. Phys. Suppl. **177**, 143–158 (2009)
67. ATLAS Collaboration, Phys. Lett. B **718**, 411–430 (2012)
68. CMS Collaboration, JHEP **03**, 111 (2013)
69. CMS Collaboration, Phys. Rev. D **88**(5), 052 017 (2013)
70. ATLAS Collaboration, JHEP **04**, 169 (2014)
71. ATLAS Collaboration, JHEP **05**, 071 (2014)
72. ATLAS Collaboration, Phys. Rev. D **90**, 052 001 (2014)
73. ATLAS Collaboration, JHEP **10**, 96 (2014)

74. CMS Collaboration, Eur. Phys. J. C **74**(9), 3036 (2014)
75. CMS Collaboration, Phys. Rev. D **90**, 9 092007 (2014)
76. CMS Collaboration, Phys. Rev. Lett. **112**, 161 802 (2014)
77. ATLAS Collaboration, JHEP **1411**, 118 (2014)
78. CMS Collaboration, Eur. Phys. J. C **73**, 2677 (2013)
79. ATLAS Collaboration, JHEP **10**, 189 (2013)
80. ATLAS Collaboration, JHEP **06**, 124 (2014)
81. ATLAS Collaboration, Eur. Phys. J. C **74**, 2883 (2014)
82. ATLAS Collaboration, JHEP **09**, 015 (2014)
83. ATLAS Collaboration, Eur. Phys. J. C **74**(10), 3109 (2014)
84. ATLAS Collaboration, JHEP **10**, 24 (2014)
85. ATLAS Collaboration, Phys. Rev. D **90**, 052 008 (2014)
86. J.M. Butterworth, A.R. Davison, M. Rubin, G.P. Salam, Phys. Rev. Lett. **100**(242), 001 (2008)
87. ATLAS Collaboration, JHEP **09**, 076 (2013)
88. D. Krohn, J. Thaler, L.T. Wang, JHEP **02**, 084 (2010)
89. ATLAS Collaboration, Phys. Rev. Lett. **109**, 211 802 (2012)
90. ATLAS Collaboration, Phys. Rev. Lett. **109**, 211 803 (2012)
91. ATLAS Collaboration, JHEP **11**, 094 (2012)
92. ATLAS Collaboration, Phys. Rev. D **88**(11), 112 003 (2013)
93. CMS Collaboration, JHEP **08**, 026 (2012)
94. ATLAS Collaboration, Phys. Rev. D **88**(11), 112 006 (2013)
95. S. Weinberg, Phys. Rev. D **26**, 287 (1982)
96. N. Sakai, T. Yanagida, Nucl. Phys. B **197**, 533 (1982)
97. SNO Collaboration, Phys. Rev. Lett. **92**, 102 004 (2004)
98. S.P. Martin, Phys. Rev. D **75**(115), 005 (2007)
99. K. Griest, D. Seckel, Phys. Rev. D **43**, 3191–3203 (1991)
100. S. Mizuta, M. Yamaguchi, Phys. Lett. B **298**, 120–126 (1993)
101. J.R. Ellis, T. Falk, K.A. Olive, Phys. Lett. B **444**, 367–372 (1998)
102. J.R. Ellis, K.A. Olive, Y. Santoso, Astropart. Phys. **18**, 395–432 (2003)
103. P. Binetruy, G. Girardi, P. Salati, Nucl. Phys. B **237**, 285 (1984)
104. J. Edsjo, P. Gondolo, Phys. Rev. D **56**, 1879–1894 (1997)
105. LAT Collaboration, Phys. Rev. D **86**, 022 002 (2012)
106. C. Weniger, JCAP **1208**, 007 (2012)
107. PAMELA Collaboration, Phys. Rev. Lett. **111**(8), 081 102 (2013)
108. S. Henrot-Versillé et al., Phys. Rev. D **89**(055), 017 (2014)
109. O. Buchmüller et al., Eur. Phys. J. C **74**, 123212 (2014)
110. C. Strege et al., JCAP **1304**, 013 (2013)
111. T. Cohen, J.G. Wacker, JHEP **09**, 061 (2013)
112. L. Roszkowski, E.M. Sessolo, A.J. Williams, JHEP **08**, 067 (2014)
113. P. Bechtle et al., arXiv:1410.6035 (2014)
114. M.W. Cahill-Rowley, J.L. Hewett, A. Ismail, T.G. Rizzo, Phys. Rev. D **88**(3), 035 002 (2013)
115. C. Boehm, P.S.B. Dev, A. Mazumdar, E. Pukartas, JHEP **06**, 113 (2013)
116. B. Dumont, J.F. Gunion, S. Kraml, Phys. Rev. D **89**(055), 018 (2014)
117. ATLAS Collaboration, Phys. Rev. D **87**(11), 112 001 (2013)
118. CMS Collaboration, Eur. Phys. J. C **73**, 2610 (2013)
119. D. Curtin, P, Jaiswal, P. Meade, Phys. Rev. D **87**(3), 031 701 (2013)
120. J.S. Kim, K. Rolbiecki, K. Sakurai, J. Tattersall, JHEP **1412**, 010 (2014)
121. D. Curtin, P. Meade, P.J. Tien, Phys. Rev. D **90**, 11 115012 (2014)
122. CMS Collaboration, Phys. Lett. B **721**, 190–211 (2013)
123. B. Allanach, A.R. Raklev, A. Kvellestad, arXiv:1409.3532 (2014); CMS Collaboration, arXiv:1502.06031 (2015)
124. G. Aad et al., ATL-PHYS-PUB-2014-010 (2014)
125. G. Aad et al., ATL-PHYS-PUB-2013-011 (2013)
126. CMS Collaboration, arXiv:1307.7135 (2013)

Chapter 11
Searches for Physics Beyond the Standard Model

Frank Ellinghaus, Kerstin Hoepfner and Thorsten Ohl

Abstract Despite its phenomenal success, the Standard Model is not expected to be a complete description of nature all the way up to the Planck scale where quantum gravity comes into play. As the collider at the current energy frontier, the LHC is in a unique position to look for signals of physics beyond the Standard Model. In this chapter, results from the first run of the LHC are surveyed that go beyond the supersymmetric extension of the Standard Model discussed in the previous chapter. At the time of writing, no evidence for new phenomena has been discovered. Instead, the LHC experiments have constrained the parameter space of conceivable models significantly. The most prominent results are on new gauge bosons, compositeness, extra dimensions, black holes, dark matter and leptoquarks. The journal publications are always organised according to the final states studied in a particular analysis. Thus they combine results for various physics scenarios. The present chapter provides the complementary view, organised according to the models for physics beyond the Standard Model.

F. Ellinghaus (✉)
Institut für Physik, Johannes Gutenberg-Universität Mainz, Staudingerweg 7, 55128 Mainz, Germany
e-mail: Frank.Ellinghaus@cern.ch

K. Hoepfner
Rheinisch-Westfälische Technische Hochschule Aachen, Physikzentrum, 52056 Aachen, Germany
e-mail: hoepfner@physik.rwth-aachen.de

T. Ohl
Universität Würzburg, Am Hubland, 97074 Würzburg, Germany
e-mail: ohl@physik.uni-wuerzburg.de

T. Schörner-Sadenius (ed.), *The Large Hadron Collider*,
DOI 10.1007/978-3-319-15001-7_11

11.1 Introduction

The Standard Model (SM) is extremely successful in describing the observed elementary particles and their interactions. There are, however, several pieces of experimental evidence for physics that is not described by the SM. The most obvious of these is gravity. The energy scale corresponding to gravitational interactions—the Planck scale $M_{\rm Pl}$—lies 16 orders of magnitude above the electroweak (EW) scale $v_{\rm weak} \approx 254\,{\rm GeV}$. In the minimal SM, with no new degrees of freedom between the EW and the Planck scale, these scales will mix under radiative corrections. The observed absence of this mixing, i.e. the exponentially large separation of these scales leads to the so-called "naturalness problem".

The proposed solutions for this problem typically involve new symmetries that prohibit this mixing, e.g. supersymmetry (see Chap. 10) or chiral symmetries in composite models. All these solutions predict an extended particle spectrum with a corresponding mass scale much closer to the EW than to the Planck scale.

In the SM, the EW scale is a free parameter of the Higgs sector and subject to renormalisation and mixing. In contrast, a complete model could generate a low scale by the dynamical breaking of supersymmetry or chiral symmetries, similar to the generation of hadron masses in QCD. Then this new scale protects the EW scale from being pushed up to the Planck scale by an otherwise uninhibited mixing of EW-scale and Planck-scale degrees of freedom through loop corrections.

In order to really solve the naturalness problem, the new scale should not lie too far above the EW scale. The LHC, therefore, is in a good position to test predictions based on these solutions to the naturalness problems or, at least, to set lower limits on the relevant scales.

Strong evidence for the incompleteness of the SM also comes from the observation of non-baryonic dark matter (DM) in our universe. A generic new weakly interacting massive particle (WIMP) with a mass close to the EW scale and couplings similar to the weak coupling constant will produce the correct dark-matter remnant density, provided that there is also a new conserved quantum number that protects it from decay. Despite of this very suggestive "WIMP miracle" (see Sect. 10.6.1), however, there is no guarantee that the corresponding particles will be in the reach of the LHC.

The perfect cancellation of all gauge and gravitational anomalies in the non-simple $SU(3)_c \times SU(2)_L \times U(1)_Y$ gauge group of the SM could find a natural explanation in the unification of all interactions in a simple gauge group. This requires the existence of additional gauge bosons (see Sect. 11.2), some of which will even couple quarks and leptons (see Sect. 11.6). However, it must be noted that the scale of this unification, as predicted by the running of the SM coupling constants, lies far above the range accessible by the LHC.

Searches for particles solving the naturalness and dark matter problems follow the established narrative of high energy physics and constitute a major fraction of the experimental programme at the LHC. However, one must not become blind for the unexpected, i.e. for new particles or new interactions that solve none of the currently perceived problems of the SM. As long as such new particles and interactions can be described by a consistent (effective) quantum field theory, and as long as they decouple sufficiently rapidly at low energies so that the SM emerges as an effective theory and provides a quantitatively good approximation, there is no reason to rule out such scenarios.

The consistency requirements for a theory of physics beyond the SM need not include renormalisability, i.e. the ability to describe phenomena at arbitrarily high energies with a finite set of parameters. Instead, there might be new physics at a higher scale the effects of which will first be felt as non-renormalisable interactions of SM fields. Indeed, effective quantum field theories (EFT) that provide accurate quantitative predictions up to an energy scale above today's collider energies are powerful and theoretically consistent tools [1]. In the EFT, any deviation of the experimental results from the SM prediction can be translated into a scale for new physics, at which the EFT must be matched to an ultraviolet (UV) completion.

The least biased description of potential physics beyond the SM (BSM) expands the most general Lagrangian compatible with the low-energy symmetries of the SM [2, 3] in powers of an inverse energy scale Λ. The contributions of operators of dimension larger than 4 will then be suppressed by positive powers of $\sqrt{s}/\Lambda$, where $\sqrt{s}$ is a typical partonic energy scale [1]. In principle, one could use the general dimension-6 Lagrangian [2, 3] to compute LHC observables as a function of all the coupling parameters in this Lagrangian and try to fit these parameters to LHC observations.

Using modern automated calculation tools, it is in principle possible to compute leading-order cross sections taking into account all 59 independent dimension-6 operators for a single fermion generation [3]. In practice, however, such a maximally agnostic approach will not yield useful results, because a parameter space of 59 dimensions cannot reasonably be constrained by a global fit to generic observables with limited statistics. Therefore one must select subspaces of the complete BSM parameter space that correspond to plausible models and perform dedicated searches. Still, taking into account the finiteness of available resources, it is impossible to cover all plausible extensions of the SM in the LHC programme by a search with a dedicated experimental analysis. Fortunately, it is often possible to recast a single analysis as a search for particles in very different models.

In this chapter, a small portion of the vast universe of SM extensions will be probed that do not necessarily solve the problems of the SM, but which might nevertheless be delivered to our doorstep in the future.

11.2 New Gauge Bosons

All observed elementary vector bosons can be interpreted as gauge bosons of an unbroken or broken gauge invariance. In fact, gauge invariance is required for the decoupling of unphysical polarisation states of massless vector bosons. Furthermore, there seems to be no theoretically consistent description of interacting massive vector bosons that is not equivalent to the low-energy limit of a spontaneously broken gauge theory [4]. Therefore, at the LHC it is natural to look for new spin-1 gauge bosons that are not contained in the SM and that are too heavy for production at previous, lower-energy colliders.

The popular paradigm of the unification of the SM gauge interactions (see Sect. 10.1) predicts a number of new heavy particles filling the adjoint representation of the unified gauge group. In most scenarios, the corresponding mass scale is too high to allow production at a collider. Nevertheless, in the absence of experimental evidence for this unification scale, an unbiased search for such particles in the accessible mass range is called for. Indeed, there are many scenarios [5] in which the gauge group contains an extra U(1) factor that would give rise to a new neutral gauge boson (Z'), which could be light enough to be accessible at the LHC (see Sect. 11.2.1). A particular example is given by models in which the SM is embedded into an E_6 gauge theory.

Among the models with additional space dimensions (see Sect. 11.4), those in which particles with SM quantum numbers can propagate in the bulk of the extra dimensions predict whole Kaluza–Klein towers (sequences of particles with increasing masses, but identical quantum numbers) of massive vector bosons with couplings to SM matter particles similar to those of the SM gauge bosons.

Models with a strongly interacting electroweak symmetry breaking (EWSB) sector where the Higgs appears as composite or partially composite (see Sect. 11.3.3), typically predict additional massive composite states with spin 1 that can effectively be described as gauge bosons. Since their masses should be of the order of the symmetry-breaking scale, they could be accessible at the LHC.

In the searches described below, possible self-interactions among the new gauge bosons can play no role yet, because the corresponding channels are suppressed by phase space and small couplings. Indeed, their nature as gauge bosons only enters through their coupling to conserved currents constructed out of SM particles.

In order to perform a search, one needs to construct a class of models that allow a consistent computation of cross sections. The simplest approach for new gauge bosons is to duplicate the massive gauge bosons Z^0, $W^\pm$ of the weak interactions in a so called "sequential Standard Model" (SSM), essentially extending the gauge group by another broken SU(2), resulting in a new triplet of heavy gauge bosons, denoted Z'_{SSM} and W'_{SSM}. Duplicating the flavour structure of their couplings from the couplings of the SM gauge bosons guarantees that flavour-changing neutral currents (FCNC) remain suppressed, as required by low-energy precision data. For the purpose of the searches, the Feynman rules for the gauge bosons can be derived from a dimension-4 Lagrangian written with an explicit mass term:

$$\mathcal{L}_{\mathrm{SSM}} \supset -\frac{1}{4}F_{\mu\nu}F^{\mu\nu} + \frac{m^2}{2}A_\mu A^\mu + \bar{\psi}\gamma_5(g_V - g_A\gamma_5)\psi' A^\mu. \tag{11.1}$$

This Lagrangian is indistinguishable from the unitarity-gauge version of a non-linearly realised gauge invariance. As long as additional degrees of freedom in the symmetry-breaking sector are too heavy to be produced in a collider experiment, the non-linear realisation provides an accurate description of a spontaneously broken gauge theory. Therefore, the models in this class are only characterised by the masses of the Z' and W' and their respective coupling strengths. It should be kept in mind that both new and old gauge bosons couple to the SM fermions, so that in general they will mix and couple to each other.

In addition to the leading-order interactions of formula (11.1), there are always interactions in higher orders of the energy expansion, such as the magnetic dimension-5 interaction

$$\mathcal{L}_{\text{eff}} \supset \frac{g}{\Lambda}\bar{\psi}\sigma_{\mu\nu}\psi' F^{\mu,n}. \tag{11.2}$$

Couplings of this type are studied in precision measurements as "anomalous" couplings among SM particles (e.g. of top quarks to SM gauge bosons (Chap. 7), or of SM gauge bosons among themselves (Chap. 4). They manifest themselves as small deviations of the observation from the SM predictions.

In the case of searches for new gauge bosons, anomalous couplings can be ignored most of the time because they correspond to higher dimensional operators that are suppressed with respect to the leading dimension-4 couplings. There are, however, scenarios with chiral vector bosons [6], in which the dimension-4 couplings in (11.1) are forbidden by chiral quantum numbers of the fermions and vector bosons, while the dimension-5 couplings (11.2) are allowed and form the leading contribution.

For experimentalists, such models (Z^*, W^*) are attractive as alternative benchmarks because the tensor coupling (11.2) produces an angular distribution of the decay products that is very different from that of the usual vector coupling [7]. They are also theoretically interesting because they allow the existence of a naturally light Higgs because it is either a pseudo-Goldstone boson, or a state related to a Goldstone boson by symmetry, or the fifth component of a gauge field [8]. In such models, the non-renormalisable dimension-5 operators are generated from loop diagrams involving only renormalisable interactions.

11.2.1 New Heavy Neutral Bosons Z′

From the experimental point of view, searching for a Z' is probably the most straightforward exercise in the quest for physics beyond the SM. The cleanest decay channels are, of course, $Z' \to e^+e^-$ and $Z' \to \mu^+\mu^-$, and one can therefore simply search for an excess of pairs of charged high-p_{T} leptons. From flavour-physics results, it is known that flavour-changing couplings are strongly suppressed (see Chap. 8), so that decays into pairs of leptons with mixed flavour are expected to be suppressed. But it must be kept in mind that, while there is no freedom in the Lorentz structure of the leading dimension-4 $Z'f\bar{f}$-coupling, the individual coupling strengths to the vector and axial-vector currents are, a priori, not determined and can differ from the SM neutral-current (NC) couplings.

11.2.1.1 A Detailed Example: $Z' \to \ell^+\ell^-$

Searches for new electrically neutral heavy resonances are predominantly performed in final states with two opposite-sign same-flavour leptons, $Z' \to \ell^+\ell^-$, where $\ell = e, \mu$. This signature with two stable high-p_T leptons is very clean and thus favoured in terms of signal-to-background ratio. In case of the Z'_{SSM} benchmark model the decay to two quarks is clearly favoured in terms of branching ratio. However, at a pp collider the cross section for dijet production is many orders of magnitude larger than the expected Z'_{SSM} signal, so that it is very difficult to pick out the comparatively small additional signal yield. This large dijet cross section even poses a problem for the search in the dielectron channel. In order to keep a reasonable level of background, the electron identification needs to ensure a small probability of a jet being misidentified as an electron, while still maintaining a high efficiency in the electron identification. Since the probability for a jet to be misidentified as a muon is very small in general, the dimuon channel is less affected.

After selecting high-p_T lepton pairs, the main and irreducible background is the Drell–Yan process $pp \to Z/\gamma^* \to \ell^+\ell^-$. Other important reducible backgrounds stem from $t\bar{t}$, diboson and W+Jets production, which depending on the actual decay channel contribute via real leptons and/or misidentified jets. The search strategy relies on isolating an excess of data on top of the SM prediction. An example of this search in the dielectron channel at ATLAS [9] using the full 8 TeV data set taken in 2012 is shown in Fig. 11.1a, where the event yield is shown as a function of the dielectron invariant mass. The irreducible Drell–Yan background as well as the $t\bar{t}$ and tW ("top quark") and diboson contributions are derived from MC simulation. The dijet, multi-jet, γ+jet and W+jet contributions ("dijet & W+jets") are derived by a data-driven method. In addition, the hadronic decay modes of the diboson and top backgrounds are not taken from the simulation but are included in the data-driven method. The reason for using a usually rather involved data-driven estimate is twofold. On one hand, as mentioned above, the cross section and suppression of the dijet production are large, so that a huge amount of MC-generated events would need to be produced. On the other hand, the probability to misidentify a jet as a lepton is not easily modelled in the simulation.

The ratio plot at the bottom of Fig. 11.1a shows no sign for an excess. The actual size of a Z'_{SSM} signal at 1.5 and 2.5 TeV is also indicated in the same figure. Even by eye, these two examples seem excluded, so that the question arises what the exact lower mass limit is on the Z'_{SSM} benchmark model, or, how small, at a given mass, the cross section times branching ratio of a hypothetical new particle decaying to two leptons needs to be so that it presently escapes detection. Both questions are answered in Fig. 11.1b, now for the case of the dielectron, dimuon, and the combined channel. The expected and observed upper limits on the cross section times branching ratio ($\sigma \times$ B) are shown as a function of the mass of the new object, i.e. new objects with $\sigma \times$ B larger than the values given by the observed limit are excluded at the 95 % confidence level (CL).

The observed (expected) limits are derived by comparing data (simulated pseudo-experiments with only SM processes) to the expected background, taking into account

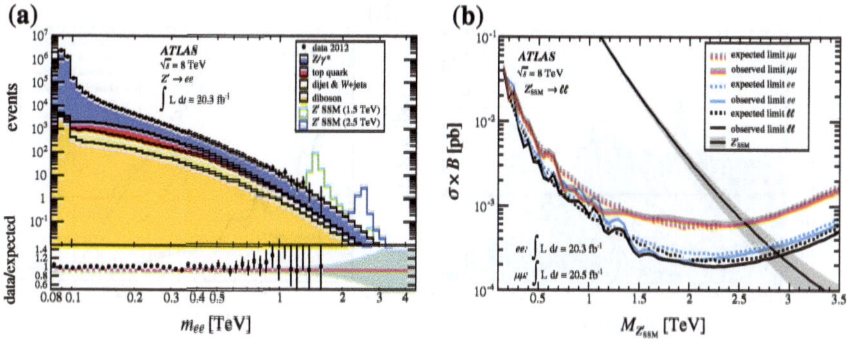

Fig. 11.1 Search for $Z' \rightarrow e^+e^-$ and $Z' \rightarrow \mu^+\mu^-$ with an integrated luminosity of $20\,\text{fb}^{-1}$ of data recorded with the ATLAS detector. **a** Comparison of the measured dielectron invariant-mass spectrum to the sum of all expected backgrounds, including the data/expectation ratio with associated systematic uncertainties. The latter are indicated by the *shaded* band and dominated by PDF uncertainties. Two examples for a Z'_{SSM} are shown as well. **b** Expected and observed 95 % CL upper limits on cross section times branching ratio ($\sigma \times$ B) for Z'_{SSM} production, along with the predicted $\sigma \times$ B for Z'_{SSM} production shown as a band. The width of the band represents the theoretical uncertainty, which is mainly due to PDFs. Adapted from [9]

the actual shape of the signal distribution as a function of the reconstructed invariant mass. As an example, these so-called templates are shown in Fig. 11.2a for three different masses of a Z'_{SSM} decaying in the electron final state. It can be seen that this narrow resonance would also be reconstructed as a narrow resonance at small and medium masses. However, at very large Z'_{SSM} masses the low-mass tail of the resonance gets strongly enhanced due to the much larger parton-parton luminosity at smaller values of Bjorken-x. The direct consequence of this off–shell production is a reduced signal yield at high mass, i.e. in the sensitive region where the SM background is small. This degradation of the $\sigma \times$ B limits at high masses can be seen in Fig. 11.1b.

Figure 11.1b also demonstrates that the combined limit on $\sigma \times$ B is dominated by the limit in the electron channel, in particular at higher masses. This is mainly due to the much better energy resolution for electrons as measured in the electromagnetic calorimeter compared to the one for muons, which is measured by means of the tracking system. Very stringent requirements need to be placed on the muon tracks in order to maintain a reasonable resolution, leading to a smaller selection efficiency for muons and thus to a smaller sample of dimuon candidates in the ATLAS analysis. Note that the dielectron and dimuon channel yield similar limits in the search at CMS [10], i.e.–the electron channel is not generally more sensitive.

For specific models with a known $\sigma \times$ B as a function of mass, the limit curve on $\sigma \times$ B can be used to set a lower mass limit on the new resonance, e.g. for the benchmark Z'_{SSM} model shown in Fig. 11.1b masses below 2.9 TeV are excluded based on the combined limit curve. This value neglects the width of the Z'_{SSM} cross-section band, which is dominated by PDF uncertainties, i.e. the value stated corresponds to the crossing point of the observed limit and the central prediction for the Z'_{SSM}.

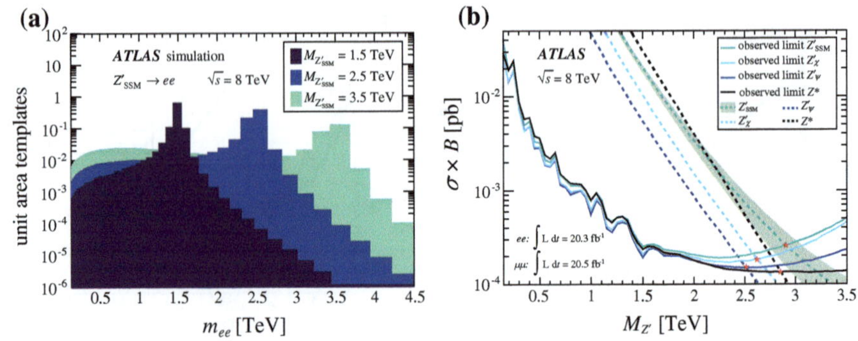

Fig. 11.2 Search for $Z' \to e^+e^-, \mu^+\mu^-$ by ATLAS, with 20 fb^{-1}. **a** Examples of dielectron signal templates for the Z'_{SSM} model as a function of the reconstructed invariant mass. **b** Observed upper limits and predicted theoretical values for the cross section times branching ratio for Z'_{SSM}, E_6-motivated Z' and Z^* bosons using the combined dilepton channel. The *stars* indicate the lower mass limits for each considered model. The width of the Z'_{SSM} band represents the theoretical uncertainty from the PDF error set, the choice of PDF as well as the strong coupling constant. The width of the Z'_{SSM} band applies to the E_6-motivated Z' curves as well. Adapted from [9] and auxiliary material

Figure 11.2b shows the observed limits and theoretical $\sigma \times$ B curves for the E_6-motivated Z'_χ and Z'_ψ and for the Z^* models as a function of the mass of the new resonance, in addition to the ones from the Z'_{SSM} benchmark already shown in Fig. 11.1b. It can be seen that at low masses the observed limits are very similar for all models, despite using specific signal templates for each model. All models predict spin-1 particles, but differences in their predicted width lead to these small differences in the limits. Consequently, the most (least) stringent limit is found for the Z'_ψ (Z^*) model, which also corresponds to the model with the smallest (largest) width. Apart from the Z^* model, the same dependence on the width is also seen at large masses, where the differences are more pronounced due to the off-shell production discussed above, which increases with increasing width. At large mass, the best limit in terms of $\sigma \times$ B is observed for the Z^*, which has a width similar to the Z'_{SSM}. This is due to its different coupling (magnetic-type only) discussed above, leading to enhanced production on the peak, and thus effectively to a suppressed off-shell production.

For the Z'_{SSM} benchmark model, Fig. 11.3a summarises the impressive progress achieved year after year during Run 1. The observed cross section limit divided by the predicted SSM cross section is shown as a function of the Z'_{SSM} mass. Thus, the lower mass limit for a given data set corresponds to the crossing points of the ratio with the line at unity, and the observed limit with respect to the SSM cross-section prediction is easy to read off for any given Z'_{SSM} mass. It can be seen that—due to the much higher centre-of-mass energy—already the small data set recorded at the LHC in 2010 yields limits [11] similar to those derived at the Tevatron [12, 13] with 100 times more data. The advantage of a large centre-of-mass energy can be directly seen when comparing the results from the Tevatron and the full 7 TeV LHC data set

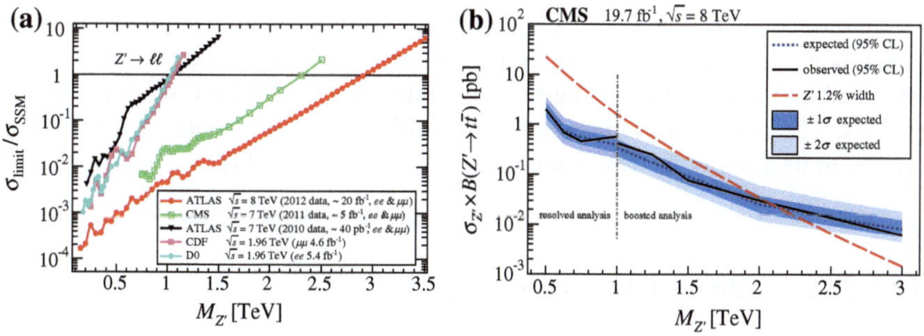

Fig. 11.3 **a** Observed cross-section limits divided by the corresponding predicted SSM cross sections as a function of the Z'_{SSM} mass. **b** Expected and observed 95 % CL upper limits on cross section times branching ratio ($\sigma \times$ B) for the $t\bar{t}$ final state, along with the predicted $\sigma \times$ B for a Z' in a leptophobic top-colour model. Adapted from [9–14] and auxiliary material

recorded in 2011 [10], which is based on a similar integrated luminosity. The further large improvement using the 8 TeV data set from 2012 [9] gives an idea about the discovery potential for the data to be recorded at 13 or 14 TeV in the future.

11.2.1.2 Z' Searches in Other Final States

New Z' resonances would also manifest themselves in other final states, but the searches are usually less sensitive when compared to the dielectron or dimuon case discussed above. One example is the $Z' \to \tau^+\tau^-$ search, which is less sensitive due to its more complicated final state. However, this was different if the new particle would not decay democratically into the three lepton generations, as, e.g. assumed in the SSM, but would happen to prefer decays to the third generation. Searches at ATLAS [15] and CMS [16] using the full 7 TeV data set taken in 2011 yield lower mass limits of 1.4 TeV on the Z'_{SSM}. These limits are to be compared with the CMS [10] (see Fig. 11.3) and ATLAS [17] Z'_{SSM} lower mass limits using the separate dielectron and dimuon channels from the same data set, which are at or above 2 TeV.

While disfavoured in terms of signal-to-background ratio as already discussed in Sect. 11.2.1.1, the $Z' \to q\bar{q}$ channel has been used to set limits on the Z'_{SSM} [18], excluding a range from 1.0 to 1.6 TeV with 4 fb^{-1} of data taken at 8 TeV. These limits are not competitive with the ones from the dielectron or dimuon channels when assuming SM-like branching ratios, but it cannot be excluded that the new object does not decay into leptons at all ("leptophobic") [19, 20].

Finally, in contrast to Z bosons, new Z' objects can decay to $t\bar{t}$ if their mass is above the $t\bar{t}$ production threshold. Consequently, limits on $\sigma \times$ B are also derived for the $t\bar{t}$ final state as shown in Fig. 11.3b. Here events are used where at least one of the two W bosons from the $t\bar{t}$ decay will further decay to quarks. At invariant $t\bar{t}$ masses below (above) 1 TeV, a so-called "resolved" ("boosted") analysis is used. The model used here is leptophobic and based upon top-colour [21, 22], where the introduction of a new strong gauge force coupling preferentially to the third generation is eventually responsible for the large top-quark mass. As can be seen, observed and expected limits agree, thus indicating the absence of signs for new physics. The lower mass limit assuming a width of the new resonance of 1.2 % with respect to its mass is 2.1 TeV using the full 8 TeV data set [14].

11.2.2 New Heavy Charged Bosons W'

The search for an additional charged vector boson—denoted W'—is another flagship analysis in BSM physics. The experimental searches are largely guided by the sequential Standard Model (SSM), which assumes the W'_{SSM} to be a heavy carbon copy of the W boson of the Standard Model with the additional decay channel in $W' \to tb$. Consequently, final states may consist of a charged lepton (typically an electron or muon) and missing transverse energy (Sect. 11.2.2.1) or of two jets (either untagged or flavour-tagged, Sect. 11.2.2.2). While the decay in two bosons is often assumed to be suppressed, dedicated searches for this channel do also exist (Sect. 11.2.2.3). Unlike the Z', the W' signal does not manifest itself as a narrow resonance, but rather as a Jacobian peak with some smearing due to detector resolution. All these searches also allow other interpretations, such as compositeness (Sect. 11.3), extra dimensions (Sect. 11.4) or dark matter (Sect. 11.5). These interpretations will be discussed in the respective sections.

11.2.2.1 Searches for W' with Leptons

Similar to the previously discussed Z' searches, the highest sensitivity is achieved in final states with leptons (electrons or muons) and missing transverse energy ($E_{\mathrm{T}}^{\mathrm{miss}}$), which profit from a high signal-to-background ratio. In practice, the signal consists of one electron or muon with an energy or momentum in the TeV range, i.e. in an energy range with very low SM background where an excess even of only a handful of events would clearly indicate new physics. The experimental challenge lies in the limited experience in terms of detector response, resolution and energy calibration in this energy range. The final-state neutrino does not leave a detectable signal and yields $E_{\mathrm{T}}^{\mathrm{miss}}$ of a magnitude similar to the momentum of the charged lepton, and opposite in direction. $E_{\mathrm{T}}^{\mathrm{miss}}$ can also be caused by instrumental effects and then has to be removed from the data. This characteristic two-body decay kinematics provides a good handle to discriminate the signal from backgrounds. The transverse mass (M_{T}) is reconstructed from the final-state lepton and $E_{\mathrm{T}}^{\mathrm{miss}}$, similar to the SM W-boson analysis.

The main background in leptonic W' searches is due to the off-shell tail of SM W-boson decays, which has to be modelled correctly including higher-order corrections. Other background contributions are from diboson (in particular W^+W^- and

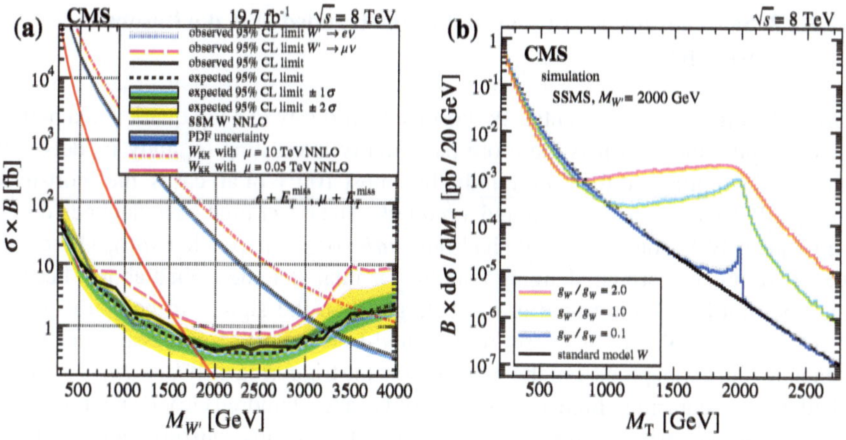

Fig. 11.4 **a** W' exclusion limits based on the full 8 TeV data set for the electron and the muon channels and their combination. The W' mass limits are derived with a Bayesian method. **b** Transverse-mass spectrum for SM W background (*black*) and expected signals for a 2 TeV W' taking into account W-W' interference with same-sign couplings (SSMS) of the W' to quarks and leptons and the same absolute value for W' couplings ($g_{W'}$) to all fermions. Different values are assumed for the ratio between $g_{W'}$ and the SM weak coupling g_W. Adapted from [23]

WZ) events in which all leptons except for one escape detection. Similarly, $t\bar{t}$ events contribute to the background.

The lower mass limit for a W'_{SSM} is 3.3 TeV [23, 24]. Using the same analysis a limit of 3.2 TeV [24] is derived on the W^* boson introduced above. The best limits are achieved when combining electron and muon channels, as illustrated in Fig. 11.4a. For lower masses, the limits are constrained to 300 GeV because of the threshold of single-lepton triggers, which are typically as high as 40–80 GeV in order to keep the trigger rate within the allocated bandwidth. Offline thresholds—applied symmetrically to the charged lepton and $E_{\text{T}}^{\text{miss}}$—are somewhat higher in order to take into account the trigger turn-on and to be in the fully efficient trigger plateau. Figure 11.4a also shows limits for a W^2_{KK}, which is discussed in Sect. 11.4.1 in the context of extra-dimension models.

Left-handed W' bosons would interfere with the SM W bosons, either constructively or destructively. The nature of the interference effect depends on the W' coupling to quarks ($g^{qq'}_{W'}$) and leptons ($g^{\ell\nu}_{W'}$) and the SM weak coupling constant g_W. The same-sign (SSMS) and opposite-sign (SSMO) scenarios assume $g^{qq'}_{W'} \cdot g^{\ell\nu}_{W'} > 0$ and $g^{qq'}_{W'} \cdot g^{\ell\nu}_{W'} < 0$, respectively. The absolute value of the W' coupling strength to quarks and leptons is considered the same. The resulting SSMS transverse-mass spectrum for a 2 TeV W' signal is illustrated in Fig. 11.4b for various coupling strengths in comparison to the SM W background. The interference is destructive for $M_W < M_{\text{T}} < M'_W$ and constructive for $M_T > M'_W$, leading to a differential W-W' cross section smaller than the SM W-boson cross section at low M_{T}.

11.2.2.2 Searches for W' Bosons and Right-Handed W-Like Bosons with Jets

Given the high mass of potential W' bosons, their decay to a top quark and a bottom quark is possible, and it is therefore particularly interesting to study models with enhanced couplings to the third family. In almost 100 % of all cases, the short-lived top quark decays to a W boson and a b quark. The exact final state depends on the subsequent W-boson decay (fully hadronic $qqbb$ or semileptonic $\ell\nu bb$). These final states are much more challenging than the $\ell\nu$ final state described above, because the SM multi-jet background is much larger than the signal. In comparison, the single-lepton final state is nearly background-free since SM processes very rarely generate $\mathcal{O}$(TeV) leptons. For the same reason, the triggering of $W' \to tb \to qqbb$ is challenging, requiring a high trigger threshold of a few hundred GeV (for comparison, the single-lepton triggers are of the order 40–80 GeV), thus constituting a lower W'-mass bound for searches in the tb-channel of $\mathcal{O}$(1 TeV).

In the $W' \to tb$ analysis, b tagging is an important tool for the identification of jets originating from b quarks and serves as necessary input along with the measured top-quark mass (172.5 GeV). For high W'-boson masses (>1.5 TeV) the two final-state b jets are boosted, and substructure techniques [25–28] become necessary.

The $W' \to tb$ analyses are performed separately for left-handed and right-handed couplings, denoted W'_L and W'_R, respectively. The W'_L couples to the same fermion multiplets as the SM W boson, and interference with the SM W boson occurs in much the same way as in the $\ell\nu$ channel. The W'_R can only decay leptonically ($W'_R \to \ell N_R$) if the mass of the heavy neutrino $m(N_R)$ is less than the W' mass. In such a case, the previous W'_L limits apply, independently of the previously made assumptions on handedness. If the mass of the right-handed neutrino is larger than the W' mass, only hadronic decays to two-jet final states are allowed, resulting in a higher branching fraction for the $W'_R \to tb$ decay than for $W'_L \to tb$. Also for these channels, interference occurs (this time with SM single-top-quark production) and is taken into account in the analyses. The $qqbb$ channel gives cross-section limits which translate to observed mass limits of 1.68 and 1.76 TeV for the left-handed and right-handed models, respectively, assuming the same coupling strength to quarks as for the SM W boson [29]. The $\ell\nu bb$ channel yields limits of 1.84 and 2.05 TeV for the left- and right-handed models, respectively [30].

A related analysis also searches for the production of a heavy right-handed W boson, W_R, that decays to an electron or muon and a heavy neutrino [31]. While the previous search assumes the heavy neutrino to be stable, in this analysis the N_R can subsequently decay to a same-flavour charged lepton (again, electron or muon) and two jets via a virtual W_R. The final state thus consists of two same-flavour leptons plus two jets. The primary backgrounds stem from top-quark pair and Drell–Yan production in association with jets. Multi-boson and single-top-quark backgrounds contribute at a smaller rate. Considering $W_R \to eN_e$ and $W_R \to \mu N_\mu$ searches separately, regions in the M_{W_R}–M_N mass space are excluded at 95 % confidence level that extend up to $M_{W_R} < 2.87\,(3.0)$ TeV for the electron (muon) channels. The observed electron limit is lower because of a slight excess in the $eejj$ spectrum

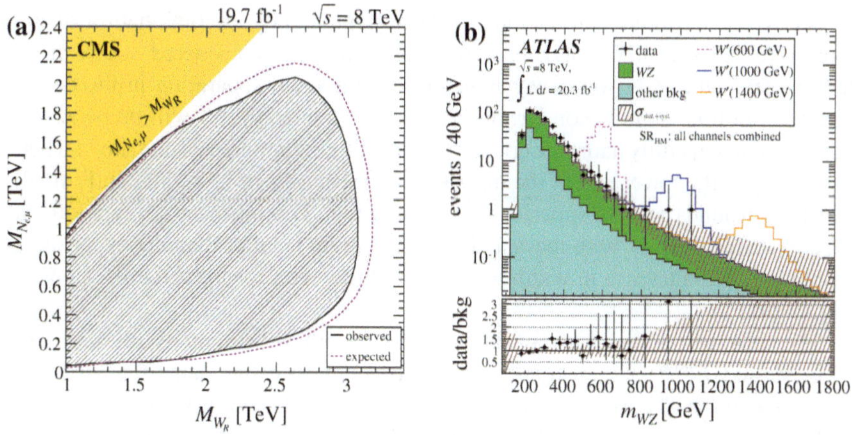

Fig. 11.5 **a** Exclusion region for the combined electron and muon channels as a function of the mass of the right-handed W boson and the heavy neutrino, assuming equal couplings in the left and right sectors and with all heavy neutrino decay channels accessible. **b** Observed and predicted WZ invariant mass for events in the high-mass signal region. An extrapolation of the backgrounds to the very-high-mass region was performed using a power-law function to fit for the SM WZ and the sum of all other backgrounds separately. Also shown are predictions for a W' signal with masses of 600, 1000 and 1400 GeV, stacked on *top* of the expected backgrounds. Adapted from [31, 33]

with a local significance of $2.8\,\sigma$ that is, however, inconsistent with the expected signal. Assuming $W_R \to \ell N_\ell$ with degenerate N_ℓ mass for $\ell = e, \mu$, W_R-boson production is excluded at 95 % CL up to $M_{W_R} < 3.0$ TeV (see Fig. 11.5a). This result is dominated by the sensitivity of the muon channel. At the same time, one can derive mass limits on heavy neutrinos that range from 1 to 2 TeV, depending on the W_R mass.

Furthermore, decays of W' bosons to light-quark (or untagged) jets are also being searched for. This is a rather generic, less model-driven analysis searching for new resonances in the dijet invariant mass spectrum on a typically steeply falling SM background. A narrow search window scans the entire mass spectrum in the search for resonances, which could be due to many models: W' or Z', excited quarks (see Sect. 11.3.2), extra dimensions (see Sect. 11.4), etc. Up to now, no hint for a signal has been found, and W' masses below 2.45 TeV can be excluded [32]. The same analysis excludes W^* bosons up to masses of 1.75 TeV.

11.2.2.3 Searches for Diboson Resonances

While the previous searches assumed a suppression of W' decays into bosons, such decays (i.e. $W' \to WZ$) can become dominant in scenarios containing, e.g., leptophobic W' bosons (i.e. W' bosons with zero coupling to leptons) [19, 20, 128–131]. The searches for such bosonic decays are somewhat generic, as are the analyses of the dijet invariant-mass spectrum, which also constrain a large number of models.

Assuming the final-state bosons to decay leptonically, a combination of three leptons (electrons and muons) define the final states to be considered, where the two leptons from the Z-boson decay typically have a small separation in the detector because the final-state bosons from the decay of the heavy W' will be boosted. In comparison to its fully hadronic counterpart, the three-lepton final state shows a better sensitivity at the lower end of the mass spectrum because of its significantly smaller SM backgrounds and its superior mass resolution. The intrinsic decay width of a W'_{SSM} scales linearly with mass and is significantly smaller than the experimental resolution. The increase in width with mass can be seen for the three W' signals in Fig. 11.5b.

The selection of candidate events requires exactly three leptons, all with $p_{\mathrm{T}} >$ 25 GeV, and $E_{\mathrm{T}}^{\mathrm{miss}}$. The invariant mass of two opposite-sign same-flavour leptons has to match the Z-boson pole mass of 91 GeV within a 20 GeV window. Events are further required to have a rapidity difference below 1.5 between the W and Z bosons from the $W' \rightarrow WZ$ decay in order to enhance the sensitivity to such s-channel resonances. The dominant SM WZ background, shown in Fig. 11.5b, is modelled with NLO precision up to high masses. A second category of background arises from photons misidentified as electrons, mainly from $Z\gamma$ production. Since no excess was observed in the data, these searches exclude bosonically decaying W' bosons with masses up to 1.5 TeV [33, 34]. Other interpretations include a simplified heavy vector triplet interpretation [33] and technicolour [34] (see Sect. 11.3.3 for further interpretations).

The hadronic version of the WZ analysis [35] slightly extends the W' mass limit to 1.7 TeV. Since the final-state bosons are boosted, new jet-substructure identification techniques [25–28] are exploited that provide sensitivity to the presence of highly boosted vector bosons decaying into a pair of quarks.

11.2.2.4 Generalised Couplings

A new heavy vector boson could potentially couple to the SM particles with non-unity coupling strengths resulting in higher/lower cross sections and correspondingly different sensitivities. The impact on the shape of the observed M_{T} distribution and on the W' width for the example of the lepton+$E_{\mathrm{T}}^{\mathrm{miss}}$ channel is shown in Fig. 11.4b in Sect. 11.2.2.1. The cross-section limits of the $\ell\nu$, the tb and the WZ searches are interpreted for other values of the coupling strength [23, 30, 34]. Examples are shown in Fig. 11.6a for the tb final state in terms of the coupling strength to left-handed (a^L) and right-handed (a^R) fermions. The lower limit on the W' mass is illustrated by the different grey shades in the third dimension.

A similar analysis was performed in the $\ell\nu$ final state [23] already discussed in Sect. 11.2.2.1, assuming left-handed couplings and expressing the varying coupling strength in relation to the SM coupling ($g_{W'}/g_W$). SM coupling strength is the basic assumption in the SSM model. The upper limit on the coupling strength is depicted as a function of the W' mass in Fig. 11.6b. The resonance width strongly depends on the coupling strength as illustrated in Fig. 11.4b for couplings of $g_{W'}/g_W = 2.0$,

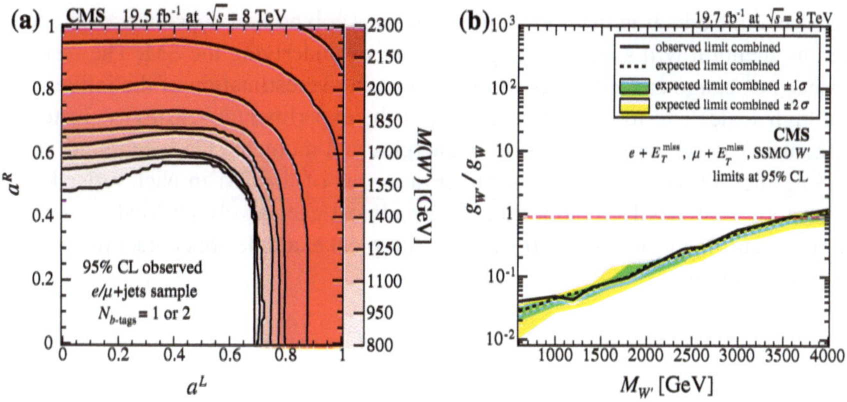

Fig. 11.6 Generalised couplings for (**a**) the tb channel where *left*-handed (a^L) and *right*-handed (a^R) couplings are studied separately. The $\ell\nu$ final state in (**b**) is taking into account W-W' interference. While the absolute value of the W' couplings $g_{W'}$ to quarks and leptons is assumed to be the same, they can have the same (SSMS) or opposite (SSMO) sign. For the SSMO scenario the upper limit on the ratio between $g_{W'}$ and the SM weak coupling g_W is shown as a function of the W' mass. Adapted from [23, 30]

1.0 and 0.1, and thus is considered in the analysis. Taking into account interference, the opposite-sign scenario (SSMO, introduced in Sect. 11.2.2.1) yields a larger cross section than W'_{SSM} without considering interference. Thus the lower limit on the W'_{SSM} mass (3.3 TeV, see above) is lower than the limit on the W' mass in the SSMO scenario, which is 3.7 TeV as can be seen in Fig. 11.6b.

11.3 Compositeness

Currently, all experimental results are compatible with the assumption that the leptons, quarks, gauge bosons and the Higgs boson are point-like particles without internal structure except spin. Nevertheless, this assumption must be tested again every time the experimental resolution is increased by a higher-energy collider. Indeed, if quarks and leptons were made out of smaller building blocks, their quantum numbers and family structure might be explained by their composition, just as the periodic table was explained by the atom's electronic structure and the hadron spectrum was explained by quarks in Gell-Man's eightfold way. In the latter two cases, however, there had already been experimental evidence for a finite extent of atoms and hadrons, respectively.

Therefore, present day searches for compositeness must start with the search for deviations from the point-like nature of elementary particles. Historically, the finite extension of protons was discovered by measuring a non-trivial form factor in elastic electron scattering. This approach can not be generalised directly, because an ad-hoc introduction of form factors with non-polynomial dependence on energy and momentum will spoil the gauge invariances underlying the SM. The resulting absence of gauge cancellations would lead to an overestimation of the influence of the high scales and to an overestimation of the experimental sensitivity. Instead, a systematic power series expansion in energies and momenta [1] is required, maintaining the observed symmetries and invariances of the SM in each order [2, 3]. Therefore, searches for compositeness are almost exclusively phrased as searches for non-renormalisable interactions [36]. A typical example are contact interactions of four fermions

$$\mathcal{L}_{\text{eff}} \supset \sum_{\Gamma,\alpha klmn} \frac{g_{\Gamma,klmn}}{\Lambda^2} \bar{\psi}_k \Gamma_\alpha \psi_l \, \bar{\psi}_m \Gamma_\alpha \psi_n, \tag{11.3}$$

where the $\Gamma_\alpha \in \{\mathbf{1}, \gamma_5, \gamma_\mu, \gamma_5\gamma_\mu, \sigma_{\mu\nu}\}$ encode the Lorentz structure of the interaction. The indices k, l, m, n encode the gauge and flavour quantum numbers of the participating quarks and leptons. Λ is a characteristic energy scale, introduced to make all couplings $g_{\Gamma,klmn}$ dimensionless numbers.

However, care must be taken when interpreting the experimental limits on the coupling strengths of non-renormalisable interactions at hadron colliders. Since their contributions rise as a polynomial of the parton energy, they counteract the fall-off of the parton distribution functions, and there can be parts of the accessible phase space where the interaction exceeds the limits from unitarity. These regions must be cut out from the analysis, or the scattering amplitudes must be unitarised.

11.3.1 Contact Interactions

There are 30 independent contact interactions (11.3) involving four SM fermions from a single generation [2, 3]. Three of these couple four leptons and are not directly accessible at hadron colliders. Twelve operators couple four quarks and contribute to final states with two high-p_T jets, while ten couple two leptons and two quarks and contribute to final states with two high-p_T leptons. The remaining five operators violate baryon number and couple three quarks and one lepton, resulting in final states with one high-p_T jet and one high-p_T lepton.

As mentioned in the introduction, while one could in principle perform a simultaneous multivariate fit to all 27 accessible operators involving hadrons, it is not practical to reliably combine the analyses in all channels required for such a fit, since the experimental precision at the high-energy limit is limited in each channel by the strongly correlated uncertainties resulting from the parton distribution functions. Therefore, a more realistic approach will single out a few of the contact interactions that are either more likely (according to theoretical prejudice) or that are experimentally interesting (because of peculiar signals).

The LHC collaborations have concentrated on models involving the coupling of two left-handed or right-handed currents, in particular the left-left isoscalar model

(LLIM) [37, 38] ($Q_{lq}^{(1)}$ in the notation of [3]). A useful complementary benchmark is the helicity-non-conserving model (NhCM) [38] ($Q_{lequ}^{(1)}$ in the notation of [3]), because of different angular distributions of the produced particles.

If the new gauge bosons described in Sect. 11.2 are too heavy to be produced on shell, their effects can also be described by contact interactions (11.3) where Γ_α encodes the Lorentz structure of the exchanged particle and $\Lambda^{-2} = M^{-2}$ is the lowest order in an expansion of the propagator $-1/(p^2 - M^2)$ in powers of p^2/M^2. Vice versa, one can make use of the fact that contact terms of this form can not be distinguished at low energies from the exchange of a boson with coupling $\sqrt{g}\Gamma_\alpha$ and mass $\Lambda \gg \sqrt{s}$. Furthermore, (pseudo-)scalar and tensorial Γ_i couple left-handed and right-handed fermions, and naive power-counting suggests that the corresponding operators are further suppressed by $m\sqrt{s}/\Lambda^2 \ll s/\Lambda^2$. Therefore one can cover the searches for the most important contact interactions by reinterpreting the searches for very heavy vector bosons described in Sect. 11.2.

One example is the reinterpretation of the search for Z' bosons in the dilepton (ee, $\mu\mu$) final states, described in detail in Sect. 11.2.1.1, in terms of contact interactions [39]. The main difference between the two analyses is the different predicted signal shape, which in the contact-interaction interpretation is a broad non-resonant excess (see Fig 11.7a) instead of the narrow resonant excess (see Fig. 11.1a) expected from a Z'. The two examples shown for contact interactions are for a value of 14 TeV for Λ from (11.3), corresponding to the characteristic energy scale of the interactions among the fermionic constituents, and for a constructive (Λ^-) or destructive (Λ^+) interference between the SM Drell–Yan process and the contact interaction. As expected, constructive interference leads to a larger excess. Lower limits on Λ are set for models with constructive and destructive interference and for left-left (LL),

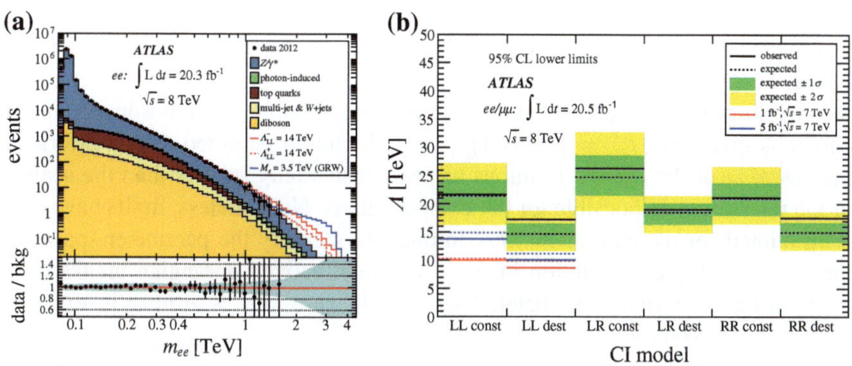

Fig. 11.7 **a** Comparison of measured dielectron invariant-mass spectrum to the sum of all expected backgrounds, including the data/expectation ratio with associated systematic uncertainties indicated by the *shaded* band. Two examples for signals expected from contact interactions and one expected from a GRW ADD model are also shown. **b** Lower limits on Λ for the combined dilepton contact-interaction search for models with constructive and destructive interference and for *left-left* (LL), *left-right* (LR) and *right-right* (RR) models. Adapted from [39]

right-right (RR) and left-right (LR) models, which define the chiral structure of the contact interaction. These limits can be seen in Fig. 11.7b for the full 8 TeV data set as well as for the full and a partial data set of the 7 TeV data. As expected, the limits are much higher than the ones on the on-shell production of resonances.

Another example is the reinterpretation of searches with a single lepton and $E_{\mathrm{T}}^{\mathrm{miss}}$—experimentally discussed in Sect. 11.2.2.1—which may be interpreted in terms of a helicity-non-conserving contact interaction [23]. Similar to the dilepton case, also here the signal is a broad, unstructured excess, this time in the spectrum of the transverse mass M_T. Unlike the previously discussed dilepton channels, in the single-lepton channels the helicity flips and, hence, no interference with the related SM process ($W \to \ell\nu$) occurs. Values of $\Lambda < 11$ TeV are excluded in the electron+$E_{\mathrm{T}}^{\mathrm{miss}}$ and muon+$E_{\mathrm{T}}^{\mathrm{miss}}$ channels [23].

11.3.2 Excited Quarks and Leptons

A generic prediction of composite models of quarks and leptons is the existence of excited quarks q^* and leptons ℓ^* [40–43]. Since these excited fermions couple to the ordinary SM fermions and gauge bosons, they can be produced via gauge and contact interactions in collider experiments and subsequently decay to ordinary fermions through the emission of a gluon, photon, W/Z boson or via contact interactions. The leading gauge interactions of an excited fermion $\psi^* \in \{\ell^*, q^*\}$ are usually parametrised by the dimension-5 effective Lagrangian

$$\mathcal{L} \supset \frac{1}{2\Lambda}\bar{\psi}^* \sigma_{\mu\nu}\left[g_s f_s \frac{\lambda^a}{2} G^{a,\mu\nu} + g f T^i W^{i,\mu\nu} + g' f' \frac{Y}{2} B^{i,\mu\nu}\right]\psi + \text{h.c.}, \quad (11.4)$$

where the strong-interaction piece is of course absent for excited leptons. In (11.4) Λ parametrises the compositeness scale and f_s, f, f' are the relative strengths of the couplings to the strong, weak isospin and hypercharge gauge bosons. A natural choice is given by $|f_s| = |f| = |f'| = 1$. It should be noted that, a priori, the masses M_{f^*} of the excited fermions can be substantially smaller than the scale Λ of the interaction responsible for the compositeness. Nevertheless, limits have often been quoted for the case $\Lambda = M_{f^*}$ in order to simplify the parameter space. On the other hand, the description in terms of the effective Lagrangian breaks down if M_{f^*}/Λ is not a small parameter. Therefore the case $M_{f^*} > \Lambda$ must be excluded from an analysis based on (11.4).

In pp collisions at the LHC, one of the partons entering the hard interaction can be excited directly to a q^* state and then return to its ground state via photon or gluon emission. Consequently, searches for excited quarks exploit final states with one jet and a photon [44, 45] from $q^* \to q\gamma$ or with two jets (dijets) from $q^* \to qg$ [18, 32].

The dijet analyses select events containing two or more jets and then combine the two highest-p_{T} objects to determine the dijet invariant mass, m_{jj}. High-p_{T}

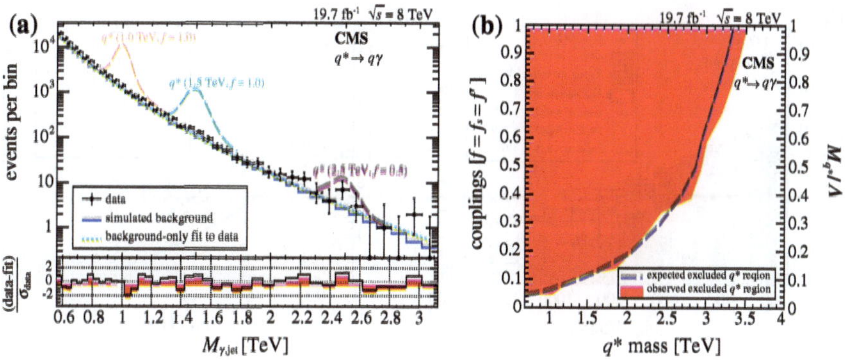

Fig. 11.8 **a** Excited quarks (q^*) would appear as a narrow resonance in the invariant-mass spectrum, here formed by a jet and a photon. **b** The excluded parameter space for excited quarks depends on their mass (M_{q^*}), their coupling (f) and the compositeness scale (M_{q^*}/Λ). The *filled area* is excluded at 95 % CL. Adapted from [44]

dijet events are, however, produced copiously by QCD processes, and along with a sub-percent admixture of additional SM processes, these processes create a smooth rapidly falling spectrum in m_{jj}. Many new physics models describe new particles or excitations created as s-channel resonances with appreciable branching ratios to final states involving quarks and gluons, consequently also producing dijet final states. If the resonance width of an assumed new bound q^* state is sufficiently narrow, these signals would appear as peaks in the dijet mass spectrum over the smooth SM background [18, 32]. In the search, different widths and shapes of the resonance are considered. As an example, signal and SM background for the γ+jet analysis at CMS [44] are depicted in Fig. 11.8a. A more detailed description of the γ+jet analysis at ATLAS [45] is given in Sect. 11.4.4 in the context of the search for quantum black holes.

The limits are a function of the compositeness scale Λ and the excited quark mass, as shown in Fig. 11.8b for the γ+jet analysis. The lower the excited fermion mass, the higher should be the compositeness scale Λ to allow such light excited fermions. Following this convention, excited quarks with masses below 4.09 TeV [32] are excluded by the dijet analysis and below 3.5 TeV by the γ+jet analysis for couplings of 1 to their SM partners (see Fig. 11.8b). The better sensitivity of the dijet channel is due to the larger cross section. Limits on excited quarks as a function of their mass and varying coupling strength (<1) are also given in Fig. 11.8b for the assumption of $|f| = |f_s| = |f'|$. As to be expected, weaker couplings lead to lower sensitivities.

A dedicated version of this search for excited quarks uses top-tagging tools when searching for pair-produced spin-3/2 excited top quarks ($t^*\bar{t}^*$), each decaying in 100 % of all cases to a top quark and a gluon [47]. The decay $t^*\bar{t}^* \rightarrow ttgg \rightarrow WWbbjj$ with $W \rightarrow \ell\nu$ ($\ell = e, \mu$) yields multi-particle final states with several jets, leptons and $E_{\mathrm{T}}^{\mathrm{miss}}$. Events are selected requiring a single isolated muon or electron, an imbalance in transverse momentum, and at least six jets, of which one must be

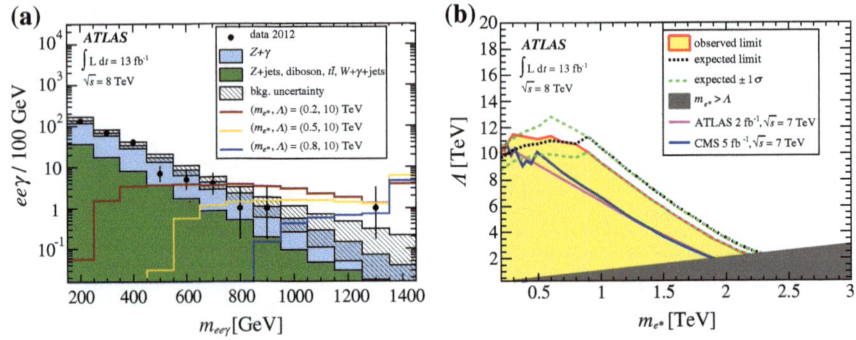

Fig. 11.9 **a** Three-body invariant mass ($ee\gamma$) in the search for excited electrons. SM backgrounds are shown along with examples of excited-lepton signals. **b** Exclusion limit of the compositeness scale Λ versus the excited-electron mass (m_{e^*}) parameter space with the filled area excluded at 95 % CL. No limits are set in the *dark shaded* region $m_{e^*} > \Lambda$ where the effective Lagrangian (11.4) is not applicable. Adapted from [46]

compatible with originating from the fragmentation of a b quark. This first search for a spin-3/2 excited top quark yields a limit of 803 GeV at 95 % CL on the mass of the spin-3/2 t^* quark in an extension of the Randall–Sundrum model (see Sect. 11.4.2).

At the LHC—in order to conserve lepton flavour—excited leptons, ℓ^*, are produced in contact interactions together with an ordinary SM lepton of the same flavour. The ℓ^* could then decay into a SM lepton via a contact interaction $\ell^* \to \ell f \bar{f}$, where f is a fermion, or via a gauge interaction. For the latter case, the following decays are possible: radiative decays $\ell^* \to \ell\gamma$, neutral-current decays $\ell^* \to \ell Z$, and charged-current decays $\ell^* \to \nu W$. Due to the missing energy stemming from the neutrino, the latter channel can only be reconstructed fully if the momenta of the incoming particles are well known, as in the case of e^+e^- colliders. Unlike in the case of the searches for excited quarks, the final states in searches for excited leptons contain multiple particles, and a potential signal is not a sharp resonance but rather a broad excess as shown in Fig. 11.9a for the example of the e^* search with two final-state electrons and a photon from the $ee^* \to ee\gamma$ decay. The leading search channel at hadron colliders is $\ell\ell\gamma$ since it has the largest cross section. At e^+e^- colliders, electrons can be directly excited as it is the case for quarks at hadron colliders. However, if $f = f' = -1$, photon emission is not possible, and searches need to rely on the emission of a Z boson. Such analyses are currently ongoing. The production cross sections for excited leptons under the conditions at hadron colliders are lower than for direct excitation, and consequently LHC searches for excited leptons yield lower exclusion limits than those for excited quarks. The limit as a function of the compositeness scale Λ and the excited lepton mass M_{ℓ^*} is shown in Fig. 11.9b for the case of the electron, using a partial data set of 13 fb^{-1} [46]. For low M_{ℓ^*}, a compositeness scale up to 12 TeV is excluded. For $\Lambda = M_{\ell^*}$ excited electrons and excited muons are excluded up to about 2.2 TeV.

11.3.3 Composite Higgs and Technicolour

The mass and vacuum expectation value of point-like scalar particles are the only parameters that depend quadratically on the physics at the highest scales through radiative corrections. Therefore the naturalness problem of the SM is confined to a symmetry-breaking sector with elementary Higgs fields.

Before the observation of the 125 GeV resonance, an attractive scenario was to avoid fundamental scalar fields altogether and to describe the Goldstone bosons required for the longitudinal polarisation of $W^{\pm}$ and Z^0 as collective excitations of new fundamental fermionic fields, akin to pions in QCD [48–50]. A narrow scalar resonance with the quantum numbers of the SM Higgs boson is then not required. Similarly to the pions being much lighter than all other strongly interacting particles, the Goldstone bosons and electroweak gauge bosons can be much lighter than the other states in the new strongly interacting sector. Before the LHC, this part of the "technicolour" scenario could be realised straightforwardly. Experimental signatures are the production of additional bound states and collective excitations, analogously to pions, ρ mesons, and a_0 mesons, called techni-π, techni-ρ and axial-vector. It is possible to build low-scale technicolour models such that these techni-hadrons are within the reach of the LHC [51, 52].

The major obstacle to building a realistic complete model was in the mechanism for giving masses to quarks and leptons, the so called "extended technicolour". Most models ran into conflict with flavour-physics results, due to insufficient suppression of flavour-changing neutral currents, if the scale of extended technicolour became too close to the technicolour scale [53]. The separation of scales $E_{\text{IR}} \ll E_{\text{UV}}$ in strongly interacting models is provided by the energy E_{IR} at which a running coupling that is weak at E_{UV} becomes strong. Thus a possible solution to this problem is to adjust the matter content such that the couplings run very slowly, i.e. by "walking technicolour" [54], also known as "almost conformal models". More recently, using intuition provided by the AdS/CFT correspondence [55], it was possible to construct realistic Higgs-less models in continuous [56] and discrete [57] extra dimensions.

After the observation of the narrow 125 GeV resonance, it must, of course, be incorporated into such models, leading to a class of composite Higgs models.

Searches are performed in various final states, such as boson+γ or two bosons decaying to leptons. The predicted new bound states, the techni-mesons, are expected to manifest as narrow resonances in the three-body invariant-mass distribution of the final state, $\ell^+\ell^-\gamma$, $\ell\nu\gamma$, $3\ell 1\nu$ or the four-body invariant mass for $ZZ \to 4\ell$, with $\ell = e, \mu$. An example of the $e^+e^-\gamma$ invariant mass distribution in a partial 8 TeV data set is shown in Fig. 11.9a in the context of a different analysis, the search for excited electrons. Using the full 8 TeV data set, a lower limit of 960 GeV is set on the mass of the charged techni-meson a_T in the $W\gamma$ decay channel, while for the neutral techni-meson ω_T most of the mass range below 890 GeV is excluded in the $Z\gamma$ decay channel [58]. The mass differences between the techni-mesons are set to $M_{\omega_T} = M_{a_T}/1.1 = M_{\rho_T}$ and $M_{\rho_T} - M_{\pi_T} = M_W$ [52]. The limits on a_T and ω_T can therefore be related to the ones on M_{ρ_T} and M_{π_T} shown in Fig. 11.10a in the

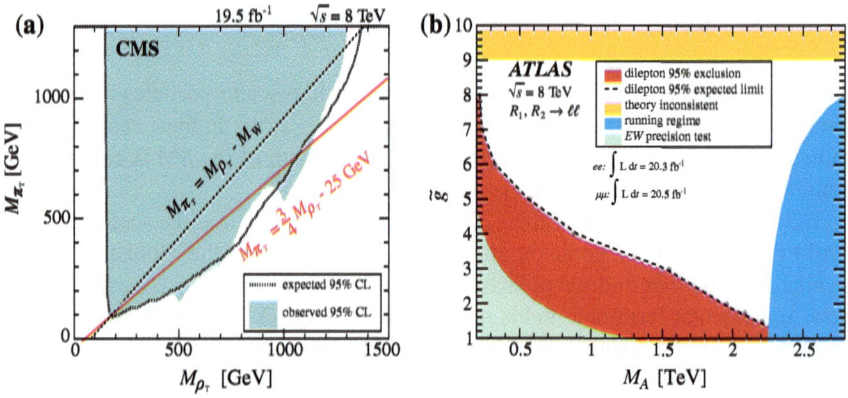

Fig. 11.10 **a** Two-dimensional exclusion limit for low-scale technicolour for techni-ρ and techni-π resulting from the CMS search for leptonic diboson final states. **b** Search for $Z' \to e^+e^-$ and $Z' \to \mu^+\mu^-$ by ATLAS, interpreted in terms of minimal walking technicolour. Adapted from [9, 34]

context of the CMS search for new diboson resonances from [34], which is discussed in Sect. 11.2.2.3. Mass limits on techni-ρ (ρ_T) correspond to 1.14 TeV for the model assumption $M_\pi = 3/4\, M_\rho - 25$ GeV.

The ATLAS search in the dilepton (ee, $\mu\mu$) channel described in detail in Sect. 11.2.1.1 is also used to set limits on the so-called minimal walking technicolour model [59–61]. In this model, new narrow spin-1 techni-meson resonances are predicted, which are produced by quark-antiquark annihilation and decay to dilepton final states via an intermediate Z/γ^* state. The analysis is analogous to the one for Z', with the appropriate signal template used [9]. Limits are set in a plane of the bare axial-vector mass M_A and the gauge coupling $\tilde{g}$ of the technicolour interaction. Via mixing, the latter enters in the denominator of the coupling $g/\tilde{g}$ of the new resonance to SM fermions. The results are shown in Fig. 11.10b. They improve substantially on the limits from electroweak precision data for weak coupling to SM fermions and for low techni-hadron mass. The technicolour coupling must not become too strong for the model to remain consistent. In addition the axial-vector must not be too heavy for the technicolour coupling to run slowly enough.

11.4 Extra Dimensions

The notion that the four observed dimensions of space-time are just a long-distance approximation, while extra dimensions open up at short distances and high energies, has been extremely popular for several reasons: it is a natural prediction of string theory, it allows to solve the perceived hierarchy problem, and it provides a fertile ground for model building beyond the Standard Model, including candidates for dark matter.

The geometry of the extra dimensions is characterised by their curvature and their size. First, in Sect. 11.4.1, flat extra dimensions without intrinsic curvature will be discussed. In these models, the scales relevant for particle physics are determined by the size of the extra dimensions. Then, in Sect. 11.4.2, warped extra dimensions will be considered, where the relevant scales are determined by the intrinsic curvature of the extra dimensions.

A generic prediction of models with extra dimensions is that the scale of quantum gravity is reduced and with it the production threshold for microscopic black holes. The semiclassical approximation leads to a thermal decay of these black holes. Allowing for deviations from the semiclassical approximation leads to a non-thermal decay spectrum that can be modelled by so-called quantum black holes. Production and decay of these thermal and quantum black holes are discussed in Sects. 11.4.3 and 11.4.4, respectively.

11.4.1 Flat Extra Dimensions: ADD and UED

If gravitons are allowed to propagate into n additional space-time dimensions, Gauss' law implies that the relationship between Newton's constant G_N and the Planck scale of quantum gravity, $M_\mathrm{Pl} = \sqrt{1/G_\mathrm{N}}$ in four dimensions, changes to

$$G_\mathrm{N} = \frac{1}{M_\mathrm{D}^{2+n}},$$

where M_D denotes the Planck scale in $4+n$ dimensions. If the extra dimensions have a size R, the strength of the gravitational interactions observed at much longer distances will be $G_\mathrm{N} = 1/(M_\mathrm{D}^{2+n} R^n)$. As a result, the microscopic Planck scale M_D, is related to the observed macroscopic M_Pl by

$$\frac{M_\mathrm{D}}{M_\mathrm{Pl}} = \left(\frac{1}{RM_\mathrm{Pl}}\right)^{n/(n+2)}.$$

Therefore, the microscopic M_D, which is relevant for the high-energy loop corrections causing the hierarchy problem, can be pushed down to the TeV scale by sufficiently large extra dimensions, i.e. $R \gg 1/M_\mathrm{Pl}$. In this case, there is no hierarchy problem [62].

In the case $n = 1$, R would have to exceed the size of the solar system, which is clearly excluded. However, already at $n = 2$, R can be smaller than 1 mm, which can not be excluded from tests of the gravitational $1/r$ potential alone. Instead, particle physics experiments are required to test the predictions of concrete models for the geometry of the extra dimensions and the fields propagating in them.

Solving the SM field equations in the extra dimensions with appropriate boundary conditions leads to an infinite series of eigenstates, commonly referred to as Kaluza–Klein (KK) towers [63, 64]. The lowest state in each tower is the SM particle, and from a four dimensional perspective, the higher states look like copies of the SM particle, but with masses that are multiples of the inverse size $1/R$ of the extra dimensions. Their interactions with the SM particles and among each other are fixed by the SM interactions. This results in a rich phenomenology that can be probed at colliders, assuming that the scale of the extra dimensions is low enough to actually solve the hierarchy problem.

The original Arkani–Hamed–Dimopoulos–Dvali (ADD) model [62] was inspired by string theory, where fields with gauge and flavour quantum numbers (open string endpoints) are naturally confined to lower-dimensional branes, whereas gravitons (closed strings) can propagate in all dimensions. This scenario predicts a tower of heavy spin-2 particles, the so called KK gravitons. The coupling of these to SM particles is small individually $\sqrt{G_N} = 1/M_{Pl}$. However, the spacing $1/R \ll M_{Pl}$ of the KK gravitons is very dense, and the contributions from the exchange of virtual gravitons among SM particles add up coherently, potentially producing an observable effect. Thus the contribution to cross sections will be larger when the density of states is larger. This is the case for a small number of extra dimensions, because the required size of the extra dimensions is smaller for a larger number of extra dimensions.

Of course, the spectrum of KK gravitons must be cut off at the energy scale where gravity becomes strong and perturbative quantum gravity is replaced by a more comprehensive theory, e.g. string theory. Since the sum over all KK gravitons is divergent, the end result will depend on this scale. A sensible choice for the cut-off is the so called string scale $M_S = 2\sqrt{\pi}\,(\Gamma(n/2))^{\frac{1}{n+2}}\,M_D \approx 4M_D$ [65–68]. The final result for the sum is then a contribution to the scattering amplitudes for SM particles from virtual KK graviton exchange of the order $1/M_D^2$. For M_D sufficiently small to solve the hierarchy problems, such effects should be accessible at the LHC.

The production cross section for individual real KK gravitons is small, of the order $1/M_{Pl}^2$. However, their decay width is similarly small and they will escape the detector, resulting in an E_T^{miss} signal. Therefore one must add the production cross section of all kinematically accessible KK gravitons, which results in an E_T^{miss} signal of the order $1/M_D^2$, that can be probed at the LHC. In contrast to the virtual exchanges, the result is automatically finite because the spectrum is cut off by kinematics. Nevertheless, one must take care to exclude regions of phase space where perturbative quantum gravity is not valid [68].

The ADD model was soon generalised to universal extra dimensions (UED) [69] (see also [70]) by allowing the SM fields to propagate in the bulk of the extra dimensions as well, leading to KK towers of particles with SM gauge and flavour quantum numbers. Appropriately choosing the topology of the extra dimensions and the boundary conditions allows one to project out states in the KK towers that would otherwise give rise to conflicts with constraints from low-energy precision experiments.

In contrast to the weakly interacting KK gravitons of the ADD model, the KK partners of particles carrying SM charges must be heavy, in order to have not been observed a long time ago. Therefore, the extra dimensions in UED models must have a size of the order of 1/TeV or even less. Apart from this constraint on the size of the extra dimensions, the parameter space of possible UED models is larger compared to ADD, because one can choose the number of extra dimensions and the boundary conditions for the particles, and one can also tune the masses μ of the particles in the bulk. Note that, while one must not add masses to gravitons and gauge bosons by hand, it is possible to add bulk masses to the fermions, resulting in so-called split-UED models [71], depending on a bulk mass parameter μ.

The flatness of the extra dimensions results in momentum conservation in the extra dimensions, which gives rise to a new conserved quantum number, the KK parity. This means that KK-odd particles can only be produced in pairs and will be stable, providing dark matter candidates. Of course, model building and experimental searches will start with the smallest number of extra dimensions possible. Since the extra dimensions in UED models are of the order of 1/TeV, a single extra dimension is not already excluded by gravity on macroscopic scales, in contrast to the ADD case.

Experimentally, searches for the production of KK gravitons in the ADD model can be separated into two classes: the direct production of gravitons leaving the detector undetected and the exchange of virtual gravitons at tree level. The latter case leads to deviations from the expected SM cross section, e.g. in dilepton or diphoton production. As explained above, in order to solve the hierarchy problem, the size R of the extra dimension needs to be comparatively large, leading to an almost continuous spectrum of KK modes. Thus, the experimental signature is a broad, non-resonant excess, similar to the one searched for in the context of contact interactions already discussed in Sect. 11.3.1. Indeed, a very similar analysis strategy is used for both searches, and the dielectron invariant-mass spectrum displayed in Fig. 11.7a also shows an example for an ADD signal in addition to the ones for contact interactions. Different conventions exist for the parametrisation of the total dilepton production cross section, i.e. the sum of SM Drell–Yan production and the additional signal contribution. These are known under the names of Giudice–Rattazzi–Wells (GRW) [65], Hewett [72], and Han–Lykken–Zhang (HLZ) [66]. Figure 11.11a shows the lower limits on M_S interpreted in these conventions for the full 8 and 7 TeV data sets for the combination of the dielectron and dimuon channels at ATLAS [39]. The limits decrease with an increasing number of extra dimensions. This is due to the diminishing density of KK states resulting in a smaller cross section, as discussed above. It can be seen that the lower mass limits, e.g. for GRW or HLZ with four extra dimensions, improved from about 3 TeV for the 7 TeV data set to about 4 TeV for the full 8 TeV data. Similar 8 TeV exclusion limits are also expected from an analysis of the diphoton spectrum, as can be inferred from comparing the diphoton results from the full 7 TeV data set [73] with the corresponding dilepton limits on this plot.

The direct production of gravitons leaving the detector undetected leads to an experimental signature of missing transverse energy in association with a jet or a gauge boson. These signatures are commonly referred to as "mono-X" and are mainly discussed in the context of searches for dark matter, see Sect. 11.5.1. An example of an ADD signal in the mono-jet search at CMS using the full 8 TeV data set is shown in Fig. 11.18a. Details of the analysis are discussed in the corresponding section. The interpretation in terms of lower limits on M_{D} is shown in Fig. 11.11b for

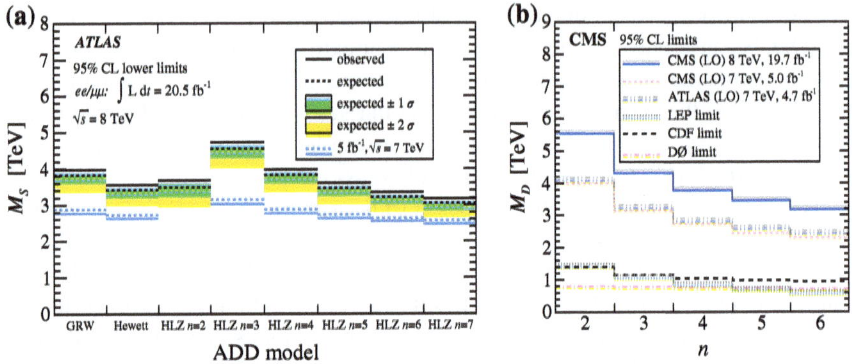

Fig. 11.11 ADD model. **a** Lower limits on M_S for different conventions used in the parametrisation of the total dilepton production cross section as well as for a different number n of extra dimensions in the HLZ convention. Results for the full 7 and 8 TeV data sets using the combined dielectron and dimuon final states are shown. **b** Lower limits on M_D as a function of the number of extra dimensions from the mono-jet searches at ATLAS and CMS, compared to limits from other experiments (also involving mono-photon final states). Adapted from [39, 74]

the CMS 8 TeV result compared to the CMS results with the full 7 TeV data set and various other experiments. The improvement in terms of limits compared to using the full 7 TeV data set and in particular to the LEP and Tevatron results is clearly visible. Searches in the mono-photon channel at ATLAS and CMS are less stringent compared to the mono-jet channel due to the smaller cross section and yield upper limits on M_D of less than 2 TeV for the full 7 TeV data set [75, 76].

UED and split UED models so far received little attention at the LHC. One exception is the interpretation of the lepton+$E_\mathrm{T}^\mathrm{miss}$ search already discussed in Sect. 11.2.2.1 in the context of the W' search. The W_KK^2 is the lowest state that can couple to SM fermions, where the superscript refers to the second KK excitation mode. Its decay into a lepton and a neutrino is kinematically identical to the W'_SSM, apart from effects stemming from the different width of the W_KK^2. Figure 11.4a shows predicted $\sigma \times \mathrm{B}$ curves for a W_KK^2 in the split-UED model for two extreme values of the bulk mass parameter μ. The resulting lower mass limits are also rather different for the two values of μ.

11.4.2 Warped Extra Dimensions: The Randall–Sundrum Model

A conceptually very different solution of the hierarchy problem is provided by the Randall–Sundrum (RS) models [77, 78]. Whereas the extra dimensions in ADD and UED are flat, the extra dimensions in the RS models are warped solutions of the five-dimensional Einstein equations, with the four-dimensional length scale $a = \mathrm{e}^{ky}$

growing exponentially along the single extra dimension y. The strength of this growth is parametrised by the scale $1/k$. As in the case of ADD, string theory allows for a setup where all SM particles are confined to a four-dimensional brane, while gravitons propagate in all five dimensions.

There are two different versions of the RS models: in RS1 [77] $y \in [0, \pi R]$ and in RS2 [78] $y \in [0, \infty)$. If the SM particles in RS1 are confined to the brane at $y = \pi R$, all four-dimensional momentum and mass scales, such as the electroweak scale v_{weak}, are exponentially suppressed by a factor $e^{-\pi k R}$ with respect to the gravitational scale M_{Pl} of the five-dimensional model. The resulting hierarchy $v_{\text{weak}}/M_{\text{Pl}} = e^{-\pi k R}$ can therefore be explained in RS1 by a moderate $k \approx 12/R$. On the other hand, the RS2 model is not a solution of the hierarchy problem and is not considered in the following.

Solving the field equations in the warped extra dimensions will again give towers of gravitons that couple to SM particles. In contrast to the ADD model, the extra dimension is not large, and the masses of these gravitons will be multiples of the electroweak scale. Their coupling to SM particles is flavour-independent, allowing searches in channels with light particles such as $\ell\ell$ and $\gamma\gamma$.

Again, as in the case of flat extra dimensions, the RS scenarios can be generalised, by allowing SM particles to propagate in the bulk [79]. These bulk-RS models have become a very active area of research, because they allow free parameters in the SM (such as Yukawa couplings) to be computed as overlap integrals of SM wave functions in the extra dimension. If enough custodial symmetries are implemented, the stringent limits from electroweak precision data can be avoided [80]. In contrast to RS1, the coupling of the KK graviton and gluon to SM particles depends on their respective localisation in the bulk. The light SM fermions are localised near the Planck brane in order to explain their small Yukawa couplings and avoid constraints from electroweak precision data, whereas the right-handed top quark, the KK graviton and KK gluon are localised near the TeV brane. In bulk-RS, therefore, the KK graviton will couple most strongly to heavy particles [81], and the KK gluon in particular to the coloured top quarks [82].

Before the discovery of the Higgs boson, variants of RS with TeV-scale extra dimensions provided the theoretical framework for realistic Higgs-less models of electroweak symmetry breaking. Today, they provide the theoretical tools for models with composite Higgs bosons, because they can be interpreted as a AdS/CFT dual of a strongly interacting EWSB sector.

11.4.2.1 Only Gravity in the Bulk

The RS1 model provides a useful benchmark model for the search for spin-2 resonances. The ATLAS dilepton (ee, $\mu\mu$) analysis discussed in detail in Sect. 11.2.1.1 is also used to set limits on G^*, the lowest KK excitation of the graviton. The data analysis for the G^* is fully analogous, except that the signal templates now correspond to RS G^* signals.

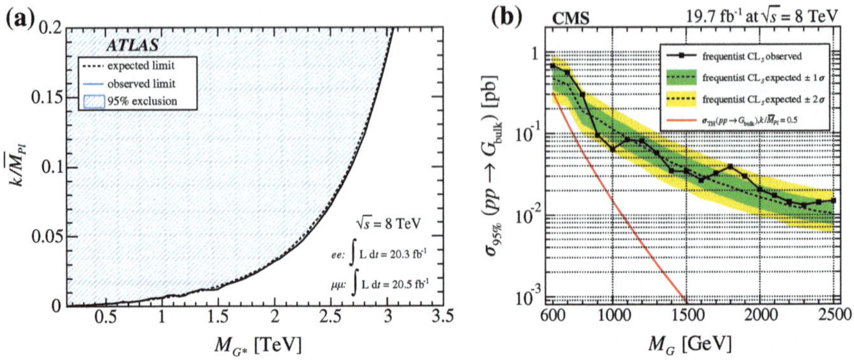

Fig. 11.12 **a** Expected and observed limits in the plane of the G^* mass in the original Randall–Sundrum (RS1) model versus the coupling strength $k/\overline{M}_{\text{Pl}}$ from the combined dielectron and dimuon search. **b** Upper limit on the graviton production cross section as a function of the graviton mass in the bulk-RS model. The limit is derived from a combination of the W^+W^-, ZZ and WZ final states from semileptonic and fully hadronic decays. Also shown is the predicted cross section for $k/\overline{M}_{\text{Pl}} = 0.5$. Adapted from [9, 83]

Despite the kinematically different decay due to the spin-2 nature of the G^*, the resulting $\sigma \times$ B limit as a function of the G^* mass is, at lower masses, similar to the ones for spin-1 particles shown in Fig. 11.2b. At higher masses it is similar to the Z^* one, which is hardly affected by the parton-luminosity tail. The predicted $\sigma \times$ B for G^* depends on the coupling strength of the G^* to SM particles, $k/\overline{M}_{\text{Pl}}$, with $\overline{M}_{\text{Pl}} = M_{\text{Pl}}/\sqrt{8\pi}$. Therefore, the lower limit on the G^* mass is a function of the coupling strength. It is shown in Fig. 11.12a for the full 8 TeV data set using the combined dielectron and dimuon channels, and is found to be 2.68 TeV for $k/\overline{M}_{\text{Pl}} = 0.1$ [9].

Similar 8 TeV exclusion limits are also expected from an analysis of the diphoton spectrum, as can be inferred from comparing the ATLAS diphoton lower mass limit of 2.06 TeV for $k/\overline{M}_{\text{Pl}} = 0.1$ using the full 7 TeV data set [73] with the corresponding combined dilepton limits of 2.16 TeV [17] and 2.14 TeV [10] from ATLAS and CMS, respectively. A combination of the dilepton and diphoton 7 TeV results slightly improved the lower mass limit on the G^* to 2.23 TeV [73].

Further searches for RS gravitons with smaller sensitivity have been carried out in final states with two jets or with two vector bosons. The excluded G^* mass ranges for $k/\overline{M}_{\text{Pl}} = 0.1$ are found to be 1–1.45 TeV for the dijet search with 4 fb^{-1} at 8 TeV [18] and, using the full 8 TeV data set, below 1.2 TeV for the W^+W^- final state decaying hadronically [35]. No mass limit was derived for the ZZ final state decaying hadronically as it is not sensitive to the small predicted cross section [35].

11.4.2.2 Standard Model Fields in the Bulk

In the bulk-RS model, in contrast to the original RS1 model discussed above, the coupling of the graviton to light fermions is suppressed, and the coupling to photons is negligible, while the decay of gravitons into a pair of heavy gauge bosons can be sizeable.

Searches for new resonances decaying to W^+W^-, ZZ, or WZ are performed, in which one of the vector bosons decays leptonically and the other hadronically [83] along with the fully hadronic final state [35] already mentioned above in the context of RS1. Newly developed techniques aiming at identifying jet substructures [25–28] are used to analyse signal events in which the hadronisation products from the decay of highly boosted bosons are contained within a single reconstructed jet. When combining W^+W^-, ZZ and WZ final states from the semileptonic and fully hadronic decays, upper limits on the bulk graviton production cross section are set in the range from approximately 700–10 fb for resonance masses between 600 and 2500 GeV, respectively, as depicted in Fig. 11.12b. These limits do not yet allow mass limits on the bulk graviton model to be set, as can be seen from the predicted cross section for $k/\overline{M_{\rm Pl}} = 0.5$ shown in the same figure.

The decay of the KK partner of SM gluons in bulk-RS has branching ratios of above 90 % to the $t\bar{t}$ final state [82]. Thus, the $t\bar{t}$ resonance search discussed in the context of Fig. 11.3b is also used to search for the KK excitation of a gluon, taking into account the fact that the resonance is rather broad. A KK gluon with a width of approximately 15–20 % [84] has been excluded up to a mass of 2.5 TeV, using the full 8 TeV data set [14].

11.4.3 Thermal Black Holes

By dimensional analysis alone, an immediate side-effect of lowering the Planck scale $M_{\rm D}$ is that the Schwartzschild radius $r_S = 2m/(M_{\rm D}^2)$ of an object of mass m is increased dramatically. If $m > \sqrt{\pi}M_{\rm D}$, this Schwartzschild radius will exceed the Compton wavelength $\lambda = 2\pi/m$ of the object. In this case quantum effects should be localised inside the Schwartzschild radius, and one expects from classical general relativity that an event horizon, i.e. a black hole (BH), forms. Since there are no other coupling constants involved, such black holes can be produced with a geometric cross section on the order of the size of the event horizon $\approx 1/M_{\rm D}^2 \approx 0.4\,{\rm nb}/(M_{\rm D}/{\rm TeV})^2$ if the partonic center of mass energy exceeds $M_{\rm D}$ [85].

Under the assumption that the above semi-classical approximations are justified, Hawking's semi-classical analysis of black-hole evaporation can be applied. Then the black hole will destroy any information about the initial state and decay thermally, without conserving any SM quantum numbers. The decay width due to thermal Hawking radiation of a quasi-classical black hole with a mass close to the Planck mass is then $\mathcal{O}(10^{-5}) \cdot m$, i.e. $\mathcal{O}(100\,{\rm MeV})$ at LHC energies. The experimental signature is therefore an instantaneous decay into all kinematically open channels with a thermal spectrum.

More detailed calculations in space-times with n extra dimensions show that the Schwartzschild radius is

$$r_S = \frac{1}{\sqrt{\pi} M_D} \left[\frac{m}{M_D} \left(\frac{8\Gamma(\frac{n+3}{2})}{n+2} \right) \right]^{\frac{1}{n+1}} \approx \frac{1}{\sqrt{\pi} M_D} \left(\frac{m}{M_D} \right)^{\frac{1}{n+1}}, \tag{11.5}$$

as long as r_S is larger than the size of the extra dimension. The corresponding black-hole production cross section is then

$$\sigma_{\mathrm{BH}} = \pi r_S^2 \approx \frac{0.4\mathrm{nb}}{(M_D/\mathrm{TeV})^2} \left(\frac{m}{M_D} \right)^{\frac{2}{n+1}}, \tag{11.6}$$

justifying the formulae derived from dimensional analysis above in the interesting parameter range $m \approx M_D$.

Experimentally, the search for thermal black holes relies on an excess of events with a large number of high-momentum particles when compared to the SM expectation. The expected multiplicity as well as composition and energy of the final-state particles in a black-hole decay depends on the model, the higher-dimensional Planck scale M_D and the production mass threshold M_{th} for black-hole production. Figure 11.13a shows the expected track multiplicity ($p_T^{\mathrm{track}} > 10\,\mathrm{GeV}$) for a black-hole signal generated with BLACKMAX [88] for a scenario of rotating black holes, $n = 4$ extra dimensions, $M_{\mathrm{th}} = 5\,\mathrm{TeV}$ and $M_D = 1.5\,\mathrm{TeV}$. The selection [86] requires the presence of at least two muons, where the leading muon has $p_T > 100\,\mathrm{GeV}$ and

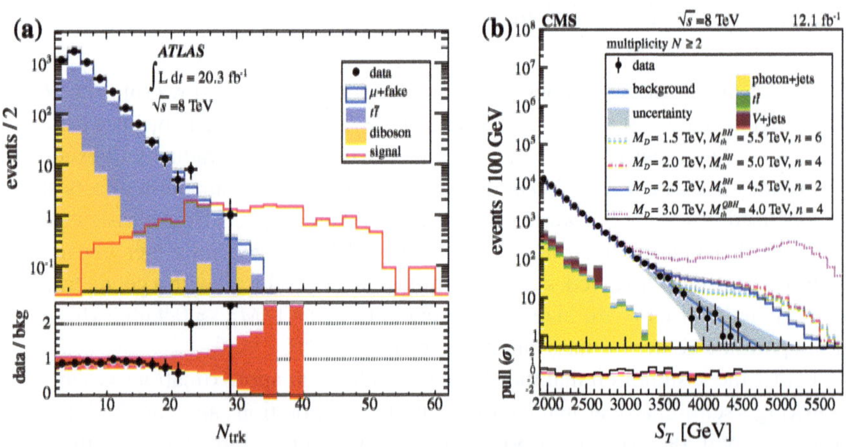

Fig. 11.13 **a** Comparison of data and expected background as a function of the track multiplicity for tracks with $p_T > 10\,\mathrm{GeV}$ in an analysis requiring two same-sign muons. The *red area* in the ratio panel displays the total uncertainty on the background. A signal prediction for a thermal *black hole* is overlaid. **b** Comparison of data and expected background as a function of the scalar sum of transverse energy, S_T, for events with at least two objects with transverse energies above 50 GeV in the final state. Examples for signals are shown for non-rotating thermal *black holes* and for a quantum *black hole* (QBH). The ratio panel shows the pull distribution defined as $(data - background)/\sigma_{stat+sys}(data - background)$. Adapted from [86, 87]

the leading and subleading one have the same charge. The good agreement between observed data and predicted background and the good discrimination power of the multiplicity as a search variable can be seen.

Another search variable is the scalar sum, S_T, of the transverse energies of all objects. The search at CMS [87] requires at least two jets and is binned in the multiplicity N ($N \geq 2, \ldots, N \geq 10$) of all objects (jets, electrons, muons and photons) with a transverse energy larger than 50 GeV. Figure 11.13b shows the S_T distribution for the case $N \geq 2$, where the sum also includes the transverse missing energy if it is in excess of 50 GeV in the event. The SM background is dominated by QCD multi-jet production, with negligible contributions from other processes in the high-E_T region as can be seen in the figure. The shape of the S_T distribution has been found to be approximately independent of N. Therefore, the background shape is determined from a fit to the dijet data ($N = 2$) in the range $1.8 < S_T < 2.8$ TeV, and the normalisation to data in each multiplicity bin is done in the range $1.9 < S_T < 2.3$ TeV. Good agreement between data and expectation is found, as can be seen in the ratio panel. Examples for non-rotating black holes generated with BLACKMAX are also shown for three different combinations of M_D, M_{th} and n.

A similar search is carried out at ATLAS [89]. It requires at least three objects with transverse momentum $p_T > 100$ GeV, one of the three being an electron or a muon. The electroweak backgrounds are obtained from a MC simulation, normalised in control regions. The multi-jet background is estimated by a data-driven method. All backgrounds are extrapolated to the region of high $\sum p_T$, the scalar sum of the transverse momenta, via fits. Also in this search no signs for an excess have been found in the signal region, $\sum p_T > 2$ TeV. The resulting expected and observed upper limits in the M_{th}–M_D plane are shown in Fig. 11.14a for non-rotating black holes generated with the BLACKMAX generator for $n = 2, 4$, and 6 extra dimensions. The black-hole production cross section increases with an increasing number of extra dimensions, leading to more stringent limits, as can be seen in the figure. For comparison, the limit from the same-sign dimuon search discussed above is also shown. It is found to be less stringent, mainly due to the smaller branching ratio for the final state with the two like-sign muons.

The corresponding limits from the CMS search discussed above are shown in Fig. 11.14b. The limit for non-rotating black holes with $n = 6$ extra dimensions can be directly compared to the one from ATLAS in Fig. 11.14a and is found to be rather similar.

11.4.4 Quantum Black Holes

Of course, the above quasi-classical treatment of black holes produced at the LHC is not necessarily justified, in particular close to the threshold when the Schwartzschild radius is of the order of the Compton wavelength. While full calculations in quantum gravity are not available, one can still derive some properties of these non-thermal black holes, often referred to as quantum black holes (QBH).

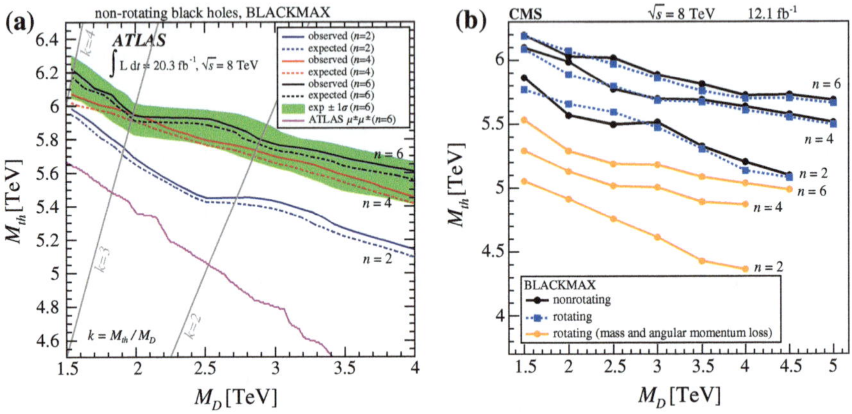

Fig. 11.14 **a** Expected and observed upper limits in the M_{th}–M_{D} plane for non-rotating *black holes* generated with the BLACKMAX generator for $n = 2$, 4, and 6 extra dimensions, from an analysis requiring at least three objects with $p_{\text{T}} > 100$ GeV, one of them being an electron or a muon. In addition, the limit derived from the same-sign dimuon search (see Fig. 11.13a) is shown. **b** Upper limits in the M_{th}–M_{D} plane for various *black-hole* scenarios generated with the BLACKMAX generator, from an analysis requiring at least two objects with a transverse energy above 50 GeV in the final state, see Fig. 11.13b. Adapted from [87, 89]

High-multiplicity decay channels with many soft particles should be suppressed, because the Compton wavelength of the produced particles should not be larger than the size of a QBH [90]. QBHs should conserve local charges and can therefore be classified by their colour and electric charge [91]. Nevertheless, global charges, like lepton and baryon number, that do not correspond to a gauge invariance are expected to be violated by gravity even outside the quasiclassical realm [92].

Thus, in contrast to the thermal black holes discussed above, QBHs decay to a small number of final state particles, most likely just two. However, QBHs can have fractional charges and may violate lepton-number and baryon-number conservation in their decay, leading to a large number of potential final-state signatures. Among these are dijets or multi-jets, one jet plus either a charged lepton, photon, or missing energy and dilepton signatures with the same or different flavour.

Clean experimental signatures are found in particular in decays that violate lepton-number and baryon-number conservation, because here the SM background is small. An example is the search in the lepton+jet final state at ATLAS [93]. Here, an electron or a muon and a jet are required, each object with transverse momentum $p_{\text{T}} > 130$ GeV. The invariant-mass distribution for the lepton+jet system is shown in Fig. 11.15a for the case of the electron, using the full 8 TeV 2012 data set. The backgrounds from electroweak processes are taken from MC simulations. They are dominated by vector-boson production in association with jets, diboson production or single or pair production of top quarks. While negligible in the muon channel, the other important background in the electron channel stems from jets being misidentified as electrons or electrons from semileptonic hadron decays ("multijet"). These

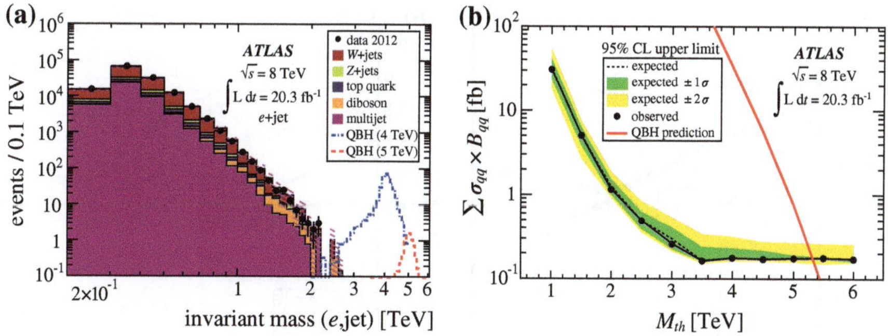

Fig. 11.15 **a** Comparison of data and expected background as a function of the invariant mass of the electron+jet system and two examples for QBH signals. **b** QBH prediction for, and expected and observed upper limit on, the sum of the cross sections of all production modes times their respective branching fraction to the lepton (e, μ)+jet final state as function of M_{th}. Adapted from [93]

backgrounds are estimated using a data-driven method. Data and expectation agree well in the electron channel, as can be seen from the figure. The same is true for the muon channel.

Figure 11.15a also shows two examples for QBH signals. These signals are based on the ADD model with $n = 6$ extra dimensions. The threshold for QBH production, M_{th}, is assumed to correspond to the higher-dimensional Planck scale M_{D}. The exact $\sigma \times \text{B}$ depends on the model used, as the assumptions made can vary in some details. It is usually assumed that total angular momentum, colour and electric charge are conserved. The signals shown here are based on the QBH model described in [94, 95]. The largest contributions to the cross section come from the production via two u quarks (uu), i.e. a charge 4/3 QBH, followed by charge $1/3$ (ud) and charge $-2/3$ (dd) QBHs, with corresponding branching fractions to each lepton+jet final state of 11, 6 and 7 %. The sum of the cross sections of all production modes times their respective branching fraction to the lepton+jet final states is shown in Fig. 11.15b as function of M_{th}. Also shown are the expected and observed limits, which are in good agreement indicating the absence of hints for new physics. Above 3.5 TeV, these upper limits are 0.18 fb at the 95 % CL, leading to a lower mass limit of 5.3 TeV on the specific model for QBHs used here.

A similar search is carried out in the γ+jet final state at ATLAS [45]. In this analysis, each object needs to have a p_{T} larger than 125 GeV. The search is targeted at production in the s-channel, requiring the photon to be in the barrel calorimeter ($|\eta| < 1.37$) and the $\delta\eta$ between photon and jet to be smaller than 1.6. These requirements remove a large part of the background production in the t-channel, e.g. dijet production, where photons can be produced in the fragmentation of quarks or gluons. The invariant-mass distribution of the photon-jet system is shown in Fig. 11.16a for the full 2012 data set at 8 TeV. Also shown are examples for QBH signals, which are produced with the same generator and settings as the ones discussed above for the lepton+jet final state. While QBHs are not resonances, it can be seen that the

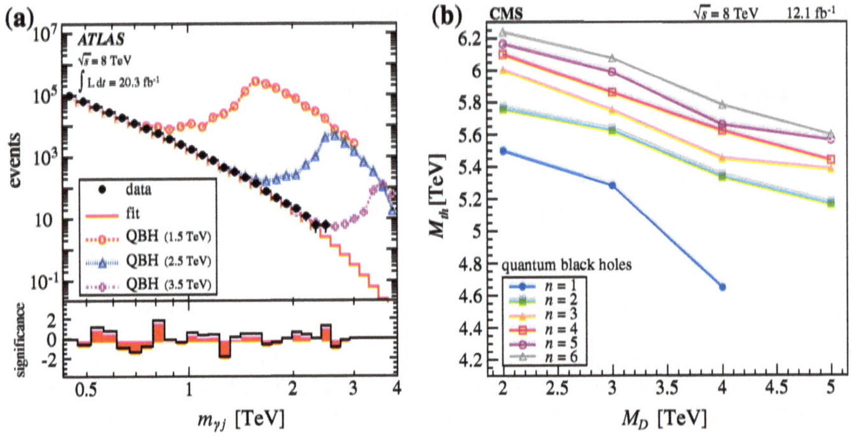

Fig. 11.16 **a** Comparison of data and expected background as determined by a fit to the data, as a function of the invariant mass of the photon+jet system, and three examples for QBH signals. **b** Upper limits in the M_{th}–M_{D} plane for QBH scenarios with a different number of extra dimensions from an analysis requiring at least two objects with a transverse energy above 50 GeV in the final state, see Fig. 11.13b. Adapted from [45, 87]

increase in cross section near M_{th} leads to predicted signals rather similar to the ones predicted for resonant production. It is therefore justified to determine the SM backgrounds by a fit to the data and simply search for an excess in data over this smoothly falling fit function to the invariant-mass distribution. The statistical significance of the difference between data and the SM background is shown in the bottom panel of the same figure. In absence of a significant excess limits are set on the visible cross section ($\sigma \times$ B times acceptance times efficiency) as a function of M_{th}. In the context of the specific QBH ADD model with $n = 6$ extra dimensions, a slightly weaker lower mass limit of 4.6 TeV is observed when compared to the lepton+jet case discussed above.

The ATLAS search in the dijet final state [32], already briefly discussed in Sects. 11.2.2.2 and 11.3.2 in the context of the W' and excited quarks, respectively, can also be interpreted in terms of QBHs. The power of this search resides in the large branching ratio (>85 %) of QBHs to dijets in the mass range considered, leading to rates that are larger by one (two) orders of magnitude at about 1 TeV (5 TeV) when compared to the photon+jet final state. The disadvantage of this analysis is the large background, which is dominated by t-channel QCD dijet and multi-jet production. This t-channel contribution is suppressed by requiring the rapidity difference of the two jets to be below 1.2, enhancing the sensitivity of this search for s-channel production. The analysis strategy is rather similar to the photon+jet case and consists in a search for an excess over the smoothly falling SM background in the dijet invariant-mass spectrum, which is determined by a fit. The resulting lower mass limit is 5.82 TeV, i.e. stronger than the ones from the lepton+jet or photon+jet searches discussed above, which use the same QBH model. A QBH model generated with the

BlackMax generator, also with six extra dimensions, yields a very similar limit of 5.75 TeV.

The lowest ($N \geq 2$) multiplicity bin (see Fig. 11.13b) in the CMS multi-jet search [87] discussed in the previous section has also been used to set limits on QBHs. The specific QBH model used is the same one as in the ATLAS QBH searches discussed above, except that the assumption $M_{\text{th}} = M_{\text{D}}$ is not used but instead lower limits are set in the M_{th}–M_{D} plane as shown in Fig. 11.16b. They are presented for up to six extra dimensions, where the $n = 1$ case is special as it corresponds to RS black holes, while all other cases are in the usual ADD context. In the RS case, the extra-dimensional Planck scale M_{D} is replaced by the rescaled Planck scale $M_{\text{Pl}}\text{e}^{-\pi kR}$ in the cross-section formulae (11.5) and (11.6) [90].

The limits are more stringent with increasing number of extra dimensions due to the increase of the production cross section. A direct comparison of the $n = 6$ case to the ATLAS results discussed above is not possible because the case $M_{\text{th}} = M_{\text{D}}$ is not presented on the plot. However, it appears that also the CMS limit would be above 5 TeV.

11.5 Dark Matter

Indications for the existence of dark matter (DM) are one of the strongest hints for physics beyond the SM. A plethora of experiments is seeking to find evidence of dark matter either through direct dark-matter nucleus collisions or through indirect searches. The resulting picture is not quite consistent, as some claimed experimental observations are contradicted by exclusion limits from other experiments. Pairs of dark-matter particles might also be produced in proton-proton collisions at the LHC, thus allowing additional searches that complement the ones discussed in the previous chapters. In addition, the LHC can also search for very light dark-matter particles. Masses below 10–12 GeV are very challenging for direct-detection experiments. Firstly, the flux of low-mass dark-matter particles is expected to be much lower, because these particles can have escaped the galactic halo. Secondly, the recoil of detector material from interactions with low-mass particles is much smaller, making it much harder to detect. Additionally, there is substantial background from neutrinos.

The most popular class of models explaining the cosmological observations contains dark matter that is weakly interacting, electrically neutral, and fermionic. Many SUSY models contain a good dark matter candidate (the lightest SUSY particle if it is electrically neutral and weakly interacting), which can be detected as $E_{\text{T}}^{\text{miss}}$ at the end of a kinematically reconstructed SUSY decay chain. Since this has been discussed in the SUSY chapter, we concentrate here on pair production of dark-matter particles without specific model assumptions. Up to now, most of such analyses consider a scenario where dark-matter production is modelled using effective quantum field theories (EFT) in which the SM-DM interaction is approximated by an effective point-like four-particle interaction. In this approach, two parameters fully determine the interaction: the dark-matter mass M_χ and the cut-off scale $\Lambda = M/\sqrt{g_\chi g_{\text{q}}}$ (g_χ

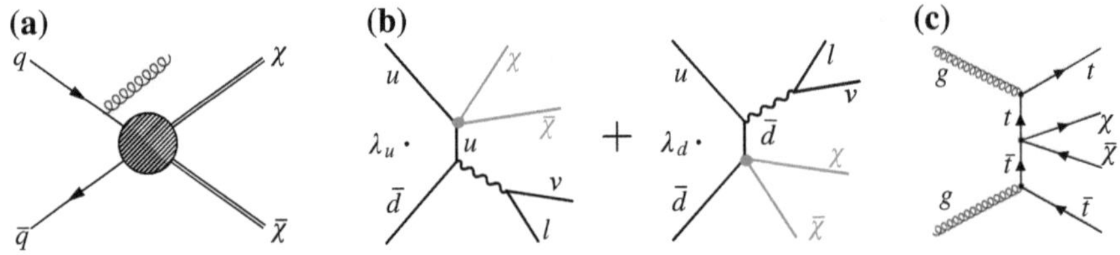

Fig. 11.17 In order to detect dark matter at the LHC, a detectable particle is needed as a "tag". This could be **a** a jet or photon, or **b** a W boson with its subsequent decay, here shown for the example of the leptonic channels. **c** Dark matter could also be produced in association with *top*-quark pairs

and g_{q} are the couplings to dark matter and SM particles, respectively). Sometimes the notation M_* is used instead of Λ. The EFT approach allows one to circumvent the need to explicitly model a messenger (or exchange) particle, similar to Fermi's description of the β decay prior to the discovery of electroweak bosons. The EFT description is valid as long as the mass M of the messenger particle is larger than the energy and momentum transfer in the partonic collision. Very recently, simplified models with real mediator particles (such as a Z') have been proposed as an extension of the EFT description [97, 98]. In these models the following parameters describe the interaction: the mass of dark-matter particles, mass and width to describe the mediator properties (M and Γ), and the couplings to the SM and to the dark matter.

In EFT, if the dark-matter particle χ is assumed to be a Dirac fermion, it has a four-fermion interaction with quarks (q):

$$\mathcal{L} = C(\bar{q}\Gamma^q q)(\bar{\chi}\Gamma^\chi \chi).$$

Here C represents the coupling constant, and the operator Γ describes the type of the interaction, including scalar ($\Gamma = 1$), vector ($\Gamma = \gamma^\mu$), axial-vector ($\Gamma = \gamma^\mu\gamma^5$), pseudo-scalar ($\Gamma = \gamma^5$), and tensor interactions. The scaling of the constant C with the cut-off scale depends on the Lorentz structure of the interaction considered. It is $1/\Lambda^2$ for vector (also denoted "D5"), axial-vector ("D8") and tensor interactions ("D9"), and m_q/Λ^3 for scalar qq ("D1") interactions, accounting for the helicity flip [99].

LHC searches place constraints on the dark-matter production cross section. These constraints can be translated into limits on the scale Λ (or M_*). Unlike direct-detection experiments, where dark-matter particles recoil against target nuclei, the dark matter produced at the LHC also allows different types of interactions to be studied, as discussed above.

Finally, the obtained limits are typically transformed to the dark-matter nuclei parameter space to allow comparisons with direct-detection experiments to be made. The latter distinguish between spin-dependent (SD) and spin-independent (SI) scattering results, depending on their target material. The sensitivity of the direct-detection experiments is a few orders of magnitude better in the SI case compared to the SD case due to the larger number of potential scattering partners.

11.5.1 Mono-X Searches

Under the assumption of only weak interactions, dark matter would manifest itself as large missing transverse energy and could only be detected via a "tag" based on particle(s) leaving a trace in the detector in order to trigger and detect dark-matter events. Following the examples shown in Fig. 11.17, such tags could be provided by radiating a single jet (Fig. 11.17a) or photon as well as through emission of a single W (Fig. 11.17b) or Z boson. These searches are commonly denoted "mono-X" (X $= j, \gamma, W, Z$), and they all suffer from large Standard Model backgrounds. The pioneering channels for dark-matter searches at hadron colliders are the single-jet and single-photon analyses, which are both pursued by ATLAS and CMS with similar sensitivities. Recently, searches based on the decay of a W or Z boson have been performed, with the hadronic final state yielding the maximum sensitivity due to the large hadronic branching fraction. Similar to the previously discussed searches, the analyses look for an excess of events over the expected SM background. However, the background for dark-matter searches is typically large, and the signal does not exhibit a resonant peak but is rather flat and unstructured. The key distributions to be examined are the p_{T}, M_{T} and $E_{\mathrm{T}}^{\mathrm{miss}}$ spectra. More and more channels are being studied, such as dark matter in association with top-quark pair production (see Fig. 11.17c).

As an example for a mono-X search, the CMS mono-jet analysis [74] is discussed. In this analysis, events triggered with a $E_{\mathrm{T}}^{\mathrm{miss}}$ or a jet+$E_{\mathrm{T}}^{\mathrm{miss}}$ trigger are selected, with the corresponding trigger thresholds determining the lower sensitivity boundary. The offline event selection consists of a cut on significant $E_{\mathrm{T}}^{\mathrm{miss}}$, $E_{\mathrm{T}}^{\mathrm{miss}} > 250\,\mathrm{GeV}$, and the requirement of a central hard jet of $p_{\mathrm{T}} > 110\,\mathrm{GeV}$. A second, softer jet is only allowed if it is located close to the first one, thus suppressing most of the large dijet background. The W/Z+jets background is largely reduced by vetoing events with an isolated lepton. The irreducible remaining background is $Z(\rightarrow \nu\nu)$+jet, which is estimated from $Z(\rightarrow \mu\mu)$+jet data. The resulting $E_{\mathrm{T}}^{\mathrm{miss}}$ distribution is shown in Fig. 11.18a, along with an example for a dark-matter signal for $\Lambda = 900\,\mathrm{GeV}$ and $M_{\chi} = 1\,\mathrm{GeV}$. In the course of the analysis, the $E_{\mathrm{T}}^{\mathrm{miss}}$ distribution is split in seven bins, which are all individually scrutinised for an excess of events over the SM expectation. No excess is observed, and the resulting 95 % CL limit on cross section $\times$ acceptance $\times$ efficiency ($\sigma \times A \times \epsilon$) for the seven $E_{\mathrm{T}}^{\mathrm{miss}}$ lower thresholds is displayed in Fig. 11.18b.

The dark-matter nucleon cross-section limits derived from mono-jet, mono-photon and mono-boson analyses are compared to selected direct-detection experiments in Fig. 11.19. Spin-dependent dark-matter nucleon limits at the LHC (Fig. 11.19b) are always stronger than their spin-independent counterparts (Fig. 11.19a). Common to all searches at the LHC is the good sensitivity to low-mass dark matter, a region of the parameter space that cannot be accessed by direct-detection experiments.

Dark matter could potentially couple differently to up-type and down-type quarks leading to interference effects [100], as shown in Fig. 11.17b for the case of the mono-

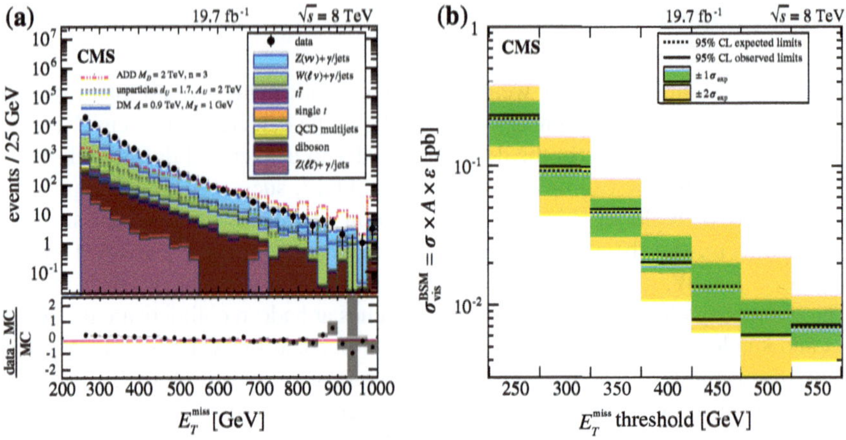

Fig. 11.18 Key distributions from the CMS mono-jet search. **a** $E_{\mathrm{T}}^{\mathrm{miss}}$ distribution after all selection steps. The SM backgrounds are shown as *shaded bands* along with signal examples for dark matter and extra dimensions. **b** Model-independent cross-section limits for new physics containing $E_{\mathrm{T}}^{\mathrm{miss}}$ above some threshold value. Adapted from [74]

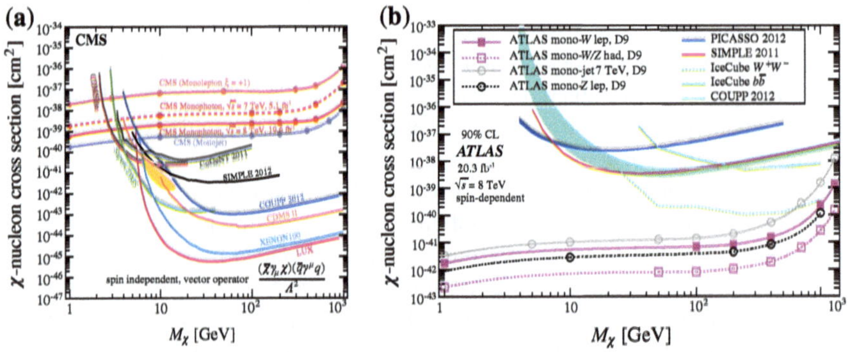

Fig. 11.19 Dark-matter results, shown for spin-independent (SI) and spin-dependent (SD) couplings. The limits are translated to the dark-matter nucleon cross-section plane to allow comparisons with results from direct-detection experiments. Adapted from [24, 102]

W channel. This channel would provide one of the few possibilities to observe this effect. While the couplings to either up-type or down-type quarks would yield a similar behaviour, their relative sign $\xi = \lambda_u \cdot \lambda_d$ with $|\lambda_i| = 0$ or 1, is most important for the phenomenology. Following [100], three scenarios are being considered in the actual searches [23, 24, 101]: a value of $\xi = -1(+1)$ maximises (minimises) the effects of interference as depicted in Fig. 11.20a. A choice of $\xi = 0$ can be assumed to either suppress the coupling of dark matter to up-type quarks or that to down-type quarks.

The leptonic version of the mono-W analysis is described in Sect. 11.2.2.1. In the case where the emitted "tagging" W boson decays hadronically—a decay blessed with a large branching fraction of ~70 %— the final state of two jets (from the W decay) and $E_{\mathrm{T}}^{\mathrm{miss}}$ (due to DM) is indistinguishable from the case where the emitted boson is a hadronically decaying Z. In either case, the two jets overlap and will be reconstructed as one "large-radius" or "fat" jet with a mass m_{jet}. A dedicated reconstruction algorithm (the Cambridge–Aachen algorithm) allows this large jet to be split into two subjets with approximately symmetric energy sharing. In two signal regions ($E_{\mathrm{T}}^{\mathrm{miss}} > 350$ GeV and $E_{\mathrm{T}}^{\mathrm{miss}} > 500$ GeV), the m_{jet} distribution would exhibit a peak-like structure in case of a potential dark-matter signal that could easily be distinguished from the flat SM background. The main irreducible background is due to W/Z+jets events with final-state neutrinos. The hadronic mono-W/Z search [101] excludes dark matter-nucleon cross sections above $10^{-42}\,\mathrm{cm}^2$ for all but the highest M_χ, as is shown in Fig. 11.19b. It can also be seen that the limits from the leptonic decay modes of mono-W [24] and mono-Z [103] are less stringent compared to the one from the hadronic W/Z decay. This is also seen in Fig. 11.20b, where in addition the impact of the different types of operators can be seen. In particular, the much stronger lower limits on the cut-off scale when maximising (D5c) interference compared to minimising (D5d) it are clearly visible.

A few analyses already interpret their result in recently developed simplified models as described in the introduction to this section. In these simplified models the mediator is no longer integrated out but actually modelled with a mass, width and couplings, and the analysis places limits on these quantities. As an example, Fig. 11.21a shows the 95 % CL upper limits on these model parameters from the CMS mono-jet analysis [74].

11.5.2 Dark Matter Through the Higgs Portal

In the previous section, using the EFT approach, the only assumption on the mediator concerned its mass to be larger than the energy scale in question. In [74], a scan of the mediator mass showed little impact on the resulting limit of the compositeness scale Λ for mediator masses ≥ 5 TeV, but variations of up to a factor 2 for masses around 1 TeV. In so-called "Higgs-portal" models [104] the Higgs boson takes the role of the mediator between the SM and DM. Such a Higgs-boson mediator does not necessarily have to be a SM Higgs, as additional Higgs bosons with different masses and exotic decay modes remain possible.

Direct searches for invisible Higgs decays such as $pp \to H \to \chi\bar{\chi}$ also require a "tag" for detection. In analogy to previously described searches (Sect. 11.5.1), the production in association with a W or Z boson offers the possibility of leptonic [105, 106] and hadronic [101, 106] final states. Another option is the vector-boson-fusion (VBF) production mode, where the invisible Higgs boson is produced in association with two forward-going jets [106].

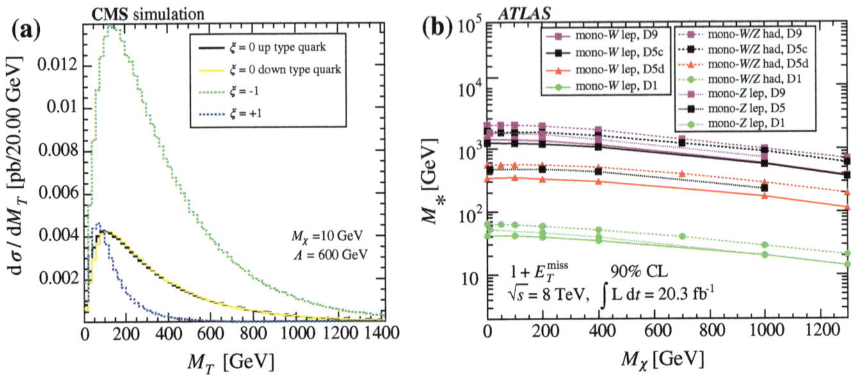

Fig. 11.20 **a** Dark matter could potentially couple differently to up-type and down-type quarks, a phenomenon parametrised by ξ. Differential cross sections are shown for four different cases: DM coupling only to one type (either up-type or down-type) of quarks ($\xi = 0$), and the cases of constructive (destructive) interference for $\xi = -1$ (+1). **b** Lower limits on Λ (here denoted M_*) where the tagging particle is a W or Z boson (leptonic and hadronic decays) for a variety of interaction types: D1 (scalar), D5 (vector), D5c (vector, constructive interference), D5d (vector, destructive interference) and D9 (tensor). Adapted from [23, 24]

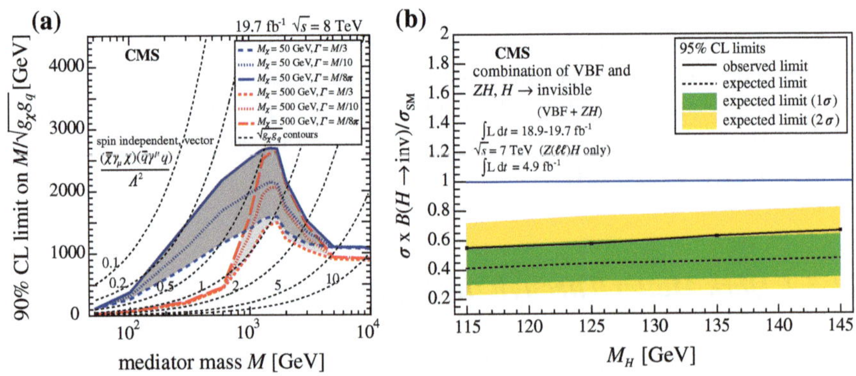

Fig. 11.21 **a** Observed limits on the mediator mass M divided by the coupling $\sqrt{g_q g_\chi}$ as a function of the mediator mass for two possible dark-matter masses of 50 and 500 GeV, assuming vector interactions. The mediator width, Γ, is varied between $M/3$ and $M/8\pi$. The *dashed lines* show contours of constant coupling. **b** Upper limit on $\sigma \times$ B for invisible Higgs bosons with respect to the SM cross section, when combing the ZH and the VBF channels. Adapted from [74, 106]

These analyses combine data taken at centre-of-mass energies of $\sqrt{s} = 7$ and 8 TeV because additional luminosity directly increases the precision. The analysis allows an upper limit to be set on the cross section times branching fraction to invisible particles with respect to the SM cross section, either assuming the measured Higgs-boson mass of 125.5 GeV or as a function of the mass of an additional Higgs boson as shown in Fig. 11.21b. The best limit in the first case is 58 % when combining the ZH and VBF channels [106]. Within the context of Higgs-portal dark matter, this limit can also be interpreted in terms of an upper limit on the spin-independent dark matter-nucleon cross section. There is no sensitivity of these models to dark matter with masses beyond half the Higgs mass since dark-matter particles are produced in pairs.

11.6 Unification of Quarks and Leptons

In the SM, leptons and quarks do not couple directly. Their gauge charges are only related by the requirement of anomaly cancellation [107]. However, many models of BSM physics, such as gauge unification (see Sect. 11.2) and compositeness (Sect. 11.3) aim to unify leptons and quarks. Consequently, they naturally predict direct couplings of quarks and leptons to spin-0 or spin-1 bosons, called leptoquarks. From charge and colour conservation, they must be colour (anti-)triplets and carry fractional electric charge. Therefore, if they exist, they will necessarily be produced at the LHC in pairs via gluon fusion. In fact, their quantum numbers are similar to those of the supersymmetric partners of quarks, as discussed in Sect. 10.2. Thus, experimental searches can employ similar strategies.

A priori, leptoquarks can couple quarks and leptons from different generations. However, the observed strong suppression of flavour-changing neutral currents and lepton-flavour violation severely restricts the allowed values of the couplings. Therefore the searches can start with the simplifying assumption that leptoquarks respect the generation structure of the SM, i.e. there are first-, second- and third-generation leptoquarks, coupling only to leptons and quarks from their respective generation.

The $e^{\pm}p$ collider HERA provided a unique opportunity to produce first-generation leptoquarks directly [108]. The experiments H1 and ZEUS have established lower mass limits for first-generation leptoquarks with coupling $\lambda = 0.3$ [108] of 800 GeV [109] and 699 GeV [110], respectively.

At the LHC, studies of model-independent branching ratios have concentrated on scalar leptoquarks produced in pairs. The searches for leptoquarks of the first two generations are in the final states $2(\ell j)$ and $\ell j + \nu_\ell j$ [111–113] with $\ell = e, \mu$. Given that all third-generation fermions are short-lived in the SM, the final states for third-generation leptoquarks are more complicated. Contrary to the light quarks of the first two generations, for third-generation analyses experimental tools are used that enable identification of the jets as either being due to a b quark (b-tagging) or stemming from a top quark (decaying to Wb), and there are many third-generation search channels being analysed [114–117]: $2(\tau b)$, $2(\tau t)$ and $\tau t + \nu_\tau t$.

An example for a third-generation LQ search is shown in Fig. 11.22a. The analysis uses the $2LQ \rightarrow 2(\tau b)$ channel. While one of the τ leptons is assumed to decay hadronically (τ_{had}), the other one should yield an electron or muon in the final state. This signature is chosen because it helps to avoid large SM backgrounds with many jets. The identified electron (muon) and τ_{had} are required to originate from the same vertex, to have opposite electric charge and to be well separated in the detector. Jets are reconstructed and eventually tagged as b jets using an algorithm based on the identification of secondary vertices. To discriminate against SM background,

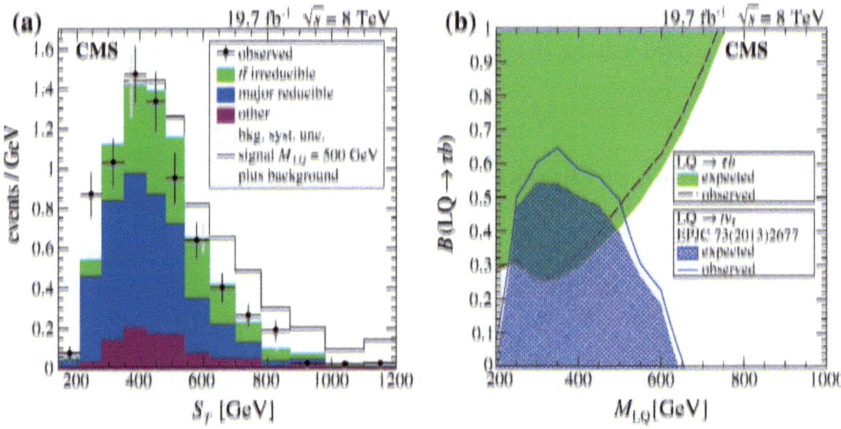

Fig. 11.22 **a** Characteristic S_T distribution defined as the scalar sum of the p_T values of the light lepton, the τ_{had} and the two jets. **b** Expected and observed limits in the plane of the leptoquark mass versus the branching fraction for the leptoquark decay to a τ lepton and a *bottom* quark. A SUSY search for *top*-squark pair production is reinterpreted to provide limits for low values of B, assuming the leptoquark to only decay to third-generation fermions. Adapted from [114]

the invariant mass $M(\tau_{had}, jet)$ of the τ_{had} and a jet is required to be larger than 250 GeV. From the two possible pairings of the τ_{had} with either one of the two jets, the pairing minimising the difference between $M(\tau_{had}, jet)$ and the invariant mass of the light lepton and the other jet is chosen. According to simulations, the correct pairing is selected in approximately 70 % of all events. After the final selection, a potential signal is searched in the S_T distribution, where S_T is the scalar sum of the p_T values of the light lepton, of the τ_{had} and of the two jets (see Fig. 11.22a). The search excludes third-generation scalar LQ with masses below 740 GeV at 95 % CL, assuming a 100 % branching fraction for the LQ decay to a τ lepton and a bottom quark. Exclusions for lower branching fractions as a function of the leptoquark mass are shown in Fig. 11.22b. A SUSY search for top-squark pair production with a $t\bar{t}$ plus E_T^{miss} final state can be reinterpreted as a search for $2LQ \rightarrow 2(\nu_\tau t)$. Assuming the leptoquark to only decay to third-generation fermions, additional limits can be placed on the branching ratio of $LQ \rightarrow (\tau b)$ as a function of the leptoquark mass, as shown in the same figure.

11.7 Searches for Long-Lived States

A large class of analyses searches for specific signatures from long-lived particles. Such phenomena could either be related to exotic charged particles which leave a distinct signature in the detector, but they could also be due to neutral particles which are only recognised through their decay products. In either case, the resulting signatures are rather exotic and require specific triggers and reconstruction algorithms for their detection.

The underlying BSM models are often inspired by SUSY in one way or the other. Assuming the lifetime of the exotic particles to be short enough, e.g. a $c\tau$ up to $\mathcal{O}(1\,\text{m})$, they decay within the detector and should be seen in typical SUSY cascade decays (see Chap. 10) if also $E_{\text{T}}^{\text{miss}}$ resulting from the neutralino is present in the final state. However, the lifetime of such particles is not really theoretically constrained and might exceed the detectors' dimensions. Ordinary SUSY searches are blind to those particles. Other extensions of the SM predict a hidden sector that is weakly coupled to the visible one. Such "hidden-valley models", invented to explain the AMS positron excess [118, 119], yield "dark photons" or "dark Z bosons", which do not interact with SM particles (therefore "dark") and hence leave no visible signal in the detector. They may decay to SM particles with sizeable branching fractions and yield electrons, muons or light mesons which then would appear "out of the blue". Also collimated jet-like structures might appear that contain pairs of SM particles that are produced far from the primary vertex of the event (see Fig. 11.23b, c).

The wealth of models results in a wealth of exotic signatures. Given their importance in the experimental BSM program at the LHC, a few characteristic signatures shall be given. Several of them are sketched in Fig. 11.23. Typical would be a secondary vertex with a displacement (several cm to m) that is too large to be treated by standard secondary vertex reconstruction algorithms. The sketch in Fig. 11.23b show two examples for a hypothetical new particle decaying to a muon pair. The rightmost pair originates from a vertex at the beginning of the muon system (so after a distance of several meters), while the leftmost pair decays in the calorimeter. In either case, no track from the inner detector can be matched, and the reconstruction can use only stand-alone muon and calorimeter information. Intermediate lifetimes can result in disappearing or stopped tracks or so-called kinked tracks (see Fig. 11.23a).

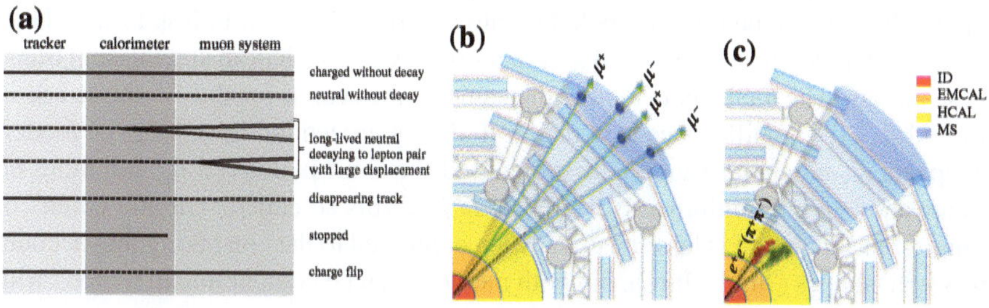

Fig. 11.23 **a** Possible exotic signatures of long-lived particles. **b**, **c** One possible signature are non-prompt pairs or collimated jets of muons (**b**) or electromagnetic clusters due to electrons or light mesons (**c**). Adapted from [120]

Lepton-like heavy stable particles would manifest as a "heavy muon" with an ionisation signature based on anomalous dE/dx deposition. Often such particles are sufficiently heavy to be mildly non-relativistic resulting in a longer time-of-flight to the outer muon system with respect to a minimally ionising particle. This provides a second handle for their detection provided the reconstruction algorithm—which normally assumes relativistic particles—is adapted accordingly. In some models, the particle may turn neutral while traversing the detector (see "charge flip" in Fig. 11.23a) and does not leave a signal in the muon system, in which case the dE/dx information remains for detection.

Signatures might also be delayed with respect to the bunch crossing due to their non-relativistic velocity. While other "normal" particles move with the speed of light and decay roughly within 50 ns, these long-lived signals might occur hours, days or even years after the collision in which they are created. The related analyses can place limits on a wide range of lifetimes, with an upper limit determined by the operational lifetime of the LHC and the lower limit given by the bunch spacing.

11.8 Model-Unspecific Searches and Reinterpretation

As already discussed in the context of Fig. 11.2b, a limit on $\sigma \times \mathrm{B}$ derived with a specific signal template can also be valid for many other models, at least approximately. It is therefore important that the experiments provide all information necessary in order to enable the theoreticians to check their model against experimental data, i.e. in order to do a reinterpretation. The signal acceptance A is one of the key ingredients, as it could differ between different models, due to e.g. different decay kinematics in the decay of a spin-1 and a spin-0 or spin-2 particle. This number A can be provided separately or it can be directly included into the limit as shown in Fig. 11.24a. Here the derived upper limit is on $\sigma \times \mathrm{B} \times A$ in a search with one lepton and missing transverse energy. The analysis is almost the same as the search for the W' described in Sect. 11.2.2.1. The difference is in the limit-setting procedure. In order to be model-independent, no signal shape is assumed and the limit is derived by summing all events above a minimum transverse mass M_T^{min}. This so called "cut and count" or "single-bin" approach is less sensitive when compared to the one taking into account the signal shapes ("multi-bin"), but both methods can be comparable in regions with a low level of background.

A different ansatz is the use of a generic shape, e.g. a Gaussian template shape, in a multi-bin limit. Figure 11.24b shows $\sigma \times \mathrm{B} \times A$ for different relative widths σ_G/m_G as a function of the mass of the new resonance, m_G. As expected, the most stringent limits can be set on narrow resonances, the widths of which are comparable or smaller than the detector resolution. These limits on Gaussian signal shapes are easy to reinterpret, but do not take into account the off-shell production due to the small parton-parton luminosity at large Bjorken-x, see the discussion related to Fig. 11.2.

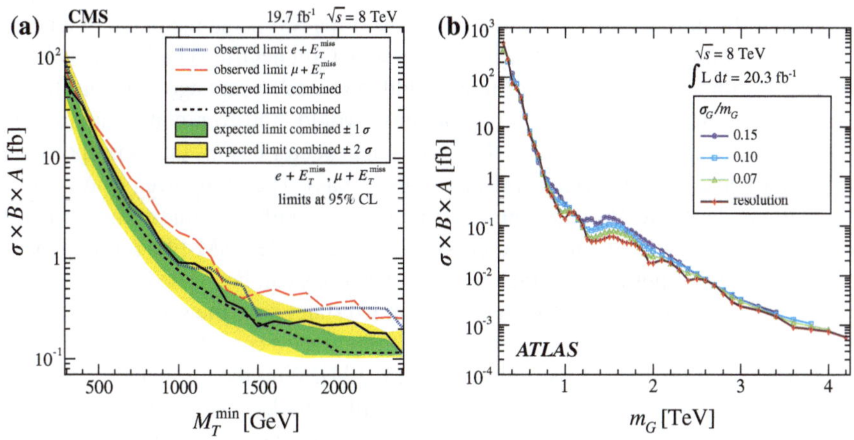

Fig. 11.24 **a** Upper limit on $\sigma \times$ B $\times$ acceptance for lepton+$E_{\mathrm{T}}^{\mathrm{miss}}$ final states due to new phenomena above a minimum transverse mass M_T^{min}. Here the acceptance includes the efficiency of the kinematic selection criteria and of the M_T^{min} requirement. **b** Upper limit on $\sigma \times$ B $\times$ acceptance for resonances with a Gaussian shape with different relative widths σ_G/m_G at reconstruction level as a function of the mass of the new resonance, m_G. Adapted from [23, 32]

Finally, the most general searches possible compare data and expected background in as many final states and sensitive variables as possible. Obviously, these searches are not optimised in terms of sensitivity. But if they are exhaustive in terms of final states, they could ensure that no obvious discrepancies between data and expectation are missed. A disadvantage of these searches is that an observed discrepancy needs to be investigated in detail afterwards, i.e. the usual practise of a blind analysis is not possible anymore.

11.9 Other Analyses

There are, of course, analyses of data from Run 1 of the LHC that have not been covered in this chapter. The available space is limited and this volume would not do the rich physics harvest of LHC Run 1 justice if it merely rushed through a list of results. Instead, the goal was to describe at least some analyses in more detail and to present some motivations that are typically left out of the journal publications.

The most important selection criterion for this chapter was a journal publication (or at least submission) of an analysis based on the full data set at 8 TeV by the summer of 2014. Of these, the vast majority has been included. Nevertheless, before concluding, at least two more published results should be mentioned that have been left out of the preceding sections.

One of the most straightforward extensions of the SM is the inclusion of a heavy fourth fermion generation. Such a generation with light neutrinos had been excluded by LEP1 with a measurement of the Z^0 width almost 25 years ago. All fourth generation fermions with SM-like chiral couplings would, however, contribute through loops to the coupling of a 125 GeV Higgs boson to vector bosons that is proportional to Yukawa couplings, but *independently* of their mass. Therefore, even heavy fermions will not decouple. Instead, their contribution rises with the value of their Yukawa coupling. As a result, the observed cross sections exclude a SM-like fourth generation [121, 122]. These constraints on chiral fermions can be evaded by vector-like fermions, where the left-handed and right-handed states have the same coupling. For example, such vector-like heavy fermions are required to cancel quadratic divergencies in the "little Higgs" class of models [123, 124]. With the exclusion of SM-like fourth-generation quarks, the search for vector-like quarks has become a very active area with many results still to come from the 8 TeV data set. One already published analysis of the full 8 TeV data set [125] has set lower limits of $\mathcal{O}(700\,\text{GeV})$ on the mass of a vector-like heavy partner of the top quark.

In the quark sector of the SM, flavour is conserved only approximately (see Chap. 8), but searches for lepton-flavour violation (LFV) at lower energies so far were not successful [126]. Indeed, the non-observation of $\mu \rightarrow eee$ and $\mu \rightarrow e\gamma$ can be related in a model dependent way to limits on the branching ratio for $Z \rightarrow e\mu$ of 10^{-12} and 10^{-10}, respectively. However, the model-independent, direct limits for this branching ratio were much poorer and—given the fundamental nature of the absence or existence of a violation of lepton flavour—dedicated searches are called for. An analysis of the full 8 TeV data set [127] improved the previous limit from LEP by a factor of two to $\text{B}(Z \rightarrow e\mu) < 7.5 \times 10^{-7}$. These analyses benefit from the large statistics available in the full data sets, but require an extremely precise understanding of systematic uncertainties. There is more to come from LFV searches in other channels and from Run 2 of the LHC.

11.10 Summary

A graphical representation of typical exclusion limits from ATLAS and CMS, as of summer 2014, is shown in Figs. 11.25 and 11.26. If results are available from both experiments, the better limit is shown. Not all results use the full data set of Run 1, since the analysis is still ongoing. In contrast to the other parts of this chapter, these tables include preliminary results that have been shown at conferences (ICHEP 2014 in this case), but are still in the process of being published at the time of writing. For this reason, actual numbers which might differ by a few percent in the final publication are not quoted.

With very few exceptions, Run 1 of the LHC has excluded all popular models of exotic new physics up to the TeV scale. In some cases, these limits exceed the TeV scale substantially. In particular the clean signature of contact interactions with

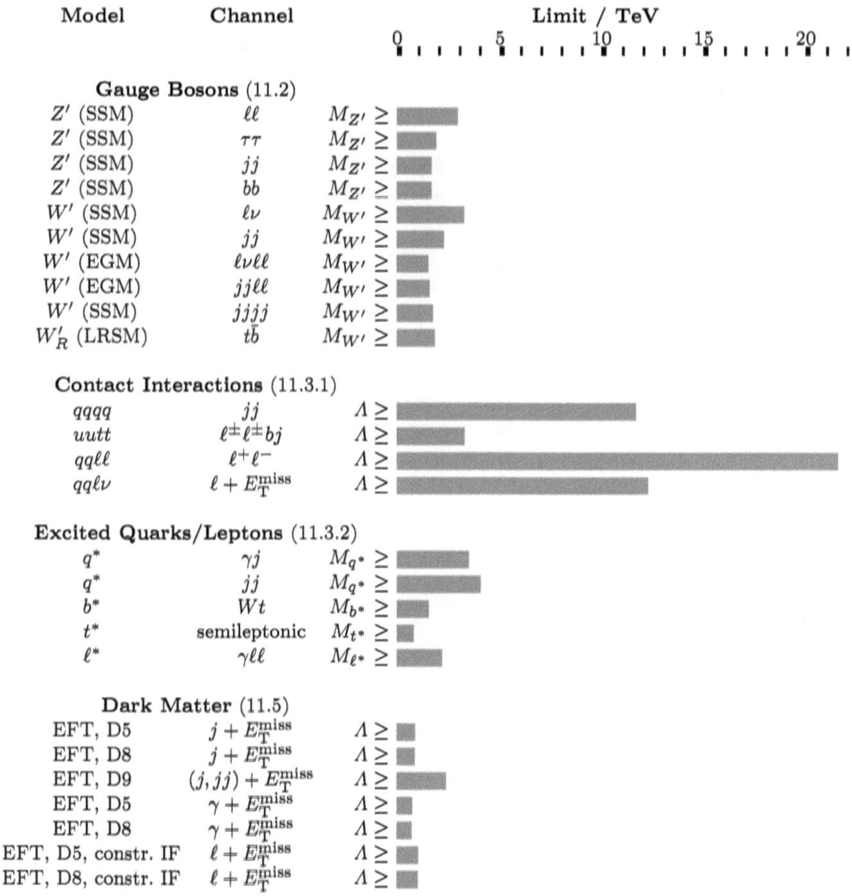

Fig. 11.25 Some exclusion limits presented at ICHEP 2014 (continued in Fig. 11.26)

lepton pairs in the final state has allowed to push the corresponding mass scale above 20 TeV.

A lot of effort has gone into the search for extra dimensions and black holes. These scenarios are now excluded well into the multi-TeV realm. On the other hand, there are cases where the TeV scale has barely been probed. In particular the masses of leptoquarks, a vector-like fourth generation and exotic long-lived particles may still be below 1 TeV.

During the upcoming Run 2 of the LHC, the almost doubled beam energy will translate into a substantially increased parton-parton luminosity, in particular for very heavy final states. Thus, analyses searching for heavy resonances can surpass current lower mass limits with $\mathcal{O}(1\,\text{fb}^{-1})$ (Z', W', ...) of the upcoming 13 TeV data or even as little as $\mathcal{O}(100\,\text{pb}^{-1})$ for e.g. the dijet resonance search. Searches at lower masses, on the other hand, do not gain that quickly and in extreme cases might not be able at

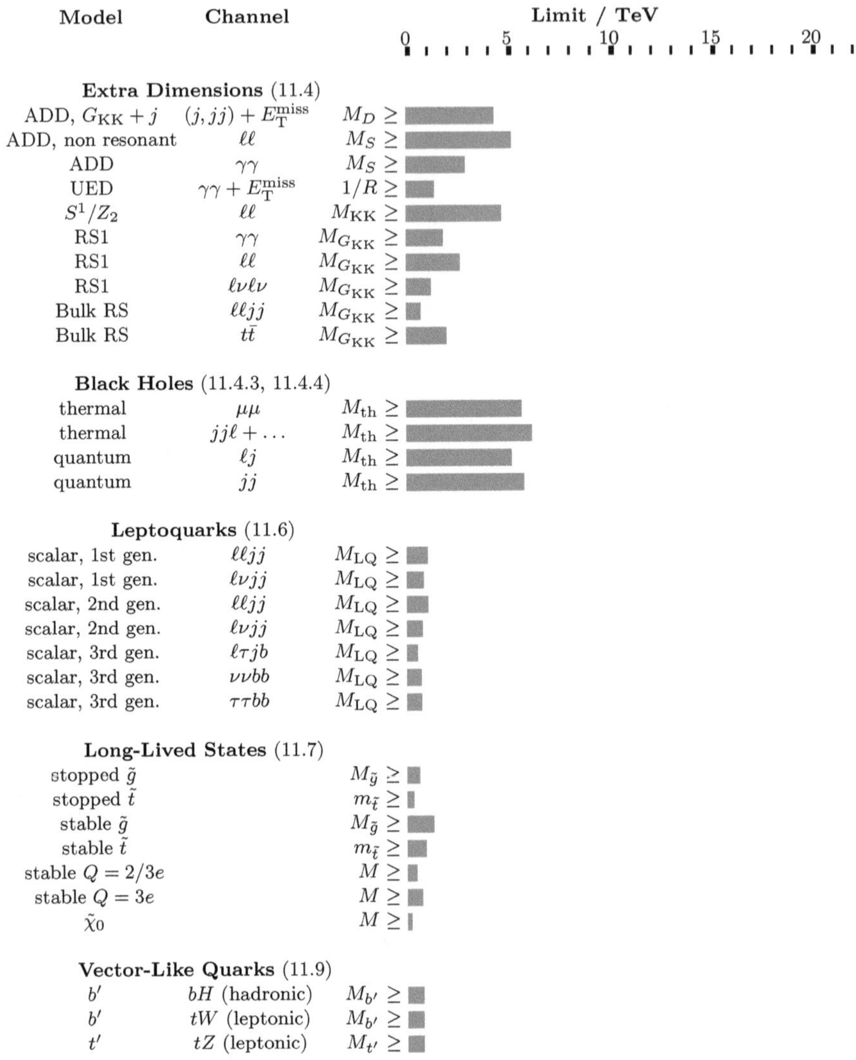

Fig. 11.26 Some exclusion limits presented at ICHEP 2014 (continuation of Fig. 11.25)

all to improve with even the full data set expected in 2015 if they are dominated by systematics. An example is the search for the LFV decay of the Z-boson. However, most of the limits described in this chapter will certainly be improved significantly, unless new phenomena show up.

References

1. S. Weinberg, Phys. Lett. B **91**, 51 (1980)
2. W. Buchmüller, D. Wyler, Nucl. Phys. B **268**, 621–653 (1986)
3. B. Grzadkowski, M. Iskrzynski, M. Misiak, J. Rosiek, JHEP **10**, 085 (2010)
4. H. Ruegg, M. Ruiz-Altaba, Int. J. Mod. Phys. A **19**, 3265–3348 (2004)
5. P. Langacker, Rev. Mod. Phys. **81**, 1199 (2009)
6. M. Chizhov, Phys. Part. Nucl. **42**, 93–183 (2011)
7. M.V. Chizhov, V.A. Bednyakov, J.A. Budagov, Phys. Atom. Nucl. **71**, 2096 (2008)
8. M.V. Chizhov, G. Dvali, Phys. Lett. B **703**, 593 (2011)
9. ATLAS Collaboration, Phys. Rev. D. **90**, 052005 (2014)
10. CMS Collaboration, Phys. Lett. B **714**, 158–179 (2012)
11. ATLAS Collaboration, Phys. Lett. B **700**, 163–180 (2011)
12. CDF Collaboration, Phys. Rev. Lett. **106**, 121 801 (2011)
13. DO Collaboration, Phys. Lett. B **695**, 88–94 (2011)
14. CMS Collaboration, Phys. Rev. Lett. **111**(21), 211 804 (2013)
15. ATLAS Collaboration, Phys. Lett. B **719**, 242–260 (2013)
16. CMS Collaboration, Phys. Lett. B **716**(1), 82–102 (2012)
17. ATLAS Collaboration, JHEP **11**, 138 (2012)
18. CMS Collaboration, Phys. Rev. D **87**(11), 114 015 (2013)
19. V.D. Barger, W.Y. Keung, E. Ma, Phys. Rev. D **22**, 727 (1980)
20. V.D. Barger, W.Y. Keung, E. Ma, Phys. Rev. Lett. **44**, 1169 (1980)
21. C.T. Hill, S.J. Parke, Phys. Rev. D **49**, 4454–4462 (1994)
22. C.T. Hill, Phys. Lett. B **345**, 483–489 (1995)
23. CMS Collaboration (2014), arXiv:1408.2745
24. ATLAS Collaboration, JHEP09, 037 (2014)
25. A. Abdesselam et al., Eur. Phys. J. C **71**, 1661 (2011)
26. A. Altheimer et al., J. Phys. G **39**(063), 001 (2012)
27. T. Plehn, M. Spannowsky, J. Phys. G **39**(083), 001 (2012)
28. M. Dasgupta, A. Fregoso, S. Marzani, G.P. Salam, JHEP **09**, 029 (2013)
29. ATLAS Collaboration, arXiv:1408.0886 (2014)
30. CMS Collaboration, JHEP **05**, 108 (2014)
31. CMS Collaboration Eur. Phys. J. C **74**, 3149 (2014)
32. ATLAS Collaboration, Phys. Rev. D. **91**, 052007 (2015)
33. ATLAS Collaboration, Phys. Lett. B **737**, 223–243 (2014)
34. CMS Collaboration. Phys. Lett. B **740**, 83 (2015)
35. CMS Collaboration, JHEP **08**, 173 (2014)
36. M.E. Peskin, in *Proceedings of the 10th International Symposium on Lepton and Photon Interactions at High Energies*, Bonn, 1981, p. 880
37. E. Eichten, I. Hinchliffe, K.D. Lane, C. Quigg, Rev. Mod. Phys. **56**, 579–707 (1984)
38. K. Lane, F. Paige, T. Skwarnicki, W. Womersley, Phys. Rept. **278**, 291–371 (1997)
39. ATLAS Collaboration. Eur. Phys. J. C **74**, 3134 (2014)
40. A. De Rujula, L. Maiani, R. Petronzio, Phys. Lett. B **140**, 253 (1984)
41. J.H. Kuhn, P.M. Zerwas, Phys. Lett. B **147**, 189 (1984)
42. U. Baur, I. Hinchliffe, D. Zeppenfeld, Int. J. Mod. Phys. A **2**, 1285 (1987)
43. U. Baur, M. Spira, P. Zerwas, Phys. Rev. D **42**, 815–824 (1990)
44. CMS Collaboration. Phys. Lett. B **738**, 274 (2014)
45. ATLAS Collaboration, Phys. Lett. B **728**, 562 (2014)
46. ATLAS Collaboration, New J. Phys. **15**, 093 011 (2013)
47. CMS Collaboration, JHEP **06**, 125 (2014)
48. S. Weinberg, Phys. Rev. D **13**, 974–996 (1976)
49. S. Weinberg, Phys. Rev. D **19**, 1277–1280 (1979)
50. L. Susskind, Phys. Rev. D **20**, 2619–2625 (1979)
51. E. Eichten, K.D. Lane, J. Womersley, Phys. Lett. B **405**, 305–311 (1997)

52. E. Eichten, K. Lane, Phys. Lett. B **669**, 235–238 (2008)
53. C.T. Hill, E.H. Simmons, Phys. Rept. **381**, 235–402 (2003)
54. T.W. Appelquist, D. Karabali, L. Wijewardhana, Phys. Rev. Lett. **57**, 957 (1986)
55. J.M. Maldacena, Int. J. Theor. Phys. **38**, 1113–1133 (1999)
56. C. Csaki, C. Grojean, L. Pilo, J. Terning, Phys. Rev. Lett. **92**(101), 802 (2004)
57. R.S. Chivukula et al., Phys. Rev. D **74**(075), 011 (2006)
58. ATLAS Collaboration, Phys. Lett. B **738**, 428–447 (2014)
59. D.D. Dietrich, F. Sannino, K. Tuominen, Phys. Rev. D **72**(055), 001 (2005)
60. R. Foadi, M.T. Frandsen, T.A. Ryttov, F. Sannino, Phys. Rev. D **76**(055), 005 (2007)
61. R. Foadi, M.T. Frandsen, F. Sannino, Phys. Rev. D **87**(095), 001 (2013)
62. N. Arkani-Hamed, S. Dimopoulos, G. Dvali, Phys. Lett. B **429**, 263–272 (1998)
63. T. Kaluza, Sitzungsber. Preuss. Akad. Wiss. Berlin (Math. Phys.) **1921**, 966–972 (1921)
64. O. Klein, Z. Phys. **37**, 895–906 (1926)
65. G.F. Giudice, R. Rattazzi, J.D. Wells, Nucl. Phys. B **544**, 3–38 (1999)
66. T. Han, J.D. Lykken, R.J. Zhang, Phys. Rev. D **59**(105), 006 (1999)
67. G.F. Giudice, A. Strumia, Nucl. Phys. B **663**, 377–393 (2003)
68. R. Franceschini et al., JHEP **05**, 092 (2011)
69. K.R. Dienes, E. Dudas, T. Gherghetta, Nucl. Phys. B **537**, 47–108 (1999)
70. I. Antoniadis, Phys. Lett. B **246**, 377–384 (1990)
71. S.C. Park, J. Shu, Phys. Rev. D **79**(091), 702 (2009)
72. J.L. Hewett, Phys. Rev. Lett. **82**, 4765–4768 (1999)
73. ATLAS Collaboration, New J. Phys. **15**, 043 007 (2013)
74. CMS Collaboration, arXiv:1408.3583 (2014)
75. ATLAS Collaboration, Phys. Rev. Lett. **110**, 011 802 (2013)
76. CMS Collaboration, Phys. Rev. Lett. **108**, 261 803 (2012)
77. L. Randall, R. Sundrum, Phys. Rev. Lett. **83**, 3370–3373 (1999)
78. L. Randall, R. Sundrum, Phys. Rev. Lett. **83**, 4690–4693 (1999)
79. W.D. Goldberger, M.B. Wise, Phys. Rev. D **60**(107), 505 (1999)
80. K. Agashe, A. Delgado, M.J. May, R. Sundrum, JHEP **08**, 050 (2003)
81. A.L. Fitzpatrick, J. Kaplan, L. Randall, L.T. Wang, JHEP **09**, 013 (2007)
82. B. Lillie, L. Randall, L.T. Wang, JHEP **09**, 074 (2007)
83. CMS Collaboration, JHEP **08**, 174 (2014)
84. K. Agashe et al., Phys. Rev. D **77**(015), 003 (2008)
85. S. Dimopoulos, G.L. Landsberg, Phys. Rev. Lett. **87**(161), 602 (2001)
86. ATLAS Collaboration, Phys. Rev. D **88**(7), 072 001 (2013)
87. CMS Collaboration, JHEP **07**, 178 (2013)
88. D.C. Dai et al., Phys. Rev. D **77**(076), 007 (2008)
89. ATLAS Collaboration, JHEP **08**, 103 (2014)
90. P. Meade, L. Randall, JHEP **05**, 003 (2008)
91. X. Calmet, W. Gong, S.D. Hsu, Phys. Lett. B **668**, 20–23 (2008)
92. R. Kallosh, A.D. Linde, D.A. Linde, L. Susskind, Phys. Rev. D **52**, 912–935 (1995)
93. ATLAS Collaboration, Phys. Rev. Lett. **112**, 091 804 (2014)
94. D.M. Gingrich, J. Phys. G **37**(105), 008 (2010)
95. D.M. Gingrich, Comput. Phys. Commun. **181**, 1917–1924 (2010)
96. R. Schnee, arXiv:1101.5205 (2011)
97. O. Buchmüller, M.J. Dolan, C. McCabe, JHEP **01**, 025 (2014)
98. O. Buchmüller, M.J. Dolan, S.A. Malik, C. McCabe, JHEP **01**, 037 (2015). doi:10.1007/JHEP01
99. J. Goodman et al., Phys. Rev. D **82**(116), 010 (2010)
100. Y. Bai, T.M. Tait, Phys. Lett. B **723**, 384–387 (2013)
101. ATLAS Collaboration, Phys. Rev. Lett. **112**(4), 041 802 (2014)
102. CMS Collaboration, arXiv:1410.8812 (2014)
103. ATLAS Collaboration, Phys. Rev. D **90**, 012 004 (2014)
104. B. Patt, F. Wilczek, arXiv:hep-ph/0605188 (2006)

105. ATLAS Collaboration, Phys. Rev. Lett. **112**, 201 802 (2014)
106. CMS Collaboration. Eur. Phys. J. C **74**, 2980 (2014)
107. D.J. Gross, R. Jackiw, Phys. Rev. D **6**, 477–493 (1972)
108. W. Buchmüller, R. Rückl, D. Wyler, Phys. Lett. B **191**, 442–448 (1987)
109. F. Aaron et al., Phys. Lett. B **704**, 388–396 (2011)
110. ZEUS Collaboration, Phys. Rev. D **86**, 012 005 (2012)
111. CMS Collaboration, Phys. Rev. D **86**, 052 013 (2012)
112. ATLAS Collaboration, Eur. Phys. J. C **72**, 2151 (2012)
113. ATLAS Collaboration, Phys. Lett. B **709**, 158–176 (2012)
114. CMS Collaboration. Phys. Lett. B **739**, 229 (2014)
115. CMS Collaboration, JHEP **12**, 055 (2012)
116. CMS Collaboration, Phys. Rev. Lett. **110**, 081 801 (2013)
117. ATLAS Collaboration, JHEP **06**, 033 (2013)
118. AMS Collaboration, Phys. Rev. Lett. **113**, 121 101 (2014)
119. AMS Collaboration, Phys. Rev. Lett. **113**, 121 102 (2014)
120. ATLAS Collaboration. JHEP **11**, 088 (2014)
121. A. Djouadi, A. Lenz, Phys. Lett. B **715**, 310–314 (2012)
122. O. Eberhardt et al., Phys. Rev. Lett. **109**(241), 802 (2012)
123. N. Arkani-Hamed, A.G. Cohen, H. Georgi, Phys. Lett. B **513**, 232–240 (2001)
124. M. Schmaltz, D. Tucker-Smith, Ann. Rev. Nucl. Part. Sci. **55**, 229–270 (2005)
125. CMS Collaboration, Phys. Lett. B **729**, 149–171 (2014)
126. S. Mihara, J. Miller, P. Paradisi, G. Piredda, Ann. Rev. Nucl. Part. Sci. **63**, 531–552 (2013)
127. ATLAS Collaboration, Phys. Rev. D **90**(7), 072 010 (2014)
128. R.S. Chivukula et al. Phys. Rev. D **72**, 015 008 (2005)
129. R.S. Chivukula et al. Phys. Rev. D **74**, 075 011 (2006)
130. T. Ohl, C. Speckner. Phys. Rev. D **78**, 095 008 (2008)
131. F. Bach, T. Ohl, Phys. Rev. D **85**, 015 002 (2012)

Chapter 12
Perspectives on the Energy Frontier

Eckhard Elsen and Christophe Grojean

Abstract The first run of the LHC propelled particle physics into the *Terascale*, the TeV energy domain and was rewarded with the triumph of the discovery of a Higgs boson. Precision measurements of elementary cross sections and couplings so far beautifully confirm the Standard Model, which is however known to be incomplete. Significantly increased statistics and higher energy reach may provide the sensitivity to shake its foundations, while other experiments and facilities may be needed to complete the physics interpretation.

12.1 The Structure of the Standard Model After LHC Run 1

The Standard Model (SM) is a triumph of the combination of the two pillars of twentieth-century physics, namely quantum mechanics and special relativity. Particles are defined as representations of the Poincaré group. Mathematically, these representations are labelled by two quantities: the spin that is quantised and takes only discrete values, and the mass, which a priori is a continuous parameter. However, the transformation laws for the various elementary particles under the gauge symmetries associated with the fundamental interactions force the masses of these particles to vanish. This would be in flagrant contradiction with the experimental measurements.

The Brout–Englert–Higgs mechanism (BEH) provides the solution to this mass conundrum. The discovery of a Higgs boson in July 2012 and the experimental confirmation of the BEH mechanism by the ATLAS and CMS collaborations has been a historical step in our understanding of nature: the masses of the elementary particles

E. Elsen (✉)
DESY, Notkestr. 85, 22607 Hamburg, Germany
e-mail: eckhard.elsen@desy.de

C. Grojean
ICREA and IFAE, Universitat Autònoma de Barcelona,
08193 Bellaterra, Barcelona, Spain
e-mail: christophe.grojean@cern.ch

T. Schörner-Sadenius (ed.), *The Large Hadron Collider*,
DOI 10.1007/978-3-319-15001-7_12

are not fundamental parameters defined at very high energy but rather emergent quantities appearing at low energy as a result of the particular structure of the vacuum.

This breakthrough discovery at the LHC surely stands out among the results of the first running period. For the first time a scalar particle has been found that assumes a vital role in the Standard Model and couples to all massive particles. In fact the Higgs particle dominates the structure of the vacuum (see Chap. 6).

The discovery itself did not come without tremendous preparatory work and refinement. The bulk of the experimental work at the new LHC collider went into understanding the initial state of any of the 5×10^{15} collisions that took place during Run 1, i.e. understanding how two protons interact. Only in lowest order can we picture this as the interaction of constituent quarks inside the protons. While first-generation quarks and gluons dominate the composition of the proton at large momentum fraction x— and are hence responsible for the most energetic collisions at the LHC—the intricacies of the initial state are only revealed when taking into account higher-generation quarks, which contribute significantly to the proton content at small x. To add to the challenge, in a given collision several partons may interact quasi-independently and thus affect the kinematic configuration of the hard interaction that is of interest. It is in this complex environment that the techniques for calibrating the jet energy scale had to be refined while at the same time initial-state and final-state radiation affects the measured quantities. Flavour-tagging techniques received a new boost in precision from the vertex detectors and also from the broad introduction and acceptance of multi-variate analysis techniques. On the theoretical side, the matching of exact perturbative matrix-element calculations with the scale of onset for parton showers was much refined. It also necessitated a better understanding of the (non-perturbative) low-energy regime of the strong interaction. All of this has been captured in advanced Monte Carlo implementations, which—surprisingly—are able to even describe many details of the final state (see Sect. 4.8).

Only with this detailed understanding is it possible to predict the production of e.g. the top quark, be it produced in pairs or singly. The top quark is an object of great interest in itself since its high mass provides its own scale in perturbative QCD and because it is the decisive agent in Higgs-boson production at the LHC. The high mass implies that the top quark is short-lived—so short that it has no chance to form a hadron. The top-quark properties thus are not obscured by hadronisation, a feature that makes this fermion accessible to experimental investigation (see Chap. 7).

Beauty quarks are abundantly produced in pp collisions. Pair-produced b quarks allow for interference experiments and thus may be particularly sensitive probes for new physics. The LHCb experiment exploits this property e.g. in the search of decays of the b quark that are highly suppressed in the Standard Model and thus are less likely to mask new physics. Observations have been made of tiny branching ratios of which probably the most striking is the measurement of the branching ratio $B \to \mu\mu$ (see Chap. 8). One of the surprising experimental lessons from Run 1 was the observation that the charm quark can also be used as a tool in the search for new physics at the LHC; charm quarks can be reconstructed and well separated from light quarks.

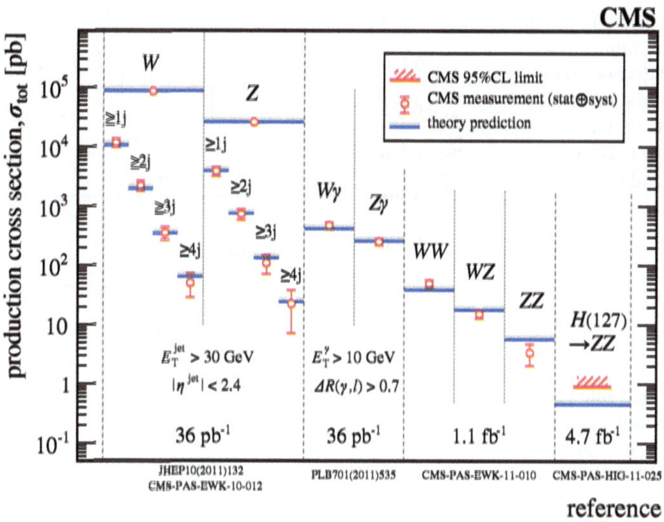

Fig. 12.1 Cross sections for processes with one or several vector bosons in the final state. The associate production of jets with single W or Z bosons is also separately indicated. These cross sections are compared to SM calculations and show good agreement even in this very specific corner of phase space. Adapted from [1]

The strong interaction of the incident protons leads to very high production rates. Cross sections that vary by more than ten orders of magnitude are accessible. The measurements have revealed an understanding of the strong interaction that can be precisely modelled. As an example, Fig. 12.1 depicts the excellent understanding of the production of heavy bosons in association with jets or other bosons. Future measurements with multi-boson final states, such as triple weak-boson production or vector-boson scattering, on the other hand, are likely to profit from the higher beam energy of Run 2 which increases the parton luminosity at high invariant masses (see Chap. 4).

This understanding of the SM cross sections is, in fact, the result of many precursor experiments to the LHC—most prominently of the e^+e^- colliders LEP at CERN and SLC at Stanford, the ep collider HERA at DESY and the $p\bar{p}$ collider Tevatron at Fermilab—but also of the progress in theoretical understanding that led e.g. to the availability of higher-order perturbative calculations. Again, as an example Fig. 12.2 depicts the history of the measurement of α_s, the coupling constant of the strong interaction (see Sect. 4.8). Currently the precision is driven by the calculations from lattice QCD (see also Sect. 5.2).

The discovery of the top quark is a particularly nice example of the interplay of theoretical and experimental progress. Data from LEP allowed early bounds to be placed on the mass of the top quark and gave very good guidance to the direct observation at the Tevatron (Fig. 12.3). The LHC data then add to the precision (see Chap. 7). This level of precision and beyond is used today, for instance, to speculate

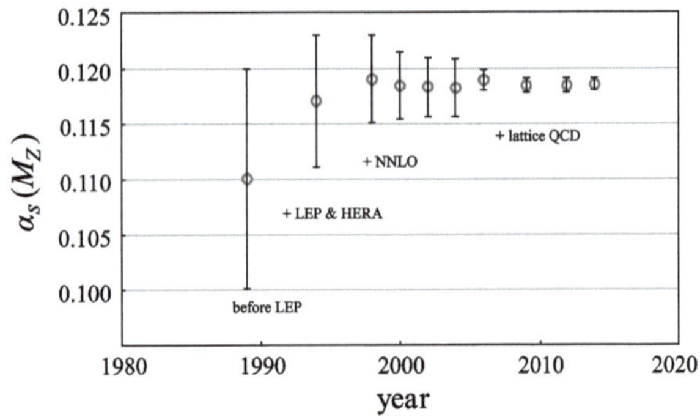

Fig. 12.2 The evolution of the measured value of α_s over several decades with indicators of the main contributions to the precision. Data taken from [2–5] and references therein

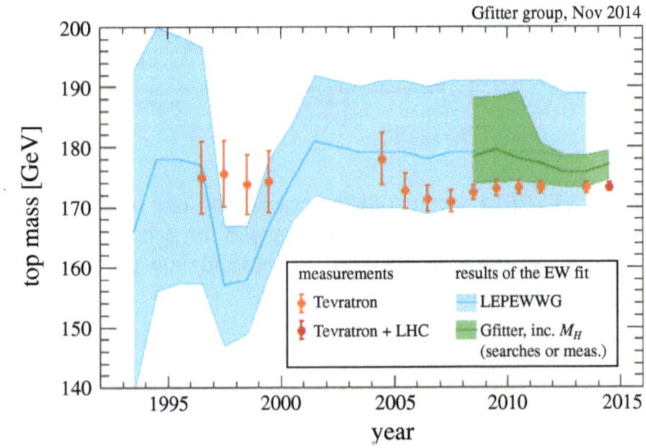

Fig. 12.3 The value of m_t and its uncertainty derived from fits to e^+e^- data without (LEPEWWG) and with assumptions (GFITTER) on the value of the Higgs mass. The values from direct measurements at the Tevatron and at the LHC are also indicated. Adapted from [6]

about the stability of the SM vacuum or to search for light stops (predicted in natural supersymmetric scenarios), the decays of which are kinematically similar to those of top quarks.

Also the measurement of the total cross section has reached a new level of precision at the LHC. The Roman pots of TOTEM and ALFA were used in various configurations of transverse beam dimensions (β functions) to explore the impact parameter of pp collisions at the largest forward scattering angles, but also down to the region were pure Coulomb interactions prevail. The LHCf experiment explored the extreme forward production of neutral particles, i.e. π^0, which aids the understanding of cosmic-ray showers produced in the atmosphere.

The aforementioned complexity of the initial state and the understanding gained in its description is taken to new levels in heavy-ion collisions (see Chap. 9). Ion-ion (AA) collisions create an initial state with much higher energy density that rapidly evolves into a quark-gluon plasma in the state of an almost perfect fluid. While the subsequent evolution of the quark-gluon plasma is well described by relativistic viscous-fluid-dynamic simulations, the pre-equilibrium dynamics and the process of rapid thermalisation are not yet fully understood. The ALICE experiments endeavours to disentangle the final state in its particle composition down to very low momenta while the other experiments are primarily sensitive to hard probes emerging from the collision due to the trigger thresholds typically imposed. Even the LHCb experiment was able to cope with the multiplicity of pA collisions and recorded valuable data at large rapidities. Heavy-ion physics is thus a thriving field addressed by all four experiments. Observations are made in minimum-bias collisions, in events with strange or heavier quarks, and with hard probes. Due to the complexity of the produced system, ab-initio calculations are rarely available. The precise nature of the initial state, the pre-equilibrium dynamics and the properties of the fluid across the phase diagram of temperature and net-baryon density are still largely unknown. The LHC provides a rich body of experimental configurations and signatures, which severely constrain the interpretation of the observations. A recurring topic is the role of flavour in the thermalisation process. In this context, the wealth of data from all experiments at the LHC provided by final states containing beauty is overwhelming. Another extreme configuration is that of ultra-peripheral collisions where photon-exchange processes relate to observations that were previously made in γp processes at lepton-proton colliders. Proton-lead collisions recorded at the LHC resemble pp collisions at low multiplicities; at high multiplicities, however, they show strikingly similar features to lead-lead collisions, an observation that has sparked appreciable theoretical work on the mechanism of multi-particle production in QCD at high energies. Heavy-ion physics has received a tremendous boost in productivity since the turn-on of the LHC.

12.2 The New Physics Landscape

During its first run, the LHC certainly fulfilled its commitments: The machine and its detectors were mostly designed to find the Higgs boson and "[they] got it!" according to the words of R. Heuer, director general of CERN, on 4 July 2012. It was an important step in the understanding of the mechanism of electroweak symmetry breaking. But the journey is not over.

Run 1 accumulated striking evidence that the Higgs field is the cause of the screening of the weak interaction at long distances. Indeed, the observation and measurement of the $H \to ZZ^{\star} \to 4\ell$ channel indicate that the Higgs field develops a vacuum expectation value (vev) that is not invariant under the $SU(2)_L \times U(1)_Y$ gauge symmetry of the SM. Furthermore, this vev seems to be the common source of the Z-boson mass and the coupling between the Higgs boson and the Z boson. If the Higgs vev is the only source of all the elementary particle masses, and if the Higgs boson is the quantum excitation above this vev, the Higgs-boson couplings to fundamental fermions are predicted to be linearly proportional to the fermion masses, whereas the couplings to bosons are proportional to the square of the boson masses. The state-of-the-art fit of the current Higgs data collected at the LHC (see Fig. 6.31 in Chap. 6) shows a good consistency with this picture: The coupling to the heaviest particles, namely W and Z bosons, the t quark and the τ lepton, is already established and so spans the key species. The measurement of other quarks and leptons, in particular the lightest particles of the first generations, will require considerably more statistics at the LHC to provide the indisputable evidence that confirms the prediction in full. Nonetheless, it is already established that the Higgs boson, contrary to all the gauge bosons, has non-universal couplings among the particles of the three different generations of quarks and leptons. The Higgs particle is not a Z' gauge boson! The Higgs boson mediates new fundamental forces different in nature than the electromagnetic, weak and strong interactions.

However, this evidence only addresses the question of *how* the symmetry of the weak interaction is broken. It does not address the question of *why* the symmetry is broken or why the Higgs field acquires an expectation value. The situation is simply summarised in the following tautology

Why is electroweak symmetry broken?
Because the Higgs potential is unstable at the origin.
Why is the Higgs potential unstable at the origin?
Because otherwise EW symmetry would not be broken.

The discovery of a Higgs boson allowed first glimpses into a new sector of the microscopic world. Now comes the time of the detailed exploration of this new Higgs sector. Are there additional states populating it? Is the Higgs particle elementary or composite? Is it endowed with a supersymmetric structure? Is it a portal to a new hidden sector forming the dark-matter component of the universe? The answers to these questions will have profound implications on our understanding of the fundamental laws of physics. Establishing that the Higgs boson is weakly coupled, elementary and solitary, would surely be as shocking as unexpected, but it may well indicate the existence of a multiverse ruled by anthropic selection rules. If instead deviations from the SM emerge in the dynamics of the Higgs, we will have to use them as a diagnostic tool of the underlying dynamics. The pattern of these deviations will carry indirect information about the nature of the completion of the SM at higher energies. In supersymmetric models, and more generally in models with an extended electroweak symmetry breaking sector, the largest deviations will be observed in the couplings to leptons and to the down-type quarks, as well as in the decay amplitudes to photons and gluons. In models of strong interactions, in which the Higgs boson is a bound state, the effects of compositeness uniformly suppress all the Higgs couplings while the self-interactions of the particles inside the Higgs sector, namely the Higgs particle and the longitudinal components of the W and Z bosons, will increase with the transferred energy. Moreover, the measurements of the Higgs couplings will also reveal the symmetry properties of the new state. For instance, it can be established whether the new scalar is indeed "a Higgs" fitting into a SU(2) doublet together with the degrees of freedom associated with the longitudinal W and Z and not some exotic impostor, like for instance a pseudo-dilaton. If the Higgs is found to have an internal structure, a detailed study of the Higgs couplings can also establish whether it is just an ordinary composite, like a σ particle, or whether it is a pseudo-Nambu–Goldstone boson endowed with additional symmetry properties, like the π's of QCD.

The question whether the Higgs boson is an elementary or a composite particle is not merely an academic issue. Electroweak and strong interactions are endowed with a gauge structure that comes with great predictivity and robustness under radiative corrections. The Higgs particle, on the contrary, brings new forces with seemingly arbitrary parameters, some of which are dangerously sensitive to quantum loop effects. In the SM this is due to the elementary nature of the Higgs boson which comes with too much freedom and arbitrariness. If, instead, the Higgs boson were nothing else but a condensate of new strongly interacting fermions, an underlying gauge structure could order the Higgs forces. A closer look at the Higgs sector discloses the weakness of the SM Higgs construction: Each parameter of the Lagrangian

$$\mathcal{L}_{\text{Higgs}} = |D_\mu H|^2 - \left(V_0 - \mu^2|H|^2 + \lambda|H|^4\right) + \left(Y_{ij}\bar{\psi}_{Li}\psi_{Rj}H + \text{h.c}\right)$$

raises its own profound questions: (i) V_0 gives a contribution to the cosmological constant about 60 orders of magnitude larger than the one required to explain the acceleration of the expansion of the universe; (ii) radiative corrections drive μ towards the largest mass scale of new physics beyond the Standard Model, leading to the so-called hierarchy or naturalness problem; (iii) the (meta)stability of the electroweak vacuum is subject to the sign and the size of λ; (iv) the Yukawa couplings Y_{ij} secretly encode the properties of quarks and leptons, their identities and their mixings, with an apparent order and structure that keep challenging our understanding of the organisational pattern of elementary particles.

With the Higgs particle included, the SM is a consistent theory that can be extrapolated a priori up to energies of the order of 10^{12} TeV, far beyond the reach of any current and foreseeable collider. Nonetheless, it fails to satisfactorily describe several experimental observations in astrophysics and cosmology that would require either a modification of the gravitational laws or the introduction of a totally new sector called dark matter. In addition, the SM is unable to explain the asymmetry between matter and antimatter since the SM *CP* violation is screened by the lightness of the two first generations of quarks and the SM cosmological history does not involve any sufficiently violent epoch to bring the universe out of equilibrium. It is still an open question awaiting the next LHC runs whether the still-to-discover new physics could transmit, via the Higgs particle, new sources of *CP* violation and whether it could alter the dynamics of the electroweak phase transition and make it of the first order, thus creating the conditions for an efficient dynamical baryogenesis.

Another striking limitation of the SM is its failure to explain why there is such a big mass difference between electrons and muons while the two particles have exactly the same quantum numbers—or more generally: Why are there three, and only three, generations of fermions that differ only by their masses? The Higgs field is the only agent of the SM that knows the difference between these three generations.

The discovery of the *W* and *Z* gauge bosons by the UA1 and UA2 collaborations in 1983 opened new perspectives: The masses of these particles called for new physics at an energy scale below a TeV to maintain consistency with the basic laws of quantum mechanics. And the Higgs boson was aptly found as a confirmation of this *no-lose* theorem. The existence of a solitary light elementary scalar field per se does not shake the fundamental principles of quantum field theory, but it questions an empirical principle governing the intricacy of the different layers in the laws of physics. The Higgs mass is not predicted in the SM—it is a parameter fitted by experiment. It is a physical observable that receives contributions from the various particles the Higgs boson couples to. These contributions tend to be very large, much larger than the physically measured value. Hence a delicate cancellation is in place and requires extreme and a priori fortuitous correlation between ultraviolet and infrared physics. This property goes against the intuition that physical phenomena at different scales decouple from each other. Based on several enlightening examples from the history of physics, this decoupling is not expected to be accidental—in the same way that the classical radius of the electron does not shrink to zero thanks to the existence of the positron; in the same way that the mass difference between charged and neutral pions is screened by the ρ meson; in the same way that the kaon mass difference is made finite thanks to the existence of the charm quark; in same way also that the corrections to the W and Z self-energies are kept under control by the very existence of the Higgs boson itself; new physics is very much anticipated to protect the Higgs-boson mass from large radiative corrections. Theorists have conceived elegant models to achieve this protection: supersymmetry, compositeness, extended space-time...—which, for now, all remain elusive. Therefore, the situation starts questioning this naturalness principle that, despite its successes, is certainly not a law of nature, i.e. a necessary condition indispensable for the consistency of the whole edifice. Accidental cancellations do happen in nature! The LHC will either observe the agents responsible for the stability of the Higgs mass, spectacularly completing the understanding of the electroweak symmetry breaking dynamics. Or it will ascertain an accidental cancellation at an unprecedented level, possibly revealing the first indications of a breakdown of the reductionism idea that prevails in the microscopic world since the experiments of Rutherford.

12.3 Expectations for LHC Runs 2 and 3

The LHC and the experiments have reacted to the challenges ahead: ATLAS and CMS have found experimental measures to cope with substantially increased event multiplicities envisaged for the runs at 13 TeV and higher. The LHC itself routinely collides the beams at small transverse separation without detrimental effects on the beam quality. This small separation leads to the concept of "luminosity levelling" where the separation of the colliding beams is reduced to compensate for the loss of luminosity as the beam current diminishes (see Chap. 2). The LHCb experiment routinely operates in this mode; the multi-purpose detectors are technically prepared to maximally exploit the usable interaction rate in this fashion.

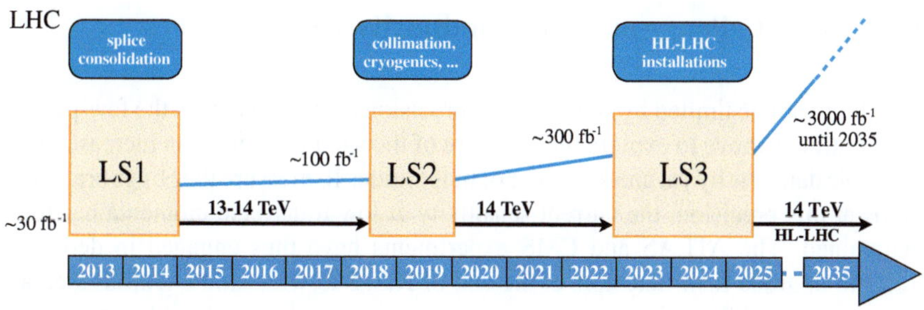

Fig. 12.4 LHC operation timeline. Run 1 covers the period 2010–12 with a total integrated luminosity of roughly $30\,\text{fb}^{-1}$. Run 2 will last from 2015 until the long shutdown 2 foreseen for 2018; it will deliver about $100\,\text{fb}^{-1}$. The total luminosity collected at the end of Run 3 by 2022 will amount to roughly $300\,\text{fb}^{-1}$

The plans for future LHC running have been laid out (Fig. 12.4). The integrated luminosity collected before the long shutdown 1 will increase by an order of magnitude in two steps. Some $300\,\text{fb}^{-1}$ of luminosity will be collected by the end of Run 3 around 2022. Together with the higher energy, this data sample will significantly increase the reach in the search for new physics. Runs 2 and 3 will also see upgrades of the experiments that support the precision requirements: improved b and c separation, better selectivity of the triggers by improving calorimeter granularity and providing track information (p_T measurement).

As an outlook for flavour physics, the energy increase corresponds to a gain of about a factor of two in the $b\bar{b}$ and $c\bar{c}$ cross sections. Here a major step forward in available statistics and thus in the physics reach will occur with the upgraded LHCb experiment, which will examine pp interactions at 40 MHz in a trigger-less mode after the long shutdown 2. Complementary input for the understanding of the beauty and charm systems is expected from Belle II that will start data-taking around 2017. While Belle II will be faced with smaller cross sections than LHCb, it can also study decay modes that contain neutral particles and reconstruct decays with missing energy.

With very few exceptions, Run 1 of the LHC has excluded all popular models of exotic new physics up to the TeV scale. In some cases, these limits exceed the TeV scale substantially. A lot of effort has gone into the search for extra dimensions and black holes. These scenarios are now excluded well into the multi-TeV realm. On the other hand, there are cases where the TeV scale has barely been probed. In particular the masses of leptoquarks, of vector-like quarks, of exotic long-lived particles or of any new particles immune to the strong interaction may still be below 1 TeV. The energy increase for Run 2 will boost the sensitivity for new physics at the highest mass scale. Small cross sections will, however, require considerable luminosity so that the experimental challenges for Runs 2 and 3 are high.

12.4 The Challenge of High-Luminosity LHC

In order to not be limited by statistics, the precision measurements at the few-percent level outlined above to explore the structure of the vacuum require an increase of the available data sets by yet another order of magnitude. To measure the Higgs branching ratios at this precision, the current sensitivity at low transverse momenta has to be maintained. The ATLAS and CMS experiments have thus engaged in dedicated R&D to upgrade their detectors at the end of Run 3 to withstand a tenfold increase in luminosity while maintaining or even improving the sensitivity at low transverse momenta. The goals of this R&D have been summarised in extensive documents of the envisaged upgrade plans, which are currently being scrutinised. The plans foresee an integrated luminosity of 3000 fb^{-1} to be collected by the mid-2030s.

The prospects of such an integrated luminosity can also be seen in the search for new physics [7]: (i) under some simple assumptions, it can be estimated that the mass reach increases by almost a factor x when the collider energy grows by a factor x and the luminosity collected increases by a factor x^2; (ii) an increase of luminosity by a factor 10 has also been estimated to translate into an increase of the mass reach by a constant $\Delta M \approx 0.07\,\text{TeV}\sqrt{s/(1\,\text{TeV})^2}$, so at the high-luminosity LHC the direct bounds on new particles are anticipated to improve by 1 TeV over the bounds obtained at the end of Run 3. These simple estimates agree with dedicated estimates in specific scenarios of new physics. The limits shown in Fig. 12.5 fully cover the parameter space of scenarios of new physics that are designed to address the hierarchy problem: Stops and gluinos of supersymmetric models above 1 TeV and vector-like partners of the top quark above 1.5 TeV generically come with a fine-tuning requirement below 1 % or call for special model-building to hide the naturalness agents into stealthy corners of the parameter space.

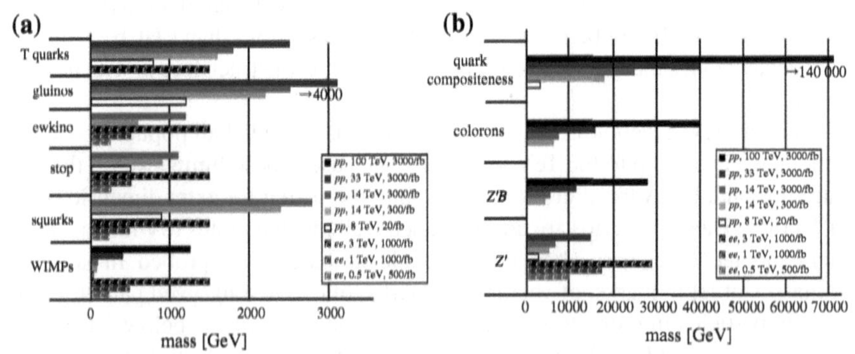

Fig. 12.5 95 % confidence level upper limits for masses of new particles beyond the Standard Model expected from current and future pp and e^+e^- colliders at different energies. Adapted from [8]

12.5 Physics Beyond the LHC

Even the ultimate LHC data sample alone may not suffice to explore the energy frontier. Despite the tremendous progress in reconstructing final states in *pp* collisions since the start of the LHC, in the end the precision will be limited by systematics, largely due to the complexity of the final state and the difficulty to fully separate background from signal events in the collisions produced by hadronic machines. The electroweak interaction in lepton colliders is both an advantage and a curse in comparison to hadron colliders. Most markedly for a given luminosity the total cross sections at e^+e^- machines are significantly smaller; the production rate is small, and known physics processes do not swamp the detector. In *pp* colliders the initial state necessarily is coloured; in lepton colliders the reaction is amongst electroweakly interacting particles. Point-like leptons also make all the initial energy available for the collision. This simplicity of the initial and final state at lepton colliders allows for better reconstruction and for the application of kinematic constraints to measure "invisible" particles.

An impression is given in the measurements of the Higgs coupling in Fig. 12.6, which contrasts the impact of the high-luminosity LHC with simulations of lepton colliders. At high luminosity the LHC will make big strides in measuring rare processes with distinct signatures such as the branching ratio $H \to \mu\mu$, while for example the precise measurement of the total decay width of the Higgs boson will be left for lepton colliders.

Among the ideas for future lepton colliders, the International Linear Collider (ILC) is the most developed one; a technical design report (TDR) is available [9]. Japanese physicists have launched an initiative to build the ILC in Japan, and the Japanese government is currently exploring the impact of such an international project. China is considering a large circular machine to host an e^+e^- collider (CEPC)—essentially a large LEP—that would reach the threshold energy of Z-associated Higgs production and that would allow for some precision studies of the Higgs decays. This lepton machine would be the first step on a path towards a Super-*pp*-Collider (SppC) [10]. At the same time, CERN is discussing an e^+e^- collider [11] as part of the Future Circular Collider (FCC) study that is targeted towards a 100 TeV *pp* collider [12]. The energy frontier in e^+e^- colliders will be addressed with CLIC that is planned for an energy of up to 3 TeV. CLIC has a far-developed technical and physics basis [13, 14] and represents the other option currently discussed at CERN (see the discussion of future machines in Sect. 2.6).

Optimum coverage is indeed achieved when using both hadron and lepton colliders, a strategy that has so well paid off in the past. Figure 12.6 illustrates the complementarity between the various future machines in the quest for more precision. Such a precision programme will only be successful if all the theoretical uncertainties are kept under control. The dashed lines on the LHC prospects in Fig. 12.6 correspond to a case where the systematic uncertainties do not scale down with the luminosity, which helps to appreciate the impact of these uncertainties on the extraction of the Higgs couplings.

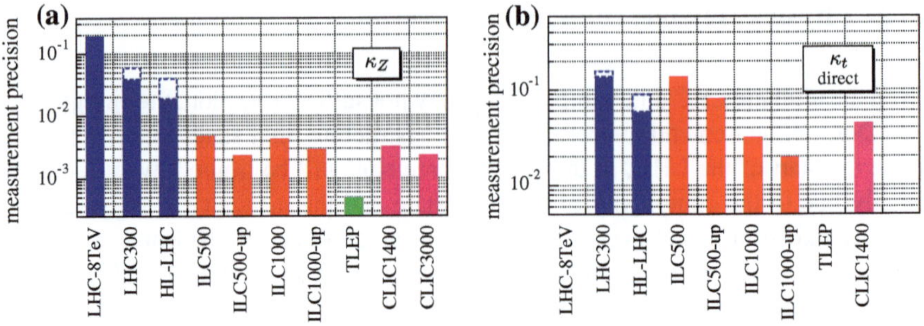

Fig. 12.6 Prospects for the sensitivity of various future machines to the Higgs-boson coupling to **a** the Z boson and **b** the top quark. It should be noted that during its first run the LHC has not observed the production of a Higgs boson in association with a pair of top quarks and thus had not direct access to the top-Yukawa coupling. Adapted from [15]

12.6 Outlook

One of the big lessons of the twentieth century is that microcosm and macrocosm are a common playground to test the fundamental laws of nature. Astrophysics and cosmology challenge particle physicists with two mysteries: What is dark matter made of? What is the dark energy causing the accelerated expansion of the universe? These two unknown components amount to about 95 % of the total energy budget of the universe. Their existence, so far, is inferred only from their gravitational interactions with the Standard Model matter.

The LHC had the potential to revolutionise the understanding of gravity by possibly observing new space-time structure via the production and decay of micro black holes. Unfortunately such black holes have not been observed in Run 1. Other more refined measurements dismissed the stirring pre-LHC hypothesis that Einstein's general relativity could bridge with quantum mechanics around the TeV scale.

The LHC could nonetheless still bring some invaluable clues about the nature of dark matter. A weakly interacting massive particle (WIMP) is known to have the correct relic abundance to match the cosmological constraints. Such WIMPs are actively searched for directly on Earth via the recoil energy produced in their collisions with SM particles or indirectly in enhanced cosmic-ray fluxes. They can also be produced directly from the collisions of two protons in the LHC that could then become the first dark-matter factory. Subsequent machines like the ILC could then establish with a good accuracy if such produced particles have all the required properties to survive the cosmological expansion of the universe from the big bang with the right density.

In the absence of direct production of new particles in Run 1, precision measurements in the sector of the newly discovered Higgs boson can play a key role in the search for new physics. The time has come to establish a clear strategy to extract the information on the origin of electroweak symmetry breaking encoded in the Higgs measurements and to pave the way for a future experimental programme.

The search for the Higgs boson has occupied the particle physics community for the last 50 years. Its discovery is now shaping and sharpening the physics programmes of future accelerators. The change in paradigm is profound: What used to be the missing particle in the SM now quickly turns into a tool both to explore the manifestations of the SM and to possible venture into the physics landscape beyond.

In fact, the LHC has considerably travelled the map of new physics options during Run 1: Rooted at the SM, mass and flavour play decisive roles that drive the remaining puzzles as e.g. established by *CP* violation and matter-antimatter asymmetry. The goal of unification remains a very attractive one; it clearly requires new physics to occur. Supersymmetry is a longtime contender for an elegant implementation of new physics although Run 1 did not reveal any candidate particles. The connection to dark matter and dark energy is still unknown. Despite these remaining questions, the progress introduced by the LHC is breathtaking and the technical prospects for the forthcoming runs are raising our expectations to further venture into the unexplored corners of the *Terascale*.

References

1. CMS Collaboration, CMS Electroweak Public Physics Results (2012), https://twiki.cern.ch/twiki/bin/view/CMSPublic/PhysicsResultsEWK
2. S. Bethke, Nucl. Phys. Proc. Suppl. **39BC**, 198 (1995)
3. S. Bethke (1998), arXiv:hep-ex/9812026
4. Particle Data Group, Phys. Rev. D **86**, 010001 (2012)
5. Particle Data Group, Chin. Phys. C **38**, 090001 (2014)
6. GFITTER Group, Eur, Phys. J. C **74**(9), 3046 (2014)
7. G.P. Salam, A. Weiler, Collider Reach Tool (2014), http://collider-reach.web.cern.ch/collider-reach/
8. Y. Gershtein et al. (2013), arXiv:1311.0299
9. T. Behnke et al. (2013), arXiv:1306.6327
10. Y. Wang, Introduction of CEPC-SppC (2014), http://indico.cern.ch/event/282344/session/1/contribution/65/material/slides/1.pdf
11. TLEP Design Study Working Group, JHEP **01**, 164 (2014)
12. A. Ball et al., *Future Circular Collider Study: Hadron Collider Parameters*. FCC-ACC-SPC-0001 (2014)
13. P. Lebrun et al. (2012), arXiv:1209.2543
14. CLIC Detector and Physics Study Collaboration (2013), arXiv:1307.5288
15. S. Dawson et al. (2013), arXiv:1310.8361

References

Index

T. Schörner-Sadenius (ed.), *The Large Hadron Collider*,
DOI 10.1007/978-3-319-15001-7

MIX
Papier aus verantwortungsvollen Quellen
Paper from responsible sources
FSC® C105338

If you have any concerns about our products,
you can contact us on
ProductSafety@springernature.com

In case Publisher is established outside the EU,
the EU authorized representative is:
Springer Nature Customer Service Center GmbH
Europaplatz 3, 69115 Heidelberg, Germany

Printed by Libri Plureos GmbH
in Hamburg, Germany